A Growing Cosmos According to Kepler's Tables

ISBN-13: 978-1535250566

ISBN-10: 1535250569

Author Peet (P.S.J.) Schutte

A Pre-Runner to A Cosmic Birth...Dismissing Nothing.

An **explanation** about the growth of outer space such as the picture above **matches every logic view we all have about the Universe**, but **does Science really provide the answers** matching our modern logic, or **are we filling in** and **compensating for science's shortfalls**. Does **reality really match** the logic of **official science**? Does the Hubble Telescope's pictures match the explanations science provide about **how Creation all started**... **where it is heading**...and **where it will end**?
What is motivating the expansion and the moving?

The book present has the dynamics to change science and that is not sensational or promotional talk. I say that with confidence because I investigated Kepler's work as far as cosmology is concerned and you may believe it or not but it seems that that is a task that was explored for the first time in four centuries. After Newton included the work of Kepler into his work science never went back to Kepler to research Kepler's work. It may astound you that any person on Earth could be of the opinion that when one read my work one will find Newton compromised Kepler's work as he did with Hook and many others I might add. For my saying that about Newton and by my claiming that Newton did not correctly analyse the work presented by Kepler, I have been rejected on many occasions in the past and by many institutions amongst the many I presume are your institution as well. The Academic world is all being Newtonian orientated and that is not surprising since Newton laid the foundation of modern physics. It must be said that it comes as a natural response when Academics bluntly ignore my work on the grounds of my negativity about Newton's view in regard to Kepler's findings. Previously there were Academics that would not touch my work notwithstanding the twenty-seven years of research that went into my work. I have only three South African Universities dealing with cosmology to turn to and their reviews of my work in the past left me in doubt about their sincerity in the performing of the reviews. The South African Academics do not attack my work because then I can defend my work...no; they just ignore the work by sending reviews that totally misses the point I make concerning my work. Every time it is on a minor technicality such as that I have to prove the Universe in outer space is not nothing, or that my view about the pendulum is incorrect or even that I misinterpret Newton. Again every one missed what I had to say about gravity, space-time, space and time, because they gave raving reviews about my focussing their attention on nothing being part of the cosmos but they shine in their failing to mention any other aspect I brought to their attention, which by the way they still understand nothing about! I truly am of the opinion that they did not read my work in the past because the comments they make is miles off on that what I say. On one of the previous occasions the nothing formed a minor a few pages in a three hundred and thirty one page book but they had no comment on any other facts or parts of the book. There whole review was as if I never wrote anything else of value in the book. Either they only understood the part about nothing or they never read past the part about nothing but they got stuck at the part about nothing as if there was nothing else they could understand. I came too this conclusion when the reviews came to my attention. I thought that it either had to be subordinates reviewing my work and they had no idea what they were suppose to review or those that did the reviewing paid very little attention. This I concluded when some of the commentating was in a review sent to their superiors and addressed to their superiors but by some accident came addressed to me directly. The information they did comment about was to my view the work of subordinates because in the manner the detail was about it shows a person with a minor ability in understanding and therefore formed a conception only about lesser part of the majority of my work. The person had a limited understanding ability of the total picture concerning the entirety of what I propose. The review does not reflect much of a broad and informed background that such a person should have and in that view it cannot be that any person that is well informed can only see such a small part of my work, as the person apparently did. From the letter what I received which was by accident posted directly to me although it was addressed to someone else to someone in the office of the superior officers and what they commented on and what I wrote was totally not the same subject. It made me realise there was only one option and that was going commercial with the most academically developed work.

This letter you are reading is my effort by which I hope to interest you in reading my manuscript in sincerity because that is the least it deserves but even that little it does not get from academics in control of Universities in South Africa. I wish to find a Publisher willing to publish my books on my behalf and there are only Academic publishers connected to Universities that would find any interest in my work. The book I wish to submit is purposely compiled with a commercial viability intension but since the theory is new and the book will be expensive to print because of the volume. With that in mind I made an effort to keep the expenses low. I offer ***A Cosmic Birth Dismissing Nothing I. S.B.N. 0-620–31609–8*** to publishers having the title of ***A Cosmic Birth Dismissing Nothing I. S.B.N. 0-620–31609–8***. To make this one more acceptable and better understood (I hope) I compromised the shorter version by adding more graphics. That I hope will render it less complex but also better informative. The book also has an own accompanying web page, which I submit as the second part of this letter. If you study the accompanying web page that I include such reading should help you to understand my ideas better by giving a platform to grasp the ideas. It would mainly help to introduce the very new ideas and the information but by reviewing the web page alone it would inform on an extremely limited scale, that which I try to bring across. The accompanying web page is condensed to suit the reader and provide some shortened briefing of the book I wish to submit. I hope that by you're reading the web page first will then entice you to read the book I wish to submit. Please keep in mind that the web page does not even cover a percentage of the work covered by the book about the content but I hope it would firstly familiarise the reader with the information in the books.

I believe that I achieved an all time breakthrough success because I can now explain what gravity is. Remember that not even Newton could explain gravity, but Kepler did without any person noticing even to this day four hundred years on. It was only that at the time when Kepler introduced the gravity he uncovered the world was not ready to realise what he uncovered. It is gravity keeping the Sun and planets in rotation and that is what Kepler formulised. From such explaining I prove the Titius Bode principle also known just as the Bode principle and that gravity comes from 10 /7 and 7 /10 in relation to singularity laws to hold a very specific relative value in relation to various different structures in the cosmos as they relate to gravity specifics. This is a law found in all planets and fragmented pieces circling about the sun, but even when all the planets equally adhere to the law, Mainstream science choose to push it off the table as inconsequential since Mainstream cosmology has no way of explaining the law by implementing Newton's physics. When I implement Kepler's cosmology, cosmology becomes ordinary simple to understand. I do prove the Roche limit from the point I traced Kepler's indicating where singularity is located. My achievements came from my effort where I separated Kepler's work from the opinion that Newton formed about Kepler's work. Behind Kepler's work hides more information and answers to unanswered questions than there are questions answered by Mainstream Science at present. Mainstream science fail to explain even one gravitational concept but I managed to formulate, define and explain all the ten gravity concepts that is a derogative of gravity. I found it to be possible through the analysing of Kepler's work and the uncovering of the singularity value. For instance from Kepler's work I can explain the operation of the Black Hole. In my opinion my explaining of gravity makes much more sense than the accepted force of Dark Age proportions, which even Newton admitted he could not explain. From my view a force is just motion applying and that is what Kepler said gravity is. Kepler said that gravity binding structures is $a^3 = T^2 k$. But singularity is hiding where Newton placed nothing and Kepler positioned the pointing of **k**! In that I followed the path to uncover singularity. I dissected **k** as a factor in the Coanda effect and found that the Coanda effect is proof of my view about gravity and singularity produces the Coanda effect, which dictates all implications in the Coanda effect. The Coanda effect is the establishing of **individual space a^3** by **applying motion T^2k** on the both sides of **singularity k^0**. Where the Coanda effect is producing gravity and such producing is stronger in a small space than the gravity produced by the Earth in that spot it pointed to the fact that there was some manner in which it had to play a part during the period of the Big Bang. As it also is in the case of the Total Internal reflection establishing singularity by applying motion to liquid produces the phenomena. By applying mind boggling mathematics solves one part of the problem where the other part is the use of

Mathematics as expressions that must be translated to the verbal used languages expressing the ideas into the normally used dialects in the manner that it says and not to please the mindset of the mathematical translator. Newton should have seen that as the most impressive mathematician yet...but he didn't... Let us investigate what Kepler introduced and see if it could simply be integrated into a "bigger" concept by diminishing the Kepler concept. Remember Newton had the attitude that Kepler's work was incomplete and therefore inferior because he added to Kepler's introduction formula what Kepler did not introduce and therefore Kepler saw no use for that which Newton found fit to introduce. It is either that or Newton saw Kepler as being completely inferior in intelligence. Still Newton saw Kepler's $a^3 = T^2k$ as incomplete and therefore he had the urge to correct what he saw to be incorrect as to be $4\pi^2 a^3 / T^2 = G(m + m_p)$. In this, the value of T and "a" are the period of revolution and semi major axis of the orbit of a planet of mass m_p about the Sun of mass m, and G is the gravitational constant. Newton added nothing but duplicated everything in his suggesting completing what he saw as incomplete. Can the symbol "*a*" be reckoned as a period of revolution as Newton suggested and therefore Mainstream science still suggests? **Is the work of Kepler incomplete?** The formula in question is $a^3 = T^2$ broken down into fragments it reads as follows:

a^3 The fact that any symbol uses a value to the third power indicates space or a volumetric established and separate unit using the six dimensions allocated to space. It is space because it is volume using the third dimension. There is no other valid interpretation or translation allowing a correct translating from mathematics to English but by categorising that space as a volumetric separate identity. It is cubic which is space. One measure a fridge or a stove or a room by the cubic measure without bringing in pi because of the volumetric content there is when using the third dimension. The fact that there is a line connecting the space and enforcing a rotating motion around a specific centre connects whatever volumetric measure the space is in the cube and the cube as a separate issue independent from the space in which the independent cubic space rotates, which brings about the circle and yet there is no need to implicate pi because the rotation brings along the circle after the cyclic completion of the rotation by the independent cubic space.

T^2 Is an indication of motion, the moving of an independent space that is holding a^3, where the space in motion will be measured as a^3 as the space a^3 that is occupying separate unit as space within other space that is travelling by using the second dimension of motion in T^2 moving the independent space from one point to another point or following a flat distance between two points and the distance and only the distance is T^2 where the distance is established as T^2 by the space a^3 but is not the space a^3 and as the independent space a^3 the space is going from point T to point T making the distance a^3 travelled T^2. It is space filled with material consisting to allow independent space within the surrounding of an enveloping space of bigger proportions a^3 being in motion T^2 and that motion T^2 is of the filled space a^3 that is taking time in the second dimension moving the space a^3 in question from one point of choice to another point of choice by which time will be established.

k^1 Is the symbol used to indicate a straight line between two points that is connecting a very specific centre with a definite beginning and a specific end position. This indication of a distance is an indication of a bigger space that is big enough to include a smaller and separate space a^3 within the bigger space running all the way from the start of k to the end of the line of k. One has to see k for what it represents because k indicate the presence of a larger space that is large enough to allow the smaller space to rotate all the way as the circle runs from T to T in the full diameter of k. Kepler introduced this absolute basic mathematical principle. It is positioning the independent space in motion in a specific relation to a controlling dynamic situated in a dominating and controlling centre. It is indicating that the space in question is in motion acknowledging the centre in control of the motion and therefore in control of the space location. It proves a larger space is holding the space in question as part of the larger space where the larger space is in ratio so much larger that the space in question can effortlessly go in motion and to full rotating motion be concluding a rotation within the larger space that k indicate. But k also must indicate therefore coming from a very small start centralising the motion of the space in question using the space that k produce from where that specific space is controlled in the realms of the larger space. Nobody before saw it in as simplistic manner as I just showed it to be and yet in the simplicity is the sensibility of it all...

When $a^3 = T^2k$ then $a^3 / (T^2k) = 1$ and k^0 also equals one therefore mathematically it is very correct to say that singularity is presented amongst many other values to the power of zero as $k^0 = 1$ proving that $k^0 = a^3 / (T^2k)$

What formed the grounds for any need by Newton to change Kepler's translations from what Kepler translated the cosmic given to mathematics and then from mathematics to English? We have to acknowledge one fact without any possibility of disputing the fact…Kepler received his acquired information spanning across the accumulated effort of two life times of work directly from the cosmos as the cosmos used the language of mathematics to explain to Kepler what Kepler was suppose to see. If the cosmos saw it fit not to include pi then one should not try and correct the cosmos because for some very obvious reason the cosmos would not supply incorrect information. It will be better to rather seek the reasons why the cosmos saw it fit to exclude pi…and that is just what I have done. The space-time that the cosmos introduced was so brilliant it took the likes of a genius such as Einstein to realise the presence thereof many hundreds of years later. Those who are in agreement with Newton's changes should ask themselves what part of Kepler's $a^3 = T^2k$ did I not translate correctly in the above explaining when I interpreted the expressed by taking the mathematical expressed to the verbal English? When I used Kepler's mathematics by my translating Kepler's work correctly I came upon answers not yet uncovered by Mainstream Science. But I had to dissect Kepler's mathematics in the manner as it was told by the cosmos in person to Kepler without Newton being arrogant enough to tell the cosmos what the cosmos was suppose to tell Kepler about the cosmos. Kepler gave the World the mathematic translated cosmic answers that Kepler uncovered long before Newton, Einstein and others got wise about cosmology... Such is the advantage of recollecting Kepler facts that it does answer many questions, which went unnoticed and therefore not spoken about up to now and some were previously never even thought about. Mainstream Science never previously thought that through any examination of Kepler's work such scrutinizing would uncover these facts that I present. Subsequently Mainstream Science elected not to ask the correct questions and in the process Mainstream Science never found the correct answers. By not asking the questions Mainstream Science could not decipher any of the decoded mathematical messages, which Kepler received directly as a mathematical message spoken by the Universe and coming from the cosmos. We all know and appreciate that mathematics is just another language and the professional mathematicians have is responsibly translating mathematics to a verbally competent language. They never attempted beyond Newton's efforts to neither read into Kepler's nor recognise Kepler's mathematical translation and thereby was unable to translate Kepler's mathematics to the other communication forms being all verbally spoken dialects. In other cases human natural study methods brought along a cultural of Academics forcing students to comply whereby the students will accept the knowledge through our inherited past which when tested by modern standards the culture driven ideas is not that highly proven. We accept the unproven coming from the past and then forcefully submit the next generation to that answers as questions already fully answered because culture demands the accepting thereof. Any one insisting that Newton acted correctly by initiating the four pi in the square…well those should especially read my book because they are in for a massive surprise about mathematics they never encountered before!

I know science is under the impression that particles grab each other as they move closer to each other, but as presumptuous as it may be on my part of trying to disprove Mainstream Physics, such a presuming does not change the truth about Mainstream science being incorrect. The question I had in hand was finding what role there was for gravity to play in the Creating process and then I had to find a method that would allow me to explain why it played a role even at the very start of Creation. Remember our view on gravity is one of contracting. With the contracting or bring together of material it cannot support any idea that having contraction while the Big Bang was in progress was much unlikely. It would seem that a Big Bang and Creation was all about expanding and pushing material further apart. In the book I present the analysing of **Kepler's formulating of the solar system without Newton's interrupting** of Kepler's work. Please let me explain what I refer too: Two of the world's most outstanding researchers dedicated their lives in researching the figures of how the planets revolve around the Sun. This was an effort in dedication not yet seen afterwards. Tycho Brahe and later Kepler made a study of outer space as there was never repeated afterwards. Kepler said $a^3 = T^2k$. We all know that a^3 is space and with the space indicated as being in the third dimension and forming a six-sided space the third dimension is unmistakably an indication that it is a volume, which by definition is presenting space. We also know from the way calculations come about by using the formula of Kepler that T^2 is the duration of a specific period of time relating to a specific centre, which is claimed by the sun. On the one hand we have space a^3 and on the other hand in direct relation to the space Kepler introduced motion coming from a centre that forms time T^2 k. Kepler gave us space-time a^3 / T^2 but no one ever took any notice. In the formula is space a^3. In the formula the space a^3 has direct relation to time T^2 If $k = a^3 / T^2$ it means that from the centre holding the gravity is space-time. Space is a^3 and the motion of space a^3 we accept as time T^2 k for the past three hundred and fifty years. Kepler also gave us so much more and that too, no one noticed. Kepler gave us gravity before Newton named it as a force. Kepler gave us space-time long before Einstein named the notion. It was Einstein that calculated and then declared that gravity is at its strongest where space disappears and from that the world concluded that the Universe is drawing flat. Kepler said it so simple. Due to the natural form of a circle there is a point in the very centre where space has to relinquish the position because of the form the circle has there is no more space to form space. All sphere are a multitude of circles duplicating as they acknowledge one such centre therefore the sphere is the ultimate circle. In a circle in the centre there is a point where space disappear, where there are no more room for space to be and since all orbiting cosmic objects form a circle all cosmic objects run from such a space less centre space to be and from such definite unseen location the circle extends space-time that control all space within that centre. If only Einstein referred to Kepler he would have seen that in the very centre where k^0 is $k^0 = a^3 / T^2$ k which means at that point where space disappear gravity control all space in motion. Where space vanishes gravity is strongest. Gravity is strongest where space is least therefore gravity is about removing space to establish an ultimate point of strength. There is no special sub atomic "graviton" sucking and spitting as it creates gravity. It is about motion and the form material takes on but although it may at first sound simple, it is a lot more complicated than such simplistic explaining may suggest at first. Gravity is about reducing space in motion by duplicating space with motion. Gravity is the motion T^2 of a solid a^3 through a fluid k or also a less dense liquid and the amount of liquid (plasma) heat in the space forms the density or supplies the liquid state of the space. The space is the second part of the formula $a^3 = T^2$ k because it is solidity a^3 in relation to moving ability T^2 in space k that allows motion. Motion can only come from two possibilities where one is heating producing space and cooling reducing space. Other than that there is not.

Another important theme is the book carries and something, which I already mentioned is about the Universe not coming from nothing and therefore outer space, cannot hold "nothing". By taking Kepler's $k = a^3 / T^2$ and using k as a line I show through using the line as an example that the cosmic Universe holds everything and all concepts. However the only thing it does not hold is also the only aspect not present in the Universe at all. That is the value of nothing or zero. Explain to yourself how it was possible to create nothing in the Universe! In as much as carrying the definition of the absolute absence of any value "nothing" in that case cannot be present because the line that light uses to flow eliminates any such a possibility. Mathematics is a means of communication about matters concerning the cosmos and as an intercultural language spanning across race and ethnicity or as a principle as such cannot have zero because mathematics indicating lines, which is about not applying the numerical number or value of nothing. That much I prove physically. Where there is any person that disagrees with this statement I challenge such a person to show mathematically where nothing in mathematical calculation as a factor in the cosmos can come to conclude a value in total other than nothing and where there was a chance for nothing ever to enter the mathematics of the Universe. If you put the starting point of a line at nothing you remove the line as an option of being something in the Universe and ultimately destroy the chance of any line being in the Universe. The line could have stared from singularity where singularity holds the symbol of what ever to the power of zero but that then retains the possibility to grow from a exponential value of zero which does not remove the single object and then the exponential zero being one will multiply by number as an established factor present in the Universe. But removing the line by replacing the line with zero will disallow any line ever forming in the Universe. You may either attempt to do it before or after reading my work but my challenge will stand since mathematically nothing replaced ether when the concept of ether was removed from space. The concept of ether was removed and replaced by a concept of giving nothing a permanent value of one. Then afterwards some parties as part of Mainstream Science brilliantly allowed the nothing they produced to replace ether ended where the nothing they could add to replace ether replacing ether as an accepted concept. To the nothing they attached a value as a concept of something able to carry a value and were able to allow the nothing-value they instituted as a concept to replace ether. Such replacing was only an idea introduced later and was never proven mathematically. Kepler gave us the relation between cosmic objects as $k = a^3 / T^2$. From the formula **k** forms a connecting straight line filling the first dimension and not the single dimension. Ask yourself the following: does Pluto hold more nothing between it and the Sun since it is further way from the Sun and with nothing being between it and the Sun or does Mercury have more nothing between it an the Sun since Mercury is closest to the Sun and being closer it should have more nothing between it and the Sun then the much more nothing there is between Pluto and the Sun.. If Mercury had more nothing between it and the Sun would Mercury not then be located inside the Sun at the very centre because there it should have most nothing. After all it is in the centre we find the sphere have the most nothing. The shorter the line is the more it will confirm most nothing and not the longer is the more the line must confirm where the most nothing is. This is a major point of review. By reducing space a line represent one can reach an ultimate reduction indicating a point where the reducing of any or all lines that form the Universe, then by such reducing can only confirm singularity, which was my first breakthrough. My realising that nothing has no part in mathematics sounds degradingly simple but it unlocks the birth of the cosmos. Arithmetic uses nothing because it works on numbers and quantities, and in that sense there can be or there cannot be any specific number including zero. But the Universe is overflowing with everything and that only excludes nothing since "nothing" as a concept or as a value cannot bring about expanding or overflowing. Try to get any academic to admit to this concept and you'll find out what a task that is! It has the same degree of difficulty that showing to Academics that according to Newton all comets must crash into the Sun whereas they do not crash ever. Newton said with the formula to gravity applying $F=G\,(Mxm)/r^2$ will completely remove the radius between the two masses keeping the Sun and the oncoming comet apart and then the comet has to collide with the sun, but that never happens and not one Academic thus far in my presence would admit to that!

Gravity is precisely what Kepler showed gravity is. In Kepler's formula **k** equals space-time or a^3 / T^2. Gravity is singularity extending and forming space a^3 through gravity rotating space T^2 **k**. Gravity is not particles pulling one another in a tug of war. Gravity is about reducing space and maintaining different cosmic sides not sharing the same sort of space while the space is spinning about singularity. This Kepler stated as $a^3 = T^2\,k$ **and while that brings about the equilibrium of** $k = a^3 / T^2$ **on the on side and** $k^{-1} = T^2 / a^3$ **on the other side.** In the beginning before and after the Big Bang (yes before because there had to be a before) gravity was about bringing across heat that was in space to material that was in another space. The only difference is that in the space where the space is filled with plasma or heat and not material, the heat was much denser then than what it is now. The gravity allowed the one part of the Universe to remain in form while antigravity causing plasma or heat to deform the form of the structure it had and that other part of the Universe became liquid turning to space.

I explain how matter is claiming heat from space through gravity. The claiming of space by gravity comes about by the implementing of the Titius Bode Principal of seven dimensions interacting with ten dimensions and produce the square root of space, which I show to be a major part of gravity. Gravity is not about material pulling and tugging on other material. There cannot be antimatter that went missing. In the cosmos every aspect that was present when the cosmos formed is still very much part of the cosmos because there is nowhere else to go, but to remain where it is during the time. Mainstream science claims that many aspects of material (matter and antimatter) and singularity were present in the cosmos during the early phases but has vanished since. Think about it clearly…What ever was in the Universe had nowhere to go but to remain in the Universe after the Big Bang. There is no place else to go! There is no chance of entering, later escaping and then re-entering again. The Universe is everything there can be. That means if singularity was part of the cosmos during the Big Bang singularity must still be in the cosmos. If antimatter was present during the Big Bang it must still be present. There is nowhere to go. Space was little and heat was massive. Heat became space as the density of heat turned to form space through a process we named exploding. All the proof of this becomes possible when reducing the straight line to a point holding singularity at the very end of the line. Reducing the point to where the line start the Kepler formula **k** becomes the extension of singularity and through extending singularity **k** applies and commands space a^3 and time T^2 as space-time where the formula then reads the mathematical equivalent of what Einstein named by using a verbal expression, which is space-time $k = a^3 / T^2$. I locate, spot and place a value on singularity where singularity is relevant to space-time forming. In the book **_A_**

Cosmic Birth Dismissing Nothing I. S.B.N. 0-620–31609–8 it goes far beyond with the explaining this letter can provide since this letter only and exclusively deals just with the fundamental basics of my theory. The cosmos has lines forming cubes and lines forming circles, which in 3D manifests as spheres. Between the circles and the cubes runs lines so the key to understanding the Universe are lines. The Big Bang was a time when the Universe was incredibly small making the running between particles that connected space with lines small. Understanding the Universe is taking the line back to its limits where such limits were during and pre-dating the Big Bang. You can reduce the Universe to where all fitted into a subatomic particle by applying maths. Behind the mathematical reducing was the reducing of the lines that formed space and particle filling space. By reducing any line to where the line will not reduce any further at that point all points land on the same spot. At such a spot all sides are on the very same side because of the singularity aspect. The spots all share one position because that is the only position there is to hold. That is singularity being one to all but it is not zero. Finding form in that point shared by all will give a value of singularity. Extend that value received to a Universal centre and bring that value to align with Kepler's $a^3 = kT^2$ and the Universe with the entire different yet unexplained phenomena becomes as easy as children schoolwork. There are suddenly no more mysteries in the Universe. Applying the new value to match the factors brought about by Kepler I managed to prove how the Roche limit works and what role the Roche limit played in supporting the Bode rule. I also show gravity comes from the Bode law applying a relation of ten dividing seven and seven dividing ten. It is only because gravity reduces the space between particles and not being some magic force found between particles grabbing all that the following phenomenon's are mathematically and principally explained:

Gravity is not being some unexplainable witchcraft-like magic force found between particles grabbing onto everything. I mathematically explained the following phenomenon:

By using Newton one cannot even begin to explain any one of or the combined effort of the above cosmic phenomenon that are all over the cosmos and which I prove that they form all the laws in the cosmos. Newtonian definition cannot even recognise any of the principles but only Newtonian science are thought to students…and Academic go about their usual ways ignoring the existence of the phenomena and religiously denouncing they're presence as coincidental occurrences. Such views do not help with tutoring students either. Newton became religiosity for the most part three hundred and fifty years and Science blames me for mentioning it. The gravity Kepler introduced is all working on a principle of indicators pointing dimensional integration and separation of space through heat densities applying different grades of space intensity. That is space-time being apart and forming densities. That means the space does not mingle, as one would expect because of the nothing value contributed by Mainstream science to space. Space is a liquid and as all liquids do, space depends on specific densities. With the specific densities borders come about. The explaining of this I manage with very simple sketches indicating the principles. I explain with ongoing proof through out the book indicating to the reader how I came to realise that gravity is not about particle pulling on each other with some inexplicable force holding an effect of matter pulling matter. I take the reader on a step-by-step voyage as I came to unravel Kepler. It is as Kepler stated gravity is motion but the motion combines a circle T^2 with a straight-line k and that it is how gravity is to be, even before Newton came up with an idea that there was such influencing going on and named the influence gravity. The one factor of gravity being either of T^2 or k would dominate the other. Gravity is $a^3 = T^2k$, which is the space a^3, that forms through the moving T^2k thereof giving the space a^3 independence T^2k from surrounding space. Gravity is space moving in a circle holding space and what is in space at a distance where that is the applying conditions. Only when such conditions are broken does space fall away and particles come crashing down to Earth. But this falling comes from a lack of motion and not a sudden jerk on the falling object. The relation in motion must reduce on the part of the object about to fall to introduce the falling action. There is no pulling but a differentiation in motion that breaks the barrier the space placed and the object the breaks through the barrier the Earth holds in place. A little science experiment such as the Coanda effect disproves the grabbing on theory. Gravity is about matter concentrating space through the spin of the proton and the reducing space by accelerating the movement of space between the two objects. A good example is a blowing fan. By establishing motion of the air with the spinning fan, volumetric occupation reduces in favour of airspeed coming about from motion. By removing the space the particles come automatically closer, and that is the principle of gravity in operation. But the diminishing of the space goes about by applying very specific rules and in the adhering of those rules is applied to everything in the cosmos.

Gravity is not <u>about particles pulling</u> each other closer, <u>but it is about depleting the space</u> the body holds and the space surrounding the material as unoccupied and occupied space. Gravity comes about as space a^3 applies motion T^2 and from establishing singularity k^0 that provides distance k to supply space-time gravity is as much part of space as the motion of space is part of gravity. Mass is the result of applying gravity by reducing space. Gravity is not the result of mass. Gravity dismisses space and by doing that the stronger gravity can have more particles fitting into less space occupied where that reducing of the space is bringing about extensive mass increases into the volumetric occupied confining more material into less space. That produces monsters like a Black hole having enormous mass and being without space. If gravity was not about reducing of space the Black hole does not make any sense. Stars reduce their volumetric size as they gain in mass by creating enormous gravity applying the sphere of influence of such a star. Gravity is the increase of heat occupied by the reducing of space in a spherical unit. From the offset of the Big Bang to the process the Universe went through development up to the very point where it is at the present, the process was about converting heat to space. I named the expanding of material through the process of overheating antigravity. Gravity and antigravity is the driving "life" of the cosmos and such a processes as the Coanda Effect and electricity can charge such "life" into being present. The Hubble Constant and expansion through heat applying is the other part of the driving engine applying antigravity with expansion where the heat transforming to space releases heat and converts the Universe to more space but less dense space. Remember how the heat came down from 10^{34} to where it now is 0 K at present? Remember how the space multiplied by expanding from the size of an atom to what we now can and cannot see as it extended beyond any measure we humans can devise to where it is at present? But in all of science this inverting connection goes lost as it is unnoticed by all. The density of heat in space surely diminished considerably since then to now. The space holding the heat concentration expanded volumetrically as the heat concentration reduces…but not one person in science can connect the two obvious relevant factors. Expansion is gravity fighting a lost battle because space is increasing as heat is reducing. Gravity on the other hand is exchanging heat through the concentration by removing space bringing about space loss with increased density of particles and therefore heat concentration. The exchanging of heat as the heat converts to space is what fuels the Hubble constant, which is the result of the explosion we call the Big Bang but is not the result of momentum. Momentum is the second form of gravity symbolised by Kepler, as k. It is the directional motion to compliment the spin T^2. The Big Bang is the result of heat expanding into the forming of space. Gravity, on the other hand is about concentrating space back to heat, and take recouped heat through to material, acting out a balance of expanding while contracting. This way gravity is applying the onset of the Big Crunch by destroying space while space is converting heat to material occupying space. The Big Crunch is coming about because the Universe is expanding where the two processes are one principle.

In the beginning singularity was present in the Big Bang. Mainstream Science promotes the idea that singularity and antimatter went on the disappearing by escaping from the Universe. But Mainstream Science never says where they went. Singularity and antimatter and all other Houdini acting stuff had no place to escape too, which leaves us with one possibility; we just have to find the new location it occupies. Everything that was in the Universe is in the Universe and will forever be in the Universe until the very end of the Universe. If it was in the Universe it still is in the Universe because there is no other place available to escape to. But that also goes for the space we now are in because either the space was there and it was filled from the beginning or the space was something else and that something else turned into space. In the totality of the Universe where the all and everything is in the Universe nothing is not in the Universe. The Universe is everything holds whatever form of material or non-material there is. That which is now part of the cosmos was always and always will be without the slightest chance to add or disappear. Going somewhere is not part of options of whatever forms the content of the Universe. Keeping that in mind one should remember that the realm of the Big Bang was forming the cosmos by gravitational influences. What caused the Big Bang is everywhere to be found in the Universe and cannot be in selected places. With realising that much and printing that concept into a realisation one should first prove that gravity concentrated matter when it formed the solids from a heat and much more pertinent is the need to prove how it came about. If singularity was part of the cosmos in the beginning it is still with us. We just have to find it by searching for it and use the characteristics it had during the Big Bang. It will still have the very same characteristics. However by the claiming that singularity vanished somehow since the Big Bang is totally incorrect and proves a lack of understanding cosmic concepts. Kepler and his formula also prove this fact. Kepler said $a^3 = T^2k$ which is the same as $k^0 = a^3 / (T^2 k)$. Using k^0 is using singularity because any symbol using the exponent of zero has a value of one and one is the value of singularity. Every aspect that was part of the Universe at the cosmic birth announced by the Big Bang has to be present up to this point. There is no room or place to which the singularity could dispose since the Universe is the only place there ever was or will be. When I truly admitted this fact and committed all my thinking to this conclusion (and not as merely accepting by thought as another fact in my mind), cosmology became a woven blanket any one can read. I prove that singularity is present within matter and in fact all matter. This is a claim not yet made by any other person. I can direct any person to locate singularity which is another claim never made by any person in the past. Singularity present space-time and where one find space one has to find time because space cannot be without time. Kepler said that much when he said $a^3 = T^2k$, which mathematically translates as space becomes motion and motion indicate relevant time. By my finding such locating and producing as well as applying a very specific value to singularity, singularity makes the process, which the Universe adopted much easier to follow. The connecting of singularity runs from the ordinary stars such as the Sun and up to the Black Hole down to the ordinary material forming our every day living in the cosmos. We have to realise that singularity is a prerequisite for the Big Bang to have formed any and all material. If the presence of singularity is not in all material in the everyday cosmos, singularity was not present during the Big Bang and the Big Bang cannot be, because what is in the cosmos now, had to be in a pre Big Bang cosmos as much as it has to be in the cosmos of the present and the future as it in all stars including Black Holes.

As I said before the motive with this letter that is accompanying the manuscript is to introduce the accumulation of my efforts into a book. This letter I hope will entice you to read my manuscript because I have grown suspicious that there are many that do not read my manuscripts or do not understand when they do read my manuscripts. I made use of the simplest of sketches all in black and white and some grey. As the book has a self-explaining title and is a condensed profile of the most basic aspects of my work the motive was to be as informative as possible.

I am not connected to the Mainstream physics in any way or form and through my views I promote in the writing of my work I found a method how **not to make friends** with those members in the ranks of the Exceptional and outstanding Gentleman of well-established and important educational background upholding Mainstream physics. Thus far I had very little success because I am not sharing much of any opinion with the Academics of the day and in that respect I have to criticize their work. Such criticizing of Mainstream views is apparently not the correct route to follow when trying to communicate with such gentlemen of exceptional standing claiming most admirable standards of the highest esteem and in every case they took the criticizing very personal. Why they tend to do that, I do not know because it is their opinions I dispute and not they personally! But with my definite disagreeing with mainstream physics and not trying to cover my disagreeing with soothing words there are no other options I have but to disagree with Newtonian disciples bluntly and not the lesser so than with Newton. As the overall extending of the work is, it is all including and that holds my opinion without reserving or diverting from that which I believe, therefore there is no short route or polishing some of the fringes of some aspects of the accepted theories to introduce some aspects about my work. To understand is to familiarise one with the entire concept. The routes I can follow through magazines by publishing articles only allow me three thousand words at a time. That is far too little to promote such a broad concept as I wish to promote. With my promoting of my view holding the complete picture it has to bring about change in science and has to change science altogether in the field of cosmology but only in the field of cosmology. This comes about because the gravity pressing us down onto the Earth is not the same principles applying when a smaller space orbits a larger space. There is two definitions applying which is a Universe apart and in that there is just some aspects thereof sharing a concept. Therefore I almost need three thousand paragraphs at least just to make my ideas logical. I wish to find a publisher willing to publish the book presented as ***A Cosmic Birth Dismissing Nothing I. S.B.N. 0-620–31609–8***.

In essence the first part of the book is an explanation with proof about how the Universe came to be before the Big Bang and before mathematics came into place. The beginning where all and when all started. It starts explaining why the triangle is 180^0 the half circle is 180^0 and the straight line is 180^0 and how all three is (as it first was equal in form sharing value before "normal" mathematics became into place) where the triangle has three lines but is equal to a line and a half circle and yet so different in shape and form. (With that I introduce a measuring standard value used by the cosmos in the same manner we use the imperial or the metric measuring system but the basic value does not use one. The Universe uses a basic founding measure it formed before it formed mathematics. From that standard and by using that standard I can explain all stars behaviour as far down to stars yet to be stars and even down to the reason why the proton is 1836 times more massive than the electron is. I can use that measure to explain and differentiate between Pulsars and Black holes, between yellow giants and blue dwarfs. I can explain fundamentally what forms a Neutron star and why does some stars have a variable change in rhythm and others going up to be frantic Pulsar stars. Moreover I can explain gravity in detail as it has never been explained before. It is only after the moment that space (the triangle) became equal to being the motion forming in the half circle that is holding the line as it is equal to the square of line it is holding (in what we know as the law of Pythagoras) and that extensively became our Universe. The Universe formed as $a^3 = T^2k$ and from that afterwards formed mathematics in a usable six sided seven to ten in relation to ten to seven dimensional Universe with atoms ranging from 112 protons down to one proton in $k^0 = a^3 / (T^2 k)$. I use the basic value too prove the sound barrier values in metric distance relating to Earth time.

The book I present has the dynamics **to change science** and that is not promotional talk. I say that **because I investigated Kepler and** believe it or not but **that is a task that was explored for the first time** in four centuries. Science **never went back to Investigate Kepler** after Newton included the work of Kepler into his work but when one read my work one will find Newton compromised Kepler's work as he did with Hook and many others. For my saying that about Newton did not correctly analyse Kepler, the Newtonian Academics Establishment at various University Institutions bluntly ignore my work in the past. Academics would not touch my work **notwithstanding the twenty-seven years of research that went into my work.**

This letter you are reading is my effort by which I hope to interest you in reading my Introducing letter ***A Cosmic Birth Dismissing Nothing I. S.B.N. 0-620–31609–8.***
 The book on offer has the title of ***A Cosmic Birth Dismissing Nothing I. S.B.N. 0-620–31609–8.***and is the actual letter I sent to various establishments.

What brings about the expanding?

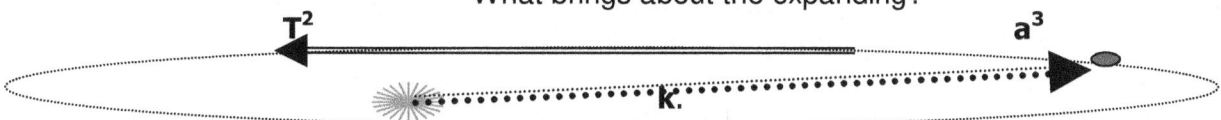

Kepler was the very first person to mathematically introduce space a^3 centre k and time T^2. Not only did he introduce space-time a^3 / T^2 but he also placed space a^3 and time T^2 in a relevancy long before Einstein did and placed gravity in space-time a^3 / T^2 even before Newton named gravity. Kepler was the person who placed gravity as the ingredient in the Universe that determines space a^3 and time T^2 and much more. Kepler was the first one that saw that gravity comprises of two factors being k or linear gravity and circular gravity or T^2 as gravity keeps space in form while all is staying together.
Since gravity also influences the space outside the sphere the space we call outer space has seven plus three points bringing about ten positions of gravity influencing space.

The influence inside the sphere also captures the space outside the sphere. Any line running through a sphere towards the centre of the sphere will reduce by half as the line progresses towards the centre. Because the material or volumetric continuance of the sphere the line will continue because at no point going down into the centre does the line sphere stop and therefore the line can never stop. Where there is a so circle there has to be a radius. The radius of zero would remove the circle and since the circle reduces but is never removed, the radius is never removed. The radius might become Πr^0 leaving only Π as a valid factor of form the circle forming the sphere does not remove in any way. One step further down the centre line would bring the reducing of the centre line to a point with the value of Π^0, and at that point singularity is in a natural position since $\Pi^0 = 1$. That must be the value of singularity because any factor to the value of 1 has the **dimension value of zero but** not the value of zero. It has a dimension with the dimensional value of zero, which by all standards is singularity. That is one of my main disagreements with the opinion that Mainstream Science has. If this argument is taken to its very roots, it changes the point where the cosmos initially started. If the Universe is nothing then that is where the cosmos started because the nothing could not end at any other point and affirm infinity. In the past the Academics charged with analysing my work never got to the point where I explain this fact because it never penetrated their realising ability that if the Universe held zero as a value in outer space then the Universe could not come from singularity but had to come from zero. They got stuck as if confused by my explicitness in my view on zero and they found that so limiting they apparently never could get passed that to read the rest where I explain my position on zero. Coming from zero means that there is no Universe since the zero would have taken the on coming Universe away before it could arrive at infinity.

Then we get to the issue of form differences. The cube is a loosely connected structure with the promise of being any form possible but the only precondition is that there must be at least six sides connecting. The six sides hold a relevancy or a responsibility to one another and provide a Universal accepted form maintaining the Universe. From the structure one can see gravity is not strongly present. All six sides support what ever are inside evenly form all sides. Why is it so important that the Universe is a sphere? It is because **only a sphere can accommodate gravity** because **only the sphere accommodates singularity** in its natural form, as I explained in the above paragraph.

The sphere is the form securing gravity since the sphere holds singularity Π^0 as well as Π which is another value to singularity, but that I explain in the book. In the centre of the sphere there is a point where space vanishes into singularity. At that point where space vanishes gravity is the strongest and that Einstein proved when he said that the Universe draw flat (singularity) where gravity is most. Only a sphere commands gravity because all cosmic objects hold the sphere as form with gravity inside singularity inside the sphere. That is the reason why one cannot find zero inside the sphere as Mainstream Science so arduously promotes. From the centre point where gravity is the strongest gravity hold the sphere true to form. At the edges of the sphere there are also point lining in 90^0 and 180^0 holding relevancy and responsibility to one another but the centre spot being the gravity point positions all the points in a location that the centre point allocate.

This means that in the cube at the point of contact between the cube and the sphere the cube experience such a contact point as if the "bottom falls out" of the cube and without a "bottom" to support objects they fall to the sphere as objects does fall to the Earth. Remember that a body "floats" in space, but at one specific point it starts to "fall" to the Earth. That is gravity and it is a dimension change much more than any force. I shall explain this last remark later on. That too is the Lagrangian system with five cosmic structures holding relevancy to the centre structure where the centre structure stands in for seven positions diverting from centre and the orbiting structures standing in for five positions in space.

Gravity is all to do with dimensional changing and reforming of forms to re-affirm alliances supporting centre. It is the reforming of space converting space to more concentrated heat.

The Universe is in the three dimensions using twelve dimensions that is visible to us and indefinite number of stages in size differences ranging from the immeasurable small to the immeasurable large where mathematics become a short fall to the next and the previous dimension.

Gravity is singularity as a factor forming space-time

Gravity is finding space-time

Gravity is proving space-time and aligning space-time with gravity

Gravity is the working principals behind all cosmic occurrences that pre dates the Big Bang period.

Gravity is the Roche limit.

Gravity is the Lagrangian system

Gravity is the Titius Bode law

Gravity is the Coanda effect

Gravity is the sound barrier

By being able to pin point prove what <u>Gravity is that enabled me to unravel the other entire phenomenon that forms gravity</u>. Each of the phenominon I mention above has one part or role in what is forming the totality that which we know as gravity.

Up to now **every one in science** is normally **acting as if gravity** is a commonly **explained factor,** which **every one knows** every aspect about **all principles that are involved in gravity** down to the smallest detail. In truth **no one in science** anywhere remotely **knows what brings gravity about** and **I used Kepler to unravel this mystery called gravity.** But no one in science will admit this fact about **Kepler being the one who formulised gravity decades before Newton came and gave gravity the name.** Newton did not underwrite or define gravity and even today the most informed in Science at best can only assert their suspicion on a rumour presumed about what causes gravity to perform as the part interlinking the cosmos but no one can go any further by explaining the concept. **Newton started this realising of gravity** but it had and still has no more substantial proof than a rumour has **and Newton admitted to it being a concept he could not explain. In Newton's ignoring to test Kepler's findings Newton missed the opportunity to find what gravity is. Since Newton every person in science also ignored Kepler and every one is guilty of missing the opportunity Kepler maid available.**

By my efforts of studying the implications that results from Kepler's finding I can now un- emphatically declare I know what gravity is. Gravity is the entire following locked into one compiling unit: Gravity is not being some magic force found between particles grabbing onto everything. I mathematically explained the following phenomenon:

Should you think this is rather a wild presumption I challenge you to spend a little more time and please think about what I say when you read about what I say in the next few pages. The first thing you should admit in private is what study did you personally so far made about the work of Kepler?

Still to this day nobody in science at present will denounce the principle of gravity as vaguely researched. Gravity has never been explained as a principle. Even when one is considering what the importance of gravity is gravity is never yet been understood, it is by now very clear that little if nothing of all objects is pulling closer in outer space. Comets are missing the Sun on a regular basis and no planet has come much closer towards the centre of the sun. There is an obvious balance in the cosmos that leaves no scope for the pulling that Newton promoted. Still everyone in science acts in a manner as if Newton's gravity idea is the best detailed proven fact and only occasionally does someone quietly admit that even Newton admitted not knowing what gravity is. No one ever come to the front and boldly state that gravity is just a rumour spread by scientist pretending to know all there is to know and knows little to nothing about what there is to know. When I do just that I am handled as if I was with the brain damage and with an infectious, contagious, catching and communicable brain disorder and best left alone at a respectable distance from society. However Newton said he had no idea what gravity was or what was the cause of gravity! Newton admitted that much when he introduced the name (not the concept). Mistakenly Newton corrupted the concept he named as gravity. Going according to what Newton introduced Newton's concept will by now have the moon much closer to the Earth than it was in the time of Kepler's studies, yet we know the moon is moving away instead of coming closer. By the same measure Kepler suggested that the space a^3 is content with the motion kT^2 as long as the motion T^2k is equal to what the space a^3 will allow. Kepler suggested motion of space remains in equilibrium as long as motion of space a^3 duplicated space a^3 by motion thereof T^2k. That is much more true than objects rushing towards one another by the pulling power of mass. Newton agreed that he could only declare gravity as a vague concept. This fact was at that time drowned by the man's stature and because of the man's position the statements were relieved from the manner of requiring the absolute burden of proof that in later Academic science became an absolute necessity. The proof that one would demand in the present milieu to day was never given to as a required demand in order to put Newton's rumour beyond doubt. Even at present when I bring the absolute proof that it is the Coanda principle that implements the Kepler formula $k = a^3 / T^2$ which translates to gravity being the motion (T^2) of space (a^3), my statement is trashed by Academics not having the will to understand what I try to say! When Newton announced a force he also admitted that the force could be anything. No one ever came after Newton and proved the fact or try to better the concept. That still underlines the fact that the force to this day can be anything. Not once could one person in the past or present provide the lack even in the present day on substantiating proof about what causes gravity as a reality by defining the very principles thereof. That includes every one since Newton as well as including Einstein and even Hawking being unwilling to take the challenge and either prove or deny gravity. Scientists can declare gravity was a factor at 10^{-43} seconds after the Big Bang but what brought gravity about or why did gravity became a presence or still remained a presence where it is still treated being tightly concealed information which all are speculating on. Even in the best and most informed circles and amongst the most educated there is no one that knows what gravity is because they all ignored Kepler and for them to ignore Kepler the price they pay is not finding the principles bringing about gravity. Using Kepler even makes the method to follow and understand Einstein's discoveries shockingly simple.

By my applying Kepler I can define gravity precisely to the point where I now can explain why the proton and the electron forms a mass difference of **1836 times**. The mass difference between the electron and the proton being the bordering edges of the atom and between the two sub atomic particles within the atom there is a difference in mass being 1836.12 times. This achievement is obtainable only when one reads into Kepler and find what Kepler did not say but meant to say when he said $a^3 = T^2k$. Kepler said it so nicely and simple, yet every body including Newton missed it because everyone is waiting for this mind-blowing discovery lurking in the subatomic structures, which is doing all the magic sucking of matter and is named the Graviton even before it's discovery. I use Kepler to explain why this aw provoking particle went undiscovered and the main reason for us not discovering it thus far is because Kepler said it is not there! I can explain why the strong forces are to the power of forty times more massive than the weak forces. Just by my studying of Kepler this became possible.

The one I offer is an effort standing outside the unit and has the book title *__A Cosmic Birth Dismissing Nothing I. S.B.N. 0-620–31609–8,__* In the letter it aims to bring a link that connects my theory to Mainstream Science without hammering out further delving into other theories, which became obsolete. In my other books I aim to supporting evidence that conclude my theory therefore this *__A Cosmic Birth Dismissing Nothing I. S.B.N. 0-620–31609–8.__* is the simplest of all of my work. All one need to appreciate the information in this book is the understanding of the mathematical expression introduced by Kepler as $a^3 = T^2 k$ and what on Earth can be simpler than that! T^2 is motion, a^3 is volumetric space = is equal and **k** is distance from a centre and this simple explanation serves the whole book.

From dissecting the formula I prove that:

I prove gravity is strongest where space is least (not where the Universe goes flat as Einstein promoted). I prove there are relevancies that are all applying equally and without such relevancies in balance there are no gravity

In the relevancies there are opposing motion where each participant provide a motion contradicting the other relevant motion. I follow on what Kepler introduced when he introduced the fact that space is half of the motion and the motion is the other half of space. The space a^3 is **= half the motion T^2 k of our six sided Universe and the motion T^2 k is the other half of space a^3.**

Therefore space cannot be if not in motion and motion is only about when it applies to individual space moving in relation to other space T^2 <k or T^2 >k. That relation is the time factor and time depends on space moving from one situation to another complexity of a situation. To shortly condense my view I would explain my view as follows: In gravity there are two opposing quantities each representing singularity within one unit. The motion of the one is getting away and the motion of the other is reeling in the one running away. From the view the onlooker has it may seem as if some rabbit is pulling some dog all over the area that is remaining in one place where the dog holds the running rabbit in one area circling around the dog. This has another angle where as from the dogs point of view the dog will love to dismiss all space and capture the rabbit.

The rabbit on the other hand would love to leave the dog at a distance where the rabbit will never again see the likes of the dog. While the space between the two is a merely a common fact it unifies their differences. Both in relevancies have to appreciate their differences by the space in the unit that is keeping them apart. The space is the factor that has to resolve the issue being the differences in motion but cannot because different relevancies sustain equilibriums. I prove that as much as there is Newton's pulling there is Kepler's running around and the running around is equilibrium of the other factor providing the running away part.

The angle science is looking at the issue science either dismisses or cannot explain the characteristics or principals, which is there none the less. The explaining of the phenomenon is quite impossible when using the pulling rope magical attachment idea in the manner science try to explain gravity. Therefore instead of dismissing the rope they dismiss all other factors present by gravity unleashing free motion but they would not release their idea about the rope. Gravity is motion between two particles that brings about mass. In the book I explain this in much detail but frankly there is not enough room to explain this in this introducing letter.

Mainstream science knows about the fact that gravity has never been defined, the Bode principal that is there in all the planets and even the fragmented planet, the Roche limit, the Coanda affect, the Lagrangian system, the sound barrier but cannot explain any of the phenomena, although the presence of these phenomena is without dispute. Yet being without dispute does not stop Newtonian science from dismissing the phenomena as coincidental. It is the explanations about what causes the phenomena that is part of the dispute but in science the way by the manner in which they defend Newton, science would rather discount the obvious phenomena that question the legality of Newton's cosmic views. I only dispute Newton as far as his cosmic principles are inclined and not as a part of general physics. In general physics the man has not got one iota wrong but in cosmology the man is way off the mark. But in order to defend Newton being obviously wrong in every dimension about cosmology Newtonian science would rather have the phenomena being there or not becoming disputed. Science fail to give acceptable explaining of such occurrences we see in the phenomena and therefore disputes the validity of the phenomena and this failing to explain the presence becomes disputing the presence thereof. I on the other hand found a way where these explaining of the phenomena took me past the Big Bang era and introduced me to the start of all starts. Science cannot get past one specific date because they do not accept or understand the phenomena, which I prove that it is the interlinking of these phenomena that started the Universe.

In such a light Scientists must somehow realise they are barking up the wrong tree with the information they have to use to do some explaining. They cannot refuse the phenomena and not realise they must have the cat by the tail as far as cosmology goes. I state once more to please remember that with this I am referring to cosmology and not general physics. There is an Earth versus a Universe of difference between the two concepts but Newtonians fail to see that because Newtonians cannot appreciate the differences thus by they're not able to understand cosmic gravity they go about blurring the understanding of gravity. If there are that many phenomena (it represents all there is in cosmology) to explain and such little ability to explain (science fail to explain even one) by using the information Mainstream science is using to explain the cosmos, then someone somewhere has to realise there is something drastically wrong in the way they present the knowledge they claim to have. One cannot be serious about science but defend your view by dismissing the validity of all unknown indicating factors presented as such. There then is some gross incorrectness in the way one reason. The Roche limit is there and no denouncing thereof can remove it from the cosmos. They may refer to evidence received from the Hubble telescope as "the star is blowing bubbles" for the lack of explaining what is occurring but occur it does. One cannot say it is some unknown gesture presented on occasions because by not explaining the pictures that presents certain foolishness. It leads to tragedies in aviation and the tragedies they are incapable to understand or explain. For fifty years they lost many pilots but still has no idea what brings the sound barrier about, or find the link gravity holds in the process we call the sound barrier. Instead they try to interpret some effect established almost two centuries ago with steam trains that is travelling at the same speed that a horse can run. No further investigation with the science in hand brought them closer to new facts! It should be a sign telling them they are going about incorrectly but it does not because tell them that because Newton did not say so. When I first came upon the amount and the totality of the unknown quantities in cosmology as well as the complacency those involved has about such unknown factors being discarded, it stirred a sense of disbelief and I decided to respond.

All principles I use in the theory I introduce with the publishing of this book. All principles I apply are part of nature. I base my theory on heat stabilizing through space using motion to produce cooling. That is gravity.

I believe some of Creation remained as some particles formed by applying gravity in motion and the lack of motion in others became the lack of gravity, which inspired overheating which then formed plasma. Plasma is the result of heat, which is also presents the fact that light is the epitome of heat. How light became plasma is rather obvious, which again I believe (within reason) I do prove. I believe heat is the destructed form of material and this information the atomic thermo explosions give us.

In the book I present the analysing of **Kepler's formula without Newton interrupting** Kepler's work. Please let me explain: Tycho Brahe and later Kepler made a study of outer space as never repeated afterwards. From this Kepler concluded that $a^3 = T^2 k$. We all know that a^3 is space and with the space indicated as being in the third dimension and the third dimension is unmistakably a cube that forms volume, which by definition is presenting space. We also know from the way calculations come about by using the formula of Kepler that T^2 is the duration of a specific period of time relating to a specific centre. On the one hand we have space a^3 and on the other hand in direct relation to the space Kepler introduced motion coming from a centre that forms time $T^2 k$. Kepler gave us space-time a^3 / T^2 centuries before Einstein gave the concept a name but no one ever took any notice. In the formula is space a^3. In the formula the space a^3 has direct relation to time T^2 If k is a^3 / T^2 it means that from the centre holding the gravity is space-time. Space is a^3 and the motion of space a^3 we accept as time $T^2 k$ and such accepting is part of our understanding for the past three hundred and fifty years. Kepler gave us gravity before Newton named it as a force. Kepler gave us space-time long before Einstein named the notion. With Newton's meddling he missed Kepler introducing gravity as $k=a^3/T^2$ space / time.

I believe that I achieved an all time breakthrough success because I can now explain what gravity is. Remember that not even Newton could explain what gravity is or where it comes from, but Kepler did that without any person ever noticing. Scientist over the years paid the price of ignorance about gravity by their unwillingness to investigate the father of gravity, which is coincidently not Newton but Kepler. From such explaining what Kepler said without Newton changing formulas on Kepler's behalf I prove the Titius Bode principal also known just as the Bode principle. I prove that the Bode principle forms gravity when using the Roche limit. These phenomena was never before explained or understood by Mainstream Science although they appear more than regularly in the cosmos. In the same breath I might add that Kepler also was never investigated.

As presumptuous as it may be on my part of trying to disprove Mainstream Physics, such a presumptions does not change the truth about Mainstream science being incorrect about gravity. After all they admit they do not know what gravity is. I am not disproving anything because they agree they do not know, which paves the way for my showing what gravity is. By admitting not knowing what gravity is they then also admit there is a chance that they can be incorrect about gravity but unfortunately mainstream physics do not see it that way (yet). The question in hand is finding what role gravity played when the Creation came about for the first time. I had to find a method that would allow me to explain why gravity played a role.

The theme of the book is about the Universe not coming from nothing and therefore outer space cannot hold nothing. By taking Kepler's $k = a^3 / T^2$ and using k as a line I show through using the line as an example that the cosmic Universe holds everything and all concepts. However the only thing it does not hold is also the only aspect not present in the Universe at all. That is the value of nothing or zero in as much as carrying the definition of the absolute absence of any value. This means the Universe is filled to the point it is overflowing which we call the Hubble constant and not there is no room to be empty. With the line that light uses to flow the lines eliminates any such a possibility of nothing being present. Mathematics is a means of communication about matters concerning the cosmos. As an intercultural language spanning across race and ethnicity or as a principle as such mathematics cannot have zero because mathematics indicating lines, which is about not applying the numerical number or value of nothing. Everything came about from singularity and Einstein proved that. From singularity nothing never had the chance to enter space. I challenge any person that disagrees with this statement to show mathematically where nothing as a factor ever entered the mathematics of the Universe. If there is any one there believing there is nothing in outer space I challenge that person to prove mathematically where nothing is a factor in the cosmos. Your attempt may either be before or after reading my work but my challenge will stand since mathematically nothing cannot be part of mathematics. Multiply whatever with zero or nothing and such multiplying results in nothing where nothing is then can establish no multiplication. Kepler gave us the relation between cosmic objects as $k = a^3 / T^2$. From the formula k forms a connecting straight line filling the first dimension and not the single dimension because k in the single dimension is not zero. It is unproven how k can backtrack to become $k = 0$. I deliberately press this point and make it an issue because that removes all the theory of mainstream science from any logical base they have in support of their views that space is, holds and comprises of nothing.

By using Newton one cannot even begin to explain any one of or the combined effort of the above cosmic phenomenon that are all over the cosmos and forms all the laws in the cosmos. Newtonian definition cannot even recognise any of the principles but only Newtonian science are thought to students. No student can have the fortune to disagree with Newton and remain a student. If the student will dare to disagree with Newton it is the end of such a students academic career. By setting this firm condition Newtonian science becomes institutionalised mind conditioning of the concepts of thought forming in physics. With my saying this I have not made one academic friend but neither have any one proved me wrong. Students are taught to accept Newton and to ignore Kepler and any student doing it the other way around will fail all examinations and other testing at Universities. Students accept Newton or they accept a ticket taking them home. Newton is an institution force fed to each following generation but saying that reserves only resentment towards me amongst Academics. According to Newtonian science space is simply nothing with no qualities but gravity separate space and space does not mingle, as one would expect if space was nothing because space does form borders. Disasters of unprecedented magnitude arise from such borders. The Challenger disaster of February 2003 is much testimony to those borders that was powerful enough to break the aircraft into pieces while the explaining contributed by Mainstream science is evidence of a shocking lack of understanding about what took place as cosmic laws were breached. I do not pretend to be of superior understanding and do not place myself on any pedestal. On the contrary the information is so simple and so easy to understand that the lack of any Academic understanding frustrates me almost witless. But academic taught culture demands all persons to miss the evidence, which is so clearly visible because academics demand researchers looking in other directions because students are forced to accept Newton's vision about Kepler's work. By the time they reach researchers status they too have tunnel vision that can only acknowledge Newton and ignore Kepler. Our not understanding laws provide a platform for future disasters occurring because it will lead to us ignoring more of applying principals that leads too space tragedies of magnitudes we have not thought of as yet. By not understanding the sound barrier tragedies have and will again come about and will increase as misconceptions become more present in the future because demand on space travel increases.

The book, **_A Cosmic Birth Dismissing Nothing I. S.B.N. 0-620–31609–8_** is about that process adapted by the Big Bang, never ended and it is still bringing over, that which is in unoccupied space to material being in occupied space. Occupied space holds matter and unoccupied space is empty of solid materials. There is contraction, which we know by the name we gave as gravity. Then there is expansion, which we gave many names being the Big Bang and the Hubble Constant or better known as simply exploding or forming plasma with all the terminology accompanying that simple idea. This I show is antigravity. Apply heat and space and a balloon lift where such lifting is antigravity. There are a balance in the Universe where gravity contracts and reforms space and heat expand becoming space and produces space.

The formula to calculate a circle, where a sphere is a compliment of multiple circles, requires the square of the radius in a square multiplied by Π. That linking r to r^2 and this allow the circles form value Π to stand directly related to the square of the radius Πr^2.

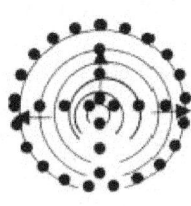

By reducing the radius in half, the size of the sphere would reduce. The reducing if the sphere must continue while the radius keeps halving every time. A point will arrive where r mathematically cannot reduce any more. The radius factor at that point then has to be in infinity $r^0 = 1^0$ since it cannot reduce more than it has reduced.

In order to build a circle we have to increase the radius. If the Universe was the size of a neutron at one stage and the Universe grew into what we now see, then the Universe extended r quite somewhat to get to where we now are. To go back in time we have to reduce the circle by reducing the radius. Keeping these factors in mind it is clear that Π are the forming the factor taken on by form and r^2 produces size. However Π is connected in form to r^2 but is only connected.

Reducing r will reduce the circle but it will not remove the circle. However if the mathematical proposition is correct and the circle starts with zero, then *$\Pi r^2 = \Pi 0^2 = 0$ leaving no circle to grow.* With that principle in mind it would be impossible to find zero in the centre of a circle. What we must find there is $\Pi^0 r^0 = 1^0 =$ singularity.

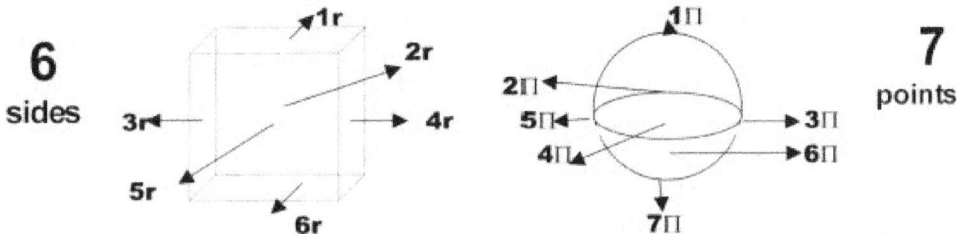

6 sides

7 points

Where space comes into contact with the sphere the cube loses one of the six dimensions it has to the more dominating seven dimension of the sphere whereby the seven dimension in equilibrium will dominate the six dimension loosely connected by r bringing about that the cube then has 5 sides to the seven of the cube. This means that in the cube the "bottom falls out" and without a "bottom" to support objects they fall to earth. Remember that a body "floats" in space, but at one specific point it starts to "fall" to the earth. That is gravity and it is a dimension change much more than any force. I shall explain this last remark later on.

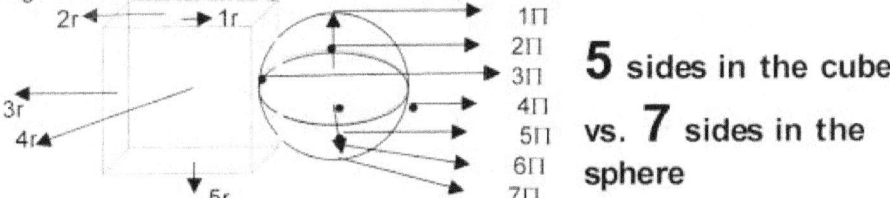

5 sides in the cube

vs. 7 sides in the sphere

What this says is that the gravity influence does not end on Earth. Although specific borders are in place in the atmosphere at various levels, the beginning and the end of such borders are in place in the atmosphere where the beginning and the end of such borders are as definite but also as vague as the gravity that forms them.

Kepler said
$a^3 = T^2 k$
but that could also be
$k = a^3/T^2$

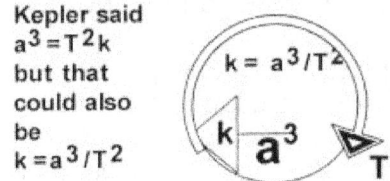

When translating Kepler's mathematical expression into a verbally spoken form of communication such as English we can see what Kepler said also read as $k = a^3/T^2$ where k is one point from a centre point that is space a^3 relating to time T^2. From a centre comes space-time

I believe that I achieved an all time breakthrough success because I can now explain what gravity is. Remember that not even Newton could explain gravity, but Kepler did without any person noticing even to this day four hundred years on. It was only that at the time when Kepler introduced the gravity he uncovered the world was not ready to realise what he uncovered. It is gravity keeping the sun and planets in rotation and that is what Kepler formulised. From such explaining I prove the Titius Bode principle also known just as the Bode principle and that gravity comes from 10 /7 and 7 /10 in relation to singularity laws to hold a very specific relative value in relation to various different structures in the cosmos as they relate to gravity specifics. This is a law found in all planets and fragmented pieces circling about the sun, but even when all the planets equally adhere to the law,

Mainstream science choose to push it off the table as inconsequential since Mainstream cosmology has no way of explaining the law by implementing Newton's physics. When I implement Kepler's cosmology, cosmology becomes ordinary simple to understand. I do prove the Roche limit from the point I traced Kepler's indicating where singularity is located. My achievements came from my effort where I separated Kepler's work from the opinion that Newton formed about Kepler's work. Behind Kepler's work hides more information and answers to unanswered questions than there are questions answered by Mainstream Science at present. Mainstream science fail to explain even one gravitational concept but I managed to formulate, define and explain all the ten gravity concepts that is a derogative of gravity. I found it to be possible through the analysing of Kepler's work and the uncovering of the singularity value. For instance from Kepler's work I can explain the operation of the Black Hole. In my opinion my explaining of gravity makes much more sense than the accepted force of Dark Age proportions, which even Newton admitted he could not explain. From my view a force is just motion applying and that is what Kepler said gravity is. Kepler said that gravity binding structures is $a^3 = T^2 k$. But singularity is hiding where Newton placed nothing and Kepler positioned the pointing of **k**! In that I followed the path to uncover singularity. I dissected **k** as a factor in the Coanda effect and found that the Coanda effect is proof of my view about gravity and singularity produces the Coanda effect, which dictates all implications in the Coanda effect. The Coanda effect is the establishing of **individual space a^3** by **applying motion T^2k** on the both sides of **singularity k^0**. Where the Coanda effect is producing gravity and such producing is stronger in a small space than the gravity produced by the Earth in that spot there was some manner in which it had to play a part during the period of the Big Bang. As it also is in the case of the Total Internal reflection establishing singularity by applying motion to liquid produces the phenomena. By applying mind boggling mathematics solves one part of the problem where the other part is the use of Mathematics as expressions that must be translated to the verbal in the manner that it says and not to please the mindset of the mathematical translator. Newton should have seen that as the most impressive mathematician yet...but he didn't... Let us investigate what Kepler introduced and see if it could simply be integrated into a "bigger" concept by diminishing the Kepler concept. Remember Newton had the attitude that Kepler's work was incomplete and therefore inferior because he added to Kepler's introduction formula what Kepler did not introduce and therefore Kepler saw no use for that which Newton found fit to introduce. It is either that or Newton saw Kepler as being completely inferior in intelligence. Still Newton saw Kepler's $a^3 = T^2k$ as incomplete and therefore he had the urge to correct what he saw to be incorrect to be $4\pi^2 a^3 / T^2 = G(m + m_p)$. In this, the value of *T* and "a" are the period of revolution and semi major axis of the orbit of a planet of mass m_p about the Sun of mass *m*, and *G* is the gravitational constant. Newton added nothing but duplicated everything in his suggesting completing what he saw as incomplete. Can the symbol "*a*" be reckoned as a period of revolution as Newton suggested and therefore Mainstream science still suggests? **Is the work of Kepler incomplete?**

a^3 The fact that any symbol uses a value to the third power indicates space or a volumetric established and separate unit using the six dimensions allocated to space. It is space because it is volume using the third dimension. There is no other valid interpretation or translation allowing a correct translating from mathematics to English but by categorising that space as a volumetric separate identity. It is cubic which is space. One measure a fridge or a stove or a room by the cubic measure without bringing in pi because of the volumetric content there is by using the third dimension. The fact that there is a line connecting the space and enforcing a rotating motion around a specific centre connects whatever volumetric measure the space is in the cube and the cube as a separate issue independent from the space in which the independent cubic space rotates, which brings about the circle and yet there is no need to implicate pi because the rotation brings along the circle after the cyclic completion of the rotation by the independent cubic space.

T^2 Is an indication of motion, the moving of an independent space that is holding **a^3**, where the space in motion will be measured as **a^3** as the space **a^3** that is occupying separate unit as space within other space that is travelling by using the second dimension of motion in **T^2** moving the independent space from one point to another point or following a flat distance between two points and the distance and only the distance is **T^2** where the distance is established as **T^2** by the space **a^3** but is not the space **a^3** and as the independent space **a^3** the space is going from point **T** to point **T** making the distance **a^3** travelled **T^2**. It is space filled with material consisting to allow independent space within the surrounding of an enveloping space of bigger proportions **a^3** being in motion **T^2** and that motion **T^2**

is the filled space a^3 that is taking time in the second dimension moving the space a^3 in question from one point of choice to another point of choice by which time will be established.

k^1 Is the symbol used to indicate a straight line between two points with a definite beginning and a specific end position. This indication of a distance is an indication of a bigger space that is big enough to include a smaller and separate space a^3 within the bigger space running all the way from the start of k to the end of the line of k. One has to see k for what it represents because k indicate the presence of a larger space that is large enough to allow the smaller space to rotate all the way as the circle runs from T to T in the full diameter of k. Kepler introduced this absolute basic mathematical principle. It is positioning the independent space in motion in a specific relation to a controlling dynamic situated in a dominating and controlling centre. It is indicating that the space in question is in motion acknowledging the centre in control of the motion and therefore in control of the space location. It proves a larger space is holding the space in question as part of the larger space where the larger space is in ratio so much larger that the space in question can effortlessly go in motion and to full rotating motion be concluding a rotation within the larger space that k indicate. But k also must indicate therefore coming from a very small start centralising the motion of the space in question using the space that k produce from where that specific space is controlled in the realms of the larger space. Nobody before saw it in as simplistic manner as I just showed it to be and yet in the simplicity is the sensibility of it all…

The Two masters did not share even the smallest part of any concept because the two does not even find any mathematical connecting.

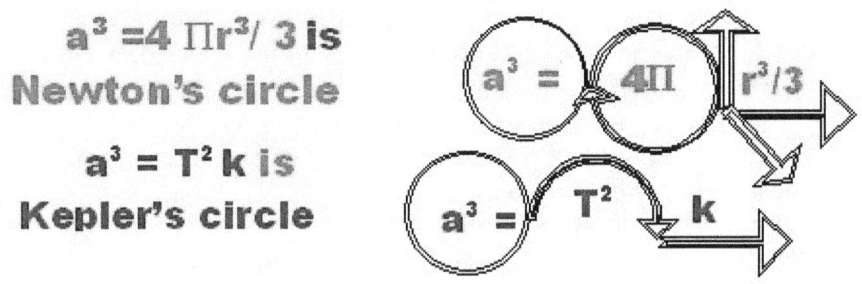

$a^3 = 4\Pi r^3 / 3$ is **Newton's circle**

$a^3 = T^2 k$ is **Kepler's circle**

When $a^3 = T^2 k$ then $a^3 / (T^2 k) = 1$ and k^0 also equals one therefore mathematically it is very correct to say that singularity is presented amongst many other values to the power of zero as $k^0 = 1$ proving that $k^0 = a^3 / (T^2 k)$

Newton said a sphere is $a^3 = 4/3 \, \Pi \, r^3$

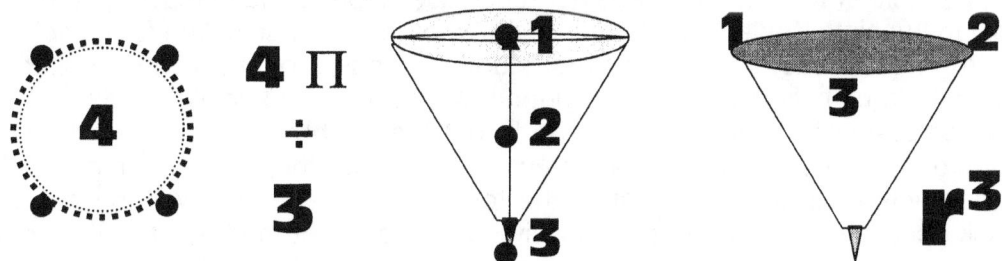

Kepler said something totally different. Kepler said firstly that in space a^3 there are two types of lines connecting space $T^2 k$ where the one is a straight line and the other is a bended line that eventually will complete the circle Those that so dearly believe mathematics should see this clearly

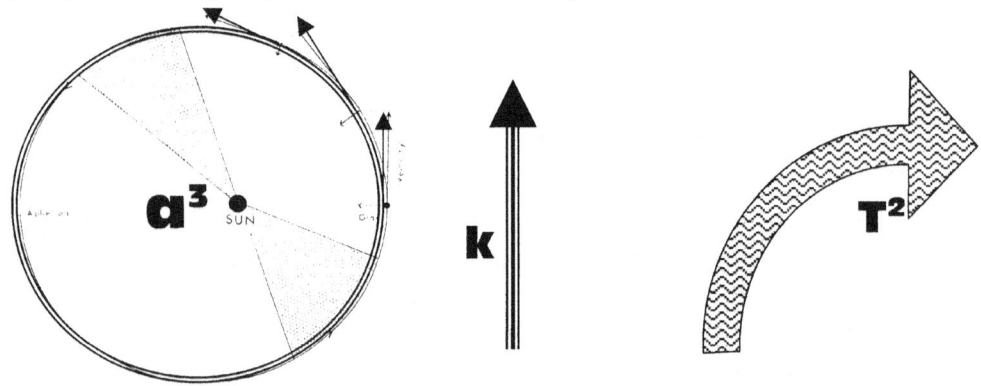

There are always in space tow forms of gravity, the one contracting as well as the one expanding and the compromise.

What formed the grounds for any need by Newton to change Kepler's translations from what Kepler translated the cosmic given to mathematics and then from mathematics to English? We have to acknowledge one fact without any possibility of disputing the fact...Kepler received his acquired information spanning two life times of work directly from the cosmos as the cosmos used the language of mathematics to explain too Kepler what Kepler was suppose to see. If the cosmos saw it fit not to include pi then one should not try and correct the cosmos because for some very obvious reason the cosmos would not supply incorrect information. It will be better to rather seek why the cosmos saw it fit to exclude pi...and that is just what I have done. The space-time that the cosmos introduced was so brilliant it took the likes of a genius such as Einstein to realise the presence thereof many hundreds of years later. Those who are in agreement with Newton's changes should ask themselves what part of Kepler's $a^3 = T^2k$ did I not translate correctly in the above explaining when I interpreted the expressed by taking the mathematical expressed to the verbal English? When I used Kepler's mathematics by my translating Kepler's work correctly I came upon answers not yet uncovered by Mainstream Science. But I had to dissect Kepler's mathematics in the manner as it was told by the cosmos in person to Kepler without Newton being arrogant enough to tell the cosmos what the cosmos was suppose to tell Kepler about the cosmos. Kepler gave the World mathematic translated cosmic answers that Kepler uncovered long before Newton, Einstein and others got wise about cosmology... Such is the advantage of recollecting Kepler facts that it does answer many questions, which went unnoticed and therefore not spoken about up to now and some were previously never even thought about. Mainstream Science never previously thought that through any examination of Kepler's work such scrutinizing would uncover these facts that I present. Subsequently Mainstream Science elected not to ask the correct questions and in the process Mainstream Science never found the correct answers. By not asking the questions Mainstream Science could not decipher any of the decoded mathematical messages, which Kepler received directly as a mathematical message spoken by the Universe and coming from the cosmos. We all know and appreciate that mathematics is just another language and the professional mathematicians have is responsibly translating mathematics to a verbally competent language. They never attempted beyond Newton's efforts to neither read into Kepler's nor recognise Kepler's mathematical translation and thereby was unable to translate Kepler's mathematics to the other communication forms being all verbally spoken dialects. In other cases human natural study methods brought along a cultural of Academics forcing students to comply whereby the students will accept the knowledge through our inherited past which when tested by modern standards is not that highly proven. We accept the answers as questions already fully answered because culture demands the accepting thereof. Any one insisting that Newton acted correctly by initiating the four pi in the square...well those should especially read my book because they are in for a massive surprise about mathematics they never encountered before!

I know science is under the impression that particles grab each other as they move closer to each other, but as presumptuous as it may be on my part of trying to disprove Mainstream Physics, such a presuming does not change the truth about Mainstream science being incorrect. The question I had in hand was finding what role there was for gravity to play in the Creating process and then I had to find a method that would allow me to explain why it played a role even at the very start of Creation. Remember our view on gravity is one of contracting. With the contracting or bring together of material it cannot support a Big Bang and Creation was all about expanding and pushing material further away. In the book I present the analysing of **Kepler's formulating of the solar system without Newton's interrupting** of Kepler's work. Please let me explain what I refer too: Two of the world's most outstanding researchers dedicated their lives in researching the figures of how the planets revolve around the Sun. This was an effort in dedication not yet seen afterwards. Tycho Brahe and later Kepler made a study of outer space as never repeated afterwards. Kepler said $a^3 = T^2 k$. We all know that a^3 is space and with the space indicated as being in the third dimension and forming a six-sided space the third dimension is unmistakably an indication that it is a volume, which by definition is presenting space. We also know from the way calculations come about by using the formula of Kepler that T^2 is the duration of a specific period of time relating to a specific centre, which is claimed by the sun. On the one hand we have space a^3 and on the other hand in direct relation to the space Kepler introduced motion coming from a centre that forms time $T^2 k$. Kepler gave us space-time a^3 / T^2 but no one ever took any notice. In the formula is space a^3. In the formula the space a^3 has direct relation to time T^2 If $k = a^3 / T^2$ it means that from the centre holding the gravity is space-time. Space is a^3 and the motion of space a^3 we accept as time $T^2 k$ for the past three hundred and fifty years. Kepler also gave us so much more that no one noticed. Kepler gave us gravity before Newton named it as a force. Kepler gave us space-time long before Einstein named the notion. It was Einstein that calculated and then declared that gravity is at its strongest where space disappears and from that the world concluded that the Universe is drawing flat. Kepler said it so simple.

Due to the natural form of a circle there is a point in the very centre where space has to relinquish the position because of the form the circle has there is no more space to form space. All sphere are a multitude of circles duplicating as the y acknowledge one such centre therefore the sphere is the ultimate circle. In a circle in the centre there is a point where space disappear, where there are no more room for space to be and since all orbiting cosmic objects form a circle all cosmic objects run from such a space less centre space to be and from such definite unseen location the circle extends space-time that control all space within that centre. If only Einstein referred to Kepler he would have seen that in the very centre where k^0 is $k^0 = a^3 / T^2 k$ which means at that point where space disappear gravity control all space in motion. Where space vanishes gravity is strongest. Gravity is strongest where space is least therefore gravity is about removing space to establish an ultimate point of strength. There is no special sub atomic "graviton" sucking and spitting as it creates gravity. It is about motion and the form material takes on but although it may at first sound simple, it is a lot more complicated than such simplistic explaining may suggest at first. Gravity is about reducing space in motion by duplicating space with motion. Gravity is the motion T^2 of a solid a^3 through a fluid k or also a less dense liquid and the amount of liquid (plasma) heat in the space forms the density or supplies the liquid state of the space. The space is the second part of the formula $a^3 = T^2 k$ because it is solidity a^3 in relation to moving ability T^2 in space k that allows motion. Motion can only come from two possibilities where one is heating producing space and cooling reducing space. Other than that there is not.

Another important theme is the book carries and something, which I already mentioned is about the universe not coming from nothing and therefore outer space, cannot hold "nothing". By taking Kepler's $k = a^3 / T^2$ and using k as a line I show through using the line as an example that the cosmic Universe holds everything and all concepts. However the only thing it does not hold is also the only aspect not present in the Universe at all. That is the value of nothing or zero. Explain to yourself how it was possible to create nothing in the Universe! In as much as carrying the definition of the absolute absence of any value "nothing" in that case cannot be present because the line that light uses to flow eliminates any such a possibility. Mathematics is a means of communication about matters concerning the cosmos and as an intercultural language spanning across race and ethnicity or as a principle as such cannot have zero because mathematics indicating lines, which is about not applying the numerical number or value of nothing. That much I prove physically. Where there is any person that disagrees with this statement I challenge such a person to show mathematically where nothing in mathematical calculation as a factor in the cosmos can come to conclude a value in total other than nothing and where there was a chance for nothing ever to enter the mathematics of the universe. If you put the starting point of a line at nothing you remove the line as an option of being something in the Universe and ultimately destroy the chance of any line being in the Universe. The line could have stared from singularity where singularity holds the symbol of what ever to the power of zero but that then remains as a possibility to grow from a exponential value of zero which does not remove the single object and then the exponential zero being one will multiply by number as an established factor present in the Universe. But removing the line by replacing the line with zero will disallow any line ever forming in the Universe. You may either attempt to do it before or after reading my work but my challenge will stand since mathematically nothing replaced ether when the concept of ether was removed from space. The concept ether was removed and replaced by a concept of giving nothing a permanent value of one. Then afterwards some parties as part of Mainstream Science brilliantly allowed the nothing they produced to replace ether ended up replacing ether as an accepted concept. To the nothing they attached a value as a concept of something able to carry a value and were able to allow the nothing-value they instituted as a concept to replace ether. Such replacing was only an idea introduced later and was never proven mathematically. Kepler gave us the relation between cosmic objects as $k = a^3 / T^2$. From the formula k forms a connecting straight line filling the first dimension and not the single dimension. Ask yourself the following: does Pluto hold more nothing between it and the sun since it is further way from the sun and with nothing being between it and the sun or does Mercury have more nothing between it an the sun since Mercury is closest to the sun and being closer it should have more nothing between it and the sun. If Mercury had more nothing between it and the sun would Mercury not then be located inside the sun at the very centre because there it should have most nothing. After all it is in the centre we find the sphere have the most nothing. The shorter the line will confirm most nothing and not the longer the line must confirm most nothing. This is a major point of review. By reducing space a line represent one can reach an ultimate reduction indicating a point where the reducing of any or all lines that form the universe, then by such reducing can only confirm singularity, which was my first breakthrough. My realising that nothing has no part in mathematics sounds degrading simple but it unlocks the birth of the cosmos. Arithmetic uses nothing because it works on numbers and quantities, and in that sense there can be or there cannot be any specific number including zero. But the Universe is overflowing with everything and that only excludes nothing since "nothing" as a concept or as a value cannot bring about expanding or overflowing. Try to get any academic to admit to this concept and you'll find out what a task that is! It has the same degree of difficulty that showing to Academics that according to Newton all comets must crash into the sun whereas they do not crash ever. Newton said with the formula to gravity applying $F=G (Mxm)/r^2$ will completely remove the radius between the two masses keeping the sun and the oncoming comet apart and then the comet has to collide with the sun, but that never happens and not one Academic thus far in my presence would admit to that!

Gravity is precisely what Kepler showed gravity is. In Kepler's formula k equals space-time or a^3 / T^2. Gravity is singularity extending and forming space a^3 through gravity rotating space $T^2 k$. Gravity is not particles pulling one another in a tug of war. Gravity is about reducing space and maintaining different cosmic sides not sharing the same sort of space. In the beginning before and after the Big Bang (yes before because there had to be a before) gravity was about bringing across heat that was in space to material that was in another space. The only difference is that in the space where the space is filled with plasma or heat and not material, the heat was much denser then than what it is now. The gravity allowed the one part of the Universe to remain in form while antigravity causing plasma or heat to deform the form of the structure it had and that other part of the Universe became liquid turning to space.

I explain how matter is claiming heat from space through gravity. The claiming of space by gravity comes about by the implementing of the Titius Bode Principal of seven dimensions interacting with ten dimensions and produce the square root of space, which I show to be a major part of gravity. Gravity is not about material pulling and tugging on other material. There cannot be antimatter that went missing. In the cosmos every aspect that was present when the cosmos formed is still very much part of the cosmos. Mainstream science claims that many aspects of material (matter and antimatter) and singularity were present in the cosmos during the early phases but has vanished since. Think about it clearly…What ever was in the Universe had nowhere to go but to remain in the Universe after the Big Bang. There is no place else to go! There is no chance of entering, later escaping and then re-entering again. The Universe is everything there can be. That means if singularity was part of the cosmos during the Big Bang singularity must still be in the cosmos. If antimatter was present during the Big Bang it must still be present. There is nowhere to go. Space was little and heat was massive. Heat became space as the density of heat turned to form space through a process we named exploding. All the proof of this becomes possible when reducing the straight line to a point holding singularity at the very end of the line. Reducing the point to where the line start the Kepler formula **k** becomes the extension of singularity and through extending singularity **k** applies and commands space a^3 and time T^2 as space- time where the formula then reads the mathematical equivalent of what Einstein named by using a verbal expression, which is space-time. **k** = a^3 / T^2. I locate, spot and place a value on singularity where singularity is relevant to space-time forming. In the book **an open letter To Selected Academics** **ISBN 0-9584410-9-X** it only and exclusively deals just with the fundamental basics of my theory. The cosmos has lines forming cubes and lines forming circles, which in 3D manifests as spheres. Between the circles and the cubes runs lines so the key to understanding the Universe are lines. The Big Bang was a time when the Universe was incredibly small making the running between particles that connected space with lines small. Understanding the Universe is taking the line back to its limits where such limits were during and pre-dating the Big Bang. You can reduce the Universe to where all fitted into a subatomic particle by applying maths but behind the mathematical reducing was the reducing of the lines that formed space and particle in space. By reducing the line to where the line will not reduce any further at that point all points land on the same spot. All sides are on the very same side because of the singularity aspect. The spots all share one position because that is the only position there is to hold. That is singularity being one to all but it is not zero. Finding form in that point shared by all will give a value of singularity. Extend that value received to a Universal centre and bring that value to align with Kepler's $a^3 = kT^2$ and the universe with the entire different yet unexplained phenomenon becomes as easy as children schoolwork. There are suddenly no more mysteries in the Universe. Applying the new value to match the factors brought about by Kepler I managed to prove how the Roche works and what role the Roche limit played in supporting the Bode rule. I also show gravity comes from the Bode law applying a relation of ten dividing seven and seven dividing ten. It is only because gravity reduces the space between particles and not being some magic force found between particles grabbing all that the following phenomenon's are mathematically and principally explained:

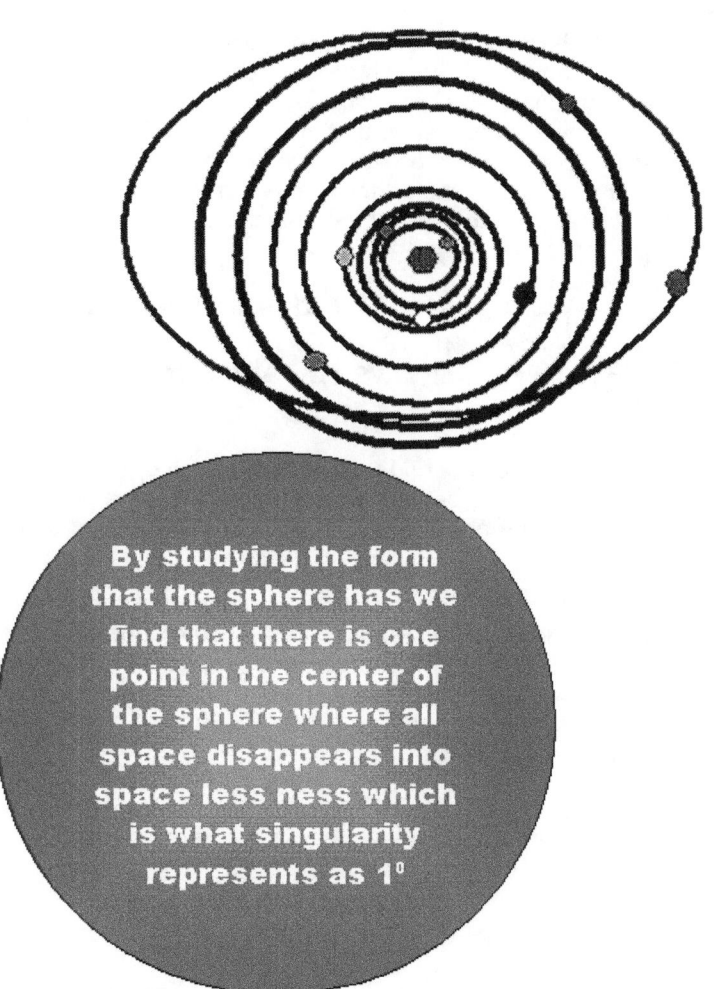

By studying the form that the sphere has we find that there is one point in the center of the sphere where all space disappears into space less ness which is what singularity represents as 1^0

The formula to calculate a circle, where a sphere is a compliment of multiple circles, requires the square of the radius in a square multiplied by Π. That linking r to r^2 and this allow the circles form value Π to stand directly related to the square of the radius Πr^2.

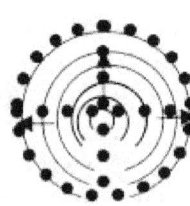

By reducing the radius in half, the size of the sphere would reduce. The reducing if the sphere must continue while the radius keeps halving every time. A point will arrive where r mathematically cannot reduce any more. The radius factor at that point then has to be in infinity $r^0 = 1^0$ since it cannot reduce more than it has reduced.

In order to build a circle we have to increase the radius. If the Universe was the size of a neutron at one stage and the Universe grew into what we now see, then the Universe extended r quite somewhat to get to where we now are. To go back in time we have to reduce the circle by reducing the radius. Keeping these factors in mind it is clear that Π are the forming the factor taken on by form and r^2 produces size. However Π is connected in form to r^2 but is only connected.

Reducing r will reduce the circle but it will not remove the circle. However if the mathematical proposition is correct and the circle starts with zero, then *$\Pi r^2 = \Pi 0^2 = 0$ leaving no circle to grow.* With that principle in mind it would be impossible to find zero in the centre of a circle. What we must find there is $\Pi^0 r^0 = 1^0 =$ singularity.

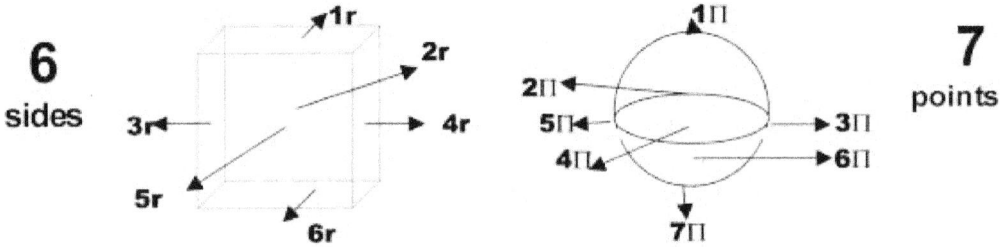

Where space comes into contact with the sphere the cube loses one of the six dimensions it has to the more dominating seven dimension of the sphere whereby the seven dimension in equilibrium will dominate the six dimension loosely connected by r bringing about that the cube then has 5 sides to the seven of the cube. This means that in the cube the "bottom falls out" and without a "bottom" to support objects they fall to earth. Remember that a body "floats" in space, but at one specific point it starts to "fall" to the earth. That is gravity and it is a dimension change much more than any force. I shall explain this last remark later on.

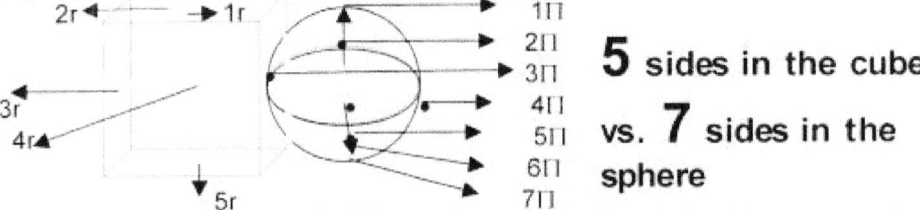

5 sides in the cube

vs. **7** sides in the sphere

What this says is that the gravity influence does not end on Earth. Although specific borders are in place in the atmosphere at various levels, the beginning and the end of such borders are in place in the atmosphere where the beginning and the end of such borders are as definite but also as vague as the gravity that forms them.

Kepler said
$a^3 = T^2 k$
but that
could also
be
$k = a^3 / T^2$

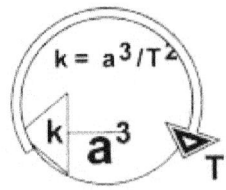

When translating Kepler's mathematical expression into a verbally spoken form of communication such as English we can see what Kepler said also read as $k = a^3/T^2$ where k is one point from a centre point that is space a^3 relating to time T^2. From a centre comes space-time

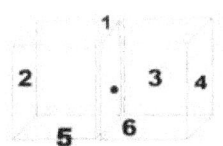

7 is the centre addition within the sphere centre

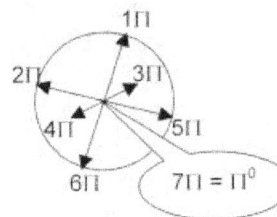

Front = 5
Top = 1
Left = 2
Back = 3
Bottom = 6
Right = 4
Center crossing = 7

Kepler's formula also indicates that a sphere is within a cube that is holding a sphere as $a^3 = T^2k = \Pi^{3 \backslash 3}$

1Π
2Π 3Π
4Π 5Π
6Π $7\Pi = \Pi^0$

The immovable three

By examining the form of the sphere we find that there are 6 points on the surface of the sphere that is holding the form at a specific and equal distance from the centre. Lines run from the centre into space at 90^0 and 180^0 angles of each other from six opposing sides. There then are six lines at 90^0 and 180^0 connecting to the centre from six points on the outside edge of the sphere. As a result of the basic shape that a sphere has there is a spot in the extreme inner centre of the sphere where the lines in 90^0 relevance cross each other and others connect by 180^0. There is also at that point a spot where all space relinquishes a position and only singularity 1^0 as form remains. At such a point we find the measure of the sphere being Πr^0 with $r^0 = 1^0$. That is where the line that represents the radius as a line disappears, as it becomes singularity r^0. After more reducing continue we get to such a point where we find only Π^0 left. At that extreme point is where space in all form disappear as the circle providing the sphere the form the sphere has, removes all possible form by going into singularity $\Pi^0 = 1^0$. Then in that area all form of any possible space disappeared leaving only the dimensions of singularity 1^0. This too, I take much further in the book but because of lack of space I cannot delve deeper into the argument in the Web page. However from such a point there runs lines that connects to space on the outside where six points on the outside points connects to the space less point in the inside. In the book I take this argument much further but for now I leave the argument at that. Those lines carry the structural straight the sphere has where the other six support every one of the six by singularity. Where there is no space there must be singularity 1^0 because the space is present although in singularity 1^0. If zero were a factor where all space finally halted in zero as the value, then zero would be able to remove the space from the centre and such removing would continue to remove the space until all space was removed. It will finally abolish all space in the sphere and it would remove the sphere. Zero removes all possibilities of anything coming about. Since the sphere is there, a zero factor in the centre cannot be present. Only infinity can be a factor from where space may grow because infinity can extend and grow into and up to eternity.

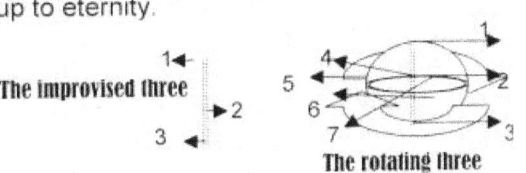

The spinning four   The improvised three   The rotating three

By using Newton one cannot even begin to explain any one of or the combined effort of the above cosmic phenomenon that are all over the cosmos and which I prove that they form all the laws in the cosmos. Newtonian definition cannot even recognise any of the principles but only Newtonian science are thought to students…and Academic ignoring and religiously denouncing they're presence as coincidental occurrences does not help with tutoring students either. Newton became religiosity for the part three hundred and fifty years and Science blame me for mentioning it. The gravity Kepler introduced is all working on a principle of indicators pointing dimensional integration and separation of space through heat densities applying different grades of space intensity. That is space-time being apart and forming densities. That means the space does not mingle, as one would expect because of the nothing value contributed by Mainstream science to space. Space is a liquid and as all liquids do, space depends on specific densities. With the specific densities borders come about. The explaining of this I manage with very simple sketches indicating the principles. I explain with ongoing proof through out the book indicating to the reader how I came to realise that gravity is not about particle pulling on each other with some inexplicable force holding an effect of matter pulling matter. I take the reader on a step-by-step voyage as I came to unravel Kepler. It is as Kepler stated gravity to be even before Newton came up with an idea that there was such influencing going on and named the influence gravity. Gravity is $a^3 = T^2k$, which is the space a^3, that forms through the moving T^2k thereof giving the space a^3 independence T^2k from surrounding space. Gravity is space moving in a circle holding space and what is in space at a distance where that is the applying conditions. Only when such conditions are broken does space fall away and particles come crashing down to Earth. But this falling comes from a lack of motion and not a sudden jerk on the falling object. The relation in motion must reduce on the part of the object about to fall to introduce the falling action. There is no pulling but a differentiation in motion that breaks the barrier the space placed and the object the breaks through the barrier the Earth holds in place. A little science experiment such as the Coanda effect disproves the grabbing on theory. Gravity is about matter concentrating space through the spin of the proton and the reducing space by accelerating the movement of space between the two objects. A good example is a blowing fan. By establishing motion of the air with the spinning fan, volumetric occupation reduces in favour of airspeed coming about from motion. By removing the space the particles come automatically closer, and that is the principle of gravity in operation. But the diminishing of the space goes about by applying very specific rules and in the adhering of those rules is applied to everything in the cosmos.

Gravity is not <u>about particles pulling</u> each other closer, <u>but it is about depleting the space</u> the body holds and the space surrounding the material as unoccupied and occupied space. Gravity comes about as space a^3 applies motion T^2 and from establishing singularity k^0 that provides distance k to supply space-time gravity is as much part of space as the motion of space is part of gravity. Mass is the result of applying gravity by reducing space. Gravity is not the result of mass. Gravity dismisses space and by doing that the stronger gravity can have more particles fitting into less space occupied where that reducing of the space is bringing about extensive mass increases into the volumetric occupied confining more material into less space. That produces monsters like a Black hole having enormous mass and being without space. If gravity was not about reducing of space the Black hole does not make any sense. Stars reduce their volumetric size as they gain in mass by creating enormous gravity applying the sphere of influence of such a star. Gravity is the increase of heat occupied by the reducing of space in a spherical unit. From the offset of the Big Bang to the process the universe went through development up to the very point where it is at the present, the process was about converting heat to space. I named the expanding of material through the process of overheating antigravity. Gravity and antigravity is the driving "life" of the cosmos and such a processes as the Coanda Effect and electricity can charge such "life" into being present. The Hubble Constant and expansion through heat applying is the other part of the driving engine applying antigravity with expansion where the heat transforming to space releases heat and converts the universe to more space but less dense space. Remember how the heat came down from 10^{34} to where it now is 0 K at present? Remember how the space multiplied by expanding from the size of an atom to what we now can and cannot see as it extended beyond any measure we humans can devise to where it is at present? But in all of science this inverting connection goes lost as it is unnoticed by all. The density of heat in space surely diminished considerably since then to now. The space holding the heat concentration expanded volumetrically as the heat concentration reduces…but not one person in science can connect the two obvious relevant factors. Expansion is gravity fighting a lost battle because space is increasing as heat is reducing. Gravity on the other hand is exchanging heat through the concentration by removing space bringing about space loss with increased density of particles and therefore heat concentration. The exchanging of heat as the heat converts to space is what fuels the Hubble constant, which is the result of the explosion we call the Big Bang but is not the result of momentum. Momentum is the second form of gravity symbolised by Kepler, as **k.** It is the directional motion to compliment the spin T^2. The Big Bang is the result of heat expanding into the forming of space. Gravity, on the other hand is about concentrating space back to heat, and take recouped heat through to material, acting out a balance of expanding while contracting. This way gravity is applying the onset of the Big Crunch by destroying space while space is converting heat to material occupying space. The Big Crunch is coming about because the Universe is expanding where the two processes are one principle.

In the beginning singularity was present in the Big Bang. Mainstream Science promotes the idea that singularity and antimatter went on the disappearing by escaping from the Universe. But Mainstream Science never says where they went. Singularity and antimatter and all other Houdini acting stuff had no place to escape too we just have to find the new location it occupies. Everything that was in the Universe is in the Universe and will forever be in the Universe until the very end of the Universe. If it was in the Universe it still is in the Universe because there is no other place available to escape too. But that goes for the space we now are in because either the space was there and it was filled from the beginning or the space was something else and that something else turned into space. In the totality of the Universe where the all and everything is in the Universe nothing is not in the universe. The Universe is everything holds whatever form of material or non-material there is. That which is now part of the cosmos was always and always will be without the slightest chance to add or disappear. Going somewhere is not part of options of whatever is the content of the universe. Keeping that in mind one should remember that the realm of the Big Bang was forming the cosmos by gravitational influences. What caused the Big Bang is everywhere to be found in the Universe and cannot be in selected places. With realising that much and printing that concept into a realisation one should first prove that gravity concentrated matter when it formed the solids from a heat and much more pertinent is the need to prove how it came about. If singularity was part of the cosmos in the beginning it is still with us. We just have to find it by searching for it and use the characteristics it had during the Big Bang. It will still have the very same characteristics. However by the claiming that singularity vanished somehow since the Big Bang is totally incorrect and proves a lack of understanding cosmic concepts. Kepler and his formula also prove this fact. Kepler said $a^3 = T^2k$ which is the same as $k^0 = a^3 / (T^2\ k)$. Using k^0 is using singularity because any symbol using the exponent of zero has a value of one and one is the value of singularity. Every aspect that was part of the universe at the cosmic birth announced by the Big Bang has to be present up to this point. There is no room or place to which the singularity could dispose since the Universe is the only place there ever was or will be. When I truly admitted to this conclusion and not as merely accepting by thought as another fact in my mind, cosmology became a woven blanket any one can read. I prove that singularity is present within matter and in fact all matter, a claim not yet made by any other person. Singularity present space-time and where one find space one has to find time because space cannot be without time. Kepler said that much when he said $a^3 = T^2k,$ which mathematically translates as space becomes motion and motion indicate relevant time. By my finding such locating and producing as well as applying a very specific value to singularity, singularity makes the process, which the Universe adopted much easier to follow. The connecting of singularity runs from the ordinary stars such as the sun and up to the Black Hole down to the ordinary material forming our every day living in the cosmos. We have to realise that singularity is a prerequisite for the Big Bang to have formed any and all material. If the presence of singularity is not in all material in the everyday cosmos, singularity was not present during the Big Bang and the Big Bang cannot be, because what is in the cosmos now, had to be in a pre Big Bang cosmos as much as it has to be in the cosmos of the present and the future as it in all stars including Black Holes.

As I said before the motive with the letter manuscript is an introducing by me of the accumulation of many of my books in an effort to place as much emphasis on explaining the combination of my work. Bu since the publishing and sending was on my account I put in an effort to keep the printing costs as low as possible. I made use of the simplest of sketches all in black and white and some grey. As the book has a self-explaining title and is a condensed profile of the most basic aspects of my work the motive was to be as informative as possible and as introducing as possible with the least costs as possible. Where the book is a letter I wrote in an attempt to promote my theory at many different Universities outside South Africa the second part of the motive is some sort of abbreviation attempt by which I try to combine all seven books into one. It shows how I came about finding the most basic aspects of my theory in the most basic form and follows a line that I followed as I progressed and developed the thoughts in formulating the concept on which I based all my other books.

I am not connected to the Mainstream physics in any way or form and through my views I promote in the writing of my work I found a method how **not to make friends** with Exceptional and outstanding Gentleman of well-established and important educational background upholding Mainstream physics. It had very little success because I am not sharing much of any opinion with the Academics of the day and in that respect I have to criticize their work. Such criticizing of Mainstream views is apparently not the correct route to follow when trying to communicate with such gentlemen of exceptional standards and of the highest esteem and in every case they took the criticizing very personal. Why I do not know because it is their opinions I dispute and not them personally!. But with my definite disagreeing with mainstream physics and not trying to cover my disagreeing with soothing words there are no other options I have but to disagree with Newtonian disciples bluntly and non the lesser so than with Newton. As the overall extending of the work is so all including there is no short route or touching up of theories to introduce some aspects about my work. To understand is to familiarise one with the entire concept. The routes I can follow through magazines by publishing articles only allow me three thousand words at a time. That is far too little to promote such a broad concept as I wish to promote. With my promoting of my view holding the complete picture it has to bring about change in science and has to change science altogether in the field of cosmology but only in the field of cosmology. This comes about because the gravity pressing us down onto the Earth is not the same principles applying when a smaller space orbits a larger space. There is two definitions applying which is a Universe apart and in that there is just some aspects thereof sharing a concept. Therefore I almost need three thousand paragraphs at least just to make my ideas logical. I wish to find a publisher willing to publish the book presented as **A Cosmic Birth Dismissing Nothing I. S.B.N. 0-620–31609–8**. The second part of the book comes from personal experience. It is about the manner I see physics forming a role in the second energy in the cosmos, which we named as "life". I include that part to indicate what I see life is and how life intertwines with the cosmos but never integrating in becoming part of the cosmos. Life is special and never overall because the cosmos is hostile to life and never friendly except on the little dot we call Earth.

In essence the first part of the letter is an explanation with proof about how the Universe came to be before the Big Bang and before mathematics came into place. The beginning where all and when all started. It starts explaining why the triangle is 180^0 the half circle is 180^0 and the straight line is 180^0 and how all three is (as it first was equal in form sharing value before "normal" mathematics became into place) where the triangle has three lines but is equal to a line and a half circle and yet so different in shape and form. (With that I introduce a measuring standard value used by the cosmos in the same manner we use the imperial or the metric measuring system but the basic value does not use one. The Universe uses a basic founding measure it formed before it formed mathematics. From that standard and by using that standard I can explain all stars behaviour as far down to stars yet to be stars and even down to the reason why the proton is 1836 times more massive than the electron is. I can use that measure to explain and differentiate between Pulsars and Black holes, between yellow giants and blue dwarfs. What forms a Neutron star and why does some stars have a variable change in rhythm and others going up to be frantic Pulsar stars. Moreover I can explain gravity in detail as never before explained. It is only after the moment that space (the triangle) became equal to being the motion forming in the half circle that is holding the line as it is equal to the square of line it is holding (in what we know as the law of Pythagoras) and that extensively became our Universe. The Universe formed as $a^3 = T^2 k$ and from that afterwards formed mathematics in a usable six sided seven to ten in relation to ten to seven dimensional Universe with atoms ranging from 112 protons down to one proton in $k^0 = a^3 / (T^2 k)$.

While Kepler and his work was considered a closed case at the same time not one of the following principles was yet successfully proven but I believe I have accomplished that goal. I first started my studies in the field of Cosmology as a spontaneous development of my natural curiosity spawned from childhood interests in the field of cosmology, which I developed even before I went to school. The studies were a reaction (I would imagine) that was part of my personal childhood development in how I was forming a personal concept of a lifelong interest that followed me into my future. At first I conducted all my earlier studying mostly on the basis that inspired me to find out more about what made the Universe tick, with no intention ever on my part to reach a point where I would be writing books on the subject. At first I was investigating cosmology on a part time basis. This went on, on and off, or the best part of twenty odd years (*as* time and *when* time would permit). Then in later life with my health deteriorating I committed myself to more intense investigation and my effort developed onto involving a study using time that is only permitted by a person when that person is involved in such a quest on a full time basis. That quest has now been going on for the last seven years in full devotion and if one includes all the years invested on my part including the twenty odd years before, part time, then the time I have spent in completing my theory when adding all in comes down to almost twenty eight years. This is to say that I did not come to realise what I am about to introduce on a light-hearted conclusion. I mention this because I wish to ensure the reader that he should have no doubt about my most sincere commitment in producing a cosmic theory on matters concerning the start and the working of the Universe during and before the Planck era. At first I began by arguing that there is a something that is blocking our progress. There is some barrier preventing humans passing a threshold whereby our understanding will pass such an obstacle. If there were any way that anyone may break through that barrier which is preventing normal research to go pre-Big Band, it would be accomplished by finding the barrier whereby the vision we use to focus would pass such a limit. If we wished on progress in our pursuit of the very first cosmic moment then we have to find and cross the barrier that blocks our view. We have to look deeper and in another direction should the desire driving us be strong enough to commit us to reach into the very birth of the cosmos. We have to rethink the strategy we use. Max Planck was one of the most brilliant men of all times and even he, notwithstanding all his personal brilliance, accomplished little. There are parts missing in what we have and that which we have at our disposal to use because if there was no such an obvious barrier then the Wise-Men involved in science would by now have found the way to break through the seal that is locking us out of the critical past which will uncover the origin of the Universe's infancy stage. I went about trying to find what everyone since Adam, (meaning all of the rest of mankind and myself) were missing throughout the ages of speculating and interpreting while philosophising about whatever we find inspirational. The obvious we saw; that was clear. Therefore I had to find a route that would lead into the not so obvious which all of us were missing, notwithstanding the best efforts of the best qualified to accomplish such a breakthrough. My efforts involved trying to accommodate that which was in the cosmos available to use by the cosmos in all phases of developing. If I had any hope of finding the answer, such an answer had to be simple because I am not very inclined to unravel what is deemed as complicated. The simplicity had to be locked in what was not yet understood about that which was in the cosmos as it formed part of the process used in forming the cosmos. My realising this brought me to focus not on that which we understand. There is not a lot we actually understand because even gravity is very poorly understood. In fact gravity is so poorly understood that there is not one person alive that can claim the prestige of understanding gravity and among the dead there is even less that can make such a claim. There are several phenomena that are presented in nature and acknowledged by science but also discounted by science and therefore not presented as accepted science. By admitting that that what we have available to us to use concerning our research of cosmology in an attempt to better our understanding of cosmology, is useless to use, then one realises that not having what there might be makes what we already have useless. It then is useless to use what there is as part of the big picture we are trying to paint because what we use is not really part of the picture. This leads one to believe that the picture of the cosmos Mainstream science is painting, is being painted without painting a full picture.

In my first attempt to understand the full picture of what science was painting I found so many colours missing there was no picture painted that anyone could appreciate. This is what made me decide to go on researching the 'unknown' in the hope it might clarify the 'known' and as the book unfolds you as the reader may agree that I was correct in pursuing the misunderstood and rejected phenomena. Finding the missing phenomena helped me to place the phenomena mentioned above in a theory where the principles also mentioned above form a part of the overall gravity used in binding the Universe. I believe what is in the Universe is not able to be coincidental because of too many influences contributing to what there is - notwithstanding the fact that this is the manner science uses when they refer to the Bode law. What is in the Universe has a role as it had a role, which is the same role that phenomena *has* had and in future *will* have. This is establishing a very new idea about the working relationship between particles and in explaining it by using Kepler's studies. Redefining the work of Kepler's views brings a new Universe to light involving new concepts that are based on old principles but principles in updating man's view about cosmology are very new in that capacity. Through that new vision I was able to come to realise what the reasons might be why Kepler never saw it fitting to include the measure of Π in his formula. I do not suggest his neglect thereof was intentional, nevertheless the formula he devised without using Π proved that there was no need for the inclusion of Π since his figures brought about a correct answer in the final end result leaving a well concluded fitting answer. The numbers he produced brought about a specific space $\mathbf{a}^3$ contained in a circle $\mathbf{T}^2$ at the distance of $\mathbf{k}$ from a defining centre thus the calculations did not require the use of Π to find a meaning. In that Kepler did not see a need to include Π. I would not go as far as declaring with absolute certainty on his behalf that he did it deliberately, however there never arrived such a necessity. It is prudent to agree on whether or not such a need is necessary, because if one is agreeing about such changing not being required a new Universe emerges. The circle that Kepler discovered came about without ever forcing Π into the frame because it is clear that the circle formation came about as a natural consequence and came spontaneously delivering an equation while he was working. In this book I prove that the reason for adding Π to the rest of Kepler's formula is unnecessary. This unnecessary addition is because when going one step further in the investigation one will find that **k** and **a** and **T** are symbolising the same value with the only difference being that each one represents a different dimension to our six dimensional or six sided Universe we enjoy. In fact I shall show that Π replaces "**a**" and "**k**" and "**T**" and that Π is the true value that should be replacing each factor as to indicate the correct value to the sides nominating Π. We humans work on a numerical base using ten as a basis where we count to nine and re-establish a new decimal numbering line by adding a nought behind the number in value. This is using the numerical basis of ten, which I suspect we took from ancient knowledge about cosmology and not from using our fingers and toes as the earliest calculating processors. In this letter there is unfortunately no room to explain my suspicion but another fact I do prove is that the cosmos uses Π in the cosmic numerical basis as a means to measure and quantify. Therefore in fact the Kepler formula should read instead of $a^3 = T^2 k$ as it does it must be $\Pi^3 = \Pi^2 \, \Pi$ where I shall show that Π represents singularity wherefrom the entire Universe sprang from Π and by forming as $\Pi^3 = \Pi^2 \, \Pi$ it is confirming that space is equal to the motion thereof. Kepler's greatest achievement was showing that the cosmos is space –time $a^3 = T^2 k$ while time is the motion of space in space. The value of Π is the primeval and most basic of measures applying as an accepted cosmic legal value that the cosmos used exclusively in the very beginning and as it does today. The measure of Π in the Universe, values particle development that brought about all development ever conducted in the Universe. Only after this stage did the rest come including mathematics and went on to freeze spilled singularity into frozen material. Reading this statement may sound suspiciously senseless but as the book unfolds the sensibility will become apparent. The full implication of such a statement will become clear when one dissects different facts coming from studying Kepler. My discovery of this fundamental basis of legal valuing ensured me again that there was no need for someone the likes of Newton to add Π in any form to the work of Kepler because Kepler discovered the ultimate Π in the Universe, the Π giving the Universe form and gravity. The concept of Π that is the only single form of all other forms available that can by duplication of Πs assemble the value of gravity. When replacing the symbols with Π the facts of the Universe become self-explanatory because the most basic form that forms the cosmos has a definitive and uncompromising value.

 But getting this far took me down roads overgrown by ignorance and which I had to uncover myself as if hacking away miles of overgrowth with a machete chopper. All of the disbelief science showed to my work in the past and their refusal to see past Newton made any and all attempts on my part as bad as they could be, strangling and smothering my attempts to announce the uncovering of the newly found insight on my part.

For decades I tried to come to terms with the inability there is in science to explain the cosmos in real terms, when using the science of official reputation. That which there is makes a mockery of science because the undisputable clues left in the cosmos makes what little correct explaining there is available, seem like a comedy of errors, when it is mixed in with all the other near Dark Age errors we still use after so many centuries that provided countless opportunities to revise the old muck. By applying current accepted Astronomy as such the phenomenon found all over the cosmos is still beyond the explaining ability of Mainstream science. This is true and it is a shame because it also is an undeniable fact in spite of the vast knowledge and progress in other forms of science taken in the manner science uses when it approaches cosmology. Cosmology truly lagged behind while the understanding and advancing of physics, mathematics and chemistry as subjects were flourishing. By comparison I saw how little there was available in explaining cosmic phenomenon and how much improvement in understanding the other departments such as chemistry, electronics, medicine etc. could offer as results were coming about from research. Even where there is a little explaining available in cosmology it turns out that such explaining is confusing to say the least and at best it highlights the manner in which science is applying double standards. For decades photographs were the only progress forthcoming as an addition to improve the meagre field in cosmology and that improvement was artificially stimulating cosmology. By providing a false impression of advancement, everyone missed what and how much was missing…To the connoisseur desperately looking for more than the obvious stirred in with some out-dated misinformation dating back to the Middle Ages, it all seemed as if it was a picture portraying the ridiculous to make the sublime look good. The pictures only proved the opposite of what progress in cosmology will represent. In truth and as such in cosmology the cover up that was hiding the lack of progress about the science of true cosmology was only forthcoming in the improving of electronic optical telescopic advances and spectroscopic progress. There were only photographs carrying beautiful pictures which pleased the less informed except the photographs did not bring progress to cosmology at any intellectual level by promoting insight.   The explaining that the photos demanded about the subject had the opposite effect of installing hope because what it did do was underline what lack in any notable progress there truly is in our understanding of cosmology and laws in the cosmos.

While such Hubble telescopic images might seem to be clear, as daylight it was more than clear there was little academic value to them. To the person in need of more stimulation than being impressed with pictures of God's marvellous Creation and the sightseeing that always accompanies such pictures, such persons always felt very disappointed. The pictures did give satisfaction to those more easily impressed, but the rest of us seeking knowledge accompanied by understanding the images left us despondent. Although they leave the vast majority in total amazement there are those less impressed about not knowing the 'why' and the 'how' in such amazing pictures. I am aware that the group I fall into may be the greater minority and the majority may only demand the portraying of the images, which is what that "easily satisfied" group demand. The rest of us rouse with anguish at the lack of information about what is known, and what lies behind what those pretty pictures are conveying. Nevertheless there can be no real progress in scientific understanding about the images portrayed by the Hubble telescope, and others, if no one is able to show the slightest clue of a deeper understanding of what is going on in the Universe. Everyone is almost breathless waiting for commentary by the most informed, which accompanies the magnificent cosmic portraying of God's Creation. When we are portraying the new images, we should also be investigating that what we see the cosmos is at that moment portraying. The lack of actual believable explanation coming from investigating by means of telescopic imaging should impress one and all, but the impressing must not be based on the colours in the images but the sensible information attached to the image investigated. It is *that* that we wish to see. What we wish to see must at least be accompanied by scientifically backed information, which provides the proven understanding coming from science. When science is employing new explanations with such photos it should also be discarding senseless baggage carried over from the past. Most images contradicted Newton and for saying that, every Academic I ever came across in the past ostracized me. That bothers me little! I know I cannot possibly be the only person absolutely discontented with what Mainstream science accepts as science. Here I refer to the out of date theorising Mainstream science still accepts amongst many others as how they suggest stars and planets are forming. One cannot promote cosmology in honesty and advocate scientific fact whilst dishing up such fairy-tale nonsense to students. Moreover I hold the opinion that amongst Academics in particular there must be many if not most that share my personal serious doubts or have an inclination to share some of them. This I say when considering the overall doubtful picture painted about what there is and what one believes there should be. I just cannot believe those forming the most intellectual group of mankind are unaware of the mismatching facts seen over the broader picture because the contradiction and lack of a plan, makes what there is so very doubt provoking. Newton dismissed the formula Kepler presented as all factors forming motion. That is where the apple cart derailed.

In honesty we have to realise that we cannot dismiss the whole formula that Kepler produced as being motion. It is so much more than just motion. It is $a^3 = k / T^2$: That is what Kepler brought into civilization for all time to come. He saw space a^3 being in isolation due to the time it uses to move T^2 claiming such space forming independence according to the lines k indicate. Let us look at the factors in more detail before we proceed with the rest of the book.

a^3 symbolises a mathematical interpretation of implicating the three-dimensional space.

T^2 is representing the period or time that Kepler suggested we should use to calculate time that holds the orbiting planet in direct contact with the space in relation to a very specific centre.

k is the space taken from the centre to the end of the line from which the planets must have grown if one accepts the Big Bang growth of particles and the affect of the Hubble constant on all cosmos material. The specific value about the centre is most important because from the specific centre gravity always applies the strongest influence.

One cannot justify Newton's dismissing of Kepler's formula as that all factors only contribute to the motion indicated because that is misleading. We all accept that the true cosmic form *would be* and most probably *is* a sphere. Everyone accepts the universe as a whole as a sphere...but why would the sphere form? What would be the reason why the original form that we devote to the Universe would take on a sphere as a natural form? Apparently our imagination grabs the sphere as form. In all natural events the gravity in that space which stands apart and independent from all other space takes on by cosmic pre-casting the sphere as form of shape … **it is because gravity chooses the smallest space to hold the strongest force**.

I am of the opinion that gravity is about dismissing space to the advance of heat increasing in such a specific and concentrated space using the concentration as measure for the heat as well as the space holding the heat in space. According to Kepler that is what he found to be true. Space a^3 will always be circling space around as T^2 in any position from the centre **k**. That is what Kepler said when he said $a^3 = T^2\,k$. Kepler indicated space a^3 will forever fight for independence and show separate individuality in remaining apart as identifiable cosmic components by means of motion. Every space will cling to independence indicated by **k** through fighting off the integrating of another coverall unifying unit by applying the motion of T^2! The problem we have to solve in this letter is what will the cosmos use to secure such independence between all particles? What sets space apart from the rest of space? First we have to admit that Kepler was the one that introduced the following.

Kepler gave us the answer to the following but no one ever took notice!
Kepler was the one that discovered **space / time** as $k = a^3/T^2$
Kepler was the one that discovered **singularity** as $k^0 = a^3/T^2 k$
Kepler was the one that discovered **gravity** is holding **space-time** relative by the measure of distancing **k** as $k = a^3/T^2$ and $k^{-1} = T^2/a^3$
Everyone able to read mathematics has to realise that Newton suggested collisions between cosmic structures must eventually come about as gravity erodes the distance separating the cosmic structures multiplied by the product of the mass of both structures from both ends. Newton said the multiplying mass of both structures destroys the distance between the structures by using the eroding force of gravity in the square. The cosmos then must end in a Big Crunch with all material joining together but that joining is not forthcoming at all...and that only indicates how much insufficient understanding there is on offer in cosmology by the educated–to-be-wise-about-these-matters. There is precious little available to explain about their field of cosmology amongst the ranks of Astronomers. So...let's us return to the beginning of cosmology before every one became oh so wise and see what there is to see.

While we are in gravity the manner in which gravity applies in our use of gravity makes us part of the Earth by mass forcing us onto the Earth as a semi unit with all other Earth belongings. Is that which we have truly gravity? By using mathematics the cosmos spoke to Kepler personally and by the use of mathematics as the medium it provided Kepler with information about the cosmos coming directly from the cosmos.

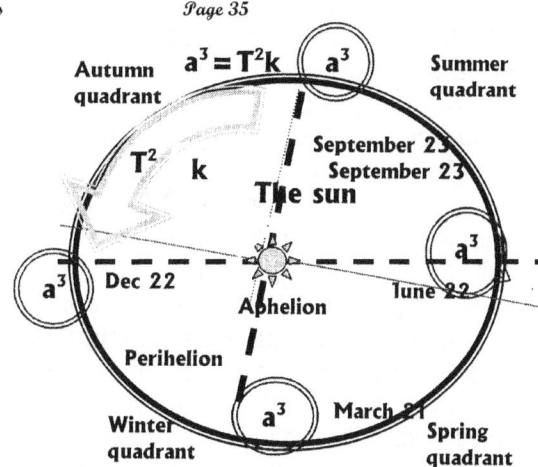

Much of the proof we use about gravity is part of our perception about gravity gained from the obscure position we have relating to gravity because we experience certain positional fixed conditions about gravity. But are our perceptions about gravity truly correct? We only experience gravity as a factor from the position we have on Earth and only while we are being forced to be a part of the Earth's totality. The cosmos informed Kepler of another gravity, which the cosmos applies much more widely and is used by nature all over the Universe.

The picture we see coming from the Hubble telescope shows why, in the perfect Universe...but can the Universe be perfect when... we see a radius between the sun and individual planets is not using a regular distance as one would expect of gravity in being a force driven by the mass and in that sense the mass is producing the gravity that always remains even because the mass doesn't alternate. As the mass is never changing on either side, that steady mass has to keep the gravity steady. But in our imperfect understanding of the Universe we find that the radius that should be constant varies considerably proving either that mass somehow adds by measure unnoticed while the structure is in orbit and later allows the same amount of mass to escape undetected; or it's the seasons adding and removing mass at will. This is an absolute contradiction to reality if mass was the factor determining the radius we find between the sun and the planets. This suggests strongly that we'd better be getting very suspicious about the idea of mass contributing to gravity. But in contrast to this, science is unshaken about their confidence in the perfection about facts they use in terms of correctness. It is well known amongst all persons that science only uses dependable and ultra reliable facts coming from sources beyond doubt. Referring to any work done by any scientist will find a remark about science only accepting facts they use to work with. It is accepted overall by all communities that in science those in science use one hundred percent accurate facts or they use no facts at all. If our view was as perfect as science would lead us to believe it then must be the Universe that is imperfect as it otherwise would not behave so mystifyingly. The unshaken confidence science uses has us believing at first consideration that the drawing of gravity should produce an even diameter positioned between the sun and the planets because of ever dependable evenly distributed gravity... but I believe there is a perfect Universe and our understanding carries the doubtful suspicions. Delving deeper uncovers even more contradictions and the level of accuracy contained by our scientific understanding then arouses more suspicion about the correctness of science. Remember Newton changed what the cosmos told Kepler leaving much suspicion as to how far the misdirection takes science. We have to correct the facts we doubt because when correcting the facts they use in science concerning our view about science such correcting brings along a better understanding and then the Universe has to become ever more perfect as one learns to understand the perfect Universe even better. But it does require an open and clear mind and it needs no culture driven preconception that should confirm interpretations about facts surmised even before they are carefully studied. It becomes obvious that Newton never gave careful attention to Kepler's findings because if he did he would have seen what gravity is. Kepler described gravity without using the name that later was given as 'gravity'. Kepler did not give the name gravity, but Kepler's studies gave Kepler the insight to coin the concept of gravity. Nevertheless it was a name and not the concept that was later named by Newton. The naming was the contribution of the Englishman. The concept that Newton later introduced is totally incompatible to the concept that Kepler introduced. What he (Newton) introduced as the force of gravity, he connected to mass, which diverts totally from Kepler's findings. With giving a name, the Englishman also changed the concept that Kepler introduced. Kepler made no mention of size or mass as part of the phenomenon that later was named as gravity, yet it must be gravity that holds the universe together. The concept the Englishman changed when he introduced what he introduced with the name he introduced. That which he introduced, he corrupted beyond recognition. The concept that accompanied his new name strayed completely from what Kepler introduced. Newton brought in something that was miss-matching what Kepler saw in Kepler's view of the phenomenon that holds the Universe true to form. The name was dominant but even more dominant and totally inaccurate was the other concept Newton introduced. In truth Newton only gave the world a name of an idea, which he then corrupted as far as cosmic physics are concerned. It is important to admit that as far as cosmology is concerned Newton gave the concept the name but *only* the name and not the concept of gravity. Newton's persuasion on matters of gravity as gravity functions between cosmic structures orbiting one another as we find in outer space is inaccurate. What Kepler saw, Newton saw differently and used the opportunity that Kepler left by not giving any name to the process he (Kepler) and Tycho Brahe worked on for two life spans. Newton did seize the opportunity to name what he, Newton, saw but that what Newton saw did not include that which Kepler uncovered. In Kepler's era the name or title was lacking but Kepler established the concept of gravity and the formulation thereof. The concept came from Kepler even before the name gravity was used by Newton to describe in the concept of whatever we today (after Newton) became accustomed to believing what the concept of gravity is about. With the help of Newton everyone since Newton confused Kepler and Newton on the issue of gravity and this confusion even begins with Newton. Gravity might not have been named but became a proven concept and factor after Kepler formulised it, which is before Newton named it. The concept of gravity that Kepler saw is about the manner in which the structures orbit because there is a space that circles around a centre and this process has kept planets secured, connected and rotating around the sun which is the same concept that is keeping the Universe secure and comes about with a process Newton later named as 'gravity'. This what Kepler saw is not the same as what Newton saw when he saw two objects drawing closer by pulling on each others mass. Then later on Newton named what he thought he saw as the force that Kepler saw but introduced another completely different concept. Kepler saw cyclic formations keeping the Universe together and never approaching each other. Newton ignored what he wished not to see but he changed as he saw fit and what he thought that should be. His experience as a young man drove him to establish a process he formulated as the process that is keeping the Universe together. In that act he corrupted as much as ignored the work of Kepler, which he also named as the same gravity that he saw as a young man. Why he chose to ignore Kepler's findings on gravity we shall never know but why the world still chooses to ignore Kepler's findings about gravity almost four hundred years after the fact I shall never know. My saying this has literally made Academics ignore me as they would avoid the plague. I am not pretending nor do I exaggerate when I say there were those in Academic institutions that questioned my mental development. Some went as far as seeing me as a joker of sorts and I have correspondence to show evidence to that fact. I know by now while Newtonians are reading this letter I have aroused the tempers of every Academic reading this far, therefore let's see what is being ignored by the Academics which I blame to do just that.

Kepler said gravity in space is about the area a^3 that would always keep equilibrium with the time T^2 it takes to travel the distance of the full circle position placed by the indicator k, therefore adjusting k as the need arrives. With k shifting in length a^3 will have to readjust and therefore T^2 will find a new relating value each time. This was the finding of Kepler and came after his intense study of orbiting planets.

Before I attempt any investigation into this matter there must be coherence in our agreeing about what gravity is. If you the reader insist that the falling of objects is the only gravity found, your further reading will convince you little. Anything we do decide upon must support the fact that it is gravity that prevents planets from dislodging from the grip the sun has on them. Gravity is not about the sun trying to catch the Earth by attracting the Earth…no, there is so much more to gravity. We must be under no illusions about what gravity is and that being the focus of our discussion and where that gravity is because we have to identify and not confuse the gravity we are looking at. We are now discussing the gravity, which is keeping planets circling around the sun, and stars around specific galactica centre. In that we do not find one example to use as proof in connection to stars coming tumbling down on galactica centres and crushing into galactica centres. If that is gravity keeping structures in orbit around specific centres we must look at the behaviour of the structures in gravity.

We have to find a reason why the planets do not reduce the radius between them as Newton suggested but we must trace the reason why it is gravity, which is keeping them apart because if anything, they are departing as they extend the radius connecting them to the sun. That is gravity because it applies throughout the Universe. The gravity Kepler found is the general gravity that is keeping structures from colliding and in that the principles are avoiding collision or on the other hand avoiding abandoning each other. It is about confirming respect for one another's independence and clearly staying at a predetermined distance while at the same time both are sharing a common space unit. That then must be the defining of gravity we have to study to find the Universal enticing gravity holding the Universe together. By close investigation one will find three factors in urgent need of investigation. There is firstly a centre that draws the object closer. This gravity is clearly a synonym to what Newton saw as gravity. If it were not drawing the object closer the object would not be orbiting around the centre and applying motion. It will draw and absorb all rotating things in its field of gravity.

The fact it does not draw the object into its ranks is because there is another gravity standing alongside this first mentioned gravity. Our recognising the first gravity forces us to accept the presence of another part of gravity. This forces us to recognise the second gravity. When saying this we are not using Newton's cosmic formula concept $F = G(M.m)/r^2$ because that can barely be what is out there happening. What Newton saw was falling. If that what Newton saw is the only gravity then whatever Kepler saw including all other parts of everything out there that are spinning around some centre must come closer to one another and connect in collisions. While that is not happening we must start to look past Newton to new grounds we can investigate. We have to go beyond Newton and admit there is more than that what Newton had us believe because it is clear that what Newton had us believe…that is not happening. That confirms the presence of the second gravity. The fact proves that everything is departing and not arriving. Even the moon is drifting away from the Earth and this information comes about from the most advanced investigation up to date, including a moon visit and the placing of measuring devices there.

Looking at the gravity intensely we find the roving structure travels in a straight line, which repeats another circle around another centre but because of the influence of a centre keeping the roving structure attached to such a centre the motion allows a circle to form by reforming motion from the original straight line to that of a partial circle. There is a centre, a connecting line travelling between what the two points establishes the specifics of a centre within a circle and the end of the circle. According to Newtonians the centre supposedly draws the rotating object closer. That is half the story.

I suggest we do some deliberation and in deliberating may I remind you THAT NEWTON'S OWN LAWS ARE IMPLIED, and again the planets disobey these laws completely!! In the modern age all evidence points away from contracting and favours eternal expanding.

The latest news says if anything is happening, investigation confirms that the lot is doing the opposite, as the lot is not coming any closer.

In our manner of considering gravity as a phenomenon we find there are three factors interacting and together the three factors form a balance, which produce and are responsible for a balance between all particles in the Universe. This must be gravity since it seems to be the glue that is holding the Universe intact. We can visually see that as the object moves in a straight line because it counteracts the pulling from the centre by a line that indicates the repositioning each time. In parallel with this it also moves in a circle. One can only interpret this action as being caused by another line just as strong but counter-directed I motion.? The circle comes into action as a counteraction that is trying to accommodate two opposing directions being evenly strong and from that counterbalancing eventually forms a rotating motion trying to satisfy the direction coming from the straight line in one direction and another straight line counteracting the first straight line. In the motion the straight lines coming from opposing values also forms an immediate circle (though only partly) but the overall complement forms a triangle. This shows a very different picture to that which Newton saw.

The lot is more evidently moving further away from the sun.

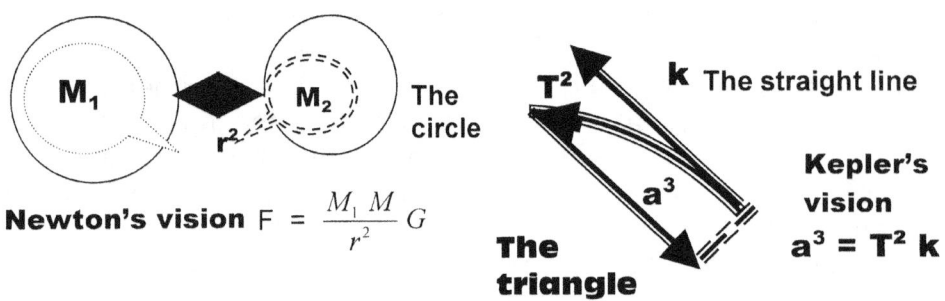

Newton's vision $F = \dfrac{M_1 M}{r^2} G$

T^2 **k** The straight line

a^3 **Kepler's vision** $a^3 = T^2 k$

The triangle

$F = \dfrac{M_1 M}{r^2} G$ This is the suggested formula confirming the behaviour of planets used by Newtonian scholars underlining the argument that contraction is coming about between all cosmic objects. What Newton witnessed, if my memory serves me correctly was an apple falling from a tree where both the apple and the tree were part of the Earth and this did not constitute - or lead to - or come as a result of - a catastrophic cosmic event happening. In the mathematical sense it does not make sense when Newton's argument is taken out and used in outer space.

 The pulling away of the smaller space. a^3

 The double counter-acting referee. T^2

 The pulling towards within the larger space **k**

What Newton saw with his falling apple was a mass influencing another mass to reduce the distance as the influencing involved motion that came about. In outer space there is another gravity where in the case of those cosmic structures in outer space there is no mass pulling each other about or pulling one onto the other. In the case where there are particles falling from space onto the Earth, that falling also results from gravity, as much as it varies from the cosmic gravity. There is another type or form of gravity different to the concept Newton introduced. That, which the concept Newton introduced, is not the cosmic gravity Kepler formulised. What Kepler introduced is a duel where both objects are clearly in an eternal compromise therefore neither party relents its position. Newton saw just the opposite...Newton saw both compromising their individual as well as each other's position. But since the mass in both cases is unchanged and the mass is the factor that is establishing the force that is used by the circle to hold the radius steady and in place, these facts point to a balance that formed bringing about the above-mentioned steadiness. In the view of science however it is the mass that either draws the orbiting objects closer or is keeping them apart. The mass does not change and since that mass of both produces the radius between both, the logic is that there has to be an even and steady radius that develops. The radius has to be equal all the time since the mass never changed throughout the rotation. The radius must be the same from any and all given points that form the rotating circle which must keep the radius equal from every angle...yet we know that Kepler proved this not to be the case even before Newton's naming and changing of Kepler's work came about.

What we see is that there is one factor that is trying to run away being a lesser space within the pulling powers of a larger space (the second factor) trying to capture and control and a referee (the third factor) is seeing to it that the even-handedness is at all times applying in the fight. That gravity which I am familiar with and know is there. In some part but not in all out representing all the gravity there might be because I cannot see the jerking, as much as I do not feel it. That is then most probably another gravity I can see and which is Kepler's gravity which $a^3=T^2k$ represents. We have a motion of pulling...yes and that is what Newton saw...but then there is another motion of establishing a motion trying to depart, leaving the centre by tearing away from the centre and thirdly there is a motion that sees to it that the balance evolves as rotation. That is what Kepler said when he saw all three factors whereas Newton saw but one of the three. The one space is filling the next space as the space duplicates the position it had in the next moving moment that brings about the next position through motion. This eventually will have confined the next point by using a circle motion, which at first was intended to be a straight line, which is stopped by another straight line. The quest in this book is to find out why the other two factors apply in outer space as only one of the factors comes about on earth under normal applying conditions.

Kepler's investigation indicates to the fact that the orbiting structure is in a motion that is going on where one strength is in a fight with a second strength and the two are pretty much matching in strength because not one of the two is very much winning the dual so no one is winning or losing the fight.

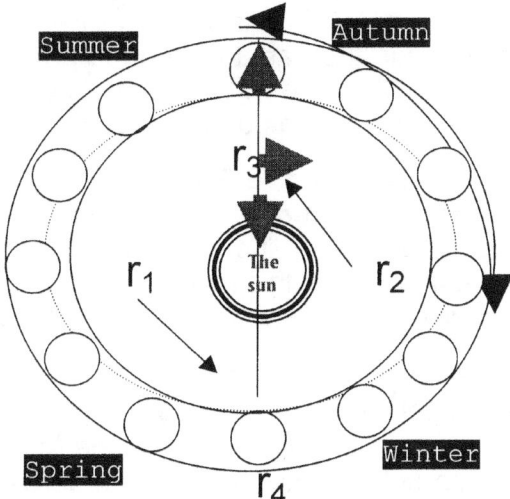

As the two factors are in a motion directional dispute there is obviously one of the two factors or strengths fighting to cut loose from the other one's grip and run off. If there were not such a force trying to escape, the first force would have a quick and decisive victory by reeling in the loser just as Newton predicted. The fleeing object and its matching fighting partner has a third party referee that allows the fight to go in a specific direction as long as there is no decisive victor.

This book, which I produced in the form of an open letter, is on a quest to find the missing two factors and I can declare with some delight and with even more certainty that I found the missing factors. By Newton's introducing gravity as a force with the formula $F = G (M_1.M_2)/r^2$ a precedent was set of gravity being a contracting force forcing distances supposedly to grow smaller. Apply Newton's view to comet behaviour. Newton insists that the sun has gravity reducing distance between the objects and while lecturers are teaching this during the day, at night they all witness how the comet follows this principle in detail showing Newton as a prophet. No sooner does the final conclusion draw near by orchestrating the final demise of the distance separating the two cosmic components when the opposite changes all concepts taught by institutions of science, the next minute out of the blue with no pre warning of the comet changing its mind, the comet defies all logic in scientific circles that apparently even included defying Newton and his logic. Because at the very point you'd think there is no chance of any return where gravity supposedly should peak because the comet is so close to the sun and due to that fact makes the collision unavoidable…then the comet chooses that very point to dart away into the blackness of outer space, missing the definite collision by miles. By the time the collision is truly unavoidable with the radius between the sun and the comet being as small as it realistically can be the comet starts gaining on the radius distance in spite of Newtonian denial of any possibility that such an event can in fact take place. The radius that should be shrinking further is instead enlarging. The radius that now begins to stretch proves Newton incorrect and it even depicts Newton as possibly being a fraud. The gravity applied that focussed on the comet reducing the radius between it and the sun was not acting predictably by maintaining the reducing of the distance until collisions come about as Newton insisted on. In our reading the Newton formula in English it says that $F = G (M_1.M_2)/r^2$ which when one translates that which is said in mathematics to a verbally spoken linguistic dialect, the translation then suggests that a force is committing the material that forms the factors involved, and forcing the material into a path that is leading to a collision. It says that the two will eventually collide because of the non-retractable mass inside each one that enforces the pulling which by the mass in each case is creating the force. The unchangeable ability of the mass and the unavoidable pulling each mass creates would bring about such a collision. The mass contributes a force making a collision imminently unavoidable. The collision is beyond any attempts of diverting any oncoming objects away from the inevitable possibility of contact. The force that mass contributes is ruling out all possible evading each other or avoiding the destruction. By enforcing a mass created force removes all chances from diverting away from the collision that is about to occur. Such a force then removes all possibilities of avoiding the oncoming collision. The force will not allow any attempt to try and bring into the equation other possibilities in as much as rerouting the approaching object and changing the course in the imminent collision that is due and in due course will come about between the comet and the sun. That which I explained is what Newton mathematically suggested with the formula. That is not what Kepler said notwithstanding so many arguments with Academics I have had in the past that tried to prove to me that the two visionaries' views were equal and the same. Well…they are **not** the same, because when we go on to translate Kepler to the verbal English the letters that come out do not even spell the same words.

Translating Kepler's mathematical expression $a^3 = T^2k$ correctly to the verbal statement in English, Kepler said that there is a **space a^3** which is **equal =** to the motion in the **time duration T^2** thereof between two specific points which is a straight line **k** that holds a relation from a centre to an end where the two ends run from the beginning of **k** to connect at the end of **k.** I might not be the smartest boy on the block but I'm not that stupid either. I know how to translate… and I translate as follows:

a^3 must have a volumetric interpretation because the third dimension is sure evidence of multiple conjunctions of dimensions put together in three sides opposing three sides having the third dimension in place. The fact that any symbol uses a value to the **third power a^3** indicates **space** or a volumetric established and separate unit. Using a cube by three dimensions symbolises a cube, a room, a space to be filled, a unit able to hold other ingredients on the inside when empty or partly filled. It is space because it is volume using the third dimension.

T^2 is an indication of something having a cubic nature other than the square forming motion that is provided by the motion the square indicates, which is where the moving object is representing a third dimensional object that is moving from point to point and it is this point to point that multiplies into the square. The space is moving as a unit from one point to another point and the moving between the points are represented by a flat square or following a flat distance between two points. The cubic space was in one instant in one place and then the second instant in the other and because time can never stand still or become single dimensional (this I am about to prove as the letter unfolds) insisting that time must always support the motion it consists of or time cannot be. It is motion that is taking time, which is motion in the second dimension moving the space in the cube.

k^1 is the symbol used to indicate a straight line between two points with a definite beginning and a specific end position. It is the location where the cube is holding space and where the space was and where the cube in space is going to be in very the next split instant that follows to which will then in multiplying form the square that indicates the time the journey took to move the cube of space from one point where **k** is indicating the location of the space to where the next indicating of **k** will shift the space being the cube pointing at the end of **k,** but since time represents the square and with **k** being the distance that proves that the **k** represents the distance the space representing the cube went to take the time represented by the square through the motion. It is the distance moving space in the cube to complete time in duration in the square of motion; therefore **k** is permitted to be in the single dimension.

There are infinitely more implications in the statement Kepler delivered than what is merely a contribution to motion and only motion as Newton was of the opinion. What is there mathematically not correct in my interpretation of Kepler's manner of translating mathematics to English and why is any changing thereof by Newton or any other person necessary in any way?

We can test any of the following symbolic values in the mathematical expression and also test the principals behind the expression in which Kepler stated them. By such testing we will find that time after time there were never any corrections in the translations required since the translation thereof was never incorrectly presented and in that a case asked for no alterations to secure the correct reporting of the cosmic information being translated. By taking the formula on face value it can change as follows: $a^3 = T^2 k$ can become $k = a^3 / T^2$

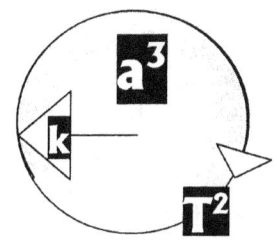

Kepler said
$a^3 = T^2k$ but that could
also be $k = a^3/T^2$

With this mathematical reality what then later formed the grounds for any individual to develop any need to change Kepler's translations from the cosmic given to mathematics and then from mathematics to English while the guilty party is renowned for his superior skills in mathematics?

Kepler translated what he found to be the cosmic given to mathematics which we humans are able to interpret from the mathematical expressed to the verbally pronounced and written but Newton still saw a need to change what the cosmos said about how the cosmos is presented and by no one less than by its own interpretation of its self structured composition.

When viewing my interpreting of what Kepler said I might have asked myself countless times what did I not translate correctly from the mathematical expressed to English after encountering a battery of Academic onslaught and resentment on my Newtonian views because after all it is directly diverting strongly from the teachings presented by Mainstream science and the diverting is not coming in a small way.

In truth from my diverting I came across very new ideas I am able to prove. By my translating Kepler's work correctly I came upon answers not yet uncovered by Mainstream Science

Kepler gave the World mathematically translated cosmic answers he received from the cosmos that Kepler uncovered long before Newton, Einstein and others got wise about cosmology...and later the wise came up with old news (old views as far as Kepler expressed their views before they, the wise were born with the purpose of coming to the conclusion that those wise men eventually did) and where the conclusions that the wise concluded brought much surprise to the world with the originality of the later Masters' initiative while Kepler said the same thing ages before…!)

Such is the advantage of recollecting Kepler facts that it does answer many questions, which went unnoticed and therefore not spoken about up to now and some were previously never even thought about.

Newton said a sphere is $a^3 = 4/3 \, \Pi \, r^3$, which is mathematically correct, however

Kepler said the cosmos told him a cosmic sphere is $a^3 = k \, T^2$ There is the two distinct possibilities which Newton saw and which Kepler saw and both are most valid. Between the two concepts there is literally one Universal difference and the two can never be mistaken as promoting the same principles. 'Ever try to answer facts about the Universe in as much as…what brings about the expanding? Kepler said the Universe plus it entire content is expanding centuries before Edwin Hubble realised what he was seeing through his telescope.

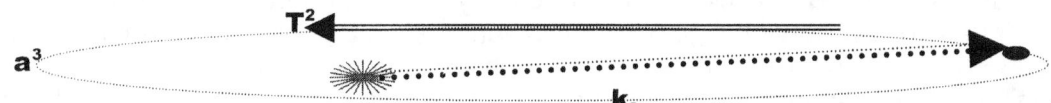

Kepler was the very first person to mathematically introduce **space a^3 centre k** and **time T^2**. Not only did he introduce **space-time a^3 / T^2** but he also placed **space a^3** and **time T^2** in a relevancy long before Einstein did and placed **gravity in space-time a^3 / T^2** even before Newton named gravity. He showed that space **k** is growing in the measure of what means the Universe attends to by promoting space-time as **$a^3 / T^2 = k^1$**. Kepler was the person who placed gravity as the ingredient in the universe that determines **space a^3** and **time T^2** and much more. Kepler was the first one that said that gravity comprises of two factors being **k** or linear gravity and **circular gravity or T^2** as gravity keeps space in form while all is staying together.

Although not one Academic has ever openly admitted to me that they as members and part of Mainstream science are more aware than I am of all the facts and doubts I point out to them, such evidence then becomes clear whenever I mention the matter to them I get more than the impression it does not come as a surprise to them and hit them like a brick between the eyes. The lack of surprise and initial doubt they should show at first when they discover the incorrectness of evidence in their theory is a tell-tale sign confirming my suspicions about their evidently knowing all this information all along. They clearly seem very agitated about every detail I show when I bring the mistakes and double talk to their attention in the hope that they may confirm my doubts. Never is there a whisper of a surprise or a hint of a suggestion that would initiate an argument carried on by the bewilderment or the astonishing surprise they should feel confirming my arguments because there is a mild complacency in their voices. My jumping them totally unexpectedly about matters they never contemplated in the least leaves them unturned. The rush in blood pressure that should be a factor on their part and part of the instant where total surprise will bring about some confusing thoughts that will inspire the unleashing of an argument in defending their holy grail should at least carry surprise in an attempt to save what they believe as being the Gospel in science and, with that, defending their honour. They lack embarrassment, which they should have in their disputing of my claim as they fight off my allegations with a countering of denial claiming foul on my part as they are in shock when finding out about any doubts. A lack of true emotion on their part is a telling sign that they also may have some serious thoughts on the quiet about any inclination presenting a flawed view of what they always thought they knew to be true. There is only that eerie dismissing of the seriousness and the lack they show in excitement that would deny or support my credibility as I present my findings. If they know about the inconsequential facts in science why is it not generally acknowledged and pronounced as a matter of fact? Why the covering up and hiding of facts that we associate with some professional criminals such as politicians. The fact that Academics are aware of this evidence in general terms of the misinformation and doubting evidence of Newton's cosmic vision, but moreover underlying this is their total denial of knowing about it and that is what is so seriously unforgivable. The fact that all Academics are aware of my evidence even before my presenting them with such evidence is beyond doubt. If that is the case then why are they forever trying to kill my viewpoint and forever trying to silence me where I am only the messenger because I bring the solution and the answer? Please note that the answer and the solution are unbelievably simple and unsophisticated. It lacks all the splendour and grandeur expected by all Academics concerned. It is because it is so simple that it went amiss for four hundred years. It is because it is so simple that it misses the grandeur that will entice them. Instead every academic accuses me of not understanding Newton while they can't show me what part it is that I can't understand and I on the other hand can't see what there is not to understand..

Newton said that it is the reducing of the distance between the objects that would bring about the un-reversible reducing that will end in a total demolishing of the radius that is between the cosmos structures, but instead we find the gravity applying in outer space is one of the instances where gravity provides an orbit circle that gravity seems never to completed as the orbiting objects follow from closing any circle that is leading into a following circle up to where the circle is completed in cyclic precision. That is not the gravity that Newton identified   although Newton admitted that there is a presence of a centre forming a point in the middle between the two objects. He was unable to know what caused or even the presence of the Coanda principle, which forms so critical a part of my theory. The formula concerning cosmic balanced gravity however leaves no room for the admitting of such a point and by not leaving a possible inclusion of such a point in his formula Newton did by such gesture in principle repeal his admission of such a centre. This had me cast doubt on what is taught at institutions of learning. It motivated me to venture back to an era before Newton came to influence science. I came to acknowledge Kepler as I came to understand Kepler. The acceptance of what I understand in Kepler involves much more reading into what Kepler said by finding what Kepler did not say in the way that he did say what he said, than reading about what Kepler said as it is written in precise detail and to the letter used in his statements. He never directly stated what he said. Again I must stress this point: When I refer to what Kepler said it most likely means reading into the part that he did not say when he was saying what he said but I accept that he meant to say what I am reading and translating from Kepler as part of what he did not say but meant to say. I have to read more with my mind than with my eyes. This comes as a result of interpreting Mathematics to the verbally expressed. I had to learn to read with my mind and not my eyes and I found that that is the manner in which one has to approach cosmology. From the first time I discovered what manner one should use if one wished to read into Kepler's findings I saw Kepler was all about uncovering the unknown. Realising that, the conclusions I drew by reading in such a way cemented my better understanding of Kepler's work, which then helped me improve my insight into Kepler's work as it increased my understanding about cosmology several fold. This helped me to realise what implications were to be found underneath Kepler's discoveries. From my realising what approach I should use, it helped me to improve my cosmic realising by using the method of reading Kepler and from that I could come to appreciate what Kepler introduced.

Only then did it bring insight and proof to me as a student of Kepler and this proof I found by dissecting what <u>Kepler did not</u> say instead of what he did say, which I now present to you with this letter, you being a superior intellectual person. Kepler said $a^3 = T^2 k$ and that correctly translates to a mathematical expression $k^0 = a^3 / T^2 k$ which in the verbal statement in English translates that Kepler said that there is a **space a^3** which is **equal =** to the motion in **the time duration T^2** thereof between two specific points which holds a relation onto a centre k^0 where from there forms **a straight line k** that is centred on the spot where space begins from k^0 **that produces k** as well as producing the circle therefore that spot $k^0 = a^3 / T^2 k$ has hold k^0 at a value of having the least space. The line **k** is centred onto a spot where space begins specifically at k^0. This point not only produces the line k^0 but represents also the space that forms the eventual circle T^2. Therefore from the centre holding k^0, k^0 leads to **k** that forms the roving space a^3, which is rotating at a distance **k** where T^2 forms the outer limit of k^0. Mathematically $a^3 = T^2 k$ will be $k^0 = a^3 / (T^2 k)$ because $k^0 = 1$. But $k^0 = 1$ also present the single dimension where all factors are a product of one. If one can locate k^0 one will find singularity. That is where gravity is because gravity is strongest where space is least. Then that suggests that gravity is strongest at k^0 because space is least. That is gravity because that is what keeps the orbiters in orbit but also that is what Newton completely missed when he changed Kepler's work. Newton failed to recognise gravity as the only ingredient in Kepler's formula. He admitted he missed this because he admitted he did not know what gravity is while Kepler explicitly showed what gravity is. Gravity is what keeps the orbiters orbiting. **$k = a^3 / T^2$** is **distance[1] = space [3]/ time[2]** forming from a pivoting centre k^0. That is a cycle and moreover it is a cycle formed **by space / time**. What Kepler said is that space is **a^3 in motion T^2 k.**

That says **space[3] (a^3/)** relates directly to **time[2]** that uses the symbol **T^2**. This is also what I refer to when I say one has to read what Kepler did *not* say when one wishes to see what he *meant* to say. Kepler introduced space[3] –time[2] long before Einstein's date of birth appeared on any calendar although Einstein is credited with the formulating of the concept of space-time and giving it a name. Going even further Kepler stated that the space **a^3** is on the move T^2 around in a circle at a distance **k.** That is what that comet we are discussing is doing. The space[3] (Comet) is circling the sun using a radius **k** to establish the cyclic time[2] as a period of continuous motion and continuous motion is gravity. That reads much more correctly and closer to the truth than what Newton predicted what according to him (Newton) was happening in space. Remember in this statement I am separating cosmic principles applying from the way that gravitational principles apply on Earth. I distinguish that which is the rule in the cosmos from what we find ourselves trapped in on Earth. The two just don't mix. I am removing cosmic physics from normally accepted physics because the gravity concerned is not the same.

The proof I bring is real however simple it may seem. It has none of the mind-blowing complexities normally associated in the presenting of investigative analyses of Astronomy. I realise the information in this book carries the arguments in a childlike manner which are very simple to follow, and for that in the past I have been blamed over and over again as being unprofessional. In my answer to that I can only reply by using another question: Are only professionals adequately equipped with minds that make them (the professionals) the only ones able to think? We being part of the human race are all thinkers. Everyone as a human being can think. Every person on Earth is a thinking thinker that uses his brainpower by exploring thoughts mainly and normally to his or her personal benefit. It is what we think about that produces the results of our efforts by which we accomplish what ever we are thinking about. I have met professional Academics that I found foolish as much as there are other cases where the so-called amateurs can credit themselves with much wisdom and insight. Albert Einstein as a patent clerk was that much but to name one. Please understand that I do not compare my achievements or myself in any way, shape or form with the likes of a Master such as Einstein although I speak my mind when not being totally in agreement with some of his or other views. My unsophisticated retracing of Mainstream physics concerning the Big Bang in detail helps to reinvestigate established principles and moreover investigate proof in the light of modern evidence. In principle I distinguish between Kepler and Newton in that Newton is one hundred percent correct concerning gravity on Earth but as far as outer space forms gravity the conclusions of Kepler and Newton do not match and they had totally different ideas about what they saw in gravity. I am in disagreement with some basic principles that science acknowledges and I divert strongly from all accepted roads Mainstream physics follow. By my doing that those who are considered and accepted as self-proclaimed members of Mainstream Physics have categorised my views in the past as incoherent. That I do not accept. I admit that my line of thought is extraordinary and controversial but only to Mainstream science and not to the standards laid down by nature. Since the concepts I follow start at the beginning, and I take Kepler at the point where modern cosmology began and in that mindset I re-evaluate Kepler's work. I start by tracing a new approach as to what I see Kepler found. The main condition of my investigation is to establish a divorce between what Kepler said and what Newton thought to add to what Kepler said. It is this divorce I create that Mainstream science finds repugnant or even in some persons' opinion repulsive. I believe the repugnancy does not come from or is not manifested in any part of my work to the letter as such, but rather what my work suggests and who is doing the suggesting. To my view in cosmology such adding to Kepler by Newton was unnecessary and it diverts Kepler's work away from cosmology. But as the generations moved on Newton became religiosity in the mind of science wherever science was taught. To students there is little or no choice in the matter since the only choice left to them is one of understanding by forcefully accepting or die an academic death since Newton is academically accepted without asking questions or raising an opinion. For the second choice, the less accepting students are greeted with a Dear John good-bye letter sending them off into the unknown sunset that such a future outside physics will bring them. That is brain washing.

From studying Kepler I saw that we have to gauge what we find in the Universe. What we find is not what we realise with our eyes but what we observe by using our minds to translate from visions coming from our eyes to our minds. We have to test the part that we are seeing much more than merely accepting what there is to see on face value. We have to not only see what other life beings blessed with much less insight most probably also should see. We must stop using our eyes in the same manner as animals do and start seeing with our minds, as humans should do. Being the superior evolved species that we are gives us the ability to read into that which only we can see and that we only can see by using our intellectual mindset. By seeing with an intellectual understanding what there is to see when we see what we can observe, we should therefore have the ability to be in understanding by looking at what we can see but moreover understand that which we cannot see. It is the same as playing chess. See what should be moved instead of noticing objects not having the ability to move on their own account. This I first found to be true about Kepler's work and when I started projecting this method of observing what the Universe is, as it scattered most previous perceptions I found that using the new method brought along answers so fast I could sometimes hardly keep up with the interpreting thereof. But as is the case with Kepler so is the case with the entire study of cosmology: One should see what there is about the cosmos which is unseen to us and then we may find so much more in the cosmos unseen to us representing that which we cannot see and that which we cannot read because we have to learn to read what is not written in light. Armed with this realising I then proceed from that point by further arguing and debating the full implication of Kepler's contribution. Kepler placed cosmic structures in relevance to one another and so does the Big Bang Theory. The backbone of the Big Bang is that relevancies apply in dynamics and such dynamics are placing all structures without any reservations independent from each other. As the Big Bang progresses all inside the Universe is in the same Universe that will always be the same, however the relations that the elements comply to bring across new relevancies with new positions to fill.　The father of the Big Bang concept is a person by the name of Father LE MAÎTRE, GEORGE ÉDOUARD　(1894 -1966) who was a Belgian priest and cosmologist. He was the first person to embrace the fact that the universe expanded from an infant stage.　His model of an expanding Universe (1927) was superior to that of W. de Sitter in that it took into account mass, gravitation and the curvature of space.　Similar models were proposed in the early 1920s by the Russian mathematician Alexander Alexandrovich Friedmann (1888-1925) but Friedman compiled various such possibilities. Lemaître argued further (1931) that the quantum theory supported an origin in the explosion of a 'primeval atom' or 'cosmic egg' into which was originally concentrated all mass and energy. As modified by A.S. Eddington, Lemaître's model provided the springboard for G. Gamow's Big Bang theory. In the wider picture of science in general a lot changed to just allow such turnabout in thought since the day of Isaac Newton. From Newton's attraction and contraction many things came into place that allowed change in the most hardened minds. Accepting facts about the Big Bang concept is quite radical. By promoting expansion the Big Bang theory contradicts gravity and our accepting of the Big Bang has to change all other concepts. By accepting the Big Bang other changes are also involved.

KEPLER, JOHANNES (1571-1630)

The German mathematician and astronomer KEPLER, JOHANNES (1571-1630) became Tycho Brahe's assistant in Prague in 1600 A. D. where he undertook to complete the tables of planetary motion Tycho had begun. Kepler first calculated the orbit of Mars. He spent much time trying to reconcile Tycho' s accurate observations of the planet with a circular orbit, but concluded (in Astronomia nova, published in 1609) that Mars moved instead in an elliptical orbit. Thus, he established the first of his laws of planetary motion. A theory that the Sun controlled the planets by a magnetic force led him to the second and third of his laws, which were published as part of his treatise on theoretical astronomy, Epitome astronomiae Coernicanae (1618-21). The Rudolphine Tables (named after Tycho's patron, the Holy Roman Emperor Rudolph II) of planetary motion appeared in 1627 and were still in use in the 18th century. Kepler also wrote De Stella nova, on the supernova of 1604 and Diptirce on optics and the theory of the telescope. The overall view followed in this book **An Open Letter To Selected Academics ISBN 0-9584410-9-X** places the true significance of his work in true contents. In KEPLER'S EQUATION is the equation that relates the eccentric anomaly of a body in an elliptical orbit to its mean anomaly. The equation is $E - e \sin E = M$., where E is the eccentric anomaly, M the mean anomaly, and e the eccentricity of the orbit. It is important as one of the mathematical relations enabling the position of a planet about the Sun, or a satellite about is planet, to be calculated from the orbital elements for any time. However this only relates to the solar system, and KEPLER'S LAWS only apply in the contents of the solar system. The three laws governing the orbital motions of the planets, discovered by J. Kepler is as follows: The first law states that the orbit of a planet is an ellipse with the Sun at one focus of the ellipse. The second law states that the radius vector joining planets to the Sun sweeps out equal areas in equal times. The third law states that the square of the orbital period of each planet in years is proportional to the cube of the semi major axis of the planet's orbit. The first law gives the shape of the planet's orbit; the second describes how the planet must continuously vary its speed as it follows its orbit, moving fastest at perihelion and slowest at aphelion. The third law gives the relationship between the planets' average distances from the Sun and their periods of revolution.

Instead of studying the true value and contribution of to Kepler's laws an Englishman going by the name of I. Newton placed his own interpretation to Kepler's laws, and in doing this, he wilfully destroyed the principle working of the Creation. Saying this I hear the alarming hooters announce Newtonian dismay. In the past my experience was that all the revered Academics lost their appetite for any further investigation of my work. That is sad as much as it is regrettable. Through Newton's tunnel vision, he applied his own misinterpretations to the correct presumptions of Kepler and through the Newtonian tunnel vision Academics did not move an inch away from repeating the same procedure. In the past it was this that had Academics shying away from me because at the point where I raise criticism of the Newtonian viewpoint I am rejected. The point where I declare my suspicions concerning their accuracy and the correctness of their theorising, which is where I should then be raising their doubts about their way of thinking, is the point where instead I raise their suspicions about my way of thinking. This is what caused the rejection of my criticism of Academic Newtonian science and evoked their criticism of my views in the past, instead of them following the logic by investigating what I said. Their rejection of self-investigating had me and my work rejected to a point where the applecart lost its wheels on every occasion. It is where Academics read my remarks and what brings (seemingly in an instant) wrath to Academics. I say this because I realise that reading my remarks or hearing me remarking about this notion brought much resentment on their part and if the reader at the present moment is a Newtonian, boiling his/her blood. It is blood boiling because I believe they see my remarks as belittling what they feel they have accomplished. This is not the case but still my remarks have the same effect on the Academic as pouring icy cold water down the back of his shirt. I mention this because I know it has happened many times before and if possible I wish to avoid this response. Therefore I ask you kindly to please be warned about the negativity you must feel towards me where you are the Newtonian and I am not. Before you lose interest in reading this letter any further please allow me to finish. In the past Academics thought me to be presumptuous and that normally became the point where all the Academics find their interest vanishes. That should not be because if Newton's work is as utterly accurate as those with faith in his work believe it is, then every aspect about Newton should stand above any and all reprimanding or any form of doubt causing a notion to reprimand. The testing of Newton's work should withstand all testing notwithstanding the person or the prominence of such a person's social or academic standing in the Academic society or even the prominence that such testing will deliver. From what I see about Kepler's work it is a flow of circumstances that lead to Academics neglecting Kepler's work and the realising of the theory I suggest is not forthcoming due to my personal brilliance. I do not consider myself to be the brilliant in any way as to be the one that can remove the verbal splinter from the eye of the Academic. Yet…if there is a splinter what else should I then do…Newton reduced the implication that Kepler findings hold by introducing to the law of gravitation. He then went about and changed it to three laws of motion. It is clear that while he formulated the laws on motion he missed the way Kepler introduced gravity as space a^3 coming about through motion T^2 and that gravity is space a^3 within space k within motion T^2. Newton also missed the fact that gravity is at its strongest where motion and space cease to be. This is most important to recognise about gravity in one of the two forms it has. I. Newton generalized Kepler's first law, verified the second law, and showed that the third law should be amended to the form; $4 \pi^2 a^3 / T^2 = G (m + m_p)$. In this, the value of "**T**" and "**a**" are the period of revolution and semi major axis of the orbit of a planet of mass m_p about the Sun of mass m, and G is the gravitational constant.

It should be clear to any person investigating Johannes Kepler and his work that Isaac Newton hijacked Kepler's work and any time there is the slightest referring to Kepler about the research Tycho Brahe and Johannes Kepler did such referring to Kepler always lead to and always include the mentioning of Isaac Newton changing the work of Johannes Kepler. It is as if the World never could acknowledge Johannes Kepler because the work of Johannes Kepler would be completely wrong and misleading if it were not for the intervention of Isaac Newton saving the skin of the less admirable Johannes Kepler. This comes in the midst of every one realising that Kepler used the information he received directly from the cosmos. I do stress this on many occasions throughout the letter because the embarrassing part is that Newton changed the work of The Universe and not of the man called Kepler. Should you reading the letter entertain the opinion of Newton and feel any urge to defend Newton you should ask the question "who is standing corrected?" Is it Kepler or is it the cosmos that gave Kepler the information he concluded? The cosmos supplied all the information by using mathematics, which Kepler then had to translate. But Newton destroyed the accuracy by altering what the cosmos said and directly by adding to that what he (Newton by name) thought that the cosmos left out. This set a precedent by Newton in cosmology and also set a trend, which was retained in all future cosmological development and it lasted in cosmology for three hundred and fifty years. In this book you are reading I am about to show that such practise should no longer be accepted in cosmology. In the process the world of Mathematics developed and the world of cosmology stood still for almost four hundred years. Faculties contributing to cosmology and feeding off cosmology improved as much as they developed, but when cosmologists see the Roche limit in action in the lens of the Hubble telescope and refer to the event as "stars blowing bubbles" being the ultimate response coming from those persons who are supposedly the Masters of cosmology affairs, then the truth of what I just said comes down on you like a ton of bricks. Everyone having any remote interest in cosmology will find they are being very disillusioned by such "official" testimony about the evidence the Ultra Wise report on. This book is about showing how great Johannes Kepler was and how enormous his work was. It will show he preceded all ideas of everyone that came later and officially introduced the novelty of such ideas. Back during the time Kepler was introducing his work the stature and the magnitude of his work was beyond any person's understanding (including Isaac Newton) and this prevailed for most of half a millennium. I do not say I am the brilliant one to uncover Kepler in the face of everyone failing that came before me, but as I am not a Newtonian such bias was not part of my repertoire and denying me the fortune of being a Newtonian added to my fortune of realising Kepler. Yet as you will notice, the work I contribute is much below the sophisticated norm of modern investigative research and the levels that modern research accomplishment demands to better the effort of the understanding ability in the splendour that investigative research work should deliver in view of our modern times. It is only pure neglect in science circles that moved science past

Kepler. Not seeing and therefore not investigating through almost half a millennium has paved a road past the inferior levels that the researching of Kepler's work holds because it was rocket science four centuries ago but the brilliance of it has faded since then. My contribution holds no astonishing flair that may add to science in general. Only failure to notice what I see on the part of those truly brilliant can explain my being able to present my contribution in investigating Kepler. Only by their passing such degrading levels of the Academic establishment in the past and the present can bring the blame for such an obvious discrepancy because any involvement in the work at such an inferior level as that which I bring cannot interest and excite a salted Academic and when thinking about it, the idea is totally unthinkable. This letter, although it is on this inferior level is about correcting this tendency and has in mind the effort to put in writing what would place Kepler in the greatness and glory he deserves. As I already said, if Kepler was wrong then the cosmos was wrong about facts and applying relevancies and tendencies in the cosmos. I yet again wish to reiterate we should never for one moment forget that Kepler received his information directly from studying the cosmos so how could the cosmos stand corrected? In spite of all the brilliance attributed to Newton nonetheless if Newton had the mind to change Kepler's work and my saying this includes all persons agreeing with such changing by Newton of the work of Kepler, those persons admit that he or she or Newton never took any time to really and truly investigate what the cosmos told Kepler. From my reading into the work of Kepler I prove gravity, the Titius Bode law, singularity, space-time, space-time relevancy, the Lagrangian system, the Coanda effect and the Roche principle, the sound barrier, the principle behind the Black Hole. The precondition for my ability in doing so is that I have to remove Newton's opinion about Kepler's work from Kepler's work. Whenever cosmology comes into question and all the phenomena, which I mentioned just now remains unexplained and by that token alone it shows to what degree did cosmology remain undeveloped. Whenever there is any mention of Newton, Kepler is never mentioned. But the reverse always applies. Mainstream physics holds the opinion that Kepler may only have an opinion if Newton can change the opinion. Kepler gave space-time, gave gravity, gave singularity, gave the Plank theory, gave the theory on relativity but no one ever found Kepler's work deserving enough to launch any investigation such as I did. I be-labour this because of what revulsion my rejection of Newton unleashed. This is one barrier much unnecessary but it has been an insurmountable barrier thus far.

NEWTON, ISAAC (1642-1727) and NEWTON'S LAWS OF MOTION
An English physicist and mathematician who developed his principal theories about gravitation, optics and mathematics between 1665 and 1666. In 1668, he made the first working reflecting telescope. Most of his work remained unpublished for long periods, partly because of criticisms by c. Huygens and the English scientist Robert Hooke (1635-1703) of his early work on the corpuscular theory of light. However, in 1684 E. Halley persuaded him to organize his work on the celestial mechanics of the Solar System, which was published as the Principia. Newton's other major work, Opticks, was not published until 1704. It contains his corpuscular theory of light, and the theory of the telescope. His greatest mathematical achievement was his invention of calculus, independently of the German mathematician Gottfried Wilhelm Leibniz (1646-1716). His profound influence on physics and astronomy is reflected in the phrase 'Newtonian revolution'. Three laws published in 1687 by I. Newton concerning the motion of bodies.

1.A body continues in a state of uniform rest of motion unless acted upon by an external force.

2.The acceleration produced when a force acts is directly proportional to the force and takes place in the direction in which the force acts.

3.To every action there is an equal and opposite reaction.

4. However there is one more law on motion that went undetected by Newton...This book is not about trying to disprove Newton...it is about adding too science more than there now is available without removing any that science already accumulated.

In this book I use Kepler's formula to either prove or to disprove the following accepted principals in cosmology and if any person in the past gave only the slightest attention to Kepler's work, many statements would have come much sooner delivered by someone else or may never have come at all. By applying Kepler's formula correctly in this letter I can either agree with or in other cases deny the following principles.

It began with NICOLAUS COPERNICUS who changed the status quo. COPERNICUS, NICOLAUS (1473-1543) was, according to the Anglo Americans, a Polish churchman and astronomer although this is just more politically inspired propaganda because his parents were both German (in Polish, Mikolaj Kopernigk). While he was completing his studies, he had realized that the Earth revolves around the Sun and not vice versa. Such a view was in that time, held to be heretical. As I pointed out in the first few articles, the Church regarded the geocentric world-view of Ptolemy as consistent with its doctrines. Copernicus set down his basic ideas around 1510 in the Commentariolus, which he circulated anonymously, because of the Islam link. In 1512-- 29 he conducted his study and concluded the observations that he needed to support his theory, while carrying out ecclesiastic and local administrative duties. In this time, he had to defend his mother in court on charges of witchcraft. In 1539, the Austrian astronomer and mathematician Georg Joachim von Lauchen (1514-74), known as Rheticus, became a pupil of Copernicus and began to spread his ideas. The published work was openly spread as the Copernican system, in spite of the life-threatening dangers connected with such a "crime", in 1543 in the book De revolutionibus orbium coelestium. However, the reality of a heliocentric Solar System was only commonly accepted, after the work of Galileo and J. Kepler. The ideas introduced developed along and proved to be correct until such a time it met a solid wall with the investigation of Max Planck.

PLANCK CONSTANT
(Symbol h) A constant that relates the energy of a photon to its frequency. It has the value 6.62076 $\times 10^{-34}$ Js. It is named after the German physicist Max Karl Ernst Ludwig Planck (1858 – 1947). PLANCK ERA. In the Big Bang theory, the fleeting period between the Big Bang itself and the so-called Planck time when the Universe was 10^{-43} s old and the temperature were 10^{34}K. In this period, quantum gravitational effects are thought to have dominated. Theoretical understanding of this phase is virtually non-existent. It is named after Max Planck (1858-1947). PLANCK'S LAW

A mathematical description of the energy radiated at different wavelengths by a black body: $E = hf$, where E is the energy of a photon and f its frequency. It was formulated in 1900 by Max Planck (1858-1947), who realized that energy is radiated in discrete packets, which he called quanta, and it formed the basis of quantum theory. The quantum of light is a photon, the energy of which depends on its wavelength.

There is one rule which is well established and which Mainstream science all agrees about. It is one aspect, which forms the very principle that holds the theory about the cosmic start together under the covering of a verbal blanket. All in science agree that it all started with singularity but I manage to go one step further where I prove that it is also where it ends, as singularity reunites space-time, which is from where Creation split in the very beginning.

Singularity is as follows: Singularity: a mathematical point at which certain physical quantities reach infinite values, for example, according to the general relativity, the curvature of space-time becomes infinite in a black hole. In the big bang theory the universe was born from singularity in which the density and temperature of matter were infinite. From singularity flows space-time.

Space-time is as follows: Space-time is a four dimensional position of the universe where the position of an object is specified by three coordinates in space and one position in time. According to the theory of special relativity there is no absolute time, which can be measured independently of the observer, so events that are simultaneous as seen from one observer occur at different times when seen from a different place. Time must therefore be measured in a relative manner as are positions in three-dimensional Euclidean space, and this is achieved through the concept of space-time. The trajectory of an object in space-time is called world line. General relativity relates to curvature of space-time to the positions and motions of particles of matter.

SPECIAL THEORY ON RELATIVITY
A theory proposed by A. Einstein in 1905, based on the proposition that the speed of light in a vacuum is constant throughout the Universe, and is independent of the motion of the observer and the emitting body. A consequence of this proposition is that three things happen as an object's velocity approaches the speed of light: Its mass goes up, its length shortens in the direction of motion, and time slows down. Hence, according to special relativity, no object can ever reach the

speed of light because its mass would then become infinite, its length would become zero, and time would stand still. In addition, Einstein concluded that the mass of a body is a measure of its energy content, according to the famous equation $E = MC^2$, where C is the speed of light. This equation describes the conversion of mass into energy in nuclear reactions within stars.

GRAVITATIONAL COLLAPSE

The collapse of a body that is unable to support itself against its own gravity. Gaseous bodies undergo such collapse if they are not hot enough for their gas pressure to balance gravity. This can happen in the early stages of star formation, or when nuclear burning ceases in a star's core. The time taken for such collapse decreases rapidly with increasing density, varying from about 100 000 years for the birth of a new star to less than a second for the formation of a neutron star. Star clusters may undergo a similar collapse if the random motions of their constituent stars are insufficient to offset gravitational effects, either during their formation or at an advanced stage of their evolution.

GRAVITON

A hypothetical particle or quantum of gravitational energy, predicted by the general theory of relativity. Gravitons have not been observed but are predicted to travel at the speed of light and to have zero rest mass and charge. A graviton is the gravitational equivalent of a photon. It is this anti-photon-being-a-graviton by just merely swapping direction and all is proved that I find not very indigestible in modern science. One of the main issues that I wish to protest by my writing this is my argument that if the Universe can be compressed back to the size it had at the point of 10^{-38} seconds after the Big Bang the daily outdoor temperatures of 10^{27} K will also come about once more. The expansion was the result of compressed space, which then formed into heat and in turn resulted in finding a Universe with all the insufficiency of space- less-ness prevailing throughout and wherever space was needed. By that it forced space-time to come into being. Space-time came about at the time of endless time duration without space availability, which brought about the period of the Big Bang wherein space growth was the converting of such heat to space. If the Universe was in a vacuum as big as being available now, then what was the temperature of the vacuum while it was empty before material filled it later? Then I presume the vacuum was there as it is now in this present day. If the Universe then employed the space of say one atom, the impression comes through that from edge to edge and from Universal border to border the space occupied was the same as one atom will claim in our present day and age. Normal gravity started at 10^{-43} seconds. The Universe was the size of a neutron or somewhere in that vicinity. The Big Bang began and GUT, or the grand unified theory, produced the attempt to describe the strong and weak nuclear forces and electromagnetism in one single mathematical theory. Somewhere before 10^{-12} seconds of counting the Universe cooled to about 10^{15} K the electromagnetic and the weak interactions acted as one single physical force. Science reckons that unification may come about at temperatures of 10^{27} K, which was the temperature of the day at 10^{-38} seconds after the Big Bang. This statement echoes my viewpoint but one has to look carefully for that to surface.

In the suggestion the presumption claims that all the space that the Universe made available at that time was the total space one atom might take up today. If that might be the case then where was the rest of the space that now fills the Universe? Or was the rest of the space we now find in the Universe and what is now explained away as the vacuum, also available back then. Did the Universe only have that one tiny hot spot it filled with huge volumes of heat? Was the rest of the space vacant out there all along during all the time running to the present date filled with emptiness standing around as a big vacuum with nothing better to do than suck on the Universe while the Universe was exploding at the speed of light. Then that statement suggests that in this hot Universe there were light-years upon light-years of vacuum waiting to be filled by the intense heat soaring in the smallest spot. If that is the case then why did the vacuum not fill in the blink of an eye by all the exploding expanding material growing at the speed of light? Was the Universe overall bitterly cold where the vacant space was locked in with one spot of the vacuum filled with temperatures so hot we can only produce it in numbers suggesting a value but never claim to be able to digest the reality thereof in the human mind? If so what happened to the natural consequence that heat flows in the direction of cold and equalises between hot and cold? Was the space available at present available then or was the hot space the only space available at the time? If so what prevented the heat from instantaneously filling the eternally cold vacuum because with the rules controlling vacuum in effect, it should have filled in such a manner in less than a heartbeat?

I believe that singularity formed space-time and space-time developed from the overflowing of space-time at the time is extending by marching onwards and outwards to this day. Space-time developed another product that everything in the cosmos has to have. It must be in such large quantities everything imaginable in the Universe has to have it and that is space using time to move about. I suggest that it is space that is holding heat in a quantity providing density and ratio to space available and in relevance to the space being available to quantify the presence of the heat and which then proves to form the time factor. The container and contained all together mixed by motion. From that very first separating of heat and space, which is what formed from singularity to produce space-time. The Universe was full… It was overflowing by the speed of light in the beginning…so where and when did vacuum or "nothing" enter the Universe as a factor if and when the Universe was so full?

The answer to that is absolutely crucial because how did the Universe decide to fill some parts with a variety of something and decide to fill some parts in between with nothing? If that is true why did gravity not prevent the vacuum filling because no gravity that came about since can beat the force that gravity had back then? This leads to another question following the previous one in asking why did gravity at the time when it was so strong with r^2 so much compromised not fill the nothing immediately as it entered with something that could absorb the nothing. At the very beginning the mass that was pulling on the mass by force was immeasurable and none quantifiable. Even more to the point is the question to be asked in how big was the radius between the materials with the immeasurable mass placed in such a little space? This is all the more important in the light that the smaller the radius is the bigger the force will become from the immeasurable mass pulling…

With the immeasurable mass that was producing the first gravity between the particles divided by an almost non-existing radius the gravity produced had to be in gigantic proportional quantities and with the separation of the radii being in the infinite measure that it was at that point then how did the Universe establish the chance to expand? It did expand, as we all are witnesses to in spite of this contraction of gravity that had to have been compromising the expanding factors. Still the expanding filled the unknown part of the unoccupied Universe, which at the time was there or was not there, and if it was there it was then filled with "nothing". If the "nothing" was not "nothing" then the "nothing" that was not "nothing" was also filling the rest of the vacant Universe which was or was not because if it was it was filled with "nothing" and if it was not then it was "nothing".

This is then taking into account then that all the reducing which is resulting from Newtonian contraction and which was going about in the space available at that time was something filled with "nothing" and surrounded by more "nothing"? With everything in the Universe being that much crowded and crammed where and how did "nothing" enter the Universe and fill the rest that was unfilled? What factors introduced "nothing" into the picture since the entire Newtonian concept finds its base on the principle that matter reduces using gravity by force which then brings about reducing or the removing of the many "nothing" between particles, which will then lead to "nothing" that has to vanish even before "nothing" can enter the space. This question may seem small-minded belonging to the mentality of a child or to that of the mentally impaired with not much factual appreciation developed yet. Please do not see it that way. If you think on those lines it will be because you do not have an answer to challenge these silly questions. Beware! Silly as they are they represent official backing by the Wise-and-Informed. If the space is "nothing" and if the space was as large as it is at present then there was no need for such a small area to fill with something leaving only the rest filled with "nothing" at first since all the space we know about was there present and by being present it was there then for the taking. Whatever filled the Universe had to start at the centre of the Universe and fill the entire Universe all over from a centre as it moved outwards filling from the inside outwards. This is a natural human instinct realisation but is beyond proving by using Accepted Scientific policy. But that leaves Newtonian science with a massive unsolved problem: where is such a centre at the present time and where does the centre produce the limits or border it apparently has to form as it expands?

By expanding there is an additional contribution to that which was there when that that was there, was receiving more of that which was there. There was therefore an increase of what there was before the addition increased that which was and by then becoming more than there previously was. The increasing has to bring an increase improving the border from where it was or must have been before the adding took place to where it is after that that was added was added. When that was less than and became more it became more when it was added too before it was at the limit, which was there before it was added to, and that limit that there was, was a limit that is the limit that I am referring to as a border being there. The cosmos is filled with unrecognised borders. The expanding has to be an ongoing filling that is at the same time expanding from the inside towards the outer limits of the Universe. Since "nothing" can enter from the outside where "nothing" is, the filling of "nothing" as a substance that would take up vast quantities of room had to fill from the very centre spot where all other filling came from. This filling of "nothing" with material has to be well mixed. The truth about cosmology is that space forms no borders but by using any Newtonian centre from where mass is attracting we must find a point where there has to be the ultimate Universal centre which is the cardinal point in the entire Universe and it is the first, the prime position to locate coming before any other concept one wishes to put forward because all concepts have to start with locating that cardinal centre. There has to be the ultimate r^2 radii located precisely between the ultimate mass drawing the other ultimate mass closer. If there was a Big Bang then there has to be the spot where from the Big Bang developed therefore there has to be such a centre connecting the past to that ultimate centre with the line of development flowing onwards to this day.

The fact that science is Newtonian proves that in the meantime Mainstream science is still of the opinion that there was the specific centre in the Universe which is nowhere to be found as it was filling the unknown with nothing coming from nowhere, but which somehow is still somewhere in the centre of all of that which is something. On the opposite side of nowhere there is an outer border in space producing a limit to "nothing" and serves "nothing" with a specific point to stop being "nothing" because that point is precisely where "nothing" ends and forms a beginning of a Universal border or a Universal end. How one will stop vacuum being no longer "nothing" was a question everyone comfortably missed to ask therefore no one ever seemed to deliver any form of answer.

One night some years ago a very close friend of mine had a meal at his restaurant and as the conversation progressed he asked me about space and where it must end. I tried to explain to him what I believed in comparing to what Mainstream physics believed but soon saw I was not gaining his understanding. Then I decided to jot it down on paper and he could read it at his leisure as he saw fit. That led to the first book written by me (in Afrikaans - my native language). What I tried to explain to Johan Boonzaier that night is that if the universe was the size of, say, even a tennis ball with only the size of a tennis ball being the very all of space available, then yes, it must take time to expand from that having the excessive heat there was back then in all the space we have at present. It then is converting heat into space bringing about the expansion. But one will find most expanding within the atoms, as the atom must grow since the Universe in all was the size of what one atom is today. The space in the atom pushed the space outside the atom but there must be plenty more to the growth. Something outside the atom contributed in its own right because there is more expanding than there can be blamed on coming from the atom. But the space then also developed as the universe developed and if space developed then it cannot be total vacuum filled with "nothing" because "nothing" cannot develop. You the reader must judge who is correct between my view that space developed with the Universe as part of the Universe and reject the official view about space being nothing or otherwise you the reader must then decide that I am wrong, but should you do that, then find a reason why the Big Bang started out small and filled all the available vacuum or what is contemplated as vacuum we have with the motion of time. When Mainstream science accepted the Big Bang as the principle that will take science into the future the view about such a Big Bang concept unlocks a different door to another view on the cosmos from birth to end. It calls for revising all aspects of the entire history on cosmology and change what dead wood needs chucking out. Most of all it was my following the lead I got from Kepler that unlocked the doors I now present to you. I claim there is no graviton as there is no gravity forming weight or forming mass. I hope the sketch helps with my explaining effort:

There is a point where the two points forming the relevancy unite in shared singularity. It comes as a result of shared motion

That part Newton saw and formulised

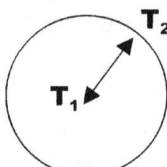

In all space-time one finds at least two relevancies where one is at the centre.

There is another part he missed. Crossing a limit of inclusion is the limit of division and such limits are in distinction by motion producing the gravity, which is parting the two objects. Motion brings about a relevancy where two positions no longer share a common point in singularity. **That is what Newton missed.**

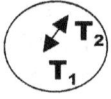

That is the gravity aspect Newton and all other Newtonians miss.

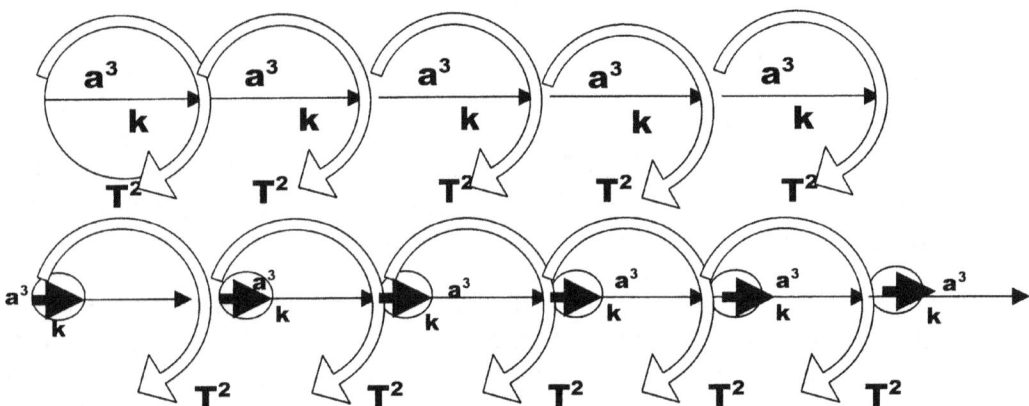

Two objects of substantial size differences are travelling at the same time but one has a space, which it has to move when it travels that is considerably different from the larger space. The larger space will produce an extending line equal to the space it moves while the smaller space will also produce a line in ratio to fit the space it holds relevant and which it has to move.

———————▶ Is the line that $a^3 k$ representing the larger space has to use to duplicate while using T^2

▶ Is the line that $a^3 \ k$ representing the larger space has to use to duplicate while using the same time constraint T^2

———————▶ ▶ The difference there is in length brought about by moving the larger space in the same time T^2 as the smaller space is what brings about mass. There are other factors too which I shall touch on as the book develop.

Mass has precious little to do with the whole affair except to be an obstacle intended to restrain the motion of the hosting space. The difference in size between the one in circular motion and the space in contracting motion must bring about that the smaller object has to move about a circle much closer to the centre because the larger space form the centre hosts it. However there is no large or small in the cosmos but only those better developed or those poorer developed. By duplicating there is more to duplicate in the better developed than in the lesser developed. When the lesser-developed space is duplicating the less developed space would hold a lesser extending from point to point forming a shortfall by distance in comparison. The motion being extended needs less extending and should therefore be closer to the centre in relation to what the better developed space would need in extending by a duplicating effort. This is the principle we find behind the sound barrier. The motion the aircraft produces forms an increase in the duplication of the aircraft, which extends the duplication of the aircraft splitting the Earth and the duplication that is producing an extension of the aircraft. The splitting does not align gravity lines with the Earth as it did before. The aircraft is reproducing more in a shorter time duration by duplicating and extending space filled by material that goes beyond the attempt of the Earth's extended of duplication by such motion.

I know this may sound barely believable but please hear me out. While we use gravity the use of gravity as such makes us part of the Earth. We see gravity as some influence or force producing mass and that mass is forcing us down on the solidness and onto the Earth. By having the mass we become a semi unit with the Earth. That is how we on Earth see gravity but when investigating gravity in outer space we must come to a basic question: Is that what we experience as gravity on Earth truly gravity? Much of the proof about gravity is part of our perception about gravity because we experience certain conditions with gravity while we find ourselves bogged down on Mother Earth. But are our perceptions about gravity truly correct? We experience mass but is the mass the result of gravity or is the mass the product of gravity. We only experience gravity, as a factor from the position we have on Earth and the conclusions we form are products of a perception we form while we are being forced to be part of Earth. It's as if we are upside down and have to decide on which route we should follow. I want to make a suggestion, which I aim to prove in the following pages. My personal being on the ground and having mass which is keeping me on the ground comes about because of the speed that I travel through space being the very same as that of the Earth.

By me not applying a speed difference I then inherit the speed the Earth places on me. But the space I use $a^3 = T^2k$ to travel and the space I use to travel through is much smaller than that which the Earth is burdened with to move and to move through. By me having a smaller space to move $a^3 = T^2k$ the space a^3 being moved k in the time it would take to move T^2 will produce less space a^3 to shift k and therefore a smaller distance k to replace all the space a^3 which is moved in the time T^2 the space a^3 needs to enable it to move k. To duplicate by motion the smaller space requires a smaller distance to shift the space but the motion will take up as much time to complete than would the larger space, though the space the larger space has to duplicate will require a longer distance to complete the total duplication of the larger space. A large space a^3 will produce a large extending k when using a^3 the same time duration T^2 when using the same time factor as that which the smaller space is required to use when under obligation to use the same time constraint. Behind this the most basic principle is hiding which allows us the fortune to be able to fly using a flying machine. It is all about motion supplying relevance and forcing on time constraints.

Because the body I have is travelling so much slower than the Earth is travelling due to my size in relation to the size the Earth has, and although I am using the same time as the Earth does to move, such a speed difference is not in the time differences it takes to complete but in the space differences which have to be completed in the same time but are unable to fill and the space is trying to crush me into the Earth where I am forced toward the centre. If I were able to penetrate the soil solidness I would reach a point where my speed as zero would equal the space I occupy.

The space I duplicate by moving from one position and placing the space I hold in the next position while keeping my space I move as it is identical in the next spot but located in the next position. Such moving by duplicating takes a certain time to move from one spot to the following spot and it will use a certain frequency that will have the same ratio in bridging the gap from one point to the next point as that which the Earth has. My speed of duplicating by motion has to be even in frequency because I am within the duplicating space, which the earth is duplicating, and as part of the space that the Earth is duplicating but the duplication of my space I do myself. But in size there is a massive difference between the space I hold and the space the Earth holds but to duplicate will take me as long as it takes the Earth. Notwithstanding this common factor the Earth has to use equal time in duplicating its massive space, as I have to duplicate my small space when we both have to share a frequency that will keep us duplicating evenly. Therefore the frequency of duplicating using the same time period will be a lot different to my much shorter frequency of duplicating space.

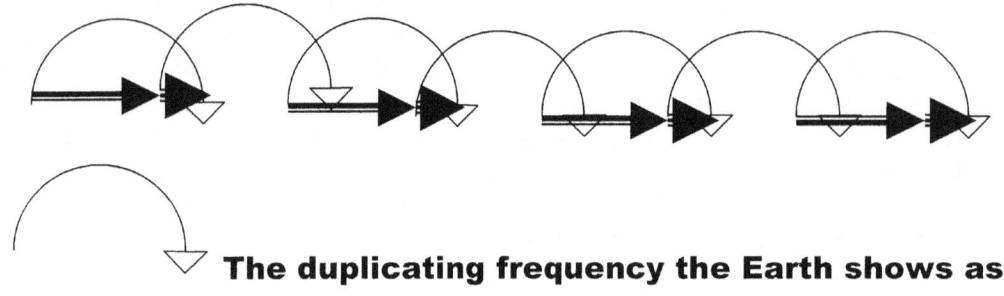

The difference is between me being in mass and me being in the correct position in the space-line the Earth has will place me in the correct position but the heat that then will surround me will fry me into non-existence. Fortunately for life the soil forms a barrier through which I cannot fall any further as to correct my location. But being where my position would have no mass would allow me to float there in that location in the same manner as I would float in water. I would be buoyant. It is because I do not harmonise the displacing frequency that I should that I have mass.

The duplicating frequency the Earth shows as

k_1

The frequency of motion duplicating my body maintains as k_2

The frequency of motion difference my body has minus to what the Earth has where that difference in motion becomes my mass.

My having weight is what Mainstream Physics uses to give me my gravity. Science purposely switched my having mass and confusing my mass with my having weight to explain what is beyond explaining. It is said that while I float in outer space in a state of suspension hanging above the Earth in weightlessness, I still have all the mass that I had on Earth. Butt in order to prove that those in science will give me a mass even in outer space whether I deserve it or not. By that token science first has to cheat all logic by reasoning in some bizarre way that I take my mass up there to where there is only micro gravity. They firstly claim that all of a sudden I take my mass to outer space and in their next argument they say I have micro gravity in outer space since my body is floating as if it is in the sea. But if I stop floating and start falling to the Earth my body and I did not gain any mass. My falling then comes as the result of my motion being much smaller in relation to the space I claim and my motion then is being less than what is required to keep me in the position I have which in I maintain my orbit when I am up there. By moving too slow I fall. I do not fall because my mass grew. But science has been proven wrong by their work without any of them ever admitting to such a defeat. All the satellites fall if the satellite motions are not reset. The satellites do not gain or lose mass. They gain or lose motion.

By amplifying either my space (using a Hot Air balloon) or by accelerating the motion that I have in relation to that which the Earth forces me to have, I will break free from my weight or mass. I shall become airborne and float as if I am in outer space. By pretending my mass can be multiplied many times over in using a process, it then is called not gravity but momentum. But motion and gravity is all the same because motion is gravity that is redirected, which then forms another part of gravity where gravity again is also only motion applying. Science maintains the argument that when I am in outer space and am no longer part of the Earth I then will only have mass. But since there is only micro gravity I will be in a state of weightlessness. My mass is what gives me gravity and while being up there I take my mass along with me. But with my mass up there I will only have micro gravity. I am floating with my mass and it is my mass that is responsible for my gravity and I am floating above the mass of the Earth, which is right down below me, but still I have micro gravity. That is true if I wish to incorporate the dubious use of double standards by separating mass from weight. The mass my body will have in a Black hole will be a billion times (at least) more than what it is on Earth. With that the Black hole destroys the fact propagated in science that my mass will be the same everywhere. That is more than permitting double standard. Because our motion is much slower than the earth is spinning we place a braking effort on the velocity the Earth has and that braking effort we accept as the mass we have. The truth is that my mass comes about from the lack of motion I have in relation to the space I occupy and has nothing to do with any gravitons pulling me down. If I increase the motion I have there shall come a point where my motion will be sufficient to pull me into the air, as I then will have the required velocity to lift from the ground. That motion being in excess of what I have and complementing the motion that I receive from the Earth counteracts the motion of gravity that is containing me. The motion I adopt then release me from the motion containing me and if motion can release me by only becoming more then gravity is my motion not being enough in the first place to keep me onto the Earth. Nowhere and at no time does my mass ever gain by having more protons that will get me back to the ground as if I am bigger or carrying more material or does my mass reduce to get me into the air as if I am smaller or carrying less material? Please note that this is my way of explaining to you the fact of bodies having weight or mass. It is not mass or the lack thereof or any means to measure occupied space within the atmosphere of a larger body that pins me onto the ground. My body is claiming space by motion in space. Gravity is the result of motion because it is in the motion that bodies have that gravity which affects them. This is proved because by adding motion the mass does get more but the body never gets bigger or hold more material, and in defiance of that statement by increasing the motion my body lifts and flies. The reality is that my body in motion has more mass being momentum but still my body lifts when motion allows my body to lift. This statement confirms Kepler that a^3 becomes more (massive) when motion T^2k becomes more (moving).

Mainstream physics admits all along that nobody, human or otherwise knows what gravity is. While investigating Kepler's work with employing much motivation and detail in order to give his work the much due credit it deserves it will also serve a valiant purpose when by the same token we try to establish what gravity is, because I believe Kepler possibly answered that mystery. We have to start with the person that introduced gravity or so everybody acknowledges. Newton saw an apple fall from a tree and he subsequently realised there is some force pulling the apple to the Earth.

Allthough he still was a student he announced his findings and became a genius on the spot. The concept he introduced as gravity gave him instant admiration from which he became the legend he is today and that reputation he gained there at that moment would last him from the day he instantaneously unveiled his mastermind, and that same genius still serves him in his honour to this day long after his death. He found that this force has to have something to do with the weight and the mass of that particular object and the mass of the Earth. There is some force pulling that apple as much as the force is pushing the apple and the same goes for the Earth because the mass the Earth has is doing the same to the apple. Between the two objects facing gravity there is a force that develops where such a force is pulling the apple on a constant basis towards the Earth even after the apple is already in a steady state on the Earth. That forms the mass and the mass forms gravity. He concluded that the mass is responsible for the pulling. Remember this observation came three point five centuries ago when knowledge and brilliance carried a much different defining than what such defining of brilliance is worth today. He realised the pulling on that apple brings about weight that brings about mass because the apple departs from its location and arrives at its end location when the falling is completed. Then he (Newton) went out and convinced all those (in science) that was in need of that type of convincing because nobody before Newton thought of what Newton thought of quite in the way that Newton thought about gravity.

Newton succeeded because he found a way of presenting science with the fact that objects move closer because of some force. He went one step further and named the force he fathered as gravity. But there it stopped! Any and all other further defining the matter or going into any possible observations of whatever magnitude concerning the topic never realised any motive to go further. Inspiration to further commitment just flew out the window as the essence to do so immediately expired as far as the rest of science is concerned. What might he have missed if he missed anything? We all fall down when we are unstable and out of balance. He never realised that balance is more crucial than brutal gravity because that part is the defining part of gravity. No one ever gave a thought about the balance part even centuries later even as we grew into all the sophistication we now enjoy. What brought about the balance that secured objects in an upright stance and supplied some form of control over the managing of a position? Any other position than being flat on the floor would have a better defining than being just at the mercy of the force gravity. Standing tall is a stance that defies gravity so there is another force other than the pulling of gravity. Admittedly the force would first and foremost have to aspire to the rules of gravity and then comply with other demands. True enough is the fact that that position would ultimately and firstly by all accounts have to satisfy gravity before any further motion could commence. Yes but then by balance motion defies gravity by changing gravity's force of pulling everything straight down towards a visionary centre between the objects. In effect this means somehow there is control over gravity and gravity does not leave objects beyond outside control. Gravity is manageable and can be controlled; we just have to find a way…

Years later someone came up with the novelty of hot air ballooning. Ballooning proved that there is antigravity but that part was missed by all even to this day. Some people speak of antigravity as if it is some mystifying mysterious concept that is so well hidden in the secret annals of the hidden Universe that only Ali Baba and his magic words can reach. Please consider the following statement. If gravity was bringing the object down, because of the effect of gravity which is what we experience as the gravitational sensation and is what we interpret as gravity by our sensation and observation, then that is only coming about by our bodies which are in a state of being dragged down. The dragging down of the body is in the direction of the Earth centre. That sensation of being firmly locked onto the ground constitutes to what we believe we experience as gravity.

When some influence brings about the very opposite affect, which then results in establishing the opposite result it deserves to be anti. In example we feel dragged down but anti will be the lifting of the body into the air. Anti will be going in an opposing direction of the motion that gravity inflicts. It will counter the influence that gravity apples. Such motion has to indicate antigravity. The counter acting of the mass dragging us down must be anti gravity pulling us up into the air above the ground. Antigravity must come from such an opposing influence that will bring about the lifting of my body. If hot air ballooning gave the object an opportunity lift, then ballooning must be antigravity. The balloonist and the entire balloon found a manner to counteract the pulling of gravity enforcing weight. The balloon can lift what gravity depresses and if Newton said gravity is the falling then later Newtonians must agree that the opposite of falling is flying or lifting. A balloon is lifting-and-flying. If gravity is pulling down objects in the direction of the centre of the Earth then flying is antigravity. Moving away from the Earth by means of motion and in particular flying is using whatever means to defy gravity where the lifting can also be the hoisting of a body by a crane. When objects lift by ballooning with hot air is blown into a massive container such balloons escape from gravity where the balloon constitutes to bring about the effect of establishing antigravity. Climbing up mountains must fall into the antigravity department because parachuting down the mountain definitely falls in the gravity department. Nevertheless it still does not answer the question of what gravity is.

Let us look at antigravity because the antigravity is releasing the object from the gravity that controls the object by an Earth fed force. The balloon starts flying when the confined space of the balloon is veraciously and violently heated in access. The balloonist shows us that in order to overcome gravity we have to introduce heat. This is the only manner in which we can defeat gravity. Even by an engine driving an aeroplane such flying can only result if an engine combusts solid fuel by creating motion as the fuel mixture is turned into heat. It is heat that makes the difference. This is the very thing that Kepler said. Expand the space a^3 and the motion T^2 will move further increasing **k**. Blowing hot air into the balloon is increasing space within the balloon a^3 which then results in providing the balloon with a larger distance **k** from the Earth centre k^0 that still holds time with in the Earth atmosphere with the Earth T^2 within the space of the Earth **k**. Using Kepler provides us with insight and the ability to see what gravity is by showing us what antigravity is (a^3 gets bigger and that will bring in a larger **k**). But moreover the larger space is enough compensation to bring about extra motion that will defeat gravity by the extending of **k.** If that is not antigravity then we can forget about Ali Baba and his magic rhymes too.

The balloon assists us to escape the Earth's hold on our body, because there has to be the force producing motion countering the motion of the Earth's gravity. The balloon shows that releasing enormous quantities of heat into an inclusive area excluding space such as that which the balloon canvas provides, which is establishing the release from the gravitated containing force on the body giving the body a means to escape by floating about above the ground. The motion is at that point breaking free from the containing gravity by moving in a specific direction, other than the direction the Earth gravity inclines the body to travel. By concentrating the releasing of heat into the balloon, the direction of motion starts to contradict the enlisting of the Earth gravity and the heat breaks the balloons confining properties while the balloon is released from the Earth as the balloon and us lift up into the air and away from our confining to the Earth.

At the point of explaining we arrive at the point where we can say what we think the difference is between the balloon floating in the air above the Earth and a body suspended in outer space floating above the Earth's atmosphere. The difference is the heat that is in the confined air per volumetric ratio favouring the heat being more in the space than what the heat is outside the confined space. If we had any method to put the required heat we need to escape from the limits of the Earth to outer space into the canvas of the balloon there was no canvas left to contain the heat. The heat is available to do the job but the means to do the job with the tools in hand is unavailable as far as we can use the balloon. By having more heat in the one area than there is in the other area it beats off the pulling of gravity. Obviously it is antigravity that keeps the balloon in the air and what keeps the balloon in the air is having a larger volume of heat per space unit than what is in the atmosphere. The balloonist shows us that by applying more heat we can defeat gravity more. Someone took the advice, because the next minute the Germans had rockets. The launching of rockets brought about the ultimate defeat of gravity but it involves almost the ultimate releasing of heat.

In antigravity we find heat more concentrated in one definitive area than the heat concentration is elsewhere. The more the heat is that we release into space the more the antigravity is that we achieve and the more release such antigravity can produce. But what connection can gravity have with heat and if there were any connection between heat concentrated and gravity, what would such connection be? The history behind Carl Benz should bring the answer but more so would be the story behind James Watt and steam although the James Watt story may not be that thought provoking because it is much less filled with the ever popular cheap thrill only sensational gossip can provide…Still both stories cover the same principles. In the Carl Benz story a housewife leaves a pot of benzene fuel on a coal stove. The pot with benzene heats up and becomes hot and under pressure. This performing of heat increase releases the heat as newly created space, which then removes the housewife with her house from the neighbourhood she used to regularly frequent as her residential address. Afterwards almost the entire neighbourhood is not there to tell the tale or ask why...

It was a stupid tragedy that brought about the end of steam and the rise of the internal combustion engine and on Earth billions on billions of human souls are in torment not to please or suffer for the advantage of coal Barons any longer, but instead they are now dying and suffering in agony to please the wishes and desires of oil Barons. How much did the world not change…While it is no longer the coal Barons shackling us in chains and telling us democracy broke our burden of slavery, we have now the pleasure of the oil Barons enslaving us with democracy and telling us to be happy because we are the fortunate slaves, there are other circumstances in which they can enslave us which will leave us worse off. All this came just because the pot of fuel created a houseful of space that was enough to remove the house from the address the house previously enjoyed. But Mainstream science neglects to appreciate this. They see the heat, they see the antigravity but they fail to add the heat, the anti gravity and the space that no longer housed the naive and rather impractical thoughtless housewife. They call the tragedy an explosion but then again everything that expands while using a noise during the expansion is an explosion. Adding of new space to the space holding the house at first altered everything that was previously proportional positioned in the space where the house was. Such exchanging of heat to accumulate and introduce more space in the process referred to as an explosion was bringing in more space that came directly as a consequence of the explosion which was producing more space where the increase in space brought disorder because the well organised material distribution and placing was before the event filling just enough of the required space arrangement that was holding every object in a prearranged order of tidiness.

Then suddenly out of the blue the space which held the house in a tidy arrangement had to accommodate more space therefore the ratio of material per space volume increased dramatically many times over in favour of the space in the balance. That part no one ever acknowledges.

However the losing of the house was not much surprising to Mainstream science back then and even today because who cares about old news. All of Mainstream science was at the time as they are today very familiar with all explosions because of wars and bombing that leads to maiming and killing and all the unspeakable monstrosities we associate with war so that the dirt poor can suffer and die to leave the disgustingly rich even richer. The poor have not the means to pay science to be clever and devise methods to save their lives so the rich do the poor the favour of paying science to find methods whereby more poor can be killed as long as the rich see it as a good investment with great capital gain on the part of the rich. Therefore science is well established in the method of creating more elaborate and destructive explosions that the rich pay them to invent. In the explosion caused by our housewife no one put up money to investigate what happened *during* the explosion but money was put up to investigate *why* the explosion happened.

That inspired an investigation in connection with the fact of the finding more about what takes place during the carnage as more money goes to finding means to create more carnage per money unit spent. At least that is why the poor were invented and that is why wars are invented. It is invented so that no money goes wasted on saving the poor people except if the poor have the money ready and available to pay the rich for medicine to enable the poor to stay alive. So science goes out and develops more fuel for carnage but fails to find out why the housewife and her house are no longer part of the neighbourhood she used to frequent. With the loss of the presence of the ignorant housewife with her house her neighbourhood and all was a normal way of leaving us with a new way of tapping and harvesting energy and untold riches which was born with the death of the absent minded housewife. But according to the mindset of science they saw not what the incident presented in space producing for to their view nothing new came about since it was just another exploding of fuel...so nobody bothered to find out how. What they missed was the part that the coal stove played in the whole tragedy. Without the intervention of the coal stove producing the heat that turned the liquid fuel to liquid heat liquefying the space that turned the liquid space into a gaseous space where the liquid space revealed its true incentive in nature by turning out as space and the newly created space that was in fact liquid space that went onto become more space, well that space was providing the one main factor in space-time relevancy. The stoves heat producing space by transferring heat leading to the expanding of the fuel as such expanding was creating new space that is transforming all other surrounding space and is rearranging every aspect that contains space or that space contains. It will bring a much different looking end. Everything about this concept is missing from Newtonian science because Newtonian science failed to investigate Kepler. Kepler said space a^3 is equal to the motion T^2k thereof and then that says without Kepler directly saying it, it says that if space a^3 goes bigger as a result of the explosion then such increasing in space will constitute to more space a^3 which has to produce an increase in motion T^2k where more motion T^2k will bring about faster displacing space. This is one small fact that Newton robbed the world of realising with his ignoring of Kepler's work.

We are now serving time in the twenty first century. One Professor once told me I must realise that Newtonian science took man to the moon and back several times and in such a view I am rather annoyingly presumptuous to criticize Newton. The Professor missed the point. I criticize Newton on what he did not give us, which he gave us as incorrect by his own admission that it is mostly guesswork on his (Newton's) part and his guessing about the facts where later that guesswork became institutionalised facts believed by all concerned to be correct and to be proven to a degree of correctness that is far beyond doubt. Newton gave us gravity but Newton never gave us the explanation about gravity. At the time Newton met strict opposition from his colleagues and peers because others felt his introduction of an unexplained force was taking Science back in time, which of course it did. Many scientists at the time accused Newton by name of dragging science back in the wrong direction of progress by introducing unexplained forces acting in a superstitious and mediaeval manner.

I went one step further by asking myself the question: If space becomes more when heat becomes uncontrolled why can space not become heat when space is under control? If space becomes more as we see with every explosion of every kind and such heat forming space releases energy, then why would space being managed not form heat being under control and produce energy. We only have to see what Kepler said gravity is. Motion gives us energy.

Where space is the least, which is in the centre of the circle, gravity is the strongest. The gravity located in the circle's space less centre holds not only the sphere together but all that is surrounding the sphere as well. It is from there in a giro action that gravity bonds all atoms forming the structure of the sphere as one unit, as well as distributing a specific alliance in shape and form. How the atoms manage that we will get to in a while, but there is a law allowing for that to take place. Gravity is the strongest in all cosmic structures holding the form of the sphere and gravity controls all around from that very centre where space is the least, therefore the more any star produces gravity. The smaller the star is as far as volumetric occupation goes, the stronger the gravity that is coming from such a centre. The less space there is the less the motion and therefore the stronger and more deliberate the motion is evoking gravity. From the centre in the middle where space is absolutely at a premium the gravity grows stronger as it draws all material.

The motion is one of confining the space to a centre by the moving or trying to move the flow of space and whatever is in the space into the centre where the space is least. Take the Neutron star and the Black Hole as an example and compare that with the sun and the answer is self-proving. I claim that gravity is all about reducing space and not attracting matter but that I explain a little later on. Therefore the matrix of gravity must be permanently located where space is the least. Looking at a sphere we find that which holds the sphere true to form is placed in the centre of the sphere, which then has to be the most intense point of gravity. Gravity is confirming the round shape without favouring any specific point. Such evenness of gravity comes from what is applying at such a centre and is in control of the surroundings. The centre that secures all of the space and material in the space holding the specific form has to be round if it is anything. That shows that in the sphere one can see that the sphere as a form is dominated or controlled from one specific location in the centre. The explanation of the reason there is control coming from the centre has a very childlike simple answer.

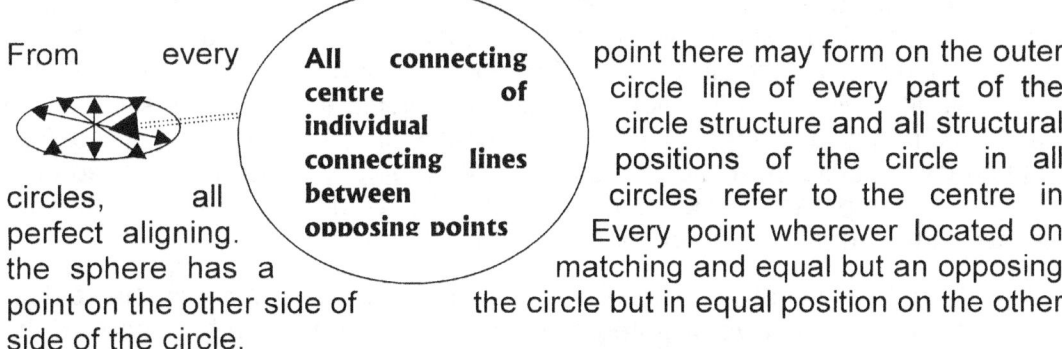

From   every   **All connecting centre of individual connecting lines between opposing points**   point there may form on the outer circle line of every part of the circle structure and all structural positions of the circle in all circles refer to the centre in Every point wherever located on matching and equal but an opposing the circle but in equal position on the other

circles, all perfect aligning. the sphere has a point on the other side of side of the circle.

Between the two controlling points runs a precise straight line connecting the two opposing points in counter balancing. When drawing the connecting line between the two controlling points and connecting such points on further edges of the circle by lines formed the lines will all cross the centre. From wherever a line may cross and from every point forming a line to the other side of the circle rim holding the connecting points there has to be a counter point located on the very opposing side that when connected by a line, such a line crosses in the centre. In the middle the centre spot bonds all sides coming from any and every direction there can possibly be. The line will run to an equal point on the other side across the same distance from such a centre and that then has to be where the strongest gravity can be located.

The Big Bang was where gravity held the Universe in the least space there ever was. To find the original gravity we therefore have to reduce the sphere to the circle and reduce the circle from there narrowing the circle down to as far as one can go. The Universe is a magnitude of spheres constructed by a complexity of circles. This is because everything sprouted from one matrix singularity. To narrow any circle down will be the same as narrowing down the Universe. In our reducing of the Universe we must first acknowledge that the Universe constitutes many spheres, which is giving the Universe gravity as a combining unifying part which is the part of the sphere giving the sphere form (or gravity) and that confirms that the sphere is a circle in many times over multiplying the positions from where gravity secures form. If we wish to go back in time by taking the Universe back down the same route and at the same time maintain some coherency we must concentrate on a single circle because a sphere is a circle by millions of possibilities linked together by just a name that changes the concept.

When one takes this accepted route in thinking that by reducing the connecting line to the connecting circle point in the centre of the lot, it must take us back in time at the same time as the circle reduces to the time during the Big Bang. During the Big Bang where all circles were as small as they can get we run into an unknown substance we came to know as antimatter. This theory is propagated according to Mainstream science but what is most surprising is I do agree with this part of the statement. All material produces gravity. I go one step further and say all material applies motion where some motion may be to contain by using gravity attributing to the contracting that leads to the reducing of their space. Then as everything in Creation has an opposing equilibrium to restore and maintain balance, there had to form another or other material that did not by our lamentable standards produce gravity because those materials produce antigravity, a concept beyond human discernment. Antigravity must be the expanding in counteracting contracting. A counter action to contracting is where expanding provides pappy to that which has no gravity. Forming pappy provides more space by losing density to the advancing of their space. Materials either have gravity by solidifying or concentrating the space they hold in ratio to the material within the space they hold whereas others lose their solidness by entertaining more space within the ratio of material to space where such material becomes liquid and in more extreme cases they become gas. Being a gas they float which gives that material a high degree of antigravity being airborne. It is however not clear if antimatter produced gravity as it did when it went to lunch on and ate up all material in the immediate surroundings. It was cannibalistic but the unanswered question is this: was it a gravity producing predator or a non gravity-producing carnivore. Did material find a comrade in their gravity forming of form or did the gravity it produced bring on the demise that subsequently followed the event as is reported by the highly informed.

The Accepted statement on antimatter reads that matter composed of anti particles where such subatomic particles that have identical rest mass to corresponding particles of ordinary matter but opposing charge and are opposing in other fundamental properties. One example given is that an electron would have a positron, which then functions as the anti particle and has a positive charge compared to the electrons negative charge. That is put bluntly in its utmost simplistic form. Unanswered and tough questions arise from such a statement. What kept the electron bonded to the atom since the protons must by implication produce expanding or by definition be repelling the atom and surroundings instead of the normal contracting or confirming of form.

What is a positive compared to a negative charge, because it is human concepts that put the directional qualities of material into a positive or a negative context as we did with hot and cold. It is human standards that humans brought about to make all human inadequacy by lamented human understanding better but it is not applied cosmos principle. If there is extracting electrons performing in the capacity as antimatter, then there better be protons by other name in service to the anti electrons, which then of course serves the anti electron in the capacity of an anti proton with an equal but negative charge to that of the proton. When matter and anti matter meet, the two opposing particles annihilate each other until one vanishes from the universe. I have to add that at the time this theory was devised the first computer games became a crazy fashion played by young and old, those wise and those foolish all alike. This game was called the packman and the packman ate up all the skulls and after eating left nothing as evidence.

The theory about antimatter has some very striking similarities to that packman game. It still does not answer the most ardent questions: What makes a positive electron different from an electron in the working place each has and can any person show such an object found in nature. Can people take a positive electron to an investigative bureau and be rewarded for such evidence? It is unwise to substitute nature with human concepts just to further mathematical equations. This was apparently presented as normal as nature was when nature developed with the Big Bang and nature then did behave this oddly just after the Big Bang came about. But one huge misgiving in this argument is declaring that everything the antimatter had as a meal vanished and even, moreover, antimatter vanished too. Where could the combination which was produced when the matter and antimatter collided go after it disappeared and did it form the by-product of antimatter science is talking about, which since then apparently vanished too. What a bloody non-intellectual fairytale, which is into the bargain one of those made-up-as-they-go-along stories, told by persons that supposedly should know better. Since there is no place other to find a location to be within than being in a place inside the Universe it is hardly possible to vanish from the Universe except in fairy tales because for one simple fact: there is no other home to have but the home we call The Universe and we have nowhere to escape to but within the walls that the Universe provide for such a purpose.

There is one Universe containing all and preserving the lot. Mainstream physics is accepting this fact. But then by the same margin they accept a principle that allows property that once was part of the Universe to leave the Universe and go somewhere outside the only Universe. They create a loophole whenever it suits them to misplace what they cannot explain readily and logically. In Creation to their and my thinking there can be no hiding of anything but in the Created Universe. This they admit and confirm although with the same breath those very same intellectuals also admit that there is another place outside of what we are able to find in the Universe. When someone comes up with the marvel where such a person can declare in all honesty that the product of antimatter or singularity escaped from the Universe to God knows where that person should leave the field of science and go for fantasy writing such as fairy tales or reporting on politician's inner deepest chastity and integrity. That is what we can find outside the spectrum of what the Universe can deliver.

With such a statement of any Universal product disappearing from the Universe alarm bells should go off in the mind of the trained and professional scientist working with such matters. Yet those in charge do not once question the validity of a statement that involves declaring the possibility that there was as now an outside of what once was part of the only place there ever can be. They can read mathematical calculations and agree on an outside the Universe without stating it in an explanation what happened to the lost and found or their ability of introducing the concept as a reality, which they claim it is. That such factor can go outside the Universe and leave the Universe by causing a Houdini vanishing act of never-to-be-repeated-again status. Science would have us believe this antimatter went into hiding in a manner that is out of the Universe. They applaud this thoughtless presumption while fully knowing that at the time they do this acknowledging that there is no other place for anything wishing for a place to be within when at first the object was within the Universe than having to be in another place other than inside the Universe. If it was ever anywhere it still is within the Universe merely because there is no other place to go than to be inside and part of the known Universe! There cannot be some factor and then misplace it as if a valid factor calculating the value can prove the disappearance and by disappearing it no longer is. If it was in the Universe it must still be in the Universe somewhere. Then we'd better start looking for it.

Another big issue is that whatever the Big Bang produced must be in equal terms everywhere. The Big Bang was a process that had the Universe act as a high-speed cocktail mixer of no repeating ever again. Whatever the Big Bang was, the most it was in the beginning was that it was one massive mixer mixing everything in it at the speed of light. With all the mixing time and time to mix there was going on with nothing better to do than mix and match the mixing was done thoroughly. That we can count on. The relevancies might change slightly and balances may change favouring opposing ends…yes and known appearances did change. But in the end all the factors must always be present everywhere throughout the Universe. By this lack of a fundamental explanation about what antimatter will look like when found, Mainstream is incredibly poorly judged by scientific standards. Those mathematicians calculating physics suggest that science should take antimatter as a cosmic fact and then in disregard of other realities they dispose of the truth by discarding its properties into the unknown. That hardly suggests plausible science by anyone's standards.

By that - Educated Scientists of High Standings are discarding even more of the old fashioned basic elementary science taught as science principles to children in schools in science classes throughout the world. One thing surer than any other fact is that matter in whatever form consists of the purest energy there ever can be. In the cosmos is, was, and will be all the material there can ever possibly be. The concepts we put forward can be faulty but nature cannot ever be at fault. Our arrangement of our ideas can be at fault, but we cannot pull a vanishing act on certain cosmic products and in doing that then dismiss the existence of such a factor or factors, which we then claim, have vanished in the further developed Universe. Our concepts of what they became may be at fault and by changing some basic principles such changing may produce a better understanding of what we think we read into mathematics. Mathematics is purely a language and mathematicians are purely translators.

Mathematicians translate from the language they read to the verbal equivalent they speak and as in all translations, certain concepts may become misinterpreted. The terminology used to explain this is "lost in translation". Mathematicians must see what there is in the translation and try to incorporate what there is available in the cosmos to what the Mathematician sees in his mathematical calculations. The Universe was full of heat and it was full of material but it was not full of free space. If that is the case then where did the heat come from and where did the heat go? Hiroshima and Nagasaki taught us many things about the horror of human nature but most of all it taught us that material is heat secured in atoms and atoms are heat tightly wrapped in a cocoon, which we named the atom. Heat in any form cannot have anti in another form. The package holding heat wrapped can unwrap as it does with nuclear atomic demise. But the anti to heat is cold and cold is space.

The undeniable fact about the Big Bang theory is the acceptance of a growing state that the entire cosmos seems to be in. With all the expansion that went on we came to the point where we now are at and in such growth. All aspects in the Universe must grow in relation to quantifiable progress in all different aspects, which takes us to that which is seen and that which is unseen and which came along as products in the Universe where everything took everything on a growing orgy by unveiling space. That is where we now are. Such expansion includes all there is, and not just with outer space growing. The dynamics of outer space alone cannot grow by leaving the growth of material behind. Should we wish to see where we came from we have to reduce that which we now see in our surroundings to apply to the measures that once applied in all aspects of the cosmos. Mainstream physics is over pronouncing the growth of space and with that suppresses the part matter must play in such growth by simply ignoring the issue. That is the reason why they prefer to ignore the evidence that material is growing, whether they believe that material is growing or not does not change the fact that it is growing in any case. Because they cannot find any reason why material should grow they refuse to admit that material does grow. This is hiding from the truth by hiding the truth. If space grows and the Universe is getting bigger then all space grows to allow the Universe to get bigger. That includes matter and space not in matter.

Space can only grow if materials that also hold space also grow within the space that is growing with the growing space. It means that stars get bigger by the cosmos growing from the Big Bang onwards and outwards to the moment in which we are at present. But if stars grow then the atoms forming the stars are doing all the growing as they secure more space within the space they claim. If Hubble saw space grow, the growth of space must include the growth of space holding material as well. In studying the Hubble's expanding theory we come across evidence that makes it clear that all material expands in a manner as if the expanding comes from the centre of each and all particles within the expanding space and the expanding grows outwards from every particle centre. It is using every star centre to grow from in all directions proportionally in all directions evenly. This leads one to believe that gravity is this securing of space in the material just as Kepler showed it to the world. It proves a connection with deliberate implications coming from every as well as in every specific centre. It proves that the centre $k^0 = a^3 / T^2 k$. It becomes apparent if and when separating Kepler from what Newton thought about the work of Kepler which Newton accepted as being inferior and all incorrect.

To find our birth we have to take back all growth that brought development in the meantime but the only way that this can be done is by man drawing the cosmos down to what man may perceive which forms man's ability in understanding. That is making the Universe small and as man grows man allows the Universe also to grow in relation and corresponding to man's ability to comprehend. We see the cosmos as a circle and we accept the circle because the circle is what gravity implement when the choice of form is coming from material that has all options to freely choose from. By taking the circle back one will follow or, put even better, we will trace the route of the cosmos to where it then started.

All stars are many circles in many dimensions, which form when all circles join into what we call a sphere, but that leaves us only with the circles in the plural. Taking the cosmos back can only lead to one point and that Kepler told us we will find singularity $a^3 = T^2 k$ which is $k^0 = a^3/T^2 k$. We can only reach $k^0 = a^3/T^2 k$ if we repeat $1/k = T^2/a^3$ in a continuing manner indefinitely. When one makes the effort to read this correctly, it says that when distance k breaks from singularity $1 = k^0$ that is then ($k^0 = 1$) $/k = T^2/a^3$ where the space a^3 produced a time T^2 equal to singularity k^0 and singularity k^0 is equal to eternity which was where all was equal to a never changing cosmos that was holding the single form into one dimensional space that included all the filled and vacant material filling in from all sides.

This is one way of looking at the issue and by doing that I am about to prove that singularity is Π. I am about to prove that not only are the planets adhering to the Titius Bode rule of seven over ten and ten over seven in relation to the Roche limit but that the Roche limit explains the very, very first instant the Universe experienced outside eternity. The atoms relate to space in the very same manner of seven singularity positions to ten points and from this motion of material interacting with space is securing material on the inside as well as on the outside. By that motion gravity comes about finding the value of Π^2. Gravity uses the relation of the Titius Bode seven on ten and ten on seven as well as the Roche factor to form gravity and gravity is always Π^2. This I see by reading Kepler's work as Kepler produced the work and introduced the work as $a^3 = T^2 k$. With this formula $k^0 = a^3 / (T^2 k)$ must also be true because $a^3 = T^2 k$ is a relevancy that has to be in relation to singularity and therefore singularity must be $k^0 = 1$. Where will we find $k^0 = 1$?

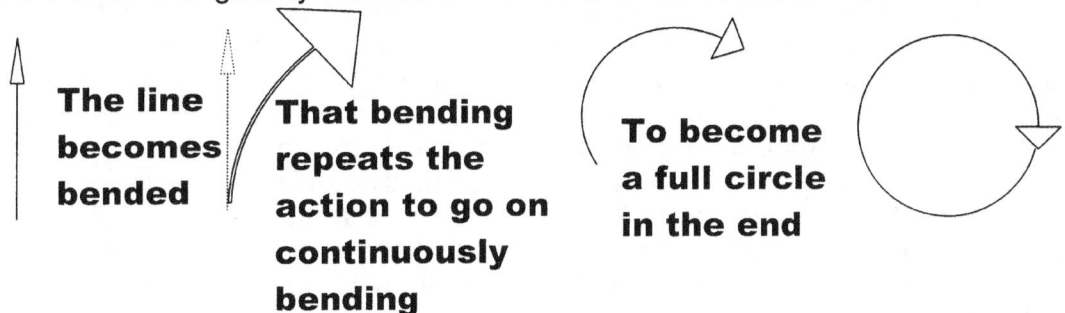

The line becomes bended   **That bending repeats the action to go on continuously bending**   **To become a full circle in the end**

All motion brings about results as the motion eventually ends in spin. Even our linear motion travelling along the surface of the Earth by sea or land seems to us as going straight but it is eventually following a circle around an axis. There are as many axes as there is always an axis. The axis provides a partition between the rotating directions that the spin of the material is securing at the location of where the axis will follow. The spin will have motion and the spin will have direction although the axis will forever instantly change the direction of the spin continuously to fit the linear part of the spin. By going straight the directional change singularity used because singularity is what it is, is continuing to eventually become the circle motion. As the direction will forever change the linear then will forever remain steady due to the eternal changing of the direction.

The linear remains linear because the linear redirects its intentional direction because of the rotational change that the linear motion always ends up doing. The line forms an eventual circle because the linear line must constantly entertain the centre.

Our gravitational falling to the Earth is a result of a circle going straight and forcing us straight down to an everlasting directional alternating circle we have as we spin with the Earth as we spin around the sun. As we fall straight down we change direction while we are falling straight down because that point we are heading to what we are falling to is changing too. But from the centre of the axis everything seems neutral. The axis does not spin at all because the axis brings about spinning motion changing eternally. That is in nature and not man-made motion.

Because the axis provides and demands direction changes to secure everything in motion around such a centre to such a centre, such a point forming the axis is beyond dimension. It has no side as it has no space and it has no motion. It cannot be detected because it does not contribute to any space the Universe has but it can be located because it does contribute to all the forming of space the Universe has. Without ever moving and because it never moves, the centre forces rotation by being in the centre as that centre is also commanding from the centre. That point allows motion to apply where such motion acts as the partitioning between objects. When the spin comes about say in a child's toy such as a top, the top gains independence producing (as long as the energy will last) an independent motion in spin but when the motion dies the independence is lost. Previously I mentioned that all circles in the plural form a sphere by duplication but never repeating opposing controlling points connecting to a joint circle that confirms all possibilities and re-ensures all possibilities. In the final analysis there is one centre one will reach in reducing every radii.

We accept that the time it takes a planet to move between two points is time T^2. Having that space a^3 in relation to the time T^2 is space-time a^3 / T^2 and that is precisely how Kepler expressed his findings $k = a^3 / T^2$. This indicates space-time that is growing through the extending of k. While it does prove the Hubble shift it underlines that that is not the gravity which we experience, because $k^{-1} = T^2 / a^3$ (Newtonian gravity) dominates by contraction where the gravity permitting expanding $k = a^3 / T^2$ is not inclined to absolutely favour contraction. Newton's gravity is totally about a decline in k. But what Kepler shows that in outer space through motion of space performing space-time But our gravity does not exclude the implications of growth because $k^{-1} = T^2 / a^3$ (Newtonian gravity) allows material growth by extending.

The gravity we feel that is dominating us which is also that which Newton saw $k^{-1} = T^2 / a^3$ (Newtonian gravity) cannot realistically accommodate growth in the Universe. This should therefore remind us living a life of splendour on Earth that we must remember that we are part of the Earth and not part of the cosmos. We may find some ability to reach outer space and remain there for a very short period but then we have to return to the Earth. The returning part is compulsory and that we must accept as we accept breathing. There are many suggestions of how we can achieve the ability to distinguish mans superiority by extensive and elaborate travelling through out the entire Universe vindicating our millions of years of being confined by the dooming gravity that the Earth grips us with and committing us to our revenge by knocking off our shackles as we cross once more yet another barrier similar to that of Columbus but infinitely more, wider and holding unlimited vastness just to secure our seemingly unstoppable ability to travel through outer space at the speed of light. Here comes the shocking part: Those that cherish the hope such inspirational thoughts may bring, those thoughts as inspiring as they may sound are no more than blatant useless daydreaming that is at best and at the worst only promoting wishful thinking.

We have as much a chance of achieving that dream of visiting the next star let alone the next Galactica as we have of never aging, never getting sick or never dying. Those thoughts belong to the mindless thinking pattern one will find in the muttering atheist, who is bent on proving the improvable by reasoning idiotically. The atheist practises a religion tempting them to think that if there is no God, they can take the role of God and be God. To do that they have to remove all barriers that divide the sane from the mentally incapacitated. Travelling through outer space on the breeze of a light bulb is just not possible to do. We are born on Earth and we are part of the Earth. Only through our attachment with Earth do we become part of as well as involved in the cosmos but that is strictly because of the surety we find with the Earth and we are secured because that is the Earth which is comforting our needs like a good caring mother should. We are not naturally part of any location other than the Earth and any visiting of other cosmic locations is artificial.

There is no doubt that such visits will be very short lived and even such a possibility is yet to be proven because from what information I can gather there will be dire consequences to follow which are to be avoided if man adheres to sanity instead of manic madness by promoting such attempts at visiting other locations. The Earth represents us in the cosmos and represents us in the cosmos on our behalf. The cosmos does not know life and the relation the cosmos has is not the relation we have with either the cosmos or the Earth. That relation the Earth represents us in the cosmos is that gravity, which Kepler introduced, while Newton saw only how the Earth jealously holds us captured by applying the gravity that Newton saw. This far man could afford speculating with his dreams because the part of science that Kepler covered was up to now just a blind spot to science. I uncovered that blind spot with the aid of Kepler.

Let's now proceed by using this information as we chase down gravity and find what more there possibly can be which it could hide. Gravity is space moving in a circle holding that space that is moving towards a centre in relation to motion. The space is identified by another space moving in the opposite direction. Between the two there is distinguishing differences and what is in space at a distance, which cannot sustain the required motion needed to maintain the gravity that is the separating second space factor that is giving independence through motion. The motion is completely different and totally harmonised holding equality by differentiating motion. The differentiation provides the equal sustained ratio in motion. If such a ratio in velocity comparison cannot be sustained the space removes as it shortens k.

Maintaining the distance of **k** from moment to moment is that requirement needed to keep velocity equilibrium sustained and velocity in ratio becomes the product and the result of gravity where that is prescribing the applying conditions forming equilibrium. Only when such conditions are broken by their inability to sustain the harmonised velocity ratio does space fall away and particles come crashing down to the Earth. Otherwise such conditions are maintained and an orbit comes about. But this falling comes from a lack of motion and not tucking each other's sleeves or pulling each other around. Performing a little science experiment such as the Coanda effect disproves the grabbing on theory. Gravity is about matter concentrating the heat in space through the spin of the proton spinning and reducing space. Such motion establishes an elected centre that houses gravity. The space holding the protons secure forms a demand on space flowing to replace space by filling from "outside sources" in order to replenish the point of space reducing.

The flow or motion comes about as a result of a need to supply space with more space as the proton diminishes space at the centre by killing of space as it nullifies the motion in the centre. There then is a vacancy forming as there in the centre is no space because there can be no motion to the space in the very centre. Because there is no space with motion that specific single dimensional spot has no part in the cosmos we know. Moving towards the centre there is a re-supplying of space equal to the number of protons, which brings on the reducing of space and accelerates movement of the space between the point of demise and the point of replenishing.

This is one part of a group effort where all factors forming the group work together to provide the required gravity. This part Newton saw not. Being the master that formulised the existing laws on motion he had to detect the consequences of motion if he carefully studied Kepler's formula. He would not have brought in the idea of a force but would rather have recognised it as a natural flow of space bringing about the duration of time. Newton did admit he had no idea what gravity is and declared that gravity is a force. On that point I disagree and my disagreeing is not on the subject of gravity being or not being. I am emphasizing my disagreeing on the force aspect because every person on Earth associates the word gravity with the word force and confuses the two in concept. Life is a force but gravity is a natural and normal flow from the start of the Universe to finally bringing conclusion to the Universe. Gravity is a natural motion of space a^3 in space a^3 and in that there is no force to be found.

One does require a force to resist the flow …yes but that resisting of the flow then becomes the force, which counteracts the flow. His view about declaring the presence of an unexplainable force brought about much rejection from his fellow academics at the time because it enlisted a vision that Physics were moving back to the dark ages at the time. No one can blame the others about the direction they saw science moving because admitting to a force without any ability to explain what brings such a force about constitutes much to the powers pagan gods, witches and other undesirable powers had at the time. However Newton did conclude that there is strength in the centre of a sphere that produces the strength within the sphere. I then decided to investigate his remark.

We all accept that the Universe uses the only true cosmic form there is, as an overall all-containing form we call a sphere. The sphere is that form, which the Universe has to be in to form the Universe and naturally the concept that immediately enters everyone's mind is thinking about it, would be and most probably is a sphere. Everyone accepts the universe as a whole will be the sphere…but why would the sphere form? If anyone in the past had stopped for a minute to think about this question such a philosopher then never stopped for a moment to write it down or to convey his conclusion to the following generations. I have heard intellectuals explain it by telling students the form is used by the Universe because it is the strongest form there is, but that carries the same value in definition as to say the Universe uses the sphere because the sphere is round. The original question then still remains unanswered because the question still stands. Why is the sphere round and why is the sphere the strongest form one may find? So declaring the sphere as the strongest form leaves the question just as unanswered as before.

What will be the reason why the original form that we devote to the Universe will take on a sphere as natural form? Yes… I have heard in Engendering that the sphere holds any point and every possible point pushed from the outside of the sphere secured by every other point the sphere has from the inside of the sphere which then is forming the sphere but that statement is as precise as it says a woman gave birth to me and not that my Mother specifically gave birth to me. It says I can be anyone's child instead of that I am specifically that persons child. It still does not reach the answer that will stop all other questions about the question. Apparently our imagination grabs the sphere as the only form of choice and that is as correct as it is true...but why...this is apparent coming from nature as natures choice to form when material is not pre-cast to have any specific form. In such an event the gravity in that space takes on by cosmic pre-cast shape the sphere as form...it is because gravity chooses the smallest space to hold the strongest force. Such a point will also establish a line we call an axis. By reducing the radius there must come a point where the ring that is in decline from such reducing is infinitely small, where it can reduce no more, where it reached its ultra limit, but at that point cannot be zero, because the point is there for all to realise but nobody to see. From the point in the centre that is no point in the actual Universe there is in one space forming a unit two points separating the unit by holding relevance and without two points there is no point.

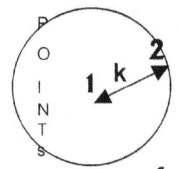 When looking at what Kepler brought into science we find a^3 being equal to T^2 by the allocating of **k**. The mention of a^3 is referring to the space filling the space that is the space at the very end of the point rotating where that point is indicating the forming of a circle T^2.

But a^3 also indicates a separate a^3 that pinpoints the allocated position of the space designated to have the smaller a^3 point out the precise a^3 that the smaller a^3 is claiming as a unit and that became the product of the motion identifying a^3 as a separate unit sharing one larger a^3 and one smaller a^3 of what all is brought about by a field invested to form the gravity. There is forever a larger space a^3 that holds a smaller space a^3 in relevance to the motion coming about in the form of T^2 and **k**. Then the relation between a^3 and the centre part of the larger a^3 there is a most relevant point being k^0.

Considering the manner in which the expression of Kepler's formula reads one may correctly be of the opinion that a^3 is in context with the broad space that covers all of the space indicated by the length of the radius which is symbolised by **k** from the centre k^0 to the point indicating the immediate border of the space **k**. Yes that presumption is very true but also true is the fact that if there was one point reserving the position for the smaller point a^3 that held a separate and independent space a^3 within the larger space a^3 which would without the smaller space a^3 not be identifiable as forming the unit a^3. If there was no such a smaller space a^3 within the larger space a^3 producing the outer limit to the larger space a^3 the larger space a^3 would have no independent relevancy in the overall totality that will distinguish such a space a^3 and to establish the containing as well as reserving position it holds. The larger space a^3 is there because of the motion of the smaller space a^3, which validates the larger space a^3 to be a factor worthy of being calculated. Only by the motion of the motion of the smaller space a^3 can the larger space a^3 claim validation and on the other end also apply independence because as I shall show later on, the motion of the larger space a^3 validates the counter motion of the smaller space a^3. The smaller space a^3 cannot be in motion if the larger space a^3 does not contribute to a larger motion of space a^3 contradicting the smaller motion by direction where both accommodate each other by motion relevancies bringing individuality without bringing independence about. Kepler said $a^3 = T^2 k$ therefore if there is space a^3 such space a^3 has to be in motion $T^2 k$ to allow space a^3 to be and have the other space within. Therefore by referring to a^3 one establishes a relation of both in the context because not one of the two would be if not for the presence of the other a^3. When referring to a^3 one refers to the larger a^3, which is containing the smaller a^3 as much as one distinguishes the position of the smaller a^3 proclaiming the area of dominance of the larger a^3 in which the smaller a^3 takes up residence in space a^3.

Kepler's formula first drew my attention to singularity in the way he formulated his formula. The most important part of his formula is not visible from the outside or from the onset of investigating and one must look for that most dynamic part covered by the mysterious coming from way within. Kepler shows us that the truth is found in the darkness and not in the light. At a point where Einstein said gravity begins we will locate Kepler's gravity beginning because space (or as Einstein referred to it) the Universe goes flat. If $a^3 = T^2 k$ is a fact then there must be a starting point where **k** starts because there is a point where **k** ends. This then will change relevancies and will mathematically equate from $a^3 = T^2 k$ to $a^3 / T^2 k = 1$ and one can be any number or symbol to the power of zero. Mind you not to the value of zero but to the power of zero $a^3 / T^2 k = 1 = k^0$. That means one has to reduce **k** to a point where **k** becomes k^0, then in accordance with Kepler's advice I proceeded...Kepler said that from the smallest space within space a^3 there is the line **k**, which is connecting in a motion covering the spaces $k\,T^2$.

The space indicated by and that is a part of space a^3 in question will run as the space-time unit $a^3\,/\,k\,T^2$. That is where gravity will form being identifiable as a unit at a specific centre from k^0. Gravity lurking in the centre at the point **k** starts the line **k** where the line **k** holds space-time a^3/T^2 secure and in form. That has to form singularity and singularity can be whatever there is a wish for as long as the wish is to the power of zero. (Singularity) $k^0 = a^3\,/\,k\,T^2$ which reads that in space-time has three sides on the one side and are opposing the first side by three other sides. If Kepler said mathematically the smallest distance between structures could at the least be $k^0 = a^3\,/\,k\,T^2$ and we all know that $k^0 = 1$ then it should be some one's duty to find that point. One must then start by accepting that Kepler also stated there cannot be nothing or zero in the cosmos since the smallest distance between two structures is k^0 which is one and not zero. I wish to introduce an argument by disposing another Academics method in his disposing of my work. Some academic found a way through which that particular Academic was able to dismiss my arguments on the grounds that the solar system was not formed at the Big Bang period or that is the information that Mainstream science is promoting.

That was a loophole he suggested because he was unable or inferior or plainly just too lazy to interpret the work I laid before him. I am not for one minute fooled by his passiveness because this is very typical of the New South Africa that everyone outside South Africa helped to create. And in addition I really can't think that he thought me to be stupid enough to be discouraged by taking his arguments seriously. In order to circumvent such a loophole I shall begin my following argument by stating that the solar system, which I am referring to, is a hypothetical one. Notwithstanding that I know my argument is solid and serious as such about all aspects in the rest of the entire argument including all other possible aspects. The following is the one part I use as a part of my argument, this part I now identify as remaining the only hypothetical possible fact in the entire argument that has a possible hypothetical truth as it stands.

There are those who avoid admitting to inconsistencies by arguing that my argument about growth of material in the cosmos throughout its entirety is invalid because the solar system was not in place at the time the Big Bang was in place. Please then keep in mind whilst reading the argument that I would like to point to the fact that my following referring to the solar system is actually referring to a similar solar system that is somewhere else and is now a part of a galactica we do not know about. That is where the dissimilarity ends. In all other aspects our solar system and the one I suggest are identical in every possible aspect. I bring this in to disqualify any academic loophole that may come about from an argument about the solar system coming into place at a later stage of the cosmic development and therefore I exclude any chance of using the counter argument that the Solar system became an eventuality long after the Big Bang was about to happen. The novelty in the forming of the solar system in the argument I am about to present is no longer an issue because I am referring to a solar system in another Galactica being precisely the same as the one we use with the exception that that solar system was around at the time of the Big Bang. I therefore hope that there are no more ambiguous loopholes that anyone may use to unfairly dismiss my point of view when the validity of dismissing my point is as simple as raising such a counter argument as the one mentioned. That counter argument of the solar system not forming a part of the Big Bang does not apply.

To avoid such a loophole again we now use a hypothetical but real solar system in space, which formed when the Big Bang took place. To them we now present a solar system that is identical to the one we know and is a precise duplication thereof. Again I say to those the argument now represents a precise duplication of our solar system and was in place ever since the Big Bang. That means with the solar systems being apart in millions of kilometres at the present time there was a time when the planets and the sun were apart by the same measure but only using kilometres instead of kilometres by the millions. The Big Bang shows a growth in space. If that is the case then there was a time when the Earth was 149 kilometres away from the sun instead of the current 149 million kilometres, which it is at present. We can reduce the distance further to fit into a billionth of a meter but I hope my statement drives the point home. But that space we reduce also has to include the expansion in size of cosmic objects because reversing the expansion shows that any argument not expanding the orbiting structures is most silly. Then there must have been a time when all the planets were between fifty-nine and five hundred and ninety metres away from the sun in comparison to the millions of kilometres we have today.

If there was no expansion of the orbiting structures in the diameter size they have then how did the planets being the size they are at present, which then was also the size they must have been back then, fit into such a space that small where that small space was keeping such giants apart. That is notwithstanding the fact that they still are with all the gravity they presently have but at the time was being apart only by the measure of 149 kilometres as is the case of the Earth? Material therefore too must be part of the growing in space. This line of argument I suppose is much below the Academic's pursuit of matters but since I am much lesser in mental standards of development than they are, such reasoning prompted me to go on an investigative journey. Light journeys throughout the cosmos and it will be sensible to follow light's travel in reverse to see where that takes us.

With objects being apart at some distances and light flowing in straight lines between them it must take light the size of a straight line to travel between cosmic objects. While I disprove this statement in future arguments I wish to stick with the officially accepted but as such a very simplistic concept for the moment. The distance light has to cover depends on the radius there is between the objects that form the total distance forming the Universe. This puts the Universe then in relation to lines forming differences in structures that are claiming independent space and space setting objects apart as such distances will be the radius standing between those objects. In reality that is what the Universe comes down to.

The objects are circles by dimensions and the space is also dimensions that are crossed by lines travelling through the dimensions. With light being a line and the Big Bang coming from a situation that was a lot more cramped for space than at present, the correct path to follow if I wish to trace the steps of the cosmos back to the Big Bang is to reduce the straight line between the structures and find where such a line will no longer be a line. I realised I had to begin at a point where I had to find the point where any and all lines end when I reduce any and all lines which will then be the same measure where such a point will show me where the line forming the cosmos started. The same procedure will apply to the material structures all being in a sphere form in our Universe. A sphere is a lot of circles forming a unit but not repeating the space claimed by the other candidates. Such a circle also applies a straight line only known by another name but still serves the same purpose. Reducing the line will lead us to the beginning of time. The reducing of the line will once more represent the point where the sphere in its role as a multi circle will begin.

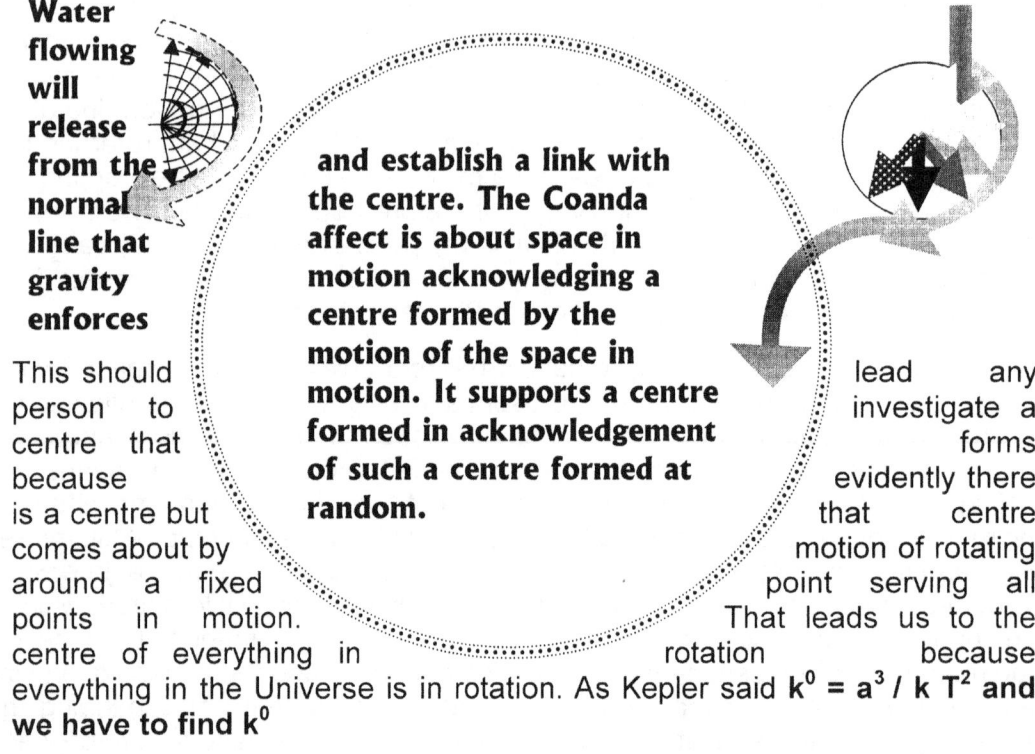

Water flowing will release from the normal line that gravity enforces

This should person to centre that because is a centre but comes about by around a fixed points in motion. centre of everything in

and establish a link with the centre. The Coanda affect is about space in motion acknowledging a centre formed by the motion of the space in motion. It supports a centre formed in acknowledgement of such a centre formed at random.

lead any investigate a forms evidently there that centre motion of rotating point serving all That leads us to the rotation because

everything in the Universe is in rotation. As Kepler said $k^0 = a^3 / k\, T^2$ and **we have to find k^0**

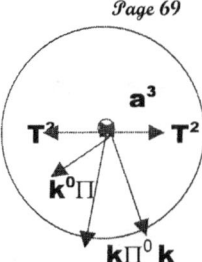

In dimensional terms, which I explain later on the value of **2k** relates to T^2.
That relation extends to the next value where T^2 relates to **k** , which relates
to T^2. The first   space   in the circle will then be T^2 **k**. From the centre
being in infinity one can realise by applying mental power the single
dimension factor not seen but present all the same. Extending that into the
3D comes six **k** and any one of the six will further extend to form a seventh
point as T^2 All this is a multiplying of $k^0 = a^3 / (T^2 k) = 7$

When translating Kepler's mathematical expression into a verbally spoken form of communication
such as English we can see what Kepler said also read as $k = a^3/T^2$ where k is one point from a
centre point that is space a^3 relating to time T^2. From a centre comes space-time

$$k = a^3 / T^2$$

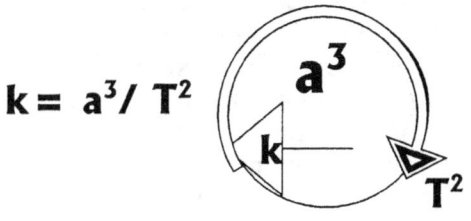

Kepler said
$a^3 = T^2k$ **but that could also**
be $k = a^3/T^2$

Locating and finding the presence of singularity

$k^0 = a^3 / T^2k$ states that whatever is, is also spinning in order to be present.

What is in the Universe is spinning. In the **precise middle** of all **objects in rotation** is a precise centre dividing the object in sectors that will **start the spinning initiation** from that centre point. Thus, the spinning object **will have a middle point**, a very specific **centre point that does not spin** and only holds Π as a specific value because no radius can apply. But also the one value such a line **cannot have is zero** because the line **is there and holds contact** to the rest of the material bringing about that **zero does not start any** line and therefore the **value of the line must be infinite**, just as described in **accordance** and by **the definition of singularity**

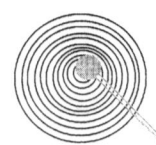

As I am introducing a very new idea, I wish to explain in better detail what I try to convey.

While the toy top is spinning one will find singularity by moving the rotating line or radius progressively to the middle by reducing the length the line has from the edge to the middle. At one point all further reducing must end but the ending cannot include zero or nothing because the rest of the line still attach the rest of the top.

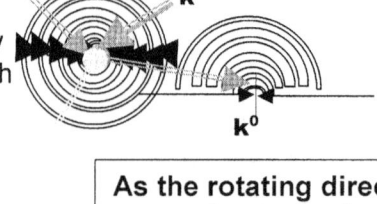

As the rotating direction moves inwards, the rings will become smaller and smaller.

That point albeit hypothetical, is also as much a reality none the less and is placed where that point **must be standing still** because every line **running from that point in opposing directions** is also in **opposing directional spin the other or opposing side.**

In considering the spinning motion in the fraction of time in the detailed instant every aspect of rotation will turn in every instant of change in time. Although the points had the same characteristics only one instant before, they oppose the characteristics it had just before and just after the very instant in which they are and to which they relate by similar points also in rotation. The fact of the graph proves my point in quarterly opposing dimensions and values,

All this was missed by science ever since Newton changed Kepler's work because Newton failed to read into Kepler's work correctly from the information he (Newton) basically discarded due to the arrogance on his part. He (Newton) concluded every factor in the whole formula of Kepler represents motion instead of the true representation it carries being space-time. He failed to study Kepler's work even though Kepler arrived at this work only when he (Kepler) received the work directly from the cosmos as Kepler was studying the cosmos. In ignorance on the part of Newton driven by arrogance he failed to notice that Kepler's work introduced for the very first time space-time as a formulised conclusion and being space-time Kepler's formula has no need of adding Π and having the formula changed from Kepler's vision of $a^3 = T^2 k$ to $4 \pi^2 a^3 / T^2 = G (m + m_p)$.

Newton tried and failed to marry Kepler's formula with an earlier concept of him (Newton) which he (Newton) received by vision when he introduced the formula $r^2 / (M \times m)$, which was his first vision of what gravity was. With that he formalised gravity but then he overstated the grandeur of his first vision by matching his concept to the Creation at large which was introduced by the work of Kepler.

In that attempt to link all creation by what he found to be condensed in the name of gravity he changed his first formula from r^2 / (M X m) to $F = G (M_s X m_p) / r^2$ which Newton then afterwards introduced accompanying his term as gravity. This was one of the most unfortunate and corrupt misleading in science that ever took place. Consider the seriousness of misguiding that came from the case of the Piltdown "Ape man" Hoax and by comparing the influence such diversion inspired the Piltdown "Ape man" incident becomes an innocent little party prank compared to the implications that sprouted from the Newton diverting of science. But it is not only Newton that is charged with the deliberate misleading in science because Newtonians are aware of the so many missing answers to questions Newtonians never dared to ask in their attempt to hide all the shared blame.

By merely diagnosing gravity being naturally pulling from the centre it is rather avoiding the question with simplicity because the question arising from this answer is where then is the centre of the universe to where the final pulling is heading? By using the gravity formula $F = G (M_s X m_p) r^2$ the centre is the vital issue directing as well as controlling every aspect involved in the formula. When this formula is mathematically translated to English the formula says there is a pulling of structures M_s $_x m_p$ towards a centre r from both sides r^2. There is the centre r^2 that is the sun. There is the centre r^2 that is the Milky Way. But that is not where the Universe stops. The Milky Way centre is not an individual end-of-the-finality-of-the-Universe in terms of coming from as we are going to. Mainstream Science declared that all the pulling is toward a centre r^2. What placed r where and what was r to achieve that much pulling power? We see that the Universe is somehow always coherent and disciplined because the rotating of structures indicates the presence of gravity.

The Newtonian formula insists on a centre pulling particles. In the formula $F = G (M_s X m_p) r^2$ such a centre is most important. Such a centre is as demanding as it is commanding and yet, Mainstream Science never came to pinpoint the centre of the Universe. Kepler on the other hand showed the precise location of the centre of the Universe. The pulling of all mass in the Universe must be towards a centre and because of Newton's introducing of a Gravitational Constant (G) in the Universe; this then demands a centre to form G. Such a factor with gravitational powers must point all gravity forces towards one centre since gravity is undeniably located in a centre of all objects. When the realising of this all out important centre arrived and Newtonians became able to recognise this factor Mainstream Science should then at the time been working towards identifying some centre before proclaiming the serious implications arriving from the presence of such a centre. Never once was there to this tome some launching of an all out search to locate this centre.

There was a silly attempt designated to Einstein because of his superior mathematical abilities to calculate the entire mass of the entire Universe but what prominence would the mass have in pulling without a precise indication of a specific location where the pulling of the entire mass is heading. Without the centre all other factors lose their validity. Where then will such a centre of the Universe be? This is what makes cosmology the shambles it is. Kepler gave us the answer centuries ago but no one ever tried to take notice of what Kepler said without Newton's interfering. Kepler gave us the ability to see what lies beyond the limitations of the visual as well as the very obvious.

When translating Kepler's mathematical expression into a verbally spoken form of communication such as English we can see what Kepler said also read as $k = a^3/T^2$ where **k** is one point from a centre point that is space a^3 relating to time T^2. From a centre comes space-time

Others like Newton and Einstein came much after Kepler and coined the phrases but Kepler formulated the concepts of gravity as well as space-time. They (Newton, Einstein and many others) named Kepler's innovations. That is very clear but only on the condition that Kepler is read correctly and Newton gossip about what Kepler is saying is ignored. What Kepler said in mathematics all the brilliant Mathematicians through so many centuries were unable to read although the coded language was written in mathematics and as it is the field of supreme speciality they have captured as theirs, therefore they should have the ability to decode what they are the masters of which they claim to be able to decipher. It is the mathematics, which the cosmos uses to allow the cosmos such communicating with humans! The cosmos said that all space stands excluded from all other space which then forms an all including unit. That is why the sun is so cold and outer space is so hot (and no… calling the sun cold and outer space hot is not a mistake on my part and neither is it due to a printing error but the sun is as cold as it gets and the outer space is as hot as it gets. We'll get to that one later on…). The two share a cosmos in which they are both apart while forming one unit. The separating part in the unit is motion driven by heat and the motion as well as the heat carrying the motion that sets the two apart while both are in the unit. I mentioned previously that there is an undeniable connection between heat and gravity. Let us do some investigating and try to establish answers.

From since the time that man discovered intelligence (if he ever did) man has been with the presumption that the sun is the hottest centre in the solar system. Later on in the more present time it came to someone's attention that the sun also holds the solar system in gravity. The Earth by its standard and dominating its sphere of which it can control with influence is the hottest centre in the space of its domain and it holds the moon centred to the Earth. The gas planets are the hottest centres in relation with the most heat and they all hold their satellites captured by a hot centre. All space structures hold in every centre there is that is confirming their independence at that point of securing independence the centralizing of the most heat it is able to concentrate and from that centre holds all material captured or controlled in the domain of what that forms the independence of the structure. I can go on and on but heat in the centre couples gravity to space-time, just like Kepler said before he was spoken for on his behalf and without his permission or his agreeing to it.

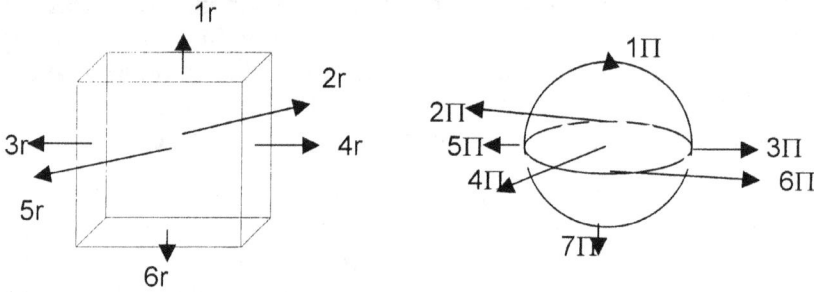

Taking the outlook from the point the sphere is holding from that centre out into space there are ten points connecting to the centre. In that are the dimensions of singularity connecting to space where five connects to space in the second dimension of singularity, and five connects in the third dimension of singularity. On the other hand does the cube show a very different characteristic, which involves only six sides (at least) connected.

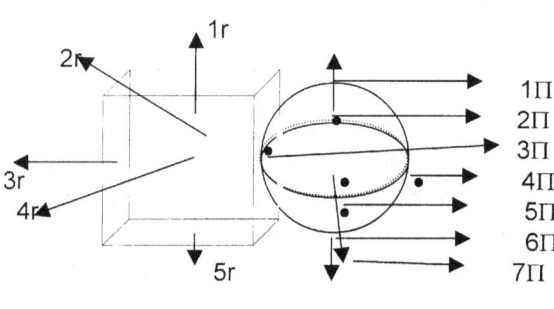

Every point securing a singularity that does not spin secure a singularity in control of the space-time that apply the motion because of heat centralising and bringing about motion through space expanding. That which is outside the sphere becomes a cube to that, which is inside the sphere

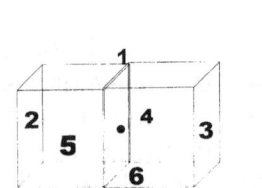
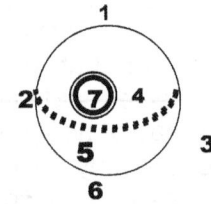

7 is the centre addition in the sphere

Kepler's formula also indicates that a sphere is within a cube that is holding a sphere

$$a^3 = (T^2 k) = a^{3+2+1} = 6$$ with the sphere presuming the position of singularity as part the of $k^0 = 1 =$ **singularity**. Einstein proved that at the point where space reduces and such reducing reaches a point where space as a factor in

the third dimension disappears into the single dimension (space going flat) gravity is overwhelming. Einstein interpreted this, as the complete Universe going flat but while it may be true that the Universe is going flat, that can only be within singularity since singularity represents the Universe as flat as it can get.

Humans' (including Einstein) interpretation of the Universe is faulty but the faulty aspect does not include the fact that the Universe is going flat but only which is the flat going Universe referred too. According to Einstein he proved that the Universe is alternating between going flat and holding space but his lack of studying Kepler lead to his spontaneous misinterpretation collected from our culture and his incorrect interpreting of what the Universe actually is. We all have a faulty perception of the Universe because not only he (Einstein) as an individual scientist but all humans throughout history have also never asked the Universe what the Universe is. Kepler did and the Universe answered using the mathematical equation $k^0 = a^3 / T^2 k$, which when interpreted means singularity placing space-time is the Universe. No one ever thought about this statement in sincerity because from a Newtonian aspect it seems silly. But rethink the silliness presented by the Newtonian Universal centre and compare that thinking about what the Universe told Kepler then decide what is silly. Newton's never acquiring the effort to do a study of Kepler's work withheld him (Einstein) from reading his very own mathematical translation accurately because apart from Newton, Einstein must be considered the second most important Newtonian ever. What Einstein saw was that space disappears and he then jumped to the conclusion that the space he saw in his mathematical equations was outer space referring to the space falling outside the parameters of the material occupied space secluded by dimensional borders. In the sphere placing the borders that the sphere holds there are deliberate and very distinctly placed edges or points forming a specific distance from the centre. The centre is also proven beyond any debate.

The centre of any sphere has to be at the very point where space completely falls away. It is at the point where all the points of line centres meet by the crossing the centre of their individual connection coming in to contact as a group. In that way one may assume that the lines connecting the controlling points on the other end are crossing on a centre point that all that is participating in the constructing of the sphere is democratically electing such a centre. Please note this conclusion very well because this forms the heart of the Coanda principle. That will put that position where the lines cross which in itself is centralising all space in the sphere at that point, such crossing point will become very distinct and controlling where that point forms in the single dimension and singularity is the single dimension. But Kepler also solves another riddle that truly got Newtonians unstuck. This, to which I now refer, is what is referred to when they refer to the Hubble constant.

The growth we see in the Universe is an adding of space in every cycle completed by every cycle, which all the protons complete. The adding is the smallest addition that can come about in the shortest period of repeating by cycle rotation there can ever be. This growth of space-time next to singularity confirms the growth of singularity as singularity recalls the space it uses to grow in the time it grows. The margin of growth will be by the extension of k in the formula $k = a^3 / T^2$. Every cycle completed in the relation to space by the initial value of k. $k = a^3 / T^2$ leaves ultimately a^1 extending as space or as Kepler chose to indicate it as k^1. But that too has to be compensated by the duration of time reducing the time aspect by the margin that the space expands. This confirms what is evident in the Hubble Constant. The further one looks at time the more time seems to race because time has the invert properties we give to space.

When one observe the cosmos one observe the night sky as one big black hole that is forming. From what we observe the night sky also fills with tiny lights here and there and in between, and the better the lenses one uses the more lights are here and there filling up everywhere. But in all the blackness and all the vastness and all the sparingly filled spaces there are three relevancies, which result in gravity. Let us acknowledge what we see from the controlling centre.

There is a position that is in motion that is forming the very edge of the outside. To be in motion the position must be in relation to a point from a centre. From the centre there must be a specific allocated space ending at the object in motion and starting from a centre that has no dimensions. The object in motion determines the one limit and the centre with no sides and no space, which is standing still in singularity, determines the other limit. By that we can see there are only one way of looking at what we can observe and that is from the outside in.

This is the one perspective. There are the others. From the outside there is a centre orbiting an unfilled space with an inner centre. The centre orbiting has to have an allocated centre with no sides because that centre secures motion that is independent from the other space surrounding but which is including the independent space forming the border.

The orbiting object also secures an individual independent and own centre but from the orbiting object the limit it holds ends at the edge it forms. Being a sphere the orbiting object secures seven positions and the larger containing sphere is ten of which seven is within the singularity dimensions within the centre, which we not observe. Immediately following that as part of that relevancy comes the containing sphere that holds space-time and another tree positions. The three positions puts a relevancy of three to the holding space that already caries seven. There are fore ever another centre that secures seven positions just because singularity chose the sphere at the value of Π and in the sphere 7 positions is made up of six sides that hold relevance to a precise centre. There are a^3 but then there are T^2 putting a^3 at a value. Then there is a relevancy named **k**, which puts T^2 at a relevancy. None of these are fixed markers because the relevancy can and does swap sides placing importance as alliances changes. When T^2 focus on another a^3 the relevancy about **k** changes and amplifies the importance of yet another space. When one applies the Coanda effect one would see just how easily new alliances come in place and secure new centres that charge new relevancies between newly established points. Going either "bigger" or "smaller" is only shifting focus on another relevancy.

I suspect this cosmic growth of all material is equal to the growth of a human hair or a human nail that presents as the duplication of cells because life takes command of what is made available by the Universe and then manipulates space-time to claim such growth by taking charge of the opportunity to use the growth to the benefit of life. But this growth constitutes multiplication from the very centre of the most inner part of the where **k** = infinity plus one.

Kepler thus gave us the answer about what Hubble found what was happening in the Universe centuries ago and centuries before Edwin Hubble's discovery. From Kepler's formula one can see that time and gravity is the same because as gravity weakens so does time reduce and as space expands so does the influence of gravitational reduce because gravity has less time per unit to control; more space per unit. Gravity is $T^2 = a^3 / k$ since the object cannot depart at any further distance between the centre and the object and is captured at that distance. Also gravity is $k = a^3 / T^2$ for the very same reason. The circular bonding T^2 of space a^3 is enforcing an orbit T^2 to gravitationally circle around a specific centre k, which indicates the gravity $T^2 = a^3 / k$ in relation to the other gravity component $k = a^3 / T^2$, and it means T^2 is a circle of gravity and k is the straight-line distance of gravity applying motion. Still Mainstream Academics ignore my statements that gravity is space in motion and motion of space is time: precisely as Kepler said. Any area to the cube is space a^3.

$$k^0 = a^3 / T^2 k$$

Mass is the result of applying gravity by reducing space. Protons are the only diminishing devices of space-time in nature but the protons remove space as it concentrates space-time to furnish material with growth of material. The more protons there are allocated to a specific space the more space the larger number of protons will diminish. Gravity is not the result of mass. The belief of mass brought about by large numbers of protons confined in little spaces is not always true. In some cases the mass produced by a large number of protons does not result in a heavy confined element since there are those with high numbers of active protons, which should enforce a large gravity. The opposite is true as the element shows as an elaborate anti gravity by being an airborne element or a gas. There are those holding protons in clusters with numbers matching heavy metals but that is categorised as gases and gases produce high ratios in antigravity.

If particles were that close and yet they expanded it is proof that gravity is not about particles pulling each other closer because then the Big Bang provided the ultimate opportunity to unite what there was instead of expanding what there is, but I came to realise that gravity is about removing the space between the body and building on that which the body already holds as well as the space surrounding the material that serves as unoccupied space whereas material holds occupied space. A fan drawing air into the blades the same sucking or pulling one experiences with gravity but in the case of the fan we know it as air that is in motion and the motion extends to affect those objects placed in the line of the air flowing. Gravity comes about as space a^3 applies motion T^2 and from singularity k gravity is as much part of space as the motion of space is part of gravity.

Fan contracting air by producing flow of space-time and not just air.

Gravity is the result of the motion of the number of protons that through such motion creates the reducing of space in the space-less centre. Gravity dismisses space and by doing that the stronger gravity is where more particles can fit into less space occupied where that reducing of the space is bringing about extensive mass increases into the volumetric occupied area confining more material into less space. Gravity is space measured over time. Gravity is the space in comparison with the time affecting the space. It is the motion of space relating to the time of motion of space. Gravity is the moving away of what fills space by extending singularity where singularity responds by bringing about space between the structure and a centre being within another and larger structure. The larger structure is considered to fill the role of the governing singularity will hold more material in less space, which is, then more material that is confined to less space.

The centre structure is reducing more space in the time factor between the moving structure and the centre structure. Gravity is the increase of heat occupied by the reducing of space in a spherical unit. If Newton only tried less to deny Kepler any recognition and gave Kepler more deserving credit of Kepler's input in the total work, he, Newton, would then have seen what Kepler had formulated gravity to be. Kepler formulated gravity as space a^3 over time T^2 in relation to a centre k. It is the space that relates to time in relation to a centre just as Kepler introduced gravity to be. Kepler said space a^3 standing is over time T^2 in motion **k**. $a^3 = T^2 k$ and to all those who try to give space-time some godly appearance with mystic properties can lose the séance-like attribute they wish to connect to space-time. Every bit of space however insignificant or however demanding forms a relation with time, which is what separates the different space from one another. It is the separation coming from time differences that distinguishes space from one another.

Gravity is working on a principle of indicators pointing dimensional integration and separation of space through heat densities applying different grades of space intensity. That means the space does not mingle, but forms layers. This is unlike one would expect from the advocating by Mainstream science of the characteristics of space. By gravity acting space becomes denser and therefore space can become a liquid and as all liquids do, space then depends on specific densities being in specific positions. With the specific densities borders come about in space. It is as Kepler stated gravity to be even before Newton came up with an idea that there was such influencing going on and named the influence gravity. Gravity is $a^3 = T^2 k$, which is the space a^3, that forms through the moving $T^2 k$ thereof giving the space a^3 independence as the independence comes about of speed differences which is motion in relevancy which is $T^2 k$. It is distancing a^3 from **k** by applying T^2 in the surrounding space and this is done by a^3 duplicating in motion when applying T^2.

The Oxford dictionary of Astronomy defines gravitation as follows
Gravitation is the force of attraction that operates between all bodies. The size of the attraction depends on the masses of the bodies and the distance between them; gravitational force diminishes by the square of the distance apart according to the inverse square law. Gravitation is the weakest of the four fundamental forces in nature. I. Newton formulated the laws of gravitational attraction and showed that a body behaves as though all its mass were concentrated at its centre of gravity. Hence the gravitational force acts along a joining of the centres of gravity of the two masses. In the general theory of relativity gravitation is interpreted as the distortion of space. Gravitational forces are significant between large masses such as stars planets and satellites, and it is this force, which is responsible for holding together the major components of the Universe. However on the atomic scale the gravitational force is about 10^{40} times weaker than the force of electromagnetic attraction

Gravitation Constant
The constant that appears in Newton's law of gravitation. It is the attraction between two bodies of unit mass at unit distance apart. Its value is 6.672×10^{-11} N m^2/kg^2 when the distance is expressed in metres and the masses are in kilograms. Although it is described as a constant, in some models of the Universe G decreases with time as the Universe expands (see Brans-Dicke theory), but there is no evidence for this.

Gravitation Mass
A measure of the quantity of matter in the body. It is measured in kilograms. Mass determines the strength of the gravitational force exerted by an object.

$$a^3 = \frac{4\Pi}{3} r^3$$

Newton saw a three-dimensional sphere holing a volumetric capacity which is a circle because the space inside has a measure in the cube equal to that of the cube.

The cube is also a three sides multiplies to give the containing measure of the inside. That is not what Kepler had in mind.

Now even three hundred and fifty years on, science still comes no closer to explaining what gravity is in contrast to the fact that Mainstream science established even more forces than the one that Newton declared at the time. If Newton only was less presumptuous about his genius and took more notice of Kepler's work he (Newton) could have seen just what Kepler said what the cosmos told Kepler mathematically about what gravity is. That effort would have saved so much misconception. But even almost four hundred years on Newtonian disciples will not recognise my personal effort to indicate to the world what Kepler said what gravity is. I have been trying to indicate this thinking by using academic channels but on grounds not related to my effort I was dismissed so many agonising times by so many academics in charge of Official policy protection.

Although it is most apparent (to me at least) that I can tell what Kepler saw and tell them that, still they the Newtonian priesthood silences me just like Newton silenced Kepler. Newtonians should have realised centuries ago that Newton and Kepler did not have the same mathematics in mind. Consider the following and then decide

In the manner the two measured the circle Kepler used a different measure of the circle than Newton did. In the case of Newton the radius r goes square while the circle Π indicates the single factor. Kepler deliberately ignored the factor Π for reasons I explain elsewhere.

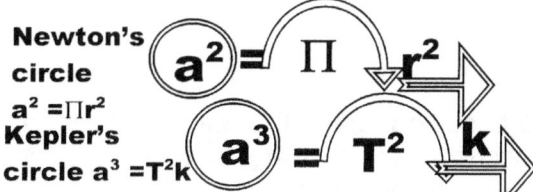

Newton's circle
$a^2 = \Pi r^2$
Kepler's circle $a^3 = T^2 k$

Kepler saw a circle because space is motion provided by singularity from a centre.

Newton wished to see a mathematical equated circle and the formula therefore was in need of revision.

Kepler intentionally stated the very opposite of Newton because the circle indicator T^2 goes square and the diameter indicator **k** remains single. That is how the cosmos relayed the given information to Kepler and that is how Kepler correctly interpreted the given information but Newton thought himself brilliant enough to change all that in favour of his ego. While Kepler formulated his ideas according to cosmic information Newton saw this effort of Kepler formulating the cosmic numbers Kepler measured as mathematically incorrect. The question Newtonians failed to ask is: How can the cosmos give mathematically incorrect information about the cosmos?
Let's look at gravity while we use our common sense.

There are three factors forming gravity. In the one scenario gravity is all about the motion of one part in space applying relevancy to another motion by dismissing space through the effort of moving through the space in a common factor. The lesser-developed point holding singularity is in an all out effort to depart from the centre spot where the gravity is vested as the strongest applying influence. The effort behind this is speed or if you wish I can use the term velocity but it will still mean it is the comparing of different motion each holding a different value but still requires

The object in motion is displacing the space that the object occupies in relation to the surrounding space and space being between the object and the centre. The object is dismissing space by motion as it is in motion through space. But there is another factor just as relative that is also applying motion. The applying of motion by the departing object only holds any relevance to the centre from which it is applying a distancing or an escaping attempt by increasing the distance between the centre and the escaping object. On the side of the escapee there is another space active, trying desperately to escape by enlarging the influencing field the escaping singularity is fighting to establish. Seen from the escapees' point of view in relation to what the escapee tries to bring in place what is important to the escapee is that the only motion there is, is the escaping object trying to escape. However much the escapee is contained the escapee is only aware of its motion as it sees all motion being only progress in a continuous charge of escaping.

Just as Kepler stated gravity is three relevancies acting in opposing as well as sustaining motion where the one action works by interlinking with the other without being aware of the other. We have to recognise the relevancies applying. It is motion performing contrasting relevancies and without all three independent actions of motion contributing as one not one act of gravity is possible. There are three relevancies applying motion in the formula Kepler left us. There is the space in which the formula presents the dimensions. The centre provides the space the dimension to move through the space as the object is trying its utmost to depart from the centre as far and as fast as it can. It is displacing the space it is in by using as short a period as it can manage. It is running off into the distance from the centre. In this motion the smaller object tries to repel the intention of uniting that the larger object has. The larger object is conditioning to unite by claiming to have the centre of space. The lesser developed singularity fights off the intended uniting with the domineering singularity while it is still part of the bigger space in which the smaller space part of.

While the orbiting structure has the effort of displacing the occupied space the material holds in an effort to produce a cooling effect on the material in the occupied space it holds, there is another centre in another object that has to be much larger than the roving orbiter and as such the centre of the centre object is removing space towards such a centre in an effort to secure the centre of remaining cool and thereby preventing destruction by overheating. From the centre object there is no escaping object but only space it is harvesting in an effort to sustain singularity in the centre of the centre object and where the strongest point of gravity is located.

While the smaller but independent space is busy with the great escape effort by putting a distance **k** between the space in motion and the centre, it fights to secure an independence from that centre. The larger space is accumulating as much space as it is contracting all space towards the centre. The independent and escaping object is moving through the space surrounding the object in this effort to put distance between the object and the centre. If it did not do that in motion it would hurry towards the centre instead of circling around the centre year after year. From the smaller object's vantage the smaller object is carrying through space displacing the space surrounding the object by motion in space. From the point the smaller object has is the space standing still while the object is applying motion to get away from the centre. The object is applying motion by dismissing the space it is moving through. $a^3 = T^2 k$

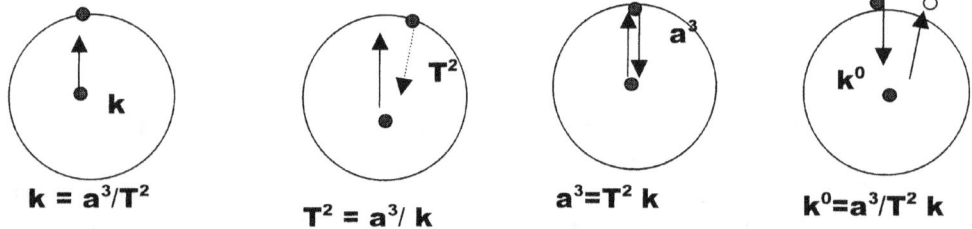

$$k = a^3/T^2 \qquad T^2 = a^3/k \qquad a^3 = T^2 k \qquad k^0 = a^3/T^2 k$$

By duplicating the space of any particle sharing space within a larger cosmos structure such as an atom inside a star or a human inside the Earth there are two relations applying.

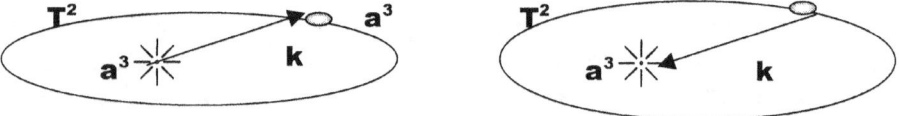

The factor **k** represents as much space as a^3 as a factor represent motion. But **k** also represents as much space as it represents motion because it represents motion.

This means that all space in motion is resisted by space in motion contradicting each other. In applying motion to space there is never just acting but always counteracting, which is precisely what Newton's third law on motion states. Because there is a motion in one direction there will be a counter motion performing a balance to the reaction on the motion and establishing a motion by a reactive motion. But also it proves Newton's first law on motion by proving that equilibrium existing of the action of space in the counteraction in space brings about equilibrium and being in motion that is in countermotion provides for the same effort as being in rest or in a stable state equal to being in rest after forming a circle eventually. Since there is equilibrium between the motion of expanding and the motion of contracting any additional motion adding to the "stalemate" existing will bring Newton's second law into action in the manner Newton described.

What this formula, which Kepler introduced tells more than any other fact is that there is no space if space is not in motion and all space there is must be about motion or else there is no space. That is establishing the fact that can only be by motion and motion is what space does to move from one point in time to another point in time by the square of such motion. If there is space the space is duplicating space by projecting space from a past through a present to a future and that means the space is duplicating what was in the past towards the future through a present. But if the space and all space there is, is not duplicating by motion it does not qualify to be in the Universe. The fact of space a^3 is about the equal = motion $T^2 k$ thereof.

The space we find is not just space but the space we find is motion of space as the space reduces to a centre by duplicating or by motion of repositioning in a rotating relation by doubling what there is in a repeat thereof. This motion, to which I am referring, is extremely apparent in the illustrated imaging of the functioning of a Black Hole where one can witness the total collapse of space toward and around a mathematical centre point forming where space and motion ends.

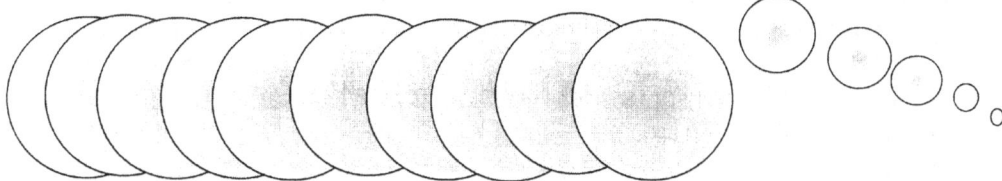

The containing structure represents duplication and in that it applies a norm set on conditions applied by the governing singularity of the space-time, which is supporting and maintaining the point in singularity. By containing the subordinate singularity the ultimate goal is to initiate the inevitable and eventual uniting of the two positions where singularity will again form one unity. The space-time first has to dismiss while the space-time fights to gain and duplicate as the one does which all singularity will do in the presence of the other that destroy space-time. This will continue to the ultimate end of all development when all singularity will once again re-unite. But in the meantime all singularity is fighting the fight of their life to remain independent, as much as possible and by what measure it may have to produce. When the two are mostly on equal motion the gravity each produce will bring about motion of maintaining independence by creating space through duplication. When the motion equality falters the space reducing of the dominating space will relinquish space by contracting space faster

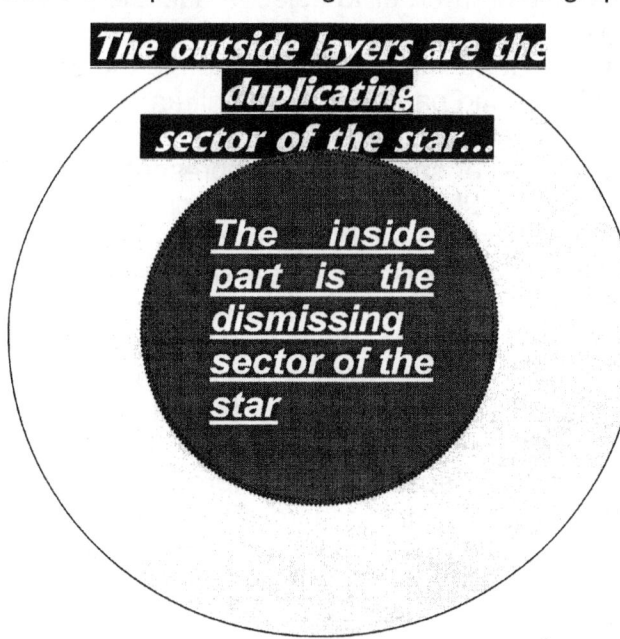

The outside layers are the duplicating sector of the star...

The inside part is the dismissing sector of the star

than the dominated space can accomplish by using the duplication of space. When the dominated space falls prey to the incorporating of such a smaller space into the domain of a larger space the fight for independence is carried on but the terms of conduct changes. As the dominating space adheres to the containing of all space within the domain of the larger space the lesser space will be reduced to the point where it may have to relinquish independence totally. By not agreeing to such terms the lesser independent singularity continues the fight for independence by an all out resisting relinquishing all individual form and to join and unite singularity. Losing independence will amount to relinquishing all forms of independent space being independent because of the independent motion it contributes to the unit of space –time within the unit of space–time. The resisting is performed by establishing individuality through the maintaining of individual space and form and in such maintaining of individual construction it confirms independence in a mode of self-protection. The self-protection is a resisting and the resisting the joining of the larger space where such resisting of relinquishing independent space brings about mass. The establishing of a mass factor is part of the resisting losing independence in a self-protection drive from early uniting with the domineering singularity. As the dominated singularity fights to remain independent it is forming resisting and when the resisting is not capitulating such resistance to capitulate its independent form, that is what we then see as mass. That will demand that the mass is brought about by all that the space contains, which is where all the atoms are individually and as a group and as a structure formed by the group where all that are fighting forms a unit and as a unit it is fighting for individual independence

In the commanding space a^3 forming the superior part of the space a^3 that holds lesser space a^3 in the unit as an independent part of the unit, which is included as a part in the same unit that normally occupies space-time as particles or at least by particles with independent space-time, the particles hold some space a^3. The space a^3 the particles hold is directly in relation to the particles the containing structure has in duplicating as an entire unit. Every particle in the unit has to fight every other particle in the unit as well as the unit as a whole for space within the time the group puts on the unit. The more space a^3 there is being relevant to the structure in comparison to the structure as a whole and as a unit that forms other relations with other units, which is duplicating in the sense of being a unit once again is relevant to the space a^3 that the structure destroys as a unit, the bigger the space is that the smaller space holds in relation to the size of all the space combined where the combination is part of the complete unit forming individual space-time as a group but still occupying space-time in the group as independent particles while in the unit. In this unit where the smaller individual space remains part of the unit but having to match to the capabilities of other space also in the unit being as part of the unit the less the smaller space will find it able to match the duplication effort compared in ratio to the entire group but still will find it able to focus a duplication standard that holds relevance to individuals in the group. Two sectors emerge in the star where the outer sector advances duplication as the main focus of the star and the inner circle is the sector that place gravity on dismissing space-time. In the inner part the particles focus much more on the dismissing of space-time in the sector as the main focus of the entire unit. The outer part is overshadowed by space-time duplication. Space-time dismissing overshadows the inner part.

The bigger the space is the more the unit as a whole will favour that particular independence to group where the individual will identify its preference in the star by selecting the group that will complement its achievement. The space will tend to associate with that group that promoted its goals. The larger space will tend to form alliances with those large enough to favour duplication and those smaller but being more solid will sink down towards the dismissing sector in order to select a suitable position it can locate in matching the smaller space needs. The more space a^3 the particle claims as an independent particle that is holding and protecting an individual singularity in relation to the space a^3 the container holds as a unit formed by the whole lot of spinning particles in the unit as one inclusive unit that relates to the space a^3 the container duplicating then is relative to the space a^3 the containing structure destroys as a whole unit. The dismissing of space-time is representing all other space on an accumulative proportionate value. It includes all particles that is spinning and this inclusiveness is serving as the combining effort of the entire group. In the group a relation exist between the individual particle and the group unit where the group is representing all particles in the group where then the group effort of dismissing space-time in each individual capacity and to the performance what each individual may achieve where such a combined dismissing manifest in a precise centre of the groups space-time. This is the point focussing the dismissing part where the groups dismissing of space-time coming from every atom would gather and present the unit effort, which is in presentation to the duplicating in motion all the atoms form as a group and in all this there must form a balance to gain gravity. As individual occupying space a^3 the atom is an individual container by own dimensions and as such duplicates space a^3 in this regard. The atom resists the dismissing of space-time in which the individual space the atom holds is also included by confirming the structural form to the atoms relevancy being k^0 in singularity that is bringing on an independent value that fits the particular k in the atom relating to $a^3 = T^2k$, which is part of the whole unit represented by the combined value of $a^3 = T^2k$.

This relevance means that without a specified container producing individual particles that control the specific duplication and destroying of space in relation to the outer space. Such a container will apply a diminishing relevancy of space-time displacing that will be equal to the displacing of space-time equal to that which a number of 112 protons can manage and no more. That number will allow a flow of space-time which will presume the displacing factor of a possible displacement figure equal to 112 protons which if they were able all are all working as a unit within a confined unit we call outer space and in conjunction with what is dismissing in the unoccupied space-time where the unoccupied space-time can withstand. We know that is a theory because the atoms in space can sustain much less than 112. In short the value of the walls serving our three dimensional Universe can sustain no more than that what a possible 112 protons gathered in one atomic cluster may displace.

Inside this outer container, which we see as outer space there are inner containers being stars that bear the direct relevancy which singularity is applying being inside stars, putting much more strain on the surviving abilities of atoms. In outer space the atom has an own relevancy of seven and the space demand on the atom is only three that the atom must maintain in order to duplicate. But in stars the containing star places a demand of the containing seven plus the space creating three in relation with the time applying inside a star, which is four. Let's put what was said just now in conjunction to the Earth. Since the Earth has no singularity demand which is that much better developed than outer space is insisting on as being a limit in the Universe needed to afford space-time self sustaining of all there is on Earth by a relevancy of k to $a^3 = T^2k$ is adequate in that which the atom normally can sustain and leaving lots and lots of space to spare, we call this state of affairs inside the atom to be quantum physics where we directly associate the concept of quantum with non-quantifiable volumes of unaccountable space in disuse. But in bigger units the space-time displacing relating to space duplication presents much more demand on atomic structures. As the demand of singularity in such units grows some relevancies within the atom come into play and I developed a formula to place such a demand in relation to singularity where the ultimate demand sets the standards. As I stated there is layers helping the development process within the star that shifts as the star progresses through development. At first the star leans heavily towards duplication. Then as the star develops the star moves across a broad range of specifically identifiable stages coming from one extreme where our Gas "planets" (which are stars in the making) are to where a tiny star such as the sun is growing towards sizeable monsters and then on to cosmic destroyers.

Throughout the variety of development there is a balance unfolding that at first supports motion by duplication and leaves dismissal for a senior partner to commit while growing the presence of a superior partner. Then the space-time development allows the smaller star to drift away from the domineering partner as space-time develops amidst all the partners involved including the material giant and the not so much giant. The growth confirms the security of individual singularity in the presence of other singularity united under an elected unifying centre singularity. From the motion that points to the stage where atoms dominate as they bring about motion with much less dismissing the stages come and go but the direction of development is always in the direction of centre singularity committing unifying in the extreme by dismissing space-time. In the end the star is totally committed to dismissing space-time by being absolutely unmovable and as it is so solidly stationary it places all motion outside the star into outer space. The aim of this development is to secure all singularity which was at first vested in every atom in independence to a shift towards and eventually including all the atomic singularity into one controlling singularity where the purpose of atoms in independence is taken over by one and all including and controlling governing singularity in the centre of the star. All the singularity that was present before is then included in a centre spot, which is not even a dot any longer, but then it returned to the spot. The Black Hole is a star as all other stars but the Black Hole completed the journey by taking all the atoms in the unit and unifying all singularity into one position that replaced all atoms and secured their singularity into one spot.

At first when the star is in a duplicating prone state, the atoms in a range of elements control and produce the drive of the star. But gradually the protons in the atomic cluster fuse into larger units and the larger proton cluster units starts to challenge the drive and eventual destiny of the star by fusing together elements with much less protons in clusters. The working process is immensely more integrate and complex than the process I describe here but this is the shortest introduction I am in a position to give. In short at first in the duplicating stages the atoms take charge of the driving of the star but as progress in development soldier on the centre singularity takes command and the star dismisses space-time while it also fuses together elements in the dismissing process. At first the singularity sustaining the atom has to group together to sustain the star unit but in time the unit becomes strong enough to sustain the star without depending on atomic independence to overcome overheating. The main picture arising from the explaining is one of a balance that controls the star throughout its development. In the sun for instance which is a minuscule small star a relevancy in the outer region might be all–favouring duplication, which is favouring duplication relating to singularity and with the atom having a sustaining displacement of favouring the electron position there is no actual danger of the atom demising. The star is still all about its duplicating mission. On Earth as in outer space the atomic difference in mass between the proton and the electron is 1836 to one. The electron is displacing the same space-time but compressing the space-time displaced 1836 times less than the proton does.

The figure mentioned is a specific predetermined ratio brought about by the dismissing of space-time in conjunction with duplicating of space-time showing discrepancies between the duplicating factor and the dismissing factor. That means the "outside" is 1836 times away from producing what the "inside" within the atom core does to reduce space-time. In the sun this ratio will be much less because the relative mass of the electron will form a relevancy number of 27 where it is on Earth only 3. That is one factor I placed on the ratio whereby the density intensity of the star space-time increases from as the star atmosphere (if you will) of the star such as the sun grows denser and gravity makes the space-time more compact. The density shifted towards the intensity the electron has and as the intensity of the density progress by becoming more compact through gravity the space between particles will match the electron at $(\Pi^2/2)(4(\Pi^2+\Pi^2)/7)=$ **55.66**. This displacement value produces gravity and it produces electricity by intensifying what is in outer space at a value of $7/10(\Pi^6)/6=$ **112.162** to what limit space-time displacement may endure while remaining three-dimensional. The number of 55.6 personifies the maximum proton displacement ability while remaining in form within the sphere. After that the sphere begins to show miss forming. It is no coincidence that all gravity driven structures have to have a molten iron core and an iron armature in the presence of magnetic inducting copper chargers producing a field can only generate electricity. In spite of the proclamation of science about the sun being only a hydrogen star the concept is nonsense because not hydrogen nor any other element except iron is able in our Universe to produce through motion the effect of space-time reducing which we either call electricity or gravity. Electricity and gravity are the same thing.

The electron position or in other words the favouring of the duplicating tendency within the sun's ability to create a working environment within the sun because that will have as a diminishing factor space-time displacement going as low as 27. This means the atom can reduce the electron space-time occupation by 27 times whereas the atom can sustain as much as 1836 times further than the electron can and the electron matches the speed of light. As the relation in the atom within the sun degenerated by 27, which means it loses but in order to find compensation in the time duration extending by the ratio, which the space reduces in the atom is left with a sustaining value of the electron plus the neutron applying space-time displacement without involving any of the neutron at all. By declining of personal atomic space in order to avoid deforming and losing individual identity which will lead to accelerating early uniting through absolute space-time demising such diminishing that will bring about the relinquishing of individual independence carried by having independent gravity securing the atomic identity which brings over the result that produces mass. That is the mass that the electron will consume in the space reducing and producing an enhancing of mass within the star. When the composition of elements within the star is such that the combined effort of all the atoms that is reducing space and thereby are improving their individual space-time dismissing. As they relinquish space-time it must be controlled because it has to remain the same factor in duplicating space-time by motion than the dismissing is demanding. By moving the control of gravity from the individual gravity vested in the motion of the individual atoms towards and in control of a centre elected spot that is holding all the invested gravity secured in the centre spot of the star, such increase of time by the reducing of space will produce a favouring of the dismissing factor much more than it will favour the duplicating by motion provided by the compliment of electrons spinning about. When the central governing singularity takes charge of gravity within the star the star goes dark as there is then less electrons in the star and it will reduce the amount of photons flowing away from the star because what we think of as light coming from a star is the ejecting of excessive electrons not needed by the spinning atoms and such ejecting of an overflow of electron production will have the star shining at night as a bright little boy shining by dismissing pebbles of light-photons into space. It is when the star gets dark that we can know the real monster. The star going dark will happen when the centre gravity will request more motion through dismissing of space-time that would light photons having the ability to escape. The proton then takes command from the electron and changes commitment within the star from duplicating in motion to dismissing by excluding all motion from the star.

When a demand of space-time displacement to the value of 56.6 protons becomes the norm the star will cease having space-time concentrated to a liquid by concentration as the star by that time excludes all electron functions and stops shining. This comes into effect when the demand on space-time duplication and reducing the reduced miss matches because of the mass overload the atom experience to space and to space supply gets so large that it destroys the heat envelope leaving the atom without a heat envelope. Only the nucleus will be able to sustain the further diminishing while the reducing of space is directly coupled to the increasing of time. The atom would shrink to such little space it will have space within the star that only the centre nucleus of the atom will take up to fit. More reducing by applying motion in creating space differentiation will leave a star with so little space the space will be insufficient to secure a position for the neutrons and the star will then have the name of being a neutron star. Going even further will find the proton rejected from the star. That is how gravity applies because it is a matter of relevancies applying between space holding and demanding conditions and space reducing in relation to insufficient motion bringing about much less space duplicated. The space duplicated brings about mass as a result. We shall again return to this topic in the new suggested theory later on after much more exchanging of information and arguing about the introduced information has taken place.

In the previous explanation it becomes clear that there are two forms of gravity applying throughout the cosmos and not just one form. Saying this I first have to reconfirm what Kepler proved that space could only be when space is in motion and the motion is in relevance to a specific controlling centre. The acceptance and understanding of these principles are absolutely vital to our understanding of the cosmos. This brings across the truth about the expression that one must not think of the heavens in the same terms as we think of the Earth. Laws applying in heaven and laws controlling the Earth by nature are one Universe apart. The Earth serves life while the rest of all the heavens are hostile to life. It even seems to us that the rest of whatever fills the Universe is meant to destroy life. Except that what is on the Earth, the rest of all created has one purpose and that is the destruction of life. If life cannot find any means of supporting life in surviving in a natural state anywhere in the rest of the cosmos out there we have to adapt our thinking about nature in considering the cosmos as totally different from what we find on Earth. We have to accept what we invaded and infested on Earth, but that invasion will be the only part of the cosmos we are likely to invade and infest.

Newton saw what physics applied on the Earth and Kepler saw what physics applied in the cosmos. Kepler saw space is in maintaining space by the motion thereof. The acceptance and understanding of this is absolutely vital to our understanding of cosmology. This brings the truth about the way we have to regard cosmology. What is applying on Earth is almost definitely not applying in the cosmos at large. We may never think of the heavens the way we think of the Earth. Heavenly concept stands widely apart form the Earth because the Earth came about to support life whereas the rest of all the heavens do not even know about life existing and is hostile to life. In the cosmos Kepler's gravity overshadows the gravity Newton saw. Kepler saw space is in maintaining space by the motion thereof. In this statement there is a balance maintaining equilibrium of space specifically duplicating by motion in precise equal duplications of the previous space that is repeated by the duplicated space by precisely copying as the following bisect of the previous copy of space to perfect in precision. I once again at this point have to remind that such duplication by bisecting is within the space-less surroundings of the proton. It is not in the Universe we see when looking at the night sky. This is what the formula $a^3 = T^2k$ translates to when turning the written mathematical code to the verbally pronounceable English. There is a balance forming equilibrium on both sides of the divide by producing $T^2 = a^3 / k$ and when barriers are broken and lines are crossed the defining ratio changes to $T^{-2} = k / a^3$ where the singularity distance in relation reduces by the time component going negative progressively. What this brings to light is that there are two points forming relevancy that indicate a separation of space although both are sharing in one space with both in the position of the identified space having motion that is balancing gravity by motion.

a^3 is space occupied by material seeking independence from the centre but that is not all because space a^3 is space holding an identifiable position all the way through the length of the line indicated as **k**. The space a^3 refers to a space a^3 within the space a^3 which all depends on where the motion draws the attention and the space a^3 will only find relevancy when motion sets whatever space a^3 one refer to apart from the other space a^3 referred to. But by identifying one both finds identification because the one is not identifiable if the other is not prominent too. The prominence comes from distinguishing both sharing a joint position.

T^2 is space in motion towards the centre of the space holding the space a^3 that is in motion, which is the space a^3 that is validating the space in motion towards the centre.

k Indicating the distance of the motion of space and in space in relation to a very specific centre. By indicating the point which **k** indicates **k** also indicate the space a^3 becoming the unit of all the space a^3 being in contact with the centre and being in space from within that centre from where **k** indicates space a^3 which through motion is distinctly not the dominating space a^3 that is in motion towards the centre but the space a^3 in motion that is differentiating by distinction separating the space a^3 that is dominating from the centre the space a^3 that is dominated by the space a^3 from the centre and through this motion relevancy the relevancy is holding all space a^3 connected to the centre.

$$k = k^{3-2} = k^1$$

It takes time for space to fill **k** in the distance. In fact it takes the distance that **k** developed since the Big Bang $k = k^{3-2} = k^1$ to fill the distance.
It also takes time $T^2 = T^{3-1=2}$ to produce the distance forming k^2
It takes space $a^3 = a^{2+1} = a^3$ to form k^3 since coming from the Big Bang

The space a^3 that is dominating the space a^3 from the centre

The space a^3 that is dominated by the space a^3 from the centre

The smaller space a^3 is distinctly distinguishing the larger space a^3 as the larger space a^3 is housing the smaller space a^3 where the factor **k** is as much indicating by length the larger space a^3 as much as it is indicating the end of the length of the larger space a^3 at the location of the smaller space a^3 by directly pointing at the position the smaller space a^3 holds. Where the larger space a^3 ends the smaller space a^3 is. The two remain as an inseparable single unit in double motion where the motion identifies the unit as much as distinguishing the separateness in the unity and always remain in absolute relevancy.

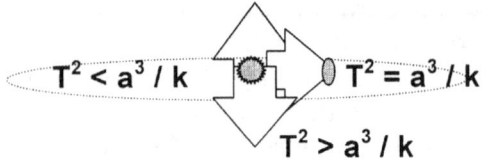

$$T^2 < a^3 / k \qquad T^2 = a^3 / k$$
$$T^2 > a^3 / k$$

Let us see what gravity is in reality when two objects perform a commune with gravity applying. Gravity applies between space occupying structures in space not occupied and a centre

If space were zero or nothing as Mainstream science so effectively teaches us then Kepler's principle formula would need the changes Newton brought about. But it is true and stands tested like no other research ever coming either before or after Brahe and Kepler's work.

From the implementing of $a^3 = T^2k$ we can see that

$$k = k^{3-2} = k^1$$
$$a^3 = a^{2+1} = a^3$$
$$T^2 = T^{3-1=2}$$

$k = a^3 / T^2$	$a^3 = T^2 k$
$k = a^{3-2} (T^2)$	$a^3 = T^2 k^1$
$k = a^{3-2} = k^1$	$a^3 = T^{2+1} (k^1)$
$k = k^{3-2} = k^1$	$a^3 = a^{2+1} = a^3$
is the same	**is the same**
as	**as**

$$T^2 = a^3 / k$$
$$T^2 = a^3 / k^1$$
$$T^2 = a^{3-1} = T^2$$
$$T^2 = T^{3-1=2}$$

It is all the same

$k = k^{3-2} = k^1$ is in direct relation to $a^3 = a^{2+1}$ is in direct relation to $a^3 = T^2 = T^{3-1=2}$. With this information staring mainstream science in the face and scream pleading at them to recognise this information they turn around and ask why can man not fly off to other galactica at the speed of light

We find that manmade structures in orbit in outer space have a relatively very short life and the corrosion up there destroys the material considerably in a relative short period. This is most apparent when comparing such corrosion material decay in Antarctica. In the South Pole articles remain seemingly destruction free for centuries whereas in the desert the heat quite literally dissolves material and even more so is the case in outer space. The heat n the desert as the heat in outer space corrodes material many times more that what is the case in outer space. That means it is not merely **k** but that what forms the concentration forming **k** that also has a strong influence.

When the astronaut is departing from space on Earth or filling Earth space it will take the departing astronaut k^2 time to reach k^1 and fill out k^3. At present and in this moment our most impressive astronautic engineers will devise an engine that would cut k^1 by say half. This achievement will come as they increase the power output say for argument sake to double what it is at present.

There are always two singularities in relevancy. The motion of T^2 seen from the centre in contraction uses the T^2 coming about as the **k** factor for the lesser space a^3 applying motion. Therefore where T^2 is representing motion to the larger **k** it is taking T^2 as the figure that represents **k** as a motion indicator to the smaller a^3. It means that $k = a^3 / T^2$ and T^2 to the smaller a^3 is the **k** factor of the smaller space soldering from point **T** to point **T** which then is the relocation of a^3 by the distance of k. $k = a^3 / T^2$ means a^3 was moved the distance of **k** in the time T^2.

It is conducive to remember that there is another part of the two relevancies applying where one is a^3 that is relevant to **k** but also there is the point where **k** has a duty to place a relation to a^3

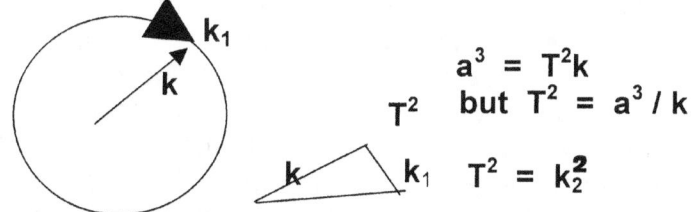

$$a^3 = T^2 k$$
$$T^2 \quad \text{but } T^2 = a^3 / k$$
$$T^2 = k_2^2$$

When the astronaut is departing from space on Earth or filling Earth space it will take the departing astronaut k^2 time to reach k^1 and fill out k^3. At the point a^3 then serves the new k will relate as much as it has to adhere to the T^2 time it takes to keep a^3 attending the new orbit T^2. At present and in this moment our most impressive astronautic engineers are devising an engine that will double T^2. This achievement will come as they increase the power output say for argument sake to double what it is at present. What we see happening in space with objects in motion translates equally to what happens to objects entering the Earth's atmosphere. There is a smaller projecting of space that changes **k** because of an altered **k**.

Mainstream science promotes the idea about particles coming into contact with the fuselage of spacecraft when they enter the Earth's atmosphere and thus cause friction that entertain heat which then rises as a result of such friction occurring. I know well I had this argument before but I cannot underline the incorrectness more of the way that mainstream science views the principle. By acknowledging this very incorrect way they see what applies when the heat blankets the incoming aircraft it disallows any further acceptance of the understanding how and what will apply in such conditions because those conditions express cosmology in detail. There is no friction between particles of the craft and of the atmosphere that is destroying the frame of the craft because it also is true that in outer space there are not enough particles there to bring about such a structural decomposing in outer space or in the atmosphere of the Earth an altitude to do it. What happens is that relevancies reapply and we know science acknowledges that material (they say some, but in fact it is all particles) reduces the space it occupies when coming from outer space into the Earth's atmosphere. In the atmosphere the reducing of space a^3 comes about because T^2 increases upsetting the ratio in $a^3 = T^2k$. When the space in the atmosphere becomes too small to allow the time it takes to enter because the distance k decreased faster than the space a^3 could compromise with the time T^2 changing from what is present in outer space comparing that to the time in to atmospheric space, the space shrinks which pushes time back into the past when the heat surrounding objects were much hotter than they are now. Time moves back as space decreases towards the time the Big Bang was present as we know the Hubble concept suggested. With the information in hand for a period of almost a hundred years and where the information forms the basis of modern cosmology since the information formulated gravity and not merely produced a name for gravity as our English friend did it is amazing that such accidents can happen and it is more amazing that no one in Mainstream physics has not the slightest idea or inclination as to why this is taking place! To top the cake with a red-hot cherry we know that our most impressive astronautic engineers are assembling a machine that will scramble the ratio Kepler introduced to a level in outer space where the ratio will be more than what the ratio in the sun is. Surprisingly they are not in the least surprised that not one object in outer space is using an excessive velocity.

Travelling is about bringing space in motion. Motion is combating heat and heat brings about expansion. Expansion is producing material as a substance that accumulates by growing into more material and producing material is about duplicating through filling vacant space. To move from one point to another point the material must release from the space it filled and fill the space it is moving into by which measure it then produces material. The lesser the material is, which duplicates in the way of being an individual unit is by taking up space in a larger space unit where the space that is taking up space forms a part of the larger and containing space. The method of filling space as forming a part of the containing space being as such the unit of such a larger space unit. But in as much as filling the larger space the smaller space still holds individuality by motion pronouncing the independence as well as the inter dependence of both individual spaces in the unit. It is the time it takes to duplicate such a large unit when comparing the two individual space units in relation of time each has taken to duplicate, that the ratio between the two duplicating will prove to be of less time duration or "slower" in time duration that the larger motion will take to complete the duplication of the larger space in comparing to the time it takes to duplicate the smaller space. On the other hand the more the space is that is in the process of duplicating the longer period in time such a process will take to perform such duplicating of space per time unit. To duplicate space per measure of space size which is having more k that holds time further from singularity by extending the space the "slower" the larger space duplication will be because of the bigger effort it takes to duplicate more space although both use the same time frame to duplicate. It is not that complicated because a lorry duplicates by motion as does a bicycle duplicate by motion but to get the truck up to eighty km per hour takes a hell of a lot bigger effort than to duplicate a bicycle at eighty km per hour. The motion T^2 requires effort to reposition space a^3 as space a^3 duplicates using time T^2 to shift space a^3 across a distance k. The more a^3 there is to extend by the increasing of k the more T^2 will be required to complete the task. The faster the duplication is the further the distance is from the centre of singularity and the longer the rotation will be in relevance. To duplicate space using the same time and duplicating a smaller space will bring about a much reduced distance from the centre using a shorter k in $k^{-1} = T^2 / a^3$.

An object can rotate in outer space as long as it can maintain a speed that will keep the object rotating in that orbit. The speed requires that the distance from the centre of the Earth to the centre of that object rotating must remain even at all times. That is the gravity applying up there in outer space

When the motion decreases and a lesser motion differentiation sets in the object can rotate in the atmosphere as long as the motion will last and it can maintain a speed that will keep the object rotating in that orbit position. But the difference is that the speed required orbiting declines. That is still the gravity applying up there in outer space

Then the orbiting object slows down to a speed that cannot keep the object in orbit above the Earth and the restriction places the orbiting object on the ground while the orbiting objects serves it speed as equal to that which the Earth provides. That is once more the same the gravity that applying up there in outer space
At that point the object has mass but by increasing the speed the object will increase the rotation speed putting more space in relation to the time it takes to orbit. One may be stupid enough to buy into the bluff that the object is still clinging onto the mass it had and that that mass is bringing on the gravity, but as I said that is only when you have a mind weak enough to buy into that propaganda.

Gravity is the maintaining of speed in relation to other motion that is either contributing to the object in motion or the influencing of such a motion.

Oh, I know they have everything confused in the red shift and the blue shift because again no object can travel even close to the speed of light. The Red and Blue shifts are all about lenses swapping relevancies and that I explain to a certain detail in "**Xepted Astronomical Mistakes**".

Singularity provides space-time but singularity is without space and therefore being without motion that takes up time to complete the duplication of space. Singularity starts at eternity and from eternity all space-time develop. The less the space is the faster the motion will be in duplicating the space because the smaller the space will be in need of duplication. But also the faster the motion is the closer such motion will be in relation to the centre as far as relative duplication goes because the bigger the extending is of the **k** in distance by measure of duplicating it applies and the less space it occupies from duplication to duplication.

Gravity is the strongest where space is the least and therefore the time that it takes to fill the space by motion will also be the most in time duration. A Black hole is altogether singularity and a Black hole is all about reducing more space into less space by faster motion dragging time to eventual eternity when space in singularity within the Black hole reaches infinity. The motion is so fast the motion reduces the space into infinity but also drags the time to eternity by the same measure. The time factor slows down so much that the light is unable to duplicate enough space in which time will allow to escape in the space that the light in the space has available. By only having the atom dismissing space-time, as is what happens to most stars in our universe such reduced dismissing will lead to more reduced contraction. That means less relative motion. In the end when the universe will draw the final curtain the final gravity will produce a speed so fast the motion will extend the time duration into eternity as it stretches the time beyond Universal limits and to achieve that it reduces the space, by collapsing all space into infinity. I refer to this action as being in the Black hole but one must remember that the Black hole is the ultimate unifying that all atoms within one certain unit can reach forming a single Unified structure. The atom's final stance is the Black hole that became a massive single atom. In our Universe however having the atomic dimensional qualities, this process of dismissing such space is found only in the atom, which achieves it by applying gravity. Somewhere down the reducing line one finds the proton is reducing space into the oblivious by increasing motion to the ultimate, but that is the proton and the star is only all the proton's accumulated efforts.

By only having the atom dismissing space-time, as is what happens to most stars in our universe such reduced dismissing will lead to more reduced contraction. That means less relative motion. The lesser relative motion will contribute to a smaller ratio in the space (not more compact but just less space used to fill) in need of duplicating. With that a shorter duration or period of time will be required to allow the duplicating to come about. The more motion that is required the more in space in the process of duplicating will come about and the further the relative duplicating will be in terms of duplication in ratio to the rest of the surrounding space. This only applies because the relative duration prolongs as the space reduces to comply with the bigger volume of space in need of duplicating.

Speeding up the motion will extend the terms of duplication produced by the motion as the space reduces to extend the time duration. In short: going faster will take longer in time because space reduces by motion duplicating more space per time unit. $a^3 = T^2 k$ – this comes down to $T^2 = a^3/k$ and that means extending k which brings about faster motion that will prolong the time duration as much as it reduces the space in motion in relation to the space holding the motion of the space in motion. Every time space halves, it will take with it the same time and therefore the time doubles through the space that duplicates. The fact that the space duplicates halves the space as much as it doubles the time within the process in duplication. As the space halves each space has an individual alliance with the time therefore as the space reduces it will prolong the time when a quicker or faster motion comes about.

Motion is gravity and gravity is strongest where space is least. If an atom is being confined in a smaller reduced space the circle of the atom will have the electron circle growing smaller which will have the electron rotate around the centre core of such an atom. The time is a fixed factor set by the occupying space but with a smaller circle to complete. The same atom will use a lesser confining space allowing the atom more space to be within. The duration the electron has to complete one cycle is the same but when the atom is bigger, the electron travels faster to encircle a bigger circle as it takes to encircle the smaller circle in the same time period. The duration of the spin that the electron will take to complete a cycle will be in the same period as when the circle was smaller therefore the pace the one electron will move about will be much different from the next electron cycle of the other atom in the other lesser confinement. Duplicating space at a faster interval will mean taking space-time back in time which will increase the direction of time to a time where singularity was starting to provide space with time, that is taking space-time back to $k^0 = a^3 / T^2 \ k = 1$ but going in such a direction involves the reducing by measure of $k^{-1} = T^2 / a^3$ to the point where $T^{-2} = k / a^3$. At a point round about singularity the gravity that the space acquires will crush the space the object claims back to the size it would have had, when the Universe was condensed to round about singularity. This cannot happen because long before it happens all space will become heat and the heat will dissolve material into photons. This is the direction we, who are captured by the Earth on the Earth are heading if not for mass forming to secure our atomic individuality firstly as an atom and then as an atom in a larger unit that is forming a group. Let's carefully look at the general use of gravity as is mostly applied between objects in consent of remaining individually separated by space and with respect honouring each other's independence.

While the smaller (planet) is in a wholesale effort to escape and secure sovereign independence there is the larger partner that is providing the centre from which the smaller object is running. The centre contracts the space it claims and from the centre the object in escaping is as much a part of space-time than the rest of the claimed space-time being the occupied and the holding part of the space unit. Both are relevant as both have a part of space in the unit forming the unit. The centre partner is providing the retraction of motion of the departing object in containing the departing. The second and centre object is retracting the space surrounding the centre object in an effort to supply the object with space the centre reduces through gravity. In relation to the centre the centre is applying motion that is reducing space and the more the space reduces the more heat surrounds the centre point where the space disappears. The space containing the heat disappears but by the space disappearing there is much heat left in the rest of the space as concentrated heat. As the space reduces towards the centre the heat level in that space rises.

The centre object is applying motion by dismissing space towards the centre as the centre applies gravity. $k = a^3/T^2$. Then there is the third factor, which is the space itself that is in motion as well as providing motion. This is $T^2 = a^3/k$. As much as the smaller object is running away from the centre, the centre is contracting all the space it claims to be space-time by diminishing the space from the centre. The centre forms a larger space a^3 that provides a flow of space T^2 which produces the time aspect that is being concentrated by establishing k being the flow towards the centre k^0 as all the space-time moves the length of k from k inwards to the centre point k^0. From the vantage space holds it finds all space-time equal that it is moving towards the centre of the Universe. The universe I am referring to is the pivotal position as the sun is in the case of the solar system. This we see with light coming towards the observer locating the observer as being in and being the centre of the Universe. I explain this statement in much better detail later on because that statement defines our improper view with which we approach the cosmos. As much as the runaway is running away the centre is contracting the space-time and as far as the centre is concerned there is no special thought going to the runaway because the runaway is all part of the space-time centre but a part which is not that much successfully contracting. The centre is tidying the flow of the runaway but not containing the flow of the runaway. The contracting is successful. It is fighting off overheating in a coming together and this is what we see as gravity. The third factor is the space reducing as it is moving and as it is moving and reducing the space by the same margin it is increasing the heat towards the centre by gravity's ability to decrease levels within the decreasing space moving the space towards the centre. $k^{-1} = T^2 / a^3$.

We have to accept that rules apply where singularity stands in regard to other singularity. Of the two one is a domineering dominator seeking control as a dominated subordinate fighting off the control by seeking independence. If no working relation is yet formed there is an ongoing fight for position between the two whereby the one will compete to destroy the other lesser developed into submission and the other will put up a relentless fight to flee and secure its independence. I was asked on occasion about my ability to prove this statement. Well, we all have eyes and we all have minds so we'd better use them therefore we all can think about what we can see. We can see there is some dominating going hand in hand with some flight to prevent full submissiveness or a fight to destroy or achieve one of the two relevancies.

In every case it is space-time in motion flowing towards the centre holding a centre spot valid as the space-time is flowing towards the centre and that is providing the motion that is affecting the others sharing the relevancies. The difference (I suppose) between space and space-time is that space is just another meaningless human concept while space-time is having a flow or a motion of a valid substance and such flow is validating the particles or objects and the space-time holding them. Seen from outer space that motion of a fluid substance is the factor that is bringing about the gravity.

Gravity is the relevance of motion of a smaller space putting a movement in relevancy of another moving space within the same space but acting as the larger space while sharing space as one unit with the smaller space and being in the same space. If any reader is in doubt about my statement then tell yourself in all honesty what a force is…but be honest while you explain to yourself what a force is…(the force Newton suggested is keeping the universe glued) and then go and scientifically differentiate in mathematical detail what the difference is between the powers of a Pagan god and a force. From my personal view a force is just motion applying and that is what Kepler said gravity is. Even if you wish to maintain the silly idea about gravity being material pulling each other all over, such pulling demands motion to initiate the pulling or the tendency to apply motion when given a chance to do so. The pulling starts and ends with motion. The answer about what gravity then is can translate directly from and in relation to the findings Kepler's work produced. In relation to the space surrounding the orbiting structure as well as the space between the centre and the orbiting structure the structure in motion is steady and motionless in concerning the motion of the space, which the centre of the largest sphere is dismissing space towards that centre. The orbiting sphere has a lesser capability to draw space towards such a centre and in that the smaller sphere applies motion in order to secure the maintaining of the lesser singularity in the effort of combating the overheating of the lesser structure.

The smaller object is applying motion by getting through space while in the larger object centre is applying motion to space-time and the space-time is providing the motion linking the relevant object to find equilibrium in motion applied. It is only in this book that I ever refer to space because there can be no such a thing as space in cosmology.

The one object tries to put space in between the centre and the object in a specific time and the centre removes the space between the centre and the object in that same period. That is making the space there is, space-time. That forms a circle and the size of the circle depends on the space relation with the period in time, which produces speed or velocity. There is space through which the occupying material moves. It is at a specific space volume during a specific time period. It is velocity or speed. If the space part is too little comparing to the time part then the time part will contribute more to the ratio and the object will decrease the distance between the centre of the circle holding the gravity applying spot and the object. It is then moving faster than the space is moving towards the centre and the space the object occupies will extend the distance that is between the centre and the object.

What Newton saw, Kepler describes best. $T^2 < k < a^3$ means that the object is falling out of the sky because the time it takes to complete a circle requires much more duplication of space within the space available to the object by the motion for the purpose of duplicating and the space available is not able to provide a large enough period of time to counteract the centres retaining the rotation by restricting such a rotation with an equal contracting.

$T^2 > k > a^3$ says that if the departing object seeking independence shows a greater motion than the distance k can provide therefore it will have to increase the space orbit in the time period by establishing the space increase in adding orbital space to establish more space to orbit. By increasing the space a^3 within the new T^2 such extending will force k to grow bigger and in so doing provide more space in which to orbit. This is mostly artificial such as one would find in the way rockets are launched but it rings true (although by the tiniest of margins) where the Universe develops by means of extending the k factor. It echoes that which we see in a normal fashion as the Hubble constant and that law describes such expanding. Even comets adhere to time old routes with cycles that are well established and as old as the solar system is. By launching the rocket straight up into the sky following the 7^0 inclinations that forms a sphere T^2 becomes k and k becomes many times the value of a^3 that finally reaches outer space. It is a case of this radical increase resulting in more space and with that in the result thereof we find that when the time of the cyclic relation provided by an extended T^2 is too slow and a larger cycle is required because of a velocity ratio that favours the object in rotation, the space between the object and the centre will increase and so will the radius between the centre and the object increase. $T^2 < k < a^3$.

When the space a^3 does not have the ability to produce the required motion T^2 or the increase in speed that is required to accelerate the speed value and the level the Earth centre demands from such an orbiter to remain in that orbit it will cease to provide the opportunity to the orbiter to remain in the orbit. The slowing down of relevancy in speed hampers all further progress of extending k, which is enabling the satellite the opportunity the Earth provides to allow such escaping to continue. The shortfall will come from k as the length of k is reduced and the deducted is in place to compensate for the short falling in the rotating motion that such an inadequate k will provide. Then the formula is $k^{-1} = T^2 / a^3$ If the rotating time is smaller than the space the centre provides from the centre to the centre end the distance k will reduce and provide a smaller space a^3 in which to rotate T^2 as to establish the required equilibrium needed to secure harmony in gravity $a^3 = T^2k$. This is what we Earthlings experience as gravity, but which is not gravity because it is a bi-product or half the result of the full complement of gravity. It resulted when some balance went imbalanced that crossed the limits of harmony. When the time factor is equal to the space cycle the orbiting structure is rotating within and hold its own in the company of the contracting motion. The lesser space orbiting should claim as much space as the centre is disposing to have equilibrium in space-time. In that the motion providing the escaping has to be the same as the motion providing the equal contracting motion. The motion of leaving is equal to the motion of staying with the circle. It is because of this that comets orbit the way they do and any thought of inter cosmic travel is completely ridiculous. To leave the sun the structure that tries to leave the solar system must beat (not only meet) but beat the gravity coming from within the very infinite of the sun's inner core where the diminishing of space-time provides the space less ness and timelessness needed for fusion between atoms. In order to leave the solar system the craft and all that the craft contains will have to fuse into one atom or dissolve into liquid heat.

By the eliminating of the motion where such elimination is coming from the centre all the space, which the smaller space is within is part of the diminishing and that includes the lesser space that is applying motion. Therefore it is the task of the smaller space to capture space and identify the captured space. When the rotation speed cannot keep up with the dismissing of space and the space dismissing is then overpowering the orbiting space, then the rebalance of gravity steps in where it will try to dismiss the smaller space in total. The orbiting structure will start its descent under such conditions and the orbiting structure will then begin "to fall". If the object is moving more rapidly than the space is depleting towards the Earth the orbiter will "lift off'. But it is all a relevancy of speeds applying placing space in relation to time. It is $a^3 = T^2k$, just as Kepler said. Where the departing speed of the orbiting structure equals the diminishing speed of space in contraction that the centre produces, an orbit of $a^3 = T^2k$ at that point holding time will come about and gravity is in equilibrium. Then gravity in equilibrium departs just as fast as the space holding the departing object diminishes and the departing velocity is the same as the diminishing velocity

When examining the illustration seen at the bottom left the motion in the top illustration indicates as one can see that the motion does encourage the seeking of independence from a centre by a lesser independent singularity. We may take the controlling object as representing the Earth gravity that is securing the object forming the role of the satellite onto a centre that could be the Earth. The reason behind this effort of the lesser-developed independent singularity seeking independence is because singularity that is the better developed and more controlling threaten it. I explain this statement much better a little further on in this letter when I get into the Roche principle. From this securing and the breaking of such gravity securing comes the sound barrier and when such securing border is broken the sound barriers represent a control that becomes invalid. The reason the centre holds the most gravity is by now very well discussed and argued. But the centre is where space disappears and where the Universe goes flat because where the centre is where one will find singularity being the singularity that becomes the governing singularity, which is forming, is forming the centre of the Universe. In relation between the three factors there is the one that is in relation by applying motion within an occupying space. In the space between the two other participating factors there is the occupying space. That space then forms a relation with the space that is roving performing the duty as the surrounding space that serves as the roving occupied space. From any of the centres the whole picture seen from the position end of the line the entire space will be filled by the motion of the space contracting. Observing such space it is apparent that the space in its role as the super container of all proportions will seem to be very motionless. We have to contemplate that the space we regard, as outer space is as big as space can get and the larger the quantity of the volume in motion gets, the more motionless it will seem. By staring at the biggest there can ever be, it will seem to us as being motionless if then only of the sheer magnitude that is involved. From the centre nothing is departing but the centre is sustaining the centre by providing dismissing of space that is flowing towards the centre where in the centre the motionless and space-less singularity kills off space in motion while the space flowing towards the singularity at the centre is replenishing the centre of space, which is being dismissed by the centre. It is space in motion as Kepler realised $a^3 = T^2 k$. Motion is independently coming about holding three positions independent in equilibrium.

The balance that Kepler noted in his formula of $a^3 = T^2 k$ is two motions applying in relevancy to each other. These two motions must form a balance to produce the balance of space being equal to the time the space is in.

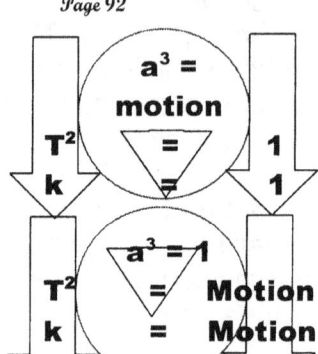

The linear motion that the orbiting structure applies is a fight to secure independence of the lesser-developed singularity where the motion is contradicting the space surrounding the motion of the second object's centre. In the centre of the dominant structure is all the gravity domination secured just as Einstein said. The lesser singularity is seeking independence from the first object and therefore tries to depart from the centre.

This independence drive to secure independence is part of any structure having the potential to produce and apply heat that will bring about that the second and lesser singularity will search to bring independence to the lesser singularity. It is how the cosmos reacts to heat coming about as reaction of heat increase.

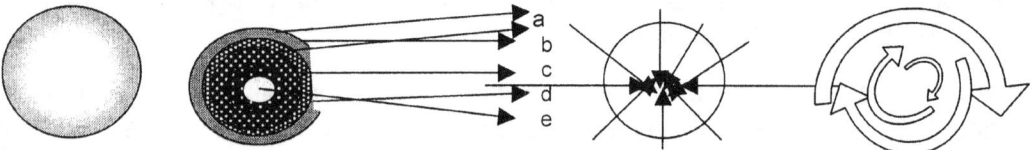

The gravity we recognise and which we consider being the " force" influencing our very existence is the reducing of space by concentrating space to form more dense heat in different layers. The gravity we find in this position is the one sector that Kepler introduced as T^2

The second form of gravity is the phenomenon, which Mainstream Science chose to name momentum but in fact is the second leg of gravity. It still holds gravity as motion but it has a directional change. This factor we normally think of as the one that Kepler introduced as k. In the centre is singularity, which at the time may seem inactive but as soon as conditions re-apply the centre can form singularity that can come alive by motion that introduces the position thereof. By electing a centre point singularity dominating can be anywhere in the cosmos at any given point selected by the motion of all the atoms in a liquid or a solid state or a gas state with the only condition the asking of relevant motion forming a centre in the application of the Coanda principle guarding singularity as $k^0 = a^3 / T^2 k$ anywhere. This phenomenon we named after Henri Coanda. This statement I shall return to and will prove a little later on. The singularity stands related to other singularity and although it forms singularity that groups together electing such a centre such a relation can only exist when relevancies to other singularity comes about as space-time or to use another term which is space in motion in relation to the individual singularity grouping to select a centre that will apply to all atomic focusing on a centre gathered by all atoms concerned.

Gravity is the maintaining of motion and motion produces speed or velocity in relation to other particles also maintaining a speed or a velocity but moreover gravity is a balance that is forming or the striving to unite singularity or aim to achieve an eventual uniting of singularity. Gravity is not a tug of war and neither is it a magical force coming from nowhere. Gravity is about half circles forming lines that produce half squares in the format of Pythagoras.

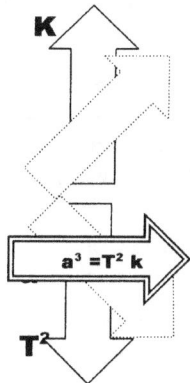

$$a^3 = T^2 k$$

Every relation formed about singularity is about surviving overheating and protecting singularity as much as preventing the loss of individual singularity by destruction. This loss of individual singularity is in the worst scenario forming the Roche limit and in the least scenario securing matter onto the earth. Mass is the product of such a resisting of losing individual singularity in relation to the singularity forming the master or the governing singularity where such an elected governing singularity takes charge of and controls all space-time from such a centre and which is from that centre totally dominating all the space-time. This resisting of capitulating of independent individual form giving identifying space by separate motion is forming a position in relation to space occupied by those atoms forming a unit, which we humans chose to call mass of the material. This does not produce gravity but is the product of gravity. Mass is the product of gravity and not the producer of gravity.

In Newton's gravity formula Newton placed the relation of the two objects in gravity in a square to each other by the moving closer of the particles. But Newton brought in mass as the principle factor, the one that is responsible for achieving gravity as a contracting force and according to their thinking, which is suggested by those most learned in science results in gravity forming. That cannot be the case because in that case there cannot be anything such as micro gravity. Micro gravity is what comes about when mass floats about. It is thought that since the floating object is out of reach of the Earth's mass the question is what will then cause the pulling of mass and in what direction will such pulling be heading. So far we have seen that gravity is motion and motion enlarges or reduces mass but mass cannot produce gravity because mass receives gravity. Since we know that which is keeping the objects afloat above the Earth is motion differentiation coming about as the two crafts harmonise but not equalise their comparing motion or speed. However we find ourselves very much unable to explain the existence of the so-called micro gravity. Consider the affect of gravity in micro gravity. It is not micro gravity we meet in outer space but it is micro mass. The pulling of whatever is up there in outer space is less than down where we normally are on Earth.

If it is the mass, which is responsible for the pulling of what we think of as gravity, then the gravity can resemble a form of possibly being micro because the mass that is producing the gravity must be the guilty party of going micro. In order to produce micro gravity one must insist on micro mass to produce such micro gravity. Since this argument is childish nonsense we have to realise the gravity up there is coming about from motion where motion produces gravity. A higher motion will increase the space between the centre and the structure while a slower spin will result in the decreasing of the space between the structure and the orbiting structure. The mass remains the same but there is a specific border that an incoming object shall not cross or the crossing will change micro gravity into gravity, which produces micro mass becoming mass. When being up there or down with us the object holds the same composition of material, which we think of as mass in the normal flow of conversation. But the mass becomes a factor not when the specific border is crossed bringing the object under the control and in the command of the earth centre singularity. If it did Galileo is wrong and if it does not Newton is wrong. All objects fall at the same rate notwithstanding what mass it has or has not. On route down to the ground there is no mass because all mass falls at the same rate. Only when being on the Earth with direct or near direct contact with the earth mass becomes a differentiating deference.

Again I stress that mass as an applying factor cannot contribute to the balance of gravity except for dragging along when being at the bottom and lying on the Earth with the motion of the Earth causing friction between the object and the earth causing what we humans wish to distinguish as mass. Newtonians, you can go on bluffing yourselves as much as you wish but Galileo insisted that all objects fall at the same rate notwithstanding whatever that mass of the falling objects might be or not be because when falling mass and all differences associated with mass differences are compromised and that discounts mass as a factor in falling. That once again confirms my view that mass is friction caused by the lack of motion contributing to duplication and therefore mass is truly only resistance. According to this explaining mass in contributing to gravity has no role to play as we all can clearly gather from evidence about performance of the objects in the increasing of gravity. Neither has mass any function in this process. It is motion that is creating space. It is about motion $T^2 k$ producing space a^3. With the increase of motion T^2 the factor k would be affected and that will affect factor a^3

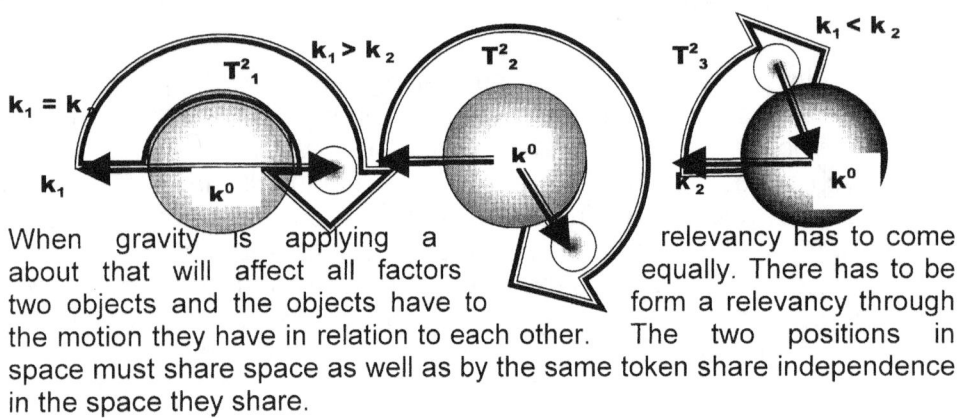

When gravity is applying a about that will affect all factors two objects and the objects have to the motion they have in relation to each other. space must share space as well as by the same token share independence in the space they share.
relevancy has to come equally. There has to be form a relevancy through The two positions in

Such a motion consists of two factors just as Kepler introduced gravity being $T^2 k$. In Kepler's formula there is on the one side space in the cube but at the same time there is a larger measure of space on the other side of the relation. The one side holds the space implicated in the relevancy to a cube but in the case of the bigger space only a line indicates the distance through which the space will measure. The distance is indicating a space that runs from the centre where at that centre there can be no motion as space runs into the contracting end where the motion stops. The space, which is indicated by the running from k^0 to the point cared by a^3 at the end of the line k can only be if such a space is in motion. Kepler said it…Kepler said space couldn't be if it is not being in motion. In that space presented by k the space a3 holds direct relevance to the time T^2 that the motion takes. That means every particle no matter how minute it may be but if the particle complies to the independence it has in protecting independent individual singularity it contributes with having gravity by the motion of $k^0 = a^3 / k T^2$. They have space-time because that is space-time. That is the space located between k^0 and k distinctly pronounced by the applicable a^3 to the tune of T^2. Having those principles has what fills the space qualifying as having space-time.

There at the end is a space a^3 identifying the space that is covering the area running from a specific given centre that is covered by the length of k. There is within one space a relevancy created by motion of space created by two factors sharing a space. The one shows a negative tendency and performs in an effort in moving away from a centre. Then there is a line that is connecting arch to the centre, which is securing the object by containing the first objects effort to bring about individuality. This produces two points and the T^2 factor accommodates the square relevancy that comes about through motion is applying. If the one k comes too soon following the second k the T^2 factor would be too small to accommodate the two a^3 sharing the spot by motional duplication and the object will fall to the centre because of k reducing. When the two points forming k is too large the T^2 will force the object into a larger orbit. If the motion in T^2 is not enough to provide k the required distance k will reduce the space holding the moving of a^3 in place. The circle T^2 in which the space a^3 moves will then reduce in size by placing the distance indicator k into a negative state or a declining measure. That is the gravity we experience but that is a small part of gravity only brought about by motion discrepancy. The object applying true cosmic gravity does not show any tendency toward mass or indicate that mass will affect the falling of the object and with the falling of the object all sizes will fall equally if all sizes and masses in that spot have the same velocity discrepancies effecting all falling objects equally. It proves Galileo correct but it also proves Newton incorrect. We see this with so many satellites and even space stations that plummet towards the Earth. The object did not get more massive and did not through adding mass to either the orbiter or the Earth begin its descent as it falls to the earth. Neither did the object become less massive and fly away from the Earth. When the speed of the object goes into imbalance such diverting of the balance occurs. The relevancy of the speed balance between the motion of the Earth and the motion of the rotating object changed to accommodate both a^3 in the T^2 that k would allow. That is gravity. That is what Kepler said gravity is when he said $a^3 = T^2 k$. What no one ever took notice of is that gravity acts precisely in the manner Kepler stated. If the motion increases the space increases and if the motion decreases the space decreases as gravity applying is motion in space forming space in motion. The more the motion is the more space is produced and the more space is affected by the increase or decrease of the motion of space. Newton's $4 \pi^2 a^3 / T^2 = G (m + m_p)$ has no part to play and it is only Kepler's $a^3 / T^2 = k$ that comes into the equation since k is $G (m + m_p)$ in any case.

When k increases all factors have to increase to compromise for the extending of k. If k extends space has to reduce because k is in direct but inverse proportionate relation. In this following argument we find two opposing forms of space where each plays its part in order to maintain a compromise done by both with mutual respect. The one we consider as the lesser trying to escape from the domineering, which is more developed and tries to contain the escapee. With k extending the lesser escaping space a^3 remains just as big as it is but forms a smaller part in the bigger space a^3 being the one in retention of the escaping space a^3 . By the bigger space having a bigger area in retention the smaller space is confined to a bigger space while remaining the same space and therefore in relevance by application is reduced in the whole relevancy where it now has a smaller part in the overall enlarged larger space a^3. But in the time aspect the completing of one cycle by the smaller space a^3 within the improved bigger space a^3 that is much bigger and is holding the much longer outer circle of the larger and more of the containing space a^3 the time it takes to circle about a longer space rim will bring about the circling around a bigger space in total using the same time that is taking longer in duration. The roving space can claim more space that will then fall into the space to be concentrated from the centre by the centre as more motion applying to the independent captured roving space will introduce that increase of space into the accumulated space shared by the factors. By introducing more space into the equation it provides a new balance that will suit all the factors in achieving the maintaining balance required. The time component will travel a wider space using the same time component but stretching the duration therefore increasing the time used per space unit gained as space holding becomes more but the length of each unit becomes shorter than previously. As the circle increases the time will be adversely affected in duration of space-time. What this implies is that one cannot have space-time and where space increases have time that is not affected by the change in the space.

We gave this forming of separate gravity coming about by means of performing individual motion a name being the Coanda effect to mention one amongst many others. The Coanda effect depends on singularity being a circle and motion establishing an independent singularity. Then the singularity cuts the Kepler formula in two parts. Evidence about this has been with us since the time of the great Leonardo da Vinci who was the first person to see the potential manipulation of space-time by changing singularity direction by motion.

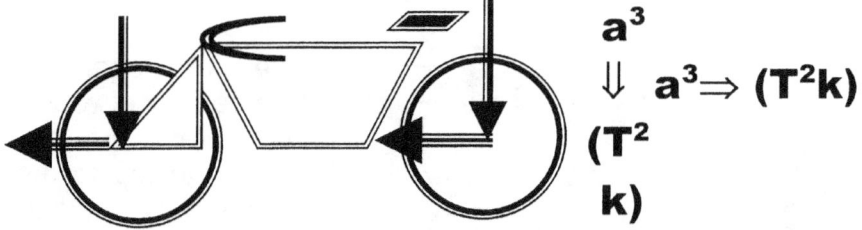

a^3

$\Downarrow$ $a^3 \Rightarrow (T^2 k)$

$(T^2$

$k)$

A low-tech mechanical human device might teach us something about the most basic rules about gravity if we pay attention to the rules as they apply. When a bicycle is motionless and free from support that keeps it erect it will fall down going straight downward towards the Earths centre of gravity. It tips over on a side as it falls onto one or the other side. As soon as independent motion other than that of the Earth comes about in a controlled manner the controlled motion alters the gravity as the motion brings about a balance that establishes another form of gravity and is in a way redirecting or channelling the motion from downward spiralling to sideways moving. If the bicycle comes into controlled motion it will redirect the gravity controlling the bicycle. However we should never forget that the bicycle as well as the way the bicycle acts is as artificial to the cosmos as life is artificial to the cosmos.

When the bicycle is on Earth within the atmosphere of the Earth, the Earth will commit the structure forming the bicycle to have a specific ratio relevancy in the space the bicycle holds in relation to the space the Earth holds in the time of motion that it takes the Earth to hold such a space. Because of the movement the Earth has, the Earth has a specific duplicating ratio with time and since the bicycle forms part of such a space claim, the Earth grants the bicycle a specific share in ratio to that which the Earth holds. That grants the bicycle structure the independence the bicycle structure has while the bicycle structure is not any part of the Earth. The bicycle holds a ratio of space a^3 in the time T^2 of motion **k** which in total is the space-time the Earth allows the bicycle to have a part in the Earth space-time.

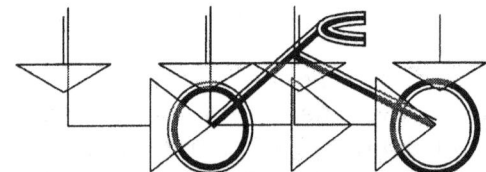

Then with the help of life the bicycle lands in a situation where it achieve more motion than what the Earth already provides. The Bicycle now has not only individual space but also autonomy in motion of claiming its individual space within the boundaries of the space the Earth provides. The bicycle still has the space during the time ratio which the Earth provides but also because of the effect of independent motion above and beyond that which the Earth supply, the bicycle has the same amount of space plus the generating of motion produces that bit more in terms of space moving threw motion –time and by that the space claimed is slightly more in the same specific time it took to fill the space while the Earth was the only motion provider.

The principle is the very same as motion which cosmic

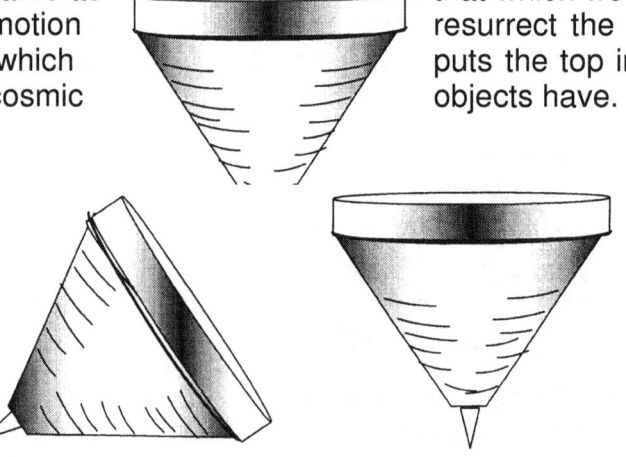

that which we find in the spinning top where resurrect the top to a life it seems to hold puts the top in a class that all independent objects have.

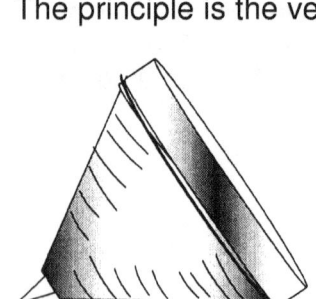

The top to the left therefore it claim to motion grants the top and individual space-

There is a space of one Universe between the composure of the top to the left and the top to the tight. has lost its motion and individual space-timer while to the right independence, time.

Every round object has a point establishing a very centre, a middle dividing one side from the other. That division determines the space from one side away from the other side. At one point there must be a point that does not fall on either side of the divide. Such a point will still be a circle, because from that side the circle divides into two sectors.

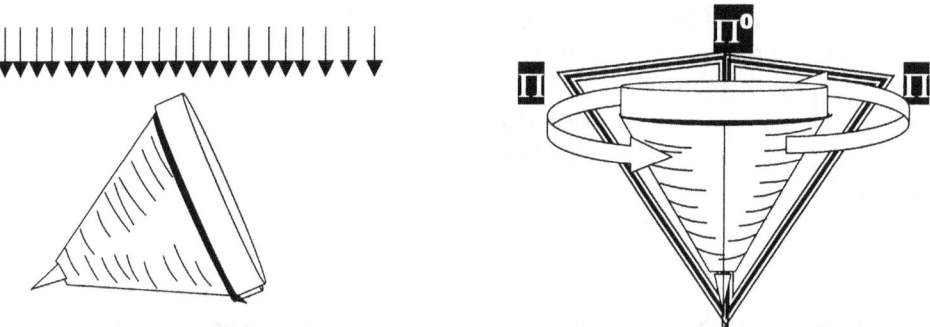

In every spinning object does not turn because it such a point becomes a born in the midst of many space outwards and from anti clockwise in all start at zero because then there is a point of infinity, a point that holds the dividing spin. However when line that cannot spin a new Universe is others. At the birth that point diverts that point the spin is either clockwise or directions. As I pointed out no line can there is no line and no rotating point can start at zero because then there is no rotation. Calculating a square involves two aspects that we think of as sides.

There is a Universe in differences between the top lying down without any individual motion and ostentatiously independent, self assured spinning top that even produce a sound to match the occasion. While without motion the top submits to the contraction lines running as the straight line holding half the value of the square being 180°. The top seems dead as it surrendered its long-term position and would eventually succumb to the Earth's gravity by relinquishing the structural independence it has. Then the motion brings life into the top and gives the top reasons to fight the Earth by fighting for independence. The top just became independent by the motion it received from the combined efforts of all the independent atoms forming the structure of the top.

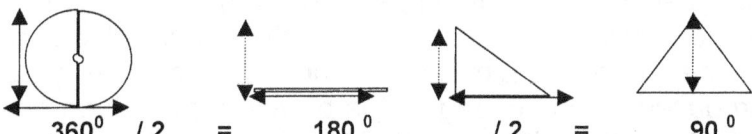

$$360° \ / \ 2 \ = \ 180° \ / \ 2 \ = \ 90°$$

The circle is a square holding a round shape, as the straight line is a square holding one side to infinity. Calculating a circle involves two aspects where the one is either the radius or the diameter that is double the radius. The other is the factor Π

It may be a coincidence that two bicycle builders were the first flyers in the air but it just might not be that big a coincidence. The bicycle represents the first phase required to fly, which is the part just before where the object must get off the ground. That is what the bicycle is doing: it is firstly getting the balance of gravity off the ground. The stability gained from motion is much more than what we humans read into it and it has even less to do with human skills

A person that acquired the skills of pedalling a bicycle has achieved the method of rearranging gravity within singularity. Without motion the bicycle falls on the spot it holds. When the bicycle is put in motion the bicycle can maintain the upright stance as long as the motion applies. When the motion stops the bicycle drops. To introduce motion to the bicycle the motion brings about a stable unsupported upright stance where balance can result from the motion the Earth enforces to the balance coming about by the bicycle using independence gained from motion of the space holding the bicycle because the gravity effecting the redirecting of the Earth gravity response comes about as the result of additional motion that is introduced to the bicycle. This is the very same process that the aircraft needs to get airborne because it replaces or repositions the singularity the Earth holds to the singularity the bicycle develops in motion. The aircraft only takes the change in direction of what the gravity is insisting on through changing direction in motion through phase one and into phase two. It all is still part of the Coanda effect. With more motion contributing to acceleration the bicycle will become airborne on condition that it is also given the advantage of a set of wings to increase the effect of creating space-time to the advantage of the motion requiring the change in singularity direction.

Standing still with no individual motion

This is where motion through duplicating changes the dimension in equation

In motion and altering gravity lines with the increase of individual motion

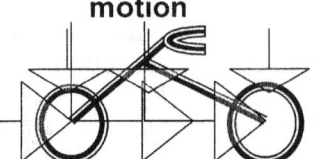

In normally applying gravity we find contracting lines running vertically as the lines connect with the Earth centre. This line form 7^0 with the centre of the Earth it connects to.

The motion of the bicycle not only extends the vertical connecting lines and not only changes the direction of the vertical connecting lines, but does both. The value added and the change in direction contributed is what brings about flying and moreover is the cause of the sound barrier.

When the bicycle is motionless the bicycle is part of the Earth by gravity applied. As soon as life steps in and brings about separate and artificial motion but still uses the support of the motion that the Earth provides it will inevitably do better than the Earth as long as the motion that life provides is not in conflict with the motion the Earth provides the bicycle becomes an object with the ability to transform the direction of the Earths domineering motion by redirecting gravity there in find the ability to change the direction of gravity.

The line of the Earths containing gravity is redirected from going vertically downwards to horizontally sideways, which then becomes separate from the vertical line running to the centre of the earth. The bicycle holds a change in gravity flow because of motion interfering with the duplicating of the lines running along the gravity line. The line running according to gravity is in conflict with the extension that the bicycle motion brings in as the line of the bicycle that implicates the gravity extends in the direction coming about from the introduced motion changing the bicycle gravity line from one vertical running line to a horizontal running line and that changes the line of gravity as it amplifies the line of motion. The line indicating the change of direction of the motion of the space holding the bicycle that is bringing about such motion will then being implicated by the Earths gravity completely redirect the direction of motion into a square in an opposing direction. From going straight down by the bicycles lack in motion when standing still these gravity lines introduce new directions that the line in gravity flowing holds. That then changes from coming straight down to going straight ahead in a horizontal direction forming a new link in relation to the normal straight down vertical and the one committed by independent motion. The relevancies about gravity changed. Every factor receives a new aspect and complete change comes about. By moving space a^3 which the bicycle holds within space a^3 which the Earth holds **k** changes to take the place that T^2 held as a new **k** connects **k** not the sun directly anymore but it now connects to the Earth centre on this occasion while the bicycle is in motion in the manner it previously connected the Earth that is performing in the motion in line with the sun centre.

The interfacing of the downward motion has to extend as the velocity increase and therefore the connecting point also have to shift promoting the horizontal direction by extending the points in distance of connecting.

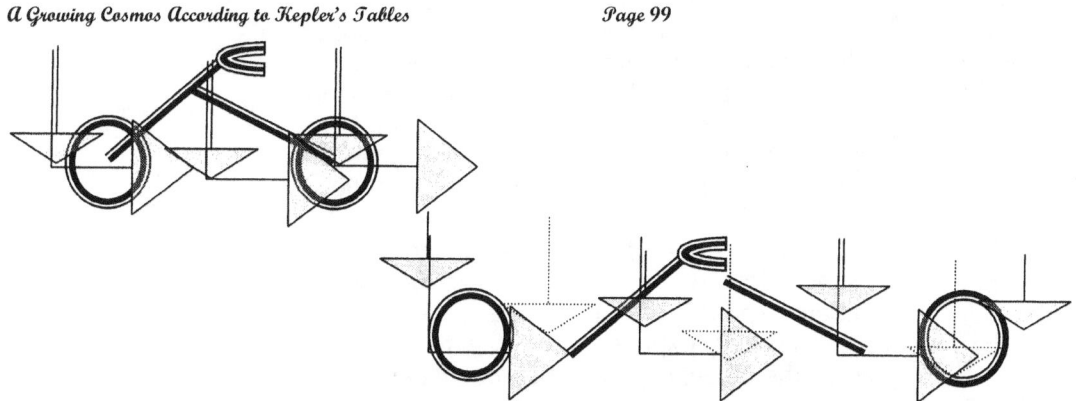

The Earth takes the position previously held by the sun by establishing the directional motion in accordance with the k^0, which the Earth then provides instead of as previously provided by the sun and while providing **k** it brings about the space moving when establishing independence that the bicycle holds in space a^3 by providing motion over and above that motion of what the Earth provides in its relation to the sun the object then takes on a directional change in motion. The change in direction also implicates other relevancies as the new motion removes the suns direct contribution as a direct factor to the role of performing as a secondary factor leaving the pivotal contribution which is the major contributing factor then to fall to the Earth in performing from then on until motion of the bicycle stops again as the major singularity centre and gives the pivotal control over to the earth.

By the motion applying to the bicycle the bicycle holds then more space in relevance per time unit lapsed and therefore there is more of the bicycle per time unit in the ratio when the bicycle is moving than when the bicycle is standing still

 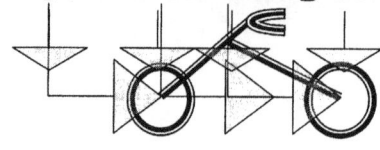

In normally applying gravity we find contracting lines running vertically as the lines connect with the Earth centre. Motion provides extending of the 7^0 establishing the centre connecting points to the Earth it connects to.

It is again Kepler's $a^3 = T^2k$ that changes the gravity relations

When the motion of the bicycle accelerates then such points that are forming the connections extend to match to motion. The motion then contributes by increasing the space factor to keep the commitment with gravity valid. The bicycle breaks its form but because it is structurally bonded. Other aspects concerning gravity have to commit to the breaking of space. When expressed extremely crudely it is put as follows but is very bluntly stated. Yet it still is the best way to explain the basics of the sound barrier. It is the compiling space holding the motion from facet going to facet by duplication together while the bicycle is in the space in motion that is part of the space holding all aspects covered by the atmosphere together in the atmosphere of the Earth. Because the conflict gravity experiences gravity first tries to break the object in motion but then extends the breaking to the connecting devices such as the sound waves in the adjoining space The atmosphere does the breaking on behalf of the object in motion since the moving space holding the object in motion as a unit shows much stronger bonding in structure unifying. We experience such breaking of space as the breaking of sound, which is showing motion or gravity differentiation.

With the Earth established as the domineering centre controlling the pivotal motion of the bicycle the Earth now presumes in the position the sun had before the bicycle created independent motion and the bicycle now applying motion that produces independence falls into the role which the Earth held on behalf of the earth and the stationary bicycle when all considered the earth and the bicycle was one unit. It still is a unit but holds independence in the unit and even more so than before. Before the motion contributed to independence the bicycle still had independence though that independence came from atomic motion that separated the structure of the bicycle composition from that of the Earth. With the individual motion coming about such motion secures the independence to a next level that will be one step away from advancing to semi self control independence when it starts flying. The bicycle motion will eventually end as T^2 but the relevancy where T^2 now will end is the drastic change that came about. Because the direct control shifter from bringing the sun in as the pivoting factor to the Earth holding the pivoting factor the gravity that is upholding the cosmic equality sense is still upholding the cosmic sense as it still applies since individual space providing motion brings on changes in the factors but not the factors implicating the cosmic law while the relevancies that are required to reposition the structure factors are still in place. The Earth now forms the centre whereas before when the bicycle was motionless this duty fell upon the sun centre. Circling about a fixed and secure centre now bringing about motion that takes the position the Earth has before the bicycle. This is a shift of one position in the gravity relevancies of $a^3 = T^2k$.

This too is the prerequisite to flying by first establishing individual space through establishing separate motion in relation to that of the Earth while it is creating motion because the motion establishes antigravity. While the aircraft is motionless on the Earth the Earth motion creates gravity and that motion also applies to the craft structure albeit that the aircraft seems motionless to us. The mass of the Earth establishes motion of space spiralling down in time. With the space being in motion the object resists surrendering the form its construction has and with that refuses the accepting of the joining with the Earth by uniting with the solid structure. This resisting action is delivering that what we believe is mass. It is where $k^0 = a^3/(T^2k)$ unconditionally. My argument about gravity being motion becomes prominent when we consider the motion versus space occupied by airplanes using wings to fly. Without motion the aircraft and all that it carries has mass or weight. As soon as the motion overcomes the space restriction the Earth enforces on the airplane the airplane loses mass and becomes airborne. It is only when applying the illogical science used to explain mass where there is no sign of mass present or that mass may contribute to the forming of gravity in any way possible. By trying to correct the incorrect holds mass and weight as a differentiating factor but that does not make sense in any case. You can try and argue till your tongue feels numb but in arguing about the matter no one will ever be truly convinced about that one can believe as being totally convinced about mass still being a factor in outer space. The protons form the centre of the atom and by spinning 1836 times more than the electron the protons dismiss 1836 times the space that the electron does. By having so many more protons in any unit dismissing space a total effort will contribute to displacing more space-time than the effort of a lesser number of protons can achieve. There is a way to detect gravity and that is in looking how we fight gravity or eliminate the effect of gravity securing objects on the ground. We know that $a^3 = (T^2k)$ locks objects in position and flying eliminates our being locked onto the ground. Let's us look at why we fly.

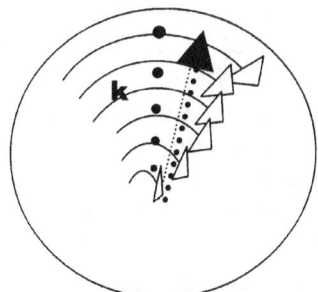

The Earth applies motion in the atmosphere of rotation but from the human position we have on Earth we have to follow the Earth by mimicking motion the motion of the earth. From our stance we are moving in a continuing straight line and the straight line we receive from the Earth is giving us the impression that we are allowing a straight line that we can see. Before investigating the principles of flying we first once again must define why the mass is keeping the object secured on the soil. The space the object claims within the space the larger object retains is not enough to duplicate using the time it holds in relation to the volumetric space retained. Not being able to match the retainer the object has to reduce the distance from the centre to secure a more relevant position with a suitable distance to relate to with the duplication of space within the retaining space. This forces the object to relocate to another point where the retainer will match the volumetric space claimed by the retained object and where the containing space is by volumetric size matching the claimed space. As the object moves down to find the correct position of relocating the object is restricted in space duplication to match that of a solid structure we call Earth or soil. The solid structure provides a restricting boundary through which the relocating or repositioning object cannot break. The effort of relocating the object as the object is smaller it is missing the required frequency of duplication be comparison with that of the Earth, the object has to follow such a search as to find the position of relocation that will match the density in duplication in relation to where the specific duplication will match the required density to duplicate the space in the prescribed harmony to the restriction of the retaining object. Because the retaining space is obligated to reproduce a much larger extending **k** from singularity that is required of the smaller space in the same situation, therefore the smaller space is in search of a position all along the line that produces the larger space in an effort to locate a comparing location to match equal duplication resulting in matching factors of space and distance in the same time experienced between the large and small spaces sharing one unit. Such a position will allow the smaller space to have a comparing spot in order to comply with equal duplication when using equal time duration as the time will allow the space then to commandeer a position in the **k** that will serve both sharing the space in the unit. By not being able to find the equal spot in the **k** that will match the space to the duration that the Earth insists on this will force the space to tend to locate while pushing against the Earth and because the soil is solid the soil will not allow unrestricted entry for the space to go in search of identifying the perfect spot of choice. This relocating and searching for an equivalent position that will match the **k** equal to fit the smaller space in comparing to a point matching the needs of both the **k** factors where the smaller space will fit into the duplication that both objects will have in space in time providing the space that will provide the mass as a factor of control. It is a difference in speed bringing about mass and not mass bringing about the restriction. In accordance with the view we receive because of the position we hold on Earth our motion is a straight line although the Earth shows a curve but in our minds that curve concerns us little people less. The Earth has motion and by all standards that life applies such motion can never exceed the motion the Earth displays. Our motion serves as an addition to the motion that the Earth provides but we cannot substitute or out perform what the Earth achieves. While flying and duplicating the concept serving mass does not enter any equation at this point. Then later on man found the magic he was always in search of; real power in the form of converting heat to space. And with this remarkable achievement man found means to break free from the establishment the Earth enforces. Man had more power available to use than just what he could harvest from human and animal muscle power.

There is very little difference between the bicycle in motion and the flying aircraft except for the motion intensity and by which the duplication of space contact will allow the aircraft to fly and leaving the other object which is the bicycle with much less space contact being without wings unable to fly. The only difference is that the aircraft produces a higher relation with space in possible motion that is contributing to space contact and space duplicating by employing a greater surface to service. In each case the aircraft wings in space contact is providing more space between the two in relevancy where the Earth forms one gravity connecting line with the object in motion being the spacecraft or the bicycle. Of the two in motion the aircraft services much more space than does the bicycle. By providing a bigger motion T^2 time factor will extend further providing the space a^3 more opportunity to duplicate, which the bicycle provides. That will increase the k factoring the case of the aircraft because the more motion will fill more space a^3 and with more space a^3 filling the totality of space a^3 will be consisting of the larger space a^3 to entertain more space-time that is in occupation than the smaller space a^3 has space-time to serve. This relation being as steady as it seems, it will prove to gain a bigger relation without providing more space–time contact. This is because the wings applying motion in itself is adding to space where the motion has more contact with space by being in motion without any additional space added in real terms in neither of the two cases. This is only because the motion and the improved wing capacity is leading on with the motion of the wing to have a larger contact of space a^3 and with more of the same space a^3 notwithstanding the fact that the quantity of space a^3 in either cases remained the same. This means the other factors in the Kepler formula being in relevancies of T^2 and k will also have to increase to comply with the balance ratio. On the other hand when the motion goes into the extreme and k proves to increase considerably the space a^3 as a factor would have to compromise visibly by reducing the space a^3 it claims. When the contact distances increase the other factors have to compensate to produce the sustained equilibrium. This comes about as a result of the wings that produce more space duplication received through increased space contact with more space because of the velocity increasing and with more motion as well as a bigger contact area more space becomes involved, which is promoted by the speed factor by discriminating in favour of the flying device to get the aircraft flying. The space ratio by duplication increases in ratio to the dismissing ratio by the combining effort of all protons within all atoms that are forming the flying machine and the machines cargo. However the fact that motion is in place as a cosmic event is totally artificial by cosmic standards of motion. In both examples the motion is artificially applied as a result of life's' extending life influencing. We must realise that the motion we find in the action of the flying machine is not cosmic driven although it is an interpretation of a normal cosmic occurrence that will take place but under much different circumstances than the manner life enforces the motion at the time the action that is brought about as a result of life's obtaining the manipulating abilities to translate to human achievement and as far as life is finding a means to manipulate a cosmic law to increase the benefit of life accepts that the action is totally artificial by cosmic standards. There can be no such natural motion where a rock starts flying because it has received from some U.F.O. outside source a set of perfect fitting wings. Judge responsively what belongs to the cosmos without life and what life can reproduce in spite of the unnatural state such duplication might be. It is most important to realise before classifying and grouping this normal physics action inspired by the intervention of life that there can be no such motion coming about on a planet without the presence of life bringing about artificial motion. One can inspect the moon all you like with the best telescope available to man but one will not see any flying object zigzagging the moon's surface. No bicycle can by its own initiative come into an upright position and start moving on two wheels. The motion is not cosmic inspired and only by seeing the difference can one have the mindset to venture into the activities applying within stars. Let's see what is artificial about life in relation to how the cosmos relates to life. Mainstream science holds the opinion that life in the cosmos comes at a dozen a penny with change repaid. What a lot of crap this idea is and I mean dirty crap. Life is alien to the cosmos while the cosmos is fiercely hostile to life being an alien in the Universe. Even on our planet life has to obey certain and very specific conditions in some cases otherwise the Earth as friendly and nursing as it is will bring about life's demise. One should try and live a thousand meters below sea level or ten thousand meters above sea level and watch as your personal demise comes to you.

Let's venture slightly away from the cosmos by trying to define the role of life as the only force found in the cosmos. Life according to my personal defining is absolute managed heat within specifically designed cosmic fibre having the ability to apply forces of a wide variety giving life power or a valid force to manipulate space-time by manipulating some motion or rules applying on motion thereof. When the body holding life is not hot or with very low intensity heat it is not with life. If the body has no motion of any sort it is not with life. If life lost the ability to manipulate gravity in the form of low electricity life has lost living. Life can create motion by manipulating space-time it occupies or which it can control or manage. The only place in the Universe known to man, who is not absolutely in all respects completely hostile to life, is this blue dot we waste for gaining money and profits. We should never confuse life's ability to accomplish with that we associate with cosmic events because an apple falling from a tree is life's manipulating motion because it needed the intervention of life to get into a position to fall from the position it took being in the tree and that is not a cosmic event. If the apple came from the outer space it would have been fried charcoal before it reached the Earth and that result is a cosmic event. That part is the part everyone in science including Newton and Newtonian disciples ignores or chooses to ignore.

We by which my referring includes most forms of life that has the ability to stand independent on Earth and from the Earth stand on Earth above the very top layer of soil holding our space in the space of the Earth. We cannot have independent excluded space if we do not fill the space we have on Earth. While being on Earth my position is $a^3 = T^2 k$ where k is because of the mass in movement standing in for k^0 by being k^{-1} Since my body duplicating is less than that which the Earth has to duplicate but is confined to the time on Earth all the same such a body will forcedly find that the distance from the Earth controlling singularity such a distance is corrected constantly as to fit into and apply with the standards that the Earth standards insist on. However with me calling the Earth space-time a force whereas it is the normal gravity flow the force comes from counteracting the flow of gravity. Being k^{-1} we are also T^2 / a^3 which is reducing us in the space we hold and that is only our mass that comes into affect as the Earth repeatedly insist that we try to reduce a^3 further to comply with the T^2 the Earth is applying and which we have to use without any further options given by the Earth. If we wish to confirm our independence of the space we have within the space of the Earth which contains the space we hold by moving through the larger then we have to produce a larger k factor by extending the normal k the normal k factor we receive from the Earth to the order of at least k^1 to find the ability to move from k^1_1 to k^1_2 which will allow us to enforce our own gravitational force in spite of the Earth's much stronger natural flow of space-time because we use T^2 to move from k^1_1 to k^1_2. So we have to improve both our independent position T^2 as well as k to accomplish motion. But that puts Kepler's formula in question. Using $a^3 = T^2 k$ and producing a larger $T^2 k$ it means a^3 must also improve. That it does by doubling the space it uses during the motion. The space a^3 becomes the next space a^3 because the motion $T^2 k$ is providing the way that will bring about the matching duplication the motion contributes. This is not that uncommon physics.

A car holds the space a^3 and is moving by T^2 through the distance of k. When the car speeds up to a higher velocity the gravity will increase on the part of the car because the distance k will increase. The mass or space in motion that remains is not able to remain even with no increase to the actual material used to move. Nevertheless the potential mass is increasing by the square of time where that is the gravity or the time by the square that increases. The increase in space is the producing of more of the same space by duplicating the same space more in the same virtual time. It is the motion providing the material a duplicate value of its mass (duplicating because of the square used by time) that then forms the increase in the material mass, which our human instincts of sensationalising prefer to call the momentum of the object. But that motion is what gravity is. The way gravity is applying is acting in the same manner everywhere but man has subdivided the concept under so many names given to misrepresent each fragment of the entire unity we divided that we cannot even find the basic principle any more.

Gravity is not a force as Newton suggested but a motion that is formed by a natural flow of space-time between space occupied and space waiting to be filled and when filling it's forming a relevancy and this applies throughout the Universe. The only force there is can only be found on Earth in the form of life. Life is the only force and only found on Earth. In spite of all absolute madness that most of the important persons in science wish to propagate they're apparently attempting to promote an even more mindless concept, which is atheism. They try so hard to pretend that life is a natural flow of normal cosmos that they go as far as to show how mindless the ideas they truly come to conclude. Without a God it means that life, which is a God linking factor, must be in abundance and if life is that plenty everywhere it is as common as stardust and then it has to be so commonly found we will trip all over other life throughout the cosmos. Until proven otherwise we find life on Earth and nowhere else and that fact is written in rock in spite of all idiotic atheistic gibber. In life we find a force different to the letter in the minutes detail to any other factor there is in the cosmos at large.

Only on Earth there is life being a force with the ability to manipulate space-time, which is placed under life's control by providing motion other than and above the motion the cosmos does provide in order to maintain and sustain space-time. We use the Universe as if the Universe was meant for us to use. We increase what the cosmos gave us to use as if the cosmos was created deliberately for our purpose and for us to use which is just as corrupting madness as is the jabbering nonsense promoted as the religion called atheism. It is precisely in such a manner that light used to accomplish moving ability to travel in from singularity to singularity. Because singularity in space is space in darkness we consider the space we see as night as dark and therefore invisible. The darkness is light outperforming visible light by duplicating much faster than can visible light. Darkness is light that breaks down and rejuvenates space much faster than light frequency can. The photons find a way to escape from the gravity applied by specific singularity points. Being in another frequency of duplication can the photon manage to secure its escapes. With the escaping that the photon does the photon can release and join the next singularity in the period being in a position where singularity takes charge and survive by galvanising a small portion of the heat forming the photon and by singularity releasing some part of the overall heat by removing space-time and forming the motion from the previous to the next infinite position in singularity in rejuvenating the point which is representing by producing space-time, which then will include the photon reassembling with the next singularity forming the space-time of the next singularity. Looking at the issue in this way we can begin to appreciate that light is the duplication of the photon by the singularity charged by the motion that provides the singularity by charging the intensity. The flow of light is about duplicating more than dismissing although dismissing does form part when the photon changes singularity. In that way the light loses intensity to the singularity that releases the light when the singularity releases the light. This process reduces the intensity of travelling light as it travels and is recharged by the singularity en route to somewhere in the future.

In contrast to the duplicating of light is the duplicating of material that is more intense and more profound. The duplication of space filled with material is the use of heat compacted in space selected from the surrounding space in the atom forming a unit, which provides the material the ability to confirm the space they hold onto the space they move into without conforming or giving up ground that is filled atomic space, which is much more than just singularity. It is singularity that is sustaining more heat than the singularity will ever require and much more than what particles ever will require. Singularity empty of material that can take charge of light can generate motion to duplicate the photon whereas material uses the heat the photon provides when the proton is clashing with the atom. The heat provided by the photon is only a part of the total heat that is required by the atom to replace the dismissed space-time that the atom needs for duplication as well as dismissing of space-time at the centre. The singularity placed in charge and inspired to dismiss has the task to make space-time redundant whereas the electron is in charge of the factor that is making the duplication of space-time, supported by space-time protected by motion in contact with unfilled space-time then even requires more than what the photon can deliver because that is why there are shadows forming the dark side. Looking at this in a clear and sober perspective we once again find a reason to believe that heat and light is the antimatter that matter ate all up and still wanted more. Material is still eating away at light as it did when space began and material still craves for more light, which is heat. We once again find a reason to believe that heat and light is the antimatter that matter ate up and wanted more. Material is still eating up all the light and then still wants more.

Man could create motion but at first such motion was far less than that motion which the Earth provides. The motion of man's ability was vested in what his muscle power could provide. But a very short while ago man grew wise to machines and the fact that machines can provide more motion much faster than could animal muscle bring about motion. By supplying machine motion it gave man extra ability whereby man extended the relation between what the object has when in normal contact with space and when extended by extra motion allowing more space to apply to the surface of the object, thus enlarging the object surface in the relevancy brought on by motion.. Then man found means to break the barrier that muscle strain held and was able to apply motion equal to that of the Earth spinning. After that eventful day then came the day man had more motion than what the Earth could provide. This is where nature and man parted their straight line common sharing of the factor **k** because the motion that man could produce placed man in a position where man was able for the first time to outperform the earth's ability of duplication by motion and thereby can go in disrespect of Earths straight line motion. This is where locomotion generated by steam concentrated gave man more than what man could tap from life and sweat. From then on man produced his very own straight line gravity in such abundance that man's gravity generating ability is no longer in harmony with the Earth's gravity ability and with man's very own straight line man could eventually leave Earth altogether. But let us get back to the straight-line man moved in as that straight line no longer followed the Earth spin. At first with man's first attempts it showed a diverting from the straight line of the Earth and we even gave that diverting a nice name as we do with all things. We called this diverting flying and flying is proving what Kepler said what gravity is in so many ways again and again. The space grows as we increase the motion with which we travel while complementing the space through which the space travels in which space we are in locomotion.

If mass was the major factor in generating gravity, the mass will play a major part in the time it takes a body to fall from any given point to the surface of the Earth. It just has to because the mass then does the pulling. Since Galileo proved otherwise the concept of mass being the producing factor in gravity comes across as rather less thought through and more than a bit silly. In flying a certain criteria must be met by involving the motion of space and as that motion is running through space in time the motion is relating to space by the scale of time. This then has to indicate that there is a restriction in all motion running through space. That we can observe by the speed of light slowing down in denser space than what the speed of light is in less dense space. When an object travels through space the density affects the motion. The slowing down by relevancy as a factor comes about by bringing into the equation where a smaller space must negotiate travelling through a denser larger space.

By being in a "thicker" density the motion has to manoeuvre through more restriction say in comparing the space-time we find as the Earth atmosphere being denser than the space-time that the Universe grants. In denser space-time there is more reference points being spots of singularity to contemplate as space distances where the singularity concertina in the time given to fill the more or the less space being contemplating. By lesser dense space there is lesser virtual singularity forming less restricting to the moving object. It is very obvious when looking at an object that is coming from outer space into the atmosphere of the Earth.   As the factor one has to consider that the space represented by the line **k** increases in the density it represents there is more space to relate to motion. It will be the same as if the space a^3 is moving faster because of relevancies re-applying to conditions changing. Then space a^3 would reduce in the relative factor presence within the larger and newly introduced **k** that comes about. Then that means that the faster a^3 travels the more a^3 will extend **k** away from k^0 and by that increase in motion.

That increase will also apply to the time relevance by reducing the space a^3. There will be a smaller a^3 because there will be more space that **k** has coming from the being in the larger **k**. A larger **k** will bring a relevance that reduces the space because the holding space increases in relation to the lesser space occupying a part of the holding space. There is always a double relation to space, which is the space the travelling object occupies and the space in the circle of time that is in control of the object by motion because the object is in motion. But this relating is affecting on both sides of space and because of that man's quest to travel at the speed of light is totally unrealistic. Those comedians hiding behind science from where they are trying to pretend to be all the wise about the Universe because of their accomplished scientists have a smaller chance of going even one tenth of the speed of light than does Little Red Riding Hood of finding her talking wolf with the ability of eating Grandma wholesale as one unit than they have of flying through space and achieving 3×10^5 km / sec. In the fairy tale Little Red Riding Hood has a bigger chance in finding her talking wolf than the chance any cosmic traveller will have to go into space flight and achieve 3×10^5 km/sec.

By establishing motion and such motion is bringing about certain contact with space by increasing motion such an increase can bring about much more space to be in contact with the moving structure and rebalance the dynamics of such space in motion. The space a^3 in motion T^2 will establish a larger area **k,** which is then contradicting the gravity of the Earth's motion in descent and this will count for a larger area present in a smaller time relevancy. While all this is happening the result is that a stronger motion line in a 90^0 directional change will come about and since the Earths motion remains the same it gives the flying structure an advantage to increase the relevancy in favour of accumulating the dynamics of the balance in space-time by space contact increasing by the contribution of more motion above and on top of that of the Earth motion.

Gravitational Constant
(Symbol G) It is the constant that appears in Newton's law of gravitation. It is the attraction between two bodies of unit mass at unit distance apart. Its value is 6.672×10^{-11} N m^2/kg^2 when the distance is expressed in metres and the masses are in kilograms. Although it is described as a constant, in some modes of the Universe G decreases with time as the Universe expands (see Brans-Dicke theory), but there is no evidence for this.

Gravitational Field
The region of space around a body in which that body's gravitational force can be felt. Within this region, other bodies will experience a force of attraction that diminishes with distance from the body.

Internal Mass
Inertial mass is a measure of a body's resistance to change in its velocity or state of rest. Inertia is a direct property of the mass of a body: The greater the mass, the greater the inertia. Although mass is formally defined in terms of its inertia, it is usually measured by gravitation.

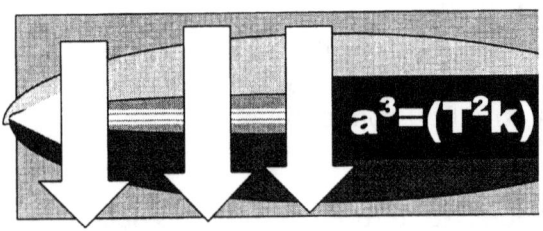

Mass is the refusing of any object to dismiss the form it has and to join the Earth solid structure. Mass cannot and does not contribute to the establishing of gravity except by depleting space through motion and such numbers of the protons in a space forming an exclusive unit.

$$a^3=(T^2k)$$

Kepler answered the question of flying and airflow dynamics before Newton gave us a name for gravity. Kepler said the motion of space must form equality to the motion of the space. When the aircraft is maintaining flight height the motion equals the mass and equilibrium ensures a constant flight height. As long as the speed and the mass are equal the aircraft will be in equal gravity balance.

Gravitational Mass.
To overcome this braking effect that the smaller object has on the larger object that we named mass of bodies being on the Earth the smaller object firstly has to transform and transmit singularity from the centre of the Earth to the Centre of the motion wherefrom a new balance sets in. This we call the Coanda effect and it works either on the linear aspect of gravity or it works on the circular

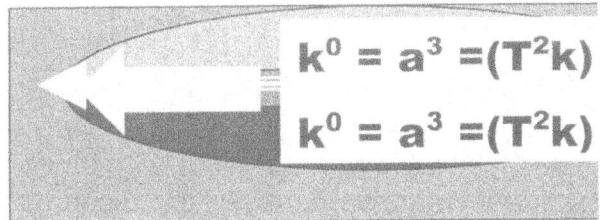

$$k^0 = a^3 = (T^2k)$$
$$k^0 = a^3 = (T^2k)$$

It is when the motion exceeds the mass the aircraft has the ability to break the sound barrier. Galileo proved that no mass is present in falling, which is also matter in the process of flight and because of that can the sound barrier become some form of constant.

It is when the motion exceeds the mass the aircraft has the ability to break the sound barrier. Galileo proved that no mass is present in falling, which is also matter in the process of flight and because of that can the sound barrier become some form of constant.

When the aircraft increases its motion, the motion changes to accommodate both space a^3 in the motion thereof T^2 that the applying **k** factor would allow. The space a^3 is a fact influenced by the Earth but directly dictated by the atoms forming the aircraft. The time factor T^2 is directly derived from the motion the earth dictates. The Earth rules on the distance that k would produce. That is gravity. That is what Kepler said gravity is when he said $a^3 = T^2\,k$. That is what no one in four hundred years cared to take notice of or refused to recognise. What no one ever apparently saw is that gravity or recognised that gravity acts precisely in the manner Kepler stated. If the motion of the aircraft a^3 increases the space increases, which the aircraft influences but also the space influencing the aircraft changes bringing in new alterations.

If the motion decreases the aircraft space relevancy increases and if the space decreases the motion changes the contact in space therefore it changes the volume of space which is reducing the relevancy of the space in motion which is gravity applying by motion in space forming space in motion. The more the motion is the more space contract is produced and the more space is affected by the increase or decrease of the motion of space. The more space that is duplicated the more space is produced but also the more space is reduced through such duplication. Seeing gravity acting in this manner does make nonsense of Newton's $4\pi^2 a^3 / T^2 = G\,(m + m_p)$ changing of Kepler's formula as it has no part to play in correcting the formula of Kepler and it is only Kepler's $a^3 / T^2 = k$ that comes into the equation since **k** is $G\,(m + m_p)$ in any case because $4\pi^2$ indicates an individual structure encircling the sun centre when one uses the cosmic relevancies which I later introduce.

We gave this forming of separate gravity coming about by means of performing individual motion which is disguised as a very well known and commonly occurring phenomena which is burdened by carrying yet another name of the Coanda affect. The Coanda effect depends on singularity forming when a solid and a liquid is in relevant motion where such motion of either the liquid or the solid or both factors has to move and such moving contributes in selecting a singularly centre point that will secure the control of the space-time, affected by such motion. The Coanda principle forms a circle and motion establishing an independent singularity in such a circle centre. Then the singularity cuts the Kepler formula in two parts where space is following motion and motion leads space. Being as human as the next person and showing as much human tendency as anyone else being human I changed the name partly that others gave to the Concept we know as the Coanda affect. I did that because of the effect it has on cosmology and by changing it to the Coanda effect I participated in human efforts because that is what humans do best. We give names best to cover our misunderstanding and distorted concepts about what we don't grasp. When humans decide they have no idea what they discover and wish to hide what they don't know about what they discovered they hide such incompetence behind a sparkling spunky and swanky new name. But with a fancy name other meaning behind the discovery gets less important and the naming becomes the accomplishment a name will scare away any one also in mind of finding out what was discovered. Like calling heat plasma when plasma is the same as heat gone liquid. Well in my case I use the name as the Coanda effect because it is a process where motion is having an effect on space-time.

The prerequisite to flying is creating motion because the motion establishes antigravity. The motion reduces the friction that mass creates and in that reduces the gravity that creates the mass. While the aircraft is motionless on the Earth the Earth motion creates gravity and that motion also applies to the craft structure albeit that the aircraft seems motionless to us. The mass of the Earth establishes motion of space spiralling down in time. With the space of the earth being in motion the object resists surrendering the form its construction has and with that refusal the accepting of the joining of the Earth solid structure. This resistance we believe to be mass. It is where $k^0 = a^3/(T^2k)$ unconditionally. My argument about gravity being motion becomes prominent when we consider the motion versus space occupied by airplanes using wings to fly. Without motion the aircraft and all that it carries have mass or weight. As soon as the motion overcomes the space restriction by defying gravity affecting the aircraft to a stand still which the Earth enforces on the airplane the airplane loses mass and becomes airborne. Notwithstanding the corrupt argument Newtonians bring in about mass remaining a factor. To prove their corruption in this argument, let those that disagree with my stating them being corrupt answer the following. On their admittance we know that mass increases as gravity in stars increases.

The more the gravity is the more the particular mass will be. But then the very opposite is true where in space there is micro gravity. Then there has to be micro mass which means by their own admission, mass disappears. It is only when applying the illogic use of mass and weight differentiating and insist on proving the incorrect correct in using a method which in any case does not make sense by any standard of arguing that one can argue about in order to prove the nonsense about mass still being a factor. Thinking about this I feel delighted that those being so very incoherent about mass see my argument about space and nothing as being incoherent. The protons spinning are supposedly bringing about the mass. The protons form the centre of the atom and by spinning 1836 times more than the electron the protons dismiss 1836 times the space that the electron does. Because there is an increase in contact with space by the body/s in motion the dismissing of space-time does not only become fully substituted by the duplication but also totally overwhelmed by the motion. It is the dismissing effort applied by the combined unit of all the atoms in the motion in relation to the contact made that tips the balance. There is a way to detect gravity and that is in looking how we fight gravity or eliminate the effect of gravity that is securing objects onto the ground. We know that $a^3 = (T^2k)$ locks objects in position on the ground and flying eliminates the flying device including its cargo being locked onto the ground. Let us look at why we fly.

The Earth applies motion in the atmosphere of rotation but from the human position we hold on Earth we have to follow the Earth in motion. From our stance we are moving in a continuing straight line and the straight line we receive from the Earth giving the impression of a straight line for us to see. Before investigating the principles of flying we first once again must define why the mass is keeping the object secured on the soil. The space the object claims within the space the larger object retains. This then is what should be overcome to fly.

When the aircraft is gaining lift the motion exceeds the mass and with that is adding heat at the bottom of the wing to create more mass a^3 added than the speed (T^2k) can create motion above. Below the wing there is more space in contact with material that is improvising to dismiss more space by collecting space compressed with heat by restricting more of the motion. On the top of the wing the motion that accelerate the flow of heat and by doing so dismisses the possibility of having more space-time dismissed as is the case applying at the bottom of the wing. In that way the wing is at the top creating an environment, which favours extensive duplicating. At the bottom of the wing we have $a^3 > T^2k$ and at the top of the wing we have motion outranking space accumulation by restricting motion therefore changing the balance on that side to $T^2 > a^3 / k$. As the speed gains, the wing will strike a balance and at a certain flight height the motion will equal the dismissing going on and equilibrium ensures a constant flight. The motion of the craft establishes individual gravity that is surrounded by the Earth gravity but the independent motion grants the aircraft some individuality and exclusivity.

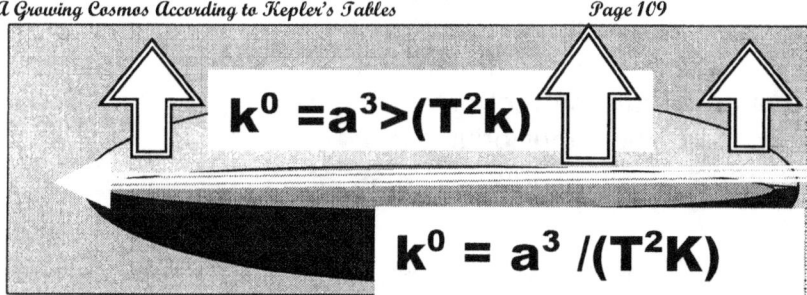

$$k^0 = a^3 > (T^2k)$$

$$k^0 = a^3 /(T^2K)$$

By decreasing motion the mass of the aircraft will tilt the balance towards favouring the gravity the Earth applies and the favouring of the dismissing factor of space-time, which then overcomes the duplicating effort of space-time by motion will contribute to the descending. In order to apply a perfect controlled landing the wing must establish additional space-time dismissing to **The establishing of independent motion of the craft secures an individual gravity and such individuality leads to the breaking of the sound barrier because the one gravity can no longer subdue the smaller motion, which is producing gravity** **allow the steady descent and the perfect landing. Even when performing the landing under the most stringent conditions the balance still rely completely on the balance Kepler gave us of $k^0 = a^3/(T^2k)$**

$$k^0 = a^3 / (T^2k)$$

$$k^0 = a^3 < (T^2k)$$

At a height of 31000 km above the Earth the mass of the wing becomes compensated only by a motion of a relevancy that comes about at 2500 km per hour. In that case the craft has to apply motion at a rate of 2500 km / hour just to create the required velocity to keep the aircraft in motion in the sky. Motion creates gravity just as Kepler said when he said gravity is about $a^3 = T^2 k$, which translates to the dismissing of space and the motion, duplication establishes a centre that controls the balance that the newly secured singularity will provide. When the aircraft stands still the sun provides such a pivoting centre but when independent motion comes about the point shifts from the sun to the Earth centre where there is a line contact between the singularity that the Earth holds which then forms a new relation in respect to the singularity activated by the independent motion of the moving body which the aircraft takes on a trip in motion and with it a position that the relevant singularity is claiming which are released as part of the minor space. The Earth provides a point from where space depletes completely within the centre of a sphere from where gravity is securing the centre spot in the form and the space surrounding the form that controls the space and time in which the independent object moves (in this case it is the aircraft). When a balance comes about between the departing object and the space reducing only then does an orbit establish a balance of speed serving time duration and space dismissing evenly. That is gravity and that produces gravity only when motion creates a centre to form a sphere.

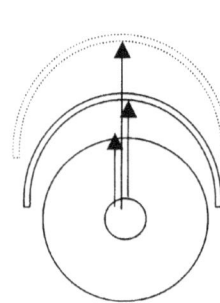

If the **k** that is applying to the distance the space a^3 being in motion T^2 requires in order to satisfy singularity k^0 the flying of the aircraft is then unequal to the motion in the previous relation that was in place between the sun and the Earth and the motion will bring a correcting in the distance **k** to put the motion in balance with space. Space will always demand a correct establishing of the miss-interpretation of the equilibrium that is needed to sustain the effected singularity because of the space-time factor.

This is very important if one wishes to understand the sound barrier. The two positions in **k** depends on the motion of T^2 in relation to the **k** the object is maintaining while the object stands in the space a^3 and related to the motion the Earth provides.

When the motion amplifies the status of the object elevates as the motion increases the relation to singularity, which is located in the centre of the Earth k^0. The more motion that applies the higher the velocity of the flying object will be and the bigger distance **k** it can sustain. If the sustaining of **k** is not required, as is the case with racing cars the space a^3 needs relative improvement and that is done by allowing wings on the car to contact more space improving the dismissing part of space-time where this contact will proportionally reduce **k** by amplifying the relevancy of T^2.

In our normal posture we are moving much slower than the Earth is spinning for reasons I shall come to later on. With our bodies not applying the motion we apply the thrust to reduce the motion the Earth has in order to achieve the much-needed equilibrium the cosmos requires.

To overcome this braking or restricting effect that the smaller object has on the motion which the larger object provides and that we named as being the mass of bodies being on the Earth as being in the role of the smaller object firstly has to transform and transmit singularity from the centre of the Earth to the Centre of the motion wherefrom a new balance sets in. This we call the Coanda effect and it works either on the linear aspect of gravity or it works on the circular aspect of gravity. By applying motion either to a liquid in the presence of a solid or supplying motion to a solid in the presence of a liquid a new point in singularity establishes a new centre elected by the motion creating the defining of the space a^3 that is in ratio equal to the motion thereof T^2k bringing dominance and the process applying the rules I shall explain a little later on.

All the principles that I make use of to explain my theory are part of nature. I base my theory on heat becoming stabilized through collecting more space using motion to produce cooling. This idea is most basic and that I admit. It may sound basic, but Mainstream science is also most guilty of departing from this most basic of principles through the employing of terminology and terminology has covered many of the crudest, most basic meaning behind the most basic principles in nature. I do not applaud a principle Mainstream science underwrites in the sense that matter in the beginning was coming about and anti matter came to destroy the matter. It is moreover the disappearing from the Universe of the dissolved by-product which antimatter somewhere not in the Universe. It is this vanishing being the result of between the two opposing materials that I strongly reject. If anything ever was part of the cosmos, it still must be in the cosmos because there is simply no other place to go to outside the cosmos. The friction that once produced the heat in the time before and during the Big Bang period is still today actively participating as mass.

Mass is the result of let us call it "stationary friction" which is the relation between two cosmic objects having motion inequality but still sharing space within space. By creating friction through the bringing about of any form of motion discrepancy between objects in any such a test performed today such friction coming about will produce heat and the heat will result in space forming. In such contact between objects in different speeds that such motion discrepancy produces to cause destruction of matter in space and heat comes about. In that the net result eventually leaves space created when overheating material no longer fills the space after cooling sets in. The cracks showing in the cooled material afterwards is a result from the overheating of the material that created the extra space and then reset the occupation to what it was before. But the material that is reducing from retracting used space with becoming colder again leaves cracks behind on the surface that proves that there was more space filled when the material was heated than what it had before the material was heated compared to the decrease in space after the material again cooled down back to what it was before the material was heated. Evidence of this is evident in all supersonic aircraft as the fuselage forms cracks in the body structure of such an aircraft. The outcome of this heat is that when cooled the material occupies slightly more space than before. The grown space then tries to fit into the area it did fit into before the heating but with the extra space it employed when it was during in the heating process it shows afterwards as cracks. This takes us back to what Kepler said. In Kepler's formula it is the extending of the distance **k** that influences the time aspect **T²** which the supersonic aircraft does by its going supersonic and by shifting **k** from the previous location the Earth prescribed to the new **k** the aircraft implicate in accordance to the **k** that comes about since the distance in effect becomes longer. The aircraft produces a new time value **T²** in accordance with the Earth time factor **T²** because of the fact the Aircraft still share space in space of the Earth with the Earth. The aircraft now has a bigger time **vale T²** in the space **a³** of the Earth using the Earth time so therefore the fuselage of the aircraft has to reduce its space **a³** it claims compensate from the extending of **k** which it does by going faster as the extending of **k** will introduce a bigger time factor **T²** that will reduce the aircrafts occupying space since the Earths atmospheric space will not compromise and the aircraft still remains in the space of the Earth.

The bigger the motion discrepancy there is between the Earth and any independent structure on Earth in the motion of the Earth and in motion within the control zone of the Earth such an object will bring about a larger pushing of the secondary space that the smaller object holds to slow down the motion relevancy of the motion that the Earth has in relation to the independent object motion which is captured by the space that the Earth controls.

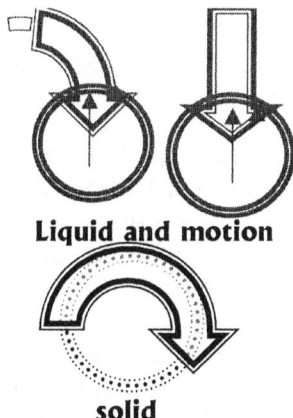

Liquid and motion

solid

All the results coming from this we also find

The object fights to retain independence while trying to slow down the Earth either as to accelerate the motion of the rebelling object or to slow the Earth down. This is all an attempt by the Earth to increase the motion that the secondary object can apply or by friction reduce the object space-time to liquid heat. There is a specific border or a definite barrier where the motion differentiation becomes so critical that the incorrect transforming of motion can have the same effect on the incoming object as hitting a solid wall. Later on, I shall show that it is the equivalent that is matching a solid wall that the object will have to break when the object falls into the atmosphere. When the object enters the atmosphere, it is through the small door there is in the area we call the atmosphere.

By having more heat per volume in ratio the material will claim and introduce new space that formed. Heat establishes space that expands. This truth science does not recognise. The claiming of more space and disposing of the space after cooling shows new space formed in the process of heat multiplying where there was no space before that which the material in the cool state afterwards cannot fill because of the void that came as a result of the material getting cold and contracting space, reducing the space as the space filled when the material was overheated. If material employs this as a basic technique today it was a basic technique back during the Big Bang. That evidence we can see when material having a heat level amplifying upwards when motion difference brings on friction and such friction brings on heat. I do not share the view Mainstream science has that when matter and antimatter came into conflict the product that came from this just disappeared without a trace of any sorts. Two opposing issues came about, but both opposing issues are still present in the cosmos somewhere in a place where we are missing the presence thereof. Material is energy and energy is indestructible. However energy can change form…yes that we all know and energy may even hide appearances. Therefore we have to search for the new form in disguise. I believe the evidence is present at this moment in our Universe and I think I know where the evidence is. I believe I can show that it is a motion discrepancy that produced matter and anti matter and we do not have to go and look for non-exiting positrons and negi-protons (if I may be excused for using such bizarre terminology but it is fitting a bizarre statement for the first one is not my doing as my brain I did not make it). But a positron must produce a negative proton and such a performing sub atomic structure cannot be possible. By changing legions the proton must then perform gravity by rejecting material or if I am correct, producing space! I am about to prove that antimatter is in fact a process where the heat that became formed heat, which forms space, and therefore space has a valid substance other than being nothing. I go to lengths to make persons see that space cannot be nothing. This is a factor that science has to accept if Mainstream physics has the will to find solutions about the Big Bang. The motion between particles in a cramped space as was the case during the initial stages of the Big Bang would have brought on friction in space we cannot even calculate.

All the results coming from this we also find in nature. The extending of **k** is as much a contribution to gravity as the retracting of **k** is. It is the combination that forms gravity by motion. The result is that in the very beginning some matter particles produced gravity in their sustaining of independent singularity by applying motion that in some cases lead to the demise of some forming space-time by converting where some compromised solidness. In order to install some form of coherency I shall for now and only for the moment call them the antiparticles which is referring to those that became destroyed. Saying this I have to immediately reprimand myself in using such terminology because by doing that I am once again bringing in our human concepts of judging some or other form to our requirements thus placing me as a part of one side or part of another side, which I like or dislike (I think it is a normal human error but as I am the one preaching against it I should be the first to try and stop such human judging and picking sides). This route the one side took resulted in plasma forming on the one side and material on the other side. This was done because there was less control that confirmed the space and the volumetric space grew. Looking at the development from a less defining point of view may say that some became soft and others stayed firm. By having some softer than the other harder ones the softer one became a liquid. The notion or defining of a liquid is very relative because as solid as the Earth seems the Earth vibrates as a seemingly liquid during an earthquake. It forms waves many meters high just like it would be a liquid like the sea. Afterwards when those in charge of damage control come to assess the damage it is hard to digest the destruction and damage because all liquid-likeness disappeared.

The fact that the Earth had more motion in its own is the indication that during the earthquake the soil served as a liquid. This becomes a reality, which because of the mould-like ability a liquid has is mostly moving away or around the solid structure that remains apparent as static. In such a manner the more solid ones through the process known as electroplating could incorporate the softer ones. Moving from one position to another will commit the softer material to coming across as the liquid part. This electroplating motion is possible since electricity is gravity to some intense extreme. Electro motion or electro flow is the concentration of gravity to the limit where we will find gravity has the same intensity in the centre of the Earth as electricity has in the open. By removing material from the less dense and electroplating that which is removed from the less dense and then galvanising that softer material onto the harder material (which by the way is a very natural process taking place all around as a corrosion) the density of the liquid will demise in the liquid sector and the material will grow in the solid sector. I believe even to this day and throughout the rest of the Universe wherever there is space such space has to have motion and space cannot be what it is without having motion. With that in mind that is space-time. Space-time is space flowing on. Where there is motion in space, the motion through all of space is carried along by time in space. The plasma is transforming to material through the motion we named gravity. By being electroplated onto material. By duplicating space in the process of establishing gravity the object does not reduce to a standard in occupying space that it had before the motion took place but by placing liquid heat into the form of solid matter the matter uses the newly acquired heat through which to cool. By absorbing the liquefied material onto the solid material it is thus freezing it into a solid to secure more material in the fight of combating overheating. In other words in the present time in our Universe gravity is freezing space to first become dense and form a liquid after which it then solidifies the liquid heat by freezing the liquid into a solid state within the substance that is the atom. However, I do prefer to use heat as the term of choice and not plasma. Plasma is confusing because of the variety of names by which it is identified. The process I just explained was the manner used by the cosmos as the cosmos came about and this is the manner that will repeat until such time as will the cosmos conclude its final motion. I believe that the first motion came about as singularity was without space and found irrepressible heat levels rising. By overheating it moved into space that was still non-existing and that had therefore produced motion to rebalance the heat. From this I also believe some material that came about from singularity overheating remained as particles forming atoms where there is this relation between the solid proton, the liquid neutron and the gas electron.

This omission I clarify later on. After this phase where the atoms formed the compromises ensuring a successful passage in cosmic development had come and gone and another phase took its place, which had the liquid heat formed as it then presented a new innovation being unoccupied space. The development of space from liquid heat such fluid was becoming gas that is space with the ultimate gravitational relevance that space can carry. This was all contributing to the lack in contracting gravity promoting expanding gravity to those particles that applied lesser motion helping the extending of space to turn into heat that again turned into space. In this, of uncontrolled release of heat performing as softer space-time such release of space-time is the destroying of singularity secured in a unit, which again I believe (within reason) I do prove. I show that on the one side singularity introduced space-time, which confirms singularity and space-time makes contact with space-time not directly controlled by singularity or that, which is directly confirming singularity. This I conclude from studying Kepler's formula. I believe heat is the destructed form of material that overheated and this confirmation the atomic thermo explosions give us. But to realise that we must beforehand find what any and all space is and we have to accept that space is made of something.

When one applies heat to an object it expands. That is primary school science. This states that more heat applied leads to more space acquired by the heated object. In sharp contrast to this is the growth in space when heat levels rise but freezing brings about the opposite result. When I freeze an object that object reduces its occupied space as it shrinks. Removing heat reduces space. That comes directly as nature responds to heat and I can prove that easily. By expanding it accumulates space to increase the improving of the size of the material. The accumulating heat for the sake of securing singularity, accumulates the heat in the material whereas the freezing tarnishes the overheating symptoms by the removal of material in unoccupied space using external matter and setting motion to the material until it contracts into a form which we see as visible heat. The heat is in the form of dissolved singularity that became material as material used it as growth. That is why by freezing it will diminish the space as to accumulate the heat absorbing into the heat into the material to maintain the equilibrium needed in space.

Taking this equation of nature to outer space we seem to confuse the natural law. With outer space as expanded as anything can get we regard outer space as incredibly cold. As heat sets in the normal flow will bring about expanding of heat into the form we think of as space that limits the heat overheating. Outer space is the very edge of expanding of space where heat cannot expand into space any more. Outer space is the limit, the epitome of expanding where heat meets space at the edge of all limits once more. Therefore being the representation of the very limit of expanding outer space has to be the hottest place there is. By applying heat to a kettle holding water, the adding of heat manifests as steam and steam is hot water that traded heat as it reviewed space. By allowing the receiving of the heat to continue the container will let loose steam in order to match the contributing of space. The manner in which heat expresses itself when confronted by overheating is to provide additional space through expanding of space. Outer space is outer space because outer space has expanded all it can. It is still expanding to the speed of Hubble's $1/H_0$ which inevitably does not only affect far-off places where we cannot be, but effects us on a daily basis. As outer space is stretched to its limit, its limit will continue to stretch but while it is stretching it has to have more than it had before in that outer space holds the limit of heats expanding possibilities. Singularity has been expanding since way back when but that means singularity is still releasing heat as space-time that turns out as space in the universal time of outer space. In outer space heat cannot expand more, therefore, except for the continual growth that benefits all singularity throughout on a continuous basis concerning all outer space.

If singularity expands when heated and there is a limit to the point it can heat, and that point of maximum expanding has been reached through the unleashing of heat, which is turning into space, we can with great confidence declare that space is the hottest space there is. Whatever expanding that is possible was done to secure the cooling and all cooling that can be introduced to bring about further cooling was performed that place is the hottest place there is just because of the sheer implication that it can cool no further as it is as hot as it gets anywhere. If that is the case then it is safe to say that galactica then is freezing cold notwithstanding our concepts of heat and space and heat in space given to us by our collective culture and not by our ability to reason. It has expanded to the maximum that it can yet we think it is cold when it is the extreme there is in heat that introduced the maximum expanding. It is the contradiction of the century - that much I do realise. At the inner core of a star all space shrinks into the oblivious but we consider it to be the hottest spot in the solar system. That just cannot be because when material shrinks it becomes cold and by shrinking into the oblivious it has to freeze into a goner. Again that is the contradiction of the century. Why will that be? The space inside the star shrunk to the minimum there can be and that tells us the space has to be cold because the shrinking took the space to a position where no space can shrink anymore.

That shrinking of no more space can only be inside the inner star and in that region where gravity is at its strongest. With outer space as expanded as nature may allow the space that grew could only grow in conditions of heat because heat produces expanding and expanding is the result of heat coming about. Space shrinks because it is cold: that we know and taking this law to the star centre it means regardless of our interpretation of hot and cold, that area in the star centre is as cold as it can get notwithstanding what our nature may tell us. Then obviously the same must apply to outer space for precisely the same reasons because it is so hot it can expand no more. We look at the hotness of space and the coldness of space but it is the relevancy to the solidity that forms the actual heat and cold limits. It is so hot no expansion can produce more space in outer space, as the outer space seems hot and quite the opposite reveals the true scenario inside the star in the centre of a star structure. That means the number of protons in motion has a lot to do with the cold and hot scenarios because where the protons are most dense the cold is in extreme. Only in the absence of space can so much heat gather in excess and the opposite is true about outer space where the least denseness found brings about the space in heat found in outer space. Our human selecting of hot and of cold and what is hot and what is not prevents us the clear vision we would have when truly understanding the applying temperature. Temperature comes about from spin and the smaller the spin density is the colder the space becomes because the more duplication produces the most cold. We think of outer space as 0^0 Kelvin but in fact it is as hot as no other place can be in the Universe. The coldest is where material is freezing solid as material does when frozen solid and the hottest is when by boiling the material is going into a gas with liquid being the intermediate position where heat acquires the space to perform as a flexible substance.

When we look at particles in outer space we see the particles being frozen. It is because there is such a severe contrast between the particles and the environment surrounding the particles and not the particles that is so frozen. The particles are in a gas state because the particles do not form a part that is part of the space unit. Hydrogen clouds of hundred of light years in diameter are a common sight in outer space. The heat we find filling space is not part of the space but like the particles the heat is a separate issue. That heat filling the space is another form of material that could conduce by diverting from space or marry the union of space by becoming more space. If it were that cold which we think it is, it would not have expanded into such a massive cloud but would have contracted forming a cube of frozen hydrogen. But as we can see the cloud expanded the gas as far as the gas can expand. That expanding is indicative of heat and has extremely little to do with gravity. If you are of the opinion that those hydrogen clouds will contract one day into forming a star, well then think again. There is just no such a chance that that will ever happen because that is not gravity. Because outer space is completely overheating the condition it has in support of the particles makes the particles appear to be in a state of freezing but the particles are counteracting the heat limit they meet. However the particles do not contract the heat because the space in outer space contracted all the heat by means of expanding the heat into what singularity will appreciate. That is not because outer space is freezing the particles it is because in contrast to the heat of outer space the particles seem to be freezing.

The atom must be the utmost coldest and the proton is even much colder because when that cold escapes it turns to heat forming space that no one can understand. When the spin of the atom allows the cold of the atom to release the heat it had it had frozen to space the atom holds but when this heat releases from the containing form of the atom it brings about much more heat than the Human mind can cope with. One may not look at the material and judge the surroundings. The fact that hydrogen remains a gas and so does helium in outer space must serve as enough proof that outer space is hot, regardless of our interpretation of the temperature gauge telling us what we wish to hear. One must look at outer space and judge outer space from the findings only considering outer space. If helium remains a gas it is hot. The removing of heat makes the centre of the Earth cold although we see it as being terribly hot. The only reason why it can seem to be hot is because it is cold and in such a cold environment the heat can gather and space can collect heat because the particles find the surroundings extremely cold.

The cold in the earth centre causes the concentration of heat by space reducing, as all cold surfaces tend to do. If it was hot the space within the Earth would expand and the space within the Earth where we think so much heat is concentrated does not expand therefore it must be cold. To gather and accumulate the space in a liquid means it became much colder being a liquid. Finding the surroundings terribly cold will allow the heat to gather and not expand but when the surroundings are hot it will not tolerate more concentration of heat and thus will expand to rid the balance of excess heat within space. Look at the sun and see how the sun turned the hydrogen to a freezing cold liquid at 6500 K. Hydrogen is in a fluid state within the sun and is colder than the hydrogen that is in a gas form in outer space. The sun is the coldest place in the solar system. That is when the protons oversupply the removing of space to produce the cold that is so apparent. By the reducing of space it can concentrate heat to a fluid state by producing the opposing cold that finally freezes the heat to a solid state. The expanding of space is a way of duplicating space without reducing space and by duplicating in the form of expanding it becomes just the opposite to duplicating by motion therefore reducing space by halving space in time. That is what gravity does. By motion space duplicates and by space halving it removes heat in space as well as by dismissing space. In all the applying of gravity space bites the dust. The density of the protons brings about space dense enough to harbour the heat in such quantities and visa versa applies in outer space.

The application of gravity that condenses space and bringing about heat by the compressing of space we apply in the way we go about tapping into the energy that nature provides. Internal and external engines combustion engines all rely on this application for harvesting motion by driving power. Compress space even today with a piston in a cylinder and then pump the compressed air into a container and such confining of space will increase the heat by the piston effort to reduce the space brought about in the container. The heat coming about inside the cylinder has no relevance to particles colliding because all compressor cylinders cool down colder because when that cold escapes it turns to heat as the heat releases from space forming a secondary form of material forming space that no one can understand when the spin of the atom allows the cold of the atom to release into uncontrolled space. This release and unification with space that heat does is the heat it had frozen because the motion of spin to space that the atom holds remains in a frozen state under the guard of the spinning electron. But when this heat releases from the containing form of the atom frozen by the spin of the electron it brings about much more heat than the Human mind can cope with. One may not look at the material and judge the surroundings.

The fact that hydrogen remains a gas and so does helium in outer space must serve as enough proof that outer space is hot, regardless of our interpretation of the temperature gauge telling us what we wish to hear. One must look at outer space and judge outer space from the findings only considered in the terms which outer space insists upon. If helium remains a gas it is hot. The removing of heat from the space that contained the heat makes the centre of the Earth cold. In our universe we see it as being terribly hot because the heat then forms a separate substance but remains a form of material (8) but that is because we see the heat and not the space derived from the separating of the heat. The only reason why the space can seem to be hot is because the space is cold and in such a cold environment the heat can gather in a much concentrated state and space can collect heat because the particles hold concentrated heat in the space separating the particles. By removing such high concentration of heat from the space that used to be expanded heat, the space then must contradict the heat by being extremely cold. We look at the heat in the space, which by that time is another form of material and find the surrounding heat in the space hot while the space is extremely cold. The cold in the Earth centre causes the concentration of heat by space reducing, as all cold surfaces tend to do. But the proton contributes to that reducing of space. If it was hot the space within the Earth would expand and explode but the space within the Earth where we think so much heat is concentrated is so much it does not expand therefore it must be cold. To gather and accumulate the space in a liquid means it became much colder when the space parted from what then is being a liquid. Finding the surroundings terribly cold will allow the heat to gather and not expand but when the surroundings are hot it will not tolerate more concentration of heat and thus it will expand to rid the balance of excess heat within space. The concentration or release of space with heat or space from heat is a direct contribution of the singularity in control of the space-time. The regard of the singularity stipulates the conducing of heat in space or the release of heat to form space by means of bisecting the occupied space.

Look at the sun and see how the sun turned the hydrogen it holds captured in its atmosphere to a freezing cold liquid at 6500 K. Hydrogen is in a fluid state within the sun and yet it is still colder than the hydrogen we find in outer space that is in a gas form in outer space. The sun is without any doubt the coldest place in the solar system. That is when the protons oversupply the removing of space to produce the cold that is so apparent in the heat levels that do not join the spell. By the reducing of space it can concentrate heat to a fluid state. By producing the opposing cold that finally freezes the heat to a solid state we find that is what matter is. The expanding of space is a way of duplicating space without reducing space and by duplicating in the form of expanding it becomes just the opposite to duplicating by motion therefore reducing space by halving space in time. That is what gravity does. By motion space duplicates and by space duplicating the material must be by dividing or bisecting - halving it removes heat in space as well as by dismissing space and in that concentrating heat. In all the applying of gravity space bites the dust. The density of the protons brings about space dense enough to harbour the heat in such quantities and visa versa applies in outer space.

The particles claim more space when heated to preserve the cold. The claim to more space produces more space and reduces more heat. Such expanding brings about cooling. When particles heat or cool motion applies in some form. Motion started at a point when the Universe was extremely hot and there was no space. By introducing motion space formed and the lack thereof produced friction that became heat that became space. It is natural and it is simple and above all it makes believable sense.

The application of gravity is that which condenses space by bringing about heat with the compressing of space. We apply the progress we have as a species in the way we go about with our skills to unveil ways we can tap into the energy that nature provides. Internal and external combustion engines all rely on this application for harvesting motion by driving power. Compress space even today with a piston in a cylinder and then pump the compressed air into a container and such confining of space will increase the heat by the piston effort to reduce the space brought about in the container. The heat coming about inside the cylinder has no relevance to particles colliding because all compressor cylinders cool down with time moving and not necessarily with the loss or release of particles. It is not only the discharging of air that will reduce the temperatures inside the container but the time flowing bringing motion about where the motion is not about particles escaping but heat escaping in the replacing of the heat density (not the density of the particles forming the material content within the container) but the space that compressed to heat will also bring about that the heat displaces through the container wall to the outside. This is bringing about equilibrium where heat will always flow from more dense areas to the lesser dense areas. This has no influence on the status of the particles on the inside of the cylinder but only concerns the density levels of the particles inside versus outside. After the pumping of air increased the heat in the cylinder which even can go to dangerous levels, will reduce back to room temperature when further pumping ceases and that stops further air movement into the cylinder and such surging of pumping air is what brings about heat stabilizing.

Mainstream physics ignored the clear connection completely, notwithstanding it being so very obvious. There is this far in their recognising of principles in natural physics not one single reference made to prove their appreciation of this matter. They are bent on particle colliding. When particles collide such collision forms an atomic thermo release and that action we call an exploding atomic bomb. What principle this argument about particles colliding ignores is that all atoms use negative charged electrons forming the atomic limit on the outside forming a definite border to the boundaries of all atoms and in both electrons from different atoms are being negative charged. In being negatively charged it means both will come out and totally reject the other. The closer they come the more violent the rejecting will be and such rejecting is the production of heat that will turn to space. The electrons repel other negative charged sub atomic structures, which the electrons are that form the outer borders of all atoms. With all electrons highly negatively charged (being as negatively charged as any possibility will allow to match the utter extreme) such electrons couldn't touch. If a train in Japan floats on a cushion of air because of equal charges lifting the train into the air, how much more will atoms repel other negative charged electrons considering how tiny they are?

The particles entering the cylinder bring with them an envelope wrapping the atoms in space that is there to distance atoms from one another. Such space formed because of and under the conditions prevailing outside the cylinder walls.

The balance at first favours the forming of heat using the space coming in, and where the space is being reduced in the containing size they are squeezed into is reducing the space from what it was on the outside. The space distribution inside then has changed considerably and reduced a great deal compared to conditions outside and with the decrease of the space distribution that space then becomes excess heat on the inside.

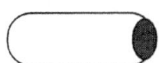  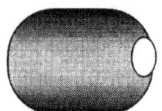 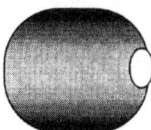

There is no correspondence between the air container and a star. The container in relevance grows while the inside in relevance is shrinking. That is pressure being in the container and the walls shrink as the pressure grows. In the star there is no containing but a time in spin bringing about space-time reducing and not space alone.

The electrons will disallow any contact directly between atoms. No force can be big enough to enforce such touching. It is because of that contact rejection electrons bring about that science has to use an overload of neutral neutrons putting them in the atom nucleus to fake a complying of charges that will eventually lead to atoms touching each other but that is through enticing a neutral stance which is enticing a positive overload for a short while. When the touching of electrons does take place the event is called a thermo nuclear reaction where heat is released in unmatchable quantities and the atoms in reaction dissolves into a liquid heat. But other than producing an artificial balanced bomb the touching of electrons will never take place because of the repelling that they would provoke amongst one another. The increase of heat by the distribution of particles in space connecting space and heat to particles is a separate issue that has nothing to do with contained particles colliding because why does it stop when pumping has ceased. This ratio of heat reduction is time connected as much as it is motion dependent. Motion reduces space by expansion as much as time contributes to space distribution by allowing the flow of heat.

This means it is not the particles touching one another in the cylinder that is bringing on the heat levels that are rising. Neither is it the particles that will eventually bring about the explosion that will follow, if safety measures are exceeded and should the pumping continue regardless of the danger rising. When the pumping stops the heat immediately starts the reducing thereof. Most important is the realising that every atom constitutes of two parts. In fact the entire universe constitutes of the two parts I am about to mention. On the inside there is a circle that contains the sphere and holds material in contact with singularity. On the outside there is heat surrounding the inner material part within the sphere and distances the inner material from the space between it and the next atom. The electron forms the division between heat uncontained and heat contained. This is why the Roche factor is so very important. There can be friction between particles in reduced space under controlled circumstances where such particles are grouped together in a unit and as a unit elects a group singularity forming the centre of the chosen form of the unit. However, there can be no friction between particles of atomic dimensions as a result of what the cosmos produces to contain a solution to this problem during the Big Bang.

The Universe separated heat from material by covering the exterior of material with heat that forms space. Some material became softer by uncontrolled overheating while others remained more solid by containing form through controlling the overheating. On the outside of all elements there is a layer that is the heat the element uses in relation to place relevancies between such an element and the rest of the cosmos. On the inside there is almost as much flexing available but we shall deal with that statement in a while. That which we think of as elements that are being a solid or a liquid or a gas is very much untrue since all elements are either solids or liquids or gas and none of them are a "natural" of any form mentioned above. It is a condition the element applies to secure a relative position under specific conditions. That is why spacecraft entering the atmosphere or which are bypassing the Mach limit such machines get covered with a heat blanket. The space surrounding the craft becomes liquid as the space becomes more intense in concentrated space that forms heat. There can be no particles in friction and even more so way up there in the atmosphere at the altitude where the cosmos meets the atmosphere just because the particles up there are so sparsely distributed in that part of the atmosphere. Above and beyond this lies the fact that all the so called air particles are very volatile and excitable by nature and they are known to turn the slightest heat into rapid motion thus establishing a scene where the particles that supposedly are in contact with the aircraft sheeting will move away from the hot incoming aircraft. If then not for any other reason it is because the particles are highly volatile and acceptingly sensitive to heat. Airborne particles are prone to motion just because it is the airborne element nature to change heat into motion and the motion comes about from their sensitivity to duplicate. No particle in the air being part of the space we call air which is in a free floating in that air can produce friction because of the volatile nature those elements have. Faced with the truth about this disinformation we have to search for other explanations that nature underwrites which forms a presentation more true to nature and will therefore be more sensible and less impractical. The craft's coming into the atmosphere produces a point where $a^3 = T^2k$ changes **to** $k^{-1} = T^2/a^3$ (the explanation is forthcoming a little later on) The distance separating the incoming object from the Earth centre reduces rapidly therefore the object starts to descend towards the centre of the Earth. We must also acknowledge the fact that there is one specific point of specific entry where this will occur more than before.

That point will rapidly increase the time factor where the incoming object crossed such a very visible border. By the reducing of distance **k** space a^3 will have to compromise in the relation of all the factors forming the equation since T^2 will very suddenly grow more acute. What happens is that the applying gravity reduces the space a^3 and the compromising factor comes about since the time factor T^2 moves back to a time where outer space was as dense back then as the density we now have within the atmosphere that then became the Earth's atmosphere. It is outer space that remained denser than what the outer space currently is. I am now referring to a process that I introduce as this letter unfolds which is by nature completely different to what is accepted by mainstream science. That which I refer to came about at a point just before the Earth established an atmosphere that grew through gravity and by measure of the Earth's gravity became separated from the atmosphere. While the gravity of the Earth contained the space surrounding the Earth in a much denser packed envelope the area not under the direct influence of the Earth governing singularity became more spacious.

The contained Earth atmosphere grew denser as the solar system developed into what it is today. As the atmosphere released from what we think of as outer space that release from outer space made the atmosphere much denser and the space above the Earth which is using a reducing time factor and that makes the Earth more compact. That established the T^2 factor to be that more condensed when one compares in ratio the density with outer space. The density at the time there was when the separation came about in outer space at the time of such parting outer space allowed objects to move away. This parting brought a barrier that is in place between the Earth and the outer space and any object coming from outer space into the Earth's atmosphere. The incoming object then would have to reduce the measure of the space the craft holds as the containing singularity set new standards applying to the incoming object with which the craft then needs to affirms its form and its status within the contained space of the Earth. The reducing will then suddenly no longer use space as the compatible factor but the focus will shift to the time factor that dictates to the space what the space can be. Such reducing comes from the switch there is in space – time where it was in outer space performing as bei**ng** $k = a^3 / T^2$ to what it has to be within the Earth's atmosphere $k^{-1}=T^2/a^3$. When the atmosphere grew apart from the outer space there are two ways of looking at the event. One can think that outer space expanded by the implication of the Hubble constant or that gravity withdrew the atmospheric space of the Earth at the time that the parting of space came about. But however you look at it there was a time when both outer space and the Earth's atmosphere shared equal density as we find it still applies on the moon and on Pluto. Then the Earth became dynamic and now they do not share any density at all. Things were overall more compact back then than at the present time and that included all things in the Universe. The space component is reducing the time component by compacting space to alter the space-time ratio.

This is portrayed by Kepler's formula $a^3 = T^2k$ It shows space as the density of space decreases. The Earth still compacts space by reducing the volumetric confinement of space $T^{-2} = k / a^3$. This we call the atmosphere. As the atmosphere becomes denser towards the Soil of the Earth there is a change in the time component. Most evident of this is when studying the pendulum. Just as we can see in the pendulum swinging, we can see that the swing reduces. Such reduction is because the space diminishes every time the arm rocks from side to side. With this there is proof that in the developing atmospheric space of the Earth the ratios change from outer space. This is proved by the pendulum arms that Galileo's experiment used to show that the swinging pendulum indicates $k^{-1} = T^2/a^3$. Furthermore it proves that Galileo was correct after all and unnoticed by science, Kepler helped Galileo prove Galileo's point. In this the net outcome establishes Kepler as being correct and the Newtonian argument of friction brought on by gases falls apart which is at that altitude where such friction supposedly should take place, the material in friction is not even present in the atmosphere. But science will stubbornly cling to the old theory with persistency that would warm any warring Field Commander's heart. Every element stands in different regard to the heat surrounding the material, which makes us consider the material to be either a gas or a liquid or a solid. The material in every element there is as such is either possibly any of all three forms. It is the way in which the circumstances are presented that the element allows the heat to gather and accumulate as the surrounding heat occupying the surrounding space. Every particle is unique in the way it regards the heat to material ratio and how much heat it uses to form either the gas liquid or solid state. If space a^3 declines then so must motion in relevance have to compensate by reducing k and limiting T^2 because space a^3 must always be equal to motion T^2k

Galileo proved space-time is functioning in the manner that space diminishes as space has to compromise to sustain the flow of time but time slows down in stronger gravity reflecting again on density changes rather than mass influences. $a^3 = T^2k$ $\frac{1}{2} a^3 = T^2 \frac{1}{2}$

As the space surrounding the Earth, which we call atmosphere, $\frac{1}{4} a^3 = T^2 \frac{1}{4} k$ **reduces in volume of space, the heat content rises as much as**
that the space holds heat having the heat rising by the same token. By becoming less, the space also becomes hotter. The ratio there is between space and heat increases as space in measure reduces.

Galileo substantiated Kepler's findings that space a^3 correlates directly to time T^2 when space a^3 compacts with the decline of k reducing the swinging arm of the pendulum that maintains time T^2. (1) The swinging arm will not relinquish relevancy, but reduces the space it moves through while it moves slower because of density rising when the distance k changes. (2) That proves that space and time $a^3 = T^2 k$ is directly related.

As the space surrounding the Earth, which we call atmosphere reduces in volume of space the heat content rises as much as that the space holds heat having the heat rising by the same token. By becoming less the space also becomes hotter. The ratio there is between space and heat increases as space in measure reduces. We have to learn to see heat where the heat in the space has two different identifiable substances. We also must see material holding space to be different from the space holding the material. We must see material to be different from the heat covering the material and compromising the space that produces the format of material in being a solid, a liquid or a gas. This changes in the state of materials holding a direct relation to the heat that also claims a stake in that space.

On the outside of all material the density provides a distance in space vowing between objects. That density also introduces heat as part of the distance of the space that is in place under the specific conditions applying. This is density because in the cold particles will be closer and when hot (3) particles will be further apart. By performing motion through pumping air into a container the pump collects particles and pushes the particles into the cylinder, which is just a cylindrical metal container. By (4) removing air from the atmosphere and squeezing that air into a container it leads to the reducing of the space between the particles *(5) in the cylinder when compared to particles outside the cylinder. (6) As the container fills the space (7) (that was space meaning it was keeping particles away from each other at a certain distance) turns to heat in a ratio to the square (8) as the compressing removes space and the material density within the space increases because the material density rising forms liquid heat. This further aggravates the heat brought in with the material in the pumping process because the compressing of space is adding to the presence of the heat by accumulating even more concentrated liquid heat..

To elements hot and cold as influencing substances are outside influences that do not apply to the core of the atom. The atom constitutes of densely frozen space flowing or liquid space and releasing (9) the liquid space into gaseous space (10) when insufficient control leads to uncontrolled expansion of such liquid heat. This is all singularity governed from all centres involved. Heat and space are influences outside the proton but we may imagine that within the proton nucleus it is bitterly cold. The heat or space will surround the atom on the outside, but has clearly no influence on the inside of the atom and therefore of the star. The star is in every sense the atom forming the star. Atoms will reluctantly compromise by reducing space but this compromise in solid structures such as the atom is will totally depend on the singularity that rules on the applying conditions. The heat or space is a state that somehow extends beyond the electron, which does not influence the proton or change the proton. (1) Only gravity coming from massive numbers of protons working simultaneously can remove space from the inner atom. The atom cannot compress but will withstand the worst pressure there can be. Remember a star cannot have pressure. Neither is there any possibility of atoms touching. Only space in the form of gas surrounding individual atoms can to a certain measure compress.

The heat in concentration or the manifesting as space is neither hot nor cold because the proton presents eternal cold. Heat is an exterior influence bringing about influence between atoms within the star.

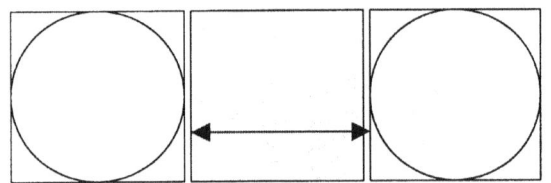

A gas will allow influences to charge as loosely connecting bonding the positioning and forming other objects occupying space next to one another. The gas will allow a lot of flexibility and compromise because in the case of gas space never becomes a premium. I should think placing space at a premium would apply when fusion comes into the picture. If a gas surrounds the hydrogen the gas will withstand as much compressing as can be induced by whatever force bringing about such compressing.

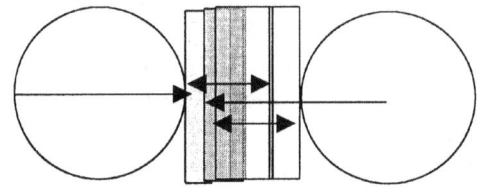

Scientists apparently in the know how of why what works express their opinion as one being that molecules run wild on the inside of the walls of the air container and bump each other while colliding and going mentally berserk in anguish and excitement. They maintain the argument that it is the clapping, fighting and general applauding of hostile, locked in particles that are in general behaving like British soccer hooligans that is going on inside the container. According to those Superior-Wise the general agony and anguish that the particles suffer while highlighting their complaints as they knock each other around ... well that is what is causing the friction, which is causing the heat to fill the container and fill the container to such an extent the heat that filled the container is flowing out through the container walls. I wonder who thought this lot up? The person obviously was a tremendous thinker with true potent ability to dream an impossible dream with not a lick of logical argumentativeness mixed into any or in all of his thinking ability.

Every particle that is of the airborne type is of the airborne type because it favours an association with surrounding itself in an envelope of heat. That makes the airborne particles airborne particles.

With that they disassociate with close proximity to others because they are airborne particles and airborne particles are what they are because they are cooking or boiling.

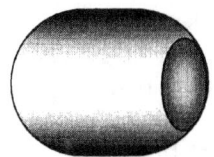

By pumping in air the molecules into an air-compressing container the molecules bring with every molecule entering a specific amount of heat from the outside that covered the molecule because all molecules entering is airborne molecules.

Scientists apparently in the know how of why what works express their opinion as one being that molecules run wild on the inside of the walls of the air container and bump each other while colliding and going mentally berserk in anguish and excitement. They maintain the argument that it is the clapping, fighting and general applauding of hostile, locked in particles that are in general behaving like British soccer hooligans that is going on inside the container. According to those Superior-Wise the general agony and anguish that the particles suffer while highlighting their complaints as they knock each other around ... well that is what is causing the friction, which is causing the heat to fill the container and fill the container to such an extent the heat that filled the container, is flowing out through the container walls. I wonder who thought this lot up? The person obviously was a tremendous thinker with true potent ability to dream an impossible dream with not a lick of logical argumentativeness mixed into any or in all of his thinking ability.

Hydrogen, oxygen and nitrogen are airborne because they are in a worse state of overheating than water is when water is vaporised. The "gases" are "gases" because in the case of "gases", "gases" are "boiled vapour" even below 200°C. Repressing the airborne elements as "gases" is the same as conforming vapour in a cylinder. I witnessed once what happened when water vapour was accidentally contained as a vapour. The house is no longer classified as a built structure. When pumping the "gases" the "gases" bring in heat because in the vapour state they are highly associating with heat. When the airborne particles enter the cylinder they are no longer able to dissociate according to will, but are forced to cluster while they will bring with their "boiling" "gas" status. This heat enveloping the elements is taking up space with particles that are enveloping them with heat as the compressor fills and that is what is compressing by compromising space where space is flexible. The solids are not flexible and cannot be compressed, not by pumping air into a container in any case. Space reduction becomes the norm needed to fill the container because the actual solidness of the airborne elements in their own right cannot compress in such a manner. That it places the burden of compensating to allow the increase in density on the space not occupied. The unoccupied space between the restrained particles deliver a gain in heat levels between all the particles and the next particle. Because of all the pumping of air which is going into the cylinder the space soon gets cramped and compressing space forms heat. That is evident in all internal combustion engines.

Since the heat is still under the specific conditions that was in place on the outside which was set by the atmosphere on the outside of the cylinder at the time when the pumping started the conditions in the air did not change by the pumping alone. As the pumping commenced the conditioning went on in the same manner, as it was on the outside and taking those conditions inwards. The air takes the same heat with when the pump takes the air on the pumping journey. The pump is taking heat surrounding the elements with the elements and the conditions they were in, in the atmosphere into the cylinder. The pumping changed little about the heat applying under the conditions as it was brought in with the elements. In contrast to such conditions that were applying on the outside there are then new rules that changed considerably. New dynamics on the inside brings about the applying of change to conditions. On the inside of the cylinder pumped with air forms new singularity elect that will enforce new controlling standards where all electrets are controlled by another set of laws. In the atmosphere where from the pump removed air the particles were widely spaced and the particles brought those conditions associated with a widely spaced atmosphere into where all are clustered in the inner confinement. In the atmosphere there were different elements all bonded by a heat holding the space between the separated particles apart. These conditions was set by the Earth governing singularity ruling space-time at that point in the atmosphere. The ratio the particles were apart was precise and specific and preset by the sun burning heat into the air or not burning heat into the air at that specific point. But the heat was the fabric that connected the material because it connected to the material by a specific ratio that set distances we call air pressure. The heat between the particles was contracted to a certain degree by gravity and was not as utterly expanded as it would be in outer space or as compressed as it would be in the container. Therefore some heat was in the space parting the particles in the conditions in the atmosphere at that very specific location. The Earth singularity brings about the gravity that determines the density of the "gluing heat" holding space between the particles in space.

The solidness in the material is actually surrounded by heat that brings with the atom seclusion from other atoms by covering the atom and the surrounding space of the atom with the heat surrounding the material. It is the density of the heat in the density of the space that specifies the conditions defining every position every particle holds as space occupied relating to space unoccupied that is separating particles and filling space between the particles. As the solid "stuff" is pumped in the solid "stuff' which is particles is covered by liquid "stuff" which is heat and together the solid and liquid "stuff" is pumped into the cylinder containing both the solid as well as the liquid "stuff". This combination forms the composition we think of as pumped air. There is coming in as much heat with the pumped element material in the form of flexible space. That flexible space is that which is containing a specific degree of heat. As there is that much material coming in as a result of what the pumping is bringing in the ratio on the inside changes somewhat. A lot of space turns to forming heat on the inside. The heat is in fact the outer part of the material coming along and is accompanying the particles as separate space but forms one unit. The particles in essence are solid when being without the heat containing the solid atoms and the atomic unit cannot compress as much as the particles cannot touch because all particles hold negative charged electrons to the outside. The electrons are all negatively charged. Excited negative electrons will not touch each other despite whatever any argument the wise in words wish to create about confirming this idea. Negative particles subatomic or not will repel one another and the repelling will become more fierce as the particles come closer. Unless scientists of high standings and other fiction writers can bring along antimatter and prove that the antimatter is hosting a positron, or what they seem to see as a positron there is no chance of electrons meeting. All electrons we know can only be what we then must call position because it is the only way that such a meeting can take place. Without the likes of a negatron and a positron meeting each other the two electrons sharing charges will not form a company. Without the likeness of such proof, (finding positrons next to negatron or whatever they will be named) the argument of touching atoms falls apart and will impress not even one clear-minded thinking individual.

We have to recognise that with the space in the form spaces come, the space that is holding heat is part of the electron status. The heat in space is a part of the un-compromised space that is not yet fully extended to the maximum as space we find in outer space. Such space carrying pure heat within the space concentration also enters the cylinder chamber. The only flexing is the space containing heat as the space accompanies the molecules entering under cover of heat. The space entering can enter as heat or as space since space and heat is the same thing. However in all circumstances there is a limit as there always is some limit to everything we find in nature.

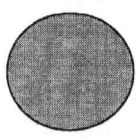

The gathering of particles is also accumulating space, which then turns to heat by reducing the available space

When this comprising of space between particles should carry on within the compressed chamber there will come a point where the heat turns from space to liquid and then the liquid exceeds the limit that atomic bonding of the container can contain. What happens next in such a container that formed liquid heat puts us all in the centre of the star. There is only one exception and that is that where we are there is no control of a governing singularity that we will find in the centre of the star. The heat turning to liquid will exceed the atom bonding of the container walls. By exceeding the atomic gravity the heat between the particles overwhelms the bonding of the elements forming the cylinder wall, which takes the space as the atomic particles that form the wall. It will supersede any limit the atomic gravity can bear in that bonding between all and every atom as they form a union that we see as the cylinder wall. The union that forms as the cylinder wall also becomes the union that forms the cylinder wall that holds the resisting imploding as well as exploding. The wall is as strong as the particles forming the wall would permit and the atomic gravity will permit the union of the wall that formed to be. The collapse of such a wall under pressure that will come after the heat in space liquefied has to come about on the side of the material bonding together to form the wall. The wall then is the weakest spot because the liquid heat will never give way once the space turned to liquid. On the inside there is no governing singularity elect to contain the heat within the unit since the whole idea of a pressurised cylinder is artificial to the cosmos. The liquid in the space will cut any and all cylinder walls to shreds. It liquefies the space in the cylinder wall. The space forms heat and the heat forms liquid and the liquid cuts the cylinder walls bringing about the explosion. If any reader is of the opinion that he or she has heard this before I advise such a person strongly to go back to Mainstream physics and find the official explanation about the material providing the pressure to bring about the explosion. Pressure has little or nothing to do with the whole procedure. It comes down to density and this process is not the same or equal to the process that applies within the star. The star is filled with heat being in a liquid form and it is not filled with gas that is true but in the star there is a singularity elect that is strong enough to contain the liquid heat. With the likes of a strong enough governing singularity, the particles as such can become small enough to fit the whole Universe into the container because some time back the lot did fit into a space that was at the time the size of the container. While saying this we also must add that this may only be possible on the condition that we can find a container able to hold such contained heat. It is the release of the particles that concern cosmology since we accepted the Big Bang theory as the correct interpretation of facts, which of course it is.

What happens inside this cylinder is most important to cosmology because through the pumping of the air into the cylinder science considers conditions inside such a cylinder the same as a star. Science holds the opinion that that cylinder is on its way to becoming a star. Such an idea is totally flawed. There is a big difference in the way the material inside the containers is contained. In the cylinder a big metal wall that is sometimes inches thick contains whatever is building up by means of pumping air from the outside to the containing inside. In contrast to this in the case of cosmic structures there is a massive pumping coming from a space less centre which does the pumping by contracting the space that prevents the escape of whatever is contained. While the contained is contained there is a centre able to contain what the star wishes to contain. That is not the case with the air container. In the case of the star there is no requirement for the walls to contain with the outer walls that are taxed by keeping it all on the inside. The inside of a star does not need a large metal jacket in order to contain what we find on the inside of the star. With that in mind it must be true that the Earth as all cosmic structures do, does not have pressure. For instance that which the Earth holds is contained from way within and there is very little chance of that which is contained having the opportunity to escape. This we appreciate but also we know that whatever is inside the metal jacket cylinder container will eventually escape despite the best human efforts of containing. The escaping is just a matter of how long the escaping process will last. Clearly the two processes are so far apart in managed control; the two processes oppose each other totally. There is no comparison between the two.

Even as far as the use of the term "pressure" is concerned, if we use the term pressure to indicate what the air container holds we are stuck with an idea of how the system functions. This is totally contradicting what happens inside the cosmic structure. With the container not forming a visible border in the star there is containing coming from the very centre and this has the opposite of pressure within the limiting walls of the container in question. The air that came in to the cylinder as a gas will turn to liquid if serious pumping continues and whenever the pumping exceeds safety limits the gas then turns to heat forming a liquid state. In the air container the density increases changing the cylinder on the inside from a gas to a fluid that becomes much more intense than the heat we think of as liquid heat produced by say an industrial cutting torch working with oxygen that is mixed with acetylene. The oxygen acetylene also forms a liquid flaming heat when the mixture is ignited but in the case of the cylinder the igniting will be spontaneous and burn through all the atomic particles. The point noteworthy is that at one very specific point in time the space becomes liquid and the heat returns back to space as it cuts through the cylinder wall bringing about "an explosion".

The exploding is actually the space that heat returns to space and takes space from times when time was much closer to the Big Bang to what space is at the present. It is a Big Bang that is coming from then to that which is presently applying. The main issue to realise is that the pumping produces a density time factor that increases and that the density increases which turns the inside from gas to liquid. The cylinder moves back in time when the pumping starts and the moving back in time realises another not containable Big Bang in the micro. It is not the oxygen or the hydrogen or whatever that fills the container that is a gas or a liquid that explodes, but it is the amount of space that turns to liquid heat that turns the container from a "planet" into a "star". Even the earth has already some flimsy liquid atmospheres in comparison to outer space. This is the only difference between planets and stars if you insist on having planets and having stars. The stage that the cylindrical container reaches when space goes liquid is re-enacting the star with the difference that with the continuous pumping of air into the cylinder there is no sustaining or governing singularity and when uncontrolled the gas will turn to heat and the heat then all goes bang. By this measure mentioned the star is different because in the star such a liquefying is a long process that the star was awaiting for one eternity. That is the difference between a compressor having a metal-jacketed wall containing what is on the inside and having a pump on the outside that becomes most apparent in comparison with the star. In the case of the star the gravity within the centre of the star is where the star is having an inside pump. By the drawing or the gravity of the star, the star really shows its worth. This is a reaction coming about from the centre of the star. Where space and motion ends is truly where the star starts to contain the inside without a possibility that the wall will burst as it goes bang.

The star does not allow escaping of material whereas the heat escapes from the container as it does from the star. When saying this we have to add that heart only escapes from young and incompetent stars such as the sun is. The star shows no relaxing in the process whereas in the event of no further increasing of pumping heat into the containing cylinder in any way possible and in the event that the pumping does not restart or continue again the process will rectify itself. Conditions inside the container will once again return back to room temperature if pumping of air into the container stops. A balance will come about leaving the heat to escape through the container walls and then release as space outside the container. However, there is no loss of materials such as atoms that rush out through the cylinder wall. This reversing of the flow of heat will escape through the wall relieving pressure without taking any of the confined particles along. This is apparent because one can feel the heat coming through the walls as conditions inside the container walls strive to become equal once again to conditions on the outside of the container wall. As the cooling goes on, the conditions on the inside of the container will again in time reach room temperature inside the cylinder too to match the temperatures outside the cylinder. Once the pumping stops, the particles colliding and causing the friction then do not play a further role. They either become very calm unexplainably and without any apparent reason or the principle taught by Mainstream physics lacks truth.

Never once could I ever find out what brought such calmness back to the material in the container without the particles having the benefit of a serious drug dosage that a paralysing tranquil medication will bring on or having a real potent Sangoma (an African witchdoctor) present to enforce a calm. After all the science that teaches us that this heat increase comes from particles colliding and that is what is producing heat inside the air pressure container, the effort in getting them still and calm again after all the excitement must be a tricky operation. What is truly amazing about the becoming calm again is that the already over filled container does not continue the friction between the particles rubbing and brushing and through so many collisions stirring up more heat which was produced within the cylinder container as a result of what science teachings lead us to believe.

When the compressor is left by itself with no further filling or any releasing of the filling substance carrying on, the temperatures inside the walls of the container stabilize as they go back to the same levels as the temperature was on the outside. When all stabilizing comes about from time moving along and allowing the continuing of the stabilizing of the heat is interrupted because there then is suddenly a motion of the air being released through the opening of a valve, such a sudden controlled releasing of the air under controlled conditions produces motion of the air as the air is relieved from the container through a release valve. The particles are unable to escape in the same process as the heat does. The heat flows through the cylinder walls and that means all the particles are still present and accounted for in the cylinder.

When the particles are released through a releasing valve the release will create a flow of air, which starts the motion of air, which then will bring cooling. The pipes through which the air flows when released will cool to such an effect that pipes can and do freeze and block all airflow. Two American Submarines were lost in this manner. It is not particles that have to take the blame but heat being released or admitted. If the cylinder is left undisturbed for a few days the stabilizing in the air within the cylinder walls will lead to re-establishing a total equilibrium and such establishing once again will take temperatures back to a freezing state as it cannot retreat heat that escaped space of the cylinder before the valve release came about. After the pumping subsides the space that is compromised then is allowing equilibrium to set in where the equilibrium equalises the particle density within the cylinder to that which is in the room temperature. The problem that results from this is that there are far more particles in the cylinder. That causes a much lower heat level or space in the cylinder than was the case when the heat was at room temperature. By stopping the flow and allowing an escape of heat the heat leaves space at levels where the levels in the cylinder are forced to be at a lower density than what was applying on the outside. This flow of air in the releasing pipes shows motion and such motion cools down space occupied. Motion is gravity and gravity is cooling of material into the oblivious. Through the motion coming from the releasing of air there is a much lower heat level in the cylinder. The motion of released air will force heat levels down when comparing the heat levels inside the container to what the comparing heat levels are outside the cylinder. The conditions will be directly in reverse as the space inside the container then is much colder by air motion than the conditions of the heat on the outside. The natural flow of air from inside to the outside will change the conditions on the inside so much in spite of the air being overfilled on the inside and the comparing lack of density in the air on the outside.

If the container and pump is left without further human intervention or human influencing any circumstances developing such absence of interfering in the conditions as far as heat distributing goes will become and remain equal on either side of the cylinder wall. The compacting of air molecules is then still much higher on the inside than what the denseness is on the outside but this difference will not produce the same heat levels inside the cylinder than that achieved during and immediately after the pumping operations. The temperature will only become affected when more motion contributes to changes in the balance. The releasing of air will extract heat from the process to a point where it will lead to freezing coming about in the narrow pipes where air is released and such airflow is the fastest. The rebalancing will go the other way as was the case when the pumping was in place but when the released air is in natural motion the flow contracting heat flows from the outside to the inside of the pipes. But since the motion of airflow will be much faster as the air is released such motion contributes to cooling whereas the motion of pumping gathered heat. Although if science is correct then the heat could have started only when particles made contact and went bumping into one another, causing friction by their bumping and dancing. It is paramount for cosmologists not only to gather mathematical proof but also to apply such mathematical proof in amongst stringent natural laws and find not only what but also why certain events come about.

The spin correlates time by supporting the flow of heat unoccupied to feed heat occupied. There is a definite correlation to establish balancing in space-time and one thing we know for sure is that if it applies today it is because it applied way back when it all started. There is a correlation between heat and space and heat in space, which is very far from being the same thing. Space contains heat because the atom is secluded including heat captured by individual singularity but space can also accommodate heat by producing expanding of space not captured directly by the control of singularity in the absorbing of heat. Heat will always flow from the hottest to the lowest region because the density of heat will bring about a flow such as water does in gravity. The mere fact that such flow of heat from hotter to a cooler region can take place makes it clear that heat will flow just like liquid does and find stabilizing as liquid does. Realising this must take our thinking back to a time when space was preciously little and heat was soaring almost beyond control.

If we tackle an issue as unimportant as a container being full of air in the correct line of thought then it might enable us to draw direct parallels from this. Such parallels might enable us to see how fusion comes into the picture. If a gas surrounds the hydrogen the gas will withstand as much compressing as can be induced by whatever force bringing about such compressing. Gas can compress like nothing else can.

By heating an element the density will deteriorate releasing heat to form extra space and the decreasing density will bring about more space becoming available within the confinement of the compressed area that is heating. But the more space does not come in the form of more material therefore the extra heat is extra space claimed by the same material occupying the space. If the confinement is such that it will not allow the growth of more space then the heat level will rise. By heating without allowing the liquid to form space then we find that liquid is even more uncompromising than material can be. The heat will grow by increasing heat levels in the liquid heat until a state will arrive where the liquid air will cut through any container that is man made. On the other hand it is true that gas will always compromise by giving away space to produce heat. With such knowledge we have to look at the functioning of stars once more but this time with a much more critical view. In fact having this realisation prompted me to look at pictures of the sun with a much more critical gaze. What I saw was, the sun being a sloshing bowl of liquid flowing as fluid once more and then I realised from what I saw that the sun is liquid. If the sun is a liquid bowl then all other stars have a liquid inside! But we also know the density levels rises extensively as the sun's space declines towards the centre of the sun. That means the liquid will become more and more dense as the sun reduces space towards its most inner centre and the sun would have heat in a sub solid state down towards the centre. This changes every conception there ever was about the inner workings of stars. One can compress a gas to a state of liquid but then the gas is no longer a gas. The gas is transformed from a gaseous state to that of being liquid. Gas can never compromise enough space to allow fusion to take place because there is far too much flexibility to bring about compromise, but liquids are quite another story. Before we come to that however, we must reclassify what we think about when we think about a star as in this case the sun.

In the process there are relevancies. If outer space is - 276^0 C or 0^0 K then it can only be because other conditions apply elsewhere. Where the elsewhere will bring about that under the elsewhere conditions there is something else at 0^0 C and that can only be valid when another object is in an environment that is allowing that something in the other environment to reach 100^0 C. In other words there has to be a scale fitting all possibilities that goes from densely deep frozen hydrogen at $- 269^0$ C to a boiling iron at 3000 ^{0} C. That is a spectrum but such a spectrum only has validity under very specific space-time applications. But if space stood alone and we gave outer space a temperature value of -276^0 C with no other references to compare, such a statement has no validity because the number could be anything just as much as it means nothing. If there is no correlation between two factors that produces a range from zero to a maximum such a scale is as meaningful as much as it is senseless. There has to be a relevancy to validate any figure in temperature used. If we pronounce anything being whatever it can only be valid if it is because being whatever requires not to be another whatever and therein lies the value of being whatever. To bring about the relevancy is only valid when allowing us some comparing between the two of different and non-equals. With the one number standing alone such a number by itself has no meaning, not withstanding whatever connections there are between the two.

With the sun being under another singularity in charge of different rules setting different standards to the markers we use on Earth the sun will have a totally different heat standard of freezing or boiling coming from the controlling singularity than we find applying in the governing singularity of the Earth. The sun we have to admit is very different from the earth and Earth standards just will not apply where the sun is concerned. Making a statement that outer space was 10^{34} degrees in the shade during the Big Bang festival is rather meaningless because what was the other boundary then? There is a question arriving from this issue which is: Am I allowed to draw the comparison that the sun was 6500 ^{0} on the rim and 18 X 10 ^{9} on the very inside at the time the Big Bang was 10^{34} in outer space? If that is the case and it has to be the case if science wishes to apply any meaning to 10^{34} in the shade at the time of the Big Bang, then the sun on the inside was an ice box at the time and during the Universe presenting the Big Bang presentation and on the inside the sun was one of the most potent freezers at the time. We then may regard the sun at the time of the Big Bang, as being a freezer storage facility, which was colder than the human imagination will allow our perspective to go. And more compelling is the fact that this then makes more sense than anything anyone said about the Big Bang or that was ever previously said about the start of Creation.

The objective of a star or any sphere for that matter is compromising by compacting the inner space as one comes closer the centre. As the star is reducing the space it claims for individual use outside the wider Universe it is also condensing such space to the inside where such space claimed is providing a reference but what reference and to what will such a reference point refer? If the sun was a gas on the inside of the sun as science declare, then there was a lot of space going around within the confinement of the sun but that is also true that there then is not much filling the in between the solid substance also going around within the sun. I have seen where comparing was made between the sun and a tire. In such comparing I wondered what was in the mind of such a Master cosmologist when thinking of the star as a tire and a tire as the sun. If the sun is an air inflated tire what then is burning? The burning is the purist liquid of all, it is raw heat flowing like the liquid it is. That is the case where a liquid fills the space because the in between space is filled with a nicely flowing liquid.

Liquids in the sun prove to be much more practical because liquid is more dense in certain ways than solids can ever be. A solid will rupture when asked to compromise because after all that is what fusion is about. On the other hand will any liquid just become more and become more liquid without giving way. Liquids did the compromising which lead to the Big Bang and after that there is no further compromising under pressure possible (if I am allowed this once to call the inside of a star pressure). That is why the strongest power engineers can use hydraulic cylinders. By using a liquid within the star too the comparable gas outside the star and the solid being within the very inner star core, we can begin to see comparisons emerging. Remember that outer space gas is hotter than the inner space fluid of the star not withstanding our perception on culture formed by schooling and with memories that our culture is shouting to us in defying our wits because the truth is quite the opposite to what our minds are telling.

When something gets hot space is added and inside the sun with all the filling of liquid and material space truly is an issue of scarcity. If it was not true that space is little in going around in the star then fusion could not have taken place. Fusion is the product of space in the demise thereof and the demise of space brings about a need for space. That is the prelude in conditions applying upon which the Big Bang followed. But with such little space going around then the absence of space makes that the space which should be available, then has to become a priority because with that little space to go around in the sun, the sun must be bitterly cold. Again I stress that heat in abundance forms space by volume and that is one thing that is very absent in the sun. If space was in abundance fusion was not possible. Heat brings about space and looking into the sun we find that whatever the space there is, there is not plenty of that in the sun. With no space it means there is a cosmic cold raging in the sun. Do not for one minute think of size of the sun that we think is taking up space because that type of space is not anywhere in the cosmos.

There is no big as much as there is no small and the space which we think of in terms of size and what size depends on is precisely the incorrect human way of thinking because that is the human instinct to measure by size in terms of big or small. If we do measure by giving size in terms of big or small, hot or cold, near or far, and we then include the importance that we shower ourselves with as we put our being the centre of the Universe in the centre of the Universe, then we set a standard judging from what we control but by putting us in a position in the Universe that makes our point irrelevant whereby we judge the cosmos. If we make ourselves important in our eyes we become part of the cosmos and that makes us incapable, as we then are irrelevant about our judgemental concepts. Having a position in the cosmos will translate to our incompetence as humans, which goes directly into our irrelevancy what we can use by which measure we judge. By placing me the human in the position that I, the human think I am in, then we give the position we think we have as forming the centre of the Universe such a relevancy by which we measure. By seeing us as life being the centre of the Universe we lose perception of what is valued in the Universe. That tendency we have we must destroy to find meaning to our thinking about cosmology. We think of hydrogen as a gas in the atmosphere of the Earth and we transform that standard by which we think to the sun.

If hydrogen is a gas here on Earth then we are convinced that hydrogen is also a gas where the sun is and then the sun is a gas structure. How much more Biedermeier can we be when we are thinking in such terms…well atheism gives such thinking of the Biedermeier manner quite a go. If we continue to repeat that line of thinking we might as well move back to the cave because such backward thinking belongs in the minds of cave dwellers. The sun is as liquid as the sea is liquid with the difference that the sea uses water and the sun uses pure liquid heat. That makes the sun a giant hydraulic pump and that removes the pretty weather system that we grant the sun from the sun. There are no winds blowing but there are rivers flowing at an astronomical pace. There then are no winds in the sun but there are rivers of flowing heat running and raging in the sun. By using hydraulic power within the sun instead of the presumed pneumatic gas as suggested by Mainstream science the rules on physics applying within the sun changes as much as day is different from night. In a hydraulic system the hydraulic power will only fail once the weakest spot in the solid breaks down. With a tough enough cylinder that will withstand all pressure pushing at it, something will give way and we know one hundred percent it will and cannot be the hydraulic fluid. The hydraulic oil will produce more fluid when overheating by acquiring more space from the heat asking for more space. However in such a case where we think of oil as a liquid the hydraulic oil is the solid substance, which we humans consider as being liquid and it is only performing as a liquid. It is not the liquid we should think of as being the substance in the sun.

We also can see that the liquid that the oil is re-enacting or is standing in on behalf of is a true liquid. Being the liquid can only be valid when conditions apply to form and not to the state of the material. If one considers the qualities of liquid then we can ask what will happen when the liquid is pure and uncompromising heat in the purest form the cosmos can provide? What happens when that which is flowing is heat unable to break down? In that case heat can only compromise by becoming more of what is containing the space and not less of the containing of space. By becoming more it also takes up more space and that is less uncompromising than what solids are. In a star such as Jupiter that has gas within the atmosphere, (let us call the atmosphere a gas this time because gas and liquid is a very grey issue) the pressure within the structure cannot apply a solid base to secure solids compromising space that will bring about fusion.

The solidity that eventually brings about the collapse of occupied material holding space is unavoidable when confronted by an uncompromising liquid and by increasing the volumetric space the liquid holds it must come to a situation where something must break. In the case where fusion is in place everything is pushed to the extent where nothing can break. In this environment we know from the characteristics of hydraulics that the final collapse will not come from the hydraulic heat. The final collapse is what happens when the solid atoms collapse from being two to being one structure and the solid singularity has to compromise independence. There are two in the space the one had previously and by fusion, the one gives way by denouncing its independence. This action can only present fusion when cold is at the limit there can be cold and the Universe at that point is as cold as it shall ever be and heat is at the very other end where heat can only peak as heat touches eternity at that very spot.

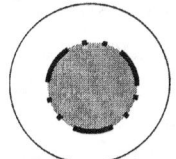 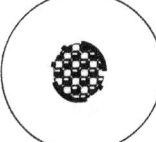 

When outer space is 0^0 K then the sun is estimated to be 6500^0 K. That is the relevancy we bring in on a scale between the sun and outer space. On the one side the gas in outer space is -276^0 C or 0 K and then the sun changes the environment to what we know the sun is. The sun in our minds is the other side of any extreme where the sun fills space. But the material within is neither -276^0C nor is a frozen liquid on the atmospheric rim where outer space meets the suns atmosphere holding a freezing 6500 K. Deep within the sun the temperature seems to be 18×10^6 K. It is only a relevancy with substance when humans relate the information to circumstances seen from the Earth. If standards applying to the sun were transformed to standards on Earth then no substance known to man including man in person would have any chance to exist because the sun harbour conditions that is beyond frightening as the conditions in the sun is hostile to life of any sort. Actually when thinking about it in a scientific manner this statement includes all conditions applying anywhere except on Earth and then not even all of Earth. By placing life anywhere and at a penny a dozen that notion put the manner science consider life and their life's ability to judge into question. When any person and in particular a scientist is assuming a role as being a judge of conditions in the cosmos such judging can only be introduced when the person in judging divorced him or her completely from they're associated with life. We have to disregard whatever we feel about the importance of life and only pursuit with what importance the cosmos regard life. We have to eagerly pursuit the sun as our nearest star after we dislodged all mind concepts with life attributing to our way of thinking. This is not just necessary but it is a necessity.

Where Mainstream science so desperately wishes to find life all over the Universe they should realise that even on Earth conditions differentiate where certain life can be and where other forms of life cannot be. Even the Earth is at some places hostile to all forms we think of as life. Life cannot be either in outer space or on the rim of the sun or in the very inside of the sun. When investigating that which forms the cosmic spectacular one thought that must never leave our minds is that whatever is out there, we have this tiny speck and that is all. With that realisation we have to come to terms with, by excluding life when judging the Universe as well as all conditions we may be familiar with because such conditions will fit in and will swing our judgement when such thinking is suiting life. With standards applying in the sun however there must be one point that is zero to give the sun an accepted range. Say the sun would be 0^0C at the point we give it the norm of 6500^0 as that will be the lowest temperature that can be found within the confined space in the boundaries of sun. This will be because the sun is a secluded, isolated Universe detached from all other universes except the immeasurable universes we think of as atoms that group together and form the single universe we call the sun. The sun holds all that is inside the universe that is the sun.

From the singularity in charge of the sun the Sun and the content the sun holds is the entire Universe and there is only that one Universe. That Universe is the sun and whatever else is there, that which is there, is there to feed the universe of the sun. Under such conditions water that is so absolutely crucial in the mind of human's would be an unknown substance holding an unknown quantity as a reference. Outer space must then be minus 6776 K if the sun wishes to use the scale we apply. But it does not work in that manner either because most of the substance that is providing life or the support life depends on therefore has no claim to space-time within the sun. All readers might be agitated by my comment about life not being able to survive in the sun, but go back and investigate what rules scientists use when they investigate the sun. All right I do admit that I am exaggerating about life being part of the sun but they (them in science) put seas and winds and even fair weather in the sun. We find winds blowing, gas clouds and such things

All conditions outside the Earth's atmosphere are completely hostile and dangerously destructive to life. If we wish to obtain a clear mindset in our discovery of the cosmos, we should not try at first to discover other life wherever we wish to think we can find life in the cosmos before we discover the cosmos. We have to set standards in regard for every Universe by setting a standard by which others following our thinking afterwards can truly apply our thoughts. The standards set can only match when all conditions are met with the criteria that any specific Universe dictates to allowing life.

We must then look for life by exactly the very same standards applying on Earth where we find life. We have to find a dot that holds water in all three conditions and we must be able to see the structure with our eyes. If we cannot see the structure those scientists then cannot boast that they are only using truths as science. They must then admit they are in fairyland and introduce their standings as such. We know water boils at a much lower temperature 100 km up in the atmosphere than it boils on the surface of the Earth and down below sea level it boils even at a higher temperature. This again supports yet again my argument about outer space being much hotter than anywhere else in the Universe because it needs less heat applied to get water to vapour. We take water as the absolute element of prominence to our survival therefore in all the scales we apply we use water as the centre value and then in one single step further we think about it as being pivotal requirement for cosmic scaling not deliberately but just because we forgot to switch off our human connection. We focus on water as the standard applying whereas water is not found in any other structure.

We even go as far as taking water to boil at 0^0 C because our thinking that way suits us in our human requirements but even that is a total abolishing of the truth. Water boils where water is located and according to the density applying at such a point where that density in space-time will have a heat to space ratio that determines the boiling point of water. True as it may be that we have little use for water boiling at 5^0 C it does nevertheless boil somewhere at altitude at 5^0 C and some places higher up at far less even less than five degrees C.

By placing the boiling of water in terms of our needs instead of hat we should read into nature throws our reception in confusion. Water boils the easiest where water needs the least heat added because there water is the hottest. We have our totally artificial manufactured perception of the boiling point of water created in lie with where we're in a centre stage y not being able to use the water we boil at thirty thousand meters in the way we wish to use boiling water we disregard such water as cold. In truth water boils at very low temperatures at high altitudes because the water is naturally extremely hot up there. We create conditions suiting our needs and then we project that centre stage we manufacturing around our needs to be the valid requirements the cosmos must have. Then we go even further to refit the entire cosmos to our human needs. This is underlining the desperate inadequacy of our reasoning about cosmic matters. We should not see how life might find the use for the boiling water but we must focus on the rules bringing about the boiling of the water.

If water takes less time or heat to boil then water must be closer to boiling point in the highest atmosphere and that can only be if water up there is hotter from the beginning before the boiling process started. The reason why water boils at so low a temperature at any high altitude must be from the more heat present in the space where the boiling is taking place. With more heat in less space less heat is required to heat more space to get the water boiling. I am referring to the actual space, which heat and material occupy, and is apart from the occupant heat, which is in a fight with material for space to occupy separately. I am referring to heat expanding to the point where the heat occupies all of the space and the space has more room to expand therefore the space took in all the heat it can use to bring about expanding. If ever life will find a way to go down into the inner earth say at a distance of 1000 km we will find that at that point water only boils at many hundreds of degrees Celsius notwithstanding the fact that the average temperature will also be many hundreds of degrees Celsius everywhere around. Conditions on Earth have such a variation, yet science has this tendency to standardise everything in the cosmos by applying constants that will fit the conditions we find applying in downtown New York on a pleasant sunny day in mid spring. If water boils that quickly the closer we get to outer space it should be a big indicator that water is naturally much hotter from the start in outer space. When the heat surrounding the outside of the molecule increases the heat inside the element has to reduce because there is a relevance attaching to the two opposing limits without the one opponent having any precise limit to show. As the outside fluid heats up, there is no breaking down of the substance because the fluid is the purist fluid there can possibly be. It is pure liquid heat. The heat will be more solid than the solid elements can be because the heat cannot give way any more than it already did at the event of forming conditions that realised the Big Bang. Just before the Big Bang arrived it took all the heat occupied or not to form heat that would be able to bring about the forming of space as a compromise to the heat that was bound to destroy the Universe. By sustaining the conditions applying even before the Big Bang there is not enough gravity left to produce more compromising from heat in a liquid form.

By heating the fluid the space the fluid holds becomes more and by heating the fluid the element becomes colder reducing the space the element holds. Space reducing is synonymous with becoming colder and becoming colder is about compromising space occupied. On the earth material will reduce space occupied that much and no more because the relevancy establishing the edges can only push the reducing that far and no more. But in the sun the conditions applying are a lot different and the relevancies can push that much further. As the elements enter conditions suitable to allow fusion one of the two factors surrounding the fusion will insist on all compromising of all space that the elements may have in separating between them or in holding the structure of the atom and in that act it supplies the biggest compromise in the relevancy between the shrinking of material as cold or the space as liquid that removes space by heating more and that is to give space over to the fluid side that removes space from the material side. In receiving the space the heat sacrificed heat for space by acquiring a sudden space.

By acquiring the space it receives a cooling that will translate to dark spots on the surface of the sun once the heat again surfaces to the top. The dark is only contrasting the light because the dark allows less density therefore has more space. There is no actual hot or cold to be found anywhere in the cosmos. In sacrificing heat the liquid obtained space and by sacrificing space the elements in fusion reached the ultimate freezing temperature it could ever achieve. The elements never became hot or never froze but the relevancies between solids and liquids. This is because mass has no role to play in the fusion process. Changing brought about conditions allowing the heating and cooling to take place without ever taking place.

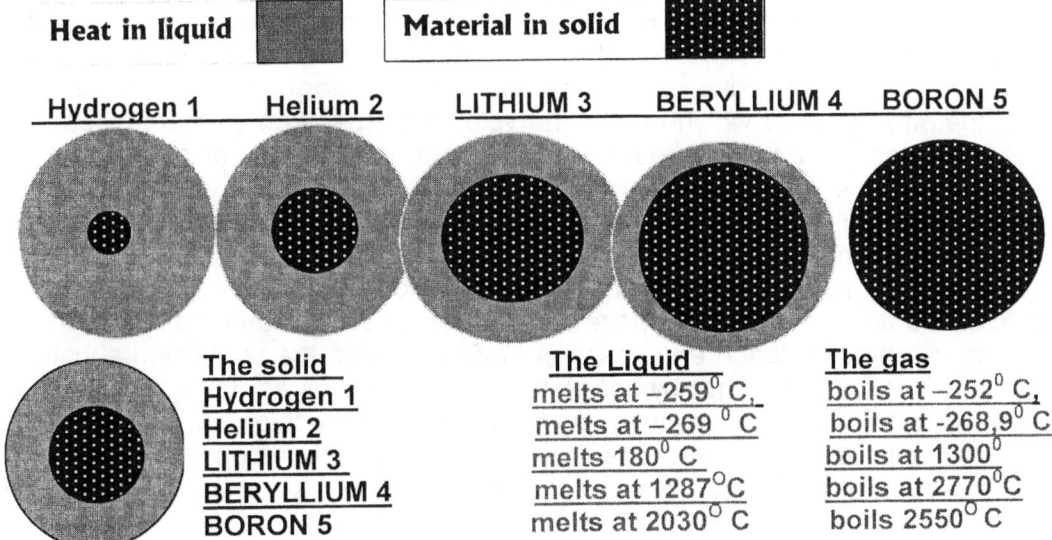

Heat in liquid   **Material in solid**

Hydrogen 1   Helium 2   LITHIUM 3   BERYLLIUM 4   BORON 5

The solid	The Liquid	The gas
Hydrogen 1	melts at -259^0 C,	boils at -252^0 C,
Helium 2	melts at -269^0 C	boils at $-268,9^0$ C
LITHIUM 3	melts 180^0 C	boils at 1300^0
BERYLLIUM 4	melts at 1287^0C	boils at 2770^0C
BORON 5	melts at 2030^0 C	boils 2550^0 C

It clearly shows that there are groups of elements with much different relations to heat than other groups have. This is such a dominant part of Creation I truly am surprised our distinguished Academics never investigated the evidence in hand of elements forming different conditions that obviously totally devastates Newtonian claims on mass and weight producing gravity.

It is believed that mass produces gravity but according to my reckoning it is gravity that produces mass and mass is only the result of gravity. Where the less space holds a smaller mass within the atmosphere that holds the stronger mass is the stronger mass creates a domain. In this domain the stronger mass will subject lesser the mass to a reduced space because of the less space the smaller mass holds. Mass increases when the mass is surrounded by atmospheric heat where the heat will increase the mass as the additional heat will add onto the mass and as the heat will influence mass. By adding heat the object becomes more massive but also more spacious. In such an event the adding of material albeit heat increases the mass. But where gravity increases the mass increase as the space reduces. That shows a complete different tendency. Therefore mass does not produce gravity as science indicate by their formulas used in calculating gravity because gravity does not increase when heat influences mass. Heat stored in motion produces gravity. Anyone not in agreement convince yourself by comparing the neutron star with the massive read giant and by your acquiring a logical conclusion that is not tainted by your opinion about big and small you will be convinced of my correctness. Science goes about in order to measure by calculation a Black hole they go and throw C^2 next to the dividing radius and throw the square onto the C by removing the square from the position the r normally has and by doing that should then present the speed of light acting as the retaining diameter. What the hell the speed of light, which is pure motion has to do with a diameter, which is merely a measure of a space distance and to top that their reconciling of the two are beyond my ability to substitute fact for imagination. With that they can the enjoy an evenings fun and games because then they can cheat enough to make nonsense of the whole lot they play with. Then they sit back and feel smart in the way they manage to cheat once more to prove their incorrect views correct because after all who will ever fly down a Black hole and return to support or deny their calculations.

The gravity applied by the Black hole is a speed, measure comparing space in ratio to time taken because all gravity is speed. Then the speed that light has is gravity. The gravity of the light can be gravity as much as it at that very same time can be antigravity. What the hell has C^2 got to do with a Black hole because you can pop whatever nuclear device far away from a Black hole and it would be at the most and at the worst very much insignificant. The light will not even escape from the gravity of the Black hole but that has nothing to do with the diameter of the Black hole except on the condition that they agree that gravity is the reduction of space and that brings about a link. But that they never even suggest. When this became apparent that the radius of stars reduces as the stars developed through progress, someone was supposed to say: hey there is a dead rat I smell. For my saying so I am the clown in the courtyard, and the Academics see me as the one with the two dead brains cells and have no more to use as spare.

What one can clearly see from the element table is that every element carries a different surrounding coming about because of different speeds and in that the proton number has no coherent reference value. Every element carries a different value of heat in relation to the atom solidness. Much more important than mass is the density factor which means the volume of space an element takes up in an overall distribution in a specific quantified space volume. It is the number of atoms that can occupy a certain volume of space that is much more pertinent than the mass erected. Airborne elements just cannot stick together and share space, as solids can. It is their qualities that carry the reason for a star performing and not the magical pulling of quantifiable man made mass that is so easy to calculate. Every element holds characteristics inspired by form and prefers placement and these characteristics make up a performance of space-time motion balance within the star. The way that an element interacts with heat in space and space is having heat as the containing substance where heat has as much a function to perform by assessing with duplication as does the dismissing of space have a function to perform. In that manner every element performs differently and to specific characteristics.   We are able to see some features of element characteristics versus the misconception science have about contained air when looking at the way a compressor cylinder is working.

In particles sharing space there is a motion ratio putting particles in a rotation sequence that seems to be a gear-locked action.

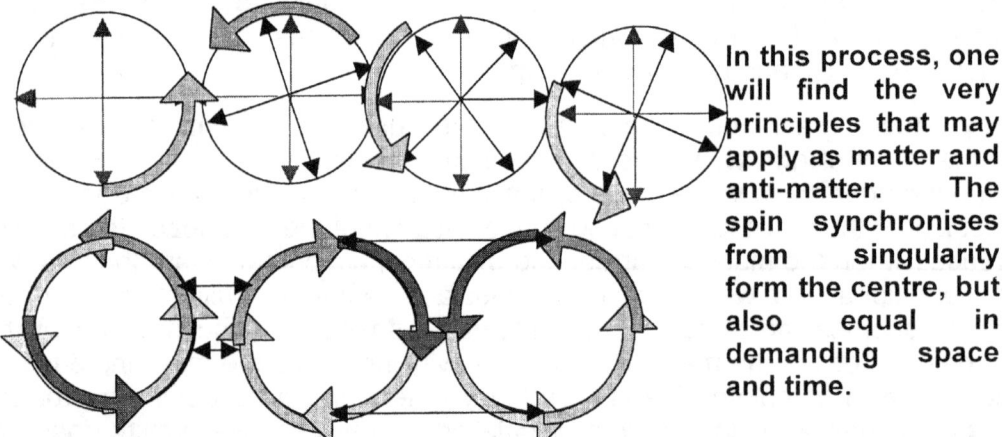

In this process, one will find the very principles that may apply as matter and anti-matter.   The spin synchronises from singularity form the centre, but also equal in demanding space and time.

At first in the beginning far back the heat was high but to have heat that high something had to be cold otherwise the concept of what is hot or cold and high or low and what is little and what is a lot in heat margins is very much meaningless.   If one takes what we have now then that being what is in place at present when compared to what was present way back only can have any meaning when comparing delivers standards and when that brings across a meaning. But what we now have was not in place back then and what was then is not with us now. If what was in the cosmos back then was being that hot then one has to focus on what is applicable now, but what is applicable now did not apply back then because what is now was looming in the background, waiting for now to arrive in the future. There has to be a ratio or a relevance that places the utmost in correlating fashion to one another. If the temperatures was 10^{34} then that which contained the heat at 10^{34} had to be at another level to produce any other field marker in reference at 10^{34}. With no zero to be found there was no 10^{34} around. If there is a zero to this day there was a zero back when things came about since nothing alone can be added or removed by science from the Universe at will and at their pleasure. The rest of whatever is and was has to remain to this day on condition that it was already in the Universe as part of what is in the Universe.

A large part of my proposed theory is that in singularity heat and space was joined as one thing in one unit. We are not aware of how or what that consists of because that with the knowledge about the margins of heat and cold joining is outside this universe. We cannot conceive what lies there because we sit within the result of what came about when space and heat parted their unified ways. That was when Creation came about and space held heat in a part being apart of space. How much ever the all-controlling Science community think them equal to God, they must swallow and accept they are not God and this knowledge will forever be beyond their inquisitive reach. Heat and space divorced the eternal marriage and space produced heat apart from space and was after that separation no longer being a part of space. The release came about when space gave birth to heat, which formed space-time in the form of structured material that remained solid or where the other second dimension of Creation lost density and became liquid heat.

At the first release of singularity creating space-time which established space-time it went about by putting all on an equal footing. The equality did not favour all and some became marginalized. Then gravity and antigravity formed and this forming brought about that some particles conformed into less dense heat, as some of the material was unable to produce gravity and secure their form. One should realise that the period, which I am referring to, took place when the electron was still inconceivable. There was no cubic universe with square edges. That came much later. Those less firm were the ones committed to motion producing antigravity. But antigravity is as much part of gravity as gravity is a part.

 Other particles formed as singularity retaining structure by their applying of gravity and constructed a valid way of retaining form as it then developed to become material by retaining a specific form that locked in heat from the other liquid sources. In the process material applied gravity as a means whereby it removed heat from surrounding space created by all singularity that was overheating. Some malformed through their motion while others contributed to contain form also by motion but the motion this time came about in the spin they accomplished around their singularity. This spin of particles through space formed gravity functioning as gravity within the space that the new particle was in as well as the surrounding space the particle withdrew heat from. As the particles removed heat from another sector it found material by which it could manage to bring about duplication in motion because the duplicating also meant the securing of material onto the form it upheld as it maintained singularity. By this application of securing form and abandoning form a trend was set where some particles was placed in relevancy and through the action of the relevancies in motion some particles could freeze heat onto already frozen matter.

Matter grew as it galvanised heat (I have to use the term galvanized because that is precisely what it is.) It is used in this sense because the process we use to bring about galvanising is the same process gravity uses to apply heat onto matter. It is for the lack of a better term or my reluctance to create one more meaningless term that forces me to use the term in hand and an attempt to prevent yet more useless naming to come into disuse. Such galvanised of heat is the precise same means used by the electromagnetic plating process but only at the rate of very concentrated gravity. The fact that we can use electromagnetic plating is proving the fact that it is a principle life may be able to manipulate but it is what nature created. Gravity, electricity, electromagnetic fields and electronics all amount to the flow or the displacing of heat. It is the very same thing with only having the concentration levels in each case and distribution of charges applied differently. This plating of heat onto space filled the already frozen solid material with more substance, which is heat that it by means of plating freezes onto and into material that then formed a sealed or frozen structure. We call this structure the protons. This process is continuing to this day as it did when creation started. Some singularity remained protected and formed material where as others was unable to bring about motion through spin or gravity but had to resolve to motion coming from expanding by overheating and relinquished form by liquefying. This statement I shall prove as the writing progresses and as I introduce more substantiating facts. To us in life this first event that ever took place is now known as the Roche limit.

One should see why did the heat come about when singularity produced space-time for the very first time. Saying this we must understand that there was two phases at least where heat came about. The first was the moving from the spot to the dot, which was a spontaneous growth but then there was the motion bringing about space. This was when the dot became gravitational or anti gravitational and each dot responded according to personal as well as group relevancies. In this second or motion heat was mainly a result of friction between created particles and exposing came due to the lack of space. If there is no space to contain the material the end result is friction between the particles. Once again the result of friction is further heating. This fact is so extra ordinary because we know that billions are spent in industry to combat friction heat. That was precisely the conditions that brought about the forming or deforming of different particles in the pre or post Big Bang second. Heat will always flow from the highest value to the lesser value. This is the concept I use for the basis of my entire theory. I base my theory on gravity producing cooling and contraction while heating produces motion by expanding and creating more space, which on the other hand also leads to cooling. By being without rotating motion cosmic objects retain heat and expand. The motion will be whether it is the motion of the unit as a whole or whether it is within the unit forcing the unit into more space. When overheating no amount of force can retain the container from becoming too little and with the heat coming about forming the expanding the space produced. With this action it will destroy the space any container may have not withstanding whatever type of material we have at our disposal and is therefore used to manufacture the container we are able to use or the container size that we can use or even the force the container in use is able to contain.

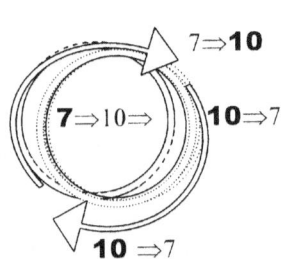

In this writing I will show how the Titius Bode principle in conjunction with the Roche limit form gravity and I show with the evidence of that how nature uses the law as a jig where it becomes a template forming space between the different planets in the manner in which they are evolving. The Titius Bode law should be seen as the process where gravity left its mark. Titius and later Bode found marks that gravity left for us in space growing but Newtonians failed to see how.

Since not even Newton could claim to know what gravity is and therefore nobody was able to read the markings of gravity this tendency was dismissed as coincidence, by those who should have investigated the process.

During Creation before the Big Bang the compactness of the particles produced motion discrepancies between different particles bringing about friction where some particles overheated and formed heat. Heat then formed space in a process that became the Big Bang where heat produced space and formed the motion we see today in the universe as the Hubble constant. In space there is different dimensions that we call sides of which there are six opposing one another. Those sides that form is forming from the heat that became space in motion but the sides play a much stronger role that is anticipated presently. The sides are dimensions and the dimensions come about since motion created the discrepancies. Let us think what the verdict at present is using current Mainstream Physics on the question why a water drop would float by forming a sphere in an outer space capsule housing people in outer space in the presence of micro gravity. Why would a sphere always form when left on its own will to capture free form?

The most important issue about gravity is that gravity is the strongest where space is the least. But when space is the least heat within that space where the strongest gravity is, is at its highest possible point. This we put to human advantage in the working procedure of the internal combustion engine. One thing that is true is that if a process is there for us humans with intelligent life to manipulate then the process is a cosmic fact used throughout the Universe in some process within the Universe. Man can create nothing and that is all man can create. Man may construct what was not constructed before, but in that effort man can only re-arrange what is already created. I suspect the process we use are the same process used by very young stars use to get heat levels rising within those young stars however the fuel that is processed is not necessarily the same. I suspect in very young stars the atomic ignition process, is similar to the process which Oppenheimer "invented" to ignite the two Japanese bombs used after the Americans got the nuclear fuel from the Germans but did no know what it was or how it worked. This nuclear manner of igniting extremely heavy elements that is found in all stars' core is mainly used in extremely young stars with a desperate need for heat and motion growth. It is reducing space to increase time coupled with non-nuclear fuel that ignites the uranium and possible natural plutonium in the core of the young star.

We all accept that the true cosmic form in Universe as a whole would have and most probably does use is the sphere...but why would the sphere form as the original form. Newton acknowledged the fact but was unable to explain it except for some vague nuance about it. If liquid material is left to a natural outcome material in a liquid state will take on the form a sphere has. While saying this we must acknowledge that all stars are only massive liquefying machines hanging in space with nothing more to do than to liquefy everything they can get hold of. With that in mind, the question about the sphere becomes extremely valid. What is that special in the sphere to secure the favourite form or form of choice of gravity in the Universe as a cosmic structure? By merely blaming gravity pulling from the centre is rather avoiding the question with simplicity because the question arising from this answer is that since we see the Universe taking on a form as a sphere then where then can we locate the centre of the universe? When applying Newton's standard cosmic formula whereby it is used to calculate the gravitational field the formula in use is $F = G (Mxm)/r^2$ the issue in the formula becomes the square radius and the radius is forming the centre whereto the mass directs its full force compliment. The more growth by which mass increases and the more mass there may come about the more the pulling motion applies by the centre towards the centre that will secure the motion of mass moving towards heat within the centre. The more mass there is bringing about the pulling force of gravity pulling that secure all gravity force within the centre by the radius square the stronger the domination of such a centre will be on the surrounding space-time. By the securing of a square radius in the centre the reducing of k will produce less space occupied a^3 but much denser space occupied bringing about a much stronger rotation of time T^2.

The piston in the exhaust/ intake position stroke

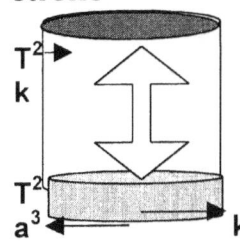

Compression / ignition

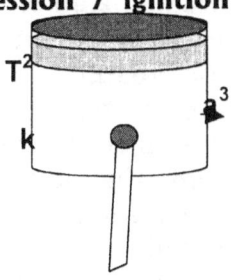

One must see the piston engine in the same manner as stars in operation or the Big Bang. By reducing the stroke **k** we find the space a^3 decreases as the temperature T^2 rises in ratio. Take this scenario and compare that to conditions applying during the period we think of as the Big Bang.

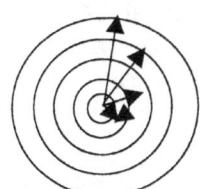

Compare the cylinder conditions to the conditions of the Earth in comparison to outer space or any star. As **k** reduces from the outside towards the centre the space represented by the declining **k** also reduces in ratio and by ratio equal in measure. However, a smaller distance **k** brings about more heat concentrated in the reduced space. If a^3 reduces then **k** must also reduce but then T^2 holding the gravity or heat concentration must increase. That is just what is happening inside the engine and inside the sphere and with the Big Bang where with the Big Bang the situation is reversed. This is what Kepler said with his formula $T^2 = a^3 / k$. If **k** gets smaller it will increase T^2 as much as it decreases a^3

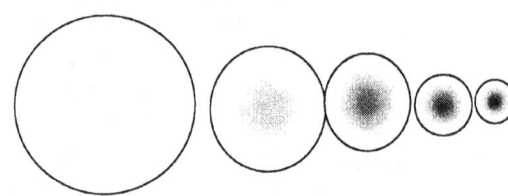

The closer **k** reduces space towards the centre of any star the hotter the star is. It is the same with the engine. The internal combustion engine teaches us the implication of gravity.

By reducing the **k** factor the **a³** factor also reduces but the heat in **T²** surges. The power computed by the engine is covered by the principle that compressing space is reducing space and that generates the heat required to ignite or contribute to the igniting of the fuel. The space has to turn liquid to command the maximum power the machine produce.

When investigating the process in all cosmic objects the gravity collects heat and accumulates the collected heat in the centre of material forming a sphere. All structures with a superior gravity collects heat and more so progressively towards the centre because in all objects with a strong gravitational field there is a collective quantity of heat towards the centre of the object. It does so by concentrating space and as such produce heat. By producing more concentrated heat it secures independence to the unit forming the individual container unit and the independence will produce motion to move away from the singularity in control. Gravity has two factors influencing space, which are a straight-line **k** and a circle going around the centre T^2. It is a balance $a^3 = kT^2$ that forms $k^0 = a^3/k\ T^2$

Gravity T^2 increases as space a^3 reduces by the reducing of **k**. Much more to the point that implicates gravity within the star would be the heat the star concentrate within the centre of the star. Such heat shows much more relevancy than the mass it holds in the sphere of the star. The more space there is between the particles forming the mass within the sphere of the star the less the gravity output is because the less motion such a star can achieve and the less space the star will dismiss. We should find more collaborating evidence when we investigate the working principle behind the eternal combustion engine.

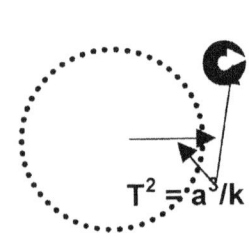

$$T^2 = a^3/k$$

A smaller factor **k** will result in much more compact area. The smaller **a³** becomes the larger **T²**. This will be from the reducing by a larger k. The compactness is the result of the density of the atom distribution increasing and by such increasing the mass or occupied space becomes denser and more intense. With that kept in mind it is quite silly to go about measuring the diameter of any star and from that detail, try to determine the "mass" because the diameter could have reduced a billion times with the motion that the inner star achieved.

The space reduced within the star centre is existing space that was the result of the left over of material, which was unable to establish gravity and reduce therefore established heat becoming space. The space that the star then moves into is that space which the star then contracts. It is existing space made redundant by motion applying the gravity the motion command. When Mainstream Science investigates stars and finds the true giants being very small as well as very massive every one acts surprised. I wonder why...? Has it got anything to do with the ignoring of Kepler as the cosmologist? I wonder why...? Has it got anything to do with the ignoring of Kepler as the cosmologist by favouring Newton as the mathematician? This question is supported by another question.

Are stars really about mass producing the fundamentals or are stars about reducing space by gravity to fit into space more particles taking up less space and thereby producing the more mass which then is the result of much denser particles holding more mass by more compact particles because of higher motion achieved? Is the increase of the density of the atoms reducing the space but at the same time increasing the mass claimed in space occupied by material? Can it be the result coming from stars with much higher motion potential producing much higher gravity?

We know that the Black Hole is spinning the oncoming space faster than the speed of light... and that means the gravity is contracting faster than the speed of light... As gravity intensifies in true large stars the space vanishes from the equation where the star becomes forever smaller. That means gravity destroys space but does not pull matter. We all are very aware which star in nature is the mighty gravity producer. It is the smallest one as far as it holds space. So then has mass the least say when gravity is generated? But it also is the one producing the most motion. That means gravity is not about space and all about motion. It seems most likely to be true. In such a star there is no more mass as all the mass turned into momentum. Later findings proved Kepler as being astonishingly correct. The diminishing of **k** will reduce space **a³** but it also will advance gravity T^2. It says so when reading mathematics correctly. By redirecting Kepler's formula mathematically: $T^2 = a^3 / k$ shows that gravity grows where space demise. Every time the diameter diminishes the gravity produced by the star becomes immensely more. The space claimed decreases as we increase the motion with which the star holds relevance in the space it occupies and which it claims.

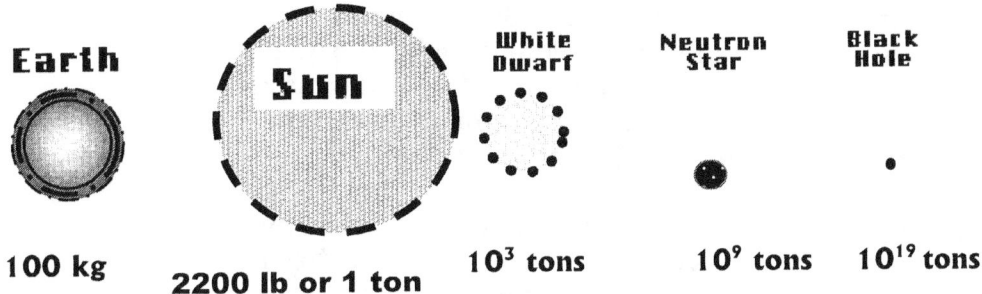

Earth   **Sun**   **White Dwarf**   **Neutron Star**   **Black Hole**

100 kg   **2200 lb or 1 ton** 10^3 **tons** 10^9 **tons** 10^{19} **tons**

The Oxford dictionary of Astronomy defines Gravitation Collapse as follows
The collapse of a body that is unable to support itself against its own gravity. Gaseous bodies undergo such collapse if they are not hot enough for their gas pressure to balance gravity. This can happen in the early stages of star formation, or when nuclear burning ceases in a star's core. The time taken for such collapse decreases rapidly with increasing density, varying from about 100 000 years for the birth of a new star to less than a second for the formation of a neutron star. Star clusters may undergo a similar collapse if the random motion of their constituent stars is insufficient to offset gravitational effects, either during their formation (see violent relaxation) or at an advanced stage of their evolution.

If mass was an issue in the applying of gravity then the mass will play a major part in the time it takes a body to release from the motion when releasing from the Earth. Mass increase when accelerated and in that then the body must actually grow bigger to have more mass. By increasing motion the actual virtual mass increases therefore motion can increase mass by gravity increasing. One has to accelerate an object to navigate release of the Earth. But in such acceleration the mass increases many times over but notwithstanding that mass increase the accelerating motion secures the release in any way. The difference in velocity that the star produce and the matching velocity the object in outer space must produce in motion to match the motion requirement of the planet's motion relating to the outer space factor such motion creates a velocity differentiation but why that takes gravity is one of the biggest mysteries I can find in the Universe. Again I must press the issue that since Galileo proved otherwise where otherwise is being that what Newton stated everyone agrees with Galileo and then completely ignore Galileo. Galileo said mass fall equal...that means mass draw equal notwithstanding mass differentiation in size or mass and while every one is admiring what Galileo said then by the same margin every one is completely ignoring what Galileo said. Newton said Mass one and Mass two draws space according to mass, which means mass has the ultimate influence in the process.

Galileo said all fall equal notwithstanding mass...Newton said mass does all the pulling and Galileo said mass has no influence on the drawing of the object...yet everyone is totally ignoring the fact that mass, be it big or small, proves not to influence the drawing of material that is falling! If Galileo is correct something else than mass is doing the pulling...and for the life of me I can get no Scientist to see what I say in the contexts that I say what I see. The concept of mass being the producing factor in gravity comes across as rather less thought through and more than a bit silly when considering the above. All evidence points directly to the idea that mass generates gravity and such presumption is most inaccurate. It is completely incorrect to think that it is mass that produces gravity since it is a notable fact that as space shrinks or reduces the stars ability to generate gravity excels. But space has   little   to   do   with   mass.

Red giant

Betelgeuse

Dia.   1400000000 km

Diameter relevancy is

35.2 km.

Before we can ever dream of understanding stars we first have to define what the purpose of stars are. What is the role that stars play in the evolution of the Universe? Even before doing that we have to define the role that the Big Bang played and what brought about the Big Bang The Big Bang was about producing space-time and by space-time developing stars release from galactica. The galactica produce the space – time from where stars arrive into the Universe.

Gravity is strongest when and where space is least. Where space is least motion is producing most relevancy and change. More gravity leads to smaller space and in such a smaller space the bigger gravity can hold more shrunken mass. This is even truer, since motion produces gravity. With the stringent reduction in space more particles will fit into less space and by having a smaller space it can manage a higher quantity of particles. But in the end the reducing of the space held by the material in the star is the result of motion contributing to space decline By producing a larger area to fill by using a specific space occupied the space occupied will have to spread thinner to compensate for the larger area commandeered to fill $a^2 = T^2k$ should T^2k increase a^3 must reduce. By this it is clear that even referring to a force applying is directly suggesting motion occurring or a tendency of a serious effort to bring about motion restrained that then is a blocking of motion occurring and that makes that mass is the restricting of motion trying to continue in a specific direction to come about. If one takes away the motion or the tendency to bring about motion then it is clear that gravity disappears. The Black Hole contributes the strongest gravity since the Black Hole places all motion in space and no motion in the star. Nevertheless the motion we see comes as a response to that whole ending of space-time is in essence the product of the motionlessness of the star. Since all stars apply motion and the Black Hole reveals the ultimate form of motion the Black Hole shows that gravity is space in motion by reducing of space towards the centre where there is space less ness and motionlessness. In the invisible centre is a point even beyond where space and motion is at its least. Without space there is no motion and without motion there is no space. Kepler tells us this as $k^0 = a^3/T^2k$.

Yellow dwarf

Sun

**Dia. 1400000 km.
Diameter relevancy
is 38 meters**

White dwarf

**Dia. 16000 km
Diameter relevancy is
300 mm meters**

• **Neutron star**
Dia. 19.2 km. Diameter
relevancy 3 mm

. **Black hole**
Dia. 9.8 km. Diameter
relevancy 1.5 mm

It is in this vain that we have look in order to find the essence that stars have as a product in the Universe. The star is a contracting space-time devouring machine that combines and unites atoms. It is about reducing. The more atoms the star unites in a process called fusion, the less space will be available to occupy by the local inhabiting atoms. The bigger the atom numbers are in proton clusters the more gravity the star may produce because of having less space to deal with. More gravity is equal to less space holding material. The purpose of the star is reducing what the Big Bang was increasing, because we are still in the Big Bang and we are still the Big Bang that is moving away from a center towards another center. The less space a star has the more gravity it has and the more gravity it has the more it will reduce space-time.

However a point arrives where the star reduces more space than what the photon can produce. At such a point the contracting motion within the star, which is gravity exceeds the motion that light can achieve. Bang goes Einstein's prediction that nothing can go faster than light and therefore the speed of light is equal to time. In the process we see stars develop way beyond what the Big Bang restricted the Universe to have. These stars are at first in a desperate struggle with light and we see that eventually they develop past the barrier that light provide. In the end the star is about motion becoming motion less by compromising space. From the evidence stars leave us we can again see gravity is motion by movement through time as much as it is motion by the reducing of time.

If one is infinitely small, all is infinitely small. There can be no space when motion is at its slowest possible speed. That forms the ultimate relevancy available and space-time or space in motion is all about relevancy. But the enormous gravity falls outside the star. In the Black Hole singularity controls matter and space applies all motion that is in fact the time factor to space occupied where the motion aspect is more commonly known as gravity. But the space less ness of the Black Hole shows that space less ness is the location of strongest gravity. It is in the place that the heat is the most, which is in that centre area of any sphere. If any one does not believe me then test nature. It means that mass has the least say when gravity is generated. According to Kepler mass in motion within space in motion and gravity is the same thing. $a^3 = T^2 k$.

The motion that the Earth resists which fills the space that applies the effort to move is the result of what ever is in effort of trying to commit motion. It is precisely what Kepler said when he said space-time is $a^3 = T^2 k$, with the only difference being motion that can also be a tendency and not only an established move. Looking at evidence we find in the cosmos it seems most likely to be true but the radius **k** distinguishes the space a^3 required and the gravity T^2 produced. $k= a^3 /T^2$. Later findings proved Kepler as being astonishingly correct. The diminishing of **k** will reduce space a^3 but it will also advance gravity T^2. $T^2 = a^3/ k$ shows that gravity grows as space diminishes. Every time the diameter **k** diminishes and the space a^3 acquired by the star reduces the gravity T^2 produced by the star becomes immensely more. In Kepler's formula $k = a^3/T^2$ the smaller **k** becomes the smaller a^3 becomes and the bigger T^2 then gets. T^2 represents the gravity that positions the space a^3 at distance **k** from the centre capturing the structure through gravity applying T^2. Looking at the illustration of the comparison between the mass and the available space in stars we have to come to a conclusion that we all are very aware which star is the mighty gravity producer. Now after some rethinking we can again ask the same question we asked a while ago.

Looking at such clear evidence one is struck by the question that we arrive at next: Has mass then the least say when gravity is generated? I would say that mass is the result of motion breaking down and therefore is a factor of and not a factor presenting gravity. The more gravity reduces space is the result from the accumulative effort of every individual atom according to proton mass (number) as a unifying effort of all the atoms in the star in accordance with mass applied. The idea preached by those that should know better is that mass is the same everywhere and is never changing. Why would there be such huge mass increases in the bigger or should I say smaller stars. Mass is supposedly another constant because a body has the same mass even in outer space where it absolutely has no mass. What would entice the material inside let's presume for instance the Black Hole stars to grow more massive if mass comes about from the pulling of one particle closer to the next particle. If it was about pulling on each other the mass of the particles could not increase through applying such a method because the honeycomb would grow bigger in size and not smaller. Even by combining the mass of two individual atoms the increase is already in the equation.

By reducing space to the ultimate in the centre of the star it happens where space is little and where gravity peaks and where the heat sour to the ultimate. As heat rises and space reduces it is surely true that the particles in such an area must take up less space so that more particles (atoms by numbers taking up less space by volume) will fit into less space by reducing the inside volume of atoms and thus in occupation totalling they then at that point in space-time are taking up less space. After all such evidence proves that more material fit into smaller space. This establishes the cosmic situation such as the conditions we are able to calculate showing what was present during the Big Bang. What was then part of the Big Bang is now prevailing in the centre of the massive star.

A star is the cosmic device that is placed in space to reduce space back to what was the case before the Big Bang event and so the Big Bang is once more repeated but this time the star is taking the Big Bang in reverse by compacting the space within the star. Within the space within the centre of the star being as little as possible the heat is as dense as possible but in that scenario more protons pack a smaller space and more protons in a smaller space accumulate more motion per space unit. An electron accompanies every proton and where the proton represents the flow of space towards the centre, there is the electron responsible for the duplication of space by motion. This brings about more space diminished as more space transforms to more heat in less space. While the star is reducing space the Universe is expanding space to comply with cosmic equilibrium. This is in place, since the heat that the star gathered through gravity will return back to the centre of space less ness and motionlessness where it was when Creation started. If the heat accumulation by motion control is not well controlled by the governing singularity we then find Super Nova events occurring.

EXPANDING UNIVERSE
Any model universe in which the space between widely separated objects is expanding. In the real Universe, neighbouring objects such as close pairs of galaxies do not move apart because their mutual gravitational attraction exceeds the effect of the cosmological expansion. However, the distance between two widely separated galaxies, or clusters of galaxies, will increase as the Universe expands.

KANT, IMMANUEL (1724 – 1804) was the German philosopher, which proposed a cosmogony, published in 1755, in which the Solar System forms, via a disk, which condensed out of primordial material. The Solar System was part of a larger system (what we would call a galaxy), and many of the nebulae seen by astronomers were in fact other galaxies, which he termed island universes. Kant was influenced by I. Newton's theories, which he termed island universes. Kant was, as everyone ells up to now, influenced by I. Newton's theories and by the English philosopher Thomas Wright of Durham (1711-86). This did deliver the ideas but still we have to be practical.

Where all motion is ending in circles, which is the result of gravity, it puts everything in circles that are evidently growing by expanding. We have to venture back into our past. As we venture back, we must take a circle along to find that the circle is representing the cosmos and reduce the circle to see what comes from there. Of all the models prepared by all the wise during the past this model was never made in reducing the straight line that runs between intervals in the Universe. That is what I did and by following the suggestions arising from Kepler's formula I uncovered a hornet's nest. The hornet sting depends on the individual person's view of what a sting is. The sphere as any circle and all circles does have two parts. One part is a line that indicates the distance between the two opposing sides being 180^0 apart. The second factor is the pi indicating the result of the square value of the other factor being the straight line which by the square produces the measure of the shape it indicates. But where as in the normal square the line matches and have angles it connects touching lines that connects one another and in the square one will find the surface of the cube in the lines. With the circle in comparison, the square falls inside the circle but the lines cross at an angle and at such a crossing of the lines in the very centre, the circle takes form from the securing of all possible corners and contains them in one point. The absence of edges hold the form value of the circle being a^3 coming about as pi cuts corners literally. But the overall prominence comes from the fact that the circle holds a cross inside the very centre as the diameter then doubles by the square and a square never crosses on the outside as in the case of the square where the lines touch at angles. With the square we refer to the lines indicating the borders of the square forming the sides that is the borders and indicate the final measure of the square or the cube.

The sides we face as in example length and breadth that becomes multiplied and form the square or in the case of the cube, the cube. In the case of the circle we use the lines crossing to the inside on the length they represent being a full line or half a line. In the case of the full line the name used for such a line is a diameter and in the case where only half the inside line of the circle applies we gave it the name of a radius. Being fully aware of the various names I prefer to use r to indicate any and all lines that forms a combination to indicate either surface or volume for either the square or the circle.

Whenever I refer to any straight line I would use r as a symbol indictor. To go back into the past where we may find the start of the Universe one have to reduce the radius because the radius of the circle indicates the size while pi indicates form. The Big Bang is all about the cosmic radius that expanded bringing about the filling of more space. By reducing the circle through the radius it becomes another matter of eliminating form because in such reducing of the radius it then becomes a matter of reducing the influence of the straight line. When reducing the circle in size one have to reduce the radius or the diameter because the pi factor is the indicator of the form as being a circle. It then becomes a process where it is just dividing the radius defined by the symbol r by halving the answer of the previous value of r every time until there can be no dividing any further and such a reducing cannot end by becoming zero because in this process of reducing nothing disappeared. The value of r reduces, although by reducing r did not vanish from the scene.

The lines just went smaller and smaller but if it was part of the Universe then or now it is still part of the Universe as it has nowhere to go but remain in and part of the Universe. One may divide by two, halving the result every time, which is a normal mathematical expression and by reducing by half the process can never reach zero. The process leads directly in infinity. Allowing zero to be accomplished through any legal mathematical equation is not a mathematical fact and I challenge any person to prove such a feat mathematically. I challenge any one of those that answered me blaming me of incoherency to that any diversion of any number can end and prove that the line is then mathematically ending in zero. The numerical procedure may become tiresome or too small for any human to make sense of the outcome or the mathematician may lose all senses of his awareness, but never can such a dividing bring about zero in the ultimate answer when the straight line is reduced eternally to end in infinity. Zero cannot divide nor can zero multiply therefore this statement eliminated the use of zero anywhere and exclude a start at zero should I wish to retrace my steps to where the cosmos started. In using the method of the dividing by reducing the answer to half the size it will forever allow such a process to continue without ending in zero because no matter how small, the next value will be dividable by two in that continuous reducing by half of the straight line will forever allow to be a value in the next place.

This stands as a mathematical fact and I do not have to prove my statement but those persons in academic positions, which portrait me as a person trying to sway facts to fit my convenience, must explain how one can multiply distances filled with zero as a concept and get a valid mathematical answer other than zero. I know that since they are the Official Academics and I am not, they can be incoherent and be blameless all in one go but they also are in a position to be incoherent and put the blame of incoherency back on me. By placing zero in as a multiple factor anywhere or either at the start or the end will remove the value of the calculation. Space cannot grow or conclude by the use of the value of nothing. If you put zero anywhere, that act disqualifies the use of mathematics. This alone is the biggest obstacle why there is so little exploring going about the space expansion present overall. In my dealings with Mainstream Science in the past Mainstream Science becomes truly all out aggressive with my trying to indicate that I am criticizing Newton and one such a fact is his way of producing zero as a result of a circle forming an end and at the same time starting with another beginning where after completing the rotating circle the final product is zero. I do not try to criticize Newton on any facts except on what he saw in Kepler's work, which he tried to retranslate but through an attitude failed to realise what the meaning was which Kepler uncovered.   I also criticize the need Newton found to change Kepler's work. Do you as the reader realise that Johannes Kepler introduced an astonishing presentation about space-time, singularity, gravity and the Big Bang in the years of his life between (1571-1630).

By applying his formula we find some answers about questions yet unanswered. I raise the questions and with much study received new answers through dissecting what Kepler introduced to science. It was only possible when taken from the mathematics, which Kepler introduced from his cosmic findings. Kepler's answers were lost so many centuries ago when some Englishman saw him wise enough change Kepler's work and bring in the alterations as he saw fit. All Newtonians followed the lead and continued his act.   I am going to show that the changes brought about in Kepler's work were unnecessary because the changes were unasked for. Such forced changes about the work of Kepler by Newton came without the modern demand on proof because such proof was back then and still is lacking **the proof** in considering to what extent the accepted norm is of proof in relation to what was required and accepted back then and compare that in the light of what the demand is on establishing absolute proof at present. Back then a man's status was guarantee of proof while today there is no comparing to what is necessary in the terms that is accepted norms today.

I have grown accustomed to outright rejection of members in mainstream physics because of my criticizing of science. Sure they can dismiss me in the light of my criticizing them on science but that they do without explaining why Mainstream Science makes such a performance about Einstein and his views about relativity, when Kepler said it far better and with more to the point explaining and with much less performing. Kepler did the explaining about four hundred years ago. By accepting Kepler one has to accept space-time, Kepler said space-time $a^3 = T^2 k$. Kepler said there can be no space if there is no time. Kepler said space is the equivalent of time. Kepler said when space came about so did time. Kepler said space is time and time is space and the one cannot be without the other. Kepler said space is motion in space in motion, which is space-time. Kepler said a^3 could only be present when it is equal (the same as) $T^2 k$ performing in the time aspect. Kepler reported on what he was told by the cosmos. Newton had no need in changing anything because by changing he brought about much misunderstanding as well. Kepler showed what time is but all Newtonians admit that not one Newtonian knows what time is. Time is the motion of heat in space where material is heat and space also is heat in some other composition. That means heat is space and space is motion filled with heat. Time is the spin or motion of heat in space. That means there is no greater all preserving Universe out there. Every point k^0 will establish space a^3 by applying motion $T^2 k.$ There is space filled with heat and the heat applying motion at the point where space is. There is the space $a^3 = T^2 k$ in motion and the motion is time. There is only space-time. There is no liberated space without the restriction of time. There is no space restricting a Universe without the liberating of time through motion. That is what Edwin Hubble saw through his lens. But where there is motion there too is gravity or antigravity whatever relevancy one wish to apply. Motion is gravity, which is motion that is committing space to form in the presence of singularity and that too is antigravity.

The man that (to my humble opinion) took cosmology into a new dimension was HUBBLE, EDWIN POWELL (1899-1953), the American astronomer. He first studied nebulae, concluding in 1917 that the spiral-shaped ones (which we know as galaxies) were different in nature from diffuse nebulae, which he found to be gas clouds illuminated by stars. From 1923, using the 100-inch (2.5-m) telescope at Mount Wilson Observatory, he resolved the outer regions of the spiral nebulae M31 and M33 into star, identifying over 30 Cepheid variables in them. This proved that such 'nebulae' were truly independent star systems like our own – other galaxies. In 1925, he devised the so-called tuning-fork diagram of galaxies, dividing them into ellipticals, spirals, and barred spirals, which he believed to indicate an evolutionary sequence. By 1929, Hubble had good distance measurements for over twenty galaxies, including members of the Virgo Cluster. By comparing distances with their velocities, as revealed by the redshifts in their spectra, he concluded that galaxies were receding with speeds that increased with their distance, a relationship known as the Hubble law. This was powerful evidence that the Universe is expanding. The dynamics of his work was so far reaching everybody (including Einstein had to revise their theories to accommodate his findings. His findings are the most disputed, undisputed observations in all of history. The HUBBLE CLASSIFICATION is a widely used system for classifying galaxies according to their visual appearance, illustrated on the tuning-fork diagram. The sequence is based on three criteria: the relative sizes of the central bulge of stars and the flattened disk; the existence and character of spiral arms; and the resolution of the spiral arms and / or disk into stars and H II regions. The system was originated by E.P. Hubble.

The sequence starts with round elliptical galaxies (EO) showing no disks. Increasing flattening of a galaxy is indicated by a number which is calculated from 10 (a − b)/a, where, a, and b, are the major a minor axes as measured on the sky. No elliptical is known that is flatter than E7. Beyond this, a clear disk is apparent in the ventricular or SO galaxies. The classification then splits into two parallel sequences of disk galaxies showing spiral structure: ordinary spirals, S, and barred spirals, SB. The spiral types are subdivided into Sa, Sb, Sc, Sd (Sba, SBb, SBc, SBd for barred spirals). With each successive subdivision, the arms become less tightly wound (but more easily resolvable into stars and H II regions), and the central bulge becomes less dominant. Two types of irregular galaxy are defined. Irr I galaxies show rather amorphous, irregular structure with perhaps a hint of a spiral arm or bar, and can be placed at the far end of the spiral sequence. Irr II galaxies are sufficiently unusual to defy assignment to any of the other types, although this category encompasses only about 2% of bright or moderately bright galaxies in the nearby Universe. The original, erroneous idea that the sequence might be an evolutionary one led to the ellipticals refers to, as early-type galaxies, and the spirals and Irr I irregulars as late-type galaxies. Colour and amount of interstellar material vary systematically along the Hubble sequence: ellipticals are red and contain little interstellar gas or dust, whereas late spirals and Irr I galaxies are blue, with significant amounts of interstellar material. The relatively faint dwarf spheroidal galaxies were not recognized as a separate type in the Hubble classification. Some variants of the Hubble classification use plus and minus signs to subdivide classes, so that Sa^+ is later than Sa, but earlier than Sb^-. The importance of the HUBBLE CONSTANT is still to this day, underestimated. This "constant" is well explained, for the first time, I might add, in this book. The Symbol H_o is the figure that relates the speed of an object's recession in the expanding Universe to its distance in the Hubble law. It represents the current rate of expansion of the Universe. This important cosmological parameter is usually measured in units of kilometres per second per megaparsec. In the Big Bang theory, H_o varies with time and it is therefore more properly known as the Hubble parameter. Its value is not accurately known but is thought to lie between 50 and 100 km/s/Mpc, recent research tending to favour values towards the lower end of this range. In the HUBBLE DIAGRAM, a graph plots either the redshift, or velocity of recession of galaxies against their apparent magnitude or distance from us. The Hubble law appears in the form of a straight line on such a plot. The original diagram, presented by E.P. Hubble in 1929, was the first indication that the Universe is expanding. The Hubble diagram is mainly used to test the geometry of the Universe, since at large distances any departures from the simple linear form of the Hubble law should show up as a curve. The HUBBLE FLOW is the general outward motion of galaxies resulting from the uniform expansion of the Universe. All motions lie in a radial direction from the observer, and the velocities are proportional to the distance of the galaxies. The real pattern of galaxy motions is not exactly of this form, particularly close to us, because of the mutual gravitational interaction between galaxies; some nearby galaxies are even moving towards the Milky Way. At large distances, however, the discrepancies are small compared with the Hubble flow. All these findings are incorporated in the HUBBLE LAW, which is the mathematical equation of the principle law that governs the expansion of the Universe. According to the law, the apparent recession velocity of galaxies is proportional to their distance from the observer. In mathematical terms, $v = H_o r$, where v is the velocity, r the distance, and H_o the Hubble constant. The law was put forward in 1929 by E. P. Hubble.

The HUBBLE RADIUS is a distance defined as the ratio of the velocity of light, c, to the value of the Hubble constant, H_o, This gives the distance from the observer at which the recession velocity of a galaxy would equal the speed of light. Roughly speaking, the Hubble radius is the radius of the observable Universe. Depending on the precise value of the Hubble constant, the Hubble radius lies between 9 and 18 billion l.y. This data is the basis on which the age of the universe depends and is the HUBBLE TIME. The time required for the Universe to expand to its present size, assuming that the Hubble constant has remained unchanged since the Big Bang. It is defined as the reciprocal of the Hubble constant, $1/H_o$. Depending on the precise value of the Hubble constant, the Hubble time is between 9 and 18 billion years. In the standard Big Bang theory, the actual age of the Universe is always less than the Hubble time, because the expansion was faster in the past. The Universe is a combination of many material formations holding positions in space. Some of such material was covered in the blanket of heat, distributing into more spacious surroundings as the material expanded from the centre flowing outwards. Hubble's constant is proof that the space between cosmic structures are departing from many centre positions between such objects and this is a trend being located between all the objects through out space but also indicating a definite growth in the radius and such radius growth follow a patter where the growth seems to flow from any such a centre point away from the centre. Without the absolute and undeniable proof coming from the Hubble constant bringing proof beyond any possible doubt in any one's mind that expanding is very much and a very big part of all Cosmic activity the accepting of the Big Bang would not be in place. $F = G \, (M_1 \times m_2) \, / r^2$ is in essence a big issue about contraction while Hubble showed the space was not dividing. The space was multiplying. The stars are growing apart and so is the galactica. This then brings in the question of space available. The discovery placed mainstream Science in a spin that still spins the daylight out of all accepted cosmic conclusions. It now even initiates a search for dark matter and all theories are in place to cover the egg on the faces of Newtonians while Newtonians only have to look at Kepler to find the correct answers.

With your accepting of modern cosmology your deciding of accepting the Big Bang principles and therefore rejecting all the other cosmic possibilities being proposed in the past as such factors coming from say the steady state Universe must be reviewed. Mainstream Science still cling to views, which still form part of other theories that now must be rejected in view of the accepted Big Bang theory. The accepted mainstream views therefore should be rejecting all the other possibilities that may lead to double viewpoint or incoherence. Then that will also bring along a new perception about cosmology that could be endorsed by the minds of the common folk of which I am a member. The changes include altering the widest picture to match the Big Bang concept and only the Big Bang concept. Such a new perception must include a new perception about Kepler. In your reading of this letter and seeing what I try to show, I shall lead you along as you will investigate Kepler and such investigations will introduce facts. The facts and views I introduce may seem as if no one before came to realise these facts belonging to Kepler. But then one feels surprised that, in the past the views were bluntly ignored, however obvious they seem to be. It becomes clear when we go about analysing Kepler's formula somewhat differently that the Big Bang theory might be exceptionally correct in the way it is presented. But at the same time the Mainstream view is incredibly flawed when an overall view establishes a universal picture. For instance Science accepts that space expands but science do not permit any view about material expanding. Science will not commit their view to the accepting that the two are linked. Try to fit the Earth as it is and the moon as it is in the present or if you wish to split hairs the material used that produces the size of the structures in their current form. The atoms forming the star must be holding space-time and when the size the atoms hold at present is projected to the past and placed in the past size Universe we had back then one will quickly realise that it was not that long ago that the two structures are holding the volumetric space that they currently do could fit an entire Universe into the space they have now. What we find in the solar system at present was representing all the available space in the entire Universe. Understanding the cosmos helps one to realise another part of Kepler for the first time. The realising involves issues that seem beyond answering given present facts because at present science is without such analysing of Kepler and therefore beyond the realising of such evidence. By scrutinizing Kepler again we find that when re-aligning such an attempt to reinvestigate Kepler and what Kepler really said. Without investigating Kepler, that which Kepler said now forms the unknown part of science. From the investigation the insight gained becomes a new vision and all the unknown becomes surprisingly clear, thanks to Kepler. But it insists on a divorce separating Kepler's ideas from Newton's ideas about Kepler's ideas. It is necessary to give Kepler the recognition as a mathematician without Newton belittling Kepler's skills as a cosmologist and a mathematician. The difference is about finding what Kepler really brought to science in relation to what Newton saw what Kepler brought to science, while Newton had no vision about Kepler's work at all. It also puts a new appreciation on what Kepler said and diminishes Newton's effort to change what Kepler introduced.

The expanding of space a^3 is the establishing of space a^3 from a specific point at the length of **k** by the applying of T^2. The motion Hubble saw was space being established by time. It is how that every spot within motion around an invisible group of dots redefines a new time T^2 releasing space-time in the presence of all aspects of the visible Universe in that region. That region is the only Universe there is in relation to that specific Universe centres. It was space-time because without space there is no time as much as there is no time releasing space from singularity by means of forming space-time $a^3 = T^2k$. Science was all impressed with Einstein's effort and some are still sceptical of Einstein committing space to time. Their scepticism runs so wide that there is still, at this time of our liberated age a group that in wisdom rejects Einstein's space-time. They refuse to connect the Black Hole as being the most poorly understood star structure to some situation that was apparently part of the Universe in the beginning. It captured the space-time before the Universe went exploding with space-time expanding. They propagate a limited space Universe wherein that era there was no space-time and refuse to connect the limit ness of space being in the beginning to the space less ness within the

space that singularity holds in a Black Hole. The problem about this poorly understood or misunderstanding about space-time brings on a concept that space-time is something we must rather use as some negotiating method to a situation that either was or will become part of the cosmos because no one knows where to place it in the current standings. We are surrounded by space-time.

I have heard cosmologists declare that we must "somehow" not think of space but rather think of space-time, but while saying that they mildly refer to something to say to please Einstein and rather nothing else. There is no corroboration between connecting space-time and reality in their explanation of what is behind the suggestion of space-time. They go about presenting space-time in such a manner as if they wish to promote an idea that will install a fashion or a trend to secure the use there of. The impression I got when I listened to those in the profession of cosmology when they talked about the using of space-time was as if an idea was promoted that was loosely connecting with something bleakly understood. It was as if they were promoting some trendy fashion amongst students without the students or the academics teaching the students seriously understanding why. It was as if they were trying to convince themselves before convincing others about their belief in and understanding space-time as a cosmic reality. They were in method more about self-convincing in their attempt in promoting space-time. It was as if to please Einstein when they were committing themselves to the promotion of space-time and their effort in the use of space-time as a concept. The way they put it was equal to promoting a fashion or the start of a new trend. During the three hours of the debate the three professors failed to give one motivating reason why anyone should start using space-time as a term instead of just space. They miserably failed to convince that they knew what space-time really was about. It is as if they were comparing the use of an expression such as "groovy" to the use of this new concept of space-time and to talk about a statement called space-time as being very similar. What was very apparent during the debate was that these people lacked understanding of the concept of space-time themselves, which is why they were unable to define it. Being Newtonians, they have not yet realised that space is the motion thereof and time is what it takes to bring such motion into a cycle where the motion of the space starts and ends at another point. Since motion from singularity is bridging one Universe to the next Universe the time it takes is in the square. That which such motion crosses irrespective of size is the bridging of space using time to go from singularity to singularity and that entails the crossing of an entire Universe. Motion establishes space-time by cyclic periods bringing about the motion of space in motion that contributes to the forming of time. They're referring to space being space-time in a fashion statement has very little convincing about their sincerity of understanding all the reasons why. This they say while the man that started cosmology some four hundred years ago introduced modern cosmology as space coming about through time. Kepler said in his formula that if there is space a^3 then that space is in motion T^2k. He declared space could only be if space was liberated and restricted at the same time by motion and the motion establishes time as a factor being as much part of space as space is part of the concept forming motion. But in the very same event the motion is the restricting of space by singularity forming time T^2. The return of space to singularity and the returning becomes a rotation forming the second part of the time factor. While space is the liberation of singularity through motion k that is part of time, the restricting of the liberation of space from singularity is time T^2 in which singularity achieves the return of space to singularity. In this comes about four cosmic laws, which I named the four cosmic pillars. 1. Titius Bode Law; 2. Coanda Effect; 3. Roche Limit; 4. Atomic relevancy.

Space is the forming of motion **k** because space is the liberating of time from singularity where the hottest part of space will find a way to move away from the rest of the cosmos and that motion forms the time component **k**.

When accepting Kepler's work as the very basis of cosmology one has to accept that space is time and time is motion and motion is heat and heat is space. The one cannot be without the other because time and space is the very same thing, it is space-time and it is $a^3 = T^2 k$. Singularity is a point where there is no space and that point can never have motion simply because there is no space to rotate about. Thinking of the cosmos must exclusively be in the form of space-time and in accepting there can be no space if there is no motion causing time. Science must review the past where concepts still hold questionable dogma by scrutinising previous concepts. If the proof that was presented then is accepted as unquestionable proven fact today to the same degree as demanded of scholars today it can again be given to scholars and only then should it be passed on as accepted evidence. But in cases in modern science where there is such evidence but the unquestionable proof was never produced in the past researchers should go back and investigate how the old dogma fits into modern evidence. It has to be investigated notwithstanding what other implications may be involved. Science still clings to previous cosmic concepts while trying hard to incorporate the Big Bang concept. Science still thinks about a mixture of theories and mix some facts about conditions applying in different theories about cosmology. If and when material is being part of the conditions presented in the steady state Universe theory to what will be present in the Big Bang all sorts of mismatching concepts arise. By trying to consolidate different theories from the past with the Big Bang rival theory such consolidation spreads considerable confusion. Trying to promote Einstein's vision and trying to bring that thought across to fit the Big Bang theory is what constitutes too much confusion. The acknowledging of the growing of space while thinking that only the space in-between structures are expanding and not the structures in their own space are expanding is a very typical of what I am referring to. Take the growth of space as one example where Science does not agree to the fact that the Earth and all in it are growing by becoming sizably bigger. The fact is that every part of the cosmos is growing rapidly. If you do not agree with my statement…well test the following: The sun at present is 1392 530 km in diameter. Let us only concentrate on the Earth in relation to the sun, as that is what concerns us humans intimately. The Earth is roughly 150 million km from the sun. The Big Bang theory wants us to believe that everything once was so small all fitted on top of a needlepoint. The sun is at present in diameter 1392 530 km. There then is a distance of 1.39 million km separating the one lot of atoms within the sun from the other atoms on the other end of the sun. Just as there is at the present 150 million kilometres between the particles that form the Earth and those forming the sun. If we are to believe the Big Bang and agree that the lot were less than one millimetre apart at one stage, we can appreciate that the distance there was between all the atoms that now forms the sun and all the atoms that now forms the Earth were one meter apart. The sun was spinning one metre from the Earth. It seems rather trivial except that it also means the sun in as much as being the space occupied by all the matter that now forms the sun was some time back reduced to a speck of what now is present. The Earth too has to be a lot bigger at present than what it was some time back. It is rather unrealistic to believe that the structures remained the same size when the space placing them apart grew from millimetres to what is now applying. Kepler brought us the insight that the distance of **k** is at present 150×10^6 but believing the Big Bang then proves that **k** once was one meter and much less than one meter. With **k** at one meter it forces the size of the sun and the size of the Earth to respond to that and shrink in sympathy. This also brings a realising that if they grow, then why do they grow because surely the idea manifesting as nothing can't grow. But even moreover is the importance of realising that material is also growing. But if so then what is material using to support and sustain such growth. It is inconceivable to place the sun as it is in the present day into a Universe that was in place just after the Big Bang. Our minds function in a rather dubious way of reasoning because if space is nothing then space can't grow because nothing can be duplicating nothing and in the process nothing adds to nothing then its nothing that is becoming more.

We then had no reason to explain why nothing can grow and accumulate because nothing is nothing, so nothing can add to nothing without any reason to explain where the additional nothing is coming from. Nothing is a concept, by which the using of nothing as a term is what makes the accepting of the idea so easy. By giving nothing a character and not a value is soothing the idea that what ever is part of the cosmos came from nowhere and that answer then must be so meaningful it better satisfy everyone regardless of reason. Silly as it seems it is also very true. The idea of using nothing to build space with must be seen by all as ridiculous as it truly is. The sun couldn't fit into a Universe of the past because the sun out grew such a Universe. The growth we find in the Universe comes from the growth we find in atoms accumulating by using gravity. We have to bring all space being k in $a^3 = T^2 k$ in realising there is growth of particles as much as particles moving apart. The growth is everywhere and not selected to the nothing part. If the sun and the Earth were that much closer back then, the sun and the Earth was that much smaller. If a sun with a diameter of 1392 530 km in diameter and an Earth with a diameter of 12756 km was one meter apart, then the distance separating the two was enough to allow the two objects to join. As we can see they did not join. That brings across that by accepting the growth of space it then brings along the growth of all space. That includes space that is holding material as much as space not holding material. It includes the space holding subatomic particles as much as the space not holding subatomic particles. It includes all space occupied or otherwise. That leaves us with one fact being the principle issue of this very book: If space is not nothing and yet space can grow, then how can space grow and what is space using to grow?

If the space we think of is constituting of as something because it not only holds material but has to be formed by something for the sheer fact that it is there in the first place and by being formed by something it then can grow as much and in relation to space we think of as nothing then the nothing has a lot of the something within the nothing because both is growing. It is an argument of complexity and it can only become apparent when Kepler becomes apparent and when everyone can see that, Kepler separates and stands apart from what Newton saw what Kepler was saying. Kepler must come into his own as a mathematician, a cosmologist and a scientist because of his tremendous achievement. If we accept that Kepler's formula is mathematically $a^3 = T^2 k$, then it must also be $T^2 = a^3/k$, $k = a^3/T^2$ and $k^0 = a^3/(T^2 k)$. What this then states is that Creation according to Kepler started at a point coming about from singularity k^0 and grew out of singularity into space standing directly and undividable related to time where the time aspect is the motion created by and continuous creating of more space.

Reading this letter will introduce facts that no one before came to realise during the past almost four centuries ago. This becomes clear when we go about analysing Kepler's formula somewhat differently. In a way for the first time it helps one to come to realise another part of Kepler. But I must press the fact once more that this realising demands a divorce separating Kepler's ideas from Newton's ideas about Kepler's ideas. We must recognise the Master Kepler as an equal to Newton and not just someone with vague ideas and the ideas he had, later needed revising by a better Master. The difference is about finding what Kepler really brought to science in relation to what Newton saw what Kepler brought to science. It also puts a new appreciation on what Kepler said and diminishes Newton's effort to change what Kepler introduced. I hope by now, you as the reader of this letter, have come to see why I say that Johannes Kepler introduced an astonishing presentation about space-time, singularity and the Big Bang in the years of his life between (1571-1630). By my reinvestigating Kepler in the state Kepler's work was seen in, isolated from Newton, I received new answers taken from the mathematics by which Kepler introduced his cosmic findings. Kepler's answers were lost so many centuries ago when some Englishman saw him wise enough change Kepler's work and bring in the alterations as he (the Englishman) saw fit. I repeat this because it is most critical to realise that Kepler saw $a^3 = T^2 k$ as space a by the sphere ^{3} in the third dimension in motion T by the square ^{2} shifting along a line k from a distinct centre k^0. This was no mathematical referring of a sphere in a drawing on a paper but a cosmic expression f space-time, which is space by motion thereof. The changes were unnecessary because the changes were unasked for, yet all the intellectuals that came afterwards and held any position of prominence in academic circles, was and remained impressed by the part that Newton changed. All academics coming afterwards were in total agreement about the changes, notwithstanding being the unnecessary changes that they prove to be, which Newton also unnecessarily changed. When reading the rest of the book, please keep in mind that it was the cosmos that spoke to Kepler directly by using mathematics and it is that what the cosmos said about itself to Kepler that Newton did not understand and then changed to what he (Newton) then understood. Newton did not just change Kepler's work but he tried to re-invent the cosmic code in mathematics which the cosmos used directly to explain to Kepler matters concerning the Universe. The Universe used the cosmic principles used in the Universe by the Universe. Then Newton changed what Newton could not interpret because the cosmos did not reveal the code to

Newton but the cosmos did reward Kepler by giving the code to Kepler. But now some three hundred and fifty years later it comes to our attention that Newton completely misunderstood what he thought he understood about what was revealed to Kepler.

The cosmos spoke to Kepler about space-time coming from singularity. Kepler gave us his findings. Any discomfort that may come when we read what is revealed must be set aside, because we must remember it is not me, or Kepler, but the Cosmos that is doing the revealing and lending us the tools we can use to decipher what the cosmos is trying to make us understand. Kepler translated what the cosmos told him (Kepler) as $a^3 = T^2k$. Translating Kepler's mathematical expression $a^3 = T^2k$ correctly to the verbal statement in English Kepler said that there is a space a^3 which is equal $=$ to the motion in the time duration T^2 thereof between two specific points which holds a relation to a centre where from there forms a straight line **k.** What is there mathematically not correct in Kepler's expression and why is any changing thereof necessary in any way? It says where there is space such space has to move. Test the following symbolic values in the mathematical expression and test the principal behind the expression in which Kepler stated them. Convince yourself about the evidence that Newton saw what Kepler saw where the translation thereof that was done by Kepler is mathematically incorrectly translated by Kepler's interpretation from mathematics to English:

a^3 The fact that any symbol uses a value to the third power indicates space or a volumetric established and separate unit which is serving an under dividable dynamically separate space being within a space. Although being apart the two in space sharing a unit can never be apart but serves as a unit by division of motion. It is space because it is volume using the third dimension. But since the space is smaller than the Universe it must be space being within space, which is within space. There a relevancy is forever present.

T^2 Is an indication of space apart from the surrounding space by granting the independent space by establishing borders through motion, an ability of moving from one point to another point or following a flat distance between two points. It is motion that is taking time in the second dimension.

k^1 Is the symbol used to indicate a straight line between two points with a definite beginning and a specific end position. The two points is valid only by re-aligning an eternal straight line to the figuration of a circle through alternating as well as recognising the control coming from such a centre. It is Pythagoras by the triangle, half the square and the straight line sharing value in the 180^0 they represent. Kepler introduced this absolute basic mathematical principle.

This leads to the question: "What formed the grounds for any need by Newton to change Kepler's translations from the cosmic given to mathematics and then from mathematics to English? The space-time that the cosmos introduced was so brilliant it took the likes of a genius such as Einstein and another few centuries of mind development to realise the presence thereof that is of space-time many hundreds of years later. What did I not translate correctly from the mathematical expressed to English? When I used Kepler's mathematics by my translating Kepler's work correctly I came upon answers not yet uncovered by Mainstream Science. Kepler gave the World a means to use mathematics in order to translated cosmic mysteries and reform such mysteries to answers that Kepler uncovered long before Newton, Einstein and others got wise about cosmology... Such is the advantage of recollecting Kepler's facts that it does answer many questions, which went unnoticed and therefore not spoken about up to now and some questions that is answered were previously never even thought about. Mainstream Science never previously thought that through any examination of Kepler's work the scrutinizing would uncover these facts that I present. Subsequently by ignoring Kepler's uncovering of decoded cosmic massages Mainstream Science elected not to ask the correct questions and in the process Mainstream Science never found the correct answers. By not asking the questions Mainstream Science could not decipher any of the decoded mathematical messages, which Kepler received directly as a mathematical message spoken by the Universe and coming from the cosmos. We all know and appreciate that mathematics is just another language and the professional mathematicians have the responsibly of translating mathematics to a verbally competent language that is understood by others that are not fluent in the language of mathematics. They did not read nor recognise Kepler's mathematical translation and thereby was unable to translate Kepler's mathematics to the other communication forms being all verbally spoken dialects. In other cases human natural study methods brought along a cultural of Academics forcing students to comply whereby the students will accept the knowledge through our inherited past which when tested by modern standards is not that highly proven. In Newton's time the accepting of accurate representation came a lot easier than what is the case today. When we use information coming from the past we accept the answers as questions that is without doubt already fully answered because culture demands the accepting thereof. Can any person in a sane mind for one second think of the fate awaiting the first year student that will question Newton's findings or the accuracy those findings represent? Let us see where the motion takes space.

The first of the three laws of Kepler concerning the orbital motion of planets is focussing on the orbiting route of planets always form an ellipse with the sun at one focus of the ellipse. This is in the total of the totality pointing to a relevancy acclimating a position that alternates the prominence of both parties in the relevancy and more specific allowing the dominant role of either on applying at any one time. The second law states that the radius vector that connects the sun to that of any and all the planets sweeps out equal areas at equal times. This is another total relevancy establishing two factors that contribute in forming space being specifically related to the time component, which is space-time as they combine in securing a definite ratio the time period in relation to the space in motion during the time applying such relating to the space involved in the bigger picture. There is a space moving through a space in a specific time period. That is space-time in any one's book. It points to the time it takes for a space to move through a space in relation to the centre of the sun. The third law of Kepler is also indicating relevancy as it states that the square of the orbital period which is nothing less than the time factor because if it does nothing else then it indicates the years in orbit proportional to the centre of the semi major axis of the planets in orbit. But it does more than that because it positions a precise location of another space also holding a^3 in relation to each other, if the larger space indicates the space factor it shows motion in relation to space where the space holds the motion every time. By positioning the lesser space the motion applies to the lesser space running through the larger space. This forms another relation where if we then were focussing on the larger space while the larger space is gravitationally centralising a position in relation to the smaller space. In this the principle is very obvious adhering to the Coanda affect where motion indicate a valid centre for gravity to focus. The Larger space shows a motion coming from a different position every time in relation to the changing position of the larger space. But motion indicates space in motion and the motion of one space is the other space being in motion or proving the existing of another space that also holds motion. The mere fact of proof of both spaces sharing one unit is the contradicting both has to each other by motion.

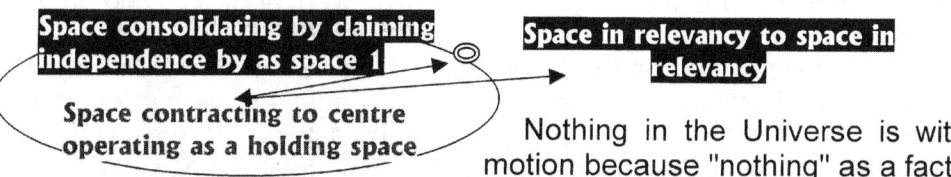

Nothing in the Universe is without motion because "nothing" as a factor is "not any" factor in the Universe. Other than nothing everything else is in motion because motion brings separation and points boundaries that establish individual space. Therefore what is in space is in motion. But as Einstein pointed out motion is a relevancy between factors of different motion in space through space. When looking at the moon on a semi cloudy night we can place the relevancy on the clouds being without motion and the moon travelling at a dazzling speed or we can put the motionless relevancy on the moon and find the clouds travelling at a large speed.

But in fact we know that there is the motion of the Earth in relation to the motion of the moon in relation to the motion of the clouds that in turn is in relevance to the motion of the moon. The relevancy of motion in relation to space filled in space not filled is an endless ratio of different objects in space in unfilled space that is at different rates. One may focus on the moon and find the clouds to be motionless but have the moon travelling at a dazzling speed. Even when we think we are motionless say when we are sleeping the fact is that we then are travelling as fast as ever before because the Earth is taking us on its journey around the sun which is going around the Milky Way. Every aspect of the cosmos has to move to be in the Universe. Every time there is motion, there is a relevancy to such motion. The Universe is in motion because the Universe is motion in motion about motion. In every individual case of the three laws of Kepler there are a larger space relating to a smaller space within the larger space. The smaller space serves as much as an indicator to the being of the larger space as the larger space serves as an indicator to the position of the smaller space. It shows space being in relation to motion of the space, which is in the space holding the space relevant through motion. It is $a^3 = T^2k$ where relevancies point to either in space in the relation of the motion. What may be linear motion in relevancy to one is of another part of the same space unit. As a whole the two factors represents space, which is the circular factor of the other indicated by linear motion that produces the border of the circle.

Any line coming straight from the planet running directly towards the sun will fall victim to the rotating action of the sun but also the combination of a group of such lines dotting the motion of the planet as the smaller object travels will prove to become the circular rotation in the end The factors representing $a^3 = T^2k$ can turn around and become interchangeable. When one candidate takes the position of the space, the other space becomes performing the motion part. The issue is that neither the holding space nor the space being held has a validity of existing without the other being a factor part of the equation. All this information above Newton blatantly missed while considering his opinion of such a standard as to bring changes to the work of Kepler.

Newton is institutionalised academic culture force fed by the generations of Academics passing by as they go from the present over to the past whereby the Academics in generation after generation introduce Newtonian ideas to the following generation of Academics. At the time the following generation fills the role of students on their way to becoming Master by promoting Newton's ideas about forces. As they progress from students to the Masters of the day they insist on the unquestionable accepting of Newton as they did as students taught by their masters of the previous generation. They were taught to learn Newton off by heart if they wished to be found sufficiently knowledgeable and intellectually competent to become the next generation of masters. All they required was to prove they understood Newton's ideas about active forces as being correct above all else. The passing on of such accepting of Newton's forces lies above and beyond and passed the accepting onto the following generation of future Academics. Providing that the student "understand" Newton by repeating Newton, is the biggest and ultimate achievement of all Academics that is presently in charge of Mainstream Science. The next generation of Academics, which is now the present day generation of students is conditioned mentally to accept Newton or die an academic death. The academic death will be in the form of cutting any further mind contamination by terminating all further study possibilities. If Newton's personal vision and his understanding on the matter of gravity and his comprehension about gravity and the way he went about proving gravity and the promoting of the understanding of gravity's accepted forces is brought into question by students with serious doubt on the matter, what would be the fait of such students? When even one student has questions about the gravity that Newton promoted while Newton admitted he had no idea what he was promoting and such a student realised the reality there is about forces in gravity by rejecting the reality of Newton how quick would Academics get rid of the poor thinker? Not one person in Academic circles would even glance at such a student's opinion. The thoughts the student will convey will be discarded as waste and he will be removed from campus. When students do not exactly and in a motorised manner repeat after the tutors what the tutors repeat after their tutors and repeat the tutors misguided culture statements in examinations wherein the students precisely duplicate definitions however invalid that was made in the past and carried to the future generation of Academics, then such students will be studying art or go finding other occupations of a none science nature. If the student so much as dares to discount Newton on any grounds of mistrust it is the student that will be vanquished from campus. Newton's is a template in which the future students are moulded and if one dares not to fit the mould because of free thought, such a student will do his thinking elsewhere. The mistakes are passed on from generation to generation and the mistake is carried on. This is not an age of free though but studying is about precondition set to students.

It is the ultimate condition of acceptance by the present day masters to all students when the students are performing the answering of their examination questionnaire. If and when the student will not accept such culture without reservations that his peers teach him, that student will fail all further acceptance and this then is disallowing the student the right to attend further classes in any of the classes given by the institution. I personally have experienced such bias from Academics in charge of institutions. By reading what Kepler said correctly so many centuries ago the effort brings all the answers to the questions academics at the present are incapable of answering. But it does not involve looking at what Newton said about what Kepler said... it is all about looking at what Kepler said. To understand Kepler one has to include the opinion of Kepler and remove the opinion of Newton about Kepler's findings. Answering the main question about gravity is locked in answering why would a water droplet form a sphere when floating in a space capsule in outer space? Kepler gave us the answer in a much simpler mathematical manner than Einstein did.

Kepler said $k^0 = a^3 / T^2 k$, which means that $k^{-1} = T^2 / a^3$ which means that by reducing gravity T^2 space will disappear. Kepler gave us the answer about the water drop forming a sphere even before Newton thought about the question. He gave us the answer three centuries before Einstein got the question wrong about his Universe going flat. Why would gravity always result in forming a sphere when gravity is left free and unhindered to capture form? Let us recollect Kepler's statements. He said that $a^3 = k / T^2$ and from that the mathematical relevancy guides one to the answer that $a^3 / k T^2 = 1$ **and** $k^0 = 1$ bringing about that singularity is $k^0 = a^3 / (T^2 k)$ which is the smallest space being in singularity, produces gravity in forming space a^3 relating to time in motion $T^2 k$. Indirectly Kepler said that mathematics prove that $k^0 = a^3 / (T^2 k)$. $a^3 = k / T^2$: That is what Kepler brought into civilization for all time but mathematically Newton saw Kepler's $a^3 = T^2 k$ as incomplete and therefore he had the urge to correct what he saw to be incorrect and changed it to be $4\pi^2 a^3 / T^2 = G(m + m_p)$. In this, the value of T and "a" are the period of revolution and semi major axis of the orbit of a planet of mass m_p about the Sun of mass m, and G is the gravitational constant. Newton added nothing but duplicated everything in his effort of suggesting Kepler's incompetence and his completing of what he thought Kepler saw but was too stupid to change and correct what he (Newton) saw as being incomplete. Can the symbol "a" be reckoned as a period of revolution as Newton suggested and therefore Mainstream Science still suggests? Is the work of Kepler therefore incomplete? Let's dissect it once again because it is very important! Newton diluted Kepler's formula by insisting that all factors only contribute as motion. By re-examining one can clearly see the finding of Newton is most incorrect.

a^3 As I said before, any symbol using a value to the third power indicates space or a volumetric established and secluded unit. It suggests the using of the six dimensions allocated to space. It is space because it is volume using the third dimension. There is no other valid interpretation or translation allowing for another translation from mathematics to English than by categorising that space is a volumetric separate identity. If it is cubic it is space. One measure a fridge or a stove or a room by the cubic measure without including a square because of the volumetric content there is in the third dimension. An aeroplane flying is a cube flying and the square part falls to the repositioning of the motion of the cubic space. The fact that there is a line connecting the space to a specific allocated centre and enforcing a rotating motion around that specific allocated centre connects whatever volumetric measure the space has in the cube to motion. The cube as a separate issue is independent from the space in which the independent cubic space rotates, which brings about the circle and yet there is no need to implicate pi because the rotation brings along the circle after the cyclic completion of the rotation by the independent cubic space.

T^2 Is an indication of motion, the moving of an independent space that is holding a^3, where the space in motion will be measured as a^3, as the space a^3 that is occupying a separate part of the including unit forming a space by opposing motion within the unit of the other space that is travelling by using the second dimension of motion in T^2. Both sharing the unit is moving as independent space from one point to another. The opposing motion combines to form a relevancy that produces gravity. Both sharing the one unit is following a flat distance between two points, but which form one point. Because the motion is coming from both ends sharing a point for that reason, time can never stand still because although there may be a single point the point refers to the square of both aspects in motion.

The distance and only the distance of movement is T^2 where the distance is established as T^2 by both aspects of the space a^3 but is not the space a^3 and as the independent space a^3 in each case of the space that is going from point **T** to point **T**. By both matching a point there is a reserved position applying on instant by both and from both the value is T^2. However when one aspect alone is considered there remains the T^2 square of motion of one single party contributing to the unit as motion making the distance a^3 travelled T^2. It is space filled with material consisting to allow independent space within the surrounding of an enveloping space of bigger proportions a^3 being in motion T^2 and that motion T^2 is the filled space a^3 that is taking time in the second dimension moving the space a^3 in question from one point of choice to another point of choice by which time will be established.

k^1 Is the symbol used to indicate a straight line between two points with a definite beginning and a specific end position. It is also the mark by which the straight-line motion is altered to accept the rotating motion. The change of the motion is directly linked to the change in direction and the change in direction amounts to the gravity strength. This indication of a distance is an indication of a bigger space that is big enough to include a smaller and separate space a^3 within the bigger space running all the way from the start of **k** to the end of the line of **k**. One has to see **k** for what it represents because **k** indicates the presence of a larger space that is large enough to allow the smaller space to rotate all the way as the circle that runs from **T** to **T** in the full diameter of **k**. It also indicates where the smaller aspect in the unit in motion crosses with the larger aspect of the unit and in that way form $T \times T = T^2$.

Kepler introduced this absolute basic mathematical principle that all others failed to notice. It is positioning the independent space in motion in a specific relation to a controlling dynamic situated in a domineering and controlling centre. It is indicating that the space in question is in motion acknowledging the centre in control of the motion and therefore in control of the space location. It proves that a larger space is holding the space in question as part of the larger space where the larger space is in ratio to the other part so much larger that the space in question can effortlessly go in motion and therefore full rotating motion. By the rotation it then is concluding a rotation within the larger space that **k** indicates. The presence of **k** is not merely a line but points to the start and the end of the space containing the space a^3 within the space a^3. But **k** also must therefore indicate its coming from a very small start that is centralising the motion of the space in question. It is using the space that **k** produce from where that specific space is controlled in the realms of the larger space. Nobody before saw it in as simplistic manner as I just showed it to be and yet in the simplicity is the sensibility of it all…this argumentatively is indisputable reasoning.

What in this is there to dispute, yet when I say these facts Academics find grounds to dispute my saying this about Kepler's saying that! Kepler gave us the answer but no one ever took notice!

Kepler was the one that discovered **space** / **time** and Kepler announced it as $a^3 = T^2k$ which can translate to as $k=a^3/T^2$. Kepler was the one that discovered singularity as $a^3 = T^2k$ that also translate $a^3/T^2k = 1$ which then is $k^0 = a^3/T^2k$. Kepler was the one that discovered gravity holding space-time relative $k = a^3/T^2$ the contracting part that Newton claimed gravity as a force is $a^3 = T^2k$ that translates to $k^{-1} = T^2/a^3$, but that is as vague as saying humans are life. If gravity is a force, then what is a force? If gravity is the product of the elusive graviton what is the graviton and where is the graviton? Mathematicians get stuck by using mathematic rules and laws. Kepler goes further by correctly using cosmic mathematics and investigating the formula that Kepler introduced intensely but without Newton interfering and telling Kepler what he (Kepler) should have found. Instead we come to a part, which takes the Universe one step further back than the Big Bang, to a time before the Big Bang to an era where no one in modern science previously dared to go before. We can reach that point by tracing what Kepler said to the time where gravity started. Newton had all the information we now use to his disposal. Instead of using it correctly Newton chose to change what Kepler said because he (Newton) did not understand what Kepler said. Newton should instead have been looking at what he (Kepler) found then he (Newton) would have seen what gravity is. He (Kepler) said that the cosmos said that space is time being space-time. $a^3 = T^2k$. The space is held in check by motion from a centre and that is gravity. It becomes more than clear that space a^3 is time by dimension T^2 and time is space a^3 without dimension **k** Gravity is a^3 / k but **k** is an addition of motion T^2. Motion T^2 of space a^3 being apart thereby forms k^1, which produces gravity. It is gravity that keeps the sun and the planets at a specific distance and apart while the planet remains in motion around the sun. It is **space-time** $a^3 = T^2k$ that keeps the space in motion and at a distance whereby the space of the sun is parted from the space of the Planet. That is gravity…what else can it be? After all it is space-time keeping the structures apart and space-time is a result of gravity. Once singularity was found the rest was simple but was finding singularity really that difficult. Not if one was guided by Kepler's formula. It is merely retracing **k** until **k** becomes k^0. It is so simple to reach singularity that it is almost ridiculous.

r /2 By dividing the radius r by the half of the value that then reduces r to a point where the left edge of the line reducing will be at the very same place the right hand edge of the line that is reducing will be. At one point the spots that formed the two ends of the line will be at the same spot where the original centre between the two points were. The two points would have moved evenly towards and in the direction of the centre by reducing all the space on both sides of the centre. Then by moving towards the centre they will at some point have to reach such a centre point notwithstanding cultural concepts favouring nothing to be filling that spot because reaching that centre point will land all the sides on the same side and because of the presence of all possible sides such presence of all possible sides removes nothing out of any further possibility.

Any further dividing will land the left hand spot past the right hand spot in the opposing half where it then will grow once again but in the opposing direction that the specific spot previously represented. All possible dividing then ends on one spot where such a one spot that represents the perfect centre point and that divide the left side from the right side and the top from the bottom and the front from the back will land on one spot. At that spot all the sides just mentioned share a location with all other possible sides. The centre that then is holding all the previous points in one spot then physically is in the single dimension applying as one spot to share a location for all sides. At such a point there is no further dividing possible. That point cannot be zero because that point represents an eternity of possible growth in an infinite number of directions available to grow. The line starts in infinity and not in zero or "nothing" as teachers teach scholars on a worldwide scale. Trace that centre while the top is spinning and one will find a centre that favours no side since the centre divide equally all sides while spinning. That centre proves to be no specific side because such a centre proves to be all possible sides. On several occasions in the past I have been accused of manipulating the argument to produce none-existing or overrated facts. That is not the case. I am not manipulating facts to create an argument as some intellectuals in the past accused me of. What I am talking about is a mathematical fact that any one can prove by calculation. By following a very simple procedure it is within any person's reach to detect the centre of the spinning top, which I am referring to, although there is no such a centre to detect, the centre is there for all to detect. A child is capable of using the two times table and the dividing by two. That is the simplest form in which mathematics can be used. It is a mathematical fact that a line will reach a point where all sides are at one spot and as such the line cannot divide any more. I have been accused of as being dubious about my arguments while it is Mainstream Science that dubiously found a way to get to zero as a mathematical starting point of any and all lines. Then they put the double standard blame on my arguments where it is I that have to prove my arguments as not being the dubious one of the two arguments. At such a centre starting point all sides share one specific spot but that spot holds all further future possible growth in any direction of all sides and since everything is in there, there is no room left for zero to be there. That point is filled with all possibilities which prevent zero becoming factor since the sides share one spot and in that sharing they are present and their presence prevent zero from becoming a conclusion. While the different sides are in one place the factor and value is one to all without allowing zero any part to play.

Tracing the centre of the Universe is still possible by any one wishing to find such a centre. The centre falls outside the accepted Universe since it cannot be mathematically accounted for but that centre is in control of every thing in its influence.

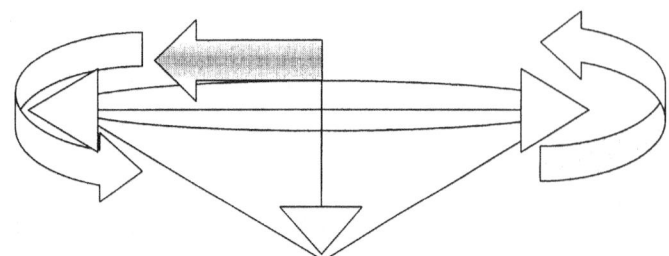

The centre changes motion to gravity by diverting the straight line to an immediate circle. By tracing the line back to where the circle is no more a straight line will uncover singularity plus one dimension. But the entire centre forming singularity is still locatable within the Universe we have.

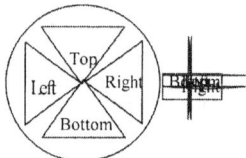

Reducing the radius r from all angles possible throughout the circle will bring about that all possible direction will eventually land on the very same spot with no more dividing possible. Yet zero cannot be a factor since the sides still hold value. In as much as holding all the value there can rise from such a spot. This is arriving at a point where more reducing will land the one side on the opposite side of the line but it will not bring about zero in the equation

What this argument further proves is that the circle reducing must then come from all points because the radius might be a line but that line represents a circle through 360^0 coming from and accounting for all possible directions. Taking that into account it is important to recognise that notwithstanding the size of a line, which any radius of any size is there is another line (or dot) eternally bigger as well as eternally smaller than the line in question. While we are in the third dimension being part of the third dimension such being in the third dimension then allows that all parts of the third dimension forever can be divided once more until the line in the third dimension is no longer part of the third dimension. When such a line leaves the third dimension it is still dividable because it might not be part of our dimension any more but it can still reduce further as part of the second dimension. By that time it has left our scope by miles but that does not mean that it ends there because from our perspective that is where it ends. But our perspective does not represent reality.

Yet, even then it can still reduce infinitely more until it has left the second dimension and then at last forms part of the first dimension. Only then when the line reaches the first dimension no further dividing of that line is longer possible. As we can never grasp what the size of a line was when that first line came about and had a size that is in size far beyond our means of understanding is the very line that came about when the first motion broke the eternal stranglehold on infinity to let out form in the space of a sphere. According to our big and small conceptions of what we perceive as large, ultra large, small and microscopic small is just mere words describing thoughts totally unrealistic in the context of what the cosmos sprang from as the cosmos moved out of the spot and formed a dot. Even by the standards of forming the dot, which we think is beyond measure, the dot still was eternally bigger that the spot it came from. The Universe exploded by measure going past any humanely possible calculation attempt. The Lines gain is size that came about was from the dot leaving the spot, as the dot and all the many dots that came from the spot formed lines. The size differentiation only between those two exceeds all limits and divides we wish to create forming borders that we can appreciate.

When looking at the circle in the conventional manner, we persist with errors brought about in culture and not by applying some significant modern logic. Take a circle and reduce such a circle constantly to where it no longer can reduce. Reduce it to a point where only form remains part of the circle because the radius has gone beyond human measure and becomes so small it is not noticeable with what ever measuring tools man may use, then what remains is pi since pi does not indicate size but indicate form, and form is all that then will remain. In any circle or sphere the size only depend on the fluctuation of r, as a component to the circle or sphere but that does not affect the form by indication of Π in any way there may be. The conclusion I drew from following this process is that from this reducing of the line no line can start at zero because when again increasing the line, that increasing attempt from zero will be a mathematical impossibility since no line can ever reduce to zero. A line will forever be able to reduce further becoming smaller but it can never reach zero because zero is not part of the scale on which we measure lines. If a line cannot reduce to zero it then cannot start at zero.

A line or spot starting at zero would therefore be shorter than the shortest line possible. For obvious reasons can no line, or any line grow or extend from zero because such a line must then quit zero and become something, thus abandoning its original value. That would mean the start of the line has a different value to the end and a line holds conformity through out. When any line is starting from point zero and it uses the factor zero, then it can never leave zero because of the influence of being zero disqualifies any possibility of growth. But when coming from singularity π^0 and the line then had to grow in all directions at the same pace the line must then become a circle π or being three-dimensional, then form a multi circle π^3 we named a sphere. Since the Universe is about circles and lines connecting circles I came to conclude that flowing from this fact is that in the Universe there can be no zero improvising as a filling ingredient for the space of a point or be unfilled space. Zero is no valid factor in the Universe. In the case of the growing sphere the value of the circle is Π, and that is where creation must have started. That gave me the clue where to start looking for singularity. One would find singularity in the value Π and the value Π will be in all things rotating in a circle but by measure one dimension smaller. As usual I am again shooting the gun before the hunt started. Lines in mathematics do not start from zero and that is no discovery on my part. It was a realisation I came too. The Universe is all about lines and the manner that Kepler pointed to the increasing of the lines by $k= a^3 /T^2$ proves growth in the composition of all lines.

UNIVERSE
Everything that exists, including space, time, and matter. The study of the Universe is known as cosmology. Cosmologists distinguish between the Universe with a capital 'U', meaning the cosmos and all its contents, and universe with a small 'u' which is usually a mathematical model derived from some physical theory. The real Universe consists mostly of apparently empty space, with matter concentrated into galaxies consisting of stars and gas. The Universe is expanding, so the space between galaxies is gradually stretching, causing a cosmological redshift in the light from distant objects. There is growing evidence that space may be filled with unseen dark matter that may have many times the total mass of the visible galaxies. The most favoured concept of the origin of the Universe is the Big Bang theory, according to which the Universe came into being in a hot, dense fireball about 10-20 billion years ago.

UNIVERSAL TIME (UT)
A worldwide standard time-scale, the same as Greenwich Mean Time. Universal Time is the mean solar time on the meridian of Greenwich. It is defined as the Greenwich hour angle of the mean sun plus 12 hours, so that the day begins at midnight rather than noon. It is closely linked to Greenwich Mean Sidereal Time (GMST), since the mean sidereal day is a precisely known fraction of the mean solar day. In practice, UT is determined by a formula from GMST, which in turn is derived directly from such observations of the meridian transits of stars. The version of UT derived directly form such observations is designated UTO, which is slightly dependent on the observing site. When UTO is corrected for the variation in longitude due to the Chandler wobble, a version of Universal Time, UT1, is derived which has genuine worldwide application. When UT1 is compared with International Atomic Time (TAI), it is found to be losing approximately a second a year against TAI. Broadcast time signals use the time-scale known as Coordinated Universal time (UTC). This is TAI with an offset of a whole number of seconds. The offset is adjusted when necessary by the introduction of a leap second, and UTC is always kept within 0.9 s of UT1. On this issue there is much more to explore than the meagrely mentioned. Time stands related to the position an object holds to a centre such an object refers too while in rotation. Kepler found for instance that T^2, which holds the orbit to a rotation specific, is directly dependent on k to value the space a^3.

By contracting the Universe is expanding and everything is based on gravity providing both actions. The universe rides on a balance and we have to locate such a balance. To prove my theory I firstly had to locate the centre of the universe. Even admitting to such a notion sounds like madness or in the least a tasteless joke, but please give me a chance to explain in more detail. I realised that my effort to locate the point holding singularity only stood any chance of success if the reducing of the line enabled me to backtrack the exploding universe to its origins. By applying some basic effort I have located the position from where all movement came and the direction it took moving forward in time…and yes, while all of that took place, I was also finding the centre of the Universe which I might add I even located at the same time. There are two standard mathematical formulas used to calculate a circle. The one uses a "r" to indicate the radius and the other uses a "D" to indicate the diameter, which is double the radius and therefore needs to be divided by a four to eliminate the Newtonian inverse square law amounting to the difference there will be between the two. This has the significance that it implicates time.

The one using the radius is Πr^2 and the other formula using the diameter is $\Pi D^2 / 4$. At the very start of my interest about matters concerning cosmology lead to investigate the travel of light through time and in particular what Einstein said about time and light. In my involvement as I progressed in cosmology I arrived at the point where I had to understand what Einstein said about light travelling one year in opposing directions and being one year apart instead of two years apart. From this I made conclusions, which resulted in my forming my personal theory about the cosmos. In cosmology normal mathematical principles do not apply that straight forward as we try to envisage. Please allow me to try and explain myself as follows. My understanding of my future theory started when I was trying to understand Einstein's view on light in motion. When light depart in opposing directions from one point jointly shared by all and the light departed will travel in a straight line 180^0 in direction to each other they are all still relevant as a continuing line.

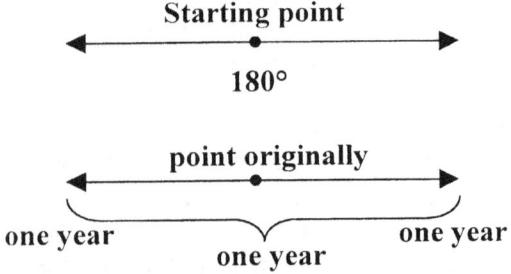

The question that nagged me for many years and on which I spent almost half a year just trying to solve the puzzle is how can light travel for one year in opposing directions and after travelling for one year in opposite directions be in two different points but was one year apart.

After one year of travel both points still are the same distance apart from each other as what both points are from the centre. Both points are from the centre just as far as each point is from where the light came from and all the points are at an even distance. The point of origin has to be evenly matching on both sides of the divide where the divide serve centre point from where the two points originated.

Under normal circumstances when applying normal mathematics the light will be two light years apart. If one could stop the light travelling to the right and have that light that is stopped, standing still, while the light flowing to the left can make a complete turnabout, it will take the light coming back one year to reach the point of origin once more. It will take the light one more year to reach the other point that is having the light that was standing still then at the time for two years running. Einstein proved that the normal way humans use mathematical thinking power is not the way rules apply with the speed of light as it is in the case we find with light. Light travelling in opposing directions for one year will be one year from the source it came and the two lights will be one year apart from each other.

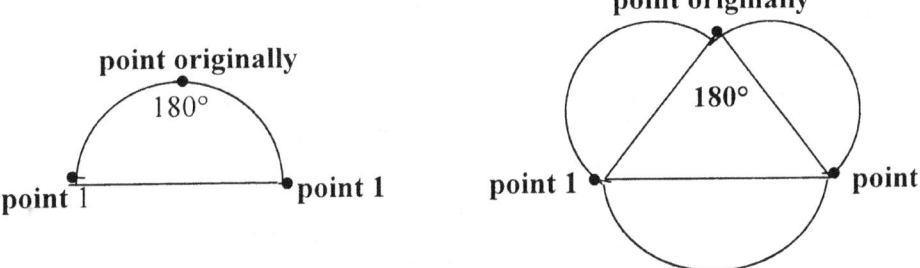

In order to give this argument mathematical logic is to put the light equal to time but this tests logic even further. Einstein's claim that this comes about because the light is equal to time did not make sense to me. Then I came to the conclusion that light became a cosmic factor before mathematics as we understand mathematics in its fully developed state. It is where mathematics started off with Pythagoras that comes into affect. If the light was time, as Einstein interpreted light in motion to be, then time was no factor to light. In that case light will travel through space in a ratio of one meaning the moment it releases from the source it is on the other side of space notwithstanding the distance the space has. It holds a factor of one with no distance to measure any restraining. That is nonsense because light is restrained by 300×10^3 kilometres of space in one second of time. Light is restrained by space-time. Light cannot be time, because light is just a simple speed ratio like any car driving or aircraft flying or spaceship launching. Light was forming distance during time duration and that comes down to being pretty fast, but it still remains speed and speed has just the same relation with time than placing time in relation with space forming distance. One should think that Mainstream Science would see from this that normal mathematics used on Earth does not apply in astrophysics but apparently that slipped past their noticing. Let us again gauge what is happening when light is travelling.

After travelling for one year the light had a distance of C multiplied by the seconds in one year to each side of the source. The points of light travelled on either end of the dividing line, which was parting the points, and securing their independence. Light had that same distance apart taken from the point the light started from. The two markers are just as far apart from each other as they are apart from the original starting point. With this in mind the use of C^2 by science might prove convenient, but it also proves with this mockery how a big farce such innovative calculation is. There is no chance of anything going at C^2 because there is no exceeding of C by light or any other particle. If that were the case light would not be present in the explosion or antigravity. The speed of light is not a force but it is a speed. It is a ratio putting space in relation to the time in the space density the speed will establish. This whole argument pointed to Kepler holding the straight line in relevance to space and time. From that point I concluded that the link must be the value of a straight line sharing a dimensional value with a half circle and the triangle. If we look at the line supposedly travelling straight we find that the straight flowing is equal to the square relating the triangle. This is completely Kepler indicating gravity being $a^3 = T^2 k$.

Look at the dimension and not the number implicated. It is $^{1+2+3}$ and transfers that to the line being the ^{1} and the ^{2} being the square being equal to the triangle as ^{3}. But it diverts very much from normal mathematics and that is precisely what Kepler's formula also does. With Kepler $a^3 = T^2 k$ and with mathematics the volumetric size of space must either be according to the measure of normal mathematics if it is a cube then three sides form $a^3 = L \times B \times H$ and in the case of a sphere the measure will be $a^3 = 4/3\Pi r^3$. This was a triangle in relation to the square we find in the half circle standing again related to the half circle. It is not standard mathematics and anyone drawing links between mathematics and the speed of light has no idea about what is involves. With that I have again antagonised millions of the most important people with which I have to share a view. I do share their view on the Cosmos but not their view holding mathematics as a standard fit all and apply anywhere in the cosmos. It is about lines carrying dimensional properties and with that we have to consider the line once again

Before coming to the mathematics I would first like to bring your attention to the practical side. I am promoting a theory in which I am able to prove there is as much contraction (moving in the direction of the Big Crunch) taking the cosmic universe back to the size it had during the Big Bang as there is expansion (moving apart by Hubble's Constant) and the contraction is as much part of the expansion.

All the difference we find is seated in the human mind. We humans set differences because we look at the cosmos by placing humans and the life we find on Earth in a pivotal centre in the cosmos instead of placing singularity in the centre and life where it belongs; only found on Earth. Einstein proved mathematically that in the presence of a strong gravity such a strong gravity slows time down. Surprisingly with that evidence being around this long nobody in science since Einstein's discovery took those statements and made any further progress from that. It seems to have been left in some drawer to dry. Science still sticks to the opinion that time did not change, not even slightly, since the beginning of the time and holds the same pace ever since the start of the Big Bang notwithstanding the implications this concept carries. Before the Earth took one year to circle around the sun and even before the sun was there a year was still the same duration of one year. How odd... don't you think ... that the only aspect in the entire Universe that is beyond change is the aspect of time? With the entire Universe including all the gravity now present and not excluding one Black Hole or dust speck pressed in such an area that was possibly the size of a lepton even then the gravity extending from that circumstances must have been beyond what words can ever describe.

When everything was that small when the Big Bang took charge, the gravity at the time was beyond light, because even today in the Black Hole the gravity is beyond the speed of light. If the gravity was that high and Einstein already proved that strong gravity slows time down, then there is one logical conclusion and that is that time was in fact at the time of the Big Bang standing still. Mathematically it is incorrect to allow gravity to compress the Universe into a spot smaller that an atom and exclude any other factors and relevancies to change.

With light not able to exceed C there is no possibility of reaching C^2.

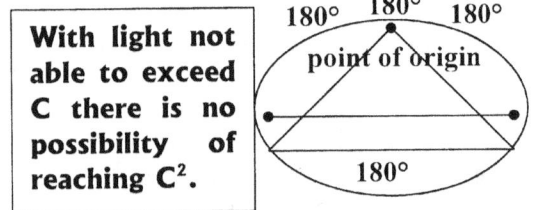

If light cannot be two years apart when travelling in two directions opposing one another but can only maintain a distance of being one year apart, light is unable to exceed C. How can light then reach the square of C?

There is no chance of anything going at C^2 because there is no exceeding of C by light or any other particle. The electron forms the limit of C but after C the space-time brakes ranks with the third dimension and accelerate to π and π^2. Light might equal gravity π^2 in space 3 becoming $3\pi^2$ but it cannot reach any value above C in the third dimension. This is the fundamental fact in cosmology and breaking this concept is reducing cosmology to rubbish. If that were the case light would not be present in the explosion or antigravity. The speed of light is not a force but it is a speed meaning it is a ratio of space over time $k = a^3 / T^2$ where space is a distance a^3 = km, k is a value 300 and time is T^2 = seconds distance of space in relation to time. It is a ratio putting space in relation to the time in the space density that the speed will establish. This whole argument pointed to Kepler holding the straight line 180° in relevance to space and time ($a^3 = 180° \, T^2 = 180°$).

From that point I concluded that the link must be the value of a straight line sharing a dimensional value with a half circle and the triangle. If we look at the line supposedly travelling straight we find that the straight flowing is equal to the square relating to the triangle in ratio. The normal manner of mathematically calculating diverts completely from that what Kepler was indicating as gravity being a sphere in motion $a^3 = T^2 k$. Look at the dimension and not the number implicated. It is [1+2+3] and transfer that to the line being the [1] and the [2] being the square, which is being equal to the triangle as [3]. But it diverts very much from normal mathematics and that is precisely what Kepler's formula also does. The fact that $a^3 = T^2k$ diverts from the accepted norm of $4/3\pi x \, r^3$ It is a clear indication that what Kepler saw does not in any way translate to normal mathematics. What Kepler saw as $\pi^3 = \pi^2\pi$ is not normal applied mathematics. That what Kepler saw, predates normal mathematics and it is our duty to investigate why that is instead of changing it to our thinking and our liking. With Kepler $a^3 = T^2 k$ and with mathematics the volumetric size of space must either be according to the measure of normal mathematics if it is a cube then three sides form $a^3 = L \times B \times H$ and in the case of a sphere the measure will be $a^3 = 4/3\Pi r^3$. This was like comparing a triangle in relation to the square. It predates mathematics to a time when we find in the half circle standing 180° related to the triangle

(180°). It is not standard mathematics and anyone drawing links between mathematics and the speed of light has no idea about what it involves. If I take what I unleashed in the past with this statement, the past is telling me again that with that comment I have again antagonised millions of the most important people with which I have to share a view. I need acceptance of my view but if my statement is not well understood I get rejected. It is therefore most important that what I say is understood. Cosmos mathematics is a standard to fit all and apply anywhere all over the cosmos. Cosmology is about lines carrying dimensional properties and with that we have to consider the line once again because what I try to introduce is not the general perception one finds in the view of science...

Let us find the smallest possible line first. We already have reached the conclusion that by reducing the line, the reduced line will eventually leave all sides on the same spot. Such a spot must be round in form. With the line being the smallest line, such a line will start off as a dot that moved away from a spot. With all possible sides being in precisely the same spot we have all possible sides onto one spot. Mathematically the spot is in the single dimension where the space is one and exponentially zeros. There the space moved over to form the dot. We now are reaching into areas only the human mind can venture by understanding and nothing more. The understanding of this concept demands our reaching the point where the mind of the animal cannot reach.

If it starts with a line that line only represents two sides being one and as such that is rather a flat Universe. The spot is not yet round because being round are requiring a shape or form and this lies beyond or before a time when any form of shape came into the cosmos scenario. It was in a period where shape and form was a part of the distant future hidden in and beyond eternity. In that time the line must have been so small it had reached a point not yet dividable in any way. If any further dividing took place such dividing would have brought growth because there then would form space between the sides going in the opposite direction. The dividing brought all there is having all sides literally on the precise same spot, and I have located singularity in just such a spot.

I came to the conclusion that the spot I found had to be singularity purely on the grounds that that spot holds only one side to serve as a start to the starting point of all directions possible. In that side is only one spot where there is only one side applicable and one dimension present. With all the factors given one can only come to one conclusion and that is that there can be only singularity. In such a case more dividing by two will land further positions on the other side of the divide. That point is serving as a position for all possible points and cannot allow further dividing as it is in the smallest line or spot there may ever be. This spot is the result of a most basic process of reduction as the Hubble constant is a most basic process of expanding during a matter of time. By reducing the line constantly the only value that will eventually remain without dispute from any party arguing about the facts is exponential zero. By only having exponential zero instead of a numerical zero and a radius as one in the square (the radius effectively becomes one holding any and all sides on one point) such a point might become any value of any significant measure implicating anything but zero as the radius. By expanding the line, it will be an evenly spaced structure growing into the most perfect round dot ever possible anywhere at the point when it starts to grow.

The reducing of the line is one dimension in six and although such reducing is representative of two indicators all the other indicators must still be accounted for too. Therefore the ring or circle is the only way to include all six sides in one aspect. In mathematics there is the formula used in calculating the volumetric inside of the sphere: $a^3 = 4/3\Pi r^3$ which holds two major components that will establish final value where as the rest is indicating ratios. In mathematics there is a line being one quantity and the circle indicator Π being the next circle indicator. Reducing the line will erode the value of Π by ratio. That will eventually lead to having a circle ratio of Πr^2 and eventually lead to Πr^0 but that is not the point where the circle ends. That is where the ratio applying factor ends but it cannot exclude the circle. The circle as a concept can still reduce when it abolishes form to the single dimension. It is not the radius that is responsible for the circle but the figure value of pi and by abandoning π only then does all the aspects fall back into the single dimension. The circle can reduce one step more when the circle eliminated r completely but the elimination of r as the factor reduced the major factor to the single dimension in Π^0. That will not reduce the cosmos to zero it will only eliminate all potential lines r^0 to potential circles $\Pi^0 r^0$ and from there the circle Πr^0 will come about as manifesting as a line but that manifesting can firstly only establish a circle Πr^2.

The only value that singularity can have although the single dimension may host the entire universe is Π^0. Pick a number and elevate it to the power of zero and in the process one may have established another point holding all points in singularity but that is not the value of singularity. Only Π^0 can ever be the accurate value of singularity while singularity will then host the rest of all the possibilities in the Universe. The first value there ever was came in the form of π. Where mathematics was still an idea in development the universe granted values of the triangle being 3 circles as π^3, which was 180° and π^2 which was half a circle also with the value of 180° and finally the straight line also being 180°. Mathematics was not yet established, but the most basic came about. Science is not taking the cosmos back as far as possible, science is taking mathematics back as far as they can but mathematics does not go all the way. Mathematics presented as numbers and symbols only became valid (as did all other aspects) later on in development. But the most basic of mathematics was in place when the spot moved on to form the dot by going from π^0 **to** π.

The reactions of those in charge of producing official policies which are responding to my argument is of the opinion that my argument is silly, but should that be your personal opinion too then test where the silly part applies. Bring the zero into the calculation, the zero that science so eagerly place in outer space and see the mathematical result. The forming of densities is once again establishing certain relevancies and when one remove one factor with a zero the density relevancy goes incoherent. By applying the distance one accepts automatically that the figure become calculated with a one. Since one is a representative of a factor that is having a value and not being without any value because as a factor it represents at least one in being part of the calculating process of the cosmos. The calculation as all calculations normally are is in order to calculate something and the something will at least stand in as one in relation to the rest being part of the calculation. When replacing the one with the nothing that science do when they say they are calculating that which is contributing to space then you can see that nothing is not what you may find in a Universe filled to the point of overflowing. But saying that the factor of one in fact represents the nothing which becomes a name and not a number since nothing is then a factor of one as it is that much the part in the calculation being calculated, then the one has to replace the zero as the fact of the factor of being calculated. You may also think that Nothing can connect a half circle, a straight line and a triangle except their sharing of a value but I try to prove that your granting of nothing is in this case a calculated value being something.

The claim becomes obvious when observing the connection between the half circle, the straight line and the triangle, which could also promote all the qualities lurking behind the pyramid. Consider the connection between 180^0 sharing three different forms all part of mathematics where each is different in form, but equal in value and then one may realise in considering the very basic in mathematics being the Law of Pythagoras on which all mathematics are focused. The triangle stands in for one factor represented by one at a value of 180^0. So does the straight line become a factor of one and the half circle also becomes one where the factor of one equals all 180^0. All three are most seriously part of shapes in the cosmos. Revalue any one form to zero and the rest too must follow and share the same value.

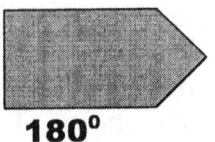   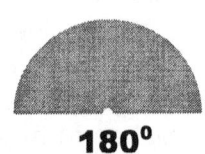   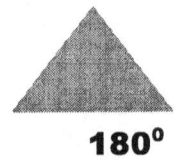

180⁰   **180⁰**   **180⁰**

The only manner in which light can move one year apart form one another while each one is staying one year apart from the source it left is when the straight line that light use to move holds 180° true to the half circle they are apart (180°) and connects the two half circles in a triangle 180°.

Only if light exceeds mathematics and become part of pre-mathematics can light find validity. $\pi^3 = \pi^{2+1}$. What we find in this sketch is what we find in the law of Pythagoras.

The Law of Pythagoras is about angles in relation to lines and not one angle can represent zero because that will reduce all the lines also to zero. The measure of angles between stars at a distance uses parsec as the indicator, but the parsec between the stars indicating an angle has represent an angle whereby one may measure distance and such a distance cannot zero because then the parsec will be equal to. Again it is multiplying the factor with the measure but if the measure is about a factor of zero, then the factor too becomes zero. That is as basic mathematics as I can present.

(diagram: two overlapping circles with an inscribed triangle, each region and angle labelled 180°; the words "to", "be", "zero." appear at left)

Put what I said just now into mathematical terminology it will be the same as saying there are 149 000 000 X 1 (multiplied by the kilometres) multiplied by what it is being measured which is 0 and what will the total come too... a full zero. 149 000 000 X 1 (km) x 0 (indicating what the km are made of) = 0 Mathematics says it. If there is something to be measured then the least value the measurement can have in relation to what is used in the measuring and as a factor that which is measured has to be one. It cannot be zero and be measured...and then we do measure outer space! It sounds as if something mentioned here is at fault. Yet I stand accused by Mainstream Physics of carrying the blame of being incorrect. It is not with my mentioning the inconsistency one should find fault but the fault is with the fact that the use of nothing is there and no one noticed! I am and neither is my work to blame just because I am mentioning it, but the blame must go where it belongs. I should think that it is by now somehow understood although I imagine the implications of the statement using nothing to that affect is not nearly accepted by all. Going back to mathematical basics it is not possible that by adding a million of nothing to one nothing there will remain one nothing and that is still nothing. Nothing cannot accumulate therefore I cannot accept anything holding the vastness of space being able to constitute nothing as the major component. If that is true why try and construct the cosmos from nothing?

Let us dissect nothing from another angle. Mercury has 58 X 10^6 km and Pluto is 5900 X 10^6 km space between the sun and the planet. The one measures about 10 x 10 times more than the other one does. The difference indicates a distance and a distance comprises of something, for if was nothing then both would have equal nothing and be next to the sun or in the centre of the sun. I repeat, the distance indicates something because nothing would place them both in the sun and moreover in the centre of the sun. Having nothing between Mercury and the sun and between Pluto and the sun, place Mercury and Pluto at the same position within the sun. By saying Pluto has one hundred times more zero in between the sun and Pluto makes such a statement laughable. Except if a learned Professor conducting a class does it and no student dare to laugh at his foolishness. If I would say Mercury holds one hundred times less nothing, such a statement will make me an idiot, but used as science makes such a statement plausible. That means the more zero or the more nothing one find between cosmic structures puts such structures further apart. There can only be distance with something concrete applying the distance between it. The problem is identifying something from nothing that defines the difference there is in science. I cannot see how nothing can become plural or more in some occasions.

When realising this I went in search of that which nothing is substituting. The issue I went in search of is what to substitute the nothing with and fill the nothing with that something, which is in place of the nothing and replacing nothing. Let us go on the interesting search of finding what prevents the Universe from tumbling in on itself. If the Universe was truly nothing, the nothing would not have the means to support the structure and the structure would disappear into the nothing that is not supporting it. The Universe is about lines forming angles and holding distance, that much we established so far. All that we know about what we now know started with Kepler's formula of **a³ = k.**

T² and that I may add was presented by the cosmos "in person" through calculating the orbit of the planets. It was rather incorrect of Newton to change this information by adding $4\Pi^2$ on the one side and G (m + m$_p$) on the other side after all Kepler got his information straight from the "cosmic" horses' mouth.

It was the Cosmos telling us humans through Kepler about the Cosmos. To change what the Cosmos told man by then telling the cosmos back what man is of the opinion of what the cosmos should be in the eyes of man is blatantly arrogant. That was what Newton's strongest characteristic was in any event and he even told the cosmos what he (Newton) as a person thought the cosmos should have told Kepler and what in his opinion was correct as how the cosmos should stand corrected when the cosmos gave the information to Kepler about itself. Newton told the cosmos what the cosmos should have been telling Kepler in the opinion of Newton. It can only be an act of utmost stupidity when man is telling the Cosmos what that person is of the opinion about what the cosmos should be telling man. This telling the cosmos instead of listening to the cosmos telling man is a trend still going on in our modern society. We run our lives by the cosmos laws every day and never notice the laws of the cosmos that we use to give us twenty first century comfort. Physics teaches us that there is only energy with no differentiation between forms of energy. The Big Bang was energy and the internal combustion engine is energy. Since there are only similarities one should then be able to find such similarities. In the fuel oil engine the engine heat up the atmospheric temperature in the combustion chamber and the fuel pushes the engine combustion chamber temperature to further levels where the equal of such levels we find on the rim of the outer edge of the sun. Igniting the fuel enables the ignited fuel to be able to push the combustion chamber temperature to the same temperature as that what the sun has on the outside edges of the sun. That means we are pushing conditions inside the internal combustion engine to conditions that is applying currently on the skirts of the sun. When the sun broke free from outer space as the sun's atmosphere released into independent space in motion such developing came from outer space at the time as the temperature of outer space dropped through expanding. The sun captured or froze the limits of outer space as outer space went on further to overheat. The temperatures we find in the engine were the same temperatures of outer space when the sun broke free from outer space at the time the sun established individuality. What applied in the Universe some time ago we humans have to duplicate to get heat released as space from fuel. That is what enables us to use as energy. If it applies to the inside of the cylinder, which is merely a container it also must apply to space in the Universe, which is also merely a container. The question to answer is how can any container contain nothing that fills a Universe?

Let us dissect nothing from another angle, as we find nothing in the presence of the cosmos where science placed the nothing as the prime pillar supporting the structure of the Universe. We all know that the distance between the sun and Pluto is roughly one hundred times more than the distance between Mercury and the sun, but both has nothing between them and the sun. This then means that one has one hundred times more of nothing between the sun and Mercury than what the other one being Pluto and the sun, has. The space filling the distance from the sun to Mercury has nothing times hundred less than the space holding nothing between Pluto and the sun. That means the distance between the sun and Pluto is as equal in relevancy than the distance from the sun and Pluto since both used the measure of nothing. If the one substituted the factor that we see as one represented by Y^0 or whatever we wish to use to represent a factor as one in the calculation with the nothing science at this stage place in place as a valid value of nothing, all laws of mathematics will go in disarray because when one multiply any number by zero it becomes zero placing both planets in the sun. When there is nothing between the second whatever then whatever must be in the centre of the sun. The distance between the sun and Pluto is **5900 X 10⁶** kilometres of space, but in that statement we take it that the one as a factor of a kilometre is present in such a multiplication. The one constitutes the presence of fact being a statement of a value. By saying the distance constitutes of nothing we have to substitute the one factor with a factor of zero. Then the calculation must read **Pluto is 5900 X 10⁶ X 0 = 0.** Including nothing as to state the presence of that part contained by the calculation delivers the total of zero. By excluding nothing from the equation space becomes something bringing in a value lying inside the realms of the infinite that must form singularity. Applying this logic to the Lagrangian system and interpreting that information to the law of Pythagoras a clear pattern comes about.

When I try to point out such in discrepancies in the thinking of Mainstream science, which they are responsible for official information, those in power don't hesitate to show me the power of authority they enjoy. I would think that Academics would honestly be surprised when someone draws their attention to the fact that they are possibly incorrect instead of getting defensive and annoyed by the whole matter. They are in a powerful enough position where they can dismiss me by blaming incoherency on my part as the reason for doing so. By the same margin as they declare my work being incoherent they use such a blame as an excuse whereby they then refuse me another opportunity to defend myself when presenting my work again and a revision about my view by simply saying my arguments and my use of mathematics is onto grabbing straws. The Appreciated Academics of Important Standings may use any phrase to dismiss me and I do not have any chance to challenge them in further debating because according to them, I have had my chance.

As much as I have searched I could not once find any method any one used to prove mathematically that zero fills outer space but my effort to draw attention to that matter brought me the sum total of nothing. Please if there is proof out there to that effect I would love to see such mathematical proof. When using Kepler's mathematics without Newton's dubious changing thereof I can prove that the same substance produce distance in space and motion in time because of the infinite and precise control we find between neighbouring structures. While Academics blame me for being incoherent no one ever showed me where I detoured from accepted mathematical principles in doing that. But not once was my statements disproved mathematically by proving mathematically that nothing (zero) can become or form distance by which to measure. No…my correspondence is merely swept from the table or ignored on the grounds of a lack of importance which they find in the subject and then because of the lack of immediate importance, such lack of importance in my work disallowed their further involvement. Please for once prove mathematically where I can locate nothing between planets because that will mean that $a^3 = 0$ and $T^2 = 0$ and $k = 0.$ That will bring about that all planets are on the very same plane being somewhere between nowhere and nothing. If space was zero or nothing then Kepler's principle formula will stand untrue and without substance everywhere in all of science. The fact that singularity spawns all natural weather phenomena is an indicator of such endless influence.

By reducing the line we come to the end of the mathematical equation of the circle but the circle does not end there. Newton did not recognise this from the figures the cosmos presented to Kepler. The circle only secures the final cosmic figure and the value to singularity where all things have equal value. At that point the half circle and the triangle and the line must start since all three having many different forms have equal value at 180^0. Only after that point does mathematics begin where all factors in 1 have the value of 1 being 1^0. In that conclusion one realises something must separate singularity from all other factors because singularity hosts all other factors but is by own initiative Π. That will be the spot of origin. That will hold the eternal spot…the smallest spot ever because all spots that ever can be was secured in a position in the centre of that spot. Because of the progress singularity follows from the single dimension singularity only allows mathematics a start at Π^0 progressing further too $\Pi\Pi^0$ and from there the line is born as $\Pi\Pi^0\Pi^0$ or $\Pi2\Pi^0$ $\Pi3\Pi^0$ $\Pi4\Pi^0$ $\Pi5\Pi^0$ where Π^0 then may form the concept and value of r. But the line starts at $\Pi^0 = r^0$. Because cosmology is singularity based and the value is $\Pi\Pi^0.$ This escaped the attention of the greatest mathematician about the work of the greatest cosmologist ever because Newton incorrectly introduced $4\Pi^2$. The introduction of $4\Pi^2$ exaggerated the value of time and removed space / time from the concept. Mathematics in cosmology does not apply pi, pi is the root value of all concepts in cosmology. The factor pi impersonates as much as it represents singularity. But we may ask why Π will come about from singularity.

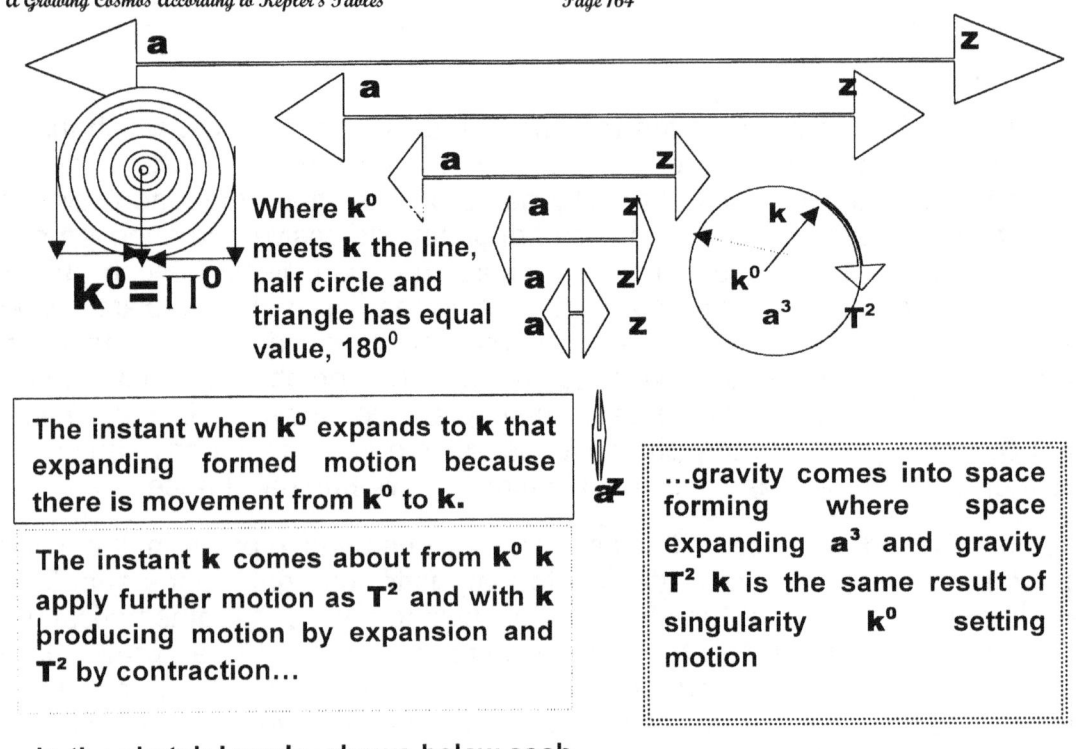

Where k^0 meets k the line, half circle and triangle has equal value, 180^0

$k^0 = \prod^0$

The instant when k^0 expands to k that expanding formed motion because there is movement from k^0 to k.

The instant k comes about from k^0 k apply further motion as T^2 and with k producing motion by expansion and T^2 by contraction…

…gravity comes into space forming where space expanding a^3 and gravity T^2 k is the same result of singularity k^0 setting motion

In the sketch I made, shows below each of the lines reducing there is a space left open

Let's go back once more and reduce the line by half every time. Then repeat the process until it can repeat no more. The reducing of the line by half every time will get to a point where all the ends land on the same position without any possibility if halving the two ends further. The points share one position and moving the points in any direction will lead too an increase of the line once more.

In the sketch I made, shows below each of the lines reducing there is a space left open between the two ends of the line that is symbolising the end of the line in reducing. In the end the two ends will share one location even by having one single point holding each one. There is no chance that I can present any sketch reducing the line to a point where the points are sharing one location literally in the single dimension. The points are there and with the points being present they may not be dismissed as nothing. From there no reducing in a natural manner can lead to nothing without changing the rules of mathematics in such reducing. But the two ends has reached a position where any further effort of reducing must bring about the start of extending because every point possible share space with every other possible point at the point of singularity where all points share one common space. By moving any of the points such moving must then bring about an increase of space once more. This also applies to the circle because the circle uses a line to indicate size. By reducing the line and by reducing the circle the reducing will end up having the ends in the same position. It is this fact of the moving of any point from that spot holding singularity that such motion will introduce space as the space exceed the previous limits of singularity.

One must draw this statement of motion back to the point where singularity is getting sides. When there is singularity there can be no sides. It is 1 (one) from all angles there can be. That one fills a space. The space it fills does not really exist in the manner we humans see space to exist. It is a spot that is there without being there. It does not visually exist because it is not filling any substance and it cannot be recognised. Once one accepts the fact of singularity that accepting of singularity then is contradicting all the things we know by not being any of the things we can recognise. There is no space. In that space there can be no motion because there can be no space to have the motion within. It is a line that is so small it is not there and the only reason why we know it is there is because of the results it left as an imprint of its not being there. We cannot detect it on the merits of its absence because it is never absent. It cannot be absent. It cannot go absent but it can never be there where it should be in the third dimension if I wish to locate it. If it was absent then it was zero or nothing but since it is there it is not there and that makes it present. The centre spot we cannot see and that we cannot detect has no sides to any side and has no place it fills because it fills all the places we cannot detect. The only way such a spot can fill space is by doubling the space it fills to become more than one place to fill. But the very instant that happens it halves the space it fill because it then cuts the space it has into two parts. Any motion from such a point in singularity lands in and on the other side of the Universe. That brings about that the point of not being is doubling the not being and by doubling the not being into being it also cuts the not being that became present into half. We have to find this spot as we find religion. It is something that we can only know is there because we cannot disprove it is there but we can never prove it to be there either. It is something seen through intellect and not through the eyes.

From the smallest ever possible dot will grow a line in every imaginable direction relating to a prospect of Π because only Π will not favour one specific direction and that puts all directions at equilibrium meaning that any form of what ever might develop from such a spot will have the end and the start being in the same position, which will also have to be a sphere as the flow outward will be equal in all directions. This is why we humans show the incentive to acknowledge this fact as we the sphere being representative of the form of the entire Universe. The smallest spot in singularity is a sphere. Please think clearly, is that not precisely the commitment we find in gravity, where gravity is flowing from singularity outwards but never favouring any side? The nature of gravity is to never end and never favour and where it seems to favour there is a valid explaining concerning singularity. This reasoning prompted me to look for singularity in such a spot because if the prime spot from which all came was a spot holding all, then the spot must hold the shortest line but more prominent it will hold the smallest form including the smallest circle or for that matter the smallest sphere. That leaves the door wide open for the advancing of any radius in all possible directions. With gravity always being in the centre of a sphere where the space is least available in the entire structure (there is not even space left to fill) one finds a flow of gravity from that centre spot outwards in all possible directions even-handedly. The fact that the original gravity will begin as a circle or will be a circle is the direction it will take when being the first spot created. All progress will be evenly in all direction because no direction will stand out or be in favour above any other direction at first.

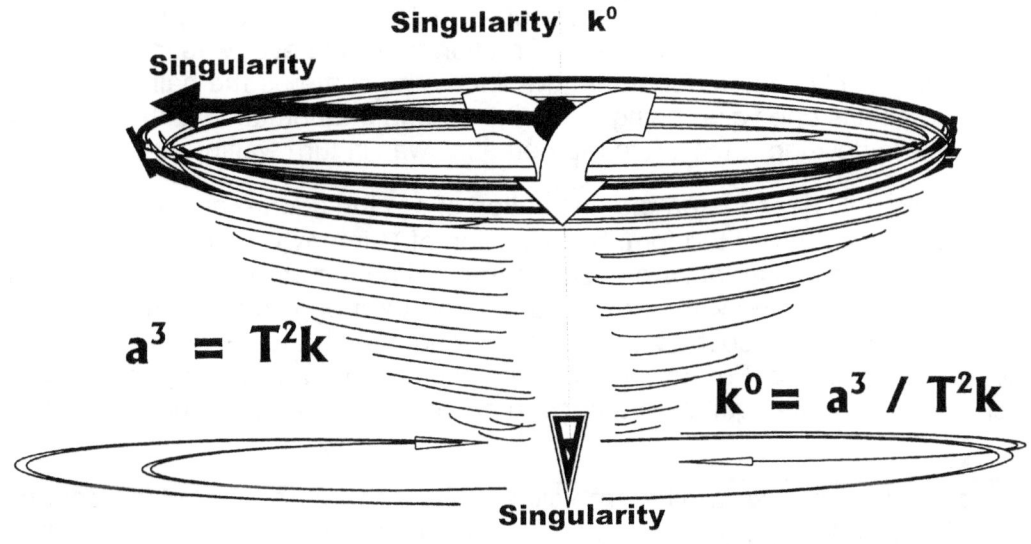

Within the circle k^0 = a^3 / (T^2k) holds gravity centred in the precise middle of the circle. By using mathematics in the way Kepler used it those rules and laws used correctly in the investigating of the formula that Kepler introduced must form the basis of cosmology. Also such intense investigation then must be without Newton interfering and telling Kepler what he (Kepler) should have found instead of Newton incorrectly correcting Kepler whereas instead Newton should have been looking at what he (Kepler) found because only then he (Newton) could have seen what gravity is. He (Kepler) said that the cosmos said that gravity is $a^3 = T^2 k$. The space is held in check by motion from a centre and that is gravity. It becomes more than clear that space a^3 is time by dimension T^2 and time is space a^3 without dimension k Gravity is a^3 / k but k is an addition of motion T^2.

The spot forms a full circle, but the line running through the circle is forever present because that is the future radius of the circle that will one day develop the circle, which is equal to the present diameter. The fact of the presence of such a possible line in such a possible circle dividing the possible circle into two parts makes the centre line equal to the half circle. The line forms the half circle but not only that the line presents the half circle as much as the line is the half circle. When referring to a circle I use the name of a circle because there is no other referring by name available but such a circle represents form and not yet space. The universe at the time was so small concepts we take for granted today was eternities apart. In the centre of the form runs a dividing line that is the form eternally but also it is eternally smaller than the form because no measure other than eternal and infinite was available to use. Notwithstanding the concept that the line was the form but also what it came from was eternally smaller than the form. This is because the line parted the form into two parts of equal halves of exact duplication. The line then is 180^0 and the half circle forming on either side of the divide is split, that formed into two separate halves leaving both half circles in equal value to each other and equal to the line parting the circle. The value in all cases is 180^0 because in singularity the two factors are the same. Mathematics as we know it at that point in eternity has not yet developed because at that stage there were only space less relevancies between the same thing split into segments that was innumerable.

The same value is of course $k^0 = 1$. In this half circle of the future, which is no half circle as yet because of a lack of space there are three future points indicating the space less ness that will go on to become space filled with something. On top of such a circle to form must be a marker indicating an awaiting boundary or future border and at the bottom of the future circle there also must be a similar marker that is no marker as yet. Between the two possible points that are not there yet is a future line running that is not there yet. Then indicating the possibility of a position to come that will bring about the half circle being a future distance apart from the future line indicating a diameter that will one day be there, a third such a marker must be established for the future. That forms a triangle with two more sides being connected by either a line being one or half pi being one. Crossing such a divide comes down to the very same as jumping over from the end of the Universe to the very end of the other side of the Universe. What was doubled so that whatever the Universe was before doubled by a not quantifiable margin in a not quantifiable number of possible jumps? Most important fact to observe is that the slightest notion of motion coming about at the time was moving in a space so small that by attempting motion just the slight inclination of motion took whatever was moving across the entire Universe. From singularity comes about that the line is the same as the half circle and is the same as the triangle and all has one value being 180^0. Space and mathematics at that point was waiting for the future Creation to develop ...therefore mathematics was in turn waiting to develop.

In any circle or sphere the size only depends on the fluctuation of r in the square as a component to the circle or sphere but that does not affect the form by indication of Π in any way there may be. The conclusion from this is that no line can start at zero because that will be a mathematical impossibility. This statement by itself excludes zero and with zero excluded one then begins to appreciate all the rest of the concepts governing corrected cosmology. A line or spot starting at zero would therefore be shorter than the shortest line possible. For obvious reasons can no line, or any line grow or extend from zero because such a line must then quit zero and become something, thus abandon its original value.

That would mean the start of the line has a different value to the end and a line holds conformity through out. When any line is starting from point zero it can never leave zero because of the influence of being zero disqualifies any possibility of growth. If the line then had to grow in all directions at the same pace the line must therefore be a circle or being three-dimensional, a sphere. Flowing from this fact is that in the universe there can be no zero point or unfilled space. In the case of the growing sphere the value of the circle is Π, and that is where creation started. That gave me the clue where to start looking for singularity. One would find singularity in the value Π and the value Π will be in all things rotating in a circle. You might wonder how does that apply to the cosmos and moreover to gravity? You cannot fit nothing into outer space because it just will not fit. If any of the factors in Kepler's formulae represent nothing, that is what you will get. The Universe will be nothing. $a^3 = 0 \ T^2 = 0 \ k = 0$ If the argument seems ridiculous it is not my mentioning such a fact that is ridiculous but the mere fact of the reasoning also becoming a recognising of an argument accepted by science making it as such ridiculous. It is the fact that one must argue about such a ridiculous matter that allows the ridiculous part enter the conversation because the trend reminds of arguing about fairies and little people and such nonsense. If space is nothing then it has a number to use indicating just that value being zero or the capitol O indicating zero. Try and indicate what is measured and calculated in space, but not by simply not thinking about the fact and therefore simply ignoring that what is measured forming the sole value of space, but put the value of nothing as part of the distance in calculation because that is what is measured. When stating the distance between the Earth and the sun place on paper what will allow the kilometres measured to represent the factor that is being measured. If represented by one being the total of one by hundred and forty nine million kilometres of nothing put that language in the International language of mathematics that spans all dialects spoken on Earth.

By reducing r indefinitely to the tune of half each time, r would become infinitely small, beyond human calculating means, however as mentioned in the case of the smallest dot holding one spot, r would become insignificant beyond human comprehension even, but never reaching zero and still Π would remain intact and dictating form. I believe one can begin too see where my suspicions are heading because the flaw comes about in the manner mathematics are practised for thousands of years. Before coming to the mathematics I would first like to bring your attention to the practical side. I am promoting a theory in which I am able to prove there is as much contraction going on in the cosmic universe as there is expansion and the contraction is as much part of the expansion. The universe rides on a balance and we have to locate such a balance. To prove my theory I firstly had to locate the centre of the universe.

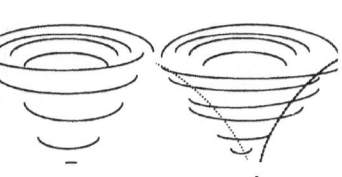

To find the invisible I had to locate singularity. I realised that my effort to locate the point holding singularity enabled me to backtrack the exploding universe to its origins

Even admitting to such a notion sounds like madness, but please give me a chance to explain in more detail. If I wish to achieve success that would depend on my ability to convince all that outer space comprises of material and as such we can locate such material even if we are unable to see such material. By applying some basic effort I have located the position from where all movement came and the direction it took moving forward in time.

We traced the line back to the spot and so we know, where it all started there was a spot. The miracle part comes in the fact that eternity shows no change. Any change, even the slightest change represents the end of eternity. With everything motionless and locked in singularity everything remained eternal because even today singularity is locked in eternity. In the centre of all objects spinning there is a space forming a divide. It divides every aspect of space spinning into sectors of space having motion as much as changing direction of initial motion. However, since it was part of the cosmos, and although it is outside the cosmos, it still is part of the cosmos. A spot holding what ever is and can be into a dimension so small it did not even have sides. Then for a reason, which at this time I do not wish to go into some miracle happened and the spot showed motion. The motion was deliberate but the motion was eternally small. The spot had one specific value, which it clung too…it had all sides on the same side and from that rotating diverting came about. The forming of Π^0 =k^0 = 1 had all possibilities available but it was only one possibility in the end. There was no space and then space expanded producing space $\Pi^0 \Rightarrow \Pi$ and from that motion comes gravity by the value of the motion creating contracting Π^2. The motion brought along Π but even Π was subjected to relevancy. From this come the most basic principles in as much as forming the ground rules of the law of Pythagoras.

When drawing a line such a line then starts of with a dot serving the spot that holds all sides equal. That means the line serving as the future radius will be equal to the half circle which is then Π. The only aspect of the point that stands in for the end of the single line forming the radius of the circle is that we then mathematically reach the single dimension. We decreased the line to where a circle being Π formed on the single dimension. This dimension also holds the circle dividing line because from that line the radius must once again generate a value and by such a gesture (because it is too small to be moving) that motion will deliver the line extending and that would form the circle that forms the sphere that eventually leads to the formation of particles. This leaves a problem to investigate.

The Roche limit is:

The region surrounding each star in a binary system, within which any material is gravitationally bound to that particular star. The boundary of the Roche lobes is an equipotential surface, and the lobes touch at the inner Lagrangian point, L_1, through which mass transfer may occur if one of the components expands to fill its lobe. It names after the French mathematician Edouard Albert Roche (1820-83).

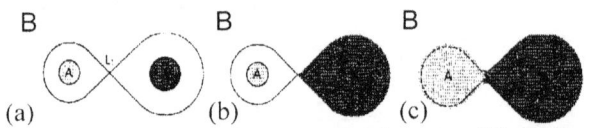

THE ROCHE LOBE: **In a binary system, the Roche lobes of components A and B meet at the L_1 Lagrangian point.**

> **(a) In a detached system, neither star fills its Roche lobe. (b) In a semidetached system, one massive component, B, fills its Roche lobe. (c) In a contact binary, both components overfill their Roche lobes and share a common envelope. Lets explain the importance of this Roche limit and how the Universe used the Roche factor to produce the Big Bang. That is where it all started...**

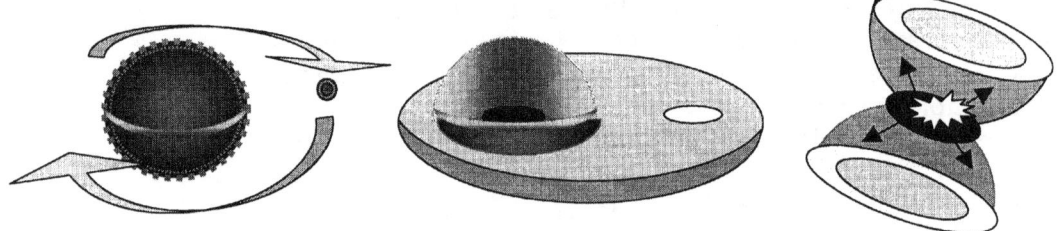

More friction resulted in producing more heat by more motion that produced more space. It served as a self-destructing devise in almost all cases but one. This cycle of heating affected all possible synchronised motion but one possibility remained viable to secure material. In all other cases the lack of space brought about heat leading onto more motion but this time the motion was directed in a single direction of securing expansion.

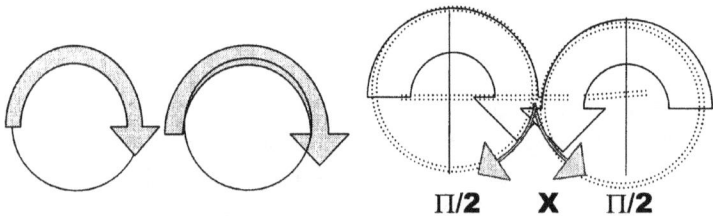

$$\Pi/2 \quad \mathbf{X} \quad \Pi/2$$

Where synchronised directional motion is in harmony with the Universe but also is in opposite relevancy to the Universe the relevancy brings about form and creates space. Some particles were reducing space this way by motion in contraction. This then formed gravity and came in as gravity. If this form produces gravity any other motion not complying with this synchronised directional motion in opposing harmony must be producing anti gravity. This centre point is present since the very first eternal instant and is present in all of nature. In every one of the natural phenomena the circular displacing of the Earth is turned into a linear motion and from the linear motion it receives rotational acceleration that generates a centre presenting singularity to take charge from that point. In all cases it is motion $T^2 k$ that sets space a^3 in relation to a specific centre k^0. Motion of many sorts establishes a generated singularity that activates space-time as it happens in the case of the spinning top.

It is easy to see why it seemed as if there were carnage and destruction in a form of matter eating up antimatter and in a sense that is the case but the destruction came about because of the lack of space causing friction that brought about heat that turned to space. This took place because space is the motion thereof. $a^3 = T^2 k$. I guess one can say that if matter dissolves such dissolving matter can become antimatter since there were only two options available at the time but I prefer to use the term heat because by any other applying name confusion sets in again. There was material producing gravity by performing motion and then there was material performing anti gravity by producing heat. This was due to the restriction where by the particles was lacking the application of motion and with heating the heat created space that formed part of material destructed or then possibly became what science refer to as antimatter. But antimatter and plasma and all other names available does not fulfil the function of establishing recognition about what one refers too. Using heat as the term is the least alien term available because in the last sense every one knows what heat is. Also in this process on the other side of another divide light became another product of heat and light is heat reflecting a sure connecting to dissolved singularity.

Singularity forming time initially formed time when time began by initiating gravity duplicating space with time in forming the Roche limit. This process occurred when material was forming space and time became a presence in the growing cosmos. This was part of a time that was an era before the Big Bang became an era. Singularity formed by the Coanda principle providing a centre Π^0 relating by Π to establish motion as Π^2 in relation to space-time Π^3 of which Π^0 controlled. But part of the centre and part of being Π four points form on the rim of the rotating circle and because of motion and because of the Coanda effect the four points establish motion and as such a relation of $\Pi^2/4$ establish four duplication points. That is how the Universe began as I describe in the **Cosmic Birth...Dismissing Nothing**

Consider what happens to a star that developed closer than the Roche limit of Π to $\Pi^2 / 4$ would allow, it is easy to see how the singularity centred grew by concentrating the heat the points in singularity brought about. Today we find that because everything started that way everything about nature comes by activating the Coanda effect in some manner where motion indicates an exciting of singularity positioned by applying motion. It is the value of the motion that brings about the time and the time "shrinks" space as the speed of motion will reduce or exaggerate the "size" of space claimed for use during that motion.

This sketch exemplifies all phenomena in nature as we use nature to us with life's advantage

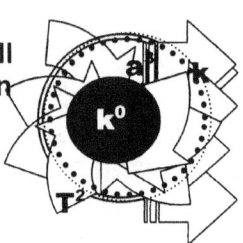

With the Universe being that small in the very beginning there were two options available for material to choose from. There was expanding because of overheating therefore becoming relatively softer or remain relatively more solid and cool off by reforming through contracting heat released by other particles. It was gravity and antigravity. One must remember that in this time when all started the slightest motion, so slight we humans can never find the ability to detect the space or the motion that went with the space, such motion took such space to another Universe or jumped to "the other side" of the Universe.

The rotating motion of liquids around a centre is what gravity is as the Earth rotates around its axis. Not for one minute should any one forget that the centre axis of the Earth is a liquid with all the heat gathered between the solid particles in the Earth centre. Take into consideration the fact that with the heat concentrated in the centre even the $iron_{56}$ in the centre is a liquid. By rotating objects in liquid (which is what the atmosphere is) the motion activates governing singularity a centre which is elected by all the atoms with the confinement of the Earth and intensifies but also localise gravity to a specific point shared by all space in that motion. It is using this principle where electricity is charged in this manner and where hurricanes generate space-time. Gravity will have a charge equal to electricity in the extreme centre of the rotating Earth. We must remember that gravity liquefies the atmosphere through compacting the space-time of the Earth from the gas it is in outer space to a liquid in the centre. The atmosphere in its natural form may display as a liquid in a natural form but can still also be a gas when water is in motion. The generating of singularity is displayed in electricity charging as it is in lightning, as it is in igniting fuel in turbine rocket engines, as it shows the ways the Coanda effect applies, as it is presented in weather and as it limits the space-time in the behaviour of the internal reflection where the flow of light is limited by the space-time supplied by the flowing water.

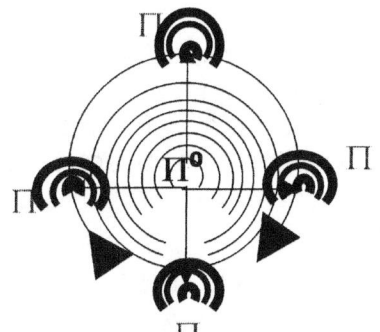

Gravity in the centre formed timeΠ^2 by Π dismissing while the four time positions started the cosmic trend of duplicating.

With every one of the four points taking form to the value of Π at a measure of $\Pi/2$ this had the result that each brought about the Roche value of $\Pi^2/4$ in relation to the developing centre. One has to remember that the star of today takes on the characteristics of the form of that era.

In this there were two options to cause material destruction, which means producing space by heat and destroying singularity at the same time or performing gravity and preserving form with the preserving of material.

Coming back to that which Kepler saw, it was Kepler that gave the factors distinguishing symbols, but the symbols hold identical value but only holds dimensional differences.

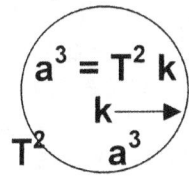

Distance k is equal to the space a^3 it is in and the time T^2 it takes to move that space around.

$k = a^3 / T^2$

$k = a^{3-2} (T^2)$

$k = a^{3-2} = k^1$

$k = k^{3-2} = k^1$

and

$a^3 = T^2 k$

$a^3 = T^2 k^1$

$a^3 = T^{2+1} (k^1)$

$a^3 = a^{2+1} = a^3$

and

$T^2 = a^3 / k$

$T^2 = a^3 / k^1$

$T^2 = a^{3-1} = T^2$

$T^2 = T^{3-1} = T^2$

It is all the same

In the past the scientific enormity of Kepler's statement passed Mainstream Physics because (I believe) that Newton at the time disregarded the importance of placing compositions in mathematical relevancies. Later on no one saw himself worthy of controlling what Newton may have missed and this blemish went unnoticed. But once a relevancy is established through investigating the relevancy from all possible sides the enormity of the concept becomes transparent. It links combinations that we now can see four hundred years after the fact.

In the Universe and throughout the Universe space contains particles by sustaining the six-sided space forming a balance. The relevancies are top opposing bottom where left is opposing right and front is opposing back. When one of the six sides moves away, due to the direct contact with a sphere forming a cosmic structure the object then makes contact with singularity. By making contact with singularity the object falls from the sky towards the centre of the sphere. It is the same therefore it is dimensions repeating to form a^6 because $a^3 = a^{2+1}$ that becomes $k^0 = a^3/a^{2+1}$. In the way Kepler presented space and time it is all the same thing that space is made of the motion forming time. With this tendency Kepler confirmed that what Kepler introduced with mathematical equations that Mainstream science preferred to ignore. But all the further ignoring will only produce more ignorance because from using Kepler observations one can read so much more into cosmology. The future of understanding cosmology can only result from accepting what the cosmos told Kepler when the cosmos told

If ever there were one scientific blunder that put science on its back exposing its under belly and brought along so many misconceptions then it is Newton's ignoring of Kepler's brilliance in the face of facts. I believe there was much loss of life, not only from space flight but also from atmospheric flying in the past because of the blunder science inherited from Newton's incorrect presumptions of Kepler's work. When saying that I have to add that in the past my saying this brought about immediate dismissal of all I have to say but I cannot ignore the truth.

Kepler that $a^3 = k.T^2$. From this we see where singularity is $k^0 = a^3 / k\,T^2$

$a^3 = k.\,T^2$ But Kepler also said

$a^3 / k = k.T^2/ k$ and Kepler said

$a^3 / k^{0+1} = k^1 = k^{1-1} = k^0\,T^2$

$T^2 = a^3 / k^1$

$a^2 = a^{3-1}$ or

$T^2 = k^{3-1}$ $k^2 = k^{3-1}$ $T^2 = T^{3-1}$ We can see why motion is the culprit forcing us down towards the inside of the earth in the process we blame as gravity. $k^0 / k = T^2 /a^3$ That is not what Kepler found $a^3 = T^2k$. Our instincts agree with Newton and we ignore Kepler because of what we feel with Newton. We feel we are pushed to the ground by some unexplainable force that is forever souring our lives. But as I have explained, it is because our motion is slower than the motion the Earth has that the earth holds our space captured in the space we occupy. Because the Earth and we share space in the same space we do not share the same space because the space with which the "earth drags us down is not the same space "we are getting dragged down with".

Due to size differences the space we have has much slower space reproducing coming from the less space we hold compared to the space we share with the Earth. The duplication of space $a^3 = T^2k$ of the Earth is much more and therefore much faster in ratio of duplication a^3/T^2k than we have. This discrepancy in space equality will tend to reduce our share of space $k^0 / k = T^2 /a^3$ because we lack the motion to equal the time thereof. We have to remember that the Earth is at $k = a^3/T^2$ in relation to the sun, but since we are the captured property of the Earth we are $k^1 = T^2/a^3$ in relation to the Earth. Reading the mathematical expression of the formula and from that translate what Kepler said we find that motion is gravity and motion produces space as motion causes space by duplication $a^3 = k.\,T^2$.

This expression translated to $k = a^3 / T^2$, which proves the systematically expanding in time relativity which Hubble discovered that is in progress.

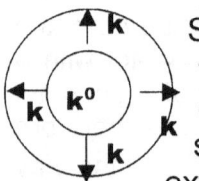 Singularity started with expanding k^0 to k or for more precise valuing from Π^0 to Π. The expanding came about from the overheating of singularity and the motion coming about with space expanding. The motion is in itself created space by extending k four times since there are four relevancies. One should remember that since there was no 3D space yet the expanding was in the flat space less world of the proton at the double square of singularity forming time related space $(\Pi^2 + \Pi^2)$. But that was not yet the Universe we grew accustomed too. This was only formed by antigravity

But keeping Π as one ($\Pi^0 = $ **1**) we keep the Universe in the first dimension.

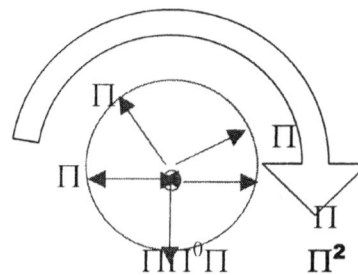

This point, which I now am referring to, is the point where Π is a fully appreciated value while the diameter D still remains a dimensional factor of one. This is the dawn of the second dimension where space was there but space was sparsely shared in some cases. It was Π^2 when Π^0 shifted to become Π for the very fist time.

By every motion that came about since the start of the Big Bang the distance **k** expanded every time the cycle a^3/T^2 completed. Our atomic structure combined produced less **k** than the Earth atomic space combined because although the Earth does not have an atomic space structure yet, the atoms accumulate their attempt to move and the concluded effort is the effort of the accumulated motion of all the atoms in the unit that produce the motion and the accumulative effort combines as the Coanda effect, but on a massive scale. As the duplication of space is a product of gravity and our duplicating is on a much smaller scale than the Earth does its duplication and while we are sharing motion or time with the Earth in the space of the Earth the reproducing of our space will lag behind that of the duplication of space that the Earth has to complete. Since we reproduce less space we are confined to space reducing by the Earth. But since our atomic units are firmer than the reducing capabilities of the Earth gravity which is space reducing we resist the reducing and that resisting gives us the mass we find we have. This reducing in comparison is the result of the earth producing an effort to reduce and even bring on the demise of our space by placing our **k** negative in comparative relation to the **k** the Earth holds. It is all about two factors that determine the question concerning the natural flow. It is the duplicating of space a^3 relating to the demise of space, which is presented by the motion **k** T^2. I delve much deeper into this aspect later on as this letter progresses with information exchanged. The two factors we now examine was the very first motion that brought any and all forms of space into the Universe.

But the equation looks far more sensibility when using the value

$k^0 = a^3 / T^2k$ forms

$1/ k^0 = T^2k / a^3$

$1/ (k^0 k) = T^2 k / (a^3 k)$

$1/ k = T^2 / a^3$

$\Pi^0 = \Pi^3 / \Pi^2 \Pi$

$1/ \Pi^0 = \Pi^2 \Pi / \Pi^3$

$1/ (\Pi^0 \Pi) = \Pi^2 \Pi / (\Pi^3 \Pi)$

$1/ \Pi = \Pi^2 / \Pi^3$

Expressing the equation by using the value singularity has instead of the symbols Kepler designated to the formula he introduced it makes far better sense expressed mathematically. By taking **k** into a negative the space will reduce the time because the space cannot sustain the demand of space growth.

$k^0 = a^3 / T^2k$

$1/ k^0 = T^2 k / a^3$

$a^3 / k = T^2$

$\Pi^0 = \Pi^3 / \Pi^2 \Pi$

$1/ \Pi^0 = \Pi^2 \Pi / \Pi^3$

$\Pi^3 / \Pi = \Pi^2$

In all my other work I make exclusively use of the value of singularity Π since it makes a lot more sense, but when I use the value of singularity which is Π then no one seems to have a remote idea what I am talking about.

$k^0 = a^3/ T^2k$ forms

$k^0 / k = T^2/ a^3$ that becomes

$k^{0-1} / a^3 = a^3/ T^2 /a^3$

$k / a^3 = 1 / T^2$

$\Pi^0 = \Pi^3 /\Pi^2 \Pi$

$\Pi^0 /\Pi = \Pi^2 /\Pi^3$

$\Pi / \Pi^3 = 1 / \Pi^2$

The replacing of the symbols Kepler used with the value of singularity the mathematic equation comes into practise.

$k^0 = a^3 / T^2k$

$1/ k^0 = T^2 k / a^3$

$a^3 / k = T^2$

$\Pi^0 = \Pi^3 / \Pi^2 \Pi$

$1/ \Pi^0 = \Pi^2 \Pi / \Pi^3$

$\Pi^3 / \Pi = \Pi^2$

In all my other work I make exclusively use of the value of singularity Π since it makes a lot more sense, but when I use the value of singularity which is Π then no one seems to have a remote idea what I am talking about.

$k^0 = a^3/ T^2k$ forms

$k^0 / k = T^2/ a^3$ that becomes

$k^{0-1} / a^3 = a^3/ T^2 /a^3$

$k / a^3 = 1 / T^2$

$\Pi^0 = \Pi^3 /\Pi^2 \Pi$

$\Pi^0 /\Pi = \Pi^2 /\Pi^3$

$\Pi / \Pi^3 = 1 / \Pi^2$

The point without movement, the point holding singularity must have a value of Π being the eternal dot but since the dot has no dimension in having form the Π that indicates the dot must be Π^0. From such a point there has to be to the side of the centre point be a point where space do start. That point will then receive a diameter but that point will have form only in being a circle. In that point there is a shift from in relevance from Π to the centre Π^0 and for the first time it brought about two separate values for Π.

The replacing of the symbols Kepler used with the value of singularity the mathematic equation comes into practise. $a^3 = T^2k$ Space created from a specific centre is equal to the motion there of in time established by that centre. Applying the relevance value of singularity the formula reads as follows $\Pi^3 = \Pi^2 \Pi$

$a^3 = T^2k$ Space created from a specific centre is equal to the motion there of in time established by that centre. Applying the relevance value of singularity the formula reads as follows $\Pi^3 = \Pi^2 \Pi$

1/ k = T²/ a³ reads the motion is in relevance to the superior space creation too slow to fill the space. When a relevancy is applying between two bodies and for the reason of not matching duplication in space by having independent motion the smaller body is unable to multiply space in ratio with the dominant space the dominant singularity will require. By the reducing of the distance factor **k** it tries to establish a point of equilibrium in motion setting an equal time applying that is duplicating space. The returning of a body towards the centre of the major body will reduce the requirement for duplicating space is unable to fulfil. By reducing the need for space duplication the individual space must find a level that will support the effort that such a space filled with those specific particles then can manage. But then the space as it then is has to increase the time relevance. Applying the relevance in value of singularity controlling and singularity submitting contact and possible friction becomes a factor. The situation will be leading to the establishing of mass. Applying singularity as a factor in value the formula reads as follows $\Pi^{-1} = \Pi^2 / \Pi^3$

$T^2 = a^3 / k$ reads that time relevancy depends on the space the distance creates in the relevancy. By increasing the motion the space will reduce by the margin of distance relating to time duration. By substituting the relevant value of singularity in the Kepler's formula reads as follows $\Pi^2 = \Pi^3 / \Pi$

$1 / T^2 = k / a^3$ reads when reducing the time as an object does when entering the atmosphere the space the object holds will reduce the distance the object maintains. Applying the relevance value of $\Pi^{-2} = \Pi / \Pi^3$. Again we see this mathematically proves that the flames we see that surround a body when the body enters the Earth atmosphere as the body is coming from outer space towards the Earth.

By reducing the space it removes heat from the density of the surrounding space and distributes that heat into particles. The motion is about removing heat from uncontrolled space situated in the neighbouring singularity and plating the removed heat onto the controlled singularity by having the removed heat joining the reducing singularity and allowing the accumulating of heat which then compensates for space gained by the plating process. The reducing singularity receives more space / heat it then converts to material. The plating would extend as the accumulated heat supplies a larger area to distribute the heat in. This is only effective in the very short term since the growth will not stop the overheating permanently. This would temporary cool the singularity. Then the expansion would once more come about, as the singularity will again start overheating due to the lack of motion, which then places some motion other than gravity in place on the space developed. I suppose in a way as the space in relevancy once more declined it would again force motion to create space once more. When the converting is done the overheating started once more and the process repeated again and again. That is time. That is space. That is space-time. The particles in a better position will slowly but evenly cannibalise the softer space-time by the growth it duplicates using the other less well place particle to develop. This action can be seen as antimatter because matter through gravity is eroding other material into more compact space. This process forms gravity in the one aspect that gravity has but there are nine aspects in all. It is motion of space moving towards a cannibalising centre of space.

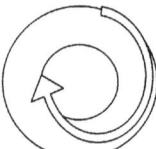 That proves that the establishing of distance **k** will produce space **a³** and set space **a³** in motion **T²** where such motion is in opposition to singularity, which means gravity or contraction is the deliberate opposite of expanding $a^3 / k = T^2$. In the beginning the expanding then also involved three more points all just outside the border of singularity but within the atom exclusivity. It extends **k** while it introduce a returning relevancy back to singularity **k⁰** by creating motion in spin and duplicating space by reducing space. This is the gravity that takes place but the reducing as such must lead to heat amplifying once more because reducing space brings about increasing in heat and with that the reducing did not solve the problem of the overheating. The expanding was leading a movement flowing into the next-door neighbour territory. It brought the seven into the realms of the ten but with the reducing singularity already made a claim **T²** on the space it went into. That space belonged to another more overheating singularity without the reducing ability because of countless factors that lead to that singularity being in that state. By reducing the space it removed heat as particles from the neighbouring singularity by joining the gravity reducing singularity and accumulating heat which then compensates for space gained by plating the reducing singularity with more space / heat that then converts to material.

Using this formulated method we can see that that is precisely what happens when Galileo's pendulum swing and the space decline because the time is being a constant. The space-time is enforced by the much larger growth in space of the Earth that reduces the substantially inferior space produced by the arm of the pendulum in the space covered by the swinging stroke of the arm. That is proof that space is reduced when time is a fixed measure as the Earth singularity will bring about and because of that we on Earth are stuck with mass.

Using singularity as a guide to find the position we are in being a subject of the Earth the formula applying to our position reads as follows $1 / \Pi^2 = \Pi / \Pi^3$. I suppose Kepler did not, by own merit, quite saw what he found. His interpretation of his findings at the time was not that conclusive but Kepler was many times over closer to the truth than was Newton with all his brilliance and the brilliance of all his followers.

In the overall view one may have the opinion that I am totally at odds with the antimatter theory which I am not. I only wish the Theorist would quantify and define the antimatter as a product belonging in the Universe. That the product he refers to is the same heat becoming eventually space that I refer to seems to be beyond any doubt in my mind. There was material producing gravity by performing motion and then there was material performing anti gravity by producing heat and with heat the heat created space that formed part of material destructed or that, which then became antimatter. But antimatter and plasma and all other names available does no fulfil the function of establishing recognition about what one refers to because using heat as the term is the least alien. In the final sense every one knows heat. Also in this process on the other side of another divide light became another product of heat and light is heat reflecting a sure connecting to dissolved singularity. Singularity carries light because light is what remained from broken up and reduced singularity

Light is the highest form of antigravity in motion. Remember that gravity can only be if there first is antigravity. Space-time can only be if there is antigravity. When light became a presence as a part of the other or the anti side the motion that motion produced not space (I lack the incentive to give it yet another confusing useless name) but the spot holding Π^0 which was producing time, which in turn created three parts of space filled with heat. In the event singularity remained space less $\Pi^0 \times 3$ on three borders $3\Pi^0$ but even so the motion remained relative bringing about the Titius Bode ritual and the product coming from that is Π^2. Light is $3(a^0) \, T^2 \, k^0$ because **k** never produced space and space never came from the motion **T²**. In this the only aspect remaining was the three positions heat or space has as well as the motion the space produces **T²** in the contracting of **k⁰**.

Kepler's formula of $(a^3) = T^2 k$ then produced another side (a) $3\Pi^2 \; \Pi^0$
   $a^0 = \Pi^0$ because $\Pi^0 = k^0 = 1$

  $(a^0)3$ space in accordance with singularity

$\Pi^2 = T^2$ the duplication of space light holds

$\Pi^0 = k^0 = 1$ which is made up of $3\Pi^0 \, (a^0) \times \Pi^2 \, (T^2) \times \Pi^0 \, (k^0) = 29.6$
Light has a relevancy to one side of $3\Pi^2 = 29.6$ and to the other side it is $3^3 = 27$. The total sustainable space of light that has the ability of motion is therefore in relevance of 56.6. Two laws came about in relation to each other that produced the Universe and still secure the Universe. It is the forming of the Titius Bode law in conjunction with the Roche limit. Mass was never a factor!

Kepler said the space from the centre running to the space isolated by time is the same. The same space is isolated and the same space is orbiting using time. Kepler said $a^3 = T^2 k$. The time used is space moved because $a^3 = a^{2+1} = a^3$ as much as $T^2 = T^{3-1} = T^2$ as well as $k = k^{3-2} = k^1$. If any one is in dispute of my statement then please show me where did my mathematics fail me? That is what Kepler's formula says and if it does not say that I wish for once that someone can explain to me what is incoherent about my use of mathematical equations. Would some one for once explain to me when during my use of Kepler's mathematical formula did I fail and what do I do wrong? So many institutions in the past declaring my verbal reasoning about the fact that "nothing" cannot be a factor being properly thorough thought and well expressed, have rejected my work. This they (the Academics) say of my using logic, which those then interpret as being incoherent on my part. That they also say about the use of my arguments concerning the use of zero where I prove that a line cannot be able to advance from the original position. That is what I cannot understand of the Highly Educated Academics. No one tries to disprove my argument as every one dismisses my argument without disproving it first.

Kepler's formula suggests duplication by motion and that makes the duplication of space a product of gravity and our duplicating is much lesser than the Earth while we are sharing motion or time with the Earth in the space of the Earth where the reproducing of our space will lag behind that of the space the Earth is duplicating. Since we reproduce less space we are confined to lesser position in space where the duplication merits. We are reduced as a unit and as unit of independent measure we restrain such reducing of our standings in relation to that of the Earth. But since our atomic units are firmer in the unit by some margin than what the reducing capabilities of the Earth is, the earth reducing gravity, which is reducing atmospheric space, that reducing of space we resist. The reducing and the resisting of such reducing gives us the mass we find we have. The atom that forms gravity is approximately forty times stronger than what the Earth presents. Therefore in the conditions the Earth presents, the atom form can withstand 40 times the gravity assault that the Earth gravity launches. This reducing in comparison is the result of the Earth producing an effort to demise our space by placing our k negative to the k the Earth holds.

The Earth is reducing our distance we have to the elected governing singularity and singularity allows such reducing of our standings in k because there is a normal discrepancy between our distance we have in value from the centre singularity and that which the Earth in its entire totality has. What I am referring to is all the atoms in the Earth that is relating to the material that constructs the Earth as a solid unit. Put another way might refer to the position the materials forming the structure have in relation to the social order the Earth insists on. In essence the balance rides on the dismissing of space-time by the group forming the unit in relation to the duplicating of space-time there is and more. The space we duplicate is less than the space the Earth duplicates and since the Earth is growing more the Earth is therefore extending the k applying to the Earth of the Earths growth in stretching k is placing the whole unit of growth of the Earth further away from singularity at k^0 centre than what we can manage in a comparable independent unit relating to the Earth effort. Through this we are naturally staying behind in the duplicating tempo and it is the staying behind that gives us mass. From our producing of space duplicating in comparison to that which the Earth duplicates in the same time factor and is duplicating at a greater pace than us our duplication involves less motion $(T^2 k)$ than does the Earth duplication by motion because we duplicate less space a^3 in relation to the motion but we attach to the same motion that the Earth allows us. Therefore our duplication in motion stands apart from the duplication that the Earth has $(T^2 k)$ to create to duplicate the relative more space a^3 our a^3 / $(T^2 k)$ is much less than the Earth's a^3 / $(T^2 k)$, which is giving us space.

When we find ourselves outside the atmosphere of the Earth a^3 / $(T^2 k)$ the odds turn much more in our favour. When we are outside the Earth boundary our synchronised singularity response has to match that of the Earth singularity growth and the body would then be a natural satellite being outside the Earth atmospheric boundaries. By being a satellite of the Earth we fall victim to the extending of **k**, which the sun permits, and the growth of the sun in relation to the space in influence allows us floating in space around the Earth much more leniency in response to the sun's growing **k**. We then change the name we use for the applying gravity we experience there in outer space then to micro gravity because we then float instead of sink. It is not our mass that brings about the reason that we stay behind in the comparative growth. If our velocity as well as our speed in growth that are in distance **k** and is dismissing T^2 in synchronising is that of what the Earth produce, we will grow in harmony and remain a satellite.

The reason for this is that we found a way where we fall outside the management of the Earth and under the equal management of the sun. If we do not manage the balance we share with the Earth, we will eventually become either space debris or we will end up as particles that are part of the Earth. Time after time, gravity reduces what the Earth increases in space. By measure of overheating the increases show a difference between the **k** the Earth produce and the **k** all other bodies sharing space with the Earth do. But by not being of the Earth and still part of the Earth such increases in the difference between the extending **k** will slightly increase. Naturally that will influence changes to the mass and the volumetric inequality there is. It is also to this reason why Galileo found that there is no mass influences of heavier objects falling accompanied by lighter objects and the heavier objects will not fall faster when normal falling reduces the **k** there is between the Earth and the different objects falling. While falling no object has a mass because of the space the falling the object is in that is unrestricted and unrestrictedly allowing the decline of **k** on par with the Earth **k**.

If an object had a mass Galileo was wrong and Newton correct. Then the pulling of the larger mass would overshadow the pulling of the lesser mass onto the Earth's mass because the larger mass has then greater gravity it can produce. But since Galileo proved to be correct Galileo that in the same token proved Newton as being incorrect notwithstanding the corrupt arguments Academics making to vindicate Newton and his corrupting of Galileo's stance. I have almost heard all the corrupt arguments before and all the interpretation as how the Academics find ways to go around the inconsistency they create but in the end all that matters in the argument is that if mass had any role to play it would play the role while falling. Since it does not influence the falling mass, it has no role during the falling. If an object had mass while falling and the mass was effectively pulling on the mass of the Earth while the Earth was pulling back by mass inflicting force that mass should produce then two objects with mass discrepancies would fall unequally. That is exactly what Galileo proved not to happen in spite of the many times I was very politely belittled in a very well mannered fashion by academics because according to their view I was "unable to understand" Newton.

Again I say as I cast all the clever and senseless arguments of the High and Mighty Academics aside that Galileo proved all mass is mass less while falling because all mass has the same mass while falling because all mass fall at the same rate which is $7(3\Pi^2)\Pi^0$. This is the factor of space displacement in the Earth atmosphere which particles has to overcome or equal in duplication or it will be the rate of their space-time demise. It is the rate that space-time displaces by dismissing related to duplicating. The duplication will grant the body in gravity confinement an independent duplicating rate as it applies a counteracting force resisting the diminishing of the space-time that the Earth produces by motion in gravity. When objects fall notwithstanding whatever mass differences there may be the **k** factor always remains the same in declining when the declining comes about from dropping to the Earth. This is because no mass is present in falling objects unless they on the ground find the ability using the **k** the Earth forcefully provides to duplicate the space created by the motion the Earth provides to the object on the Earth. This fact is the reason why dinosaur skeletons have grown that much bigger that they seem that much larger today than what the true size of dinosaurs were during their lifetime while they were walking on Earth. Once the skeleton is buried and forms a part of the ground in which it is buried, the skeleton takes on the same growth that the soil composition receives from the Earth expanding **k** by extending gravity because we have to remember the skeletons are referred to as skeletons but in truth that is the very last thing the stoned fossils are. As it became stone from bone it grew as part of the rock formation of the Earth while it caste the carbon qualities aside. The distance **k** is forever growing by increasing both time T^2 and space a^3. That makes the rock fossils still growing even after death. Being part of the Earth increases the relevancies applying.

From this relevancy we can see there is a clear issue to be made that at the point where singularity was presented by k^0 it is where k^0 breaks into space a^3 through motion (T^2k) forming matter from which time (T^2k) develops. This is distantly linked to an eternal k^0 but still is changing **k** all the while in dimensions through growth. That means time (T^2k) is as much a product of space a^3 as space a^3 is such a product a^6 when incorporating singularity at such a point. $k^0 = a^3/(T^2k)$ which then in our case only apply to all material which is within the limit space-time confinement in duplication thereof by the rate of $7/10 \ \Pi^6/6$. The value that singularity applies to keep the Universe in a six -sided 3 dimensional Universe is $7/10 \ (\pi^6)/6$ but I am getting to that explaining in a short while. From singularity k^0 time (T^2k) is bringing motion (T^2k) at the release of space a^3 from singularity k^0. Singularity k^0 is forming the six sides of space a^6 or then $7/10 \ (\pi^6)/6$. I also have indicated that even in the six dimensions a^3 is claiming to be in, the one half of the six sides is the very same as what T^2 represents or as that **k** represents. What will solve the problem is retracing **k** and from there we will see how time is aiding space and thereby is giving space the full compliment of what we find to be eternal space.

To trace **k** is to recognise **k** for what it is because **k** is k^3 or k^2 which is the same as a^3 and T^2 and those are just another form of **k**. **k** can be one line in six forming the cube with six sides as we all know but also **k** is the radius or the line running from the centre of any and all circles including all spheres to bring size to the circle or sphere. One has to bring **k** back to the value of k^0 to see how space develops in time from singularity. $k^0 = a^3 / (T^2 k)$. When reducing the circle in size one has to reduce the radius or the diameter because the pi is the indicator of the form as a circle. This divide is possible by reducing the **r** until there can be no further dividing. Such dividing cannot end in zero because no matter how small, there will forever be a value in place. Our mass comes about from the fact that we have a negative relevancy k^{-1} (being a factor less than one) and falling into the 0.9 bracket of the comparing singularity to that of the Earth holding 1. One should not judge this by mathematical laws but rather see it in " less logical" terms of singularity applying cosmic laws.

It has never mathematically been proven that nothing is a factor in outer space notwithstanding Academics blaming me of incoherency when I suggest this fact should be mathematically proven. In overwhelming contrast I am about to prove the building blocks used to construct outer space. What Mainstream physics whish to convey is that in $F=G(M_1 X m_2)/ r^2$ the G factor as being the gravitational constant can be replaced by zero, but not one of them ever thought about that. Then that will bring about that $F=G(M_1 X m_2)/ r^2$ forms $F=0(M_1 X m_2)/ r^2$ and the answer would remain the same. Any one with the least bit of knowledge will see the answer has 0 be zero. Take $(M_1 X m_2) / r^2$ and substitute any of the factors with zero and the result coming about has to be zero. The factors in the equation have to have any and all the elements at a value of at least one. Only if **r** was a factor of one can gravity bring about any mathematical equation that is developing from this argument and which is giving it coherency. That means the mass on both sides must have a factor of one being a limit, which does not allow such further reduction of **r** and any further reducing of **r** beyond the limit will not be tolerated. Only if r = 1 then r^2 can be 1 and mass can be apart. Like it or not but believing in the Big Bang must also bring about the accepting that the cosmos moved apart somewhat by some measure since the start of the Big Bang and the shifting used some means of control. The fact that r brought increase by r increasing in between objects sharing a Universe in the space separating the different particles, the friction causing mass then produces a problem that was solved already. About a century and a half ago Roche found just such a limit.

There is a limit to the radius that is tolerated between two independent cosmic objects. But before we try to find why the Roche limit intervenes as it prevents objects crossing the limit singularity places on the boundary we have to replace the Mainstream idea of having a value of nothing in outer space as a mathematical fact. With such a zero in place in space it nullifies the Roche limit presence as a cosmic law by upholding the senseless idea of zero being a calculation applying in outer space. It hampers all my explaining about how the cosmos introduces space – time and when singularity brought about space-time by introduction of the Roche limit as a starting law by placing that at the point where the cosmos actually began. Once again I was confronted by zero becoming growth. There is a huge hole that needs filling when bringing into a relation any forming of an alliance between a cosmos coming from nothing and filling with nothing and a cosmos growing spontaneously through balance shifting prominence.

Mathematically the fact of applying nothing as a value applying in the cosmos is not a strong and convincing argument. The minute one brings in zero as a multiplying factor forming a definite value working into the calculations of the cosmos, growth disappear. If growth was not a factor, the zero factors could be involved with some form of maintaining stability and where then further growth will accept the responsibility of zero. The closest encounter worth noting we ever had with this law in the modern age of news and Television was the Shoemaker-Levy 9 incident during the previous century, a bit more than a decade ago. At the time and even in the present no one drew any similarities but after completing this book the reader should find why I could draw such similarities, which there is between this incident and the Roche limit. Even the phenomenon called the Sound Barrier becomes clear when applying the Roche factor with the laws governing the influence of singularity.

At first there was singularity holding the entire cosmos with what ever is and will be in the Universe captured and contained in an area that to this day cannot fit inside the Universe we see and we appreciate. The best of all is that there was only one spot, which became innumerable dots but at first before the beginning began there was one spot with singularity in that spot. That spot had the great total of Π^0 but it could have been what ever you choose to use as a symbol as long as it is to the power of zero and not to the value of zero. There was no space therefore there was no motion but because there was no motion there could not have been space.

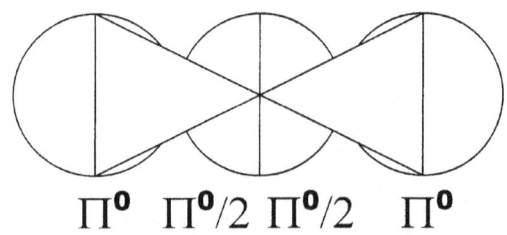

$$\Pi^0 \quad \Pi^0/2 \quad \Pi^0/2 \quad \Pi^0$$

Let us start telling the story as it was.

At the very first sign of any of the sides departing from the centre shared by all, all other points must also show signs of a willingness to depart. There will be one point where r still is one coming in as a factor but pi moves out from only being a factor of Π^0 **= 1** and at that point pi will become a full factor of Π.

The single dimension is a dimension covering everything into the dynamic of one. This brings about that **k = 1 a = 1** and **T = 1**. That is the first dimension and the first dimension is a dynamic of one being the result of a dimensional 0. $k^0 = 1$ $a^0 = 1$ and $T^0 = 1$. The factor **k** was at no stage zero. Only the dimensional factor is 0. The extending that **k** was capable of was zero but **k** as a factor was never zero. The factor of **k** was never zero. The factor **k** can never indicate zero as a point from zero or point to zero because just one zero will dump the entire Universe into zero. The Universe moved from the spot π^0 to a dot π being a multiplication of dot π. That is why by having two π it forms π^2 and with time differentiating the value then becomes $\pi^2/4$. In four sectors of gravity applying gravity then become $\Pi^2/4$ **X 4 = Π^2**

Then motion became a factor. Motion created space as space gave room for motion and space gave room to motion, which is what the Universe is. The room it gave became liquid plasma or heat and the motion formed space growing or space in containing. With motion space is growing as space however the reducing of space is affectively securing and establishing the maintaining of singularity. That brings about growth by means of containing. Then through the relevancy of motion change came about but change involved gravity.

In some way I guess there is room for improvising slightly and find grounds to establish some basis to try and incorporate Newton's idea but that is only to find some connection and not to incorporate Newton.

In an effort to begin explaining, we place potential opposing parties that came about from expanding in the relation to the very frozen instant it expanded by positioning the one mass or Em and across the space separating the two will be the other mass which is also a mass or then form Emtoo.

With no line possible to part the two because the two are taking up anything that may qualify as a line, there had to be another dot that formed since the Universe has many dots that formed lines. But let us not to get confused and lost in the range of possible diversions but let us stick to two dots. One dot was next to the dot next to the dot, but as I said we stick to one dot next to the second dot. M X M / r^2 is the first step gravity began with. That leaves us with a huge problem in as much as when r = 0 then r^0 = 0 and 0 dividing any value will leave 0 as the answer. If the particles were inseparable at the start it must bring about that gravity would not be forming since the distance will not permit any dividing.

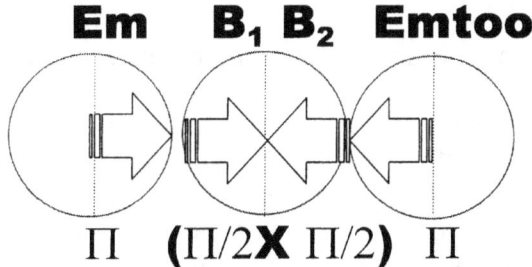

Em   B₁ B₂ Emtoo

$$\Pi \quad (\Pi/2 \textbf{X} \ \Pi/2) \quad \Pi$$

We return to the fact we established before that Em and Emtoo was divided by r and then r had to be one since r could not be zero. Such a centre would then carry the same value as Em and Emtoo. That means whatever value Em and Emtoo receive has to go in equal measure to r with Em sharing half of the divide and Emtoo sharing the other half of the divide. By allowing the distance separating the particles to be zero, the particles melt into a unit. Again this is Mathematics and not my incoherency as some Academics chose to interpret my work or rather then to find grounds on which to dismiss my work without bothering them with the effort of reading my work carefully. Let me run through the argument one more time because I have been insulted by Academics in the past telling me I am bending mathematic rules with my applying double values to try and produce some argument. While they are blaming me it is they whom are being guilty of blatant misrepresenting mathematical laws but by deflecting blame onto me as being incoherent, that incoherency they pass on to me to use to blame me. Please judge my arguments in line with finding correctness in the argument and not just to establish a loophole through which to find an escape root as to avoid reading the work. Please think carefully while you examine the next couple of sentence because this is what max Planck with all his mind-boggling brilliance missed.

A   B₁   B₂ Extending into the distance

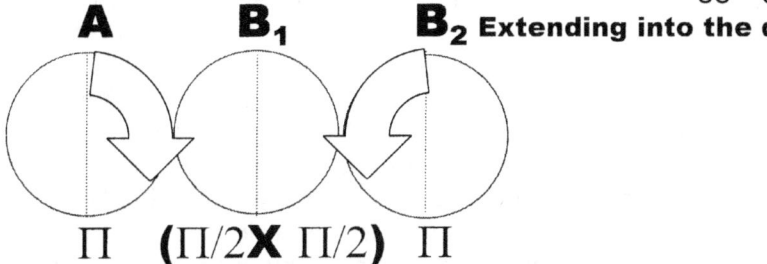

$$\Pi \quad (\Pi/2 \textbf{X} \ \Pi/2) \quad \Pi$$

We have established the fact that Em and Emtoo was divided by r and then r had to be one since r could not be zero. Such a centre would then carry the same value as Em and Emtoo. That means whatever value Em and Emtoo receives has to go in equal measure to r with Em sharing half of the divide and Emtoo sharing the other half of the divide. This is the only manner by which the division could have taken place

At first the two particles moved apart from the spot, as they became dots, which by then they became neighbours as much as associates and formed from at first being an inseparable unit to the then later divided associates after the split separated their sharing of a spot. We know that at least two dots formed because there are many more than just two dots that remained to become our all inclusive as well as the visual Universe. Let us name the dots because that is what humans do best if they do not know what to do with what they have to do. Let us call the one dot Em and the other one dots next to Em we then call Emtoo. Between Em and Emtoo there wasn't "nothing" because Em and Emtoo were separable.

By their being separable, but still remaining infinitely close, we would naturally be inclined to think that the separation value should be nothing or at least zero. But putting zero in that place is a mathematical excluding procedure leaving future mathematics excluded. With m multiplying m_2 and then dividing $\div$ r with r being zero (r=0) such a procedure will leave the lot at zero and with that much of a total forming nothing then that much nothing (or is it many nothing), the nothing is going nowhere. That means although we think the space between the two parts are nothing it is our concept whereby we are placing such a value in that location because with our flawed position we are looking at the cosmos we rely on corrupting the statements that we make about the non-existing space. We know well that the space has to be at least one to be a future factor.

Because the three points existed on equal terms in singularity sharing a same spot the coming out of singularity forming dots will enforce that equal value comes to all. The expansion had to produce Π due to the reasons given in a similar explanation elsewhere in this letter. That means the circle becomes Π, the diameter becomes Π and the distance setting the structures apart will also become Π. This is what the coming from one point brings along. Only when being part of the second dimension can there start being separate values. While the form was still being in the single dimension from the one side of the form the dots had to establish identities apart but not separated yet. The one circle had a factor of $\Pi^0 = 1$ and the centre had to have a value of ($\Pi^0 / 2$) extending past the very next object but also cutting such an object into a square double half value that was going to come about as soon as the other dimensions came into form. In the relation at present Em is extending towards Emtoo by means of establishing a valid r and Emtoo is establishing a valid extension to Em by using r and this leads to two valid values for r being ((7+7) /10) and (10/7). The values I give here I shall explain later on because the full understanding in the correct context requires much more introducing with many more facts introduced and established.

The only definite place one will locate zero is in between the starting point of the lines going in opposing direction in the position the lines hold before there was the least of directions applied, but that is only because there is no such a position, not because any line is coming from there. As I have indicated and positioned Em and Emtoo the two points may share a position but separation is forever a possibility and for that reason should there then be no other reason we then have to put a dividing possibility at a value of one ($r^0 = 1$). By dividing it means the shared dimension is $r^0 /2$ which then is half ½. The two lines are still one holding the opportunity of parting as an option but have not yet parted and therefore are on the very precise same spot from which came duplicating the next dots.

Being on the same spot does not mean being inseparable or being the same. Everything that is now in the Universe was part of Creation before Creation started because when the dot moved from the spot all of singularity that now is present in the innumerable dots was present in the single spot as well as being present in our Universe. Every dot that was were and is because nothing can remove from the Universe but the rest which was in the spot formed a dot and every dot is part of our Universe. It only means that everything was sharing a spot. The line coming from every dot being there is already there because it already has the choice of going in any and all opposing directions in spin direction and when it starts running it will place filled space in that location not yet present but also holding a factor of one since it will become filled in the future. This is because the space at present is filled with a line and the line is sharing an equal value that does not double the value yet. Where this space is, is now already filled with a line without the line exactly being there in accordance to every detail of the three dimensional standards we apply at present. However it may be the line has to have had a start. The starting became the line running and by running the line is filling space. That means with the line there it filled the possibility that a line could form and not with a line not being in place at all. It is again taking the r separating Em and Emtoo on its factor value of one and not our human visual accepting value of zero. One may not discard any future possibilities of growth by giving those possibilities a value of zero. A line might form or space may form where the line later may form. But by using a value of zero in that spot one then remove such a spot and all the future potential values that may fill such a spot in the future. We humans tend to dish out a value of zero where ever we do not visually are able to find a value at that precise second and in doing that we also place such a value as a running obstacle into the future. But our habit of doing that is proving to be a human shortfall because with our shortsightedness we think of the here and the now in excluding possibilities while we should think of the future by including possibilities. With myself personally not being on the moon and with the great likeliness there is at the moment that I may never be on the moon does not remove all potential future possibilities and with that my personal possible future ability to reach the moon whereby putting my future chances at zero such issuing of zero is removing of all chances there is. The zero will have me never ever able to be on the moon. While I am alive there may be an infinite but still applicable possibility that by some extremely remote chance that might occur however infinitely slim such a chance is still a possibility however remote such a possibility may ever may be,

By disregarding a positional value as zero we exclude such a position from ever being possible. On the other hand when giving it a factor of one we include such a position as a future possibility. When reversing a line we might find a better idea of what is in place and where it is in place. Gravity is according to official sources a force without limits going past and through borders and has an unlimited reach. It seems to remain even and this is conflicting with the flow of perceptions about mathematics. In as much as showing that r is serving in a factor value, as one such a value has to form a limit where Em and Emtoo has divided sides of r and aligning the discovered singularity produces the Roche limit as such a dominant factor in the cosmos. With my effort in retracing a simple line that retracing helped me to find an explanation about the Roche limit and that is a feat not yet done in science. The Roche factor is next to singularity the second most basic foundation in cosmology and is the starting point where singularity spawned into dimensions. As it is fundamental in all cosmic development and with that it denounces the gravity principle introduced by Newton as $F = G (M. m) / r^2$

The formula $\mathbf{F = G\ (M_1.m_2)/\ r^2}$ is unable to explain the principle discovered by Titius and later by Bode and in contrary to all statements to that effect made by Accepted Science policy makers the Titius Bode principle is not coincidental. In fact it is one of the four most adhered and important cosmic pillars holding the cosmos structural in place. From the two examples mentioned above comes gravity. In past few pages I proved how one could arrive at the facts that prove how the Titius Bode Principle leads us in the direction of the origins of the solar system. But before we can accept the influence of the Titius Bode Principle we have to return to the aspect of nothing being what is what the Universe is built of. We first have to deal with "Nothing" and as such dismiss nothing from science. I wish to introduce the fact that the reader should entertain the idea that the Universe are made up of pockets of space built by gravity where gravity cements the layers in blocks of 7/10 and 10/7 in relation to the Roche factor of $\Pi^2/4$.

This measure cannot be possible in a Universe where the manufacturing material is nothing and lumps of nothing onto the bargain. "Nothing" in the Universe is coincidental; "nothing" in the Universe does not apply. Where mathematics places objects and objects meat with lines nothing disappears. Nature subscribes what applies in the cosmos. Should any principle not match that which an accepted theory translates to or nature principles has to change to fit what the Philosopher introduce in support of the new theory in order too get it accepted, then the theory does not apply. If one cannot place what is in nature today in the cosmos at the very first instant of cosmic birth, the cosmos then had to change later on to accept that which the theory introduces and frankly, that is not possible. The cosmos is unchangeable. To imply that this or that was formed later on, or this that developed from, such presumptions are totally irrelevant.

The content of my work holds a new view about Cosmology, which I have been working on for the past twenty-seven years and exclusively for the past seven years. I always had a problem with the idea that space constituted of nothing, while I came to realise that lines mathematically couldn't start at zero because there is no evidence of zero as a factor in mathematics. Should you disagree with my statement about outer space being mainly formed by nothing the question in need of answering is this: What will the length of the shortest hypothetical line imaginable be and moreover, what would the total overall length be in that case? I once again come back to this idea after I introduced my idea about how the Roche limit came about and why the Roche limit forms the absolute partition in the cosmos between objects with variably similar dimensions. But seeing this can only become clear if the notion of nothing forming outer space is demeaned to the nothing it represents. By coming to the point where the line cannot possibly reduce more than it already reduced such a point that is holding the shortest line is precisely where the Universe started.

At that point the Roche factor came into prominence, but so too did the Bode law come into prominence because the one intertwines with the other. But to get to that we first have to abolish nothing as a valid concept. If one cannot trace zero at the start then the start must be filling the Universe and those in doubt whom are persisting on having zero must prove where zero later found the opportunity to enter the cosmos. The shortest possible line (hypothetically) must be so short it must have an initial and the eventual ultimate point sharing the same spot. The two points must be one with all the other potential separations being in the future and only then can further reducing of any line not occur. As I said before I say again: if any or all lines used zero as a start, the zero part would not count, because with the slightest growth the zero will either continue extending its current value of zero. Using zero is the same as going nowhere all the way into eternity. It is either that, or if otherwise the composition must change into some other value at a specific point. But that is invalid because what inexplicable reason, would come about to bring such doing. It cannot then remove zero.

The line will continue repeating the make of it in the same way as what it was before without change and that fact will bring about that the line then will only start. By changing such composition the line may regard what it was before as something and this is still totally unacceptable to science. There can be no changes to the line from what it is when it is flowing from what it was at first. At a point past zero where the line then will start forming an infinitely small spot. Even by that measure such arguing eliminates zero as a valid factor. I press this point about the dot moving from the spot in urging the reader the understanding this concept because there is such a point from where the Universe is ruled, but I have to get acceptance first about such a fact in my attempt to underline the fact, I have to convince the reader to abolish four or five thousand years of accepted and practised mathematical culture and that is no easy feat. Notwithstanding more than five thousand years of thinking we have to eliminate zero as a valid factor from mathematics.

In applying the most basic method of taking the line back as far as possible bring us to a dot Π and going past the dot Π till you reach the spot Π^0, which is actually still a spot Π^0 that is infinitely smaller than the dot Π but since that is only a notion of something being outside our Universe we will have to call it a dot Π since the next value carrying is the spot Π^0 is already taken by singularity Π^0 and we are passing that to a spot where Π^0 is single and is infinite in the extreme infinite Π^0. Because of the equilibrium that will stem from such a position the dot is the most balanced form there can ever be. The spot Π^0 is in infinity plus one but, however small it then might be, it still is not zero. Zero dismisses the position zero claims by vacating all possibilities of such a position being filled leaving the place full of nothing. Zero ultimately means not existing and then that point, which misses zero by one single eternity, holds nothing in such a start from such a point that does not exist.

Taking the line down the line by reducing the line proves that the line must still have a start and an end that disqualify zero as any of the two points or all the ends from front to back must be zero at all times. There is no other option as to have no line at a value of zero or a line starting from infinity, which is a valid number and running all along the line in a uniform rate to the line's end.

The smallest line has a beginning and an end at the very same spot located in infinity, and infinity may be beyond any possible human scope, though infinity is still not zero. Infinity puts the start and the end at the same spot, but in that it does not remove the line and all possibilities of a future line from the spot holding the line, as zero does when using zero as such a starting position. Infinity may constitute of something we do not yet understand, but we may not define our human misunderstanding of being zero just because infinity is not present in our minds and therefore by not sensing a value we disregard such a value as nothing whereas if it is visibility nothing then that does not mean it is nothing but in being potentially there it qualifies the point that it is one. It is the same as a person hearing a dog bark and investigate. When not sensing what the dog was barking at, the person turns around and disregards the barking as the dog is going on about nothing. The dog's reaction was not the indicator of the nothing because ultimately the dog sensed something. The man's wits let him down and his wits produced the nothing. The dog will not bark about nothing because then the dog will not bark at all. The fact that the dog barked produces a possibility of something being out there, which the dog is getting annoyed about. The man's inability to detect what it is that the dog is sensing becomes the nothing, but that "nothing" does not exclude the possibility of something out there being worthwhile to investigate by someone with better senses.

From the onset my approach to cosmology proved to be somewhat unconventional but through the abandoning of the accepted, it enabled me in locating the precise location of singularity that forms the connecting basis of the universe (and this I say with some degree of confidence). There are two locations but I shall first concentrate my explaining effort on the prime singularity. Singularity did not vanish into the unknown after the completion of the Big Bang development but is in a place science incorrectly valued and classified incorrectly and in that, there is something hiding which we named as nothing and through that became an obstacle that is hampering our recognising of what is the truth. If singularity was or is where the beginning is we have to go back and see just where such a beginning was.

It is also true that where infinity hides singularity is at a place where we can only detect nothing as we cannot detect the position or location of singularity by using our meagre senses we have. But since we are the only part of creation that is (presumed to be) blessed with wits we have to locate such position with our human side and not with our animal side. We must bring in something that the atheist cannot find because that which is in control of the Universe by appointing value to singularity from outside a point in the Universe understanding such a concept goes beyond the animal and will also exceed that which the atheist can relate to. Understanding that the concept that the control of the Universe comes from points that is not part of what we perceive as the touchable Universe is far beyond the capabilities atheists have. If they had such abilities they would not be atheists in the first place.

I cannot accept that the Universe started at zero and neither does anything else in the Universe start at zero. My excluding the possibility of zero includes that the Universe is not filled to the top with nothing and neither is "nothing" part of outer space. The universe is about lines allowing light to flow from one point to another point and in following that line it has to continue in the line as the line has to represent something. The Universe is all in relation about lines indicating distances between cosmic structures.

The cosmos is in short about lines connecting points in space being apart. It is about a line starting and continuing from such a start. But science advocates their opinion that such a start of a line flowing between any and all objects that can hold zero because according to them the Universe is full of nothing. The Universe and outer space is a container filled with nothing or that is how science thinks it is at the present moment. However it is the only available container and there is no place to place anything that was previously part of outer space but at present needs to be discarded. If it was present previously it is part of the Universe at present. What was previously part cannot be discarded in some rubbish dump out of eyesight. No place can be created to become a make shift storing facility in order to substitute for the overflowing of the Universe at present. No place can serve as a holding area where to something could be released and there is no emptying of what ever unwanted to be put that filled it before. We must then accept from what is not in the Universe meaning that that is absent at the present time was not in the Universe at the time during the start. Everything, which we find that is at present in the Universe in accordance with our observation of the Universe which is present in this present time and is according to science still part of the present can vanish because it then still must contain the same nothing and must have that same filling from the start present. If it was nothing it still must be nothing and that same substance being nothing is what it also used to grow and by using nothing to grow brings conflict in the conception that forms because how can nothing accumulate it as it grew because it filled outer space with nothing growing from and growing to nothing. Is that true? If such a presumption is true then the filling of the Universe could not go anywhere if one has to presume it started off from nothing and from there it kept filling with nothing since what ever was in the Universe at the start had no place to escape to or no place through which to escape.

That is only applying if it is nothing filling the Universe at large. Can nothing grow as much as a line is growing from a start of nothing? The answer is that such lines not only indicate a distance but since the Universe came from such a small space as science propagate with the theory of the Big Bang then all particles in the Big Bang Universe were rather cramped for space when the Universe started from that small line between particles and is now the same line but is now so big. In the past everything seemed being so small and showing that the space between particles seemed then to be awfully short at the time during the start of the cosmic concept. It was short but how short was it? Did it start off as nothing? Is the line starting at nothing as science wishes us to believe? If it does then all lines must start from nothing so we better investigate this trend with the start of a line. In this following I show my argument with which I hope to prove the counter part of what science believes. Later on in this letter I am about to prove that which science sees as nothing in space and in material is the very location of singularity. But lets return to the start before the confusion came about with space being nothing. The start has to start with Kepler because science in the new era started with Kepler and not with Newton.

I have to belabour the nothing for the last and final time because from what I am about to present cannot be presented if Academics dismiss my presenting of my work with their bluffing everyone with nothing being used in outer space. If there is one still persisting on nothing being used in outer space, such a person should either by now be convinced about my reasoning or that person will find no benefit in any further reading of this letter. In that case please donate the book to someone more intellectual and therefore more presentable to the obvious. Kepler's finding cannot stand true if the cosmos is nothing and by persisting to the accepting of nothing then Kepler becomes a discoverer of nothing, which he absolutely was not.

The value of Kepler's space he indicated as a third dimension a^3 does not depend on indicating a structure a^3 that is in rotation T^2 but only needs one position having a constant of some sorts. Any point where k may indicate a position one will find a value matching a^3 and the matching location will fit T^2 at that point. That is the relation there is in the solar system between all planets and the sun. The sun always indicates the centre and the planets always indicate the rotation. But $a^3 = T^2 k$ is only producing a relevancy of three dimensions that is equal to two plus one dimension.

Let us take it from a point where the sun provides a centre k then that centre k will provide a line from the centre and the line k will provide three spots in a formation that produces a structure by the square T^2 of the dimension. That means every single point that k indicates there are three positions a^3 implicating sides of a double dimension. $k = a^3 / T^2$. That is what Kepler said. There are three dimensions a^3 between any two points T^2 flowing as time from the centre of the sun, which is indicated by the line k. The implication of the relevancy produced by the use of the formula $k = a^3 / T^2$ brings about that when dividing T^2 into a^3 there is k left. The fact is that a^3 is a three dimension (^{3}) of single k (^{1}) showing one or T^2 is two dimensions of k being the one dimension it means that k is a part of space a^3 or T^2 which is time. It is the same thing in a double dimension or space being a triple of k then k is one factor and k cannot show a position of zero. If $k = 0$ then there is no possibility of $k = a^3 / T^2$ because $k = 0$ then $0^3 / 0^2 = 0$. That does not make sense.

Mathematically space cannot be zero because those being of the opinion of space being zero or nothing must first prove mathematically that space is zero. I have tried to convince the Super Educated by using that line which is more than correct and being more than correct brought me nowhere. Those in charge of serving Academic policy decided that nothing in outer space is a proven and established fact proven and accepted by those with prominence and who ever comes afterwards has no reputation to produce whatever truth there might be to produce. Moreover they then must prove mathematically how zero can grow through the Hubble constant. That too says nothing. Those in charge do not have to prove anything about nothing because what ever they say, notwithstanding how incorrect it may be, is being accepted as fact just because they are saying so and that makes all my principle efforts not applicable, because they disagree. With their thoughtless denouncing my challenge on their correctness goes wasted with nothing still being the norm. I cannot prove anything notwithstanding that Kepler proved the lot there is to prove. Kepler said space could only be space if space is in motion and therefore how can nothing move? If k cannot be zero then k could not start from zero. With $k = a^3 / T^2$ no point can be zero because k shows space a^3 in the duration of the time T^2. Then the next thing I know is that through the inspiration of Newton, Kepler is not accepted and Kepler's formula is not even disputed, it is blatantly ignored by misrepresentation. Even if $a^3 = T^2 k$ is about motion we know that nothing cannot move and therefore only if $a^3 = 0$, $T^2 = 0$ and $k = 0$ can outer space be worth nothing.

We use nothing not as a value to measure by but to avoid what we wish to disregard when the effort to trace and determine becomes too stringent and tiresome to further investigate and not to valuate. In this aspect lies the difference there is between arithmetic and mathematical science where arithmetic can have position such as zero since arithmetic excludes the cosmos calculating numbers only. The nothing we see and that nothing what we made, we made that the nothing we find but the fact that there is a visible and measurable distance between the structures which we may appreciate as being there proves that the distance is there, it is separating structures and by that is bringing in the factor of one which we might not be able to explain but as such we are able to see. It is the way we try to disguise our inability to detect which produce the nothing we then use as a value, but still we substitute the nothing in applying arithmetic with the name as nothing and then the names used becomes the factor of one. Cosmology is not about numbers because no one can calculate the number of stars in spite of ridiculous Critical density attempt. Cosmology is all about lines and angles positioning objects, and in those lines there features no zero. The cosmos is about better or in other cases lesser development by extending of **k** as the barometer of singularity development in space-time.

No line can be zero long and forming a position of zero degrees in relation to another object. Doing that shows we use our culture to hide our inadequacies behind just one more misconception. Let us find a place where zero does apply. A man may have that many oxen or so many sheep and even this amount of wives, (in Africa) or not have any therefore having then a total of nothing, but there cannot be nothing between the sun and its orbiting structures. The having and have-nots are part of arithmetic. Light will indicate a line flowing between the sun and whatever planet, following dot after dot from infinity crossing infinity to reach the next infinity and thereby proving the existing of the possibility of something going about by a straight line. Any straight line in is relation to other straight lines and will be valid under the law of Pythagoras where the law is in as much as obeying the rules of trigonometry. At any and every given or imaginable point between the two points forming the line, the line can be interpreted by something just larger than the epitome of the infinite small line and up to the size just larger than the size used by the line.

 Regarding the possibility of zero there is no possibility of a straight line not forming in space. If there is space, there can be a straight line. Kepler said space is the motion thereof. That means space is immediately following another straight line by motion thereof. The mere fact of two spots having different positions in space gives the two dots different values. If the line has the length of zero and is the line separating the points that it is not present, which puts everything represented by the two objects and the separation between the two points holding the ends of the line apart, outside the Universe we have to use. We gave a name to something we identify as being representative of something we gave a name to such as nothing and not the fact of zero as such. The nothing is a name and what the name nothing represents. Then the triangle where one angle is the zero it then is no triangle because all other angles are dismissed at the same time.

Mathematics converts the values of integrating lines according to Pythagoras and arithmetic is about numbers to be added or subtracted. By mathematically excluding zero from cosmology a new Universe opens to the human mind. With the distance between the sun and Pluto being roughly one hundred times more than the distance between Mercury and the sun, the distance must hold something more than pure vacuum filled with nothing except one atom hear and there occupying the vacuum between whatever object we speak about and the sun. If space supposedly comprises of nothing how can nothing then become plural forming more or be multiplied by a number as to indicate a growth in something not even existing. As the one becomes one hundred the one cannot substitute a value of nothing but then must be part of something. If the one substituted the nothing, all laws of mathematics will go in disarray because when one multiply any number by zero it becomes zero placing both planets in the sun. If Pluto was one hundred times closer than it is at present was it then one hundred times nothing closer? In Mathematical term using a mathematical expressed manner the words used then translates to being mathematically expressed as $100 \times 0 + 0 = 0$. That is the expressed factor what we read into mathematics! By allowing the three hundred a value carrying the value of nothing then nothing must form one making that which is between Pluto and the sun not nothing but representative of something we gave a name of nothing to as we would name someone George or Jack.

The nothing is a name and what the name nothing then represents, such representation has to be something as would George or Jack. Allowing this concept to apply this argument then follows mathematics to the letter and in precise detail. With Pluto and the sun being apart that being apart has to have one of something a in place of a value forming the being apart from each other's cosmic position one time where the one is a factor multiplied by the many ones we find in that space standing relative to other space regarding whatever the space becomes what is that we think is between the sun and Pluto. That factor cannot stand in for the value of "not being" one, which is the same as nothing. As that is because one cannot take the place in the position that zero secures. By excluding nothing from the equation space becomes something bringing in a value lying inside the realms of the infinite that must form singularity. As the zero becomes a dot, something else becomes clear about the dot. Looking at the night sky we find darkness overwhelming the space in relation to the stars bringing across light. In another one of my books I show that we are unable to see darkness because we consider the darkness we see as nothing which represents no visibility but we can see darkness and we do so see darkness very well therefore the darkness we see must be light we see. That excludes nothing on another term.

From the onset my approach to cosmology prove to be somewhat unconventional but through the abandoning of the accepted, it enabled me in locating the precise location of singularity that forms the connecting basis of the universe (and this I say with some degree of confidence). There are two locations but I shall first concentrate my explaining effort on the prime singularity. Singularity did not vanish into the unknown after the completion of the Big Bang development but is in a place science incorrectly valued and classified incorrectly and in that, there is something hiding which we named as nothing and through that became an obstacle that is hampering our recognising of what is the truth. If singularity was or is where the beginning is we have to go back and see just where such a beginning was. I cannot accept that the Universe started at zero and neither does anything else in the Universe start at zero. My excluding the possibility of zero includes that the Universe is not filled to the top with nothing and neither is nothing part of outer space. The universe is about lines allowing light to flow from one point to another point and in following that line it has to continue in the line as the line has to represent something.

The Universe is all in relation about lines indicating distances between cosmic structures. The cosmos is in short about lines connecting points in space being apart. It is about a line starting and continuing from such a start. But science advocates their opinion that such a start of a line flowing between any and all objects that can hold zero because according to them the Universe is full of nothing. If the Universe in as much as outer space is a container filled with nothing at the present moment, and there is no place to place anything that was part of outer space previously to substitute for the overflowing and there was no emptying of what ever filled it before, then it could not get rid of what was in the outer space when it first started with what it started off with. We must then accept from what is not in the Universe meaning that that is absent at the present time was not in the Universe at the time during the start.

Everything, which we find that is at present in accordance with our observation which is in this present time according to science part of the present because it then still must contain the same nothing and must have that same filling that was present from the start. If it was nothing it still must be nothing and that same substance being nothing is what it also used to grow and by using nothing to grow brings conflict in the conception that forms because how can nothing accumulate as it grew because it filled outer space with nothing growing from and growing to nothing. Is that true? If such a presumption is true then the filling of the Universe could not go anywhere if one has to presume it started off from nothing and from there it kept filling with nothing since what ever was in the Universe at the start had no place to escape to or no place through which to escape.

That is only applying if it is nothing filling the Universe at large. Can nothing grow as much as a line is growing from a start of nothing? The answer is that such lines not only indicate a distance but since the Universe came from such a small space as science propagate with the theory of the Big Bang then all particles in the Big Bang Universe were rather cramped for space when the Universe started from that small line between particles and is now the same line but is now so big. In the past everything seemed being so small and showing that the space between particles seemed then to be awfully short at the time during the star of the cosmic concept. It was short but how short was it? Did it start off as nothing? Is the line starting at nothing as science wishes us to believe? If it does then all lines must start from nothing so we better investigate this trend with the start of a line. In this following I show my argument with which I hope to prove the counter part of what science believes. Later on in this letter I am about to prove that which science sees as nothing in space and in material is the very location of singularity. But lets return to the start before the confusion came about with space being nothing. The start has to start with Kepler because science in the new era started with Kepler and not with Newton.

I have to belabour the nothing for the last and final time because from what I am about to present cannot be presented if Academics dismiss my presenting of my work with they're bluffing everyone with nothing being used in outer space. If there is one still persisting on nothing being used in outer space such a person should either by now be convinced about my reasoning and if not that person will find no benefit in any further reading of this letter. In that case please donate the book to someone more intellectual and therefore more presentable to the obvious. Kepler's finding cannot stand true if the cosmos is nothing and by persisting to the accepting of nothing then Kepler becomes a discoverer of nothing, which he absolutely was not.

The ether might now be regarded as unnecessary, since it is recognized that electromagnetic radiation can propagate through empty space but it is the empty space that replaced the hypothetical ether that brought along the misconceptions I am fighting as hard as I can. When cosmology was in an infancy stage Mainstream Science then realised there had to be a conductor to conduct gravity as a force. Ether is or was a hypothetical medium found in space that was presumed to act as a conductor of the force of gravity. Electromagnetism was the presumed conducting force through which the force flowed. Some how back then Mainstream Science had the sense to foresee that electro magnetism was part of electricity and electricity needs conducting to flow. If magnetism did not need a conductor, electricity would either flow without challenge of resistance or electricity would not find the ability to flow at all. Ether is a hypothetical medium once thought to permeate all space, through which electromagnetic radiation supposedly travelled; formerly spelt aether. On the basis of this supposition, the Earth should move with respect to the ether, and it was predicted that the speed of light would vary when measured in different directions.

This presumption is based on another presumption namely the time in which we are and which applies to us every day, that is a cosmic standard time. The presumption was and is that the time applying on Earth as years, days, seconds or whatnot is equal everywhere from Mars To Magellan's cloud and every where thought may reach. The tests were based on the idea that time was similar through out the Universe. Please I whish to put one thing straight and remove whatever doubt my arguments might provoke, but I am not trying to re-establish or re-institute ether. My fight is on the incorrect view of the standard unified Universal time that we can set our clocks by from here to whatever stellar system there is fifty billion light years away. The way the experiments were done in the 19th century (e.g. the Michelson-Morley experiment) which failed to detect any such variation in speed was as flawed as the ether theory by its own merit. The ether tests using the time as was done is by indication as if time could be measured by a clock watch simply because the standards of measuring used in the principle experiment was the true indicator applying to time wherever time was to be measured. The way in which the measuring was done typifies the cripple manner in which science does not understanding the concept of space-time being space which cannot be if not in motion through time doubling as the second space component.

The ether is now regarded as unnecessary, since it is recognized that electromagnetic radiation can propagate through empty space but that statement alone mesmerises the calculations that electricity requires. If nothing is able to conduct the flow of electrical current the current resistance must be zero because nothing simply cannot resist a flow. That makes the actual flow of current as measured by the heat displaced from one point to the other point beyond any limitation or measure of any sorts. The volts must be eternal which is just another name for a concept if we use an infinitely large measure to try and pass on an idea of the flow of current so big it goes beyond understanding. Using human terms available belittles the measure possible for the current that might come about when there is no resistance containing the flow of electrical current. The volts will burn holes into a Black hole while the amps will fry the Black hole to smoke without resistance coming about to the flow of electricity. Again it boils down to the fact that science trash my work as incoherent and not worth reading but in the mean while every time I return to the nothing issue I find words lacking of what there is to express how incoherent the argument about outer space and nothing is. The first mistake was the presumption of space being empty. Space can be less dense but space cannot be empty.

Mainstream Science is forever so concerned about mathematical proof, except when it suits those in charge not to insist on such proof. As much as I searched I never once could see how any person brought about calculations to prove that outer space is empty or outer space contains nothing. I challenge all concerned to show mathematically that there is nothing in outer space. I do not promote the idea of ether as it was first called or I do not think either is part of the cosmos. I do think there is something in outer space or outer space constituting of some form of material that we cannot detect because such material is the basis, the new ingredient, which formed all material including heat and space. The Big Bang was the ultimate nuclear explosion which all War Lords of the Warring Empires in the world today dream of.

The War Lords I refer to are those who bomb nations to dust because they promote peace and democracy and Human kindness and Brotherly love in such a manner while they go robbing them blind of their oil. They kill those children they say they save from brutal dictators while the dictators killed far less innocent woman and children in a lifetime than the big saviours do in a month. We all know who they are because all commercial oil funding flow through the banks of those nations. Let's get back to cosmology. Then through some investigating testing proof came about that ether does not play a part in the conducting of the flow of electricity. Those tests had to reflect on the flow of light as well. Light is also just electricity because electricity can produce light as much as light can produce electricity and light flows through space. In the events that followed the disproving of ether being present in outer space led to a belief that a vacuum came about. The vacuum was present in the minds of those explaining what was present in outer space and the vacuum then transformed to the void and from there the void went out to outer space, which then formed space. But the only void I can detect was in the heads of those placing the void in outer space. The critical density proved that no void could be in outer space just because a factor indicating the possibility of a critical density was detected. However there was a realisation that the required density was not enough to prove Newton correct but it did not prove that much "nothing" was filling the vacant space between the thinly spread material either. In outer space there is less density of space than in the more compact atmospheric space.

The density I refer to is that of material that would be obtained if all the matter contained in galaxies were smoothed out across the universe. Although stars and planets have densities greater than the density of water (about 1 g/cm^3), the cosmological mean density is extremely low (less than 10^{-29} g/cm^3), or 10^{-5} atoms/cm^3) because the Universe consists mostly of virtually empty space between galaxies. The mean density of matter determines whether the Universe will continue to expand. By accepting the density parameter, as the ratio of the mean density of matter in the Universe, which is standing in regard to the critical density of matter means that there has to be a constant of the distribution of material. That cannot be because material is definitely no evenly distributed anywhere at all. That which accompanies the material is what proves the density factor. That is released heat that becomes space and the space, requires another associated material of compromised singularity to bring about the eventual collapse of the Universe. A value of Ω (where Ω symbolises the density parameter) will bring the accepted ratio to ensure that the collapse is immanent and Newton can rest in peace in his grave as he then was ultimately eternally proven correct beyond question by his followers. It is if the Newtonian presumption is presumed that the Universe must collapse to vindicate all who believed in Newton and with Newton being correct and gravity is contracting the Universe. The value of one or more will ensure science that the Newtonian concept they use will apply and will bring such a gravity collapse. It is also anticipated (correctly) by those Super-Educated that the opposite apply. On the other hand will a value of less than one bring about expansion forever and the Universe will drift apart with space forever growing.

Off course the whole concept I just now mentioned is flawed and corrupt beyond any effort of saving and not even Einstein's best efforts could manage to bring a rescue about to save the face of all the Newtonians involved. Notwithstanding that not even Einstein was able to correct Newton's incorrectness.

Most surprising with all the geniuses they have, it is still the way the mathematicians go about arguing. In an effort to prove Newton above all logic, those bent on proving Newton say there is space and somehow the space contributes in the effort of gravity. The space with least protons contributes by producing the gravitational constant. What will bring about the gravitational constant where the gravity is somehow elastic because Kepler's **k** does flex from season to season? Why then would nothing flex? Why does the Universe not expand into oblivion since nothing means not having any restriction at all present as much as there being nothing to bring on friction tension? Why would Kepler's a^3 apply in combating such unrestricted expanding and what prevails when material is absent? The fact that any density of any sorts is present which secures space proves the fact that "nothing" cannot ever be present. The nothing is like all other things. The nothing is a relevancy brought in to fit the needs of man and to support some or other concept that those in charge of producing official policies wish to promote in such an advent. Not one of those in charge of producing official policies ever tried to prove mathematically that space is nothing because in space there is density and density of whatever kind destroys the concept of nothing.

We can fill or empty a container. The container does not mean we are filling the container with nothing. In cosmology emptying any container can go back to the era where the sun started forming a cosmic independence because it is in any person's range to reduce the stroke of **k** back to what the outer space was when the sun started to burn space. We have used this method for ages on end to tap into energy sources. By our using of fossil fuel of whatever means must involve a process of pushing time in a container back by reducing a relevant **k** that will alter the space a^3 and that will produce a time running to earlier applying conditions in space –time by prolonging the duration of timeT^2.

The internal combustion engine works on just that principle and so too does the turbine engine. By reducing space the space becomes hotter and the heat forms the workable energy that we humans found a way to tap into. In our quest to harvest energy we take the environment within the cylinder walls in the engine back the conditions as it applied in outer space when the sun established that very first thin border line that parted outer space from the sun's inner space as the sun declared independence from the Milky Way. That is when the sun became a star. We use this cosmic development technique by retaining such development. The manner in which we do it is to place air in a large area, which we reduce by motion into a very small area thus changing space and converting the space into usable heat, We then introduce into the reduced space fossil fuel and mix it well with stored heat that we pushed back to a time that dates when the solar system awoke from the Milky Way heat blanket. That was long ago.

We place this either solid or liquid fossil fuel and place it in a very small container that is terribly hot and well secluded from the container we think of as the Earth atmosphere. The container that is burning our mixture is small. By method of harvesting a combusting mixture it is pushed even smaller. By the reducing of the stroke or the **k** factor of combusting the fuel we have a process where we turn the fuel into heat and with that much heat the space can duplicate space in favour of antigravity. In that manner we push time back billions of years in this very small container. By returning time to where the sun started its quest to individuality from the Milky Way we then can harvest energy created millions of years ago and which nature stored in solid fuel. By reducing the space a^3 contained to predating the arriving of the solar system and introducing fossil fuel by method of igniting it, the process enables us to create motion we can use in the present day to the advancement of life's wishes.

The piston in the exhaust/ intake position stroke

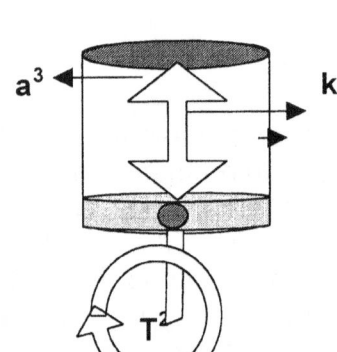

The internal as well as the external combusting engines drive on the same principle as the Coanda effect and it represents Kepler's formula precisely. In analysing, the engine we find the principle of space which Kepler introduced applying in the same way.

By decreasing the stroke, which will be the distance, **k** the area **a³** diminishes considerably while roaring action of the crankshaft **T²** produces the time motion keeping the sequence in ratio.

Compression / ignition

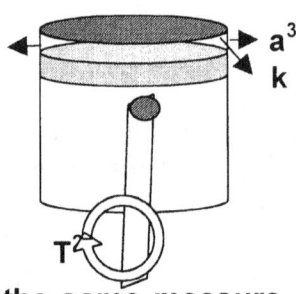

By the same measure that the distance reduces, space demise in ratio but the heat concentration represented by **T²**

The reducing of the stroke of the connecting rod will have precisely the effect on the space where the reducing will come into effect as the reducing of **k** produces a smaller combusting area. The cylinder with piston and rod turning will also adhere to the same formula Kepler proved. The area of the cylinder **a³** depends on the connecting rod length **k** and the concentration of heat **T²** that is representing the gravity aspect applying increases in ratio. It once again proves $a^3 = T^2 k$. The only difference there is, is that we turn the prominence of the different factors around to suit our needs in accomplishing the harvesting. But the factor most important is that the motion increases the heat density as the space reduces. **T²** increases by doubling or tripling as space declines in volume. $a^3 = T^2 k$ is an undividable unit wherever space, time and distance form an interlinking action. The focus must be on the temperatures rising as the stroke reduces the space. This implication is what space-time is all about and that is the strongest force preventing any space travel.

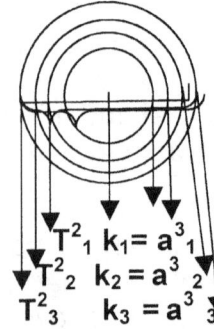

$$T^2_1 \; k_1 = a^3_1$$
$$T^2_2 \; k_2 = a^3_2$$
$$T^2_3 \; k_3 = a^3_3$$

The space on point from **k** towards time is space. Space in the six sides is time directed from space by one dimension. The motion of time provides the three positions another three to form the universe as we see it. By motion at the point of singularity the time forms the space doubling in value from 3 to six. Take away time and space collapses and takes away space and time disappears. It is not possible to have time without space or space without time. By destroying one the other will disappear. Where singularity meets space and **k** changes from one (k^0) to k^1 space as well as time comes about and time is the compliment of space where space is result of time coming from singularity.

In this the vital part is the release of heat in producing space that establishes a linear motion, which produces a circular motion. If not for repeating this cosmic action, no motion would have come from the whole episode. It is heat producing linear motion that turns to circular motion that produces the energy we tap. Whatever way we look at the working of artificial or manmade machines, the Coanda effect is ever present. There is a centre point we call a driveline. There is a distance k we call a stroke and there is a rotation factor T², which we call the revolutions or the machine speed. The rotation personifies the cyclic rotation we find present in the completion of a circle and all rotation from an engine and an electric generator / electric motor down to the mathematical expressed graph use Kepler's symbolic space-time. It further lays waste to Newton's view that by rotating a value of zero work is completed

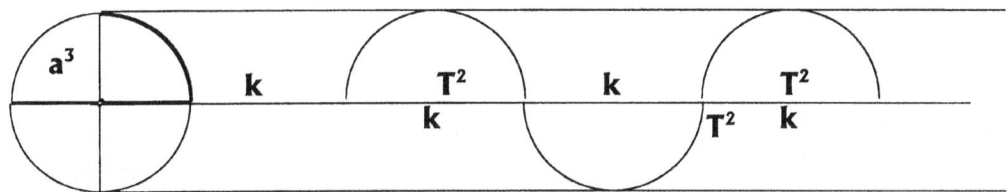

From the graph, one can establish the link in the circle's rotation around a conforming unit being singularity.

When realising the error of science in they're accepting that a value as zero is presented and is legitimate in mathematics, we then have to take it further. One can establish from that that the circle does not employ zero as a value after the completion of one rotation therefore $F = G (M_1 \times M_2)/r^2$ is invalid, one has to return to Kepler's $a^3 = T^2 k$ and establish a value from that.

Saying that one therefore has to admit that the smallest spot has to hold space because the most insignificant dot can transmit light and being able to accomplish that, one must accept it then too has to carry a value of something. If that spot had the value of nothing that would mean that such a spot was not there to begin with. If it is holding space-time then one should return to the original formula indicating space-time in as much as $a^3 = T^2 k$ where $a^3 = \Pi^3$ and $T^2 = \Pi^2$ as well as $k = \Pi$. Being space-time and time to space it has to alternate positions and that can therefore only apply to **k** where Π will indicate a relation to the space-time in question or the relevancy to singularity being $k^0 = \Pi^0 = 1$

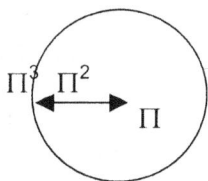

Time is always a displacement of space in relation to the implication of singularity, and comes about between two points in space relating to the centre of singularity as positioned by **k**, either to the value of **k** or to k^0.

There is no larger **k** or smaller **k** but when one factor changes other factors has to compromise and it is this compromising that places cosmology above normal Mathematics since nothing in the Universe is static enough to use general mathematics.

$\Pi^3 / \Pi^2 = \Pi$ or $a^3 / T^2 k = k^0$. With this fact established we then must return to the value as indicated by singularity, being Π. In this we find that $\Pi^3 / \Pi^2 = \Pi$, and Π is a stand in for Π^0. This brings about the value relating to space-time relevancies as a formula consisting of $\Pi^3 / \Pi^2 = \Pi$ or $\Pi^0 = \Pi^3 / \Pi^2 \Pi$ in various forms and relations. One must also keep in mind that there are always four time factors relating to the universe from any point holding singularity, and since every point in the universe contains singularity in what ever form, every spot in the Universe comprises of four time points initially extending to the next spot by means of $\Pi^2/4$, which we know as the Roche factor. By rotating space-time the atom forms as an identifiable independent cosmic structure.

It is always about relevancies applying differently because the mass of the atom can be higher on Jupiter than it is on the Earth or lower on Mercury than it is on the Earth. However the atomic relevancy will always remain in place except in stars where the star already abandoned the implication of certain factors that no longer can apply since the motion, which the star established exceeds the limitation that those factors carry.

From such a relevancy there then must be four different values relating to singularity and since the atom has a relevancy of $(\Pi^2 + \Pi^2) \Pi^2 X \Pi X 3$ that then also must be true.

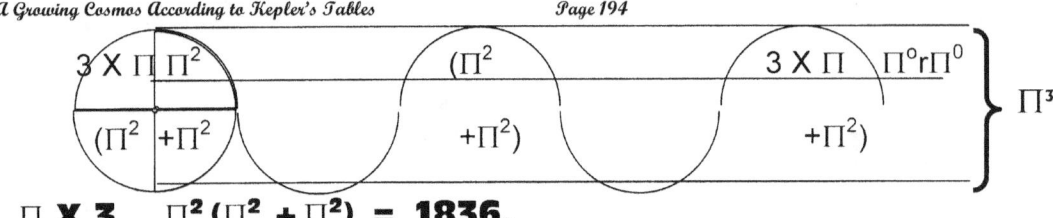

Π **X 3** $\Pi^2 (\Pi^2 + \Pi^2) =$ **1836.**

In our ability as we found a way to reduce space and energise singularity the possibilities we can achieve seems to be endless. Our generating energy or tapping energy goes by the means of taking singularity back one stage to a time that goes back as far as we may go.

$r/2 \bullet r/2 \bullet r/2$ dividing r reduces r to infinity but not Π as Π remains stable, protected by the rotation of matter forming a circle around singularity .

When one starts reducing singularity from where we stand in the Universe such reducing is physically endless.

0.9 ⇒ **0.09** ⇒ **0.009** ⇒ **0.0009** ⇒ **0.00009** ⇒ **0.000009** ⇒ **0.0000009** ⇒

Taking into account the behaving rules of singularity, it is important to recognise that notwithstanding the size of a line, there eternally is another line (or dot) eternally bigger as well as eternally smaller than the line in question. This is gauged from our perspective because we can never achieve singularity. We can never grasp the size of a line that forms the utmost or the least of possibilities in size and therefore size belongs to the human mind forming conceptions of big and small, but it has no place in the cosmos at large. This concept not only applies to size, but also to all limits and divides we wish to create that is forming borders, which we can appreciate. When looking at the circle in the conventional manner, we persist with errors brought about in culture and not by applying some significant modern logic. The reversing of the circle radius is not alien to nature at all.

An observation coming instinctively to mind one may recognise is that the form reminds rather explicitly of natural phenomena such as hurricanes, water whirls and even the shape most commonly favoured to express the cosmic object referred too as a Black Hole. The similarity may be more than coincidental. Let us consider the statement in the reverse. In our calculating of a circle we apply two formula methods. The reducing of the line can go as far back in history to where the line does predate mathematics because the line came well before mathematics arrived and it predates the numbers we use to apply in mathematics as it even predates positional relevancies we use in trigonometry.

This means it even predates directions in space therefore it totally predates human perspectives and goes further back than what human perspective can go. It goes back to the point where only the most basic mathematics being the size the squares and the shapes can take the human mind and that I am afraid is further than the human mind can go. There is no person that can truly explain more than just accept that a straight line and a half circle and a triangle could all carry the same value of 180^0. This fact makes the mathematics one are able to use in cosmology rather different from the mathematics one may use to design an aircraft wing or a large hanger to store the huge aircraft wing. Mathematics going around in normal everyday use does not apply that straight forward into the field of cosmology since cosmology carries laws mathematics never heard of and was in place long before mathematics was concluded. This description was possibly what confused everyone to this day.

The one Mathematician will use an r to indicate the radius and the other uses a D to indicate the diameter, which is double the radius and therefore needs to be divided by a four to eliminate the Newtonian inverse square law amounting to the difference there will be between the two. The one using the radius is Πr^2 and the other formula is using the diameter is $\Pi D^2 / 4$. The factor that mathematics normally allocate to the circle which is carrying the square is given to the radius or the diameter which indicates points in relation to singularity running from singularity to the edge or the border of the circle and that circle is implicating the factor of time in cosmology. Every mathematician during so many centuries has missed the chance to observe that Kepler's T^2 proved to be time and time is the gravity that is applying at that point in space-time. In that the process goes beyond what mathematics may deliver. This misunderstanding is about time not being a fact but a factor or a relevancy brought about while spreading much confusion.

Time predates mathematics. Time is motion of space ($a^3 = T^2 k$) and therefore time can never go single. Science put time at a single value as $t = \sqrt{(1 - (C^2 - V^2))}$ but having time as t is the same principle as showing a photograph of an event that happened in the past as time taking the onlooker back in the past where the image froze time to the single dimension and the image has to rely on some other person using that persons imagination to interpret the photo. However we look at it, it is the interpretation that we make of the picture we see in the photograph that takes on the role that t is suppose to have and it then is not the photograph that claims space-time in the third dimension. The photograph holds space a^3 in the motion of time T^2 because the photograph is constructed from material that is part of the Universe while only the image we find in the ink is presenting the part we consider as time in the past by the image we have. Not even the ink, but only the picture printed with ink is what Einstein's time formula t represents. By motion of space the image we find to hold our perception of time was destroyed one moment after it was established.

Time is duration; it is motion of space filled with particles running from one specific position to another position or from one point to another point. It takes time, it fills space with motion, presenting time. There is no one cosmic Central African Time in cosmology where an American clock in Washington sets all time through out the Universe. Every time k changes through motion time changes. This makes that every time space changes through motion time establishes many different cosmic time zones. Time is created by motion and motion is heat spinning through space. Any dragster car driver will tell you that the time he or she experiences while driving is much longer in relation to what the clock tells. The motion brings about time discrepancies Academics are unable to explain this. The time exaggeration the driver endures is not physiological but is physics in the truth, as science does not yet appreciate. The acceleration enhances the space duplication that produces less space per occupied volume in more time. The sound barrier is proof of this. The Coanda effect is proof of this. Momentum gained or lost is proof of this. Momentum is the increase of linear gravity affecting the mass by accumulating space occupied in time duration that is increasing through motion applied to an individual body compared to what they can manage under normal conditions when the Earth secures time to what is valid on the surface of the Earth.

Let us reflect for once the moment on the truth of every aspect about the Universe at the time in the Universe before the Universe was formed. What ever you may think of as being part of and in the present in the cosmos in the present cosmos now being seen to be a part of the overall and all-inclusive cosmos was not yet in the present term thought about but still very much under development in the far future. It all was locked in a spot, which was one eternity away from the first dot. Whatever it was, was not yet but it was confined to a spot that was so small it did not as it still do not fit into the Universe we now know. The growth that came about depended on the resistance to growth that the growth responded to by not resisting. It means the longer the waiting was the more severe the outcome was and what wait can be longer than eternity which was the length of the wait. However drastic the need was to grow there was no space to grow in so with no space the confinement was the natural while the growth was the unnatural.

That is very much in contrast to our thinking now in the present where the growth in outer space is the natural flow of space-time or gravity and the resisting by containing that produce mass, which is the force to resist and remain independent and any force is part of the unnatural. Then before the release of space into the Universe in waiting on independence forming, it was no option while at present fighting for independence is a struggle all matter will eventually lose in the end but until such time all matter is fighting the struggle for independence rigorously. Just before space came about a time had to come when the need for growth became as strong as the need to remain confined. The need to remain confined was helped by the fact that there was no space to allow space to form space. This may sound as if I wish to become poetic but it is not because with every mention of space it refers to yet another condition through which space formed and those conditions we will never be able to imagine.

Giving it complicated names would be the normal way of acting but I do not whish to confuse myself with adding more names to the congested naming already in place. The only solution was to invent something new such as space in order to break the dead lock, balancing the Universe by creating space and starting Creation. Remember eternity persisted therefore everything was in equilibrium. There had to be some introduction to change the status quo because eternity and time standing still had the Universe in a state of not changing for many eternities. Motion became the tool and what has more motion than light? Light is motion where all motion faster than light presents the killing off of motion. But motion is the altering of relevancies by changing locations and positions. By motion the material strike independence from whatever it is moving away from. The independence coming from motion applying brings about space differentiation in space differences between independently active areas. The important issue is to

realise that with cosmic birth came a change about that was contradicting what was natural.

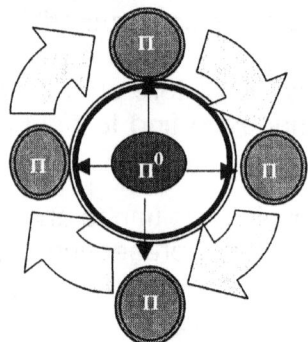

By moving from 1^0 to 1^1 and from $1^0\Pi^0$ to $1^1\Pi$ requires space. Yet such moving does not leave the realm or the domain of singularity. The motion is still within singularity because moving involves forming a relevancy between heat and cold between infinity and eternity, between space and time and most of all producing what will in the far future develop into a Universe that can even be a host for life albeit on a very small spot for a very short while in relation to the vastness space has and the duration cosmic time has. This where time started and time remains at this edge of forming space by motion from singularity that cannot move because it has no space.

Motion increases space as motion brings about space .

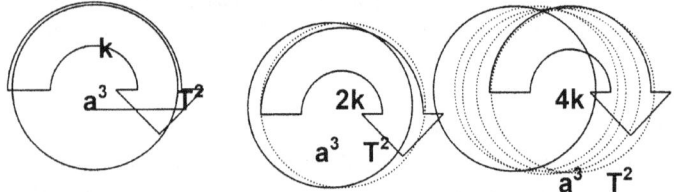

The object seems to be motionless but in reality the space occupied is evenly distributed in relation to the centre. As the motion changes the positions of points in space-time and although it seems motionless it is changing by motion. That brings about the factor of time.

By motion coming about the motion doubles the positions space would occupy when at a reduced motion and the space halves during occupying duration while holding the position because the space occupied is reproducing twice instead of only once. The space then is reduced by duplication, as **k** increases and by that space **a³** occupied captures more space **a³** through motion in less time **T²** bringing about more space distributed in lesser time duration.

By increasing motion the space that the object captured is increasing the validity of singularity connected to the space-time in question. By moving the object remains in time duration but space-time increases. Since one may either place the relevancy on space occupied or time experienced with the increase of **k** it will increase space-time **a³ /T²**. But since our observation stands related with space remaining the same and time at a constant we find nothing changing. However, we on Earth do not experience the true cosmic effect of gravity in balance.

Let's reflect on the moment of birth concerning every aspect about the Universe at the moment of birth. We are discussing a time when what ever you may recollect in your memory is something that still is something yet to be. The Universe was locked in a spot that at the time mentioned still had no space and no place to have space where that place or space forms part of our Universe in 3D which we now know so intimately. Then the dot came about asΠ^0 enlisted Π to perform duty. This brought along growth but the growth depended on the resistance there was to growth. The growth was equal to the resisting of growth and that brought about gravity as cooling coming from retracting and heat was material releasing and expanding. What brought on such resistance must have been the fact that there was no space to grow. There was no space to grow into or to progress the distance between points in singularity in relation with one another.

That means there was no space yet to become space and that is a concept we must accept as much as we accept that a line is represented by 180^0 as is a half circle and a triangle notwithstanding every thing we see as huge differences. Everything we now see originated at that time. The only solution was to invent something that will break the deadlock the Universe found it in. Creation came to a solution by creating motion, which is not yet space but is a change in relations between particles where the altering of the position meant the creating of space during time. Motion became the tool to use in all further cosmic development from that moment on. All that motion actually is the altering of relevancies by changing the manner in which they apply from moment of change to moment of change. Motion became the change in direction, which then progressed into forming space and the duration, or the distance it took became the time where they formed a relation where the one cannot be without the other. The motion established independence but not yet space because the independence created space-time and that is what it still is. The space-time is only created differences in relation to changes about relevancies that apply through motion acting as such.

This effort totalled in material capturing more space through motion, which in truth is the other half of gravity that brings about space by singularity fighting for independence. Mainstream Science Academics refused to acknowledge that gravity is space applied in time through motion of space. By failing to salute the truth Science went about making such a mess of the concept behind cosmic gravity. By not looking at Kepler they opted instead not to recognise gravity stated as such by Kepler and refused to put it down as part of gravity. Instead they use another name in identifying the second part of gravity as being momentum. Any motion and all motion alter space and change time.

Time and space is interlinking and is the very same but the one is following the other. It is what Kepler said. If mathematicians cannot read mathematics it surely says much about those holding the profession. $a^3 = T^2 k$. It is precisely what Kepler said when he translated what the cosmos told him using the language of mathematics. Space will increase as time duration slows down $k = a^3 / T^2$. Space and time is so much interlinked they are the same. There is no time constant as much as there can be no space constant. There can be no constant. As space expanded the Universe came about. However, one look at that confirms that the expression constitutes to what we find when we use the term momentum. .First space and time freed from eternity slotting in a position being next to eternity in a place that was just between infinity and eternity and that is where we still are. Therefore that is why such space and time seems to us as eternity, which is coming from infinity. Then matter occupied space through the duplication of space by using time. Space excluded heat from space by having excluding occupied space forming separate time. Some particle overheated and created space and others formed gravity and conserved space by freezing space.

Take a circle and reduce such a circle constantly to where it no longer can reduce. Reduce it to a point where only form remains part of the circle because the radius has gone beyond human measure and becomes so small it is not noticeable with what ever tools man may use, then what remains is pi since pi does not indicate size but indicate form, and form is all that then will remain. I believe one can begin too see where my suspicions are heading because the flaw comes about in the manner mathematics are practised for thousands of years. Space is though to be nothing because that means man thinks about space as a standard fit all issued everywhere that came about when time came about. Before civilisation taught man to read and write, even wind was part of magic. Today we know wind is part of heat forming space in motion. Then it was thought that space cannot increase and winds were ghosts blowing their breath.

If there is any suggestion of this thought being ridiculous then how ridiculous is it to pronounce space as nothing. Nothing and ghosts are more or less similar therefore scientifically amongst the wise and informed of the day, little has changed since then and now. More seriously wind is space holding more heat than other space holds heat in relevancy.

All the four pillars of the Universe depend heavily on the form of Π and if not for the liquid of the neutron and the form of a double Π (to the square which indicate gravity), the Coanda effect would not have been able to establish the atom in the form we now find the atom. The importance of the Coanda effect in duplicating space-time while producing a diminishing thereof in the form of Π is an indication that Π serves as a mould and form to the entire Universe. Winds are as much antigravity returning reduced space back to the ranks of increased space. Before coming to the mathematics I would first like to bring your attention to the practical side. I am promoting a theory in which I am able to prove there is as much contraction going on in the cosmic universe as there is expansion and the contraction is as much part of the expansion. The two factors are inseparably the same. The Universe rides on a balance and we have to locate such a balance.

To prove my theory I firstly had to locate the centre of the Universe. The failure of Newtonian science to locate the centre of the Universe is an obstacle they never noticed. It should be a fact they have to predominately first establish where to locate the centre of the Universe . With gravity pulling everything in that general direction and finding such a point should show where the lot is heading to where all contraction will eventually lead. Identifying that precise location is a far greater problem to investigate than is the critical mass density factors a devastating problem. This inconsistency to point where the contracting should be heading proves to be the Waterloo of science because science has no idea where to position such a centre. If we backtrack instead of fast track the contraction of the Universe we should be able to find the point of the beginning of everything. It is because of where science position the end of the Universe some thirty odd billion light years from where we now are that I concluded the centre of contraction must be allocated. Closer to home we must search for the point of gravity where the gravity is the strongest as it must be in the centre of the Earth.

The Universe limits run from the Earth centre equal in all directions since the Earth is connected to singularity by gravity

$$(10 + 10 + 1.9991) / 7^0 = \Pi.$$

and when drawing this map that is in progress about the cosmos the allocated centre must be where the Earth now is.

That was what inspired me to locate my centre of my Universe. Even admitting to such a notion sounds like madness, but please allow me the opportunity to explain in more detail. I realised that my effort to locate the point holding singularity enabled me to backtrack the exploding universe to its origins. By applying some basic effort I have located the position from where all movement came and the direction it took moving forward in time...and yes, even time as such. Gravity is the dimensional changing of space holding r as reference in the cube as to the sphere holding Π as the reference. In order to generate spin that is producing time in matter occupying space, therefore creating dimensional change, Π has to be a factor indicating the possibility of spin because by implementing Π the circle sides will follow one another without establishing separation. As soon as motion takes gravity straight, singularity will reposition the direction changing the direction of motion by 7^0. It is this turning of motion by redirecting the continuing of motion that sets the critical time within the proton connecting to singularity. Instead of r being a line, gravity will inevitably be Π, which is the form value of singularity. That is this 7^0 redirecting in the square of space, which is ten on both sides of singularity and time is that what we find to be the Titius Bode law of 7 / 10 and 10 / 7 in relation to the Roche limit of $\Pi^2/4$ which is producing the gravity of Π^2. However the reducing in it is going from ten that is on one side and is crossing over the figure of 1.9991, (which is singularity on both sides of the Universe) and coming into contact with another 10 while turning 7^0 that we find to form Π. In all being the total forming on both sides of the Universe it is $(10 + 10 + 1.9991) / 7^0 = \Pi$. The answer must be in finding Π, and thereby locating singularity. If singularity is in affect the original point of the cosmos birth, the reducing path we should follow will indicate the whereabouts such a point must be. That is where cosmology diverts from mathematics.

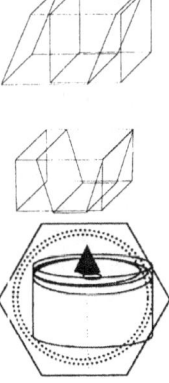

Anything occupying space in the cube will apply r and by r I mean just a distance not using Π because Π serves as a form indication while the collective product of r will determine form as well as accumulative dimension total. Notwithstanding the name used confirming the shape or r named as length width or height, it is all just a straight line bringing about the cube with all its other names that may find attachment to specific form but nevertheless still remains only a six-sided cube with connecting lines applying different angles changing in some cases. The normal perception is that any circle growing spontaneous would grow by the radius, which is r. In mathematics that may be true but it is not true in nature. In nature that cannot be the case because, r is an indication of a straight line. By growing with the aid of a straight line from the centre to circle the influence that that would have on the circle would result in many circles following one another and not a continuous growth.

In the normal applied mathematics there are two standard formulas used to calculate a circle. The one use an r to indicate the radius and the other use a D to indicate the diameter, which is double the radius and therefore needs to be divided by a four to eliminate the Newtonian inverse square law amounting to the difference there will be between the two. The one using the radius is Πr^2 and the other formula using the diameter is $\Pi D^2 / 4$. By implementing either neither produces results therefore such a lead will bring one no further than the understanding that person has. However one looks at the mathematical expressions and Kepler's formulating of space-time and we find there is an exceptional difference between the two scientific uses. When investigating Kepler's formula one do find it appreciably differs from the normal Mathematical equation such as we find the normal allocations to be $a^2 = r^2\Pi$ and $a^3 = 4/3\ \Pi r^3$. In the normally used mathematical expressions such equations tend to concentrate on the volumetric aspect. In the case of Kepler's expression it is something else that wants to surface. It is totally another idea that is coming to mind. In Kepler's formula a^3 stands to symbolise the third dimension and such a third dimension becomes equal to two other dimensions grouping and sharing value to equal a^3 efforts. It is not the circle of the rotation because with such a normal circle the radius is in the square and Π evaluates the form.

Here in the calculations Kepler received from the cosmos there is no mention of a factor Π, which one would expect to be somewhere applying since the circle is Π and Π is the circle and the two are inseparable. But not in Kepler's a^3, where there is no mention of Π at all. The fact that there is a radius of Π used to replace r of some sorts used to indicate a position, which cannot hold the square as it normally does in the case of the normal equations. In the mathematical equation the factor indicating the position of the circle edge has the square value being called the radius or in some cases the radius doubles and which then is the diameter, and the circle indicator is Π. But in this event the formula value will bring about a square value to the answer one receives. It will bring a value to the surface of the circle. In Kepler's formula it specifically does not. I am not the first one that brought Newton into disrepute. Before I did the cosmos did. The comets with they're not colliding did, and so did Roche and Lagrangian principles. Hubble was another one and it becomes apparent that every one that made a study about matters in the cosmos was in some disagreement about Newton. However no one in the past had the audacity to confess that they are in disagreement with Newton.

By Newton's effort to improvise on behalf of Kepler, Newton made a statement that Kepler never made. In all honesty nature reacted strongly against the claims Newton made on behalf of Kepler and not about Kepler's work but about Newton's modifying of Kepler's work. In short: how can a comet sail past the sun time after time without colliding and still apply a contraction in the manner which Newton suggested by the one claiming a freezing grip on the other? This strongly contradicts $F = G (M.m) / r^2$. How can five structures as the LAGRANGIAN POINT form around a centre structure while the centre structure keeps the five in position at equilibrium? It is so clear that all of the cosmos is rejecting Newton's improvising and rejects just as strongly the contradicting of the formula used to incorporate what we think of as cosmology in $F = G (M.m) / r^2$

In the event where I refer to r, more terminology of indicating a line than it is referring to a radius of a circle.

We find this proof in what we see in the Roche limit. In the Roche limit the extending of the radius does not commit r in any way but produce singularity by form of Π. It is the sphere reducing the other sphere by one applying material robbing through gravity contraction. The one sphere heats the singularity of the second sphere by some form of electrical crossing space-time without acknowledging a radius of sorts. It does implicate singularity by measure of $\Pi^2/4$, but that refers to singularity bridging tie and it has no hold on a radius. The mistake science made in the past in their studies of the Roche limit was their trying to implicate r as a radii factor.

Applying the Roche limit brings about a divide to the value of $\Pi^2/4$. This places singularity at a relevancy in a divide with singularity. When there is insufficient dynamics to place the correct space-time in order to bring about the minimum relation in space-time dominance, such dominance comes about where the one would secure a proper **k** factor with the minimum **a^3** space and the **T^2** time of the dominator will capture and increase the gravity charged by the dominant which then turn out to be antigravity to the lesser body. In the manifesting of the Roche factor the growth of gravity of the dominating singularity turns the space-time of the dominated partner into liquid heat, which is equal to antigravity of the sub structure. This process repeats the very first cosmic action there ever took place but since this action was part of the cosmos it will remain part of the cosmos until the end. In order to establish space-time singularity has to give up the privileged position of capturing and securing what ever may follow from such a release. From that singularity the dominant factor will release space but it will release time as it intensifies the heat density in the released space.

The heat becomes the time factor but then forms a sector of space-time and will lose the inclusive qualities singularity has. Before such a release as which I just mentioned singularity space in heat is forming time as a unit. We will not be able to understand this relevancy and such a ratio because such a unit falls outside the Universe or domain that we have to secure in our concept of reality and use to form our concept about our Universe. That position singularity holds is very much not part of our Universe though it controls our Universe by laws and whatever forms or represents singularity. That makes our concept of singularity fruitless and singularity is very much beyond our explaining. But the space-time released forms a limit we find running between specific borders in law as underwritten by singularity. There is a heat or hot which is liquid but is still material representing the heat and then holding the liquid is space representing the cold that places the "border on the other side". We humans regard the heat, which we measure because we cannot measure the space holding the heat that forms the limit in the cold the space-time can reach. Space forms the cold and what space contains is the heat, which limits the cold from expanding by accommodating the heat into the space. The representing of the two and the establishing of space-time is also representing the demise of what forms singularity. In singularity within the star we find that singularity reunites heat and space into something we are not aware of but that we can see diminished all space as it took heat to whatever limits heat can achieve.

This we find in place in outer space and because of the contraction we are unaware of, we find outer space to be cold while outer space is in reality so hot it is exploding and has been exploding since the Big Bang commenced. The exploding process we named after Edwin Hubble. In contrast to this we find in outer space heat at the very ebb in space in outer space at the highest pinnacle space can achieve with the time that developed as far as time did develop this far. Hot and cold loses unity when singularity releases the two aspects and singularity unravelled as it has in outer space again join the two aspects where singularity is totally unravelled in becoming space, which is representing heat. Before the unravelling space and heat was one but when Creation started it was in favour of heat whereas outer space now is space and heat still contained in a unit that is demising all the time but is favouring the heat aspect. The separating of the two and the establishing of space-time forms the demise of singularity but as heat levels demise space levels increase.

By space-time expanding singularity is in the demise. On the other side in forming material singularity vested the counter of this action by forming motion that demises space to increase heat and reinstall heat into singularity eventually where such singularity is maintained by material in gravity. The motion of expanding in outer space **k = a^3 / T^2** establish singularity favouring recouping of heat by increasing space producing more space and the motion in the atom $k^{-1} = T^2 / a^3$ favours the recouping of heat by destroying space But in all cases it is singularity in control of cosmic law to secure the maintaining of singularity.

If we wish to believe in the Big Bang and we wish to accept the factor of singularity then we have to accept that there was a period where there was no space anywhere at all. We can back track the space to a point where the space is no longer space on the precondition that the space is coming about only from the motion of space. If more heat comes to such a centre the centre will produce more motion. The motion will produce more duplication of space and the duplication of space is the gravity we experience as a contracting direction of motion where as heating is the expanding of space through motion. But it had to have started with a space less motionless dimension-less Universe wrapped in singularity. The differentiation coming from motion is a dimensional barrier that changes many aspects in cosmology. The dimensions came about as the Universe came about and each had its individual introduction period. Space and time parted at $(\Pi^3)^2 = 961$, material formed identities at $\Pi \times \Pi^2 \times \Pi^3 / 5 = 192$ and $\Pi^2 \times \Pi^2 \times \Pi^2 / 5 = 192$ where space either had material or had heat without material and space separated from heat and matter at space holding $10/7\pi^2/2(\pi^2 + \pi^2) = 139$ material $7(\pi^2 + \pi^2) = 138$ and space having liquid within $7/10 \; \pi^2/2(\pi^2 + \pi^2) = 136$. This is suggesting that these are meaning this was the first time liquid became part of the cosmos while all were still part of the same unit as the Roche principle would suggest $(\pi^2/2)$ as well as the $(7/10)$ and the $(10/7)$.

By my mentioning this I am not only once again jumping the gun but it is more a case of my being half way down the race track before the gun fired the start to the race. This, which I mention is ahead and almost at the end in the direction of where I am heading... I am on the one hand forced to do this because I am unable to find a connection between what I try to introduce and connecting that to what is available in science for me to connect it to. Let me try and once more somehow locate a connection between what I suggest and what Mainstream science finds acceptable.

Material seems to get glued to the earth by some force, which holds the name of gravity. Moving such a gravitated particle needs some drag by motion. The secret of lessening the effort in applying motion to the object in need of shifting is reducing the drag that is not being dragged at all but is all about motion that is not in motion and hiding behind the name of mass. Let us find this drag in nature and work from there to find a better natural understanding of being in a solid state on ground or a liquid flowing down into the ground or a gas floating above the ground. It is accepted by science that water can rub together and form static electricity. Those Super-Educated teach us that that is how lightning is generated and the best of all are that those Super-Educated then sit back and feel pleased in sharing their vast insight and wisdom. Can you believe respectable men say this while still feeling blameless about their view and acting so absolutely shameless!

The scientist that thought this one up had some or other big problem with his hair and thought that rubbing his hair with a plastic comb will be the same as water rubbing against water. I cannot believe that science can indorse such hogwash and hogwash it is! How can water form static electricity in vapour in the atmosphere by rubbing water to water because lightning is the product of heat that has expanded to gas then by motion of wind and cloud again concentrate such heat that before this expanded into gas. When the water vapour crystallise the vapour then again condense and form water drops but to condense the water release the heat that kept the vapour apart. The heat in turn then also condense back to heat being more intense in a smaller space as the heat is going into liquid (the most common and widely present example of heat to form liquid heat is in the form of electricity) and comes down by gravity as lightning in the turbulence of heat. The liquid heat forms lightning and as water does, the heat condensation we see as lightning flows down to the Earth in the form of transmitting electricity. The electricity we refer to as lightning has more in common with wind than it has with electricity but that I explain on another day in another book... Let's remove gravity and find mass.

It is well documented that heat forms more space when space explodes and space becomes more than what it was before. The space, which now is the matter, is occupying less space before the event of exploding came about when the heat was under natural singularity control and spinning in the atom at the speed of light. But when the heat releases from the control that the atom singularity has binding the heat to motion at the speed of light, such release turns heat into space and then heat forms additional space extending the relevancy and slowing the motion down This expansion of heat into less dense space is a method we gave the name to as exploding … While the action is a well documented fact for many, many centuries Mainstream science to this very day never realised space creating and exploding is directly or even indirectly connected!! By heating material there is an introducing of space because the space needed after heating becomes more than what the space would have required before the heating started. By heating the material with a sudden burst will bring about so much space available it brings along the destruction of matter in the position and form it holds. This destruction we know as an explosion. Because culture gathered through many centuries of steady science development left us the name we use for the process of exploding as an inheritance package. When the explosion occurs there is abundance of heat released and by turning the heat into space (which they then call shock waves as if that serves at all to explain the action that occurs).

The advancing of the newly formed space reconstructs the position layout the matter holds in all the immediate surrounding space. With the knowledge and countless demonstrations brought about by war and other destruction, this knowledge is edged into our minds to the same extent as getting dressed or eating. With everything having an opposite and a counter action the opposite must also apply. There has to be a removing of heat from the condensed space by the condensing of space. In that case the reduction of space bringing about the forming of condensed heat, concentrated by a removal of space as the reconstruction of matter. The removing of material from one side and replacing it onto the other side By removing space from space in concentrating space such space that is removed materialises the condensed space to liquid heat and then further into solid material by way of the electron condensing space further than what the atom can achieve.

The electron serves the liquid neutron as a gateway into the universe of the atom. Where matter removes heat from uncontrolled space to reconstruct its element worth and element position in value, the reconstruction is the direct opposition to the deconstruction of matter by explosion, therefore the relevancy changes and with the relevancy changing the result therefore must become reversal to the explosion. If the Big Bang was antigravity in exploding of uncontrolled or compromised singularity, then gravity is the constructing of un-compromised singularity confirming the compromised singularity onto the realm of the un-compromised singularity and extending the un-compromised singularity to the devastation of the compromised singularity. In such a manner space-time does not waste but is re—affirmed by control.

In outer space an object floats. On the moon nothing solid will float but there is no liquid air either. Those not familiar with this statement must think about what will be the difference there is in outer space of space in outer space and space in the atmosphere of the Earth and why objects entering the atmosphere suddenly acquire the ability to heat up and burn out. The reason why outer space is a gas is because the density applying in the space in outer space and material holds very little liquid material in comparison to what we are use to in the Earth atmosphere. Outer space is not colder but hotter, much hotter. In space all objects are very loosely connected and move quite freely about. This occurrence reminds one of a gas because in a liquid there is much more density in the matter relation and when solid the matter is as close as can be found.

Density comes as a result of a cold environment (not a cold atmosphere) Therefore the conditions in outer space form a gas and a gas is the hottest of the three conditions there are available to substance material. Comparing the likeness with anything we can compare to with our vision of what is on Earth, we must move to something we all consider to be a natural in all three forms, one being solid ice (very cold), two being liquid water (less cold) and three being gaseous steam (very hot). Conditions in outer space come down to steam because there is much space between the particles bringing about more space and less in the density in the space bringing about that there is a lot material. In the following example I use water as the subject because of every person's familiarity with water however the example rings true with any substance we may choose to inspect. By introducing heat to water, water changes from being a solid we call ice where there is much more material in the ratio between space not filled and material filling space whereas with a liquid substance such as we call water there is slightly more space unfilled between the solids than there normally is in other solids being apart but much less space as there is in the substance we call gas and the gas form of water we call steam or vapour. The scenario does not apply directly that much to water but I hope I'll manage to bring the point I wish to make across being what fills the space which fills and what constitutes what the solid / liquid / gas relevancy there is. The state in which material is in form being between material molecules is not written in rock but can change as situations change. It is space that is filling the entire overall space in relevancy to denser or less dense space filled at that point sharing space-time.

By introducing heat to water we change water from a solid (cold) to a liquid (less cold and more hot) and with the introducing of much more heat we get the heat to become a gas such as it is in outer space. By introducing even more heat to water we get clouds forming. By introducing even more air heat to air we get clouds moving, and with more heat added it is moving excessively, where the movement in fact displays a density increase. The motion provides a density release of sorts.

There is a density increase in the atmosphere during the storm and therefore the wind can then uproot large trees. Hail falling in whatever intensity can strip leaves but it cannot uproot trees and rain falling cannot break tree trunks. Yet those responsible for deciding what the scientific know-how must be and other superior members in Academic circles hold the opinion that air born particles such as oxygen, nitrogen, hydrogen, helium which are all extremely mass less particles, can have a density of superiority by such means that by wind blowing the particles the particles will collide with the trees and the intensity of such collisions between the air born and air driven particles will gather sufficient momentum as to remove the tree in totality of trunk, branches, bark and all from the soil holding the roots of the tree.

To suggest that something as light as say oxygen and nitrogen can blow down a tree with the quantities present in such a density as one find in the space we call our atmosphere proves how little science are able to think! With more increase of density in the wind we find spiral motion adding to the in lateral movement. With wind circling it has terrific density because in such a form it not only uproots trees but also takes on houses and much of what man can build. The wind blowing in excessive motion is not blowing molecules at a faster pace. The density of heat increases as the motion condenses more heat into less volume of space. The extensive motion produces the same qualities of destruction and material damage as water in a river causes when the water rages and the damage by wind is the same (almost) as the flooding that the water in the river can produce. When the density increases by adding motion much of the increase goes along with vapour, that is a form of air that is thick with water, (I distinguish between the terminology because why not only use one word, either steam or vapour. After all it is the same thing!)

In clouds we find lots of vapour but we find little water. The difference between water and vapour is that vapour has more unoccupied space and less space filled with water material in ratio. The thick density is there, but the air is so thick the vapour and the air combines to form a gaseous liquid we can see as a cloud. Remove the heat in the cloud (which the cloud needs to be if it wishes to be being vapour) the water returns as rain or as hail. On the other side the heat separating the particle excels in motion that becomes space also turns liquid by motion of space and the space forms a blanket that (by motion forming space) it may become so dense as to rather remove the tree with all the tree holds than to allow space between the particles to part. We also see such dense liquid space remove as a liquid and produce a form of heat we named lightning,

The heat that I refer to is different to the idea we have about the heat that makes you sweat but a much more concentrated in spin value such as electricity has liquid heat. It is all part of Kepler's formula where space density is the product of the motion and space-time is all about spinning motion. The liquid forms when the heat concentrates with a much more formidable spin that produce a higher electrical charge and comes as pure electricity in the manner it is produced. As the liquid heat we call static electricity contracts in becoming denser in measure that uses less space the static receives motion and the motion becomes lightning. The heat is forming static electricity because of the lack of motion present and that static is in between the vapour that liquefies forming electricity which is just plain old liquid heat separated from the vapour water and by letting the vapour form a denser water drop and allow the water to separate from the liquid, the water forms a specific density because the motion lets the water vapour crystallise to water. Then in that case and being next to more vapour the water forms the solid although in our mind set water is a liquid but still… it then is a solid in the form of water we named rain. The liquid which was the water did not vanish but became liquid air which we call lightning. The question about density increase always comes from material being more prominent and more abundant in such a space. With the increase of density it always accompanies the increase of heat and the discharging of heat.

If one would think that it is vapour in the wind that increased to such extend that the wind can uproot trees, then why does hail with such a lot of solid water not uproot the trees. In order to judge the logic try to imagine how much water is needed that is converted to spray and is forming a steam or vapour. Then think how much of that is needed to form a substance with the veracity that is required to uproot a large tree? There is a world of difference between windstorms and hailstorms because of the abundance of electricity or more bluntly phrased heat in spinning motion. Windstorms having the ability too uproot the trees have a very sticky substance between the molecules and the more sticky evidence there are, the bigger the ability to cause damage. This sticky substance can only be liquid air or liquid heat. The air substance shows a bigger resistance to part or create space than the tree shows its willingness to remain secured to the soil by its roots. The substance can be sticky to the point where it breaks braches that will require an effort of many hundreds of Newton meter to break.

Even spraying the tree with water which man created artificially by pressurising the water would hardly break the branches and even less hardly uproot the tree during windstorms. If it is done, the water flow will be enormous but I have seen trees uprooted many times without one drop of water visible. The only logical remaining supplier of such a density increase must be the heat in the wind becoming denser. In this there are changing relevancy dynamics, which I then introduced as equal to the substance found in atoms. Let's look at relevancies applying according to singularity distinction.

By accelerating gravity such acceleration can reach a point where space-time diminishes to the size of a photon as time accelerates to the speed of light. Establishing a Coanda effect basis for singularity becoming established in an iron confinement where copper dismisses the space –time electricity changes as gravity is in the atmosphere of the Earth to what gravity is where the Earth dismisses space-time in the inner core.

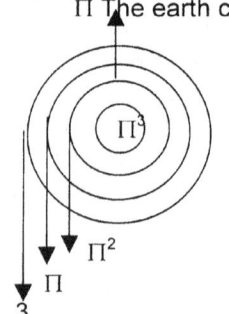

Π The earth core

$\Pi^2 + \Pi^2$ The earth outer crust

Π^2 Densified particles within the earth's atmosphere

Π Gasses and water vapour within the atmosphere

3 Heat separating molecules within the earth's atmosphere

When a structure is part of a solid it holds the value of ($\Pi^2 + \Pi^2$) which is the proton number and that number I dedicated to the solidness of the soil and everything within the soil structure. The formation of ($\Pi^2 + \Pi^2$) is uninterrupted from the centre base of (Π^3) which we find in the very centre of the Earth through all the way to the solid top of the Earth soil. Everything attached but not being part of the Earth where the Earth then holds ($\Pi^2 + \Pi^2$) and forms another separate value but is still resting on the Earth soil. Nevertheless, I gave the bottom value of the neutron as much as forming a moving liquid would have, being (Π^2). The Earth holds a proton value of ($\Pi^2 + \Pi^2$) and the structure that is solid but has an individual and independent solid form that is unattached to the Earth carries (Π^2). Being (Π^2) the object is in the realm of the solid but which through possible motion still has to find independence from the Earth through achieving separate motion and construction. This would be a rock being loose from a mountain or a tree or even life being on the ground or which then is equal to the proton part which is valued at ($\Pi^2 + \Pi^2$). When it is not struck to the ground and has an ability to apply motion independent of that of the Earth, which makes it float in the atmosphere I gave the second neutron value of (Π). Water vapour in a cloud formation will be Π but when forming rain in condensing to a semi solid the water drops will then become Π^2 by becoming relatively solid to what it was before. As Π^2 it will move to ($\Pi^2 + \Pi^2$) which is the ground. This has all to do with space-time duplicating versus space-time dismissing. When the object is in total suspense and away from the Earth all the time being part of the atmosphere such a particle is valued at the electron value of 3. Water forming vapour will have a relevancy of 3 to (Π) where 3 would carry the relevancy value of the electric charged air and (Π) would then take the relevant value of the vapour. When the vapour condenses the heat will turn to liquid as lightning and change the relevancy of heat from 3 to 3 (Π^2) where then the water will become a solid form of liquid being ($\Pi\Pi^2$). One should take note that relevancies can change as quick as motion would allow such change from where liquids can be as dense as solids (Π^2) or being a liquid (Π) or a gas (3). It indicates position and is not a tool to use as calculated formula or as measuring devises.

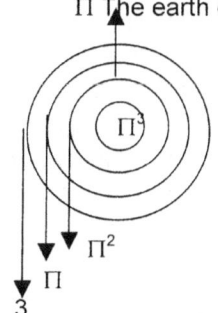

Π The earth core

$\Pi^2 + \Pi^2$ The earth outer crust

Π^2 Densified particles within the earth's atmosphere

Π Gasses and water vapour within the atmosphere

3 Heat separating molecules within the earth's atmosphere

Changing from solid to liquid or to gas is re-establishing a relation between particles such a relation is also between a solid proton $\Pi^2 + \Pi^2$ or a liquid neutron $\Pi^2\Pi$ or gas 3. In each dimension, the light displaces space-time that is more concentrated and therefore influences the projection path of the light's future position in space-time to unoccupied-, occupied-, and densified space-time. Motion is what causes gravity. Motion is the contributor to gravity or the replacement of gravity and motion comes about when heat concentrates in space. In every development it is possible to see what implications in changes came about where the Roche principle was either $(\Pi / 2)^2$ or $(\Pi^2 / 2)$. The Roche limit is therefore a joining space or a separated space.

The very first of every aspect that became the Universe started when the very first action came about and that action we now call the Roche limit. As singularity expanded motion brought about a relation in singularity that reflects a value of space shared by motion singularity shifted from one Π^0 to innumerable formations of Π just by applying motion by overheating and gravity resulted in the motion of contracting to form Π^2 where neighbouring Π started a rotating relation and all along Π stood in for the values coming about. But because the space shared is about seven points standing in regard to ten points over the Roche factor the square of Π formed the divide as well as the joining of such a divide. But, at the time and even the represent we find where singularity forms space – time space is very much at a premium therefore singularity is halved.

The discoverers discovered all the singularity that forms a unit also produce a singularity taking charge of the combined effort and this yet again is named after the discoverer because no one at the time had the foggiest notion of what. In the case I just mentioned the discoverer was an engineer by the name of Henri Coanda. Henri Coanda discovered without him or any other person ever realising that what he discovered is the effort singularity in single produce can remove and form a unit in an allocated centre from where the motion of gravity takes charge to establish a space with boundaries set by motion. In this the Earth as just another cosmic structure is no different and all the atoms in the confining space of the Earth elects a centre wherefrom gravity takes charge as space is dismissed. At such a point singularity-forming gravity shifts the collective singularity to govern by producing gravity at $\Pi^2 = 9.8$.

This position is what Newton saw as cosmic gravity but cosmic gravity it is not. It is a bit of localised gravity that has no cosmic importance at all. But that is not where the centre of the Universe is at...no it is in the centre of any product of singularity that charge space-time and by charging space-time it calls upon the use of the Roche limit as it introduces contraction in Π^2. But since gravity is motion only half of such gravity takes charge on any particular side of the Universe and since the Universe we discuss at this point is the accumulating effort of all the centres of the Universe that group together to form the Earth. When a body is within the atmosphere of the Earth half of gravity applies as $\Pi^2 / 2$ and between separate bodies the value is gravity which is time divided by the factor forming time which is four $\Pi^2 / 4$.

There is another system called the Lagrangian system where a centre aligns with 5 other points to form a six-pointed relevancy.

LAGRANGIAN POINT is another result flowing from the singularity position and most directly coming from Pythagoras principal and connecting that most basic mathematical in conformation of Kepler's formula.

As singularity parted two very much opposing factors rose from that action where each formed a relevancy in relation to the other as well as a relation to the third factor which was singularity controlling the positions. The two particles that came about in a fight for space, was round and the space required a square to fill. In one case the round could remain round and stay solid by applying motion in moving about an axis and in the other case the other one had to apply motion by expanding and relatively grow softer as it deformed and lost shape. There were many more than the two that stood against each other forming the required motion but only two points in singularity apply a relevancy matching each other. Each one in such a confrontational position had to form a match between two particles where the two were one Universe away as two separate Universes matched.

The one was in a position to apply gravity and kept the form remaining true to structure and shape at the expense of the other where the other miss formed by overheating. The one factor, which remained solid, we now call material and the other got the name of plasma but to spare confusion I use the term heat, because all things are either one of two forms of heat. The one form is heat frozen solid in a unit of heat and the other is liquid in a fluid heat. But in all events it is heat-matching space that is holding the heat that is in a unit, in another unit. The softer compound became cannibalised by the solid firmer structure and demise came to the liquid singularity in order to save the solid singularity a sure fate of destruction. There are today two factors, which still is in opposition still as relevant as the first motion. There is heat and there is matter, which is a Universe apart although they keep a position next to one another. The crucial factor is the relation every element of every atomic grouping has. It is the form it holds by number of protons grouping in clusters. Being secured by an electron outside, a neutron fluid and a proton solid centre is the characteristic all atomic elements hold.

With the electron, neutrons and electrons in equilibrium there is a matching of representation in the unit it is in holding. The cluster forming is by itself the unit into another unit that is much more securing to form. The Lagrangian layout plays a cardinal role but is far too time consuming to explain at this point, however I do explain it in my book Matter' Space In time : The Hypothesis. The relation the atom form in cluster grouping of the protons absolutely dictates the relation the atom accepts with the heat surrounding the atomic structure. The Lagrangian proton layout will eventually dictate heat/material relation applying in every case of every structure of every atom. Since any star is a cluster of atoms, and any cluster of stars are several groups of atom clusters, the groups of heavenly bodies will assume the characteristics enforced by the characteristics of individual atoms.

Kepler brought in a formula that came through the dedication of two lifetime studies and the formula reads that space is equal to time in motion. Mathematically it reads that $a^3 = T^2 k$. This formula brought Newton's claims into dispute because what this formula said was that geodesic space is not confined space and while laws apply in the confined space of Earth it does not apply in geodesic space. From Kepler one can see the precise moment the Cosmos started. It is Mainstream science that is hindering the accepting of the formula that is dampening the human understanding of the Cosmos. Kepler said without saying that the first moment was when singularity $k^0 = a^3 / T^2 k$ produced space through motion and we know motion is the product of heat releasing space. That is what Kepler interpreted when he declared that $a^3 / T^2 k$. Understanding the cosmos largely includes our realising that heat and cold is the very same thing. It happened at creation when heat brought about creation. The moment, the instant heat came about, cold also was invented and the maximum heat there was, was also at the same instant the minimum cold there was. By creating heat it created cold that separated cold from heat. The birth of heat produced the birth of cold because the same thing was born at that instant. Heat and cold is the same thing bringing borders we human's whish to create but such creating only furthers our inability in performing worthwhile understanding. But like planets heat and cold are human creation and have little interpretation in actual cosmology. Our being part of the Universe, depict us with a realising of borders and boundaries but also prevent us from seeing the full picture as observers of the Universe should.

"When did the cosmos start?" This is the question every one is in search of since Biblical days. Even at present the remaining unanswered question is this question every race and every culture is in pursuit of. I would say it had to have started with space but science has the cosmic time linked with time. The truth is that there could not have been space without time and there was no time without space. The first moment came when **k** moved a way from singularity **k⁰** to establish space. Creation was when **k⁰** introduced individual entities **a³ / T² k** and went about moving the different entities apart. It was when the confinement was broken and particles appeared for the first time. Understanding this comes hand in hand with accepting that there is two forms of structures other that elements confined in an atom.

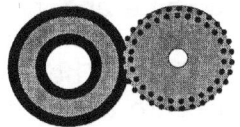

The heat surrounding elements is not a match that fit all and is a separate issue in every case with every element. The heat /space /material relation gives elements their characteristics they are recognised by. Such characteristics changes as the heat /space/ material relation changes and alters the presumed characteristics

Hydrogen 1	melts at -259^0 C,	boils at -252^0 C,
Helium 2	melts at $-269\ ^0$ C	boils at $-268{,}9^0$ C
LITHIUM 3	melts 180^0 C	boils at 1300^0
BERYLLIUM 4	melts at 1287^0C	boils at 2770^0C
BORON 5	melts at 2030^0 C	boils 2550^0 C
Carbon 6	melts at $804\ ^0$C	boils at 3470^0 C
Nitrogen 7	melts at -210^0C	boils at -195.8^0 C
Oxygen 8	melts at $-218.8\ ^0$C	boils at -183^0 C
Fluorine 9	melts at -219.6^0 C	boils at -188.2^0 C
Neon10	melts at -248.59^0 C	boils at -246^0 C

Melts (meaning that the element becomes a liquid) at -259^0 C. **Hydrogen boils (meaning that the element then becomes a gas) at -252^0 C, This does not concern the basic atomic element but only applies to the space on the outside of the atom**

This is the matter and the antimatter that science is referring to when science claims the devouring of matter by anti-matter, according to my opinion. Some particles became matter and others became antimatter or plasma or heat or just what you wish to call the by-product that came about from the friction that brought about the heat that led to the expanding of the space. Those forming units we call elements had established characteristics relating to space-time where each one holds an individual identity according to the number of protons the element has in one cluster unit. There are elements. The elements presume in the role of being solid and form as units the solidity of solid materials. Then there are liquids. Amongst those elements mentioned, we regard some of them as being natural "liquids" and others being natural "gasses", however they are solids notwithstanding what our culture call them. There are those we consider to be mainly gas or liquids and only then there are the state of being "frozen gas" (as we regard outer space to be) but such presuming underlines our mistaken culture we have.

The two forms I have just mentioned being gas and liquid are the same and the only difference there is, is the state in which elements can find form being between the two forms of having liquid/gas between elements or not one of the two which in that case we then call the state in which the elements are being a state of solids. The form of being liquid or gas or even solids alternates and the changes come about with more or less heat being part of the density factor. The form of heat being more (when in the form of gas) or less (when in the form of liquid) or absent (when in the form of solid) establishes and reforms the state in which the elements cluster together or have heat in a specific formation parting the atoms. Nevertheless liquid and gas is the opposite of the very same thing that came about before the Big Bang event took place and represents a period at the time the cosmos was still forming the second form other than atomic solid elements.

The forming of liquids came from the friction between solid elements that were in a position without unoccupied space-time. This forming of a form other than being atomic clusters took elements into having the opportunity of choosing between another forms, other than solid atoms forming clusters, which we think of as elements. This is a very much identifiable period as it developed in the cosmos as the cosmos developed and went through a specific development period. This period went on up and until a time when the Big Bang took place. The Big Bang is a period after the forming of space-time came about and normal cosmic growth started influencing the development direction of the cosmos.

The Big Bang commenced when some particles went liquid (or softer) which took liquid space and turned a space-less liquid space into a spacious gaseous space by adding more time per cycle to the relevancy. But elements found in nature are not naturally solid, liquid or gas as the element table wish to teach us. They do not form unbreakable columns that one can take anywhere and it will match everywhere. The element table is a human concept that is valid on Earth and not a cosmic law and then again when thought about it realistically it is not even valid everywhere on Earth. The heat or space that brings form is an ultra blanket that wraps the element into an enclosure to suit the conditions applying at that point. We then identify the conditions atomic elements may be in, in such a thin margin that it only applies under certain conditions which is mostly artificially created by humans in laboratory conditions at a very precise pre-arranged heat range under a very specifically stipulated environment. These results will change notably and have a varying outcome when tested in different climate conditions through out locations on the Earth.

One cannot put the characteristics these elements will have and prescribe the same characteristics to conditions applying in the sun or on Jupiter. All elements are all forms and it depends on the conditions permitting what form will apply at that point. Water boils at very low temperatures on Mount K2 and at very high temperatures down at the bottom of Stilfontein Gold Mine, which are several kilometres straight down vertical. To the Human mindset we form, water must boil at 100° C or 210° F because that boiling temperature suits man and human grew accustomed to that idea. It is as far from the truth as thinking about outer space being cold or the sun being a hot gas structure.

Forget about the example we find in water because it seems if any one talks about the three materials all concerned immediately thinks about water as one structure that can freeze and can boil when it is not flowing. That is not even an example because water is not a true substance but is a compounded combination of volatile elements forming the most outrageous concept the cosmos could think up. By combining some of the most volatile elements the cosmos created the least volatile substance known to man, which is water, that in the combination of CO_2, where the elements combine space-time to from in that combination, a substance killing fire. But the three forms water personifies are the structures compiling the universe and prove stages of development. Water on the moon is ice. Water on Venus is not even a thought that will ring true and water has never been on the sun. Water is to us a vital means having an ability to sustain life, which to our concept is to sustain the most critical of just about all substances. We even think of water as only in the way it is used as fuel to get us away from the future possible planet we might be visiting. As long as we create a Universe with life as the centre of all focus we will find a miss-fitting Universe we can never understand and what ever we think will never apply anywhere but on Earth. We have to start our focus on singularity in charge of whatever space-time we investigate. In the beginning there was singularity. That makes singularity the centre from before the beginning of all times. Singularity is still the centre on which our focus must be. Singularity came first but also staged the centre.

From singularity came material as a solid substance. Some rubbed against others forming heat in some cases. From elements disposing form comes liquid, a softer substance that will allow some form to match other less flexible forms. Then comes gas when singularity loose control and turns space to liquid, which loses density as a result. Again we can see a pattern coming about. But it involves Kepler more than any Mathematics. It puts distance **k** in space at point a^3 between **k** and k^0, which provide distance **k** with two points bringing about time T^2. $k^0 = a^3 / k T^2$. Once Kepler showed relevance existed where no one could denounce such a relevancy. Not even Newton could because relevancies in mathematics show clearly an interaction that changes and reapplies. Yet Newton did by putting **k** at a value of T^2 and put that at zero.

A circle has no point serving as the start and therefore there is never an end and cannot form zero. Singularity brought about Π from the instant when the point took up more space than what space was available. The space less ness brought k into an equality of T^2, which was then equal to a^3, and that equality still performs as 180°. The result was that π formed and Π has no particular start or end. It cannot show a start because there is no ending spot to singularity. In all cosmic wheels there is no specific point that indicates where the circle motion first started at any given point. Therefore after one cycle it may end precisely at any point where the immediate start starts. All starting points or ending points are human concepts but has no relevance on the truth of the state of affairs where we wish to divide our life in seasonal changes. There is only relevance reapplying as motion contributes to cyclic reoccurring. Therefore there never can be a cyclic zero as Newton believed. All singularity would form a duplication of another singularity that forms a ring as relevancy to the inside and a relevancy to the outside but the value remains Π at all points in all stages.

Singularity forms a divide Π that is sharing value since there will always be a possibility of yet another line in the realms of singularity lying between the two lines in question. By reducing the size infinitely to either side of the divide is what we humans create. Boundaries therefore are human and as man made substances it does not belong to the cosmos outside the influence of man and must be discarded. The understanding of this insists on one vital precondition. We have to abolish our instincts about things being big and small, high and low, hot and cold, tall and short. All the human measures are truly unfitting to the cosmos. The biggest of whatever there is in the cosmos is beyond human detection and will be classified according to human standards as nothing because in human standards we are unaware of the largest there is. The same goes for the smallest there is because man will never even have any ability to detect the smallest in the cosmos.

The biggest and the smallest are human inventions. Those are measures made by man. In the cosmos the tiniest object is the key to that particular Universe. By going smaller one is going bigger. By going within the atom we go outside and meet the Universe. In singularity we find the cosmos because every singularity not only represents the cosmos but also in fact is that Universe. No mathematics will ever measure the thickness, because as Π is standing still it cannot have a width at all. The moment a width appears which one can measure or calculate; the line will become part of the factor forming the divided and not the divide. The instant when space connects, the spin direction will produce the partisanship of space and spin. Any form of space including even in the smallest state, will produce a favouring of direction. Through such motion gravity comes about because the motion is gravity and that produces the time aspect T^2 thus thereby changing the direction by rotary motion. The minimum is Π^0 going onto Π. It is the spot forming dots.

The moment there is an area there is a measurable rotating brought about and no longer a non-interfering divide that is there where Π^0 forms Π. Such a line holds space in a position that runs far beyond the boundaries and limits of the three-dimensional. Another factor of such a line would be that the radius (let us substitute the radius r with the using of Kepler's **k**), **k** would be immeasurably small. The factor **k** cannot be zero because infinitely close to that first **k** is the start of the third dimension where time plays the part as the fourth quarter. The presence of **k** is undeniable and recognisable yet it is not visible. The fact that **k** is there albeit stripped of any influence, disqualifies it from being zero and therefore not being there. With **k** already beyond any measurable space, leaving a^3 as a factor of one and not being able to pin any volumetric measure to that one **k** will have to be to the power of 0 being k^0. It is exponentially zero but zero it is not. In Kepler's formula $a^3 = T^2 k$ the area a^3 would be one because of the dimensional non-existing of measured sides in any direction. If $k^0 = 1$ and $a^3 = 1$ the only alternative T^2 could possibly have is also one. The factor of T^2 identifies the time in the formula and when the formula indicates time as one. The time component must therefore be eternal. Only time in eternity does not change. In all human calculations an end as well as a beginning has to enrol the whole concept being calculated and when that is not possible the calculation as well as the way in which it is calculated becomes suspicious.

I know somewhere there is vast numbers of highly intellectual mathematicians that will disagree in all sincerity about my next statement but nevertheless I have to state it. Mathematics has to use a beginning and an end to bring about coherency. That secures fixed values and implying numbers in relation to determine size differences or quantities. However, since that may be and however clever it would make the mathematician appear, it is an irony and it focuses on stupidity. Such quantities and numbers are only deceptions and have no place in the cosmos since there are no size differentiations or numerical number quantities to calculate when applying the cosmic standard rules.

Gravity bends light is a statement very synonymous with Newtonian cosmology. To determine the gravity such determining formula requires a diameter. To find a diameter of a star one has to measure such a diameter by means of looking at the light. The gravity bends the light, so no true measure can be reached by using the bended light to measure the gravity, which is bending the light in the first place. One cannot use an unreliable measure to measure that, which is enforcing the unreliability of the measure. In that case what Kepler introduced makes most sense of all. Such was the case with Kepler's findings where each provides separate applicable relevancies and Newton's dispute thereof. Newton saw no reason except for his denunciation of what he thought was Kepler's inferior mathematical skill in Kepler's choice not to use the Π while Kepler never thought of Π because the cosmos never gave Π in the formula whereby the calculations were done. But behind the cosmos not including π naturally is a Universe filled with reasons.

Newton observed a need for the more commonly used mathematical formula applying when the calculation of the sphere comes about is $a^3 = 4/3 \, \Pi r^3$ where it places one third dimensional but lesser factor in direct relation to another third dimensional relation and all that in relation to the form that is applying. By using the normal mathematical equation $a^3 = 4/3 \, \Pi r^3$ it is definitely not what Kepler said when he produced his formula. Kepler allocated the square to time and by the same margin allocated the cube to space in the same formula. This means that mathematics do not find representation in the normally used mathematical expression of $a^3 = 4/3 \, \Pi r^3$. In mathematics we find a deliberate lack of the time factor when using the equation $a^3 = 4/3 \, \Pi r^3$. This must prove the differences in what Kepler found to express and what Newton thought Kepler tried to express.

However when using Kepler's manner of measure in calculations concerning the cosmos there is no criss-cross matching of dimensional accumulation. If I am reading the situation correctly Newton saw Kepler's mathematical skill somewhat below the dimension of Newton's genius and that spurred Newton on to bring changes to Kepler's formula but in doing so many things went missing through incorrect translations and miss interpretations of the use of such a genius. Kepler was not referring to a mathematical space on a flat sheet of paper, which was what Newton saw. Kepler produces a value linking space to gravity being time of space in motion in as much as calculating space in the geodesic forming a motion and measuring time in the geodesic of space in motion and bringing about factors in formulas through figures that geodesic space informed Kepler about. It is the motion or time part Newton did not notice.

If Kepler was mistaken then the Cosmos was wrong about the cosmos and if Newton was incorrect it was about mathematics that Newton then was wrong about. Kepler places time in the square directly in relation to space in the cube in association where time shows two distinct qualities. But thirdly this relation applies to a second space flowing in contrast and securing a three dimensional cross point that gives every aspect in individuality. That is the part Newton missed. The one factor is time in the circle rotating while the other is in the linear or the straight line implicating the position that the other would have. But in all cases the straight line responsible for the factor **k** will lead naturally into a circle T^2 and the circle forming T^2 will grow into forming a larger **k,** which will become a larger T^2. The cycle shows no end and that disqualifies mathematical calculations, which demand a specific start and a specific end. Any starting point demands motion because the space that requires identifying can only be identified if that space is in motion. This proves that time or gravity (with time and gravity being the same thing) is the method that the Universe uses to control the dimensional growth of the Universe. It is $\mathbf{k = a^3/T^2}$.

The space we find between the planets is representing the space we find between the sun and Oort cloud. It is the same space and what we find between the planets we will find way beyond the planets all the space links to the sun centre $\mathbf{k^0}$. That space we will find outside the solar system in the regions between the solar system and the next star is the same space and still is controlled by the same centre. That space is filled with the same filling and it comprises the same substance. To see Kepler's finding only applicable to the Planets is so Newtonian. It is as thinking of gravity as the ability of a towing rope that can be used for pulling all innocent material with an invisible rope. Space is everywhere the same and holds the same value because space started out being the same substance everywhere. That is what Kepler introduced. Tycho Brahe and Johannes Kepler were the first cosmologists.

Official astronomy defines Kepler as follows being The German mathematician and astronomer KEPLER, JOHANNES (1571-1630)

German mathematician and astronomer became Tycho Brahe's assistant in Prague in 1600 A. D. where he undertook to complete the tables of planetary motion Tycho had begun. Kepler first calculated the orbit of Mars. He spent much time trying to reconcile Tycho' s accurate observations of the planet with a circular orbit, but concluded (in Astronomia nova, published in 1609) that Mars moved instead in an elliptical orbit. Thus, he established the first of his laws of planetary motion. A theory that the Sun controlled the planets by a magnetic force led him to the second and third of his laws, which were published as part of his treatise on theoretical astronomy, Epitome astronomiae Coernicanae (1618-21). The Rudolphine Tables (named after Tycho's patron, the Holy Roman Emperor Rudolph II) of planetary motion appeared in 1627 and were still in use in the 18th century. Kepler also wrote De Stella nova, on the supernova of 1604 and Diptirce on optics and the theory of the telescope. The overall view followed in this book **Matter's Time in Space** places the true significance of his work in true contents. In KEPLER'S EQUATION is the equation that relates the eccentric anomaly of a body in an elliptical orbit to its mean anomaly. The equation is $E - e \sin E = M$., where E is the eccentric anomaly, M the mean anomaly, and e the eccentricity of the orbit. It is important as one of the mathematical relations enabling the position of a planet about the Sun, or a satellite about its planet, to be calculated from the orbital elements for any time. However this only relates to the solar system, and KEPLER'S LAWS only apply in the contents of the solar system. The three laws governing the orbital motions of the planets, discovered by J. Kepler is as follows: The first law states that the orbit of a planet is an ellipse with the Sun at one focus of the ellipse. The second law states that the radius vector joining planet to Sun sweeps out equal areas in equal times which as it says refers to time and not the circle. The third law states that the square of the orbital period of each planet in years is proportional to the cube of the semi major axis of the planet's orbit. The first law gives the shape of the planet's orbit; the second describes how the planet must continuously vary its speed as it follows its orbit, moving fastest at perihelion and slowest at aphelion. The third law gives the relationship between the planets' average distances from the Sun and their periods of revolution. Instead of placing, the true value to Kepler's laws I. Newton placed his own interpretation to Kepler's laws, and in doing this, he wilfully destroyed the principle working of the Creation. Through Newton's tunnel vision, he applied his own miss interpretations to the correct presumptions of Kepler. Newton reduced the implication that Kepler's findings hold by introducing the law of gravitation. He then went about and changed it to three laws of motion. I. Newton generalized Kepler's first law, verified the second law, and showed that the third law should be amended to the form; $4 \pi^2 a^3 / T^2 = G (m + m_p)$. In this, the value of T and a are the period of revolution and semi major axis of the orbit of a planet of mass m_p about the Sun of mass m, and G is the gravitational constant. The major aim of this book is to correct these misgivings of Newton. I shall return to the statement about $4 \pi^2 a^3 / T^2 = G (m + m_p)$ In all instances of measuring the distance the orbit travels around the sun as the space displaces or space covered by travelling in the time it is covered and dividing such a ratio one find the distance of the orbiting object from the sun in relation to the other factors form one or very close to one.

What Kepler saw was the cosmos telling Kepler that the circle goes square and the straight line stays single when calculating cosmic cubical measurement. There is a very good reason for this because $a^3 = T^2 k$ is in fact $\Pi^3 = \Pi^2 \Pi$. There is no room in which to add Π as a form indicating symbol. The cosmos uses Π in that manner as a standard measure to begin with.

The table is just about numbers but about that it is much, much more than just numbers. The numbers paint a picture and tell a story. It is the only time ever that the numbers first was produced and concluded the formula applying arrived at. In all other cases numbers are the result of a conclusion brought about by the use of a specific formula. Kepler produced a formula from the numbers and not numbers from a formula. The numbers brought about answers for which we fail to ask the correct questions. By seeing a mathematical circle one miss the total picture and the story the numbers tell. The figures explain dimensions working in conjunction and together they combine dimensions where the picture behind the story becomes a colour spectacle in comparison to the mathematical grey that a mathematical circle produces. In the case of the solar system it is relevancies carried from the sun and the sun is the governing singularity representative for the entire solar system. This is about relevancies applying throughout the Universe and not just locally. This balance is much, much more than what the figures say. It underlines and it explains gravity as a life form indicative of the cosmic life in the cosmos and is very different to what we consider our life to be. In the argument Kepler made he had hidden much more facts into one formula than what I think even he realised. Well, it is much more than that the Accepted Policy Protectors Of Science ever came to realise. He officially formulated space-time, he officially coined not the name but the origins of the Universe being the Big Bang and he was the first to put the speed of light in relation to cosmic development…and all of that with his rather simple formula. He said the space a^3 not the circle (a) or the circumference a^2 but in the circle a^3… where such a circle represents a factor in the third dimension. The formula he compiled was not rather but very specific about the area being a third dimension area and to prove it beyond doubt he placed it in the relevancy of the formula in a ratio of presenting the third dimension in space. He said a^3 is equal to T^2 **k.** That specifies space as motion. Newton and Newtonians came afterwards and played with mathematical toys as to challenge their mental capabilities but brought little new ideas to the table. Newton introduced a $4\Pi^2$ to indicate the presume circle on the one hand and on the other hand he brought this lot equal to $\{G (m + m_p)\}$ which he then presumed to be the general Universal gravity constant (G) and the sum total of the two

structural masses. Newton saw a ring circling around a centre having $4\Pi^2$ to indicate such a ring outside a centre and he positioned $\{G (m + m_p)\}$ where the two mass factors combined the gravity effort in the general grand gravity constant in space.

What Kepler saw was more of a dimensional nature than the practical mathematic symbols and values. On the one hand was a value to the third dimension, which equalled two-dimensional values one the second dimension, and one to the first dimension.

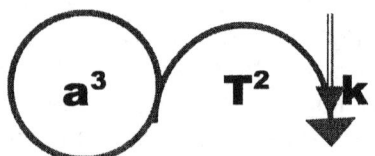

Planet	Period T years	T^2	Distance	Space a^3	Ratio
Mercury	0.241	0.058	0.39	0.059	0.983
Venus	0.615	0.378	0.728	0.381	0.992
Earth	1.000	1.000	1.000	1.000	1.000
Mars	1.881	3.54	1.524	3.54	1.000
Jupiter	11.86	140.66	5.20	140.6	1.000
Saturn	29.46	867.9	9.54	868.25	0.999
Uranus	84.008	7069	19.19	7067	1.000
Neptune	164.8	27159	30.07	27189	0.999
Pluto	248.4	61703	39.46	61443	1.004

At the first glance Kepler's formula seems to be numbers and positions applying between the sun and specific but different planets in the solar system.

I have had so much resistance in the past from all Academics but that is not what I see what Kepler saw. With what I saw what Kepler's saw I shall trace that back even as far as to the centre of the start of creation. In their eagerness to calculate the Mathematicians calculated a formula to measure the circumference a^2 of a circle being Πr^2. I have seen an Astro physics examination question paper where they use $4\Pi r^3 / 3$ as the formula to calculate the sun and other stars' volumetric space! Not one mathematician was for one second in doubt about the manner they may interpret the radius with light and gravity bending each other and all that!! They formulated the measuring procedure of the circle being in the third dimension that will show how big the volumetric space is of a sphere at a^3 being measured with the procedure being $4\Pi r^3 / 3$. Then some Mathematician and an Englishmen of Importance in academic standings came onto the idea of gravity. Being a mathematician the Englishman placed the Universe at the feet of mathematicians. He saw circles where Kepler saw three dimensions interlocked in an ever intergrading relevancy where one feeds on the other and feeds the other factors forming circles where the one circle grew from a straight line forming a circle which comes down to the interaction of dimensions feeding and being fed by one another. Newton saw pulling and shoving where Kepler saw space in motion forming the interaction of dimensions.

Kepler knew time had to be somewhere as something and then covered it by pronouncing the circle cycle as space in motion that is responsible for the time aspect. What then is it that Kepler saw as he formulated $a^3 = T^2 k$. At the normal flow of time it takes the electron a certain time to spin around the atom. That is in short space-time. The atom uses space a^3 and the atom is a certain length k that forces the distance the electron has to travel in one cycle period T^2. The atom a^3 space connects the electrons k to go about circling around the space a^3 at the time of gravity T^2. The relevance k produce to support a^3 is to point T^2 to two positions the electron will be in the duration of one specific time. The electron travel will be cyclic and periodic in relation to the space the atom holds. The space stands related to the gravity with which the Earth reduces space and with the space and speed with which the atom travels through space.

Einstein fathered the perception that light travels as fast as time flows but I disagree with that idea. Nevertheless the perception is there that the speed of light is as fast as travelling of any sorts can reach. We accept the electron's travelling speed imitates the speed of light as much as it is permitted by gravity to do so. By this imitation the electron come as close to time as it can ever come (that is in accordance tot the general persuasion of science). The electron rotates around the atom nucleus indicating an atomic border of some sorts. From the centre one can draw a line pin pointing the position of the electron during the duration of a time period. In this time will indicate movement of the electron through space. The time indicated must be T^2 should space be in the third dimension a^3 and singularity will connect through k.

The one position in space will place the electron in time where the electron then will be below...behind and above the atom sub with particles the which shares electron frequency. The k indicates and position implicates

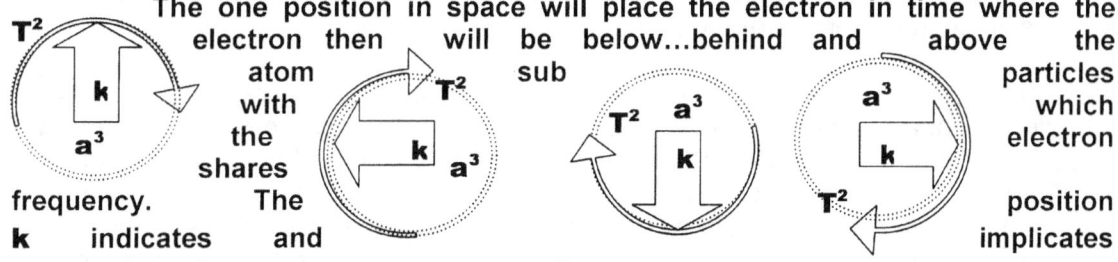

the electron T^2 in the environment that a^3 establishes.

By linking space a^3 to singularity, which produces time T^2 it will have to indicate the influence of singularity through the single dimension connecting of k. In the relation only k will be representing the single dimension factor since that places the Universe in space and time. Should one place the time factor as a cosmic relevance, but presenting time as t in a single dimension role the two dimensional time have to disappear with the three dimensional space that also will disappear.

If not for any other reason then simply because the moment space holds one position a double time unit, eternity sets in and space by dimension disappears. That is what Kepler said when he said $a^3 = T^2 k$ and that is what Newton missed when he corrected Kepler's $a^3 = T^2 k$. The moment motions end space and time falls into the single dimension. In that space and time must disappear. We all can accomplish that task by taking a photograph and print the image on paper and call it still photography. Then the paper will hold time in the square while the paper is in the cube all indicating individual and complimentary k pointing to individual and complimentary singularity producing space-time. The image returned to the single dimension while the photographic paper serves time in space. That freezing of motion and bringing the image and not the motion into the equation is what makes the image of motionlessness while the electron cannot stand still.

By taking the line to k back to where the line or k cannot reduce further $k^0 = 1$ establishes such a value where k then finds a position in the single dimension. In that case, a^3 also is equal to one and so is T^2. In fact, k still has to produce a line and we find that k represents a^3 to the full as well as T^2. The point where k forms the most slightly distance the area a^3 establishes a value outside the single dimension because T^2 adds a value. The one position in space will place the electron in time where the electron then will be below…behind and above the atom sub particles with which the electron shares frequency. The position k indicates, implicates the electron T^2 in the environment a^3 establishes.

By taking the line to **k** back to where the line or **k** cannot reduce further $k^0 = 1$ establishes such a value where **k** then finds a position in the single dimension. But in that case a^3 also is equal to one and so is T^2. In fact **k** still has to produce a line and we find that **k** represents a^3 to the full as well as T^2. The point where **k** forms the least or the most slightly distance the area a^3 establishes a value outside the single dimension because T^2 adds a value. The fact that T^2 comes in as a factor in the presence of the first sign of **k** appearing, it indicates the start of motion taking a^3 from one location to another specific location. It indicates the travel of the planet during a month or a day or an hour. Trying to freeze time will place the electron in two positions because the electron then is from …(a point visible in time) to … (another point visible in time) where it is seen as two individual points in that moment one wishes to freeze time. It is not the points of electron visibility that forms time but the space occupied by time being between the two points showing a position of the electron. The electron cannot freeze, because if it does, there is a nuclear bomb about to begin rapid expanding which no one appreciates when in closeness of the location.

It does not indicate a circle except at the end when completing one cycle. T^2 is the distance in time a^3 that will take **k** from indicating one point to indicating another point. The formula points to a referring of the very time space was indicated by position location and time. The astonishing part is not as much the way Kepler formulated his formula to cover the movement and the position of the electron in relation with the rest of the atom, but the brilliant way the mathematicians neglected to see the fact. Kepler saw a three dimensional a^3 something in a specific position in time T^2 relating to a specific density **k** of the atom. With space in a cube as it cannot ever be otherwise the time too has to be in a square because placing time in the single dimension of **t** the time then becomes part of a single dimension such as one may find in a photograph picture. One can justly use the same formula to implicate the electron taking time to complete the distance between two points indicating the area from the centre of the atom. Allow me to establish by crude illustration time used to space produced in relation to space using time within the atom, which is in motion other than circling the proton and thereby creating gravity through motion. When we view the atom we see an electron spinning as the electron forms the atomic boundary.

The simplistic concept science holds on the Universe and cosmic travel is rather less to very little thought through. Mainstream science is of the opinion that the future space travellers are very much comparable to the adventures Columbus endured. It is about seeing new places, meeting new faces and an all around adventure. They are quite to the letter of the opinion that travelling around the Earth by ocean as Columbus did and travelling through space is very much a similar journey. Columbus set out to travel across a sea but Columbus and his crew was at the same time remaining on the Earth. The Earth forced time and gravity onto the travellers. On his way he came across some islands where he was able to allow his crew some shore leave. There was life on the islands breathing the same air he required to sustain life. To leave the island he only had to set sale and be off if the winds were blowing. This was in the time of Columbus that was today's *news-braking-space-travel living-on–the-edge* where the most developed minds participated on achieving the most inspiring mind provoking futuristic travel thought up by an exploring human mind. But let us face it, breathtaking it may be, it is far from a journey to the moon. Travelling through space on an Intergalactic hopping one needs to go much faster than sailing speeds will allow. Let us study the following concept in more thought through detail on our part. When we glance at an object travelling from the Earth to Mars we see the sun rotate during the motion the object travels by. The sun might turn a few degrees but those degrease is immense when standing in relation to the motion of spin we find in the material atoms forming the travelling objects. We must remember direction has changed because our travelling now includes cosmic laws and not the laws of King and country.

The cosmos is not to the outside where we glance at the sky because we are the outside or the final limit of the cosmos we form. We are where our Universe ends because we are in the centre of our Universe. What we see that we think of, as the Universe is an array of light travelling through space. To our outside or away from us it is the light forming space-time that we have contact with and not the space we think we see. The space we see is light telling us to form a concept about what time tells us and not what space brought us. We see the history of space but the history of space is written in the ink of light on the paper of time and we are reading about the history of time but we use the ink of light to see the writings. It is telling us of space, while our minds know the space we think we see changed so much that we cannot see the space that changed. The space we are in contact with is to the inside. Space-time to us is towards singularity, towards the atoms forming our Universe of which we are the outer edge.

When we look at an object we see it in a specific colour. We think the colour we see the object in represents the colour the object has but in truth it is the last colour the object relates to because the object is rejecting the colour that we see. That we associate the object with by appearance is completely wrong because the object rejects the colour we see and by it is represented by all the colours it associates with which is not one colour we see. The object we see associate and cling onto all the colours there are that falls outside the spectrum we see. That in spite we still connect the object to the one colour the object rejects. The object rejects the colour because from all the colours there are, that specific colour the object is not. The very same argument concerns outer space. We see the sun as a star shining bright but with all the light it deflects the sun has to be as dark hell. On the other hand with the night sky as dark as it is it is keeping all the light to itself because it is not passing out any light to us. It is keeping the light it has under control and that makes the darkness we see being the light it is holding back.

The relevancy would apply as long as I hold a motion relation of equilibrium in relation to the motion standard set by the Earth governing singularity. The motion sets the trend of gravity applying and sets the deviation in motion equilibrium or the maintaining of such equilibrium. That is where the factor **k** produces its input to relevancy applying. If motion equilibrium is sustained **k** is duplicated by precise measure. The motion in double value holds on space with the other part valued at Π^0. However, when gravity does not set equal motion reproduction the relation will change where the Earth demands one value and the independent space in motion will have to contribute a motion higher and on top of the motion that the Earth already provides.

The growth we find in the Universe is from the point where singularity charges space-time and space-time extends from the centre to the outside. The atom cannot disappear, as it cannot vanish. The space we think we see by the light coming to us has already disappeared and has already vanished through the interfering of time. After all what we think we see is energy by heat that expanded. That what we think we see to the outside is the "outside" we believe we see forming outer space. It is the light that came about after the Overheating sparked a Universe out of complacency forming the light to be. **THE COSMOS IS NOT OUTSIDE IN OUTER SPACE; IT IS INSIDE, INSIDE EVERY ATOM. THE ATOM CANNOT DISAPPEAR, AS IT CANNOT VANISH.** The cosmos ends where you are located, and the reset...well that is just space-time. To us everything but us is all just energy. Yet it can be relocated back to singularity. Travelling through space has this same relation because where I am I am the border of space-time. My relation with singularity applies only because my singularity relate to the singularity in outer space. My singularity will prove its worth in space-time by the relation in time through space it has with singularity points other than those under my control. My singularity, which is in my control is in relation with other points of singularity not under my control and by that I am in fact under control of those my singularity stands relevant to. If the photon and therefore light was travelling at the speed of time, (t= C) the photon has to cross the entirety of space at the very instant it enters space. After all when light is time, time cannot restrict light, yet it does to the value of C. Time can restrain light therefore light cannot be time. Time is motion because while in motion set by the Earth motion the point I would hold would place my general or my governing singularity direct in relation to the gravity of the Earth.

When there is motion of equality between the Earth and a satellite orbiting in gravity equilibrium to that of the Earth there is **k** applying as well as a related k^0, which proves to duplicate an extending of **k**. In order to be either k_1 or k_4 depends on the motion difference there is in the mutual gravity applying motion.

In the atom **k** forms a relation of space a^3 to distance **k** connected to the double proton time factor T^2. In this case **k** represents the atom spinning and the atom represents the Universe.

Due to factors from the Big Bang and the fact that from that all singularity connects the same, the more the **k** that connects the atom **to the sun centre will increase** by momentum, the more will **k** within the atom reduce. Such reducing will affect the space a^3 that will affect the time T^2 period. The period of time will increase because the time duration will reduce.

The reducing of time and the reducing of **k** results in the reducing of a^3. That is because there is more duplicating of atomic space by motion of the atomic unit as a whole.

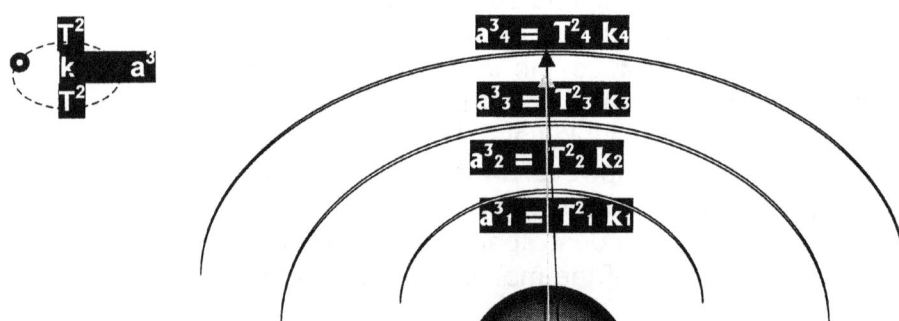

To move the already rotating object from k_1 to k_4 requires that **k** duplicate by extending on top of the required rotating speed and that extending must be linear in relation to the circular motion gravity set the equilibrium conditions of duplication to. It is when the duplicating produces an extending factor as a compliment to what already are in motion that the relations that is applying changes somewhat. By increasing the motion to a forward motion many compromises must be accounted for since many relevancies change drastic. The higher the forward velocity is, the more drastic will the changes be.

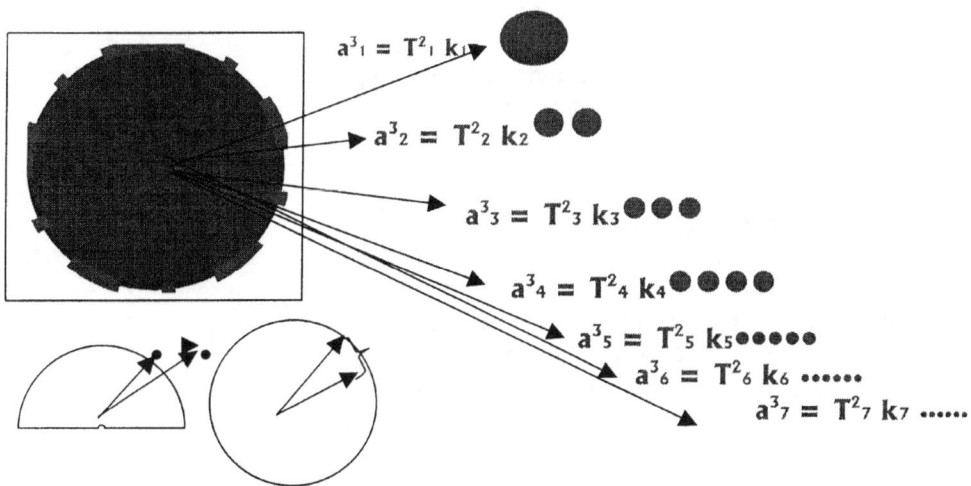

$$a^3{}_1 = T^2{}_1 k_1$$

$$a^3{}_2 = T^2{}_2 k_2$$

$$a^3{}_3 = T^2{}_3 k_3$$

$$a^3{}_4 = T^2{}_4 k_4$$

$$a^3{}_5 = T^2{}_5 k_5$$

$$a^3{}_6 = T^2{}_6 k_6$$

$$a^3{}_7 = T^2{}_7 k_7$$

When **k** duplicates by remaining in equilibrium the rotation value duplicates to the exact same measure that is repeated. The circle **k=k**

The duplicating requires a repositioning of the aligning of **k** from a certain position to a more fore ward position in relation to and in that **k** will also have to extend a value when moving from **k₁** to **k₂**

Route the electron follows

There is **k** that forms the distance between the proton and the electron while the electron is spinning T^2 around the proton k^0. While all this action is going on, we think of the atom as being very still and satisfied with being a small part in a lump of metal we call iron. It could be any element but I use iron just as an example this time. The lump of iron is as motionless on earth as anything can be while being.

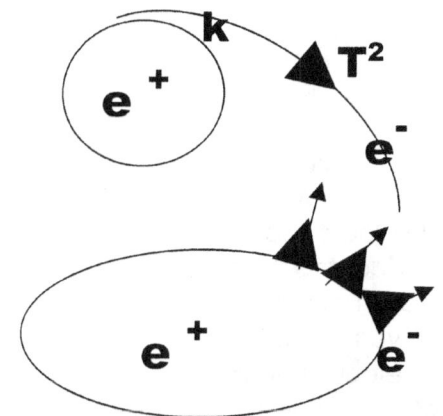

With the oval form the electron orbits in, proves that a second relevancy is present where the electron motion is in another relevancy with the proton centre and that relevancy also applies a **k** that places the motion as one that runs straight in defiance of the centre in order to secure independence of such a centre.

If the proton starts to move at the pace the electron does which is apparently C the neutron would have to compromise space-time to allow the proton to go faster. The electron is directly related to the proton by the neutron and therefore the motion the proton has will directly influence the motion of the electron. The electron will never allow the proton to catch the electron because that would destroy all space the neutron holds and the neutron forms all that is the atom. When the proton moves with the electron in front the space has to reduce.

The atom working becomes real to our thinking if we place the space the atom claims as a^3 and the motion of the electron as T^2 while the distance of space the atom has as k. By motion of the electron does the atom form a bonded unit precisely in the manner as the Coanda effect must have. The liquid neutron takes the electron spinning around a spherical proton, which personifies the solid, but roundness and that serves all the requirements we are looking for to establish the Coanda principle.

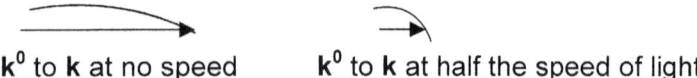

k^0 to k at no speed k^0 to k at half the speed of light

As k^0 now applies a relevant motion k in distance has to comply with motion k^0 produced by changing relevancies at a specific rate other than just circling about the atom and the compromising of k in motion had to endure.

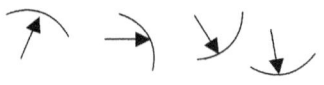 Indicating a very much- reduced k^0 to k at travelling half the speed of light speed while sustaining the atom.

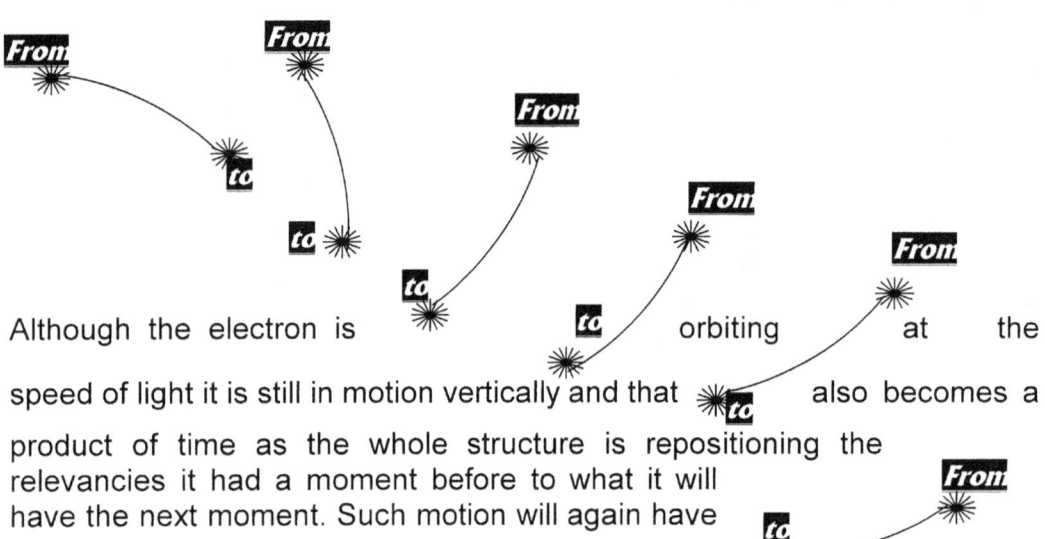

Although the electron is   orbiting   at   the

speed of light it is still in motion vertically and that   also becomes a

product of time as the whole structure is repositioning the relevancies it had a moment before to what it will have the next moment. Such motion will again have

an influence on the relation in the position the   electron forms with the rest of the Universe while the lump of metal is now travelling as a spacecraft destined to other galactica. It is if we use the logic those intellectuals calling them Academics show and those Super-Educated that advocate how we may travel to far away galactica while we go on skipping the nearby galactica that is only two to twenty million light years away. Since the electron is duplicating by motion the motion links the electron to a time constant. The time constant is linked to the speed of light but time as such, is part of the speed of light. The faster we take the electron to go straight in the motion man produces, the less time there will for the electron be to circle around the atom. If we make k bigger in relation to increased motion, the smaller will T^2 produce a usable space.

However these are many relevancies that require compromises to remain in equilibrium because the Universe insists on equilibrium above all else. To allow the one factor to expand another factor has to reduce. That is what the whole Big Bang event is all about. It is about expanding producing compromises. By moving forward there is a time differentiation coming about. It is a factor of time to space $T^2 = a^3/k$ and as k increases the unit space-time it has to reduce the object in motion space-time. It is all a question of rotating timing just as one would find in gears rotating in synchronised motion. The reducing will be in space but as I said there is only space inwards and light to the outside. If one wishes to apply motion reaching the speed of light, the space towards the inside must reduce to the size of a photon and the space to the outside must incorporate the entire global

Universe we so desperately but incorrectly whish to incorporate into our personal standards. Our Universe is the total compliment of our atoms.

The motion provides a circle in which the atom claims space. The slower the motion is in which the bigger the circle is that space. The motion claimed space where the two common unit but of the two steaks a claim to the unit.

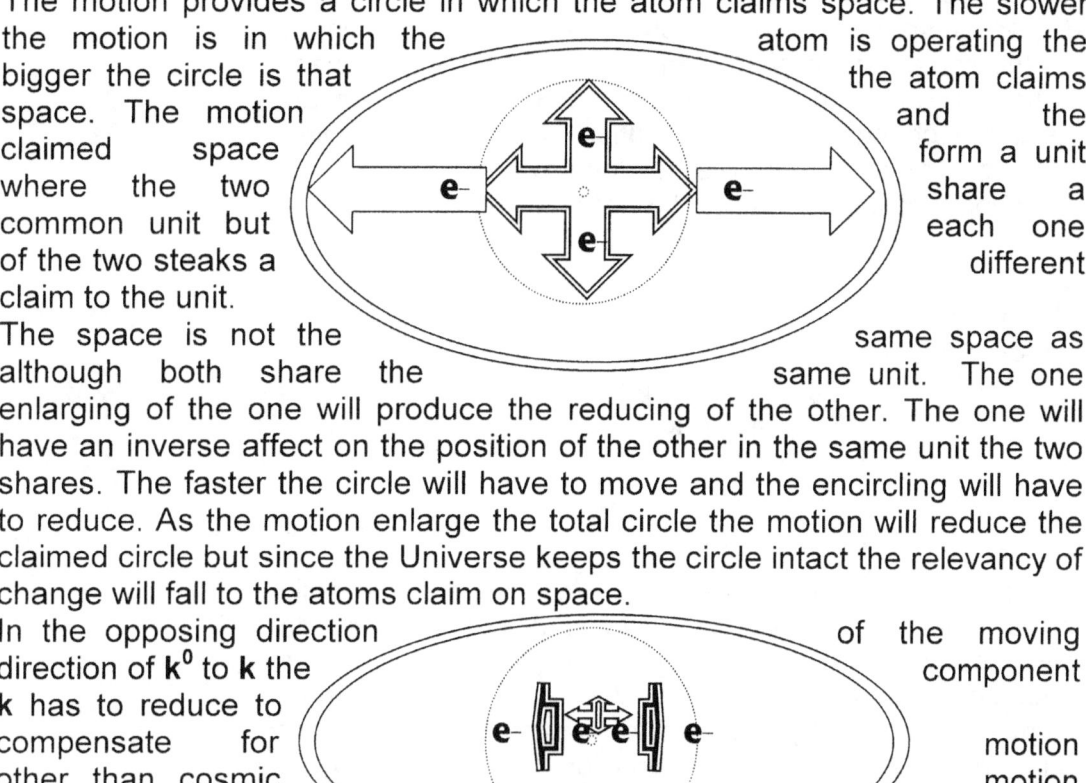

atom is operating the the atom claims and the form a unit share a each one different

The space is not the although both share the same space as same unit. The one enlarging of the one will produce the reducing of the other. The one will have an inverse affect on the position of the other in the same unit the two shares. The faster the circle will have to move and the encircling will have to reduce. As the motion enlarge the total circle the motion will reduce the claimed circle but since the Universe keeps the circle intact the relevancy of change will fall to the atoms claim on space.

In the opposing direction direction of k^0 to k the k has to reduce to compensate for other than cosmic coming about and reduce the distance of the moving component motion motion thereby between singularity and the electron. It means that k^0 to k has to reduce in distance between the atom centre and the new atom borders that form but in the compensating other factors has to compromise for such changes brought onto the reducing the line formed a k^0 to k. With the line k^0 to k forming the indicator that indicates the space between the atomic centre and the electron forming the atomic border.

When the electron is in rotation behind the proton with the proton moving at the speed that electron normally move the heat will compensate for space-time relevancies returning to what was about when the Big Bang became a fact.

However now we change our perception in allowing the atom to move while everything about the atom is moving as before and this brings about the second relevancy of independent motion of the electron in relation to motion around the proton.

This is because moving the atom faster in a straight line will reduce the electron time to complete the circle and that reduces the space the atom occupies. The relevancy linking everything is motion. Even the electron and the atom is a result of the manner in which motion brings about gravity by studying the Coanda principle and applying that principle to the working of the atom set up we find the manner or method of the forming of the atomic bonding as it clearly indicates the working of gravity by motion.

That means the second motion coming from the object travelling through space which arrange its atoms in the same time duration as that it would have used while being stationary on Earth pushes the electron circle out of the normal sequence that electron had while being in a steady state of motion on Earth and under the control of Earth time.

The principle, which I described in the motion of the spacecraft travelling through time, and space in space-time also apply in the same manner in the Coanda effect. It is the same motion producing gravity by applying a space relevancy that changes in the same duration as when not in motion but through motion the space changes to, a new motion that comes into play. In the motion producing a relocated centre a new centre will play its part in the compensating motion that will force upon the relevancies a new space-time dispensation

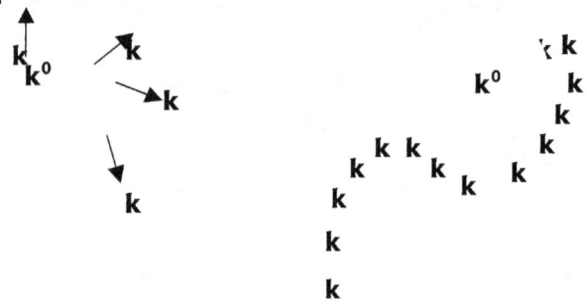

By applying motion the flow of the liquid around the round solid which represents as much as rein acts singularity the motion bring about a new space-time as the factors has to compensate for the motion and the motion in itself bring about a new controlling k^0 in the position elected by all atoms forming part of the relevant motion. The motion introduce a securing of the position that is elected by all the atoms and is forming the new part that will serve as a controlling singularity by substance producing the motion or the substance of the solid securing the position of the newly elected controlling singularity. By applying motion the Earth gravity is interrupted by the space producing motion that through motion take charge of the space and turns the vertical flow of Earth space into a circular motion although only for a short while. This is how electricity is charged but the motion then involves the reducing to the flowing electrons to compensate as much as producing speed of light. Other elements and their characteristics are also involved and in the charging of electricity the Coanda principle personifies the gravity that comes about in the condensing of space-time around singularity.

By reducing the space towards the centre in the using of the Coanda effect re-insures the correctness of the Big Bang theory. A newly established centre will create a newly controlled Universe and the motion of Iron $_{56}$ will force space towards the newly established centre. By forming the flow of space in relation to copper the space will break down and form the inside we find the space to flow on the inside of the Earth centre. We have to remember that the relevancies, which applied during the Big Bang is present in a minute capacity on the inside of the Earth.

It is not only the relevance applying from the centre to the electron but also the electron holding an allegiance with the centre in the precise manner as planets do in relation with the sun. It is because of the dual relevancy that the electron stubbornly clings to the newly elected centre before being overpowered by the Earth providing the controlling centre. It is all about relevancies attaching centres that formed when creation came about.

The time that Newton froze on paper in his establishing of a single t that is representative of time is effective in remembering the viewer of an event but that cannot be the event that is part of the present any longer. If we reduce the moment to a snapshot the picture we focus on can only be what we see for a very short instant and then forms that which was how the event occurred during the time from where the camera shutter opened T_1 to where the camera shutter closed T_2 and the time frame T^2 was then during the open period of the camera shutter. But as soon as the shutter shuts, time moved on and another T^2 formed leaving the image taken as time never to repeat again.

Afterwards the image we see as the picture represented **t** and when looking at the picture, the looking of the picture became an event during a specific T^2 that went from where one is taking the first look to where one is looking away from the paper carrying the first dimensional image of an event gone by and that is at that stage a representation of **t** in another milieu of $a^3 = T^2 k$. The **t** in the single is when mathematically presented as only **t** indicating a mathematical single flat dimensional view of time and is then correctly applied because it represents a reminder of a four dimensional event $a^3 = T^2 k$ that went single dimensional because the moment in the fourth dimension was then frozen in a single dimension on paper with the paper being part of space-time while the fourth dimension $a^3 = T^2 k$ soldiered on and time will always be representing T^2 as Kepler stated.

In $4 \pi^2 a^3 / T^2 = G(m+m_p)$
$a^3 = T^2 k$
$a^3 / k = T^2$ but
$k / a^3 = 1 / T^2$
$k = a^3 / T^2 =$ singularity
$a^3 / T^2 = G (m+m_p)/4\pi^2$
and $a^3 / T^2 = k$
then $k = G(m+m_p)/4 \pi^2$

All Newton's changing was possibly done with good intensions but even that I doubt. The end result however was in some cases far from good, as it does not do such great credit to Newtonian insight into cosmic affairs. Only Kepler and only Kepler unaided without the intimidation and interfering of Newton can explain the Coanda effect. I grant the fact that the Coanda effect was discovered before Newton saw himself fit to change Kepler, but only Kepler can explain the Coanda gravity effect when Kepler is without the attentions of Newton.

But I showed that $k = a^3 / T^2$ and Newton's claim is that $a^3 / T^2 = G (m + m_p) / 4 \pi^2$
Time is in the square, and that is allocated to space having a cube. Kepler said gravity is $a^3 = T^2 k$ at a time even before gravity got a name. But reducing the dimension of time to a single **t** one will find the ability to mathematically design the paper on which the photo image will be printed in time T^2 using space a^3 in the third dimension to apply the ink in the third dimension. Printed on the paper is an image that is not part of space-time while the ink used is space-time and the paper is space-time. The ingredient of ink on paper all hold different values since the image has a value we as humans grant the image to carry such a value. The image, the ink and the paper all hold different relations but the image only relate to thought in our mind. The image has no k indicating only references with and to what forms part of a realistic different singularity coming from the Earth centre and connecting to individual atom groups, forming individual as well as group space-time.

In the past no one even thought of placing the Coanda principle in line with gravity. By following the Kepler's formula we find that the Coanda affect, in fact is gravity.
Gravity is moreover the Coanda principle than gravity is any other variety of forces or concept of contractions. The Coanda principle is gravity. It is the how all gravity is charged and distributed. That forms the basic principle of Cosmology.
The Coanda affect is proof of the functioning of gravity inside the atom. It proves that motion (T^2) of the neutron establishes a centre in line where the compliment of material forming the atom will secure a controlling singularity that is governing the entire atom. That forms the centre of the Universe. Singularity then finds a position at the distance of (**k**) and such motion claims the space (a^3), which is the atom by construction from a centre within that motion (T^2). The motion (T^2) creates a centre at the

line of (**k**) and a centre of the space (a^3) the motion (T^2) establishes a gravity field all along the lines and at the distance of (**k**) in the space (a^3) that the motion (T^2) created.

In the Coanda affect we can read how the atom became the Universe. The proton is the substance performing as the solid on both sides of the Universe, The neutron being a liquid is what establish the gravity, which helps the singularity secure space-time by heat compensating for overheating. With the neutron forming a liquid and the proton providing the spin, the neutron established the space that forms the atom to the inside of the electron.

With the facts well established in the Coanda principle, much more of Newtonian visions come into question, one being his all-to-famous view about gravity and on the establishing of gravity. Gravity is precisely what Kepler first said it was. The Coanda effect is the best prove of that and from the Coanda effect comes a sound explaining about the sound barrier. Gravity is created by motion of space within space forming motion that is served by points. Without gravity space would not be there at all. $k = a^3 / T^2$. The proof is there in every atom. Every atom manifests the gravity being all over and that gravity can be everywhere. The establishing of a space providing gravity by motion proves the validity of the Coanda effect in the producing of gravity by the atom reforming space by motion using the same period of time to apply the motion.

Kepler said $a^3 = T^2 k$

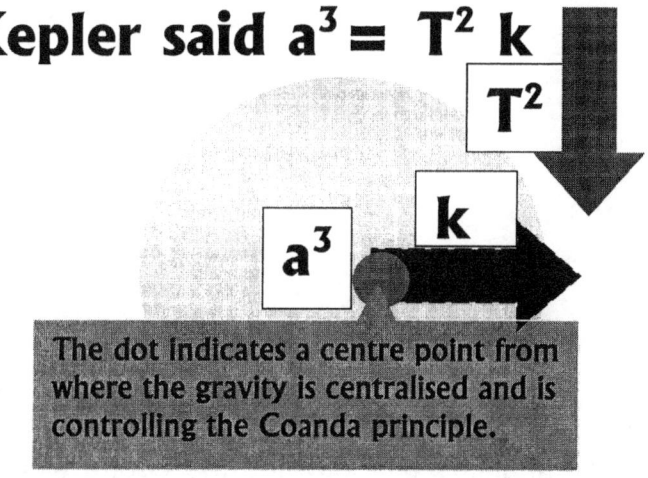

The dot indicates a centre point from where the gravity is centralised and is controlling the Coanda principle.

---

a^3 forms the space the atom claims while travelling in the Earth spinning all the way and travelling with the Earth around the sun

k positions the electrons travel in the space relating to the space the atom holds while travelling with the Earth in the Earth around the sun in relation with a specific position **k** will indicate in relation to the sun.

T^2 is the time it takes the electron to be relevant to the position the Earth places on T^2 while the Earth captures the space of the atom by providing the space for the atom to be within while the Earth travels from one point T_1 to another point holding T_2 in frequency to the atoms T^2 relating to the Earths T^2 in perfect harmony with the sun having another T^2 relevant to all the other factors we call cosmic particles. Big or small it is only about cosmic particles holding space in time in relevancy.

The motion of the liquid which the neutron is proves to be the time (T^2) aspect because as it increases it claims the space (a^3) in the at a distance k of time (T^2) that the running has increased. The faster the motion is the stronger is the gravity that the motion generates in the space it claims by the gravity it generates

---

The atom creates a unit wherein the electron establishes space and from the electron to the centre we call the atom but when the centre of the atom brings about a relation where such a centre has to comply with changes in space holding time such motion redefines the space the atom claims. This is the way that motion creates space independent from space within space. The following movement of a continuing stream of atom forming a circle relations with atoms in a solid structure forms what Kepler introduced as space-time $a^3 = T^2k$. The position **k** will establish, as a point will come into affect as soon as motion or time T^2 sets any space a^3 apart from the rest of space. That rule applies to all motion on Earth. That rule applies to all motion in the entire Universe and that rule is the Universe. That is what happens to an aircraft and that is why an aircraft will go through the sound barrier. The aircraft is setting a new filled space $(3\Pi^2)$ at the position of 7^0 diverting from singularity and the independence will allow a motion of Π^0, or Π to 5Π within the boundaries of Earth's gravity, but as soon as the limit of 5Π is shattered so is the sound barrier and a new limit sets in place being the Roche factor by half $= \Pi^2 / 2$. The complete formula comes to $7(3\Pi^2) = 207$.km per hour, which is the maximum velocity any object in free fall, will hit the Earth.

That for instance makes sound part of gravity.

The Coanda effect is most probably the strongest suggestion that my theory is correct. I have a theory where I show the singularity coming about be reproducing a circle motion that is resulting in the creating of the circle centre where the circle enhances or motivates the centre singularity to produce gravity point that secure the hold of singularity in the form of applying gravity through motion. There has to be motion either by space or by material and the flowing water on this occasion is the liquid and the liquid falls into the solid round space that holds the gravity part of the cosmos. The flowing water is producing the motion and such motion represents the space where the water is flowing creating motion, which is creating gravity

We are very aware of the Coanda effect where the water takes another route other than the normal way water will fall when using the gravity the Earth supply. Adhering to the Coanda effect the water flowing will invest in a new centre holding the flow of water related to the space the flowing water will establish thus proving that the motion k in the time duration T^2 create a new space unit a^3. The natural thinking is acknowledging Newton's mass pulling mass and taking the shortest route to flow but the Coanda effect denounces this very prominently. As the water flow around the bowl it will divert in direction by producing a longer circle to flow and eventually loose the grip it had secured on the round bowl. The restraining of the circle forming stands in relation to the velocity of the liquid or the solid that supplies the motion. By clinging onto the surface of the bowl the water tries to flow upwards and stay realigned to the centre it created by motion producing such a centre. It conforms Kepler $k^0 = a^3 / T^2k$ in defiance of Newton's $F= G (M.m) / r^2$.

The circle forms in defiance of the natural flow that the Earth gravity will enforce that is according to Newton. The water secures a longer route around the round bowl as the water clings to the bowl by running around the edge of the bowl and even tries to run up the bowl circle. When motion is introduced to the bowl, as a car tyre would do the clinging becomes much exaggerated because a wheel of a car can produce 25 mm layer of water clinging onto the wheel at high speeds. This is even where the clinging water supports the mass and the motion of the entire car. The Coanda gravity keeps the water in place where the circle of water surrounding the entire tyre can withstand the force of the car's entire mass, as well as the mass that comes about as a compliment to the motion that is exaggerating the initial mass of the car. The situation renders high speed travelling in wet conditions to be very much unsafe and it is well known to all drivers of motorcars. But in essence this is confirming the Coanda principle as the Coanda principle confirms Kepler statement because as the wheel has more motion T^2 through maintaining a higher velocity it will secure more liquid k onto the solid tyre structure a^3 and thereby establishes more space a^3

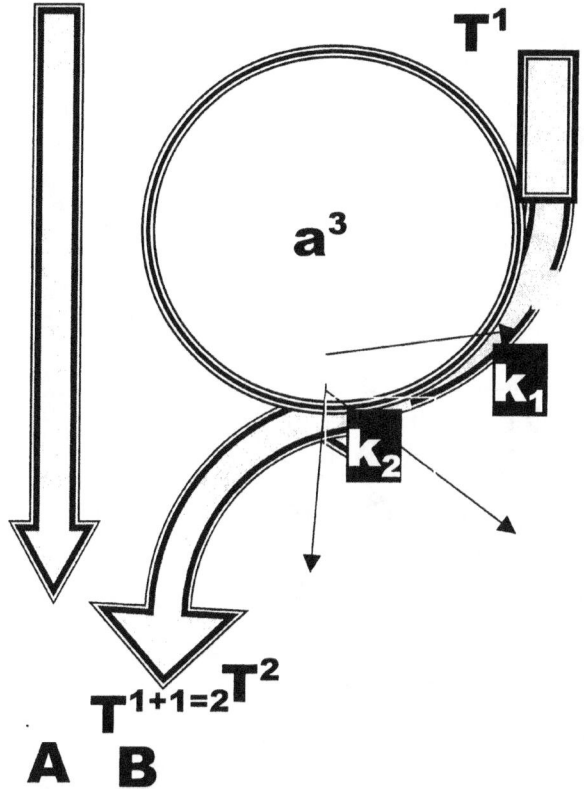

T^1

a^3

k_1

k_2

$T^{1+1=2}T^2$

A   B

Normally water will run down to the centre of any gravity point, as A shows. By allowing the flowing water to come into contact an object of a specific form the flow will divert (B) from the normal line and follow the contour of the object presented.

The Coanda effect is proof of gravity coming about through space forming motion. In the case where water diverts the normal directional flow the space that translates to the motion is deflecting singularity with the flowing water charging the motion. In the centre of the object having the round form, singularity is duplicated and by transferring Π to form Π^2 and the motion of the water creates a line of gravity that pushes the flowing water to follow the direction that the newly gravity applies to the water. This again proves Kepler's statement of **k** $= a^3/ T^2$ that specifically states that space (in this case the object transferring singularity to a new position within the round object) and with the motion of the water redirects the gravity flow of the water to new space in new time. Only Kepler can explain the phenomenon but only when Kepler stands alone, correctly interpreted and divorced from Newton's opinion about Kepler's statements.

For that to take place there is one condition that has to come about. The form has to be circular to influence the movement and therefore there has to be movement.

However creating a circular singularity is one aspect of the Coanda gravity alteration but there is more establishing of such a singularity. By increasing both more a^3 in supplying wing contact that creates more space duplication as well as increasing motion T^2 by producing a turning momentum. This is very apparent in the motion of a propeller through which singularity is established as it then forces the singularity to recognise new linear singularity relevance by producing a neutral **k.** We call this flying**.**

Let us again review the location where I traced singularity. Looking at a motionless top which children spin in amusing their playful minds, the top is a solid structure with every aspect of the top being the same. The antigravity comes from the motion we deliberately invest in the top as we bring about a spinning motion to the top from where the top then finds liberation by motion performing as artificial antigravity, which are strong enough to provide a release from the Earth's confinement. But it can only be possible if motion applies in a spin and no other way. Throwing a ball horizontally brings about different influences coming from singularity, which I explain, in another book "**Xepted Astronomical Mistakes"** .

But the Coanda effect also is more than only that.

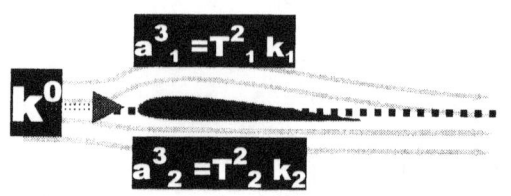

This all confirms that translating Kepler's mathematical expression $a^3 = T^2 k$ correctly to the verbal statement in English Kepler said that there is a space a^3 which is

By moving linear the wing becomes the **k** in Kepler's formula and the earth is $\mathbf{T^2}$. That is the reason why the aircraft can fly. By creating an uneven $\mathbf{k^0} = a^3$ on top of the wing compared to the $\mathbf{k^0} = a^3$ below the wing can lift or descend. It is a balance going unbalanced.

equal = to the motion T^2 thereof between two specific points which holds a relation to a centre being **k** a straight line.

The ball has to spin even when travelling in a straight-line to secure direction control. Even by shooting a round ball through a vintage cannon will not establish direction control by force. The producing a spin secures a point that is creating the dynamics of singularity and that singularity in charge will strive to create independence for the space a^3 that is (=) now spinning T^2k. The motion must come from the way of spinning and the spinning must establish a centre. Then only can the top move about an individual axis and that axis grant the top independence and freedom from the confining gravity the Earth apply to secure the tops position as part of the Earth. So we find motion establishing a centre and from that centre the motion in the top secures the space that is the top releasing the space that is the top from the space the Earth grant the top when the top does not apply motion This spinning top is using the principles we find in the Coanda effect. This way the Coanda effect is the generating of electricity, is using the principle in the crank and connecting rod action of fossil fuelled engines and is every motion we conceive. That all are the Coanda effect in living proof and in acknowledging Kepler's formula of $a^3 = T^2k$. It shows the top having control over space claimed from the Earth by applying motion and thereby defying gravity confinement. The motion establishes a centre k^0 from which there is space claimed a^3 by rotation motion T^2 relating to the solidness of the ground **k**. It is once again $k^0 = a^3 / T^2k$. That is the circular effect and then there is the linear effect.

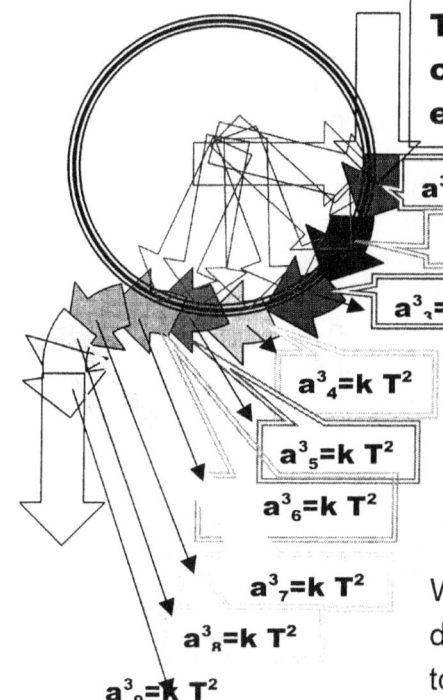

$a^3_1 = k\ T^2$

$a^3_2 = k\ T^2$

$a^3_3 = k\ T^2$

$a^3_4 = k\ T^2$

$a^3_5 = k\ T^2$

$a^3_6 = k\ T^2$

$a^3_7 = k\ T^2$

$a^3_8 = k\ T^2$

$a^3_9 = k\ T^2$

There has never been a better and clearer explanation about the Coanda effect.

Every motion that the flowing water establishes in relating to **k** from the centre such motion introduces new space and the space is a continuing of direction changing in relation to the centre prevailing.

Every time an a^3 comes about claiming a new space spot for the flowing water the motion of the water determines the spot in space created.

When the water fills a new spot a^3 it is the directional change of singularity running from k^0 to **k** that produces the two points of **k** between T_1 and T_2 forming T^2 creating small instances of new relative values where $a^3 = k\ T^2$ is equal to the motion within the boundaries set by singularity. It proves Kepler correct. It proves that space is the duplication thereof through motion.

Space is created from one position to another position and the duration it takes to complete the distance is time. This the Coanda effect proves as water flows past a round object and the contact the flowing water makes diverts from the normal route the Earth gravity will enforce. Gravity is the very same but it is the recalling of the space by creating motion in the space. By duplicating space through motion in relation to singularity the flowing water diverts from the normal route. By recalling the space it is also reducing the space because it is counter acting the time expansion provided. That then is clarifying the reason why gravity will always on the limit be stronger than light. At a point it slows the time component down to being on the limit and of the time light takes to move. Gravity is motion that is going way past that point and the motion gravity produces is past where the limit of light is. Gravity as motion, can be much stronger than light which too is just motion. At a point gravity slows the time component down to such that the space reduces faster in that time than what light can produce motion. This must be time because the Black Hole contracts light back into the star. That is also precisely what Kepler introduced as.... This is why the Coanda effect applies. $a^3 = T^2\ k$ then being $k^3 = k^2\ k$ and this is showing that the space k^3 is equal $=$ to the motion $k^2\ k$ of the space k^3 seen form one specific point.

$a^3 = T^2\ k$ then $k^3 = k^2\ k$ and this is showing that the space k^3 is equal $=$ to the motion $k^2\ k$ of the space k^3 seen form one specific point.

In the past no one even thought of placing the Coanda principle in line with gravity. By following Kepler's formula we find that the Coanda affect IS GRAVITY. Gravity is moreover the Coanda principle than it is any other variety of forces or concept of contractions. The Coanda principle is gravity. It is the how all gravity is charged and distributed. That forms the basic principle of Cosmology. The Coanda affect is proof of Kepler's statement. It proves that motion (T^2) establishes a centre at the distance of **k** and such motion claims the space (a^3) from a centre within that motion. The motion creates a centre and a centre establishes a gravity field. The motion of the liquid water proves to be the time aspect because it claims the space in the time it is running. The faster the motion is the stronger is the gravity that the motion generates in the space it claims by the gravity it generates. Even Einstein's square of light is not the square of light but refers to gravity being equal to the speed of light.

In the formula MC^2 I have been protesting that the speed of light cannot double and by that I seriously have doubt about the formula. All I can see is that Even Einstein had no idea what to his formula presented to the world. the has no feature about harvesting the speed of light in what light represent. The Einstein formula is depicting the Coanda gravity and it works on the fact that where the neutron places the electron, the space-time displacement goes to gravity at C which is the speed of displacement of space-time the electron is charged with.

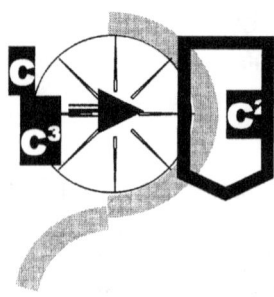

The C^2 value Einstein came up with has no bearing on the speed of light, but is equal to the gravity displacement the neutron has when entering the atom. It shows the stopping of the flow of space – time and the atoms overheating in the process when the two atoms touch and stop one another's space-time flow bringing about cooling. The C^2 is the gravity at the point of the atom presented as a result of the neutron centralising the gravity C^2 at the rate of the electron C^2. What Einstein saw was gravity at the point where gravity equals the speed of light, but it has nothing to do with the actual flow of light going square. It is the flow of space-time. Going nuclear takes the atom back to what happened at the start of the Big Bang.

Space is created from one position to another position and the duration it takes to complete the distance is time. This the Coanda effect proves as water flows past a round object and the contact the flowing water makes, diverts from the normal route the Earth gravity will enforce. Gravity is the very same but it is the recalling of the space by creating motion in the space. By duplicating space through motion in relation to singularity the flowing water diverts from the normal route. By recalling the space it is also reducing the space because it is counter acting the time expansion provides. That then is clarifying the reason why gravity will always on the limit be stronger than light. At a point it slows the time component down to such extend the space reduces faster in that time than what light can produce motion. This is why the Coanda effect applies.

For these phenomena to occur some independent singularity spot has to prove their independence by overshadowing the Earth's gravity. Two conditions have to apply. A round object **k** has to be in place allowing the establishing of a space a^3 within the space, and then movement T^2 has to establish borders to such a space. That proves Kepler correct. And that proves gravity can come about at any given point by establishing singularity independent from the space it holds, the movement will generate heat which will energise gravity setting borders and confining that singularity in $\mathbf{a}^3$ by and confirming the singularity energised with $\mathbf{k}$ producing the boundary of motion $\mathbf{T}^2$. That was what Kepler said before Newton's interfering...he said that space comes about by the motion of time from singularity $\mathbf{a}^3 = \mathbf{T}^2\,\mathbf{k}$.

As the example of the atom in motion through space showed motion changes relevancies. But in that it changes the relevancy of space-time affected by the altering of the motion and not applying relevancy of singularity. The Coanda example shows that motion establishes singularity, dominating space-time, controlling motion and the direction thereof. That means the motion establishes the space-time and the position as much as the direction of the flow of space-time. Motion creates space-time as much as space-time is supplying motion to form space. The only way to enable that to become a reality is that motion creates space as much as space follows the direction of motion. That is what Kepler said when Kepler said the space is equal to the motion thereof $\mathbf{a}^3 = \mathbf{T}^2\,\mathbf{k}$. There is no solid $\mathbf{a}^3$ Universe but all interrupted by positional changes that recreate the space in the and according to the new direction singularity will create as singularity allows space to flow by motion by fragmenting space into time sectors.

By any object applying motion such motion is reducing the time the object is occupying the space it holds as the occupier of a position in that space. By motion the object occupies more space and by occupying the space it holds, the motion creates the reducing of the space in favour of the time coming about and duplicates the object reduction of such an occupation of space in virtual size.

The faster the object will duplicate such occupation by motion the more the space will reduce in favour of the time remaining the same. That means the duration of the time will produce smaller space but longer occupied space. Some pages ago I gave a specific example using an atom in motion in space-time by moving through space-time. Within the realms of the Earth the space tends to remain constant while the time duration changes accordingly. Motion is space duplicating and comes from the point where singularity contributes to space-time. Singularity is at the point where Mainstream Science now view that that point holds nothing.

The growing in size of the space occupied is a result of relevancies coming about. Since $\mathbf{k}$, moving towards $\mathbf{k}_2$ is growing by the increase of $\mathbf{k}_2$ the position $\mathbf{a}^3$ holds in relation to $\mathbf{k}_2$ and $\mathbf{k}_1$ will result in a larger $\mathbf{k}_2$ running from $\mathbf{k}^0$ to $\mathbf{k}_2$ through the previous point of $\mathbf{k}_1$. Therefore $\mathbf{k}^0 - \mathbf{k}_1 - \mathbf{k}_2$ is overall larger in ratio by measure of $\mathbf{k}^1 - \mathbf{k}_1$ was, therefore in ratio the space unit grew as $\mathbf{a}^3$ remained the same. But that is the virtual position. In reality the space $\mathbf{k}^0 - \mathbf{k}_1 - \mathbf{k}_2$ is a unit, and that unit grows by the Hubble constant and by no other means.

Now the same principle comes into effect, which takes place in the sound barrier and which takes place as the spacecraft enters the Earth's atmosphere. But the Earth holds its atmospheric space as a controlled unit. Outer space holds its measure by time. Therefore the object extending the motion has to comply with changes coming about. We know that the unit forming the craft, holds the unit by space to size taken from the accumulated compliment of all the atoms taking part to form the unit. The two units in relevance are also in dual because the object in motion holds relevance to a larger space than it did before. But seen from the role the space has, outer space holds the same measure by the same means. Outer space does not compromise because as far as outer space is concerned, outer space remained the same. In addition the space the object unit claims by atom occupation is much inferior to the gravity that outer space insist upon, therefore to adapt to the new relevancy, the spacecraft atom unit has to adopt the new relation.

The spacecraft atom unit has to reduce the claim it has on the space of the totality in order to compromise for the bigger overall space in time in relation to a larger space in time by the craft repositioning its time it has with a larger velocity. Therefore the compromise is that the lesser applying gravity will reduce the space it claims and such reduced space compromised will translate to heat in liquid taking up the discrimination.

The very same scenario occurs when the craft enters the Earth and since the Earth's gravity is totally dominating the craft's gravity, the craft's gravity has to adhere to the Earth's gravity and the craft has to compromise in size. In the case of the Earth atmosphere versus the craft's, the extending of the **k** factor comes about as the extending of the **T²** factor since the compactness of the Earth's atmosphere translates to more space per measure compared to the craft's atomic space claimed. But all this might seem to us to be in the outer whereas in truth it is in the inner Universe of the Earth. From the centre of the atom the electron will align in a position that it will maintain in relation to such a centre. As the relevance of the increase in motion outside the atom reduces the space occupied and by holding synchrony to singularity in the centre the electron will use the same duration to cover less distance since some of the distance is increased on the outside of the atom. That means the electron will take the same time to complete the rotating circle but since there is less space to cover the motion would be relatively slower because the motion is shorter. The duration of space duplicating or space interrupting at that speed favours the interrupting part more than the space part in relation to the situation where the relative motion of the particles were stationary.

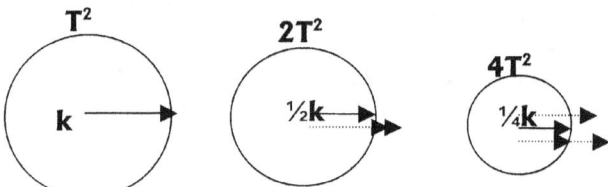

By going faster the duration of lost space is filled with liquid heat because the motion that reduces the **k** factor also takes the moving object back in time going closer to the conditions that applied at the Big Bang. Time slows down as space occupied decreases with singularity dynamics coming closer to the position singularity holds. The time does not in affect slow down as much as the duration of the time it takes to produce the interrupting of space. The duplication will be more, which is taking longer in relation to the time in the space that is viable. But one has to break time down to $((1836)^3)^2$ multiplied by the speed of light, which is C in order to establish the frequency.

By producing motion to the object that relates to a change in the space-time relevancy changing as the motion applying alters the relevance in time producing space. As motion or gravity doubles the atomic border distance of the electron position would half whereby the space being occupied will be the quarter of that which it was before. The linear motion will establish a centre singularity k^0 just as the circular motion does but since k^0 comes about from **T²** time changing the relation and where **k** must be responding **k** will no longer set a **k** confirming the relation of the border but **k** will reduce the relation to compromise for the changes that alters the k^0 position at every alternating point there is of establishing space once again. Where the object in motion has an elected k^0 that holds a relevance to k_1 the faster motion will produce more duplicating and that duplication will stand relevant to more space outside the atomic accumulated space. In that manner k^0 of the elected atom unit that is going faster must reduce its claim on space occupied because the space not occupied is more, but if the relevance is placed on the other side the space remains the same and therefore k^0 holding the claimed space has to reduce. The motion coming about is also gravity increasing because as I shall show later, gravity is purely motion and gravity is not a force of the magical kind. By altering the space as the space reduces it allows the electron to remain in harmony as the space duplicate according to gravity.

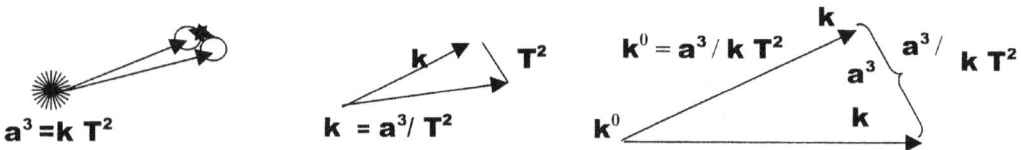

Motion is space duplicating and comes from the point where singularity contributes space by motion. Singularity is the point Mainstream Science now view that holds nothing as a value. The only definite place one will locate zero is in between the starting point of the lines going in opposing direction in the position the lines hold before there was the least of directions applied, but that is only because there is no such a position, not because any line is coming from there. The two lines are still one holding the opportunity of parting as an option but have not yet parted and therefore are on the very precise same spot. The line coming from there is already there because it already has the choice of going in any and all opposing directions and when it starts running it will place filled space in that location because the space was already filled with a line starting and not with a line not there at all. When reversing a line we might find a better idea of what is in place and where it is in place.

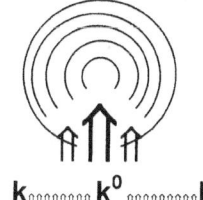

$k_{nnnnnnnn}$ k^0 $_{nnnnnnnn}$ k

With singularity placed in infinity within the centre of every rotating object every atom and its relation to its surroundings including other atoms form space-time diverting from the point holding singularity as far as rotation goes because every object holds three relative positions in as far as where it was, where it is and where it will be in relation to singularity providing time. I shall elaborate this a little.

When the line came from singularity and expanded into space forming space and by using Kepler's formula we can see two dynamics coming into place. There was singularity remaining at value holding six positions in relation. This is a how Kepler saw the involvement of space-time in relation to singularity. But there is another involvement. When we observe the formula Kepler introduced we must see where the formula holds value and by doing that we find three factors that is equal in relation to forming a relation.

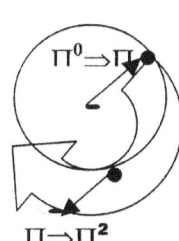

$\Pi^0 \Rightarrow \Pi$

$\Pi \Rightarrow \Pi^2$

There are two ways of looking at this issue. The one is looking at it from the centre that is keeping the rotating object honest or there is the rotating object forming space in relation to the centre and placing the centre in the centre. It will always be one taking prominence to the other and where Kepler introduced the formula it is indicating motion producing gravity which is gravity that is keeping form outside the sphere. Gravity is motion but the motion we see is much different from the gravity we experience while we know it has to be the same with only relevancies changing.

In the one formula Kepler introduced the relevancy suggests that space is producing motion to becoming space filled by motion. $a^3 = T^2 k$. That is an indicator used by material forming a circle outside the sphere to accommodate the motion of the sphere. Clearly such an indicator would not use the direct value of singularity because it is not in direct contact with singularity. In this relation I showed where the factors are the same although the symbols used suggest otherwise and that made me realise to search for a symbol that will produce equality. There are three sides in space and the three sides are moving in the direction forming the next three points the previous three points will form. What may seem to us being as realistic as man can be, we see a solid structured Universe boxed into a nice cube. But instead we have a hugely flickering Universe, which is only relevant when viewing space-time from singularity. From singularity there is motion as three markers, which is moving to fill the position the three other markers where the space will be and are pointing at. That is space filled by motion. The space is a direct result of the motion coming about as the outside of the sphere has six sides establishing the space. Then what about the sphere? We can only see the differences there are between calculating the cube and calculating the dimension of the sphere when using the manner that Kepler calculated. By turning Kepler to a more precise stance it shows that the calculating of that space is just as much different from the manner that Newton used to calculate the sphere. There are distinctly no comparing in the calculation methods used in either one of the methods in measuring the sphere versus measuring the cube and that should show that the answers Kepler arrived at do not indicate the normal mathematical calculation methods normally used.

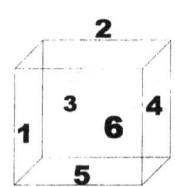

Outside the sphere we find six sides where three sides oppose three sides. The six sides place a specific dimension but they do not define specifically impenetrable edges. The sides are vague and undefined but from the sides one can introduce borders that will come about by connecting angles on the outside using 90^0 as connecting angles.

There is front 1 to back 2. There are left 3 to right 4. There are bottom 5 to right 6

The practise clearly points to another method because Kepler did arrive at and where Kepler found answers and in fact the formula came about from the answers he arrived at. His work did not need the changes that Newton introduced to find the calculations that came to form the formula. The calculations produced the formula and not the other way around. This is bringing the crux about cosmology to the foreground. It proves that the mathematics applying to cosmology is not standard maths used where mathematicians are in the designing of a high-rise building. In the cube there are six measurable sides.

The layout of the sphere demands a totally different perspective since the sphere on the inner parameters are about specific borders in the stringiest control from the centre and the length of the edge calculated from the centre is in measure precise to five other opposing borders. From such a precisely located centre there are no margins of flexing the radius which will alter the points forming the edge or border of the sphere other than adhering to precise cosmic principles concerning gravity. Gravity is the strongest where space is the least and that is in the centre of an evenly space sphere where such a centre produces the gravity in precise measure to accommodate every possible point in six opposing sides at any position on the edge of the sphere.

$[k^0 = a^3 / T^2 k] = 7$

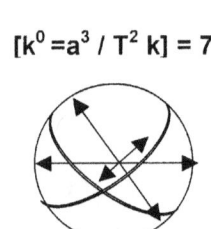

Edges come from solid structures within the space having the six sides in relation to the sphere and from the material in the sphere there are six points securing one another by 90^0 and 180^0. But in the centre of the sphere where all the pointers connect as well as cross there is one point all points refer to when locating every point holding a relative position. That then must be one. This still would not have made much sense if it were not for reapplying the Kepler formula but this time to suit the connecting in the sphere.

This still would not have made much sense if it were not for reapplying the Kepler formula but this time one must use the Kepler formula to suit the connecting sphere. Since a centre from a centre in which the centre is part of a sphere defiantly controls the planets in the sphere, which we named to be the sun. In the cube we find the one side provided the other side space but the sphere goes beyond that. The solid structure of the sphere not only relates but also places the one in connection with the other and this is directly placing the relation to a centre where such a centre must be one. That suggested to me that centre that mathematically according to Kepler there must be a centre that will provide a connecting. $k^0 = a^3 / T^2 k$. From there I divided the line $k / 2$ that brought me to singularity and from singularity I could value singularity and find the connection that the cosmos uses to bring about cosmology in the form of $\Pi^0 = \Pi^3 / \Pi^2 \Pi$.

But as one can see, I also realised gravity is relations of motion applying in two factors. There is no separation of the two factors acting as one but both have different applications and values in the unit. It was what gravity was because this action prevented expanding whereas the centre commanded the borders and assisted in the collapse of such borders. This is the result of singularity having three parts acting as one but giving three distinctions in application. Gravity is as much part of dismissing space as it is about making contact with space in time. Since the connection comes about as a circle, the connecting points will relate to Π as the value. Due to the spinning nature of such a point with all surrounding the point will be alternating direction favouring change every instant of a time frame used and in that the value to such a point can only be Π because of its constant changing. Using **r** would specifically oppose another **r** from every angle because the use of **r** will bring about a static relation to the previous and following instant and therefore it will cancel the constant spin flow. By reducing the line to its maximum possibility one end with Π being the minimum but that Π is actually Π^0 which can also be $\mathbf{k}^0$ or $\mathbf{a}^0$ or $\mathbf{T}^0$, which all indicate positions in singularity. Only when forming a value past singularity does independent identification come about. When the atom formed, that atom applied a relevancy of ten positions where seven positions are included in the atom spinning and three positions are part of the exterior of the atom spinning but all the positions relate to singularity but as space flight taught us, such relevancies can change when an object is within the space boundaries of a larger structure or roaming free in outer space. Within the boundaries of the atmosphere where the sphere border touches the space borders the space borders hold six positions and the sphere holds seven points. But at the precise place where the points make contact with the sides one side falls away in favour of the point it connects to leaving five sides relating to seven and where one of the six sides takes control in removing one of the cubical sides by replacing that side with a sphere point position the object then becomes directly controlled by singularity positioned in the centre of the sphere. The object seems then to fall from space and enter the atmosphere becoming a shooting star. What the Coanda effect proves above anything else is that gravity in control of space-time comes about from a centre and such a centre can be created by motion applying to a liquid in relation to a solid. That means there is undisputedly a flow of space-time towards a centre and the centre has to diminish the space-time reaching such a centre to create the flow and therefore the control from such a centre. That's the one pivot of gravity.

In the cube we find that the one side is providing the other side, which is the side directly opposing each other with space by duplicating space through motion of space but the sphere goes beyond that. The solid structure of the sphere does not only relate by opposing sides mounting opposing borders, but there are no precise centre such as the case is with the sphere being in precise contact with the controlling centre. By having precise cross referencing on each other makes the sphere superior as there are then also placed in relation a precise centre where six points on the outside are relating by one in connecting with the other. The opposing sides run a connection in a precise duplicating of the one onto the other as the cube does but goes beyond that and this is directly placing the relation to a centre where such a centre must be one centre to every possible crossing line running from any given point to the directly opposing point through such a centre position. That suggested to me that there is a controlling centre in all spheres that can mathematically, according to Kepler, form. There must be a centre that will provide a connecting and by reducing such a connecting line from both sides will be where I then will be able to find $\mathbf{k}^0$ or singularity as suggested by the formula translated from Kepler $\mathbf{k}^0 = \mathbf{a}^3 / \mathbf{T}^2 \mathbf{k}$. From there I divided the line $\mathbf{k} / \mathbf{2}$ that brought me to singularity and from singularity I could value singularity and find with that the connection that the cosmos uses to bring about cosmology in the form of $\Pi^0 = \Pi^3 / \Pi^2 \Pi$. The way I concluded the value of singularity as Π^0 going to form Π I have already explained.

But as one can see I also realised gravity is relations of motion applying in two factors. There is no separation of the two factors acting as one but both have different application and values in the unit. It was what gravity was because this action prevented expanding although gravity is about expanding as much as contracting. The result is that singularity has three parts acting as one but giving three distinctions in application in two opposing variations where both points in motion are controlled by the centre. Gravity is as much part of dismissing space as it is about making contact with space in time by duplicating space in using time. Since the connection comes about as a circle, the connecting points will relate to Π as the value. Due to the spinning nature of such a point with all surrounding the point will be alternating direction favouring change every second and in that the value to such a point can only be Π because of its constant changing although such changing is remaining the same. Using **r** would specifically oppose another **r** from every angle because the use of **r** will bring about a static relation to the previous and following instant and therefore it will cancel the constant spin flow. By reducing the line to its minimum possibility where all ends on one end it will be with Π being the minimum but that Π is actually Π^0 which can also be $\mathbf{k}^0$ or $\mathbf{a}^0$ or $\mathbf{T}^0$, which all indicate positions in singularity. Only when forming a value past singularity does independent identification of distinct value differences come about. When the atom formed that atom applied a relevancy of ten positions where seven positions are included in the atom spinning and three positions are part the exterior of the atom spinning but all the positions relate to singularity. But as space flight taught us, such relevancies can change when an object is within the space boundaries of a larger structure or on the other hand roaming free in outer space. Humans are taller in space than they are on Earth. Within the boundaries of the atmosphere where the sphere border touches the space borders the space borders hold six positions and the sphere holds seven points. But at the precise place where the points of the cube and the points of the sphere make contact with the one another's sides the domination of the seven contact point, we find in the sphere, including the controlling centre will dominate the six sided cube and will bring about that one side of the cube will fall away when making contact with the sphere in favour of the point it connects to. By one side removed, this is leaving five sides in the cube in relation to the seven sides in the sphere where the relating to seven eventually stand in for one as the Lagrangian system spawns from this. Where one of the points in the seven-sided sphere takes on the six sides, the seven will take control in removing one of the cubical sides by replacing that side with a sphere point. This is when the sphere centre takes control of the position. The object then becomes directly controlled by singularity positioned in the centre of the sphere. The object seems then to fall from space towards a centre point and enter the atmosphere becoming a shooting star. What the Coanda effect proves above anything else is that gravity in control of space-time comes about from a centre and such a centre can be created by motion applying to a liquid in relation to a solid. That means there is undisputedly a flow of space-time towards a centre and such a flow contributes to space being liquid. The centre has to diminish the space-time when space-time is reaching such a centre to create the flow that produces the motion contributing to gravity being in place and therefore the control must come from such a centre where space is dismissed by the lack of possible motion. That's the one pivot of gravity.

Since the Coanda effect shows gravity is control of space-time by motion flowing towards a centre, which then allows controlling and managing the space-time within the realm of such a centre that then also prove and explain the one part of gravity. The part it explains is that space reduces space by increasing time towards a centre and that is established by motion and the lack of space establishes a lack of motion in that very centre. Because there is no space in such a centre there can be no motion of such a space in the centre and since there is no motion of space in such a centre there can be no duplication of space in such a centre, which then brings about the killing of all space before and after the space can form.

A larger object will allow more time to duplicate the bigger space it has in the same time duration than would a smaller space find a need to duplicate a smaller space in the same time duration. But in the cosmos there is no big or small and only relevancies produce links with and to singularity. By having less space while having to produce the same time the illustration of the atom once again enters the equation we think of. By having s smaller space to duplicate than the larger partner a shorter k is needed when the same time T^2 is enforced. But as Galileo's swinging pendulum proves the space diminishes to allow the time T^2 component that the Earth enforces on the smaller object to be adhered to by the smaller component. Since the time does not match the duplication required to sustain the k factor and the time factor cannot change the space of the smaller object has to reduce in relation to the diminishing space of the Earth forming the atmosphere. Therefore the smaller space seeks to find a position where the k component will match the time component because the space is offered as compensation in any case.

We must not think of the Earth but we have to think of the unit of all the atoms within the Earth unit that elects a centre where all the atoms establish a governing singularity. All objects are part of the elected election but at the same time they have a proxy vote in relation to the totality of the ability of all the electrons to duplicate space-time while the protons are dismissing space-time. In the larger unit is a smaller unit, which is a unit in individual ring but is also part of the collage of atoms forming the larger unit. That then comes to a selection of relevancies applying to establish a pecking order by the units all forming the larger unit in relation to each other and in relation to the unit forming the umbrella. Mass is this pecking order by proxy or gravity within the control of gravity supreme.

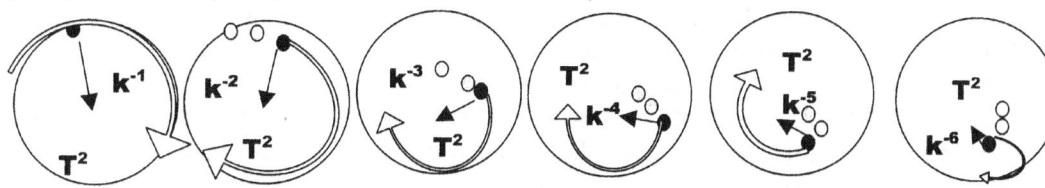

Because the smaller object holds much less space the duplication of the lesser space being part of in a lesser relation than the main object and because the time factor enforces the duplication period to match the unbalance in the unit duplicating makes that mismatching in the unit between members of the unit reign supreme. This will lead to chaos and the cosmos is about order in duplicating by precise repeat. Therefore something in the applying ratio has to give in to allow the major relevancies to remain in place. Since a^3 representing the lesser unit as part of the holding unit in the holding unit, the lesser space has to match the unit in effort of reproducing space by duplication and in that find the relevancy whereby the lesser space can apply to the conditions it requires in the unit from the unit. The time factor is enforced therefore the space duplication will be lesser in proportion taking a period to duplicate space that is relevant to the larger unit. The lesser space must obey the time it is in and standards set by the larger object forming the holding space unit. By such differentiation of duplication in applying the same time a new relevancy comes about where the new a^3 will bring along a reducing position **k** because the time factor T^2 has been set. The relevancy between the lesser units will have to define a position with the diminished **k** holding the lesser unit in a position that the Earth enforces. Since the space that motion reproduces is smaller in relation to the Earth, but the Earth enforces the same time value. The relevancy of the joint time factor in value will deplete by enforcing **k**, to establish a new position but not in a straight line because all factor changes will then only be carried by one factor. In this way the diminishing of the lesser space position produced will help the cyclic time factor to decrease the lesser singularity position with the distance that grows smaller that serves the governing singularity and the lesser unit.

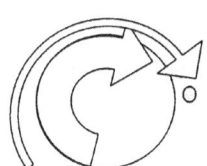

When the object is released from the atmosphere of the dominating space, this very same gravity $k^0 = k\,T^2/\,a^3$ ratio will still be enforced since it is not the law of the Earth prevailing but it is the law of the Universe applying. Outside the atmospheric borders the Earth no longer have the means to remove one of the cube sides that form the lesser object space and where the cube reinforces position by keeping the rotating object in position floating above the Earth.

Newton did not see it that way. Newton did not see relevancies. Newton ignored the vital role motion plays in gravity. Newton placed one relevancy between the objects as to serve both objects. But there is so much more to the relevancy because the motion applies from both ends in regard to both ends.

This is the very same principle applying when there is a lesser object such as a planet in relation to the sun floating in orbit with the sun. The minor object then received an individual $4\Pi^2\,a^3/\,T^2 = G\,(M_s + M_p)$ What I am trying to say is by using Newton's symbols which he added to fill the picture he saw helped nothing. What he did was to repeat what Kepler said by duplicating what Kepler said. He only extended what Kepler said. The value of **k** is what he suggested as G $(M_s + M_p)$ and $4\Pi^2$ is representing **k** by implicating gravity to the full time value of **4.** Later I shall show that the Earth or any free orbiting structure in relation to the sun is $4\Pi^2$ and in that sense the relevancy of all stars committed to fusion has a relevancy value of 7/10 multiplied by $4(\Pi^2 + \Pi^2)$. If we think of the statement as Newton put it, it comes down to the manner how the Earth as a planet connects directly to the sun with having the Earth only as a relevant mediator and not as a controller. It is because Π is the basic basis which the Universe applies as a value. Because π is already the principle the use of π was included as the base value and that is the main reason why the cosmos did not care to involve the use of Π when it produced the formula which was unveiled to Kepler.

The only value that must apply is that the distance from the centre holding the factor **k** holds space and time valued. The Earth then sets the speed of motion or the gravity factor demanded by the Earth of the orbiting structure orbiting the Earth. After the glitz and glamour that accompanies a launch of a rocket into space, the celebrated factor of such a launch is placing the object as a sun controlled factor that after the launch became independent. Such independence would have to rely on the gravity that the orbiting structure produce to validate the sustaining of such a relevancy in continuing the independence as the object puts itself in the role as another satellite with semi solar factor to encircle the Earth as a satellite. To sustain such motion the object must be in a position to charge the required amount of heat and offer the heat as payment to the centre of the Earth and then set off the orbiting object. Remember never, for one second, to forget that the launch and the orbit are manmade and directly resulting from the intervening of the will of life and are therefore as far as the cosmos go totally artificial. Where such orbiting took place in relation to cosmic law such an orbiting object had to secure just the right amount of heat to establish the velocity it must have in order to bring about the gravity it needs to stay relevant with the motion of the Earth. In order to maintain the freedom received by the orbiting structure that structure must apply a time in motion in relation to the centre where this orbit speed must match the distance from the centre in harmony with the Earth.

If the object is going to slow it will retreat by falling towards the Earth centre and when going to fast it will apply more distance taking the object further away from the centre. It is the motion that the object has in relation to what is coming from the motion of the Earth dictated by the centre in the Earth. As it should be clear to every one, one find the gravity that is then applying to be a relevancy as speed or motion is a relevancy where space in motion relates to a specific duration of time from point to point $a^3 = T^2k$.

With the term micro gravity, such a term is placed in meaning to bring the idea across that the cosmic rules does not apply any longer in outer space because gravity as we know gravity does not apply and therefore mass does not apply because there is no mass to generate gravity or if it then applies, it applies in a very diminished sense. It is documented that an astronaut can pick up something like an object that will show an Earthly mass of about four tons of mass when on the Earth when he is working with that object in outer space. The condition to this lifting is that the person then must have contact with a device that is in control of a much more dominant singularity. The result then is that again there is a precondition attached. Such a precondition is also a relevancy of some sort. Bonding is one of the contradicting ideas. This time a much larger object must secure the position of the astronaut such as the spacecraft when he does this lifting of the four-ton object. The larger object then produces a controlling point serving as the elected singularity governing to all the space-time that is forming the structure unit. This is the result only when the objects are not being captured in the space claimed by the Earth but is still captured by the motion the Earth brings along.

From the centre in control of the Earth this motion will apply to space duplication in relation to the sun without the Earth applying boundary control over such motion in that specific space. The only control coming from the Earth centre is the speeds forming the gravity or gravitational position the orbiter has to secure. I guess one may say the objects are in the region represented by the ominous G (using Newton's formula) as in the Gravitational constant that is everything but a constant or has any Universal equal application. The bodies at that position relates to three independent markers in singularity where every marker is duplicating space in relation to what the sun demands and what the Earth provides but where every object still responds to the individual dismissing of space according to individual singularity protection. It is all about relevancies applying attachments of which mass is but only one implication of such a relevancy

In all units forming in space, there are six sides in relevance with three sides in attendance held by singularity in relation to relevancy of space-time and three sides coming about through motion or repositioning singularity space-time. Before singularity overheated that brought space-time about and placed space-time as a form of heat in a position to heat we have to presume that all space-time that formed in the form it did, was there already present but it was only still in the theory part of Creation. The development thereof came about as gravity came about and formed relevancies by applying the motion of the three sides. The motion established relations by three in three with three in forming a balance to indicate relevancy with singularity. It is most likely that it is ten points each in independent singularity where each is forming a relation. The ten plus one point nine plus ten is in relation to seven which forms pi, but that means there are always ten points in seven to pi. Only when such relevancies exist is where the relevancy establishes a partnership or connection. Gravity can only come about as a relation of motion differences between points occupied by singularity in maintenance. The motion of space-time inside and outside between atomic particles establishes gravity by motion of space-time flowing. Then a moment arrived where **k** developed from k^0 to **k**. This brought about a revolution of cosmic proportions and this was the only time that such an expression is not exaggerated. Matter divided from singularity as matter claimed space. The growth of **k** had to produce a^3 which is an interpretation of the space **k** will bring in place. The most tiny and slightest of growth established **k** coming from k^0 to **k**. We can never bring about any concept by which to understand this. By establishing **k** that very second a^3 also came about. But through Kepler we can see how singularity achieved space to come about. It came through spin.

10 Everything is space-time by confirming space in establishing time

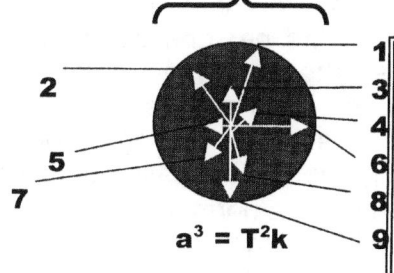

$$a^3 = T^2k$$

When the Universe was in the beginning with the entire cosmos still in a single dimension there were no limits as we know limits to form in the Universe we use and no borders indicating limits because after all it is the single dimension where there is only one dimension holding so much diversity. The borders were part of development because we can witness the legacy of such borders in the present day holding the 3D in place.

Remember that space was at a premium like Creation afterwards never can repeat again when the first moment arrived for the very first time. But in contrast matter had to expand as heat surged in search of space to Create cooling. Because of the overheating, space was a desire but the heat had no means to expand to find a form of release from the eternal grip of singularity being space less. Kepler said that $a^3 = T^2 k$. Space broke away from singularity by applying spin. Gravity is spin. When **k** extended it secured a^3 being space but through the spin of T^2 the space separated between particular points holding singularity, to establish individual singularity. It secured individual space from space by rotating space. Through the rotation came boundaries, which we now consider to be particles in time. It is the time (from T_1 to T_2) that it takes **k** to swing into a different relevance to a^3 space. It established movement in the area holding the least space. The relevancy started when **k** claimed space by motion from singularity. That is what Kepler claims. What it does say is how Kepler incidentally forgot to improvise for the claim of a circle but accidentally forgot about including $4\Pi^2$ on the one side and G (m + mp) on the other side. Kepler placed the growth **k** directly relating to the area a^3 separating as the spin T^2 provides the space. Kepler showed with his formula what gravity is. Gravity is the least space (singularity) k^0 claiming space a^3 through spin T^2 throughout the entire distance of k. Kepler announced space-time, the Hubble constant, the Big Bang and other later cosmic developments. He said space comes by the motion thereof.

Our need that we have to form the understanding of this notion that space produces time to duplicate space by dimension is all inclusively to underline what space-time is as much as it is critical and demands our full attention in forming our understanding concepts we have about the Universe and Creation at large. My believing Kepler then tells me that the space we have is set in a solid state by the motion that the atom brings about. But way down in the atom where singularity takes charge every action is the vibrations of pulsating protons, flickering to set motion to the immovable singularity. The atom finds the way to draw "flat" and re-establish space-time again. By taking full control the singularity that is elected by the compliment of atoms begins to govern as it unifies all individual atoms into one structural compliment. A star such as the Pulsar is an atom that by now is in a position where the size of one star is fast becoming. That is the direction the Universe is heading one atom.   That reflects what happens in the final stages of the Universe before singularity once again steps in and takes control as it does in the Black hole. At first singularity spawned and in the end singularity will grow to capture and remove all spawned uncontrolled space-time to get the Universe under direct control of singularity once more. The pulsar expands and we see that as light streaming from the star.

This provides space by motion expanding the heat and increasing the space. Then motion produces contraction and the star starts a spin that removes whatever heat/space there was taking it by spin back to singularity. We might observe a broken stream of light as flickering and it may occur let's say twenty times per second. Inside the star one such cycle of flickering will last say one thousand years. I am very conservative in an attempt not to shock the reader witless but in other books I use much more realistic periods in time duration. The time on the outside where we are located and the time duration which the pulsar froze when the pulsar developed individual space-time from that of the Universe has a differentiation in period duration of billions of years. As space a^3 expands, time T^2 has to change since k enlarges. Being in the star reduces the time so much that if the pulsating cycle is one cycle every thousand years then by the same standard will be that the time on the outside where we are located in is having the duration of twenty thousand years per one pulsar cycle. Remember there are possibly forty flickers per our one second. It takes a thousand years duration for the cycle to end and the full duration from the start of a flash to the end of a flash or the start of a dark faze to the end of a dark faze is twenty thousand years in one faze. The star was shining for twenty thousand years or it was not shining for twenty thousand years making one cycle duration becomes forty thousand years. The time going on in the star is Π by10^3 years, which is a normal relevancy in a star cycle of that development. Then the other side of that Universe take charge and all the factors repeat. But since we find ourselves part of a six-sided 3D universe our set of time frames per space Unit that is largely out of synchronise with such a massive star the star is in harmony with our inner atoms. Such a star may embody our entire galactica many times over into one growing singularity as far as material occupation positions a star. Therefore, time on our end is pretty slow to the pulsar as the space frames takes a much quicker duration to complete. We see the flash at a pace of forty flashes per second but to realise the other side we have to place ourselves in the position of the pulsar. We have to place our minds in the time applying in the star.

If the pulsar takes forty seconds of our time to complete one period or faze and that phase takes twenty two thousand years in the time frame of the star to complete we must present a pretty motionless picture to whatever is looking at us from the inside of the star. At such a rate, it is no wonder we find us in a solid Universe, which we know that that Universe, has to flicker but the flickering is so fast and uninterrupted, we find it to produce a solid state of affairs. When the sun broke free from the Milky Way, it secured individual space-time. That also happened when the Earth secured an individual atmosphere. But the capturing of the space by singularity also accomplished the freezing of the time duration that applied at the time when the Earth or the sun or any star captured space-time.

Space increases as time in duration decreases $k = a^3/T^2$ and by becoming individual it holds space and time as part of space a^3 time T^2k. Even if the star had the ability to break down all space-time and reposition all space time at a rate of forty times in our second then in one second of the time relating to such a star we would still go very slow. The well-developed star would then find one thousand years worth of time in every second we have. To argue that the star and us use the same time component in duration is quite thoughtless and laughable. It is pretty Newtonian to think of space-time in that way. One cannot alter space and not time while in the same breath science promotes the idea of space-time. What happens to space forces the reverse to happen to time. It is precisely what Kepler said when he said $k = a^3/T^2$.

There has to be some difference to time when space changes, since it takes the light coming from such a star many millions of years to reach our part of the Universe and there was some Hubble development since then, when the light was sent on its journey to now, where we are. We see it to be one second but the inside of the star sees us as moving much more likely only once every million years because our moving about might take us a twenty four hours to complete our cycle. Compare that in the relation it takes the pulsar to complete twenty-one and a bit times multiplied by many thousand years in one second and multiply that with thousands of years in each cycle. When looking at us being on the other side or the outside we travel so slow that in the view of such a pulsating star from on the inside of such a star we, on the outside of such a star have gone solid. Our time duration seems never ending from the position the developed star has because our second is a matter of thousands of years compared to what is coming from the inside of that pulsating star. Those on the inside of such a star see us standing without motion for one million years and then in that million of their years we do the travelling that we do in one second of our time. To them our Universe froze solid and that is what happened. That means by using a ratio of one million years to one second of our time we freeze the moment in our time to a standstill for the duration of one million years. If we did not freeze space-time we would not have had the fortune of being solid in the 3D that is producing the six dimensional Universe captured in space-time we enjoy. We are motionless because we are taking such a long time to apply motion. Our motion remains relative to the motion of the pulsar and what we see in the star, the star sees the very opposite in our behaviour.

The space the pulsar has relates directly to the time it takes to replace the space but not only that the space-time inside the pulsar is relatively in opposition to what we find space-time to be. When saying this we must not stray too widely away from the flickering because that flickering is also inside every atom that forms our Universe. The motion creates our space on the outside much slower because we (our atoms) create much of the space-time we enjoy but is using the same heat to manufacture the space-time we use as frozen material. But to get the heat as concentrated as it is in the pulsar it takes a million years to concentrate the space to the amount of heat used to produce motion in such a dense environment. The pulsar froze a small part of the Big Bang space-time as the rest of space-time being outside the star developed away from what was space-time when the pulsar star came of age. When we look at such stars we see a Universe whose stars froze in space-time to our benefit. It is the converting of such an amount of space from the amount of heat that makes up the duration differences because the slow part of the star is in motion at the speed of light and the really fast part of the star is way beyond the speed of light. If space as well as time did not change since the Big Bang to where we are, the only scientific explanation about the star since the Big Bang up to now would be to think of it as magic. But since magic is not part of science, science better adapt ways taking the thought of magic out of Newtonian science where science starts to understand and explain what is understood. The time it is taking the pulsar to create space is in our view instantaneously because there is in truth comparatively very little space to create but there is enormous heat involved and in our view very little time to recreate the space. Motion duplicates space and the more space being created by motion applied taking up time the more solid the space will seem. The truth we find in the Universe is not what we are looking at but it is in the Universe that we cannot see. The Universe is where singularity comes together or where singularity parted to form what we see. But what we see is the extending of reality and not reality.

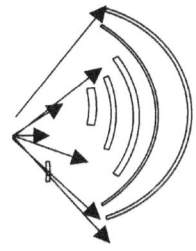

Although space and time is the same the one is the opposite of the other. $k^0 = a^3 / T^2 k$. Space is the same but the inverse of the time there for less space constitutes to more time. Looking at the pulsar we see a little space to duplicate at the inside where the Pulsar is much closer to singularity. From the inside looking at us the space will seem eternal and the time it takes to duplicate the motion will have an eternal relevancy just because we see the pulsar to act instantaneously.

The pulsar is much more Universe because it holds much less of our Universe and is much closer to the true Universe being the singularity it claims standing in relation to the singularity I claim. But it is the singularity within that Universe and not the motion providing space to be within the creation. Because the Pulsar is flickering so fast we have the fortune to be slow enough to be solid by space and motion. The space becomes the motion. The heat that is directly released from a pulsar will be able to destroy large regions of the part of the Universe we occupy by cutting the space-time into cosmic threads. Still, we also must remember not to regard the pulsar much alien from us, because within every atom we compose of and that forms us as well as our Universe hide all the aspects that governs the Pulsar. The pulsar only used a huge number of more protons to get it unified in the singularity it holds than we have at our disposal. But as time (not space) changes bringing about different space, we are heading in the direction the pulsar froze space-time. It is only a matter of having other aspects in our atoms, which in our Universe takes prominence by importance compared to the Pulsar's Universe. It is the measure the singularity sustaining the drive of the Pulsar has gained by destroying space-time and duplicating heat through light reflection as it incorporates all singularity within the confinement of the star as a unit forming eventually only one atom the size of many galactica combined. With that much singularity focused into one unit the relevancies of the Pulsar has to change considerably from the relevancies governing our little yet-to-be developed planet.

If we observed the Pulsar as equal to our space-time, and used matching space and we used matching time duration as well then the time in durations between us and the Pulsar would have matched we could have had the attitude that all things between the Pulsar and us are on par. Since we can see that the applying gravity producing space-time in the Pulsar exceeds the entire space-time the Milky Way generate but the space claimed is much smaller that the sun claims there has to be many discrepancies and none lesser than time and space equalities. The Scientific way of thinking being practised by the Super-Educated wants to take our Universe and the Pulsar Universe into the same arena. Doing that ignores the space aspect since the space is obviously different to what we have. With the space reduced by that much, we see the time as the reduced factor which means we have an inverse view on the practicality of space –time towards the inside of a star. Our perception must always dictate our observation and not the other way around. That is what brings about the relevancies we have to acknowledge.

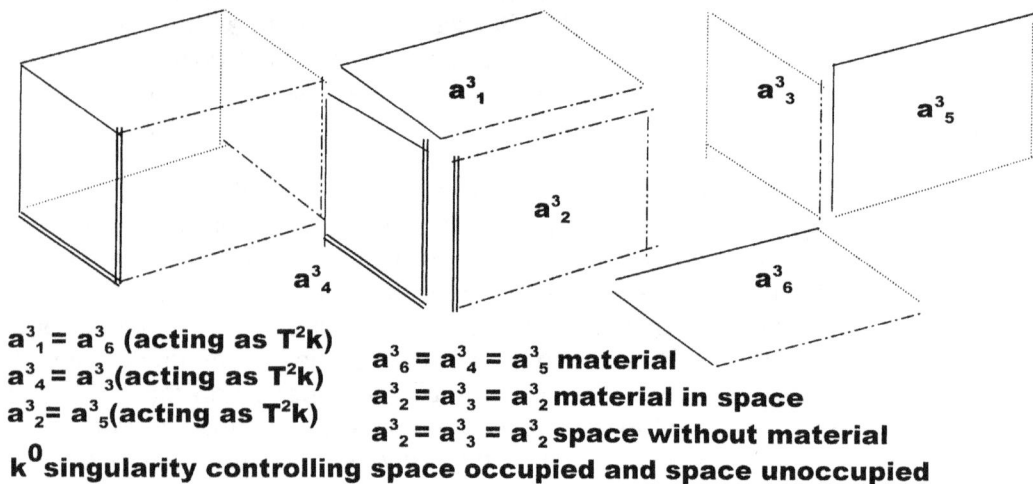

$a^3_1 = a^3_6$ (acting as T^2k)
$a^3_4 = a^3_3$ (acting as T^2k)
$a^3_2 = a^3_5$ (acting as T^2k)

$a^3_6 = a^3_4 = a^3_5$ material
$a^3_2 = a^3_3 = a^3_2$ material in space
$a^3_2 = a^3_3 = a^3_2$ space without material

k^0 singularity controlling space occupied and space unoccupied

Space developed sectors through time applying differences as singularity changes the universe from **k** to **k** through a^3 and T^2. The factor **k** positions the centre of the universe and then sets rules applying in that Universe as far as setting space in time by applying space-time. Space = a^3 and time T^2 is coming about from singularity k^0 pointed by **k**.

Space-time $a^3 = T^2 k$

According to Newtonian statements, it is said that the Titus-Bode law is no actual law. It is said that the law has no theoretical basis, but it shows how orbital "resonance" can lead to "commensurability". By following the guidance of Kepler the Titus-Bode law is most certainly a principle law affecting every aspect of gravity. It stands miles above the Newtonian denouncing of such a law as being coincidental. How the Newtonians would have the lack of cosmic vision to dilute such a law to the level of being coincidental goes beyond my understanding of the Newtonian mentality.

If the sun held a relevancy of 10 relating to seven with one or two or three of the planets…well yes that might be coincidental, but when it shows such a relation with all of them where all of them includes planetary fragments being between the Earth and Mars making such a coincidental claim on ten structures perfectly distributed then any claim on coincidental is more than ducking the truth. To honestly be honest about the finding of scientific truth and discarding such evidence then as coincidental is being unfaithful to the search for truth. Coincidental can be one planet holding such relation and on a stretch maybe two but when all of the planets shows this precise tendency without skipping one or even giving the fragmented planet material a miss that truthfulness then shows it holds the ten to seven ratio in extensive regard and becomes some aspect that is shouting to be investigated since the law of average outlaws coincidental ness and then exceeds the being coincidently by miles.

From singularity twelve points rises in forming a relevancy and puts a relevancy on space and on what we find to be time. Since two of the positions in singularity never concern us, where we in space and being part of 3 D, it is only worthwhile concerning us with ten positions. The moment space cracked the motion confirming the crack was also the cause of the crack.

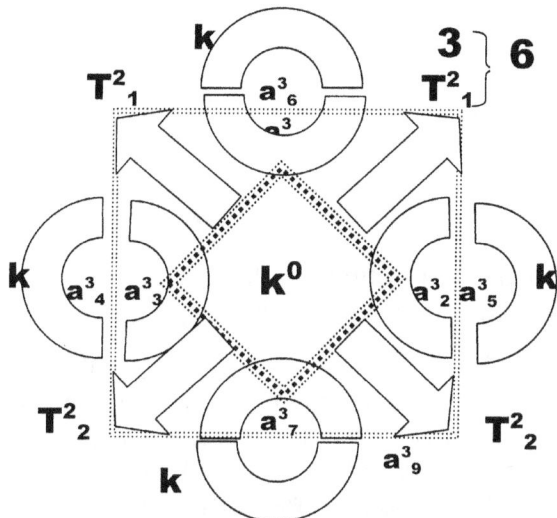

One cannot remove motion from the forming of space because the forming of space is the motion in which the space is then formed. With the motion space duplicates what singularity establish but singularity cannot be discarded as a factor. Therefore singularity is the referring factor to the other six sides forming space in motion.

While considering you as level minded person, who all scientists that is truthful should be, and being an Academic pillar they have to be level minded under every circumstances life may introduce them to. They then should judgemental to the truth in facts consideration without any form of self-righteousness due to their standings in life. Never ever should they think about lesser beings that are in reality so far down the social ladder they are hardly aware of such little people? They must always consider everyone in terms of other persons to be a participant equal to an opinion than may matter and distance them from any form of bias. While being all this in the same breath they must be big enough to consider the smallest opinion and never blow away such clear evidence because the candidate is much unlikely to have an opinion of importance. Their judging should be honestly considered as gross clear dishonesty.

Then if one takes the importance of the cosmos who is the actual presenter of facts being so clearly presented by the cosmos requires investigation by those claiming to be non-bias. To go and dismiss this certainty as coincidental because it does not fit into a Newtonian Universe is stretching the truth to beyond the accepting norm. Science should always remain an open field that avoids the path of being a one way thinking street while science should not try to focus on one person even if he is Isaac Newton. They should rather focus on an image of such a created spot or dot where all came from, because there is no image of any person that will uphold cosmic laws. The cosmos shows evidence and in such evidence we must seek the truth. We must research and investigate such obvious evidence where we have to go beyond the obvious until we find singularity hiding outside the Universe we see, feel and experience. Such a singularity divide may be beyond our viewing capabilities but it is still a point although that point is beyond the visibility being only where we can reach mentally in reading singularity by using intellect. Singularity is elected wherever singularity is needed. The spinning top that uses the Coanda principle is the indication of that.

There is an influence generated by the spin of the top that keeps the top upright while the top is spinning. The line is generated but the line is far from magic. The line is where the centre of the Universe is. The Universe is the particles that fill the top. The particles in motion generate motion by electing a centre from the centre of every particle in the top. Such an elected centre becomes the centre of the universe as far as the top relates to a Universe.

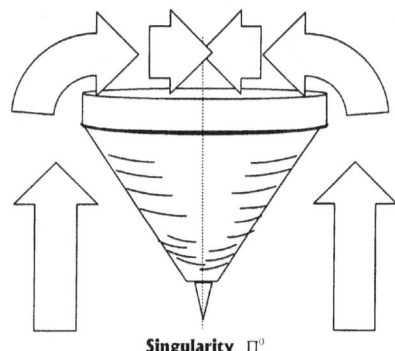

Singularity π^0

By rotational motion the top creates such a line and by generating the line the line charges gravity. The gravity is what drives the top as the top and as long as the top spins

When the top is spinning it will spin most of the time standing upright as if it is a tin soldier. Applying an effort of impact the top will start to spin at a high velocity thrusting from side to side. The effort will seem as if there is a fictitious wall that the top wants to climb. The top will try to lift as it bends over to one side and then changes sides bending over too the other side. One does not need any imagination to see the top is trying an utmost effort to lift itself from the ground on which it is spinning.

The line dividing that the top created by motion is a product in the cosmos which is the centre part of any particle and that runs through every particle, no matter how large or small it is or even if it might be beyond our vision. Such a small line might be so small it is not even noticeable to the cosmos, it might be in the 3D but nevertheless it is large enough to part the cosmos into sectors. It splits the biggest there is into particles and we are not even able to notice the precise location of such a split. It divides a star as massive as a Pulsar into bright and dark periods, yet it is so small it still remains invisible. When observing singularity from our stance, in truth there is no top or bottom that we living in 3D can see because there is no large or small that we can see. Standards measures borders and sizes are all man's creations because man is part of the Creation. We shall have to use a general conception brought about by intelligence when we observe the cosmos. One's intellect tells one about such a spot, but that is all because that spot is on the other side of the Universe (quite literally).

On the surface at first glance the top is an ordinary piece of dead wood that is machined into a sloping shape. With a sharp needle point SA Universe rides on the point

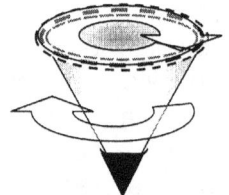

When life attaches a string around the piece of dead wood we use as a toy and throws the top with a jerking motion, the top will land on the needle sharp point at the bottom and the top will start spinning. The top received motion which then with the motion is where the top (for a short while) become a dynamic participant in the established Universe.

By motion that creates gravity applying from a specific established centre one can see how the Universe came about, and that the centre of the Universe is elected where ever motion duplicates space to the ratio of space-dismissing. The centre of the Universe is like the Universe; it is not a fixed feature but is created as motion centralise space to create gravity by time applying.

There are two factors present in gravity. The one factor introduces duplication of space-time by motion and the other factor represents contraction by reducing of space-time. The duplication is presenting space a precise or as near a precise replica of the space that was available the previous instant and it is the reproducing that science thus far never acknowledged.

Without giving the recognition that the Coanda principle is due, science in general science will never rise from the slump of the Middle Ages and science may even fall back into obscurity, even with the brilliance of persons such as Max Planck. The way gravity is formed and the way the Coanda affect not only presents gravity but also becomes gravity will be lost. It is how gravity produced the atom. Some solids that were also protons became liquids, which today we call the neutron.

The higher the spin of the solid (or the liquid) is, the more gravity there will be. The more gravity by motion there is the more space-time is secured in solidness. A spinning wheel secures liquid water onto a solid tyre to the extent that a car runs on an inch of water at say one hundred and what... kilometres per hour. The gravity that forms around the spinning wheel, secures more solid water than what ice can provide in density. One cannot drive a motorcar on one-inch thick ice but one can drive the motion car on one inch of water that is secured by the motion of the spinning wheel. Take this phenomena back to the atom and one would see how the atom formed before the Big Bang became in place.

The whole process of gravity comes to the surface when Kepler's version of gravity is used. The neutron providing the spinning motion manifests the motion T2. Since the neutron produce the space for the proton to have, the proton and the neutron establish a symbiotic relation of equality. The motion of the proton uses space to dismiss while the motion of the neutron uses space to duplicate the space in which the neutron is as well as the space in which the proton is dismissing the space in which the neutron is. To secure a balance or a favouring of any one of the positions, the electron ads or removes it's favouring of placing a dynamic to the atoms form.

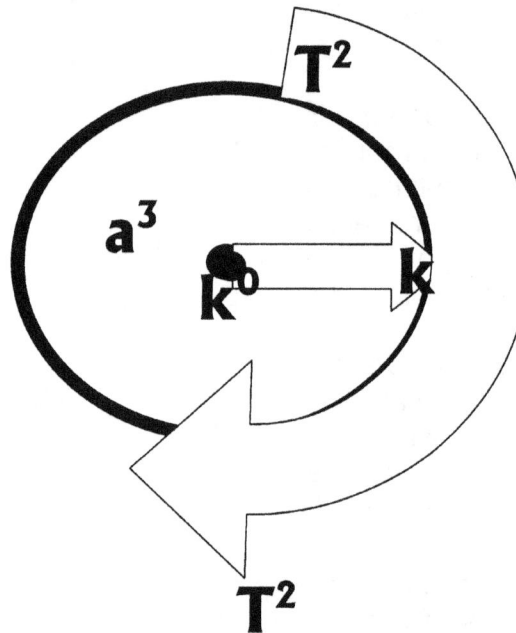

Having that we find that all stars go through a developing that ranges from being all about liquid heat to becoming liquid material as the neutron is and ending as space solidness by dismissing all that the neutron may provide in duplicating space-time. In this we trace the beginnings before the Big Bang.

From the centre of the dot there is a top and a bottom spot. From those points there is connection with four quarters. That produces six connecting points that are all aligning to the centre.

Motion is the time it takes to reform space from being three sided in half the Universe to a six sided six dimensional Universe extending from singularity by seven which includes singularity and claims another three parts in singularity under the motion of space. Such motion is time since the duplication of space in motion produces the time the motion takes to confirm the space then formed.

Space a^3 is both cyclic and linear in one unit $T^2 k$ by time

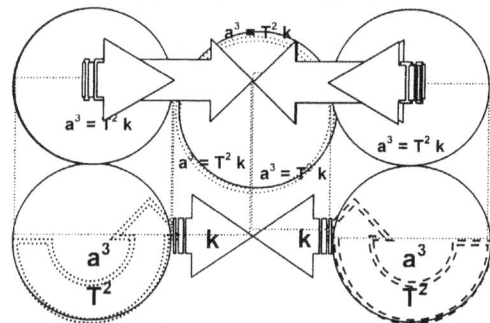

There are seven points in the sphere as Kepler's formula indicates.

$$(a^3 = kT^2) = (a^3/kT^2 = 1) = (k^0 = a^3/kT^2)$$

$$k^0 = k\, T^2/\, a^3 \ .$$

Since any value to the power of zero is representative of singularity it is from singularity that the first line of material that did not dissolve. It therefore remained in direct contact with singularity. In the position singularity holds gravity allows space to disappear thus presenting all material with the sphere as form. Singularity connects to seven points representing singularity in the sphere. Connecting to the seven is the space going into motion a^3 or the space coming from motion $k T^2$ but since that space is coming about through motion only three sides apply . The total is then the seven being in and being part of the ten, which include the non included three.

In this singularity idea there can be no sides and without sides there can be no drawing showing the explanation by means of illustrations. However, I do that to try and bring some understanding across from an unexplainable concept. Where I do just that, I ask for your forgiveness because being human means I have no capable means of performing an explanation. Yet I am forced to do just that and I have to allow the transfer of what can only be in my mind formed by the way of sketches, well knowing the implication that such act is not allowed. Let us again go to the beginning of creation.

The Coanda principle indicate that the gravity described in the previous page is generated by motion of liquid in relation to a solid anywhere motion can produce gravity. There is no mention of mass because mass is a derogative of the gravity which the motion creates. A centre is formed where the surrounding space-time forming the one group is relating a position from the "centre point". That forms one inclusive relevancy between points within the gravity field. The gravity field is holding "back" and "front" running through "the centre" where the other line is relating from "side" to " side" running through the "centre point". The fact of the line in the centre is that "it is there", but we cannot see it. Try as you may, no one will be able to calculate the very position that forms the lines, but as they change all particle characteristics, the lines are a reality as the spin of the matter is real. Being too small to hold atoms, the space holding such a centre line is no space at all and with that knowledge we may presume then therefore what ever the line constitutes of must become part of singularity, where singularity is a spot in the centre with two lines crossing the spot at an angle of 90^0. That is the basis of singularity, and since all the positions still relate to a centre of a circle, forming a part of a spinning circle, Π must form the basic value. The second major reality that one has to recognise is that the only way singularity was broken was by motion. The only way motion can come about and break space less ness is by establishing heat which establishes expansion and the

Universe became a possibility and later a reality by expansion. The heat swell into space and the space swelling is the motion that produces the gravity we find visible in the Coanda principle. The space at first was presumably filled with material because the expanding could only be material. The Coanda principle alters time and establishes with such alterations to space-time a new Universe with borders and all. By introducing motion it sets a new time standard by which the space created will apply a newly generated gravity.

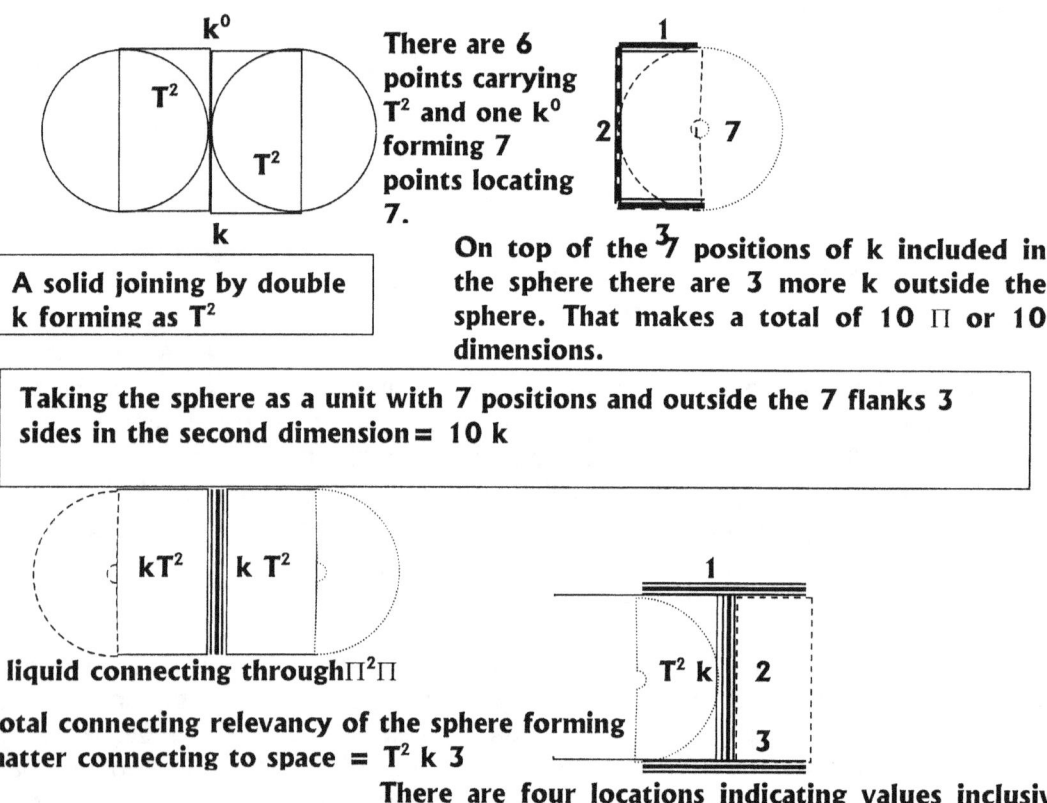

k^0

T^2

T^2

k

A solid joining by double k forming as T^2

There are 6 points carrying T^2 and one k^0 forming 7 points locating 7.

1

2   7

On top of the 7 positions of k included in the sphere there are 3 more k outside the sphere. That makes a total of 10 Π or 10 dimensions.

Taking the sphere as a unit with 7 positions and outside the 7 flanks 3 sides in the second dimension= 10 k

kT^2 k T^2

A liquid connecting through Π²Π

Total connecting relevancy of the sphere forming matter connecting to space = T^2 k 3

1

T^2 k   2

3

There are four locations indicating values inclusive of the space of the atom. That is the base of gravity in the atom and in the atom is the base of all gravity.

There where singularity grew to proportions that singularity without the help of space-time can once again control space-time by eternal power. The formula $t = \sqrt{1 - (C^2 - V^2)}$ is an appreciation of how little concept there is in Mainstream science about space-time. Time can never be single because to be single time then has to fall back into singularity. Time is motion and when motion ends, there space ends that the motion brings to time.

One such a relevancy is the sphere.

The sphere has six edges relating to one another at all times

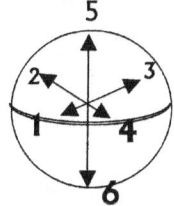

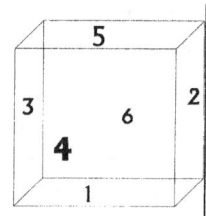

The cube has six sides in three pairs relating to one another at all times

Then connecting the six sides is a centre from where the control comes about that places these edges at specific related points and the points in return puts the centre at the precise centre

The cube has six flat sides loosely connected at corners where the corners prove even weaker connecting points than the flat sides convey support to the structure as a whole.

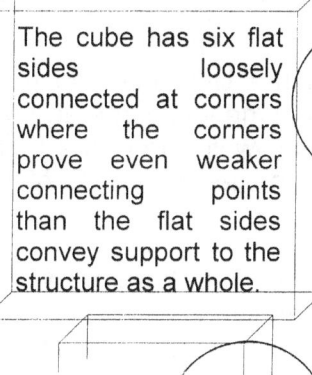

6 - 1 = 5

7

At all times the sphere has six precisely controlled edges connected by a supporting centre that is in such a position the six plus one in the centre is in immediate support of any or all of the points at any time. When touching one point the point reserves the strength to its disposal that is given by all seven points, which are backed by the entire structure. Try beat that for form strength and that is why a sphere is the ultimate form that provides structural strength.

Where the sphere makes contact with the cube the sphere loses one dimension to the sphere. Because of the absolute domination the sphere has in form and in control coming from a centre the sphere removes one of the six sides the cube has leaving the cube with five sides in relation to the seven the sphere has. That is another factor that gravity shows. This explanation also concerns the Lagrangian form.

The matter in motion between eternity beginning and eternity to follow and in that will hold a time related position of eternal motion until eternity ends motion and cannot be standing still in an eternal position while having space not moving. The lines and the centre spot hold an eternal position as the position never changes. Therefore the circling matter is either in time coming, or time going, but can never stand still. That is the reason why time will forever relate to a square of coming from as much as going too, but never becoming one. Being one time then must be standing still relating to singularity in the single dimension of $t = \sqrt{1 - (C^2 - V^2)}$. Time in a relevancy of t will be eternal one, and all forms of what ever matter can or will take on where that then is holding space forming time can never enter that line or dot. When time stands still space comes to an end.

In that manner we know that that was the way particles formed combinations just after the arriving of moment-Alfa. Singularity brought the Universe but also singularity brought the divisions between the many Universes that followed the immeasurable many Universes that came after the flooding of Universes to follow the leaders. At this point mathematics renders it useless. Every slightest point in space became an opportunity of establishing a Universe with most different functions and ingredients there might form. This is apparent from the fact that it still takes place at the present moment by motion attaching new singularity through duplication and through duplication release previously attached singularity from serving the purpose of duplicating by motion.

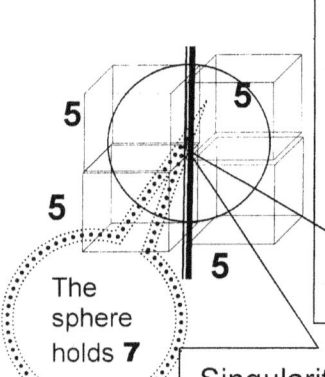

5   **5**

5

5

The sphere holds **7**

Where the sphere and the cube meets there is this most robust form with seven supporting the one point that is in contact with the cube and this point will always completely dominate whatever the cube is resisting. That is not where it stops. The time factor that supports the sphere claims four quadrants in total, which means the sphere holding seven relates to five times four totalling seven in all.

Singularity holds the eternal one or Θ^0 or Π^0 or whatever value there is in the idea one wishes to attach to the notion of the concept of the original.

Then one tenth less, which is one fraction of the square of space is the other side of the Universe and there singularity is a value of one minus one fraction of the square of space making the value one and one singularity measure (-.1-.01-.001-.0001- .00001) deducted from another one which that one being deducted from sits as singularity being part of the same singularity but on the other side of the world…and on the other side of the Universe.

Remember there is no Universe towards the outside because there is no outside but image. The Universe runs towards the inside or away from the inside and every singularity is an individual Universe, only one Universe away from the next Universe. Since the Universe starts and ends in infinity and that ends all definitive values of big and small, which is merely human appreciation of what cannot be. It is a relevance of what came when and that is all. Everything past singularity is space created time driven temporarily substituted by the unreal. There is, was and will never be one fixed solid Universe one can touch and smell, but the Universe is timely created space by motion of duplication in time delay. Once motion stops, time stands still and space falls into a black, Black Hole of eternal space less motionless reality where all the created concepts of space and time are contained in reality of eternity. That also is not religion but it is physics. Time can only stand still in the Black Hole of empty space.

There was the spot that became lots of dots. The dot had no borders therefore there was no separation and still we know there were more than one in a group of one. When Π^0 moved to form Π the evidence of this move is very present in the cosmos at present and one can find such evidence all around us.

The overall picture resulted in a ring or circle due to the release of from motion by all parts and all rings hold Π to secure the form. The only form that existed then was Π and therefore even today the borders use Π to indicate positions. But in the single dimension such definitions were far from clear and the only distinctions came from securing singularity in preserving the position of singularity to apply gravity and thereby absorb all anti-gravity. But anti-gravity could not control expansion by counter acting contraction through gravity so the overheating continued forming non-existing borders. The borders appeared in some material that was infinitely solid just as Einstein predicted because this took place before light came about and therefore before the speed of light. That which we refer to at this point even pre-dates light and therefore light at that point was excluded as being part of the cosmos. The cosmos formed a partnership with one side overheating forming antigravity by expanding into space through the applying of the overheating. In the relevance which the Universe is all about there is another side and the other side formed gravity or contracting of the expanding space.

This says it all and yet every person with a position of influence in science is missing all there is to see in Cosmology! Greatness in Cosmological terms is not in size, but the measure goes by intensity of density and lack of space. A smaller (a^3) results in a larger T^2 where (a^3) is the space the object holds and T^2 is the sizable gravity the cosmic object has. The suggestion confirms Kepler and disagrees with Newton. According to Newton, the ultra gigantic Red Giant Betelguese should be formidable when applying gravity whereas we all realise that the Black hole is the true undisputed giant! The red giant is sloshing around like the bowl of liquid heat-soup it really is while the Black hole unleashes gravity to the point where it devours even the smallest photons in the largest waves thereof that we can imagine. By taking the diameter, as the means to measure is clearly no solution in a method to calculate the gravity of any given star because it solves not one thing. They go on to even circumvent this failing.

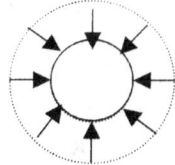

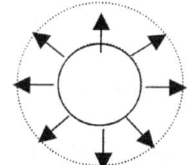

Gravity is about reducing space

Expanding is all about heating. Heating takes up more space and gravity reduces space.

The Academics change their approach by applying the usual radius r forming the square in the use of the formula in those formulas, which Newtonians devised to measure gravity. However, instead of having the radius holding the square as is done in normal mathematics when calculating the gravity, those practising astro physics then get really mathematical. Instead of squaring the radius they bring in the speed of light in such a place within the mathematical equation and put the C under the dividing line in the formula next to the radius. By them using the C to indicate the speed of light as C they place the C in the formula to bolster the radius value and that diminishes the size the star has so many fold. By bringing in C they reduce the star because the radius then gets bigger by the multiplying of the speed of light. The star suddenly gets reduced by the factor that the speed of light produces and then they go on disrupting the truth even further by applying the square that should fall on the r in the normal calculations to the C that indicates the speed of light. The star then reduces by the square of the speed of light while the radius, which should carry the square value suddenly remains in the single. That is supposedly their way to put a measure there that somehow has the means to bolster the gravity. The speed of light is the worst or best form of antigravity depending on which way one looks at matters.

Light is the strongest or the most intense antigravity there can be and to throw that into the Black hole by the square to hide the insufficiency of the methods applied to calculate gravity is once again another cover up to hide the Newtonian ideas, which are not functioning in cosmology. By producing C^2 in an attempt to bolster the gravity figure they supposedly are able to calculate in the gravity of a Black Hole and placing C^2 in conjunction with r symbolising the radius. This blatant further method of corrupting the corrupted is just the way not to improve the incorrectness, which their theory quite deliberately brings about as a measure to determine gravity. Inadvertently they did confirm that gravity is a speed ratio, but their confirmation of such a fact past them by without their noticing it. We can take the proof of my statement about the incorrectness of using the radius to measure gravity one step back and see how the method applies to the Neutron star or the Pulsar. Then what about the measuring of the gravity in the Neutron star by using the diameter and how do they explain the Neutron star, in principle. The Neutron star will either be stronger in gravity than the Black hole when also using the C^2 method, which is clearly not the case or it will be pathetically weak when not using the C method. If the Black hole has a diameter of 10 kilometres then the Neutron star has a diameter of twelve kilometres. By using C^2 in determining the gravity in the Neutron star the Neutron star suddenly have a larger gravity than the Black hole has or instead is much weaker in gravity than the moon is. It is either that it is stronger than the Black Hole or the neutron star is so weak it has less gravity than a comet.

This ridiculous scenario developing just shows how little grasp mathematicians have about cosmology. Yet they and those are the persons denouncing my thinking and me. They (the mathematicians) should keep to building dams or skyscrapers where the use of mathematics is useful and is appreciated and stay out of cosmology. Nevertheless, I know that for this remark they will get back at me as they usually do. It is precisely because what we can learn from the incorrectness of mathematics that we can determine that mass is the result of gravity and gravity is not the result of mass. Gravity that forms restricted motion as gravity applies in and becomes mass due to the gravitational restriction but mass certainly does not bring about gravity by some magical intervention as it then is applying as a pulling force. Gravity brings about mass but mass does not produce gravity as Science wishes to advocate. For instance, there can be no mass factor in a galactica that is generating gravity because all galactica never shows any singularity restriction. Gravity creates mass but mass does not establish gravity. $k=a^3/T^2$; Mass comes about by the reducing of **k** in the case where the Roche limit has been bridged and singularity reels in the dominated, or claimed space-time. Any object that can slip past the Roche limit value of $\Pi^2/4$ times the diameter of the star, the star would reduce to something no more than mere heat and treat the reduced objections as space-time. Any object that can slip past the first barrier or safe guard then becomes the dominated and is a $k^0/k = T^2/a^3$. When reading the formula, such reading translates the formula to objects turning the equilibrium around by becoming time-space $k^{-1} = T^2/a^3$ because the time factor at that point has to reduce the space factor since the duplication of the space within the time set does not match. The object lost its independence for one reason only and that is that it could not match the gravity speed set by the dominating singularity because the lesser structure could not produce the independent gravity to secure sufficient heat in its core to drive the lesser object. The distance depends on the position that the orbiter developed space-time $a^3 = k\, T^2$; The space depends on the distance the space developed from the centre and the speed the space moves around the centre. $T^2 = a^3/k$; the speed the space orbits around the centre depends on the distance of development and the size into which the space developed. In this view our standing on Earth makes us part of the Earth space-time. The space-time established by the Earth governing singularity has put a claim on us to dominate the independence that makes us as much captured properly as we are part of the electing atoms that elect the governing singularity in place. Singularity running negative or singularity declining motion places the incoming structure under the governing singularity control by diminishing the space value through enforcing the time aspect when the relevant **k** reduces to incorporate all the space-time the Earth will grant. That is described as $k^0/k = T^2/a^3$. What this verbally means is that since the speed we have representing all those on Earth and which we all then inherit as being part of the unit the Earth, we secure through motion, we all contribute to a decline of direction compared to distance or space we occupy and that what is also declining. The factor **k** presents in relation to singularity k^0. As all independent atoms each have an independent singularity, which stands related to the elected singularity, the independent to elected **k** is in declining growth and this is because our motion T^2 lacks space and thereby space a^3 declines through the deteriorating of **k**. If by some fluke we gather heat and we increase our velocity the formula which is time-space $k/k^0 = a^3/T^2$ we take on then may turn about as we will just start flying.

In the light of all this above-mentioned facts Mainstream science still promotes the idea that gravity pulls material closer and even more so in bigger stars. But that gravity pulling can only come from the accumulative effort of every individual atom according to proton mass (number) as a unifying effort of all the atoms in the star in accordance with mass applied. The idea is that mass is the same everywhere and is never changing. Why would there be such huge mass increases in the bigger or should I say smaller stars. What would entice the material inside such stars to grow more massive if mass comes about from the pulling of one particle closer to the next particle. If it was about pulling on each other the mass of the particles could not increase through such pulling. Even by combining the mass of two individual atoms the increase is already in the equation. By locking the two into a unit, should not change the mathematics when calculating the total mass because $1+1 = 2$ whether the two share one unit or two units. There cannot be any mass increase in a star because all material is within the unit the star holds. By fusing atoms the star cannot become more massive using that specific method since it gains no further mass. The two hydrogen atoms was there all the while and then adding one oxygen atom can at best be equal to if not less massive than one neon atom. Though by using this principle the mass of the star cannot grow because the star does not produce mass or work to further the mass it has. The star reduces the mass, which is the thinking of Mainstream Science and if we stick to the accepted views that was formulated almost half a millennium ago, we find that those in Mainstream Science hold the opinion that the star diminishes the mass it has by burning mass as does a coal stove in the old days. Today we all find we have progressed to the era where we are fortunate and blessed with electric stoves that do not reduce mass to allow a fire to consume the mass the stove gas as was the case with stoves in the Newtonian era. Unfortunately the principles behind the process of driving stars and new stoves using electricity has not yet reached cosmology where the Mainstream Scientists find their theories and therefore stars in cosmology still consume material to operate just as coal stoves did in the time so many centuries in the past when modern cosmology theories was brought to book. By fusion protons will only join without further raising the mass they have apart or combined. If the particle has a mass as two units, and the units join in volumetric occupation, they still have the same mass. We can see that some facts about Newtonian science are too astonishing to be real!

The above facts are part of Accepted Science and accepted facts, but my theory about gravity being a result of dismissing space to the advancement of compacting matter is not accepted through all my trying to introduce my ideas to Accepted Science. Where space reduces all space reduce. Occupied space might reduce less in relation to unoccupied space, because the atomic individuality has a resistance factor of being forty times more than the gravity applying. One hundred pounds in mass will be equal to a mass of one ton in the sun. This does not come about because the mass grew more but it is the result of the overall space being much reduced and as a result it is having a bigger attack on the atomic individuality. The forty times difference there is between the atomic mass fighting for independence and the gravity attacking such independence grew much closer. Therefore the space reduced somewhat and that includes the space granted to material. One cubic meter being one ton on Earth will hold ten thousand tons of material in a star one class more developed than that which the sun is. In more developed stars the figures rise above human comprehension. But it is so clear that the space diminishes as the mass becomes denser. By compacting matter the space reduces in the same process. I have been trying for years to get any professor to admit to that by producing the rest of my theory but with little success. Our understanding of any star development must lead to our understanding the Big Bang, because a star is the reverse of what the Big Bang was.

The heat available during the Big Bang explains the lack of space at the time. Space is heat expanded and heat is space contracted. The one is the reverse of the other. To expand is to bring about excess heat and to contract is to produce gravity by eliminating space. The Coanda effect backs my argument. Where gravity is applying the strongest the heat is the most ancient rated and therefore the available space is the least.

As the space develops from the Big Bang this view lends itself to favour space. Seen from the developing one must also see the space that the star reduced in size while space in outer space then remained static. It is not only outer space shifting but by compressing (for the lack of a better word) the star is reducing and the reducing is accumulating heat material from which the star material is growing its singularity. Therefore the star froze space as much as outer space grew beyond the size the star captured while one is losing what the other is gaining. All three concepts are correct in the final analysis. It is a matter where one wishes to place the distinction that brings about any concept. That too is one reason we never were able to understand gravity because gravity never favours any one side and therefore do not produce the type of relevancies we humans need to understand. Gravity is the motion producing differences and is dishing out even-handedness.

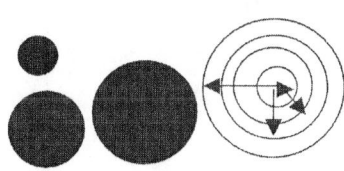

Looking at the affect of gravity it shows the precise quality of no distinctive point, as gravity never seems to end at a point but flows all over affecting all that holds a position in its sphere of influence. The gravity coming from China meets the gravity coming from America at no particular spot but intermingles without distinction.

But we humans can only understand by applying distinctions.

In the sketch above the circle to the right would come about from a straight line r growing influencing the appreciation of Π, but to influence Π would lead to a breakdown in r as Π and r are different entities. The circles to the left shows a continuous growth by extending Π every time a cycle of π^2 completes and since Π is the same part as what the previous Π was, only extending that immeasurable of an infinite quantity which is many times smaller than what we can understand each time, the circle will truly continue without any signs of a break. In the context of dimensions one find coming from the centre $Π^0$ an established eternal flanking of Π to six positions since $Π^0$ forms the centre to the six sides and all six sides are not having a diameter yet must apply Π to indicate specific value. What I try to say in this elaborated effort is that where **k** extended for the very first time when creation started from Π^0 to the edge of 3D where Π begins it had a certain distance. Such a distance is much beyond our understanding nevertheless we may never ignore the distance forming every time. We humans are incapable of ever measuring that extending growth but be as it may, it is there. Such expanding is beyond human thought or comprehension but it still had enough value to raise a Universe. By allowing this expanding of **k** personified as Π to continue uninterrupted as one flow with a continuing growth of Π being a line of that might be ever so small, but as it is still productive it will follow one line upon another line producing a cover of the full area of a^3 by forming the time factor $\mathbf{T^2}$ or for that matter gravity Π^2. Gravity π^2 will be flowing constantly through out $\mathbf{a^3}$. The factor $\mathbf{a^3}$ will be improvised by the singularity measure of Π^3 and the time factors $\mathbf{T^2}$ and **k** will then be in singularity terms $\Pi^2\Pi$. Yet all is controlled by the dot from within the very centre, which I am referring to, where rotation must end or start space depending from what vantage point the relevance is placed.

Singularity cuts by dividing without contributing or participating in segregation. It divides without any form of favour that it shows in forming by favouring sides. That is singularity not having a dimension of space and not having a dimension of time, or a radius connecting the rotating distance to **k**. In the explaining earlier on about the matter of gravity it was apparent that **k** developed into $\mathbf{T^2}$ and $\mathbf{T^2}$ was serving a bigger space $\mathbf{a^3}$ as **k** interacted with the larger space $\mathbf{a^3}$ by intergrading $\mathbf{T^2}$. Singularity holds from the beginning the point where $\mathbf{T^2}$ breaks down and $\mathbf{a^3}$ stops to exist and has never relinquished such a point. It is a point, which we find in all-rotating objects. Every rotating object holds a centre from where the rest of the rotating direction will differ at any and all given points to all other rotating points. That is what provides all in the Universe with independence and with individuality. Not one point is exactly the same, but in the very middle, in the very centre no one can draw, measure or see is a point not in motion. Although the point is beyond detection by our looking at the spinning top, the point is - the point is there all the same.

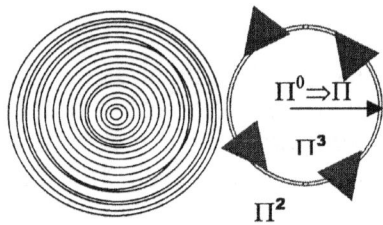

The very centre forms an eternal divide that will not allow what is on the one side to present an influence on the other side. It divides spin. It divides direction of spin. It divides all rotation from the outside that one may detect and such divide is there because at one point spin will run to the left coming from the right and just immediately next to that point must run a direction from left to right.

 π^0 in the centre runs an axis line that forms the division of rotation. No one human will ever be able to indicate the precise line, but such a line must exist because of our logic telling us about such a line. In the centre one will always find one more line smaller than the outside but forever also always bigger as it is towards the inside.   The law of Pythagoras is seen by science as a double square forming on two sides of a triangle and the combining of the two squares then becomes calculated as the root of the third side of the triangle. That is correct but again Mathematicians are satisfied with half the story and never seek the rest. The law of Pythagoras is also that what Kepler saw that sides relate to one another by line and by the opposing square. Singularity comes about first of all by Pythagoras where the centre line claims division. By the centre line holding the centre divide in the single dimension the square comes about from both ends of the singularity circle. Then the square of Π forming two points on both sides of the centre line is equal to the centre line becoming equal to the centre line and being in the square. In

that the centre line provides the square to which the half circle attaches. In locating singularity we must return to the spinning top.

Coming from a unit where time as motion divides the circle into equal sectors one unit will form two parts in the square holding singularity as an equal divide. Resulting from the rotation, space forms that positions in a centre, an elected governing singularity charged to secure a dominating and controlling centre... through which comes a point where the sharing of singularity of the entire unit positions the gravity in charge. By that, not one of the two particles can ever be in the space the other claim during the time of the occupying claim. Since it is sharing the mutual singularity that sharing will bring about an eternal bonding. Although such eternal bonding is in place, the unit tries to escape the limits it shares with the other half by claiming the space of the other half and never reaching the claimed space. In the centre forming a line **k** the line **k** is just a point referring to singularity in proton as well as particles we think of as neutrons. One must appreciate the fact that Pythagoras is mostly about a line **k** if one wishes to use a symbol forming a square, possibly T^2 that is symbolising circular motion. In relation to space, which uses another possible symbol, as an indicator in relation to the three sides of the triangle possible symbol used is a^3.

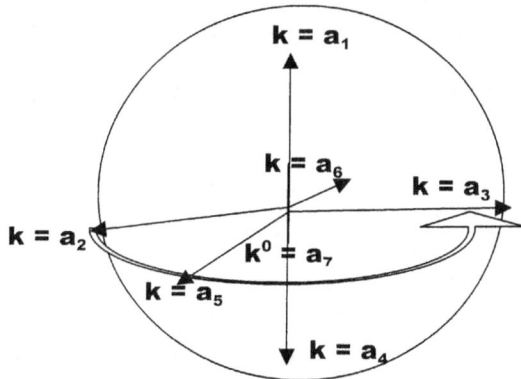

Kepler translated that from singularity k^0 came *the three sides by three sides* we receive by motion next to eternity. In outer space outside the sphere the relation is only three sides relating to three opposing sides being $a^3 = T^2k.$ In the sphere the space is secluded by another dimension enforcing a boundary from which the inner material cannot escape freely. The boundary of the sphere has also six sides where three opposes three other sides because singularity maintains very strict form there is one connecting point holding gravity in a part where space disappears. In the centre a seventh point forms as $k^0.$

That is where the Universe started and that is where mathematics started although the rest of mathematics formed some distance down the line. The Universe is mathematics because three dimensions prove equal to two dimensions being equal to one dimension. Although that personifies a sort of mathematics we can only accept but will never full heartedly understand. There are the three-sided triangle a^3 being 180^0 and in relation to the square T^2 being 180^0 where they then are equally related to form the line **k** also being 180^0. In the six sided Universe it does not apply but where dimensions originate (inside the atom core belong the protons) this is true and therefore space a^3 can go flat T^2 because the space then become equal to the square of time as it moves out the Universe by reducing **k** to the point where **k** becomes k^0. In the Universe that we occupy being six sided the method is revealed in order to duplicate space by motion.

The fact of form proves that the sphere captured all sides that can possibly influence the sphere. The sphere therefore holds $k^0 = a^3 / T^2k$ within the boundaries designated to the sphere. When a body is placed in a location on the outside of such spherical borders, that object seems to float in any direction. There is no control one can establish that will secure movement in any specific direction of preference except by releasing heat to counter act the required motion in a specific direction of choice. We all have seen what happens to any object that comes into the border area of a sphere. The object suddenly is motivated by motion to follow a specific designated direction and the motion leads the object to move towards the centre of the sphere. It is as if the support of the six opposing sides forming the cube has lost one side where the sphere makes contact and took over the control and movement starts in the direction of the Earth centre. The support of one side is literally removed by the centre of the Earth. In the centre of the Earth there is one point where space goes flat and that is the actual position that Einstein saw when Einstein claimed the strongest gravity is drawing space flat. The motion of the object also starts in that point and from such a point gets a direction. There is no pulling on the object but there is removing of space by the centre of that specific point leading the object and the space it is in as well as the space it carries to move to the centre spot. The centre spot can only become activated by the spinning motion of the space occupied by a rotating object that becomes independent from the surrounding space by applying a spinning motion. From that point the spin influences as it produces space-time that stretches the influence of gravity (space in motion) from singularity spinning all the way to where the borders of the sphere form. The spinning motion is activated by a centre that is activated by the accumulating effort of all the atoms within the structure, be it a sphere or whatever, that is in motion and driven by the centre singularity. But in such a very centre the revolving structure it self is motionless in one specific spot in relation to the space spinning around it.

However when the cube comes into contact with the gravity of the sphere the sphere extends that boundary and removes one of the six sides from the cube at any and at all particular points. By removing the space of one of the sides of the cube at a point where the cube and the sphere hold a sharing point and where the cube makes contact with the dimensions of the sphere, the sphere brings contact to that point coming from the sphere centre where we locate singularity, and by superior dimensional contact the object within that point will fall or descend or move towards the centre of the point where gravity is the strongest and space is the least. Gravity is the strongest where space disappears and that is in the centre of all structures holding a sphere as form and has motion and is therefore by motion granting a status of representing the central motionless singularity. That Einstein proved but his lack of studying Kepler as an individual Scientist withheld him from reading his very own mathematic translation accurately.

What Einstein saw was that space disappears. He, (Einstein), then jumped to the conclusion that the space he saw in his mathematical equations was outer space or the space falling outside the parameters of the material that occupied space, which is secluded by dimensional borders. In the sphere the borders the sphere holds are deliberate and very distinctly placed edges forming a specific distance from the centre. The centre is also proven beyond any debating. The centre of any sphere has to be at the very point where space completely falls away. That will put that space at that point in the single dimension and singularity is the single dimension.

 Without me trying to teach a Master and fall into the same trap as Newton did, I do have to make a suggestion as I see fit. Newton's observations are very correct except for one condition. It is space that should relate to space. It is the space captured by material that must be in relation to the space influenced by material. It is not space occupied versus space occupied, but space occupied reducing space unoccupied.

$k^0 = a^3 / T^2k$

$a^3 = T^2k$

It is space standing relevant to singularity in the formula $k^0 = a^3 / T^2k$ standing in opposing the other space. It is the occupied space going to the place where it will be in the position that is not in direct contact with singularity control. It is space-time $a^3 = T^2k$. The borders extend within the atom as well as beyond the atom. The proof of Pythagoras being part of Kepler is as obvious as creation itself. The factors of Kepler and Pythagoras verifying each other may not be that much part of natural mathematics but the cosmos is not that much part of natural mathematics either.

Mathematics has a place on Earth and some application in space travel. Take singularity for instance. Singularity was with science from a time before science was science and yet not one mathematician ever noticed singularity while little more could be that obvious. Understanding all I said about space-time being connected intimately is so very important and even more so all conditionally to the fact of accepting that all individual particles in the universe use motion and therefore spin.

The sphere holds six sides relevant to one centre point in relation to its form that forms a unit and all other forms hold six sides in relevance as does all other shapes and forms. All forms have to have at least six sides indicating different exposures to the Universe. Where gravity has a free choice, gravity always chooses the sphere. As I shall prove elsewhere, gravity is the strongest where the form produces the least evenly distributed space. The first condition for gravity is even-handedness throughout the sphere holding the applying gravity and the second is to have most or the strongest gravity located where the space is least. That is where gravity then has a position in the very centre of the sphere and from that centre the gravity produces all the edges or borders that the sphere consists of. In the case of the sphere this factor makes the sphere much more dominating than any other form does. From the centre point controlling all sides is gravity and with gravity applying control the sphere has seven sides to the square in any other possible form having at least six sides.

The rotation or spin ends on the border where k^0 becomes k and steps away from singularity into the very first point proving to form 3D as a factor of T^2 where time comes into place

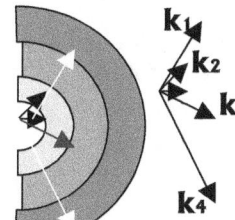

k_1
k_2
k_3
k_4

Every time k reduces (shortens) it changes. It is more than merely the mathematical consequences changing. It is the Universe changing and when entering such space-time, the newly relevant motion will insist on all aspects within such a Universe to change. It is either because the density of heat surrounding the concentration in the space and sharing with whatever is in charge of duplicating promotes duplication more easily with a lot more heat available or contains the time it takes to duplicate because of more heat available.

Or another scenario is that space in need of duplication is less. It effects the time considerably and it does it to the extent that we experience gravity in the manner we do on Earth because of that. That is the reason that the proton is 1836 times more massive while applying the same displacement that the electron does.

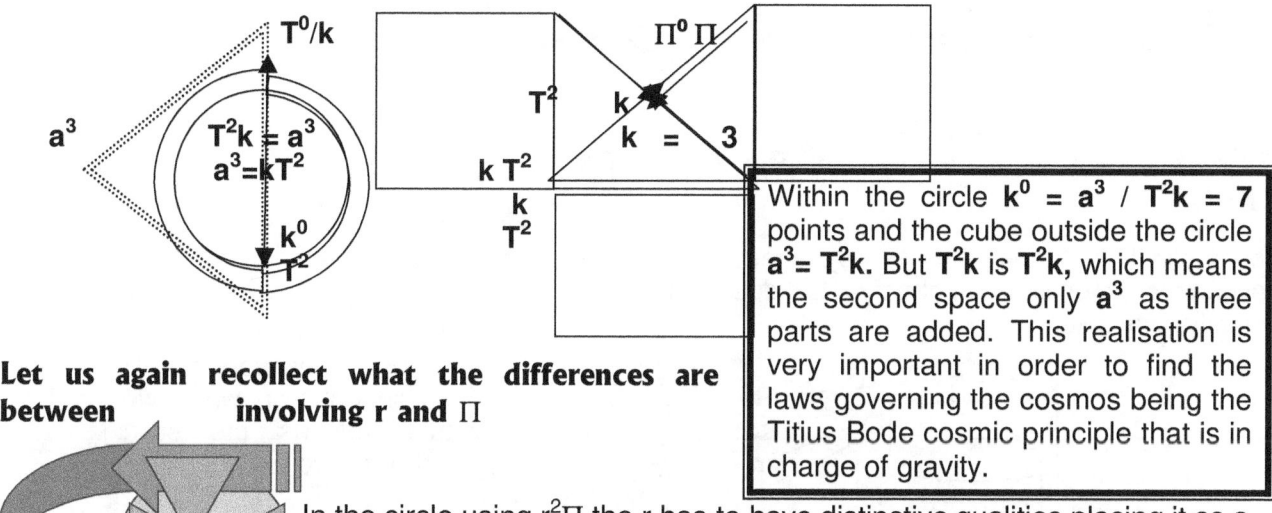

T^0/k $\Pi^0 \Pi$

a^3

$T^2k = a^3$
$a^3 = kT^2$

k^0

T^2

T^2 k
$k = 3$

$k T^2$
k
T^2

Within the circle $k^0 = a^3 / T^2k = 7$ points and the cube outside the circle $a^3 = T^2k$. But T^2k is T^2k, which means the second space only a^3 as three parts are added. This realisation is very important in order to find the laws governing the cosmos being the Titius Bode cosmic principle that is in charge of gravity.

Let us again recollect what the differences are between   involving r and Π

In the circle using $r^2\Pi$ the r has to have distinctive qualities placing it as a factor apart from Π. Where the growth shows no separate distinction but a continuous flow from the precise centre to the precise edge the flow would become in relation with Π depicting the circle and Π replacing r as reference to any point on the circle.

By using r as a distinction in the circle division is possible but by using Π there is no distinction possible making it a solid flow. Any object being in outer space floats and such floating is seemingly random with no specific detectable interfering favouring a movement in a particular direction. Such a devise is depending on influences not in our scope of detection. But then the object comes closer to the Earth and reaches on specific point where the six dimensions that influence the object in the cube suddenly changes. At one point one of the six dimensions fall away as it disappears and the object quite literally falls to the Earth. It changes a stance from floating to falling. It is the point where $k = a^3/T^2$ becomes $k^{-1} = T^2/a^3$. It is where the Universe swaps end. While the object remains in outer space the object is floating but that floating stops at one given point and then the falling of the floating object starts. The support of one side disappears and the centre point of the sphere takes over the control. At that point the object is under the influence of one centre point in the sphere where the sphere in this case is the Earth and we also know that in such a centre point one will always find the strongest or the controlling gravity.

In using the formula Kepler produced and finding gravity in the formula one will get a total of seven factors inside the sphere and three outside the sphere.

Since gravity also influence the space outside the sphere the space we call outer space has seven plus three points bringing about ten positions of gravity influencing space.

The influence inside the sphere also captures the space outside the sphere. The space outside the sphere is 10.

The reason why the electron receives a value of **3k** is very incorrect from our stance since the proton shows the three dimensions we find our Universe to be representing our space. Our side receives the values but do not apply the values.

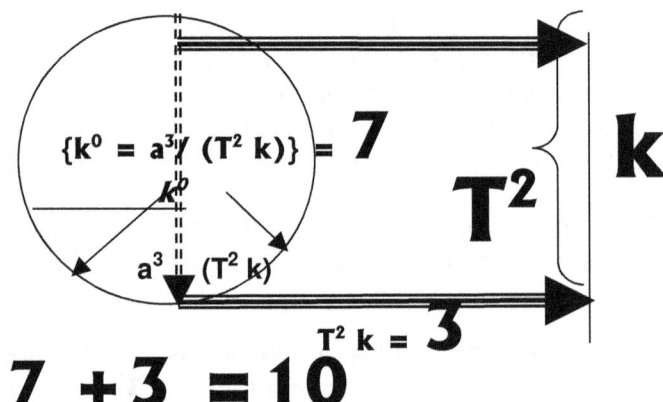

$$\{k^0 = a^3 / (T^2 k)\} = 7$$
$$T^2 k = 3$$
$$7 + 3 = 10$$

From singularity where there is either **k** being Π or Π being $Π^2$ the lines are just **k X 3**. That is why I prefer to use r as a symbol to represent as line because man needs many options and names but nature use one where it may be necessary otherwise it is a repeat of one.

Pythagoras shaping the cosmic beginnings

Totalling electron =3

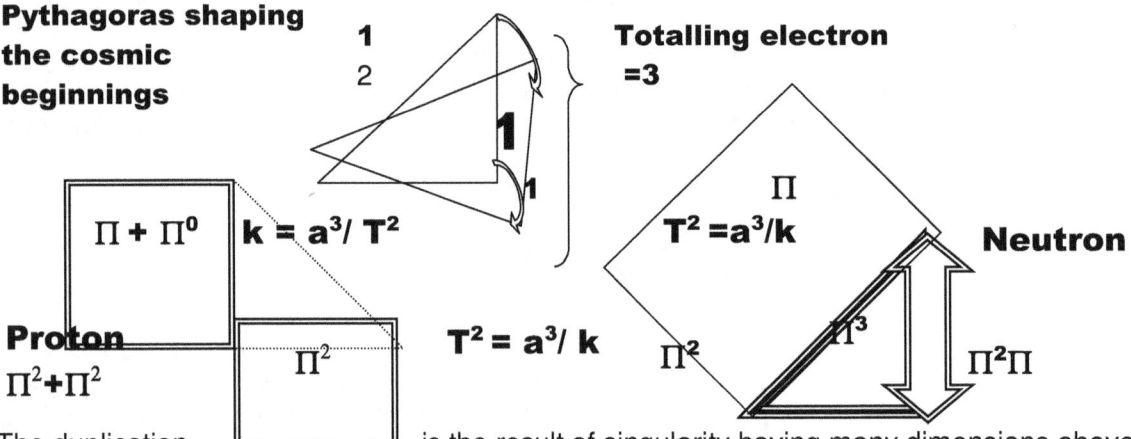

$$Π + Π^0$$
$$k = a^3/ T^2$$

Proton
$$Π^2 + Π^2$$

$$Π^2$$

$$T^2 = a^3/ k$$

$$T^2 = a^3/k$$
$$Π$$
$$Π^2$$
$$Π^3$$
Neutron
$$Π^2Π$$

The duplication [ ] is the result of singularity having many dimensions above and beyond the dimensions we observe. But due to that it now is possible for us to live in a (seemingly to us) solid Universe.

100 (10 X 10)
10
1
0.9   (1 − 1/10)
0.99   (1 − 1/100)

The dimensions running into the Universe stands in support of what we think of as the solid Universe we experience.

$$Π + Π = 2Π = Π^2.$$

The cube can have square corners; the corners are of less importance because it may come in whatever form there may be. In contrast the sphere adheres to precise measure and behind this principle is all that forms the Universe. The cube has six sides connected loosely and can change form just by changing the relevancy between one side (or more) in relation to the distance brought about by the other sides. The sphere being a complex circle stands related where the sides has to apply precise measure in equality. This becomes a law because in the precise middle one will find the strongest gravity as that gravity holds the object in form and true to form. If there is even gravity spread in all directions the form must be a sphere and the sphere insists on seven points relating to sides or borders.

1r   1Π

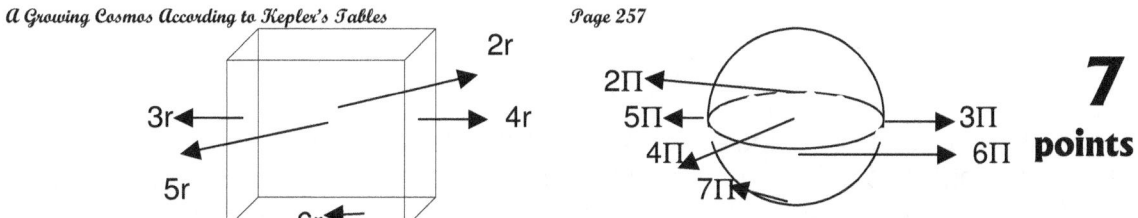

7 points

In the sphere, which I am referring to and that one find in cosmology there can be no radius but only the extending of **k** from the centre point k^0 that then with motion becomes T^2. From the centre k^0 in six opposing directions there are precise located points that is crossing a centre and is relating to one another by the square but the factor is remaining **k** because of the unity the matter holds in relating to space. The factor has to remain **k** since the motion takes on the square as T^2 and because of this implication there can be no radius in the square or to the third power. In the event of using r the individuality of r as a concept will bring about a specific ring at any specific point where every such a ring will define every time as one ongoing circle. Such a circle has no validity in the Universe since the circle indicates an absence of motion. In the cosmos all space has to have motion. Since gravity is motion of space the use of a radius that will indicate a square is principally flawed. One cannot use such a definitive line as r would be because such a line will have to cut through atoms at some points while running from the centre to the edge. We have to take such a line as **k** and that gravity is **k** that becomes T^2 and T^2 is securing another space a^3 boundary that a new T^2 becomes a new **k** in relation to another object influencing the space that will again become T^2. It is a never-ending, always continuing cycle of interactions. That is the effect one gets from gravity. Gravity includes and joins all aspects within the field but such a definitive line will have to exclude some and include some because of the definite points and rings developing. The gravity influence we are in search of is something like a woven cloth covering the area and covering all in the area. It is like a silk blanket covering all the aspects. That means the value **k** is running into and past the entire surface as if the surface is one consistency without different particles. That is what motivated me to look for gravity as being heat concentrated in space reduced, because everything is space, that consists of heat and if gravity is condensing heat, gravity then will be all including. Gravity is all-inclusive therefore it is not possible to draw a precise line that would form a precise ring when using r as an indication of gravity and not cut some atoms in parts. Because where r is used there will always be an atom disallowing the precise positioning of the circle. The circle continues on a solid basis holding a dot in all spots as a positional reference and not r. In every sphere there then are the seven dots relating in precise dimensional and positional equality forming equilibrium to the centre spot as well as to one another by 90^0 and 180^0 implicating the dimensional positioning. Therefore the sphere holds 7 spots with six being dots and the cube. This point then serves as $k^0 = a^3 / T^2$. The cube holds $6 \times r^2$. I use the symbol r to define the idea of a line whether the line might be used as a radius or to be indicating length breadth or width. I do that to avoid the pitfall of using names to hide truths. Where space comes into contact with the sphere the cube loses one of the six dimensions it has to the more dominating seven dimensional sphere whereby the seven dimensional in equilibrium will dominate the six dimensional cube that is loosely connected by r bringing about that the cube then has 5 sides to the seven of the cube. The space surrounding the sphere takes on the shape of the sphere and not the other way round. The sphere revolves and in accepting the form, the cube is the lesser form. The rotation of the sphere allows one to presume that the form of the sphere is the most dominant of the two choices that allows rotation.

In the centre of a sphere there is a definitive position where the strongest influence of gravity is located. It is the centre of the sphere where the space is the least. From that centre point gravity extends in keeping the edges of the sphere perfectly true to the form singularity has being Π in every aspect. Also such extending continues beyond the specific edges of the sphere where it influences the space surrounding the sphere. We know this by another name given to divert every one from realising that no one has a clue why that phenomenon is applying. We call it the Coanda effect where the process shows that even liquids submit to change of form.

The condition for such a submission is that singularity that is carrying k^0 is in place and committing the influenced space. Only by applying singularity through using **k** does the Coanda effect apply. With the sphere being defined by singularity to confirm singularity with the use of **k** as form the sphere is influential enough to remove one side of the square cube, which is, loosely connecting sides to form the cube. The cube is very loosely formed and the sphere is very differently controlled by singularity therefore the sphere is able to dominate the space in the cube by removing all the sides it is in contact with. With the removing of the side the cube in form looses one supporting dimension and therefore will not be able to secure what is in the sphere to the form of the sphere. That is gravity. It is reforming space to the requirements of singularity by reforming form of space through motion.

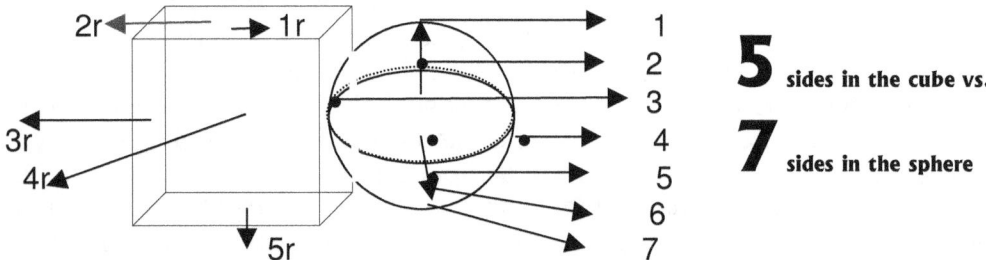

This means that in the cube at the point of contact between the cube and the sphere the cube experiences such a contact point as if the "bottom falls out" of the cube and without a "bottom" to support objects they fall to the sphere as objects fall to the earth. Remember that a body "floats" in space, but at one specific point it starts to "fall" to the earth. That is gravity and it is much more a dimension change than it being any force. Then at any given point the space outside the sphere holds five points where as within the sphere, there are six points relating to the centre spot of the sphere. I shall explain this last remark later on. That too is the Lagrangian system with five cosmic structures holding relevancy to the centre structure where the centre structure stands in for seven positions diverting from singularity and the orbiting structures standing in for five positions in space.

Gravity is all to do with dimensional changing and reforming of forms to re-affirm alliances supporting singularity. It is the reforming of space converting space to more concentrated heat. The Universe is in the three dimensions using twelve dimensions that is visible to us and indefinite number of stages in size differences ranging from the immeasurable small to the immeasurable large where mathematics become a short fall to the next and the previous dimension. There are always 10 in positions running smaller and running larger. Even to us thinking we are at the edge, or since may think we are in a centre that is because we use light as an information source. By our use of mental thought instead of light to obtain information we will find that the Universe are infinitely bigger than what we see and infinitely smaller than what we see.

Gravity is the dimensional change of space taking space from 10 to Π^2. This happens by means of applying the Titius Bode configuration of space adapting form through the seven dimensions interlinking ten dimensions to reform the concentration of the space to heat. In this manner gravity is "building" space by motion of space.

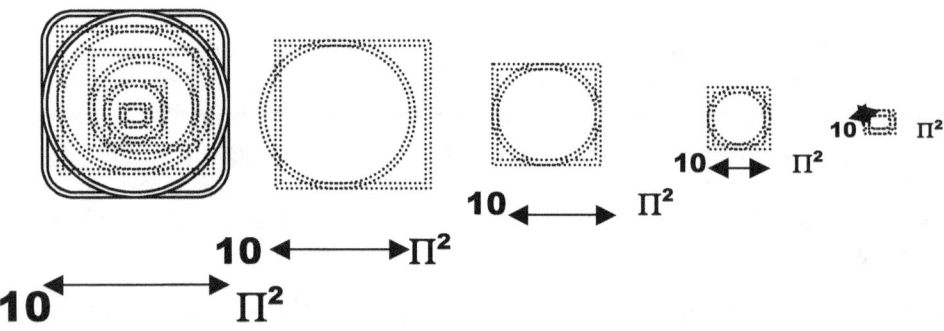

From the gravity "building" space through motion the motion of the building leaves an imprint which is detectable in the sequence the Titius Bode law saw in the number arrangement of 3; 6; 12; 24; 48; 96 etc. The incorrect application of the Titus Bode law lies in subtracting the figure of 3 from 10 leaving 7. The other way of reasoning is to add four each time to the first value of three starting with 3 and so on. One has to see the Titus Bode as two relevancies in the unit bringing across the building of space-time. The true significance of the Titius-Bode law is that it points directly to a circular growth of 7 in the sphere leaving the marks as it grew in stages. The 7 relating to 10 is a precise derogative of the Roche limit or the Roche limit is a precise derogative of the Titius Bode principle because the two systems interlink.

The principle is an indicator of what gravity is and of what produces gravity when material **(7)** moves in space **(3)** and through space **(10)** to leave the imprint of gravity applying on the space growing between the planets.

The sequence the Mathematicians found was undeniably correct. There is a defining relation between all the planets and even the fragmented one shows this tendency too. It is the way space relates with material but because it did not fit, Newtonian thought it was dismissed. One should see the Titus Bode law as a vibrating ripple effect leaving marks about those times when collecting heat manifested in material by growth every time it forms a cyclic pattern in the growing space.

If it applied only to one or maybe two planets I can imagine there might be some inclination to doubt the obvious but when all the planets show the same cyclic pattern in all of them such evidence just turns things very much against a Newtonian principal dismissing. Also if this was the only doubt that the cosmos cast against Newtonian principles then some denial was excusable. But the further studies go the more Newton seems to be wrong. At what point will science eventually come to their senses and look beyond the hundreds of years of student brainwashing.

There is a Universe parting Physics that apply on Earth to Cosmology. In cosmology one work with space-time $a^3=T^2k$, whereas on Earth one can work in a fixed and general space with a fixed and general time. However, in cosmology there is nothing fixed and there is nothing general except the four pillars Creation was built on. Using Newton, however one cannot even begin to explain any one of or the combined effort of the above cosmic phenomenon or the four pillars that are all over the cosmos and forms all the laws in the cosmos. Newtonian definition cannot even recognise any of the principles but only Newtonian science are taught to students. No student can have the fortune to disagree with Newton and remain a student. If the student will dare to disagree with Newton it is the end of such a students academic career. By setting this firm condition Newtonian science becomes an institutionalised mind conditioning of the concepts of thought forming in physics. With my saying this I have not made one academic friend but to the same degree did any one in the past proved me wrong. Students are taught to accept Newton and to ignore Kepler and any student doing it the other way around will fail all examinations and any other form of testing at Universities. Students accept Newton or they accept a ticket taking them home. Newton is an institution force fed to each following generation but saying that reserves only resentment towards me amongst Academics.

According to Newtonian science space is simply nothing with no qualities but gravity separate space and space does not mingle, as one would expect if space was nothing because space does form borders. Disasters of unprecedented magnitude arise from such borders. The Challenger disaster of February 2003 is much testimony to those borders that was powerful enough to break the aircraft into pieces while the explaining contributed by Mainstream Science is evidence of a shocking lack of understanding about what took place as cosmic laws were breached. Let us now start to timely investigate cosmic laws as presented by the cosmos.

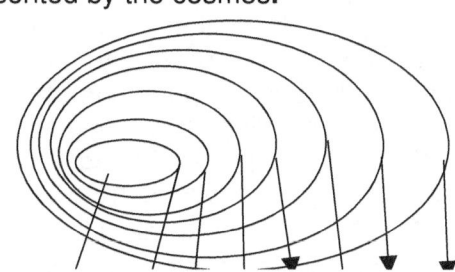

Planet	Mercury	Venus	Earth	Mars	Ceres	Jupiter	Saturn	Uranus
Bode's Law dist.	4	7	10	16	28	52	100	196
Actual dist.	3.9	7.2	10	15.2	28	52	95	192

The TITIUS BODE Principle Outside the sphere

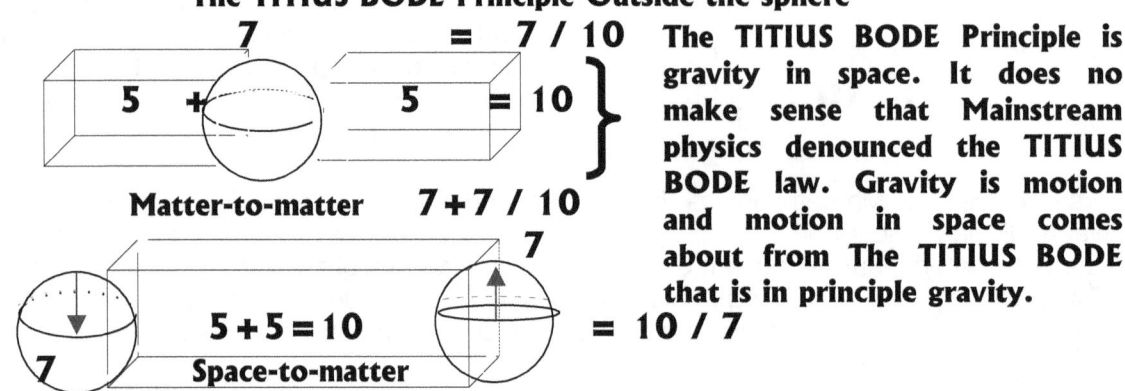

The TITIUS BODE Principle is gravity in space. It does no make sense that Mainstream physics denounced the TITIUS BODE law. Gravity is motion and motion in space comes about from The TITIUS BODE that is in principle gravity.

Investigating Kepler is also the understanding of Kepler and that no one did since Newton named gravity and defined what he saw gravity was but gravity is undeniably Kepler's discovery. There is a Universe inside the sphere and there is another Universe outside the sphere. One has to see it from the point singularity holds $k^0 = a^3/T^2 k$ and $a^3 = T_2 k$. Both represent a view from singularity but the two are not the same and neither do they share a principle. That is the relevance bringing about the gravity Kepler discovered. What would Kepler discover that Newton did not discover. Newton named gravity but Newton also missed the chance to discover gravity because Kepler discovered gravity and with all the laws on motion that Newton introduced Newton failed to recognise the work of Kepler as that Kepler introduced gravity. Newton did not see what Kepler introduced as relevance between participating object in motion forming sequence. The gravity Newton saw and the gravity Kepler saw is not the same gravity but since Newton meddled in Kepler's work by changing Kepler's work Newton then being the master on motion should have seen that while we on Earth uphold motion in serving the time factor T^2, which produces the space such a space factor will compensate for this inconsequence in the formula $a^3 = T^2 k$. Kepler saw gravity forming a relation that becomes a square when the two bodies interact between the different positions the bodies hold in space that is relevant to the motion each structure performs. That too was what Titius and Bode both observed. The Titius Bode and the Roche limit is only a part of the complete gravity process where gravity forms with motion placing relevancies. Mass does not bring about gravity but gravity produces mass by performing as a resistance to motion. Saying that we better find a means to distinguish the concept we have about more or less particles per unit forming more or less mass, and mass forming the gravity contracting resistance that the unit upholds. In our case we are descending or falling which gives us a negative distance growth. Normally the Kepler's formula read that $a^3 = T^2 k$ therefore $k = a^3/T^2$. This is normal where the planet is orbiting the sun in a regular pattern and does not bring about a decline to the factor **k**.

In the situation we are in and being captured by space-time control we have **1/k** because we are within a reducing or growing smaller than the **k** the Earth introduces, so we submit to the **k** of the Earth. Reduce **k** as it happens in the gravity Newton presented then k^{-1} would produce a smaller space a^3 if the time T^2 component of such motion relevancy is forced to be equal to both notwithstanding what differences there are in space a^3. Newton had available at the time the findings of Galileo and it was Galileo that showed by using a pendulum, that the swinging pendulum arm will reduce the space to compensate for the establishing of the time. By compromising space a^3 the pendulum can manage to uphold the time T^2 component because the pendulum is visual proof of $a^3 = T^2 k$. In our case the **k** attached to our position in space is in decline or negative which is the formula taken in another relevancy of $k^{-1} = T^2/a^3$. That is precisely what is happening. We are reducing the space between the Earth centre and us (our singularity) because we are travelling too slowly in relation to the Earth. Again Kepler showed himself being correct and not only that but he proves his brilliance.

Newton saw in Kepler's work what all other people ever since Newton see, that gave Kepler's work the smallest attention and Newton was the only one that gave Kepler's work the smallest degree of attention. Kepler calculated the movement of the orbiting structure around a centre sun. Newton did for starters not interpret this rotation as gravity. Secondly he saw what clearly is space, as motion. Therefore Newton was unable to see that Kepler said space in motion is gravity. I challenge who ever is of such an opinion that Newton made an in depth study of the integrate meaning of the work of Kepler, to prove that. Newton made a serious observation about the work of Kepler. Newton saw a circle forming by a planet that is in rotation, running around the sun. To correct this Newton added $4\Pi^2$ on one side and on the other side he introduced his version of what he saw gravity to be. Newton as a person failed to see in Kepler's study any gravity applying. If Newton did recognise any gravity aspects then he, Newton would have concluded that forming another interpretation by adding is very unnecessary. Newton did not act incorrectly when he added $G\,(m + m_p) = 4\,\pi^2$ but he merely duplicated what he then neutralised as a fact. He brought in his vision he had about gravity and that the two structures in motion is combining a time related value of $4\,\pi^2$. He duplicated what Kepler said and by duplicating as well as neutralising the duplicating he brought misconception and he covered information, which I then uncovered. Through my uncovering, I came to the conclusion that what Newton said what Kepler saw is not what the cosmos told Kepler.

The cosmos told Kepler of two aspects coming in place and forming a relevancy about two axes which is related by a space-time unit $\mathbf{a^3 = kT^2}$.

The understanding of this aspect is most crucial and all the condemning of my work in the past, was a result from bias formed with my criticizing Newton. I feel I cannot overstate this condemning of my view just plainly because of favouritism about Newton. When a circle of cosmic proportions complete a rotation. The circle does not neutralise the combining effort by forming a rotation. Much rather is the circle the result of the square $\mathbf{T^2}$ forming and with that reducing the extending line to one factor. Kepler did not find two particles in space pulling at each other and bringing on a force $F = G\,(M.\,m)\,/\,r^2$. By placing this factor on the other side of the changed concept duplicated what is a common fact but in doing that it neutralised all concepts forming about Kepler's research.

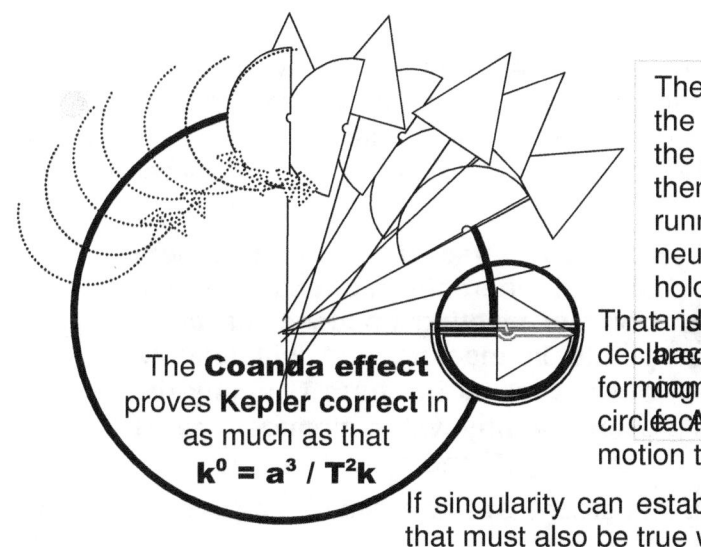

The **Coanda effect** proves **Kepler correct** in as much as that

$$k^0 = a^3 / T^2k$$

The first axis is the one every person knows. It is the straight line forming running from the sun to the planet and that has a symbol value of **k**. But then the is a second axes value coming about running in contrast to the first and that first line neutralising the second line forms another factor holding the square value. The planet is departing

That is stretching k very opposite what is Newton declared when planet around reducing circle motion forming angerabout therefore being one traised as the circle factor circle came about from forming a double motion that is equal and opposing.

If singularity can establish space by creating motion the reverse of that must also be true where motion by space through another space entices singularity. That statement the Coanda effect proves

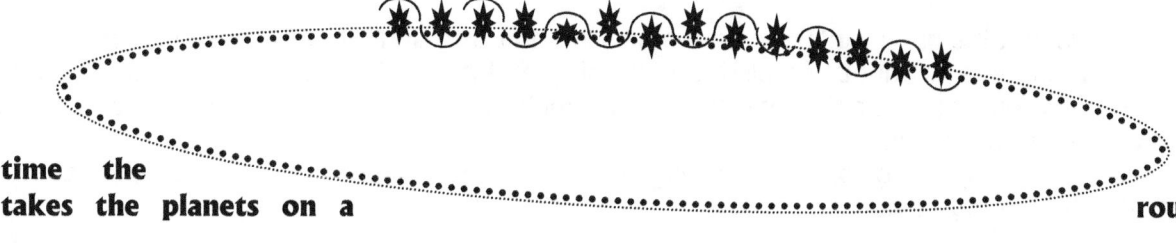

Every

time   the   **sun**

takes   the   planets   on   a   **route march**

around the Milky Way they travel through space-time and that does not come to nothing, since every aspect in the cosmos re-align totally in one rotation and in every rotation. What was never will be repeated in precise detail and what is in the past changes the future but never repeats the past. That can never add to zero accomplished because that is space-time. It is space in travel. It is gravity. After every rotation everything in rotation grew by the Hubble parameter. How on Earth can that constitute to nothing...only a Newtonian will make sense of such double talk

Kepler already introduced the square in the time factor, but the square does not divide the gravity, it is the result of the combining of such relevancies doubling. As humans, it is very human to look at everything with a single-minded point of view. We have to see that where an object holds seven from one singularity, there is another singularity, which is concerned with 10. There is always a doubling of one where one in singularity holds seven, but the second part of singularity holds 10 whereby it is unified by relevancy. The doubling of the affect of the distance between the two cosmic structures is a result of the combining of the motions applied by both sectors bringing about the Titius Bode configuration, but it is the result of changing perspectives from both sides and that brings about gravity as a interacting of seven over ten and ten over seven on both sides of the relevancy ends.

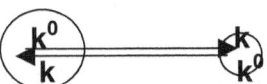

| The duplication of the relation **k** has forming gravity from one point to the other point is the gravity coming about. |

This confirms that T^2 is gravity and is Π^2. The circle having four interacting relations with each other forms $4\Pi^2$ and by Newton's claim that the planets circling the sun is $G(m + m_p) = 4\Pi^2$ leaves no doubt that my argument about gravity and Kepler is correct.

SPACE = Π^2 = 9.8696 = Space and time in a dimensional implication.

The space to the outside of the gravity ring holds a dimensional relevancy of 10. The space to the inside holds a value of Π^2 in relation to the top. In one case the seven stands to ten through motion of reducing by spin and the reducing of ten then becomes Π^2 as the space reduces. It is an ongoing process carrying form from as wide as one wishes to take it down to as small as it gets when entering singularity. The seven and the ten has very intermingling relevancies of relations swinging from top to bottom and from bottom to top.

The Titius Bode law is an extending dynamic deriving from the law of the gravity dimensional factor where the space factor in a square of ten relates to a matter factor in the square by half (half since nothing can be in two places in the universe simultaneously) of the matter factor of Π^{7+7} or the square of space (10) relate to the matter factor of 7. The Titius Bode is a relevancy of space reducing by implication of the square going to a circle.

In every sector the directional flow will provide a distinct meeting of Π linking r to Π^2 and this allow the time component in the rotation.

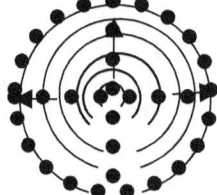

As the meeting of r points to a very distinct different r in direction such a point of meeting opposes the other points in meeting and will lead to destruction of the form Π in any the event of any value changes by Π changing Π^2 and r.

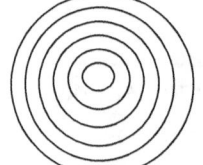

Keeping these factors in mind it is clear that Π^2 are the choice of gravity and not r^2.

What this says is that the gravity influence does not end on Earth and although specific borders are in place in the atmosphere at various levels the beginning and the end of such borders are in place in the atmosphere where the beginning and the end of such borders are as definite but also as vague as the gravity that forms them.

If we go back to the spinning top might it might reveal some factors about motion that activates singularity that is forming borders by motion.

Every quarter provides a distinct value that indicates the progress of the flow of time from the one point Π to the next point Π.

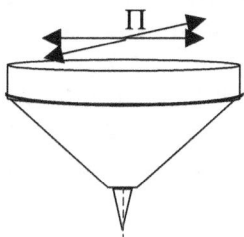

Any changes occurring in Π will lead to an unequal triangle providing two different values to r and will alternate the link between r and Π^2 bringing about different form (Π) and time (Π^2). When singularity forming the lines of the triangle is not in equilibrium the triangle will destroy the matching of half circle.

It is motion that brings about 3D as much as it is motion bringing about gravity Π^2

Gravity comes as a result of interaction between spaces maintaining singularity of different independence. Space is created by motion thereof, forming dimensions which is involving Kepler's view totally. The material uses seven contact points to singularity and the material including space the material influences through the motion, is filling space a^3 or release of space filled a^3 and holds ten positions.

The figuration found in the planetary layout and has taken the name of the Bode or the Titius Bode system, such a system is the mould that remains in space as part of space after the concept establishes the relation and as such is then detected by observation and growth leaving the mould in space as the departing of planets from a space concept in relation with the centre. It is a manifestation of the evidence what Kepler brought from his investigation of the cosmos. Gravity is where space vanishes and as such space is in the centre of the sphere. The sphere controls not only the space included in the sphere but also the space excluded or on the outside of the sphere.

In the sphere or the including part of the sphere, there are seven points $k^0 = a^3/T^2k$, or $k^0 = k^{0=3}/k^{0+2}k^{0+1} = {}^7$ in relation with one another at all times, which totals seven all included. Then there are three more in space and three more at the same time coming from motion $a^3 = T^2k$, which then also is $k^{0+3} = k^{0+2}k^{0+1=6/2=3}$. Adding the centre or sphere, the total is worth seven and any one side of the Universe you wish the sphere would move to a total of seven factors or dimensions fill with three more when the space is in motion filling the relevancy on one side with seven dimension characters and on the other side with ten dimensional relevancies. This role changes to a mirror image on the other side of the Universe where ten relates to seven.

If gravity is motion T^2 the process may sound exceptionally silly and simple for Masters such as the Academics of distinction fighting to create fusion and Space Whirls and lots of other very impressive goodies but it is very complicated to repeat. By motion T^2 space comes about, forming a^3. But the motion T^2 will mean a crossing to the other side of the universe since singularity divide the Universe into sectors.

Material produces the dismissing or the concentration of space by applying the motion. Surrounding all elements are a layer we call the atmosphere and even Pluto and the moon must have the atmosphere because they have gravity. In this the relevancy of ten to seven forms this layer and it results in forming a circle, because the combining of the motion duplicates the singularity factor Π, forming from that gravity as Π^2

THE PROCESS PARTED USING THE ROCHE PRINCIPLE

By establishing motion and creating motion, singularity quadruples to 4Π in rotation. But since the rotation is motion duplicating the space established as four times the value of singularity the motion divide the space coming about by halving such space by the dimension, which is putting a square root over the quadrupling of space. In this comes about the direction gravity takes the universe. The expansion is always double the square root but the square root is neutralising the expansion and that brings about that the neutralising of the expansion creates a contraction that seems dominant to us but it is not. The contraction is doubling the expansion by halving the effort of the expansion.

Space duplicates as space moves. The motion that is becoming another duplication is the result of singularity being unable to move. Since singularity is eternally motionless singularity cannot shift in a straight line. The only way singularity may achieve a motion is by re-directing the continuous duplication of space by re-alignment. That is what the Coanda effect proves. The singularity is unmovable since it is motionless but it does shift position as we see in the manner that the top spins. Our position we have on the Earth does not allow the spinning top complete freedom in which the top can spin. To the spinning top there is no other options to be but to be within the Earth atmosphere using the Earth solidness as security which provide the stability for the top to perform but to use the space the top claims from the Earth for rotating purposes. The duplication by employing a singularity allows the rotating or the continuing of time to be flowing in a fluid like manner but we also know that there are other functions involved. When a train and a fly go in a direct head on collision, it is obvious that at some point both objects must come to a standstill.

This is because there is a re-arrangement of particles layout at structure re-arranging, especially as far as the fly's interest in the event is concerned. To find such re-establishing of material positioning there had to be some stopping and applying different motion to different particles to create the re-arranging of the structure of the fly's body.

We know that when two objects go in a head on collision between say a train and a fly at the moment of impact the fly stops the train as much as the train stops the stops the fly. Only in the very next motion does the rearranging of construction begin. This is because singularity has to demolish space-time reposition the centre forming singularity that every atom will reapply the next position they will hold to in relation of the motion that determines the previous and the following position of singularity relevancy.

After installing singularity at the next position the singularity then re-institutes the space-time in relation to what the space-time was just before. Only if there is an interruption by another singularity of more dominance in proportional displacement such singularity then will it stop the singularity on route to fill that position in the direction where it is heading. As that position is then already filled will it force the atom to place the singularity of lesser displacing qualities to seek another position to occupy in relation to all the surrounding singularity it forms a relation with. This all is the result of singularity being immovable the singularity must attach to a new focus point whereto the focus of the atom having the singularity will shift. In that it is not the shifting of the singularity because such shifting is impossible.

It is the motion of the space-time generating another alliance with the same singularity (singularity everywhere is the same singularity) but finding in motion of space a new point, which the motion by moving generates as it excites singularity in the specific. The Coanda shows only rotating will find a new point to excite by motion of centralising space ands generating singularity from which s the motion generates space. That is why there can only be space-time anywhere. The motion or the momentum it has and that momentum continuously creates the Universe or creating of space is the value of time. After focussing a new point to install singularity in motion the space-time has to duplicate in relation to the new position that came about by the motion bringing new points on which to focus. One may never exclude the thought that it is one singularity having many points where each holds an individual Universe.

In the event where the point has been filled and cannot find a vacant spot in time the singularity will not match the time component and will turn the space-time it then cannot duplicate into heat, which will then form a crack or a tear or show some form of material deformation. By duplicating space-time in relation to a new allocation the singularity had to diminish the space to time being motion then shift the point of focus and then reinstall the space from motion into the new found position that may or may not bring new relevancies about where every atom indirectly form a new alliance with every singularity in the Universe since all the singularity maintain a relevance but all remain the very same.

That is the same action we detect in the Pulsar where the space breaks down, relocates and then re-establishes and floods out into outer space. But the pulsar is only getting in the single atom stage while us observing the Pulsar find our position is one other Universe away from the pulsar. We can never reach the pulsar as much as the pulsar can never influence us directly. The controlling singularity remains to the most specific sequence there can be and the duplicating of singularity that is applying motion by producing space to move or time in releasing space in order to use time to reinvent space from inside the proton that is inside the atom. It is like letting a strobe light flash $((1836)^3)^2$ times per second and watch the content of the Earth move in that strobe light.

Because of the fact that the motion will introduce a new space coming about such a space has to be within limits. The maximum limits or minimum limits depending on where you place the relevancy will be half the value of singularity dividing singularity as much as duplicating singularity. Gravity is about factor holding singularity and placing relevancies by putting relevancies in place. The motion creates new space $a^3 = T^2k$, but the motion duplicates the space created as well the motion producing space $T^2k = a^3$ where the singularity stands overall related to singularity changing into space and into motion $k^0 = T^2k = a^3$. The duplicating of space by dividing of space reduces space by half after it quadrupled on both sides of the divide.

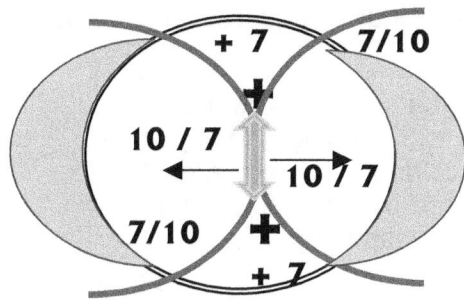

Gravity is motion in space and motion through space. Gravity is a relevancy between space travelling and the time it takes the space to travel. Gravity produces or reduces space during a certain period of synchronized spin of material in motion. Gravity is $a^3 = k\ T^2$ where it then becomes $k = a^3 / T^2$ In the light of this all other explaining fails the test of accuracy. It is no force because mass depends on gravity and gravity does not depend on mass.

The Titius Bode principle is a relation where space is the ten factors and material is the seven factors. By space being diminished by material, one relation comes about and where material dismisses space, another relation of seven to ten comes about. Gravity is the motion where space conservation is applied by motion control. By applying motion in sequence to space movement Universal harmony is installed. By applying motion at a faster pace than space conserve motion, the motion that is also gravity, controls the space by motion duplicating space at an even rate. In that event gravity is applied in the manner we recognise the working of gravity and space.

This is not possible to have any strobe with such sequence I know, but it serves as an indication of what applies because if it was possible we would have been able to see photons move about where as now we can see the photons flicker, while in reality the motion is so solid while flickering it sustains the Universe as a solid.   In all of this we find gravity as the controlling factor holding the lot together as follows:

Matter in relation (part of) to the total dimension of space.
(10 / 7) \ (7/ 10) = 2.04
1.4285 / 0.7 = 2.04
Taking from both orbiting influences

SPACE DIVIDED INTO TIME
(7/10) / (10/7) = 0.49
.7 / 1.4285 = 0.49 Taking from both orbiting influences

SPACE MULTIPLIED WITH TIME
7/10 / 7/10 = 1 and   10 / 7 X 7/10 =1 Therefore not influencing change
THE PROCESS PARTED USING THE ROCHE PRINCIPLE

10 / 7	$(\Pi/2)^2$ **The Roche influence on Titius Bode**	
7/10	$2.04 \times (\Pi/2)^2$ =	5.033
$(\Pi/2)^2$	$2.04 \times (\Pi/2)^2$ =	5.033
10 / 7	5.033 +5.033 =	10.066 **from both objects**

7/10

7/10 / 10 / 7= 0.49 + 0.49   **on both sides of the divide**
0.49 X 2 =.98

10 / 7

$$\frac{10/7}{7/10} = .49 \qquad \frac{10/7}{7/10} = .49$$

.49   + .49   = .98
.98 X 10.066 = 9.8 =Π^2 TIME SPACE = Π^2 = 9.869 TIME

Gravity is in place when space is in motion producing a duplication of lesser space than time will form a sequence. In the other scenario overheating or space expansion or antigravity will come about when gravity cannot sustain space duplication in the preventing of duplicating to bring about reduction. With space remaining at an even duplication without adding heat space will become lesser dense and such thinning will also increase space as antigravity, which is the same thing as overheating, and space growth comes about. The motion secures, prevents or supports space in motion by forming harmony in frequency to the motion, which is in truth duplicating space. One side of space is the duplicating of the other side because time is eternal at singularity. It is the double motion applying that performs the gravity between the objects in rotation. But the way Kepler introduced it diverts drastically from the way Newton introduced gravity. The moving away in relation to the coming towards forms the square that we see as the rotation. The rotation does not bring about an accomplishment of zero as Newton suggested but it brings about a square, which Kepler introduced.

On the inside there are the seven markers of which singularity is the focus point in the centre of the centre. The markers are representing one aspect of space, which for argument's sake let us call it cold. Then there are three more markers on either side being part of the space but not captured in the space. It is space in motion by the influence of the motion of the Earth.

That is, is space ....ferent relation with heat. We also realise that space is heat and heat is space. In differentiating we then have to call one cold, which we know is incorrect and the other hot, which also by the same margin is as incorrect as the previous name given. There is not hot or cold. Yet for the sake of being humanly inferior we are forced to be incorrect in order to assert our incorrectness. The one limit has to be the furthest position away from the next and opposing limit.

Let us refer to this as cold. Where electricity is heat and is hot there then has to be cold to limit the heat. There is no possible manner to separate hot from cold and yet gravity does just that by concentrating heat into cold space. The unseen cold and therefore unnamed, unrecognised and unappreciated space, which forms the basis of the space, has the ability to move independent from the hot part, which is heat in space. We cannot separate the two forms of space and yet with the aid of gravity the Earth turns the shorter **k** which represents a lesser (volumetric) space by gravity in reproducing a smaller area where the very end limit that was in place one moment before the Big Bang, are to be a bitter cold. On Earth we named the coldest regions the Artic Pole and the Ant Arctic Pole since we have nothing better to call it. Our way of reasoning the difference there is between the hot equator and the cold poles has to be the sun shining on the equator and not shining on the poles. The argument is that it is the sun that is shining at such and such an angle that makes the Earth middle sector hot and as the sun cannot reach the poles and that, not reaching the arctic regions, makes the pole regions cold. What a lot of baloney that lot is because the sun is far too big to have the Earth's roundness restrain the heat distribution of the sun from the centre the top and the bottom of the Earth. In another book I explain the dynamics of space-time generating heat This is all grand but that is beside the point and belongs to another book, nevertheless I will explain some aspects very quickly

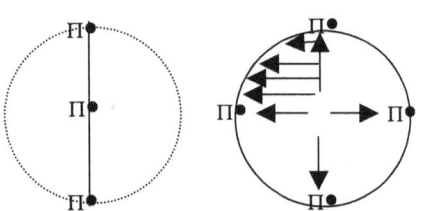

Points seen from the side form a line that never moves and is a line in singularity holding three points

Space-time is allocated in progress from the line singularity offer. That space-time consists of heat and can therefore store more heat, which strands in contrast to the cold than singularity presents. By providing more space-time at the equator there is more space-time to allocate to heat being there

There is a lot more to say on this issue and I do just that in other books. With the size that the sun has it will fry the Earth right through the very core the Earth has. The discrepancy we find in the equator regions compared to the poles is all about reproducing space as cold and reproducing heat in cold, which at the poles are much less to bring about through motion. If it was all about centrifugal forces flinging mass and thereby bulging the Earth as I have read some Academics announce is the case, we should also have all the massive particles such as the iron, copper, lead and so on swinging at the centre on the outside of the Earth core. It is about duplicating the cold, which is space without much heat, and by turning that it requires a fluid such as that which heat is to be within the space required to turn in. The seven can never totally separate from the ten, but by singularity being the same but being on the other side it is withdrawing space-time altogether. See it as seven (let us think of that as the cold basis of space) spinning or turning in the ten (which then will represent the hot part in the cold basis) and the ten is part of the seven but the seven is not part of the ten. The third factor is the axis around which hot as well as cold will turn. Therefore when reading the next page please envisage a cold base turning in a hot and cold space. The purpose of this is not to define whether the argument is correct or not but it is to help **the reader** gain understanding of the process principles involved. But motion also converts space to relate to space by changing relevancies through motion matter is in relation (part of) to the total dimension of space but is not the total dimension of space.

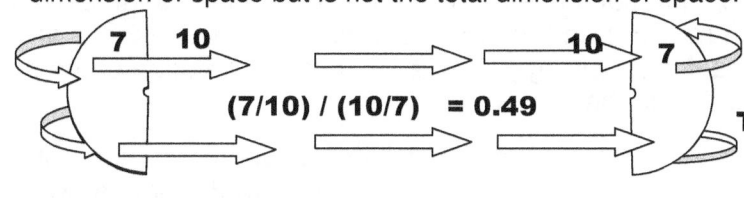

$(10 / 7) \backslash (7/ 10) = 2.04$
$1.4285 / 0.7 = 2.04$
Taking from both orbiting influences

$(7/10) / (10/7) = 0.49$

SPACE DIVIDED INTO TIME
$(7/10) / (10/7) = 0.49$
$.7 / 1.4285 = 0.49$ Taking from both orbiting influences

THE PROCESS PARTED USING THE ROCHE PRINCIPLE

$10 / 7$	$(\Pi/2)^2$ **The Roche influence on Titius Bode**	**Crossing the**
$7/10$	$2.04 \times (\Pi/2)^2 = 5.033$	**singularity divide and**
$(\Pi/2)^2$	$2.04 \times (\Pi/2)^2 = \underline{5.033}$	**activating the Roche**
$10 / 7$	$5.033 + 5.033 = 10.066$ from both objects	**principal $(\Pi^2/4)$**

SPACE MULTIPLIED WITH TIME

The crossing of the divide will implicate singularity on both sides of the divide bringing about the Roche factor
$10 / 7 (\Pi/2)^2$ The Roche influence on Titius Bode $7/10$ $2.04 \times (\Pi/2)^2$

$7/10 / 7/10 = 1$ and $10 / 7 \times 7/10 = 1$

Those factors being equal and therefore equal to one is not influencing change. The space that the motion establishes creates a relevancy of seven factors in space while the direction of motion involves another three dimensions or points, which is incorporating the other singularity in the unit. While the motion is at the same time moving out of ten points in relevancy and only occupying seven points the very opposite comes about through the same action being duplicated. The motion turns ten points to seven by moving from ten and filling seven points through the motion ending before the next cycle starts.

SPACE DIVIDES INTO TIME

On the one side of the Universe
$7/10 /10 / 7 = 0.49$
on the other side of the Universe
$7/10 /10 / 7 = 0.49$
And on both sides of the Universe
$.49 + .49 = .98$
$(10 / 7) \backslash (7/ 10) = 2.04$
$.98 \times 10.066 = 9.8 = \Pi^2$

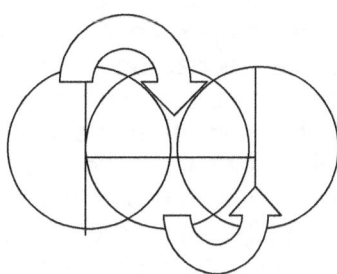

The value science use for gravity is 9.81, which they measured to much detail but the moon has a lot to do with the recorded difference coming about.

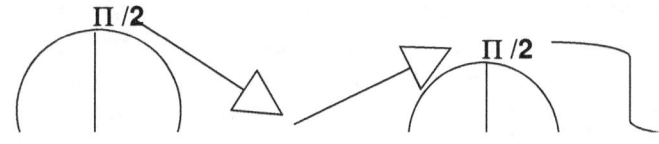

$.98 \times 10.066 = 9.8 = \Pi^2$

TIME SPACE $= \Pi^2 = 9.8606$

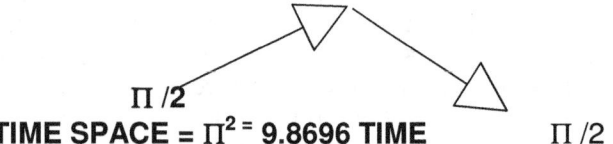

TIME SPACE = Π^2 = 9.8696 TIME $\Pi\,/2$

This is the prime element; the state where everything started. It started at the time when mathematics was still to be invented by nature and only singularity had a value of Π and a reference of 180^0 in all directions in relations to other positions singularity established. It is at this point where everything other than **singularity was $\Pi^0\Pi$ becoming Π going on to be Π^2 through motion forming Π^3.** One must take into account that gravity is different motion of particles forming a relevancy about duplicating space and dismissing space in relation to the effort of the particular and specific element. The motion differences in the motion between two particles bring about relevancies. It is a seven factor standing in relation in motion to a ten factor of which the seven then is included and part of the ten factor. The four time factors applying gravity as time is on the edge of the sphere in relation to the centre line forming singularity in the sphere.

Time involves a slightly different aspect than the explanation I give above. When the sphere is spinning there is a line running from the top where we place the pole North to the bottom where we place the pole South connecting the top, the centre and the bottom. The three points in this line does not spin but is as motionless as singularity always is. This line represents singularity from where the Universe formed the first dots and the dots formed a dotted line. It still is singularity to the value of Π^0. Just next to this line we find Π and alongside Π we find Π and from this we find space-time. Since singularity does not move, space-time is generated and the generating or establishing of space is the breaking down of space and rebuilding space to a new alliance with another sector of another Universe.

Space is totally destroyed and the rebuilt by duplication but the rebuilding is done with changes in direction as to always consider the control singularity has on space-time. This rotating is done in form and is done while considering relevancies in relation to a unit that forms about singularity. As much as singularity is breaking and building the form in the circle forming the edge of the sphere in space-time, by the same measure is the centre singularity charging motion within the ranks of the orbiting singularity. Establishing a line holding singularity in the centre, singularity is as much a task of controlling space-time as it is charging the sphere space-time with motion. From the centre the rotating body holding singularity is as much space-time as the rest of space-time from the line in singularity forming in the centre the centre singularity is charging by changing motion. Yet the orbiting singularity in reality is as much the same part of the Universe as is the centre singularity and both maintain the Universe by being the Universe each to its own Universe. By not having motion the line confirms its part in the original Universe before the Universe was the Universe.

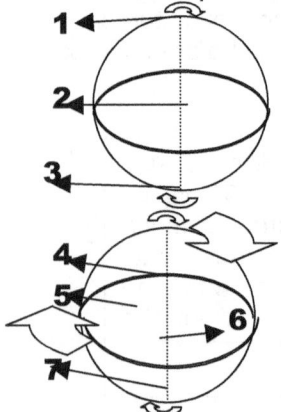

By not having motion the lines also have no space as the space extends to form space forms space and the line includes serving the three points to the outside. Where there is no motion, there is no space and where there is little motion there is little space. The only space the line may relate to can be a point that is on the border of the sphere that is crossing singularity and connecting the two edges on either side of the sphere that is forming the sphere. That means the line from one point holding singularity to another point holding singularity that line will cross the centre line which gives the line in singularity valid space-time to control. Singularity does not have the ability of motion therefore singularity does not hold space. Singularity is also eternally indifferent to motion and motion can excite singularity but singularity cannot be shifted by motion. Three points form the line.

It is not possible for these points to have motion or to produce spin since spin requires the refurnishing of singularity transfer or displacing space-time from one point holding singularity to another point holding singularity. Where space-time or space by motion is active there the motion carrying space has to be transferred from singularity without motion and space to singularity being without motion or space. As the space transfers from one locality to the next locality there some energy has to be transferred with the space that the motion generates to a new location and such relocation is going about with the transfer of space-time. However, the motion excites singularity but does not shift singularity. Time shifts singularity because it is the exciting of singularity to get singularity to hold space for a period that forms time. Time is the flow of heat through space in space.

Space-time is conducted from one point to the next point just like the flow of electricity is conducted. The conducting medium however is not similar. Should motion be required, there has to be a transfer of space-time from one location to the next location. This transfer is not that immediately instated but is a process that has to be developed. It takes time to break down space-time, relocate the position generating the space-time and then reinstate the energy that will produce space-time. The Newtonian view is that mass shifts but the truth is incomprehensibly more complicated than purely shifting mass.

There is a breakdown of singularity attachment and then there is the motion transfer of space-time followed by the rebuilding and re-instituting of space-time that the motion generates, just like the motion in the Coanda effect is responsible for generating the pulsing of electricity. Generating electricity is in fact a process of creating and re-establishing space-time but only at the level of C by creating gravity at the level of C. Every point holding space is singularity being charged with receiving as much as distributing space-time at points that receive or let go of the next space-time it is charged with. The space-time, the point is charged with becomes the space-time it is displacing the charging of. It is four points in receiving as much as the four points are in distributing

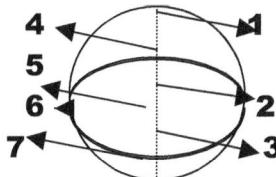

Every time motion takes place, the centre line holding 1,2 and 3 stands still. There is little generating going on and reducing to a point where there is no generating going on. On the outer edge of the rim however there are four points that do shift. The points shift from one location to the next location by generating space-time.

Those four points are the measure of conducting time by redistributing heat as material or in a state to receive material. We live in a Universe of time being from **10/7 (4(($\Pi^2 + \Pi^2$)** down to **7/ 10 (4(($\Pi^2 + \Pi^2$)**. The use of the four in front of the proton relevancy indicates the time aspect. Involved in the equation. Space-time involves the incorporating of motion in relation to singularity by distributing space from one sector to another sector. There is no wild theorising necessary of antimatter devouring matter when nature presents the most adequate explaining in the simplest form. But by using mathematics one can never reach the true origin of the Universe because such beginnings were even before mathematics. How do we know this? Because that part of mathematics are still with us and present in mathematics we currently use.

On the one-side space forms and in the forming process of space two relevancies come about. The seven departs in the direction putting the ten it is moving to in relation to the seven it is coming from. **7/10** Then by the same token and in a separate action which from our 3 D point of view is the same action the ten is losing the seven as the seven is departing from the ten. **10 /7**. All this is happening while the crossing is all concerning singularity moving from one sector of singularity to the other sector of singularity which is **(Π/2) X (Π/2)**.

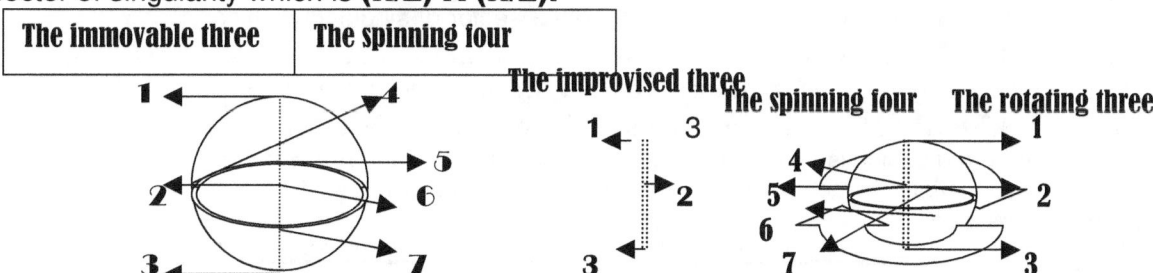

The relevancy forms part of the duplicating and dismissing displacement of space-time we call gravity. In that we are looking at relevancies and no precise specifics. However this is the manner in which the Universe was built block by block. As it was but is no longer only form that applies in the Universe but concrete measurements also come into play therefore even the relevancies may apply in different relations as they switch over to compensate for other factors alternating as they are coming into prominence. The lesser developing sphere orbits the dominating sphere and between them there are definitive relevancies. The centre circle singularity line of three is unaffected by spin which I shall call the immovable three. However the immovable three holds such a stout position as far as the centre sphere is concerned. In relation to the orbiting circle the centre line is part of the building and destructing process that manifests as duplication as the centre singularity maintain domination and control over the orbiting structure, but also it has a major part in the motion building of the sphere in orbit and moreover building by generating the singularity line that generates the lesser and the orbiting sphere. In that relation the centre sphere reflects the centre line to serve the orbiting sphere by supplying the reference needed to establish motion in the orbiting sphere as one Unit. In that there is an undisputable reference of seven orbiting the centre and the centre providing three as a reflection of the seven which in all accounts for ten relating to the four which also is spinning as time and in total forms the seven taken in relation from the orbiting ten.

Then there is the shift in space-time in the centre sphere component that holds four in orbit around the centre singularity line. As part of the motion is the singularity of the lesser-developed sphere, which is also part of the generated space-time in relation to the centre sphere because the singularity line came about at the same developing phase as the directly linked space-time of four points came about. In that it is the dominating sphere holding four orbiting and as well as the orbiting singularity that is related to the dominating sphere as merely space-time that carries three. Albeit that the orbiting singularity line is singularity, it is related to the dominating singularity line in the same manner as the same as the space-time in orbit forming the better developed sphere.

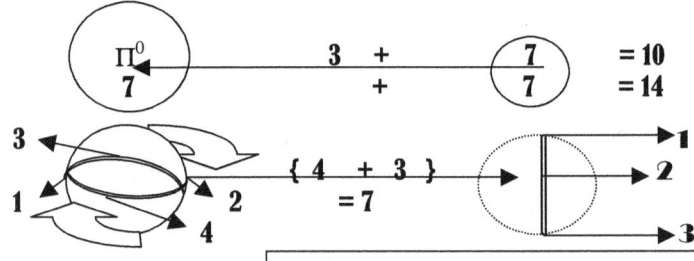

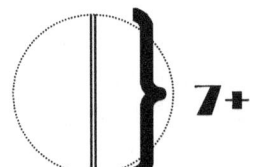

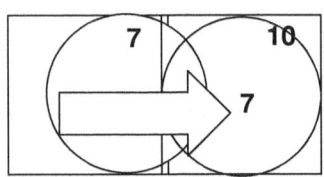

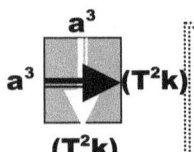

This forms 7 as a unit standing in security as singularity to the orbiting sphere. Within the orbiting sphere there is the line of three included that also forms a relevancy to singularity in relation to the space-time of the orbiting sphere.

This then puts the fourteen in relation to ten, which gives 1.4. From the other side there are the seven orbiting which includes the three reflected bringing the total to ten in orbit. That ten stands related to the centre sphere of seven and as it stands related it is a division of 10 / 7 = 1.42

7/10 (Π/2) X (Π/2) 10 /7

On the one-side space forms and in the forming process of space two relevancies come about. The seven departs in the direction putting the ten it is moving to in relation to the seven it is coming from. **7/10** Then by the same token and in a separate action which forms our 3 D point of view is the same action the ten is losing the seven as the seven is departing from the ten. **10/7**. All this is happening while the crossing is all concerning singularity moving from one sector of singularity to the other sector of singularity which is **(Π/2) X (Π/2)**.

Realising that singularity k^0 produces $a^3 / (T^2k)$ in Kepler's Universal findings of $k^0 = a^3 / (T^2k)$ then one can see why gravity applies a sphere when form of free choice is an option. This statement has very and many far-reaching implications in how our Universe came about as it chose forms when it developed.

$T^2 > a^3 /k$ **k** reduces Gravity contraction comes into play	$T^2 < a^3 /k$ **k** increases Antigravity brings about expansion and "growth" comes about

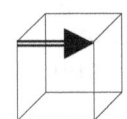 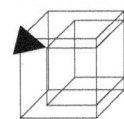 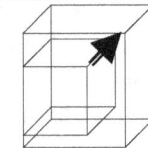

Gravity only will be possible when the reducing of space will produce more space for heat to ensure the reducing of heat and prevent overheating which will contribute to expansion. There will forever be growth.

$T^2 = a^3/k$ $T^2 > a^3/k$ $T^2 < a^3/k$

The growth will either be by expanding to secure more dense particles or the increase of space will produce the increase of space by reducing density. If the time effort $T^2 = a^3/k$ can duplicate space from singularity no increase in space will come about. If that was the case, then the status quo in space-time will be in place and time will duplicate space to a precise eternal value. When time **reduces space** to a lesser value $T^2 > a^3/k$ gravity applies reduction and space through movement becomes smaller than it previously was. That is the way the Earth perceives gravity. When **space produces larger expansion** than what time can duplicate $T^2 < a^3/k$ by the extending of **k,** the growth of space will be larger than what time can manage and space will increase. This applies when a spacecraft launches into outer space and astronauts become taller. All three systems are in place and Universal growth comes about in such a manner.

Within the circle incorporating a sphere as the formula $k^0 = a^3 / (T^2 k)$ shows it holds gravity centred in the precise middle of the circle. By using mathematics in the way Kepler used it, it shows that those rules and laws, when used correctly in the investigating of the formula that Kepler introduced, is a testament of the cosmos. When using the Kepler formula it shows a space-time relation that must then form the basis of cosmology. Also while investigating as we now are doing, such intense investigation then must be without Newton interfering and telling Kepler what he (Kepler) should have found. The formula should be applied directly instead of Newton incorrectly correcting Kepler whereas instead Newton should have been looking at what he (Kepler) found because only then he (Newton) could have seen what gravity is. He (Kepler) said that the cosmos said that gravity is $a^3 = T^2 k$. The space is held in check by motion from a centre and that is gravity. It becomes more than clear that space a^3 is time by dimension T^2 and time is space a^3 without dimension **k** Gravity is a^3 / k but **k** is an addition of motion T^2.

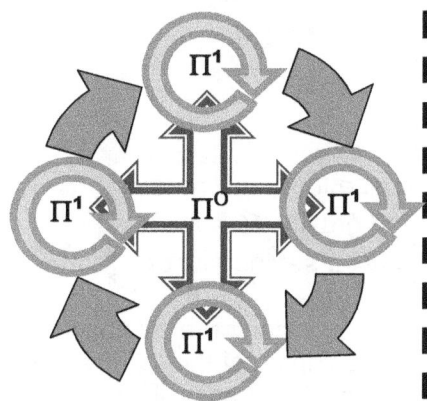

The major issue is to find where the cosmos started by using Kepler. Let's go back once more and reduce the line by half every time. By

All rotating object has to be round to rotate. From the ends of the Unit rotating there will run a line running horizontally which turns with the top and as it turns with the top, the horizontal line is crossing another line that is running vertical but is not running at all because the three points cannot turn. The picture on the side is a picture showing the rotating object from the top as one would look down onto the line in singularity that does not turn. That line in singularity is representing the Universe because it is representing eternity as much as it is eternity. It will never disappear because it is never there to begin with. The line cannot stop turning because the line can never start turning. The line is absent because it can never hold space, yet the line is always there because any motion may charge the line into presenting the centre of the Universe. To find the centre of the Universe is to reduce the line because Kepler said $k^0 = T^2 / a^3$.

repeating the process some other aspect comes to light every time. What happens to one line is also applicable to all lines. It is cutting every distance there is in the Universe by half every time. The process we refer to at this point I might add is not in the space we use and know as outer space but in the inner, inner space where the proton is and we cannot use, so please do not confuse the space with outer space. Then repeat the reducing process until it can repeat no more. The reducing of the line by half every time will get to a point where all the ends land on the same position without any possibility if halving the two ends further. There, all possible points share one position and moving the points in any direction will lead to an immediate increase of extending the line once more.

In the sketch I made shown below of lines reducing I left space between the two ends of the line that is symbolising the end of the reducing that will share one location even by having one single line. There is no chance that I can present any sketch reducing the line to a point where the points are sharing one location in the single dimension. The points are there and with the points being present they may not be dismissed as nothing. From there no reducing in a natural manner can lead to nothing without changing the rules of mathematics in such reducing. Any further effort of reducing must bring about extending because every point possible share space with every other possible point at the point of singularity where all points share common space.

From where all possible points landed on the same spot any further moving of any point must then bring about an increase of space. No matter what motion comes to that point when moving such an act will result in the increase of space between such a point and singularity. At that specific point further reducing will bring redundancy of the line, so further reducing is no longer an option. Since all the reducing that did not bring about the redundancy of the natural line that there are covers everything that is going to develop from the line, the fact of zero that is not included is a proven fact. With the reducing zero was never encountered and zero therefore cannot then enter the process because the process fill without the need for zero. This also applies to the circle because the circle uses a line to indicate size, such reducing of the line and by reducing the circle will end up in the same position. It is this fact of the moving of any point from that spot holding singularity that such motion will introduce space as the space exceed the previous limits of singularity.

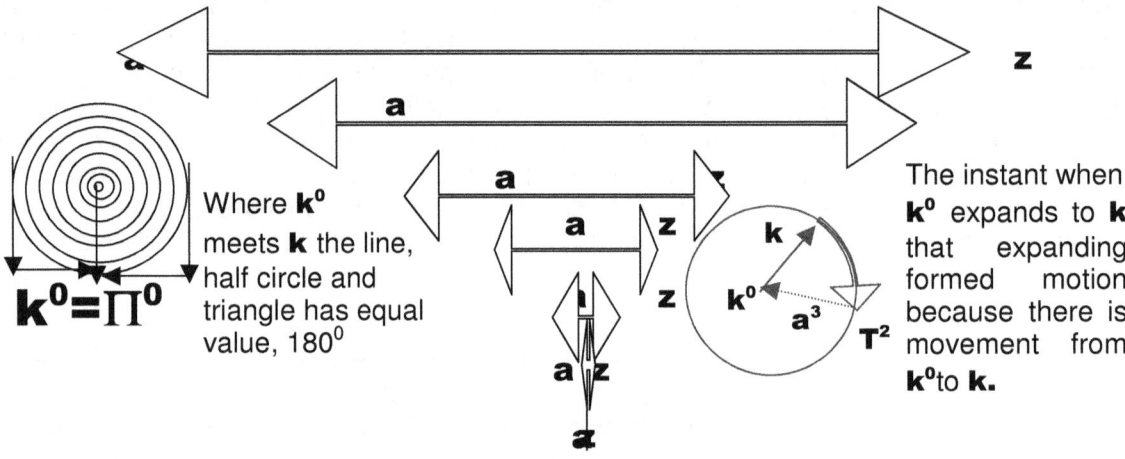

Where k^0 meets k the line, half circle and triangle has equal value, 180^0

$k^0 = \Pi^0$

The instant when k^0 expands to k that expanding formed motion because there is movement from k^0 to k.

The instant k comes about from k^0 k apply further motion as T^2 and with k producing motion by expansion and T^2 by contraction...

...gravity comes into space forming where space expanding a^3 and gravity $T^2 k$ is the same result of singularity k^0 setting motion

This process is a natural normal occurrence everywhere in nature without any person ever noticing. We see this so clearly in the spinning top. When the top is spinning such spinning of the top creates a centre and the lines start reducing space in the direction of the centre of the spin. The centre establishes a balance in space-time where at a point it finds partial independence from the dominance of the Earth's gravitational motion that is depressing the tops movement to a standstill.

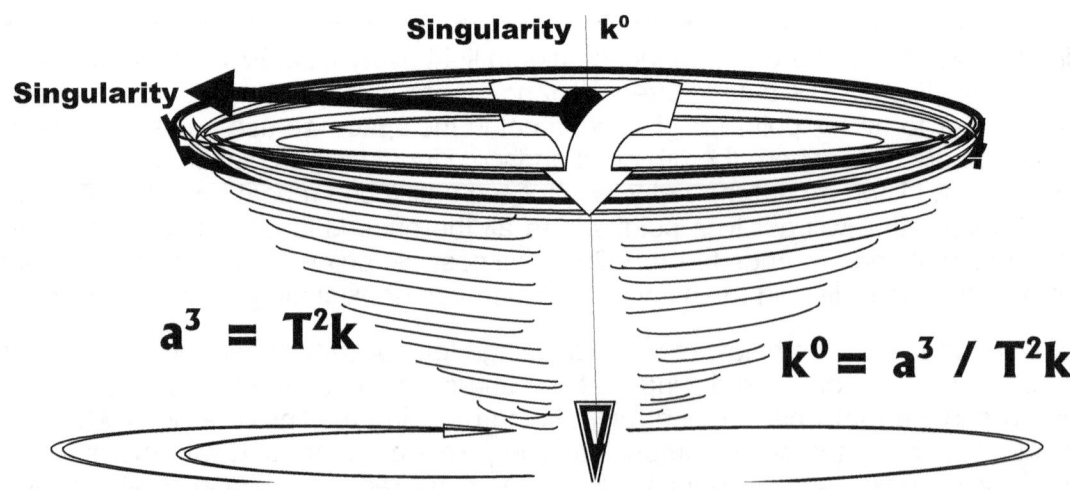

Singularity k^0

Singularity

$a^3 = T^2 k$

$k^0 = a^3 / T^2 k$

Singularity

When the top is spinning there is an obvious centre point established. This point is also the application of the Coanda effect. It is establishing a "non-existing" but also eternally circling centre that is allocated but cannot be located. That point secures a centre by motion, because the motion establishes a centre $k^0 = a^3/T^2k$, which means that centre establishes individual singularity. The space created a^3 must be equal to the motion created around singularity $a^3 = T^2k$. The only part is that no such point exists in our human reality, but such a point can only be realised as much as such a point can only be understood. That point where directions change while the spin continues, is without space a^3 which is k^0 but on the condition of motion $k^0 / k\ T^2$.

One must draw this statement of motion back to the point where singularity is getting sides. After all, that is where creation started. When there is singularity there can be no sides. It is 1 (one) position away from all angles there can be. That one fills no space but all space flows from there. The space we allocate to that spot does not really exist in the manner we humans see space to exist. It is a spot that is there without being there. It does not visually exist because it is not filling any substance and it cannot be recognised. It is clearly holding a position in a location where the location where such a locality is found is beyond this Universe. Once one accepts the fact of singularity, that accepting of singularity then is contradicting all the things we know by not being any of the things we can recognise. In singularity there is no space. There can be no motion because there can be no space to have the motion within. It is a line that is so small it is not there and the only reason why we know it is there is because of the results it left as an imprint of its not being there. It is not recognised by its absence because it is never absent. It cannot be absent because it is eternally present. It cannot go absent but it can never be there where it should be if any person wishes to locate it. If it was absent then it was zero or nothing but since it is there it is not there and that makes it present. The centre spot that we cannot see and that we cannot detect has no sides to any side and has no place it fills because it fills all the places we cannot detect.

The only way such a spot can fill space is by doubling the space it fills to become more than one place to fill. But the very instant that happens it halves the space it fills because it then cuts the space it has into two parts. That brings about that the point of not being is doubling the not being and by doubling the not being into being it also cut the not being that became present into half. We have to find this spot as we find religion. It is something that we can only know is there because we cannot disprove it is there but we can never prove it to be there either. It is something seen through intellect and not through the eyes or light transfer. It is a point far beyond light. It is in our being and not in our vision. Most important about this is to confirm what liquids are and what solids are in space. Because of control of heat, the confirming of the control of heat or the lack of control of heat we find the state matter occurs in.

Flames are a liquid heat. Smoke is a liquid heat. Vapour forming compressed space is a liquid heat. Rocket thrust is liquid heat. Air and atmospheric space is liquid heat.

The Coanda effect is the manifestation between solids, liquids, heat and motion. The fact of being solid or being liquid is locked in a relevancy that brings across a comparison. Hydrogen in Jupiter may just be ore solid than iron is on Earth. However, hydrogen on Jupiter will always be a liquid compared to iron on Jupiter. The cosmos puts a huge difference between solids and liquids. In fact the Universe divide on that perception. Both are extensions of singularity but it is the extension that brings across the distinction because in the one case gravity forms part of the inside of an atom and in the next extension that extension allow gravity to bring about the motion of space which is gravity.

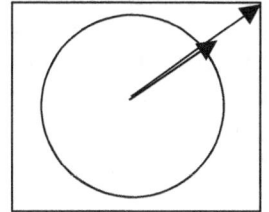
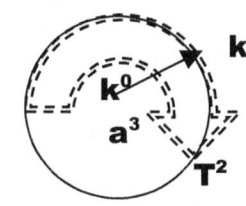
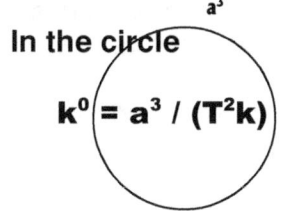

In the circle

$$k^0 = a^3 / (T^2k)$$

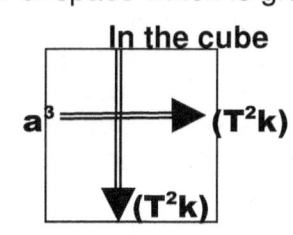

In the cube

$$a^3 \rightarrow (T^2k)$$
$$\downarrow (T^2k)$$

The big proof coming from the Coanda effect as explained above is that space flows through time as a product of the creation by motion op space-time. The opposing of a^3 by (T^2k) on the "other side" produces the **six sides** we now came to use. But that means gravity is where space disappears within spherical or circular structures and not in outer space.

$a^3 = T^2k$ therefore $1=a^3 \setminus T^2k$ and is the same as $k^0 = a^3 / T^2k$ which means that
$k^0 = a^3 / a^3$ or $k^0 = T^2k / T^2k$

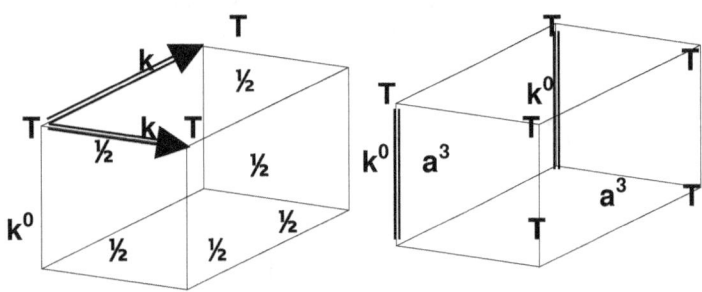

Gravity (cooling by contracting) and anti gravity (heat bringing expansion) is space created by the motion of the space in a line running between two points.

Space in the cube a^3 is the space in the square T^2 by motion through a straight-line **k**. That is mathematically what Kepler says. Space a^3 is formed by the square T^2 going double in creating a straight-line **k**. That means space a^3 is the doubling through motion T^2 by a straight line **k** and time T^2 is creating space a^3 by implementing motion T^2 using straight-line **k**. By the motion T^2 of singularity k^0 space a^3 comes about in a straight-line **k**. Applying motion T^2 is the forming of gravity or anti gravity.

In a solid structure the structure select a point of forming a centre that will refer the function of such a centre onto a centre which all the atoms sharing the unity of the space of the group holds which by motion separated the isolated space as a separate Universe from the rest of the Universe. That is solid structures and stars form an excluding unit with one singularity at the centre which dominate as much as control all singularity within the unit. In such a unit an object loses independence when the object does not bring about independent motion to confirm the singularity as being apart holding individual space-time outside the units space-time confirmed by every atom in the Unit. All motion of the unit will presume and any motion will be affective above and beyond the motion of the unit operating as the larger structure. When the dominated body finds any ability of producing motion such motion will serve as an attachment of the larger singularity in motion by spin. By it having individual spin, such accumulating of the group effort is secured by the larger objects being the accumulated unit's motion and find security in that motion. But since the structure is finding security in motion by adopting the combining unit it has to adapt to the combining unit space-time by establishing a measure of duplication of space in line with that which the combining unit produce to sustain the combining unit gravity. This connection comes about since the object does not have an independent created singularity it can sustain.

From this point that is there without being there that point now has to create a point that is there where we can see such a point start. The only way to do it is to double on by adding and not by multiplying. It has to be in two places at one time but the time is filling the one place while the space is filling the next space. Therefore it cannot ever afterwards be in two places or on both sides of the Universe at one time because it is already filling both sides of the Universe and as such cannot again repeat such action. In space there is forever the divide singularity produce.

If one is in any position of judging location, one part of the six-sided Universe will fill space and the next part will fill movement. We may find one part filled and the second part empty but in truth there is a divide which have one part filled when another part seems empty only because it is going to be filled by the space filling the part that is now seemingly filled but which is going empty as time develops. Time always develops space across the divide of singularity and singularity is always forming the divide. Should one wish to name such a divide, one may place space on the one divide and time on the other divide making all there ever can be space-time. That space then can move from the space side to the other side being the time side or it is the other side being in opposition without filling space that is forming space whereby moving from the space into the next space it is filling space, forming space. It is holding space that is not there and only by applying time or motion it can duplicate such a space and make that space being there valid.

Space is either filled or waiting to be filled whereby when waiting to be filled, time validates the existence of such space. The one side is space not filled and on the other side it is space only filled by the space not filling one point duplicating such a point and then only filling the point through the motion that excludes the point from being on either the one or the other side of the Universe. Without the one side that is going to fill the other side and without the other side that is going to halve the one side not filled into two sectors of half the standing of one not one side is possible. Even if the following space seems to us as filled, to the space in motion every particle finds the space in which it is, as being filled and the space it is moving to, as unfilled or empty. To every singularity the Universe is only filled by what that specific singularity fills the Universe with. All the rest are empty space.

All singularity points are one singularity point and singularity does not apply motion because singularity is without motion. It is the space-time that transmits space as motion but since all singularity is identical and one, the transmitting of the space-time is not extending as far as the singularity but only as far as charging singularity. Again I have to insist that when one think of space-time and space motion one has to exclude any comparing that with life. Singularity is charged by motion and it is motion that sets the space-time. By charging singularity, division comes about while singularity remains one spot with immeasurable dots, but it is motion charging the singularity into dots. Singularity is as much a dot as it is a line.

The only fact that the line ⎪ is there is because the line is not ⎪ there. By duplicating ‖ the line not being there the line establish the fact that it is there but by duplicating the line the line at the same instant becomes half of what it is at the time it is duplicating what it was not. This is coming straight from the horse's mouth. The Cosmos told Kepler that $a^3 = (T^2k)$. That is it. That is gravity. That is how the cosmos unveiled the cosmos. The cosmos said that from singularity k^0 came space a^3 with the motion establishing time T^2k. If it was this simple why make it so terribly complicated just to please the playing of the games by the mathematicians. The moment space realised a^3 such realising presented itself as motion. Translating Kepler's statement from the mathematical equation of $a^3 = (T^2k)$ to verbally spoken English can only be translated as "space moves" and "space is motion" and "space becomes motion". It is space filling as motion or motion will complete space. There is no other translation one can draw from this except by altering the concept through incorrect adding of facts the Universe never entered or stated in the formula the Universe unveiled to Kepler.

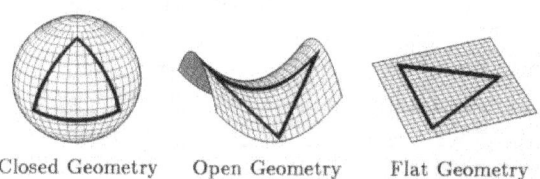

The ⊽ line can only duplicating motion to a^3 or but never duplication it T^2k can be no motion be if the line fills the space it is not filling when the line is the filling of the space the line does not hold by applying the filling of the space of the line. Then there will be either both as we now wish to see the Universe. By the therefore insists on relevancy because without relevancy there and no motion means no space. The strongest proof there is about this is the manner in which the Coanda principle applies the reproducing of space taking shape from a round object and involving motion to produce such duplication.

In the general relativity the geometry very closely connected to the distribution of matter. In the space of two dimensions, (as if the universe is like a flat rubber tyre) Euclidian geometry applies so that the sum of the eternal angles of a triangle is 180^0. If a massive object is placed on the rubber canvass the sheet will distort and the path of the objects moving on the sheet will become curved. This is in essence what happens in general relativity. Well missed by science is that the water drop proves the point the best. But not only does the sphere acclimate this concept, but all of matter interacting with space producing time does the same and will result in the same. The issue the brilliant mathematicians miss in every argument about they're manner in which to theorise and claiming the use of mathematics there has to be a centre point they can produce to secure the precise location whereto space will disappear. It is fair and fine to talk about how to calculate gravity but showing the precise location to where the space will flow of the Universe, they have to indicate that there is the point of a precisely located centre of gravity.

Closed Geometry   Open Geometry   Flat Geometry

It is fine and fair arguing about sides curling in one direction on the other direction and pronouncing a flat rubber sheet, but if there is gravity, then one has to admit that gravity comes from a centre! So where is the centre in the theory where gravity applies control? The curve of the sphere is relating to a specific centre. This is diverting strongly from the theoretical picture we have that the cosmos is always presented in. Any one with a clear mind will immediately dismiss the flat Universe Einstein tried to sell us. Our Universe is 3D in six sides as every one can see. That makes the curvature of space-time not a very magical awe inspiring unrealistic threat to the human senses but it is very common to find and it is very common to understand. It is derived from singularity by duplicating singularity in forming the Π and π personifies the sphere.

With the roundness in which Π is formed and by securing Π a centre is created. This is evident in the Coanda effect where gravity comes about in such a manner. From the centre there then is control coming, which resembles the same principle in the spinning top. Much more important worth noticing is that that is what Kepler said. A centre forms when k^0 establishes Π. From k^0 a centre forms holding a^3 valid by allowing motion T^2 to secure independent space-time, which is a bowl normally holding a fluid we know as pure heat. Then there is the motion of the liquid $\Pi^2\Pi$ securing the space by producing the motion and thereby establishing a centre k^0. In the example we see where the Coanda effect establishes just this principle by having water flowing around a round bowl, the motion of the water forms the liquid part $\Pi^2\Pi$ which generates the space Π^3 or a^3 which validates singularity Π^0 or k^0.

The motion of the water becomes the liquid part $\Pi^2\Pi$, which generates the space Π^3 or a^3, which validates singularity Π^0 or k^0. Looking at what happens in the Coanda effect we find a centre that is activated by motion of liquid since the centre takes charge of the space within the bowl. That is k^0 that produces a^3. Then on the fringe of the space motion, the space-time becomes part of the space holding the centre which can be seen in a formula coming from Kepler's formula which is $k^0 = a^3 / T^2k$ and when using relative cosmic values read as $\Pi^0 = \Pi^3 / \Pi^2\Pi$

The Curvature of space-time is as common as space-time

The Titius Bode principle   **Half the Roche principal**   **The Lagrangian and Titius bode principles**

The Sphere.   The Black Hole.   The developing galactica

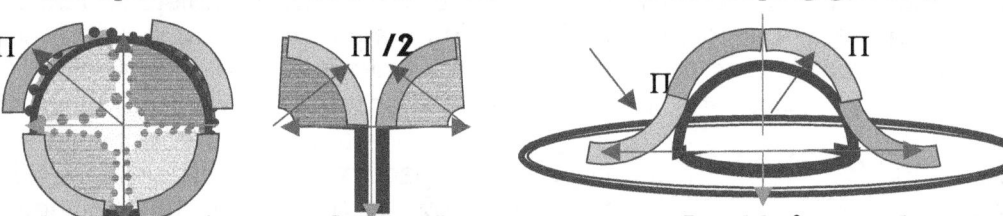

Bubble or normal Π   **Inverted Π**   **Double inverted Π**

The reasons for this applying I explain explicitly in detail in other books

With singularity being where it is and in a common place as it is the curvature of space-time also becomes a common factor but not yet commonly distinguished because singularity changes the profile to suit space-time.

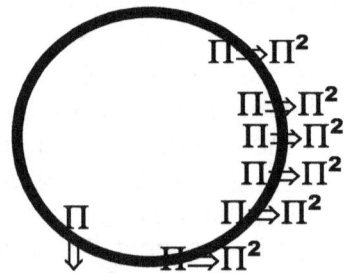

The Coanda principle which in fact should be seen as a law because it is that strong is the principle of gravity duplicating with the motion that provides such duplication a relation of the particles having the seven factor and such a factor of seven produces through motion another three dimension. This total that material fill while in motion is ten and when ten crosses the line of singularity to duplicate the seven in the other side of the Universe the crossing cuts singularity in two as much as it puts singularity in the square.

Π^2 But it involves the motion of concentrating space to be or hold fluids around solids that
Π may or may not move. In this must be a solid, a round basis Π, fluids concentrated in
$\Downarrow$ space and motion applying to one or all of the factors.

Π^2 The extension of Π is well received as a dimensional implication to matter holding seven positions from singularity and space having four quarters through out the rotation of singularity forming the centre to the five dimensions (one side lost to the cube's six sides connecting to the five remaining sides) making the total sides facing space from the point holding singularity at any given instant at a value of twenty (4 X 5 = 20). Then adding the singularity cross of Π being (1+1) = 2 the relation becomes 22/7. This is crude because in more precise calculations it becomes 20 = 0.91 + 1 = 21.91/7 = Π

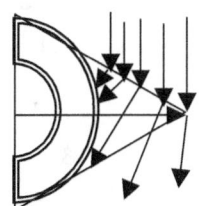

From what this says, it is validating Kepler's formula again. By bringing about Newton's claim distorts facts, because if some distortion must become evident, it should progressively be evident. The Big Bang did not happen yesterday and the progress from then to now must indicate that the Friedmann universe or the Walker universe is taking place.

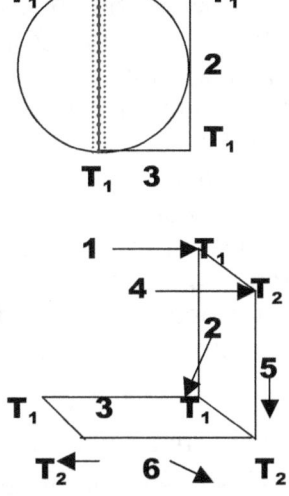

7 Singularity establishing

Π

Π^2

The relation forming the duplication of singularity is a duplication but applies as a dimensional forming of Π and placing 7 in relation to 10 forming Π^2

The liquid applying motion forms the 10 disciplines. No motion leaves no Coanda as well as no gravity because gravity is motion that duplicates singularity.

Conditions that prescribes the enactment of the Coanda effect is that the one surface has to duplicate singularity by establishing Π as a form. The round surface Π will bring about the shape of singularity Π that becomes enticed by the action of the motion of the liquid or of the solid or the motion of both around Π, which then establishes and confirms singularity by form. The next factor is the presence of liquid. Air or atmosphere is liquid and water is liquid. Heat is liquid. The third factor being just as important, is the motion establishing Π^2 by duplicating singularity as singularity becomes relevant through the applied motion that produces gravity from the singularity spot that provides the form.

Time or spin or rotation, call it what you like but it is the moving from T_1 to T_2 using time that provides the dimension of depth to the dimension of distance between dimensions

In the motionless Universe there will be on point in time and that point will represent k more than anything else. Every point being T_1 will only show the extending of **k** from singularity to that specific point. The fact that T1 indicates no motion brings the universe to a stand still and to a flat Universe.

It is that which give **k** the coming from the first dimension and by only extending from singularity it the forms two more positions becoming a^3. $k = a^3 / T^2$ but remove T_1 to T_2 from he equation and only **k** remains $a^3 / T^2 = k$. By the effort of spin or motion the universe becomes the three-dimensional object it all seems to be.

Motion is what **gives space stability** and **security** in **six sides** where **three sides** are **opposing three sides**. This Kepler shows so very clearly in his statement $a^3 = T^2 k$. If there is anyone out there that cannot see this and miss the interpretation such a person cannot read mathematics.

That too forms the answer about the question concerning the Titius Bode gravity implicating of cosmology. The seven sides are linked by rotation nothing changes because there is a steady linking to the inside centre of the sphere. But it is to the outside that this rotation brings about dimensional complications. There are five T_1 points moving to five T_2 making contact with five moving points. The moving non fixed points is the point before reducing by five to the point after reducing by five that bring along the ten points in stead of the five to one point as it is the case with the Lagrangian system. Two points relating to seven points coming from by continuing in the same direction it is going to remaining seven points as going. That means in matter there are five times two points relating to seven in a moving constant and seven fixed rotating points

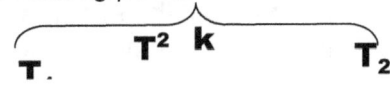

k_1 k_2 k_3 $T.$ $T^2 k$ T_2 a^3

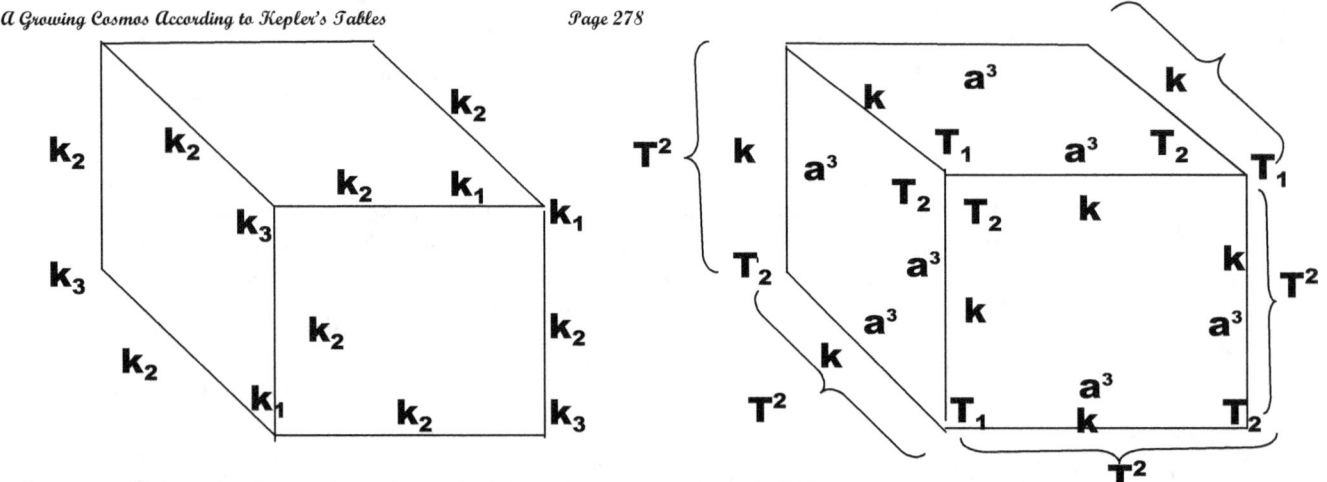

$a^3 = k. T^2$ As I indicated previously it is all a dimensional differentiation.

$T^2 = a^3/k^1$

$a^2 = a^{3-1}$

$T^2 = k^{3-1}$

$k^2 = k^{3-1}$

Einstein said gravity is the strongest where space disappears and even he, the Master, misinterpreted his mathematics. The space mathematics show is not the Universe we see but space located at a point where space departs from singularity applying the value of k^0. It is a point within all atoms secured in a dimension smaller than the spin of the proton. It is where $k^0 = a^3 / k T^2$. It is at that very point that the one side is where space disappears and on the very other side motion gives space dimension stability holding size sides in form. It is beyond the micro cosmos at a point not mentioned yet or named yet.

If it was not possible for space to use time in providing space, for space to move to in duplicating, the Coanda effect was not possible. But the Coanda effect is only applying gravity in a way we are not used to because we see our motion as being in a straight line following the curvature of the earth. In our way of thinking about gravity, it is that we go down wards by not going downwards. This then, according to our misconception, comes about as a means of pulling, which is very incorrect because mass is the result of the lack of space we must duplicate, to still remain in the cosmos. By cosmos law we must move to fill the space we are moving towards and by our not applying we create the mass we are in. As we are part of the Earth space we are duplicating in time with the earth because we have to comply with cosmic law. Instead we double the space by standing still of space as motion insist of replacing the space we have. By not being able to duplicate the space as we move onto the next space our motion creates we establish mass as a means to cheat gravity out of the space we should be duplicating. Mass is the effort we have to bring about since we cannot duplicate our space with motion. The duplication we are using to double our space stems from the protons up to the stars where singularity is providing such duplication.

The Universe we know shows space having six sides and even that Kepler proved as he proved space in spin $T^2 = T^{3-1}$. It is the same, therefore it is dimensions repeating to form $a^6 = (a^3 X a^{2+1})$. This becomes a factor of space forming six sides when Π forms Π^6 in relation to the six sides of space ($\Pi^6 / 6$) in the presence of gravity supplying motion of 7 / 10. That is outer space $7/10\Pi^6/ 6$ forming the element space-timer limit of **112** in which we find all the known elements.

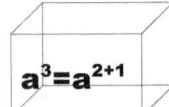

Our six-sided Universe is about space-time. It is space in motion and the motion is placing relevancies that are applying the time. Time even to human standards at present is the positioning of objects after some other arrangement of positioning took place and repeating by duplication brought about by transmitting space as the transmitting or the moving and the pace of the moving of the space is repeatedly duplicated to the previous order it was.

One part of space is time, which is the pace of the duplication of the space duplicated or the spin of heat and the other part of time is the dismissing that is establishing the flow of space towards a dominating centre. If there were no space moving or space to move within or available to move towards, then the next time period will find no space to fill or find itself unable to fill the vacant space and if there were no space to fill the space with space being the filler, the Universe will collapse back to singularity where all is without space being without motion thereof. It is what Kepler said it is, being $a^3 = T^2k$

Space-time in motion and movement are all the same things only separated by dimensions and dimensions are formed space, where the dimensions become space being in motion and the space is motion by contraction or by expansion but because time is almost eternal at k^0 our perception of the universe we are in is a stable and steady eternal structure. Gravity is motion and motion creates space to the third by the third in the third that interacts with one but establishes ten.

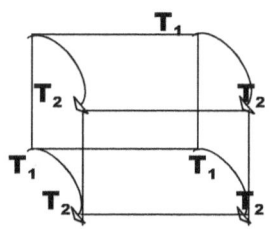

In this very manner is gravity the very same as speed where gravity is the moving

The reason why man can never fully create the complete 3D is because of mans' inability to recreate motion that we find in time. The duration it takes any one point T_1 to move to any other point T_1 will have at the time of the arriving of T_2 produces the 3D Universe we are in. By such means does the Titius Bode changing five relating to seven to ten relating to seven bring about gravity or time or moving singularity. Use the name you like but it is

Space is created from one position to another position and the duration it takes to complete the distance is time.

Gravity is the very same but it is the recalling of the space by creating motion in the space.

By recalling the space it is also reducing the space because it is counter acting the time expansion provides. That then is clarifying the reason why gravity will always on the limit be stronger than light. At a point it slows the time component down to such an extent, that the space reduces faster in that time than what light can produce motion. But gravity and antigravity must be seen in the speed relations that come about from the process. Gravity is speed or velocity applying. It is space in volume in relation to time in motion. It is $k = a^3 / T^2$

When the object is stationary on the Earth a certain value applies k in relation between the object and the Earth to the object. The space a^3 surrounding the craft influences the craft to maintain the time T^2 that the Earth applies. By departing to outer space the k influencing the space a^3 in the time T^2 changes as the time T^2. This is the result coming from the changing of k changes and since k changes all relevant factors change.

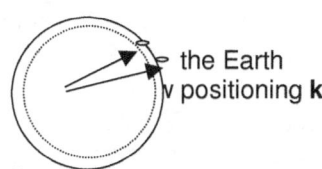

the Earth
positioning k

Atmosphere

As the craft extends the k influencing the space a^3 the craft requires the extending of k brings about a bigger time T^2 component. There is a slight change in the space occupied by material that composes a lot of fluid such as the human body but the time T^2 is the big factor that changes. T^2 is the gravity applying and it is all clear that the Earth gravity dynamics do not apply. The object then secures a gravity depending on the k it receives from the sun. In that manner the six sidedness of space apply and not the Earth's sphere having a seventh point that produces a new k and separate gravity

However the ratio of $k = a^3 / T^2$ does not end there. By shifting position k also produces an extension of increase on the formula.

The k that science sees that positions an object in relation to the centre of the sun is not the only k applying. The line is k and there are many lines forming. There are the lines forming k by motion.

The only way k can extend is by revaluing T^2 to produce an individual gravity T^2 in the space a^3 that the occupier occupies. By increasing the heat that supplies the cosmic structure individuality, the structure then receives a new gravity T^2 because $k_1 < k_2$ in producing $k_1 = a^3 / T^2$ compared to

By motion increasing or decreasing gravity changes values. This is very evident in the Coanda effect. It is also very apparent in the sound barrier as well as the re-entry of objects entering the atmosphere. An object entering brings about that the k of outer space changes to the k of the Earth. The Earth k secures a new time component T^2 but such a time

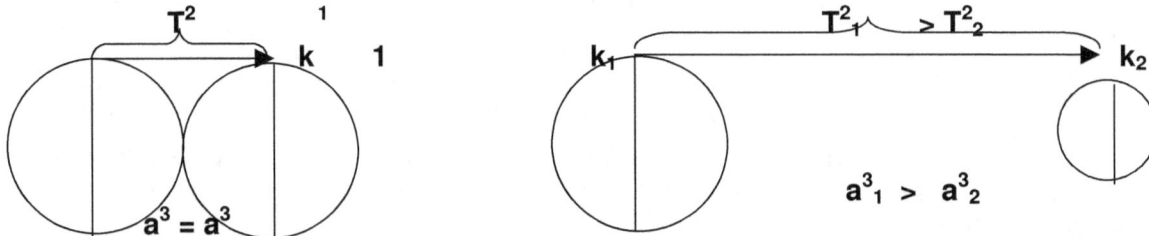

This also comes about and influence travelling in space. There is no unlimited speed that can be achieved once an object is in space. By increasing the time T^2 or decreasing the movement k such changes will bring changes on all factors and will demand revaluing of all aspects. This we find by using Kepler's formula $k = a^3 / T^2$. This is what happens to objects entering the atmosphere. There is no particles touching which introduces friction and such thoughts are senseless, however there are friction brought about by motion restriction.

There is a time component present to activate one singularity to the next singularity as singularity is generated or excited by motion of the space moving as it shifts by the space it holds in motion, which generates the singularity to control the space. Considering when a body enters the atmosphere there is a tremendous re-adjusting going on about in the governing singularity that the group or unit elected and that spirals down to individual atomic singularity and therefore all the atoms forming part of the craft must undergo space-time re-adjusting to match the singularity that the Earth has in time related space occupying. There is not enough material in that area of the space we call the atmosphere limit to bring about the friction required to unleash the heat we see develop from such an entry into the Earth's atmosphere. The flames surrounding the structure is the area a^3 that reduces because of the changes brought on by the new time component and the new k factor changing the relevancies between the factors. Space travel is limited and the distance to time endured is gravity. For us to be able to leave the confinement of the solar system will require anti gravity equal to the gravity in the centre of the sun. It is the same requirement there is when we wish to leave the earth centre.

By looking carefully at Kepler one can see that space comes about from matter moving by which space is created through the motion causing time to become an integral part of space. There is no chance of wiping out time or turning time in reverse, because by fiddling with time motion moves space back to space closer to the realms within singularity. Changing motion is taking gravity back to singularity and that reducing of k will lead eventually to conditions that prevailed before the Big Bang event. Gravity starts where gravity started in the very first instant. The very place where k left singularity and stepped into the 3D k extended ever so slightly but never more influentially. The point is where k formed a line outside singularity and went from k^0 to form k.

The length of the extending of k^0 to k is so small it is beyond any manner of human conception or mathematical comprehension but so it is vital to the Universe. It is the result that comes in the form of space we call the Universe and the Universe comes from that action. With that action of extending k by the very utmost, utmost slightest of margins the Universe we know came into being what it is. The factor k went from k^0 to k. By the extending of k space a^3 came into place. But that space only came about by T^2 producing motion. By k extending by motion, space came into place. Space came into place only through the motion of gravity and antigravity. The motion T^2 produced brought k the independence which a^3 secured. The factor is documented as $k = a^3/T^2$. Seen from the point singularity holds every aspect there is in the cosmos including other singularity dots is space-time. It is space generated by motion attached to the centre or governing singularity. From where k starts expanding k produces space a^3 through time or motion T^2. It is $k = a^3/T^2$. To one point holding singularity that point of singularity holds everything there is and there is only the space-time extending by **k** producing **k = a³/T²**. There cannot be space without motion. There cannot be motion without gravity and there cannot be space without gravity. Gravity is motion producing time forming space capturing gravity. It is a confined unit belonging to every singularity extending.

Accepting gravity to be motion is half the story and that is very much literally half the story.

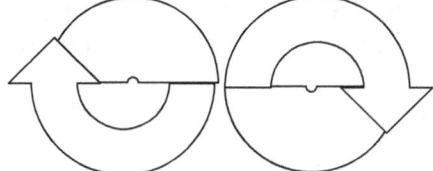

The heat brings about expanding singularity from a one sided affair to filling a volumetric Universe. But all of it is a relevancy where ten positions will sacrifice individuality and compromise singularity in order to secure two positions in singularity that holds a relevance of one.

By using the Titius Bode principle of seven on the one hand relating to ten and all ten coming about to sacrifice their position in order to save one vantage point, material as well as space has to cross the border of singularity and fall in the other side of the Universe. The centre point holding singularity still forms the divide but all other points have a task to perform. Securing the centre singularity and maintaining the centre singularity brings about securing all singularity and maintaining all singularity. It is six relating to six where three is on the one side and three is on the other side. There is space formed and three is motion in space formed.

That which I refer to at this point being ten to seven is material in unoccupied space. There always is a relevancy of singularity being relevant to singularity and one takes charge of seven while the relevant partner is taking charge of the other singularity (7) in the unit plus the three in motion between the two in relevance. There is another relevancy in place that only takes space occupied by material into account and in that confirms the sphere as forming space occupied. In this second relevancy the same principle duplicates once more because of the direct attachment to singularity and then there is the space of ten duplicating twenty in a dimension above. The first dynamic involving the Titius Bode principle was relevant to space filled with material in relation to space providing the motion and thereby filling the space.

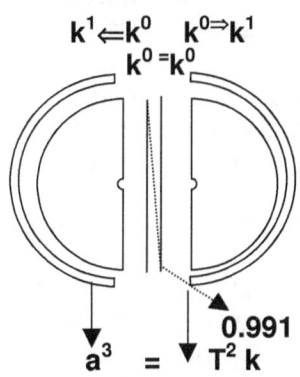

$k^1 \Leftarrow k^0 \quad k^0 \Rightarrow k^1$
$k^0 = k^0$

From the relevancy of the overheating which is bringing about space by creating motion there are positions taken in space occupied and controlled by singularity as well as positions influenced by singularity. There are those on the one side forming ten and then the eternal divide and there are ten on the other side also involving the one and the infinite. On the one side there are $k^0 = a^3 / T^2 k$ and when that adds it is seven points. Then there is another three either being space or being motion creating space a^3 or $T^2 k$. It is holding either but not both and will establish duplicating once more.

0.991   Singularity eternally running to the inside

$a^3 = T^2 k$

It is seven relating to five, which is duplicating five in either sides of the Universe by involving the Roche limit. The second one is duplicating ten that relates to the line representing singularity as one and the line reducing towards singularity. The second dynamic involves the curvature of space-time by producing through duplication the singularity dynamic of Π.

On the one side of the Universe ten positions form a web with one and a range of dots smaller than one (0.9991) to form the protecting of singularity. On the other side there is the same number and that number forms a unit 10 + 1 + .0991 + 10 = 21 that forms the space effected by the seven forming material and forms the motion of the Titius Bode principal. The factor of space that is surrounding material (always in the sphere to protect singularity by applying gravity) brings singularity back to the original value of Π by maintaining Π² in the motion and space required to maintain Π. This produces the dome covering the sphere forming the circle mimicking singularity. The atmosphere is as much part of the sphere as the sphere is part of gravity forming an extension of singularity being the eternal Π.

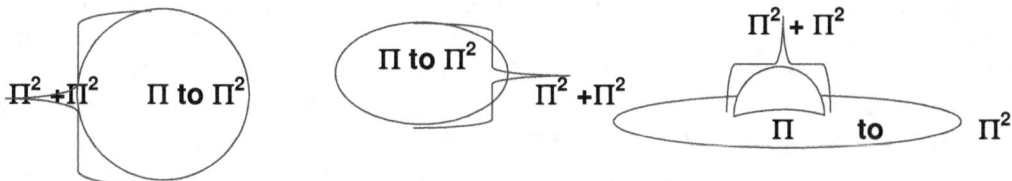

The atom once was as round as any sphere but with the enormous development of space-time and the massive favouring of space in relation to material the dome form grew flatter. This does not change the singularity forming Π **to** Π^2 in any way. We can see this very same tendency in almost all galactica of substance. The centre still holds a dome or a sphere while the edges grow flat with gravity growth reconstructing the original form. It is important to realise that because motion establishes space as well as control space where the ten factors is representing the square of space. The square will always grow much more than the seven factors do in the relevancy.

We can even follow how the proton relevancy reapplies in the shape that the galactica form or as the galactica takes on a "normal" neutron form. The control of motion producing time is essentially locked in singularity where time starts the motion as it is establishing space. From our position outside singularity the motion creating space is the point, which is where the slowest time can be when not being eternal because there time is periodic and not eternal which, singularity is. Singularity cannot shift because singularity is the first dimension and as such immovable. Therefore anything not eternal is temporarily created and beyond being permanent although I agree that we are seeing the Universe as structurally solid. The universe we see as solid is solid because of the motion being the flicker or stroke relevancies and as time duration slows down as space expands, space relatively remains longer in one position of duplication than in the other position of resembling positional correlating. By it going out of frequency we see the space as space only because it is relative longer remaining in space and that is what confirms the 3D qualities of the space we enjoy.

That only comes about as our perception. In stead singularity creates space and time and by allowing space to reduce from singularity the reducing is placing the motion establishing space to reinforce as well as secure the immovability of singularity. It is providing singularity with permanency while singularity is removing the permanency from the creation sector to the power of three to the power of two. Where the time excels beyond our measure is within the atom structure. That which we think of as a mass difference in the atom, is a reducing of space in stages and a reducing of space is an increase in the duration of time by motion. The proton reduces space 1836 times to the electron and that we see as mass. In fact that is space-time differentiation because all mass is just space-time differentiation. Things go even more reduced as the proton sheds all space to enter singularity.

As the one proton moves into singularity and allows space to disappear, the motion brings about the other proton to move into the place of the proton that disappeared. At that point time is next to eternal because the very next spot time becomes eternal as it enters the immovable singularity, which is eternal. This is like watching your nails grow. After time one can see the nails showed growth but the growth is so time consuming we find the growth of the nails to be next to eternal. By the motion of the proton, the proton is doubling the next proton that disappeared as it connected to singularity. This motion at that point is 1836 times to the power of three to the power of two more time consuming than is the speed of light because the electron is representing the increase of space-time displacing to the speed of light and at the point where the proton becomes singularity the motion is so much time consuming the time created to indicate the motion presents us with a permanently secured structured. It seems as if it might be $((1836)^3)^2$ times the speed of light but explaining this is far too time consuming at this point. This period of space relative to motion being $((1836)^3)^2$ is the motion securing space in the Universe and is from our position eternally fixed. We see the proton as vibrating because the proton is securing the permanency of singularity. Any light shining permanent is actually flickering and the flickering that is so fast gives us the impression that it is permanent. Anything less permanent will seem broken into fragments when holding comparison to what we find to be permanent and at the same time anything flickering $((1836)^3)^2$ faster than light becomes solid.

At the present moment science is looking at the speed of light as being three hundred thousand kilometres of space being displaced every second of time on Earth. This again is a speed. The speed of light is a form of gravity that has gone anti-gravity. It places distance in relation to time but according to science they place the emphasis on space where the emphasis should rather be on the time factor. That space we are looking at when we look into the night sky and that space which we think we see when we look into the night sky is not the space factor what we truly see when we are looking at the larger cosmos forming the geodesic space-time. What we see in outer space is light bringing images of space gone by, therefore it is time we see because the space out there we can't see. There is no more left of any of the space, which we think we see. We can see in the time that light brings and misinterpret that information as space. It is however the light coming across time, which we interpret as space that we vision as space but we stand to be corrected because it is in fact time, we see. It is a light holding no space a^3 because it is pure motion T^2k that is coming across the vastness of space. When saying space it is just a linguistic cultural expression because we know by now that space is time and motion of space. What we see the light representing in the image it brings across, is the space but the space is since long gone as it disappeared from the Universe. In the meantime all that remains from the time of the space is the flatness of time delivering the light that represents not the space of back then but the time now forming an image of the time then as it was time that was crossed and not space that was crossed in a flat dimension of time. The light is about space. Yes, but the space is history that can never repeat and such history is not time because time is what it took to carry the light and not the space from there set in when to here set in now. The picture of space we see is about space but not space. It very well is time because the light is the image of space where the motion of space only left the image of a vision of light representing what the space was while light by duplicating an image but it is carrying the image of space gone by. The space in motion is in a completely different location. History is once again repeating as it is duplicated through the effort of light. The light is duplicating an image through time (it is not the space duplicating but the light coming from time of the space that is duplicating) and that is not the same thing. In the sun on the very inside, the displacing tempo might take three hundred thousand years to displace one kilometre of space or said in another way, it takes the photon three hundred thousand years to travel one kilometre. The space creates time and if the space is dense the time that creates the space must be slow moving. Einstein declared that time moves slower in larger gravity. In this it takes light an effort of about ten million years to escape from the inner core of the sun to the outer ridge. When space reduces time expands and when space expands time reduces. $k^0 = a^3 / T^2k$ therefore the bigger k will bring about a bigger T^2 since it reduces a^3 in the process. On the other hand will a bigger a^3 bring about a longer k and thereby shortening T^2.

From what we read into Kepler it says that time contracts space at a specific value and that value remains a unit that brings about a rule. The factor k moves one specific distance creating a very specific space a^3 that takes a specific time T^2 to duplicate. This ratio applies throughout the Universe but the relevancies of the factors as the factors stand is in sequence with singularity extending. That is why the atomic conditions applying inside a Black hole is almost unrelated to atomic structures in the sun or on Earth. The applying conditions demand different requirements on space-time brought about as the space-time is influenced by singularity at that point. The atom changes space-time to comply to standards the singularity allows and with what standards are prevailing within the borders that singularity establishes. It is because of this reason exclusively that stars vary so much in the way they present and react to space-time. If it is motion that contributes to space-time differentiating in the form we call gravity and gravity is exclusively coming from within the atom it then is highly advisable to find the properties and influence of gravity within the atomic relevancy. I elaborate this aspect much more in another book of mine called **A Cosmic Birth...Dismissing Nothing I.S.B.N.0- 620- 31609**

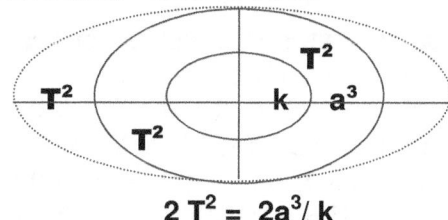

$$2 T^2 = 2a^3 / k$$

That is the relation there is . At the point of cosmic birth the cosmos is Π^3 as Π extends from Π^0 becoming 3D through spin. Where spin ends space ends and gravity ends. It is a unit undividable. Linking Π^3 to Π^0 is the motion Π^2 brought about by spin or rotary motion.

The proton forms the first line of contact from singularity into space-time. That is the manner in which the atom sets space-time as well as rules about space-time.

There are four time sectors in the Universe coming about from singularity and singularity fills every one. In the one sector there is a proton Π^2 connecting to singularity Π^3, which is connecting singularity in the form of Π to singularity in the form of Π^3. In that there is the Kepler formula $\Pi^3 (a^3) = \Pi^2\Pi^0 (k T^2)$. Then there is the other proton filling the second opposing quarter of time implementing the same procedure with the same result coming about. In the third quarter is the first neutron connection following an identical path but linking to a proton forming the space-time. In the fourth quarter the motion divided even further by splitting the motion as it conforms space-time from **3 to Π,** which in the end is the motion of Π^2. In all instances space a^3 is confirming singularity k^0 in time T^2 except in the one that produced the Big Bang. The k^0 confirms T_1 to T_2. Singularity is and remains Π therefore Kepler takes on Π but the dimensional impact remains the same. This connecting happens on both sides of the Universe. Time has four parts to fill space and maintains singularity by criss-crossing it fills singularity by connecting directly space-time through the proton to proton $(\Pi^2 + \Pi^2)$ or the proton to neutron (Π^2) or the link of the neutron to electron link $(\Pi X3)$. The proton removes space by doubling space to a point of the oblivious. The space the proton duplicates is so little and the duration so long that eternity devours space into the oblivious. By the proton circle reducing as it constantly duplicates, the duplication becomes so fast that the period it takes to duplicate the space becomes eternal in the centre circle. The atom takes space where space reduces into the electron as a factor of 3, where the neutron further reduces the space and extending the time by the factor of $\pi^2\pi$ and in the end the proton diminishes the space-time to the measure of $(\pi^2 + \pi^2)$. Adding this total confirms the mass difference between the proton and the electron at 1836 times. The proton has a mass of $1,673 \times 10^{-27}$ which is $1836,12$ times greater than the electron's mass of $9,109 \times 10^{-31}$. $(\Pi^2 + \Pi^2) \times (\Pi^2) \times (\Pi X3) = \mathbf{1836}$.

Mass comes about as a result of one component duplicating space in a motion that is inferior to another more prevalent duplicating of the space holding the inferior duplicating space as a unit. The difference in reducing from the electron that is holding a displacement rate of space-time that is equal to the speed of light and reduce space-time from there even further to the dismissing of the proton where in that there is a difference or a reducing of space-time or if you wish, a stronger restriction in the flow of space-time to the tune of 1836 times. But in all that it is exclusively about motion speeding up as space reduces from the point of entering the atom to the point of departing from the atom onto singularity. The entering and the reducing of space-time produces time relevancies. However we look at it as mass but mass is only a product of restraining the moving of space-time, which produces resistance. That resisting is about restraint to conform in shape and change the independent space-time to what form is needed to comply with the solid structure. It is a fight for dominance and liberty where mass is that restraint of further motion. By reducing the space-time the lesser singularity is claiming singularity independence by offering reduced space, which will result in promoted time with heat being the net result. That heat is filling the space, which should then be entered by the independent space in motion if the motion of duplicating is brought closer in a relation to what the matching tempo requires.

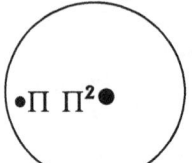

It is the mass that space generates where the space has to reduce the size by becoming more intense and concentrates space-time to the time of 1836 time more when entering the point of the proton being on the verge of singularity. In single dimension seen from one aspect, with single dimension contacting the edges forming the sphere it will still keep the seven positions because the sphere remains a unified structure though apart because of singularity. In the core of the sphere the proton connects in alliances as $\Pi^2 + \Pi^2$ with the solidity of the neutron holding Π^2 as a second forming a π value. That brings the atom unit of π to a number of seven.

It is clear that the density of material in motion is $\Pi^2 + \Pi^2$ but since that is $\mathbf{k}$, which extended we know that that extending cannot sustain the initial speed. Since the speed is reduced the space in motion will value less. Taking these atomic relevancies into account, we can detect what relevancies brought about the atomic Universe of $(\pi^2 + \pi^2)(\pi^2\pi)3 = 1836$. The first substance that formed from singularity was solid and if that was the case the contra substance would then be a fluid with less motion filling more space taking a shorter time duration in duplicating. The fluid substance that then formed was one less than the proton in motion which makes it slightly more in mass since it duplicates more with less space that then has to form $\Pi^2\Pi$, which has one Π less from $(\pi^2+\pi^2)$ which is resolved becoming a fluid like substance relevant to the first solid substance which is the proton. The loss of the one Π then became the factor claiming more space that is holding less substance. In this fluid state the neutron has more duplicating of the substance than is required of the proton. That what we find in space we also must find in the atom because the cosmos is not keen on inventing but is passionate on duplicating. This fact will also apply to space-time in many forms. That means investigation must prove the same results and what we find in the atom then also has to present in the cosmos at large.

From the centre to the outside there is a connecting of **k**, running the length of space-time in relation to the T^2 that brings about the liquid or neutron form. The fact of the liquid neutron is to produce duplication by repeating **k** as $k = a^3T^2$ whereas the role of the proton is about relinquishing **k** by $k^{-1} = T^2/a^3$. In the centre of all the atom's space is relinquishing a position through the dissolving of heat by means of maintaining singularity. But through it all, another singularity forms in the very centre of the structure, claiming the position where space is the least available that binds the singularity of all the atoms sharing space as a unit. With that evidence I realised there is a connecting of singularity and that connection is electricity. Later on I prove that electricity in its most intense form is gravity, which we will find in the very, very centre of the Earth. In the cosmos all objects form a sphere. Some solids do not seem to be a sphere and space is no sphere, but the truth is hiding in the way of connecting. At the centre connects T^2 forming the base of the solid. At any one specific given point forming the surface of the sphere is another marker holding the connecting relevancy of **k**. When there is no sufficient heat to form space that will part **k** from the other holding of **k**, the two will combine in a solid joining connecting as 2 **k** that translates to T^2.

In the way space and the sphere connects, the sphere will have 7 points holding a relation to 3 points not within the sphere forming the 10 that creation started with. This will mean there is a division forever, and such a division may run smaller as it runs everlasting. With fluids connecting it is simple to recognise the sphere, as **k** will indicate the point-to-point location of **k** as the form of the sphere. By gas forming the connection there are the three points of space being apart and not forming **k**, but still holds a relevancy to T^2 through the value of **k**. The gravity applies as much to material as it applies to form installing form. In this way stars are spheres while the stars are just one more cosmic atom and all rules apply as much to stars being just cosmic atoms as the rules apply to atoms being just individual stars. Remember that big and small, hot and cold, tall and short or any other boundaries we humans can see so clearly are no valid boundaries to the cosmos. In the cosmos a star filled with atoms and with the compliment of atoms working in conjunction, all the atoms eventually unite and combine as a unit. In the very end of their developing of the star, the selected and developing singularity will again, as it was before the beginning of time, combine all singularity in one structure fitting and holding all singularity of every atom within the star unit. The Universe will end with a united singularity and the Universe started with a united singularity where from there is no size but there is just space-time. The rules in the cosmos are the same applying to all in the same manner. The relevancy of the atom being $(\Pi^2 + \Pi^2) \times (\Pi^2) \times (\Pi X3) = 1836$ remains but the structure of the atom relevancy demise as space demise by prolonging time in the growing star. It might be beneficial to compare the atom and the Earth in composition. At the spot where $k^0 = \Pi^0$ becomes $k = \Pi$ space comes about in the form of $a^3 = \Pi^3$ by the measure of $T^2 = \Pi^2$. That puts space a^3 or Π^3 in relation to three sides and it will forever be three sides coming about from one point. $a^3 = k\,T^2$ where $k = T$ and therefore $k^2 + k = k^3$ or $\Pi^3 = \Pi^2\Pi$

The sphere was the first to come about and only after the sphere could not produce the gravity required to suppress the overheating forming the antigravity did the Universe try other options. The double proton formed but by that the rising heat excelled. The proton deformed where one became a neutron and the other became space in heat. With the demise of many protons favouring the forming of other particles and substances such formations united. Then the atom came in to form...did the atom clusters form...did the material clusters form...did the antigravity apply enough to allow overheating to bring about expanding where the relevancy will produce one softer and one firmer structure...from which the antigravity expanded into space. When the heat turned to apply antigravity that eventually produced space, as we now know space to be the Big Bang was happening. But before the Big Bang there were some mighty jolts, cracks and jerks announcing the Big Bang to come. It was the proton $\Pi^2 + \Pi^2$, which afterwards connected to other protons which became independent Π^2 connecting yet again to another Π^2 particle that with help established further antigravity which then overheated and expanded to Π^3. One can see there were many stages to produce what we now have. However most important is the fact that if the laws that once implemented the stages of developing all periods in the Universe then those laws must still apply and must still be present in our Universe in the same way as what it was when Creation started.

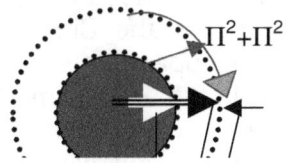

In the circle $\Pi^2\Pi$ which consists of the atmosphere the space surrounding the rotating object will also extend by Π as the concentration of the spinning motion draw or drag on past Π^2 extending the influence of Π^2 by the value of Π. Through very clear evidence about this one can see as the Coanda effect. This extending

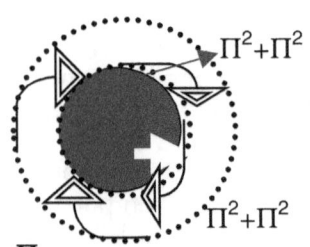

Π

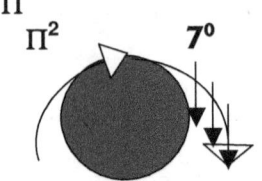

With the circle being Π^2 Π the Π^2 will reflect the circle in the square with Π forming the extending of Π^2. This is an extending of the six Π forming in alliance with the centre Π. This produces that any extension of 6 forming material one further extending goes into space and relates to a seventh dimension.

"How many dots w[...] was undividable solid and yet it did group together to form every sub atom particle located in the 3D. From what is now we should be able to trace what was back then when Creation was still fresh. Let us dissect how our Earthly atmosphere brings about duplication or dismissing in relation to singularity boundaries applying on the ground and in the atmosphere.

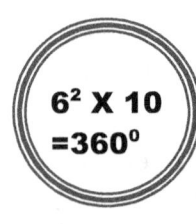

The extending of Π will not end immediately but will carry to the surrounding space the circle influence through rotation. The influence immediately above the circle will have the biggest influence and reduce gradually as the value of Π reduces in the leverage that the space has on Π and a gradual but definite change from Π to r will affect the extending of Π progressively more. The decline of Π will follow the same contour of the circle at 7^0. Every one of the dimensions indicates an individual significance as I shall show later and the increase into space runs by 7^0.

Individual singularity and governing singularity and group singularity enhancing the gravity every time singularity find an accumulation of influences. With our ability to look at Kepler's in a mathematical sense it is clear that from singularity comes space by three duplicating space in time by three. $k^0 = a^3 / (T^2 k)$. Very clearly the dimensions produced space and produced more space by applying time and gravity as motion. The space comes about as three and the time coming from singularity that is standing still to the motion, which we call gravity, three duplicates three. The universe came into position by deploying dots supporting other dots and some dots remained dots while other dots went on to become dots of hybrids as it was supporting dots through claiming dots of lesser density and pass that on to dots with larger density. We must also see that space within particles stands related to singularity, which through material also implicates space, which does not hold material as dense as we recognise material to be. It is in this that three comes into contact with three that is in contact with three plus one. That gives a total of ten and that brings about that singularity contacts the Universe in formations of ten. It is one of the most crucial properties we have to recognise to understand the cosmic relevancies.

I do not see that the Creation started with one particular or another specific relevancies but brought about that all particles grouped and as relevancies dictated different circumstances so the dynamics changed to accommodate the Creation requests. But laws were in place to guide the development according to precise lines to follow. It is however quite clear when looking at stars advancing the development that there is the tendency to reverse the sequence followed during the Big Bang. Looking at stars and in particular Neutron stars we see sufficient contraction of space-time where such reducing of space-time brings about a full scale demising of the Neutrons in the space-time the Neutron star claims. Then as a final phase in the development of all stars when conclusion comes to singularity development that ends the final development, there is no space-time left within the margins of the star. Then as the star progresses to the final development, we see the dismissing of the proton by ejecting the double proton out into outer space. Following the trend we see as the

Black hole does it, it does seem that a method of producing came into place. Remember the star is the Big Bang in reverse.

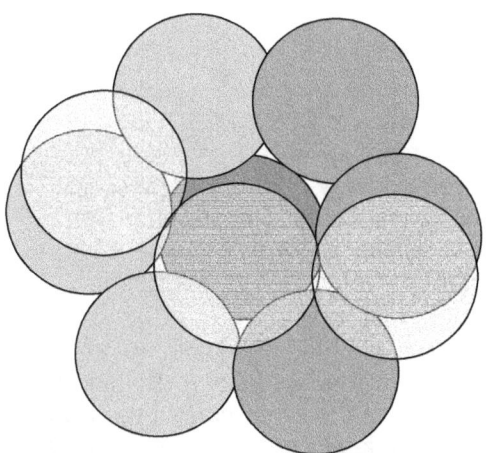

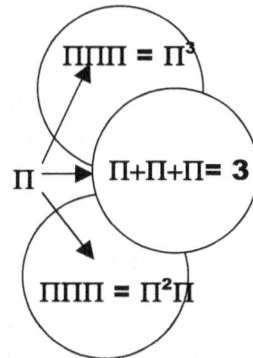

$\Pi\Pi\Pi = \Pi^3$

$\Pi \rightarrow \Pi+\Pi+\Pi = 3$

$\Pi\Pi\Pi\Pi = \Pi^2\Pi$

Space formed as motion came about through singularity overheating. Singularity k^0 produced motion at the point where k^0 became **k** and a^3 became T^2 by motion duplicating space. But according to Kepler a^3 is equal to **k** relating to T^2. Using that, I have already shown how three is a duplication because $a^3 = T^2 k$ and this shows that every aspect of creation used three Π to create space in time through motion and space Π^3 secured by time $\Pi^2\Pi$ applying motion and space using time to form space $\Pi^2\Pi$ **X3.** This formula is the atom, which is the universe because every secluded atom forms a Universe apart from other Universes. $(\Pi^2+\Pi^2)\ \Pi^2\Pi$ **X3**

Matter formed where matter had to have $\Pi\Pi\Pi = a^3$
space to occupy since matter was to use in some space $\Pi\Pi\Pi = T^2 k$
in the space outside the atom $\Pi^0+\Pi^0 + \Pi^0 = 3$

therefore $\Pi\Pi\Pi$ **met with** $\Pi\Pi\Pi\Pi$ **to form the proton in** $\Pi^2 + \Pi^2$ because the matter is within the space it holds and another Π^2 employs Π as a representative of singularity. This then placed the seven positions of singularity as the ending of matter and the three squares ($\Pi^2 +\Pi^2$ **and** Π^2) of singularity as the limit of material. The last $\Pi\Pi\Pi$ became k^0, k^0, k^0 and that became the space producing heat without occupying matter in order to allow heat to be restrained inside the dome singularity provides. When I refer to an atom it includes all unified cosmic structures holding an excluding formation such as an atom or a star or a galactica does. This is where **k** defines many but as a whole also one a^3 / T^2 determining space within time holding space within time to the value of unifying the lot in one cosmos container. There is little difference in the cosmos in comparison with for instance a lead atom or a giant star or a large galactica. It is all space confined to time extending singularity because in the cosmos there is no big or small. From that the effect of gravity as a restraining on the exploding of space came into effect.

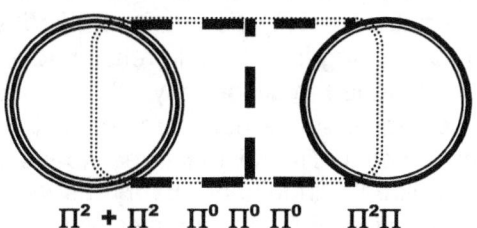

$\Pi^2 + \Pi^2$ $\Pi^0\ \Pi^0\ \Pi^0$ $\Pi^2\Pi$

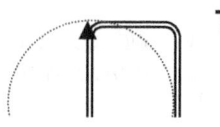 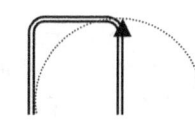

TOP

It is all about relevancies applying the relations gained and lost through relations. If one place $\Pi^2 + \Pi^2$ on one side then $\Pi^2\Pi$ is related form where $\Pi^2 + \Pi^2$ is in the other side of the Universe being on the other side of the relevancy. Then $\Pi^0\ \Pi^0\ \Pi^0$ will again relate to the other two factors forming the "outside" of the other two being the "inside".

But when the Universe was in the single dimension, all values were Π, therefore every value related to $\Pi\Pi\Pi$ forming three of the same that was very different because it

MIDDLE

$$T^2 \longrightarrow k \quad \text{and} \quad k \longleftarrow T^2$$

BOTTOM

Singularity | Dividing | Singularity

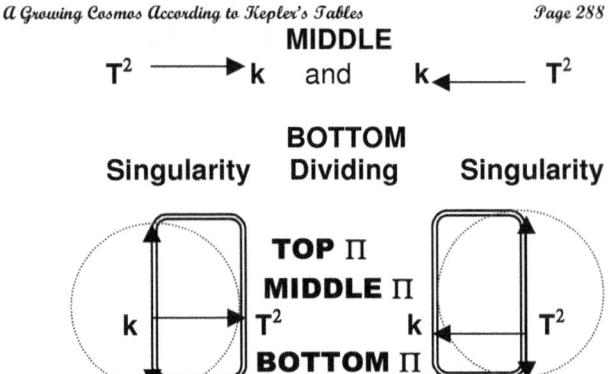

TOP Π

MIDDLE Π

k T^2 k T^2

BOTTOM Π

Singularity | Dividing | Singularity

The names I use in TOP, MIDDLE and BOTTOM must not be viewed as sides but merely as terminology using names to implicate divisions. Direction depends on positions and positions form a value only when the observer forms part of the cosmos and not part of the observing.

The universe divides into two separate issues because of singularity. Nothing can be in two places at the same time where as all the rest in the Universe has to confine to the law applied by singularity. Objects can only be in one side of the universe holding three parts or in the other side of the universe holding three parts. From the totality three will be a double with six sides to show, but that forms 3D. From singularity it is flat with three sides forming on either side of singularity.

During the Big Bang two things happened. Particles all overheated. Heat established motion wherever overheating applied. By overheating, material enable the securing or claiming of more space. Only by overheating or increasing heat can material claim more space. In order to supply **k** with any reason to grow into where **k** then becomes the fibre of material, there had to be a way fitting natural processes to do such extending. The precondition of material to grow and claim space is to accumulate heat, and that must have happened because we can trace the excessive heat there was even today. If **k** grew, the temperature had to rise. But all temperatures were the same in singularity. Singularity is homogeny in all areas including heat. Singularity can generate heat on one condition and that is by producing more spin. The Coanda effect is vivid proof of my statement. But to generate heat it must spin at a higher rate and by spinning singularity then produces more heat. That is in motion and all proof still exists that singularity had and has no movement. Spinning the top will bring the top to become more existed and then attempt to elevate the motion of the top to separate from the Earth gravity. In singularity there was no motion yet and there was no heat yet. Still, to get **k** to extend, the temperature had to rise. Let us have a good look at gravity.

As the relevancy between the particles promotes overheating or applying antigravity (overheating) to the responding cooling or applying of gravity, the one repels material into space-time while the other is collecting material into space-time. The one being uncontrolled singularity loses material and that ensures a model of preventing further escalating of overheating while the other gains material as it removes what the other produces as a result of overheating and thereby the controlled singularity sustains and prevents overheating being a model of fighting overheating. By gaining the one prevents overheating and by losing the other part prevents overheating. The one principle we named the Hubble constant where overheating produces space and the other one we called gravity where gravity is demolishing space, but both phenomena is at present dominating the flow of time in the Universe and will do so until eternal relating again comes about.

At first when material presented one side of the Universe, matter had three sides to show. Matter had to have space to keep matter somewhere in some part of some universe and that made up three positions. Between the two universes **k** and T^2 placed a value but since only singularity applied any values the value therefore was **k** T^2 where $T^2 = \Pi^2$ indicated time coming from 7/10 in relation to 10/7 and $\Pi^2/2$ (proof of that is somewhere in the book) and   **k** = Π valued by singularity. When space-time developed 3D the dimensions falling outside the sphere becoming space-heat formed as $\Pi^0 = 1$. The electron holds a relevancy of 3 relating to the Neutron being $\Pi^2\Pi$ and the three keeps the electrons in different universes relating to separate or individual singularity. Every aspect of cosmology is influenced by singularity and singularity carries the value of Π in many disciplines. The ratio started at the start and that ratio still applies.

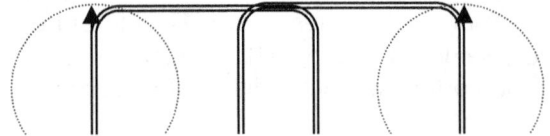

The relevancy between the two particles secures individual positioning between the opposing particles, which positions the material so that sufficient space secures

$$\Pi^2 \longrightarrow \Pi \quad 3 \quad \Pi \longleftarrow \Pi^2$$

Singularity   Dividing   Singularity

There then came a tendency to remain solid or become softer and more flexible. That is evident by the simple deducing of logic. If there was no space available and everything was solid something had to give in. We have a solid state (freezing at $?^0C$) and we have a liquid (boiling at $?^0C$) therefore liquid is the addition to a solid where the solid state forms a liquid after a certain relevancy has been bridged. It is not the solid element of whatever element that becomes liquid because all elements are natural solids. After conditions have been bridged and heat surges the liquid engulfs the solid to protect the solid from expanding beyond borders. Some had to apply a more fore giving form to allow shape or form to come about. We still find evidence in this since some elements hold heat relating more rigged while other comply much more willing to adapt and comply with heat deforming the structures. In this there are matter that grinded the other antimatter into heat and destroyed most of the antimatter that destroyed form to space not connecting to form demanding singularity any longer.

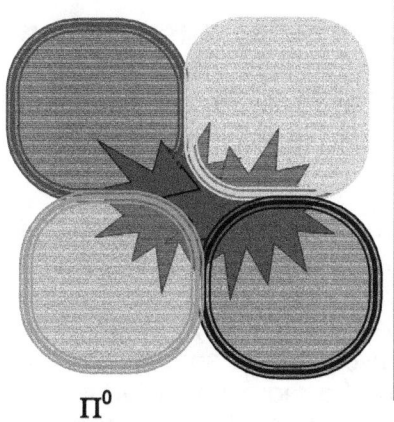

Kepler show that motion and space is the same therefore all space involves time and all space evolved through time. By repeated duplicating space grows using time or motion of space. The Big Bang is the proof. But through time motion places discrepancies and that also brings synchronised harmony. That could not have been in place when the very first motion brought about the very first space. With all space coming into birth from singularity all space was filled as some space was spinning less, which had to bring about more space developing by heat expanding and other space contracted by reducing the space through material motion.

Π^0 Π^0 Π^0

$\Pi \Pi^2$ Π^3 $\Pi \Pi \Pi$

Singularity split the Universe into two parts that under no circumstances can ever meet. The one side of the Universe performs a balancing act to the other side of the Universe that duplicates but never double. The dot started overheating while the dot remained cool by activating gravity, the dot duplicated forming a sequence while the dot redefined the position in control.

Space
$\Pi\Pi\Pi$

     $\Pi\Pi\Pi$ **Matter**

$\Pi\Pi\Pi$

Time

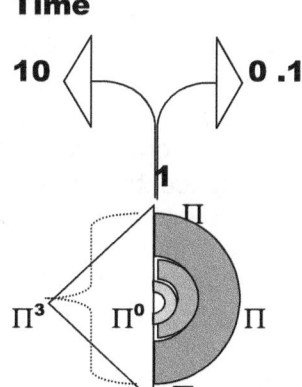

With the first dimension came matter, but also came space and came time splitting the universe in segments of matter relating to space filled with matter and time influencing the spinning matter.

Taking the queue from the numbers line that runs in opposing directions Π^0 going larger as well as smaller. But the centre takes a value of one. It is a private choice preferring $k^0 = 1$ or $\Pi^0 = 1$ but that splits the Universe into two part, being smaller and being larger in relevancies.

It is apparent that one cannot substitute the correct formula used to measure the area of a circle by using $a^3 = \Pi r^2$ because if k is the diameter then the formula must be $k^2 \Pi$. But k cannot be Π because in Kepler's formula k takes the value of the radius. In that case what will the value be of T^2? That places the formula outside the normal use of mathematics practised in the normal sense of $a^2 = \Pi r^2$.

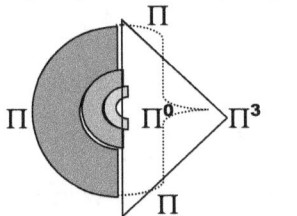

Space

ΠΠΠ   **Matter**

Π   ΠΠΠ

ΠΠΠ

Time

If k is the middle being $k^0 = 1$ then $a^3 = k^0 = 1 = T^2$. When time is in a shift freezing then $a^3 / T^2 = k^0 = 1$. In order not to overstep my limits by changing valid formulas I changed Kepler's formula to $R^3 = T^2 = 1$.

But the book being written in Afrikaans the R stands for Ruimte meaning space and T is time. From that I deducted that the space used in a specific location will equal the time meaning the density of the heat in space. That brings the proof that space equals heat and space is the same as heat. Heat deforming or exploding is the equal to the space created. Also it confirms the substitute between Kepler and Π is correct

By the same token dismissing brought on reducing of space-time where no space was available in any event. With no space available and a demand for space as big as the Universe self in size some had to go soft to allow others to remain solid and that set the course for the next stage to last the one eternity to come. There were to groups then formed and the two groups are still today present. The one group went on to form two groups and that we call the Big Bang event where space and heat separated for the third time. It was all about securing space by demand of a requirement for space. The proton and the neutron came about and one softer applying motion by duplicating part came in relation to another dismissing space to remain solid part. But further collaboration was required since the demand on space was far from fulfilling to satisfy the need that was developing. Proton /neutron clusters grouped as the successful cluster drove the unsuccessful particles to form more heat in the separating of space and the heat still went softer. As the dismissing proton clusters demanded space-time to stop allowing further overheating and the duplicating draw heat around the sphere the neutron formed as protection from the overheating. The heat formed space, which the neutron duplicated and passed on to the proton to dismiss as its part in the sustaining of singularity. The proton cluster $\Pi^2+\Pi^2$ entertained the space producing neutron $\Pi^2\Pi$ to engulf the sphere which now is called the atom with heat that relates as three from where the neutron taps heat to duplicate as well as send some of it on to be dismissed by the protons. The accumulating heat in the third relevancy then turned to uncontrolled space producing from that the Big Bang. In the 10^{-42} to time zero that is accounted for a development took place that took many eternities to take place. Today we find secluded space that we call atoms covered by space we call space and the secluded atoms group to form secluded space we call stars in between space we call space.

There then formed material $\Pi\Pi\Pi = \Pi^3$ and there formed time (matter in relation to heat surrounding material) $\Pi\Pi^2$ and then there formed the space to fit this lot into $\Pi\Pi\Pi = 3$, which with the singularity including Π makes the total value space-time in motion has being 10Π relating to the Π^3 bringing about gravity or contracting Π^2. The cosmos formed in equal part forming from the same substance, which might in some sectors be more or be less concentrated but remains equal. The entire cosmos resulted from one basic form of material (π^3) that formed motion ($\pi^2\pi$) as well as space $\pi\pi\pi$ to fit the lot but initially it was all the same ingredient. The substance forming the entire Universe came about from a natural occurrence: heat in solidity contracting heat in solidity that caused friction. It formed not antimatter but a natural occurrence: heat and it did not vanish from sight it became a natural substance: space-time. The results are the gravity and the anti-gravity we know.

Why would a water drop floating in a space capsule in space, in micro gravity always form a sphere when left to capture free form? We all accept that the true cosmic form would be and most probably will be the sphere...but why would the sphere form as the original form when matter is not pre-cast to have any specific form and therefore take on by cosmic pre-cast the sphere as form? In the past, all intellectuals looked passed this question but in this concept we find the origins of the Universe.

We know gravity is there in the very, very centre, but the qualifying of gravity as a force lets the process of investigating science a bit off the hook. Things become rather simplistic in the modern age when all else is so highly investigated but gravity is merely defined as a force influencing matter. Why would we find in space, where there is supposedly nothing, something we named micro gravity and what would bring about micro gravity when gravity is not present? What would cause gravity up to one point, which is in the gravity region of the Earth and one step away from such a point where there supposedly is nothing; we find a phenomenon, which is micro gravity? Answering this question will take us back to the origins of the Universe.

By freezing **k** such a value came about and set the Universe into a concept other than being the one unified lump of not being anything of sorts. When creation came about and when the spot moved pi from the spot π^0 to the dot π heat came about, but being heat, it froze solid by motion. As everything is heat, motion is what freezes the heat solid. The freezing action brought on **k** and **k** brought on the freezing of space through spin. By the extending of **k** did the Universe obtain space-time a^3 / T^2. However, it had to produce **k** by freezing **k** into existing from where k produced a^3 / T^2. This we have to understand about fusion. Fusion is freezing liquid to solid and space into eternity. Fusion is about creating a freeze in the deepest of heat and where all is engulfing in heat surrounding all, from that freezing must come about what is to apply the freezing we call fusion. It is this first action ever that has to repeat to establish fusion. It is in gravity attempting to secure the most heat under the prevailing conditions where gravity eliminates the most space to establish a freezing centre in creating fusion. But that does not solve the indicating action.

How did the Universe liberate material and heat space from singularity because with singularity comes an unchangeable eternal condition that is non-changing-everlasting in all conditions and aspects that is remaining in absolute equilibrium. This equilibrium maintains because all development extends form precisely in a detailed equal equilibrium throughout. Think about what brought the cosmos out from the eternal rest it was in. The eternal rest still maintains and is therefore our detection. What inspired the eternal rest the cosmos was in and inspired change to the state of eternal rest? What evoked change? That is the question the Atheist will never be able to answer but that too is the most basic and ever-lasting fundamentals of the Universe. What changed in this split second start before the official start? I do not wish to ponder on this matter in the letter I am writing at this minute, as there are other books where I delve into this matter. From the deep freeze of creation came the Hot Big Bang and the 3D Universal displacement came about with the relatives being $10 \div 7(4(\pi^2 + \pi^2)) = 112.795$ and then a second one established 3D by introducing the six to seven sides Universe at a density point of $7 / 10\ \pi^6 / 6 = 112.162$. There is of course a lot more information about this establishing of the Universe than what I mention at this point. The question is what made the Universe freeze, to form the Universe in space and through time. It had to start with a specific reason applying, which brought about space-time. Once the process started there was no stop to it, but there is no chance that the initiation of the start was spontaneous by nature. With everything being in one spot, all within that spot was in a state of eternal rest. While all remains the same and nothing changing, what brought on the sudden change of everything, shocking everything from and out of the eternal rest? With $k^0 = a^0 / T^0$ all stood still in singularity and that factor is still with us controlling and generating the cosmos. It is there for all not to see and for every one to establish.

The effect coming from $k^0 = a^0 / T^0$ and about $k^0 = a^0 / T^0$ is beyond denial. The part of proof being beyond denial is because it is still there for all not to be able to see it. That centre spot in the rotating of objects is up to this point of cosmic development as incapable of having space-time and can only secure space-time as it was capable of at the start and as it will be capable of up to the very end. In only $k^0 = a^0 / T^0$ being present, it was as stable as it still proves its stability. There were something external and from the outside of the Universe that was outside controlling the $k^0 = a^0/T^0$ Universe that set the lot into space being motion. That control coming from the outside of the Universe is still a provable fact; which I am about to prove in this very letter. That is the one aspect natural physics will never answer. What forced the cosmos out of the stability of singularity as $k^0 = a^0/T^0$. There are two controlling cosmic principles in control of it all. The one is gravity and the other is the light performing as the example of the epitome of antigravity.

There has never been any explanation offered about gravity being able to bend light except mathematical calculations producing the factor prevailing and the principle produced. Einstein declared that large objects producing massive gravity could bend light. Has any one got the answer to the question about light bending? If light can bend what is light then made of? If light is what science portraits about being a simple line running along a very flat piece of paper one might just find correctness in such an arrangement but that is if light is that simple to explain. Taking more facts into account, there is a lot more to light than such simplicity can ever cover. Light can bend if light was a solid with the flexibility of a taint of liquid mixed in between. But if you think of the broad spectrum of information that one stream of photons releases when coming through my eye about a Universe being so wide in its entirety it will take me several lifetime periods just to inspect the Universe close up where the Universe is hiding information about the most complex parts there is out in the Universe. It once more shows how simple mathematicians make concepts to portrait the most complex issues in the most meagre sense. Human abilities are meaningless when brought in line with the totality of the cosmos, therefore what we perceive as massive and what we perceive as little is in cosmic standard amounting to about the same. It is our norm we create bringing on our true human incompetence in realising limits prevailing in cosmic standards. If large gravity can bend light all gravity can bend light by affecting the flow of light. Gravity does not bend light because light is not a solid that can bend. Yet again it is relevancies in space-time in differentiating velocity that change because gravity is velocity differentiation between two cosmic principles filling space-time.

By dismissing the space through which light travels such dismissing must lead to rerouting or redirecting the path light will displace space-time. Also by duplicating space and duplicating photons in the photons travelling across space the photons are duplicating as they travel by changing frequencies of singularity. The path they then follow, will affect the harmony and therefore the flow direction of light. The closer the light travels or pass the centre core of any object the closer it comes to the main gravity within the structure because the less space there is available the stronger the pressure of gravity is and that alone will bring on a greater effect as the more reducing of the space conforms to density in heat. Gravity is bringing space in relation with time through applied motion. It is about space becoming reduced or redundant because of motion either by space moving or by matter moving in space-time through which it is moving. The duration of the time in relation to the space affected brings about the gravity applied. Gravity is space in relation to the centre while at the same time it is the centre coming in new relevancies to space surrounding the material structure. There are always two components to gravity in space being $a^3 = k T^2$. It is k in the way the centre stands in the space changing in relation to a centre a^3 but at the same time it is a centre T^2 changing relation of that centre to another centre k applying control. It is a constant re-matching prevailing centre influences on space occupied by material and space not occupied directly by material.

It is the time the space takes to bring about new positions in space occupied by motion and through motion that takes a certain duration in time to move from point to point. Gravity is motion of space towards and in relation with a centre and the time is the period such motion takes while that centre is attempting to produce space by doubling space through motion leading space away from a specific controlling centre to another specific controlling centre. The time it takes to complete such an attempt provides space the opportunity to double its status. Moreover, the time stands affected by this motion material creates to duplicate space-time by generating singularity and activating different locations holding singularity. Gravity is speed and speed in space in motion through time duration. Gravity is motion combating heat expansion and supplying space with space producing through motion providing the space the opportunity to expand while remaining in relevance.

At the start the gravity part invested heavy in material. However in space light came about since there is no gravity in space. Light is the attempt to establish motion not controlled by any centre and the time it takes to establish space between such a centre of space control and the light finding an ability to dispense the space by reactive motion. Gravity produced space by allowing as much as producing overheating with a feeble attempt to combat the overheating. But what is the meaning of heat if there is no cold to set the standard for the heat to become a value, which is then related to the other end, where such another end must be the limit in cold. The moment singularity produced space-time heat distanced from the cold factor. Space produced a cold base to have heat within. How did singularity part the shared principle of being the unification of heat and cold as a unit? Singularity froze in applying gravity bringing about particle separation within singularity. Heat and cold parted to produce frozen heat by atoms forming and captured heat by unleashing uncontrolled material that ended as space in time.

In the investigation of light and gravity as well as objects and gravity, the mathematical rule of the invert square law must apply without question. However, according to the observation of Roche factor that is not the case. From what one gathers through the Roche limit implicating two orbiting structures, the opposite is applying. One must accept that although **k** proves to be an indicator it is also much more when complying the thin influences brought about by singularity in the values carried on by singularity.

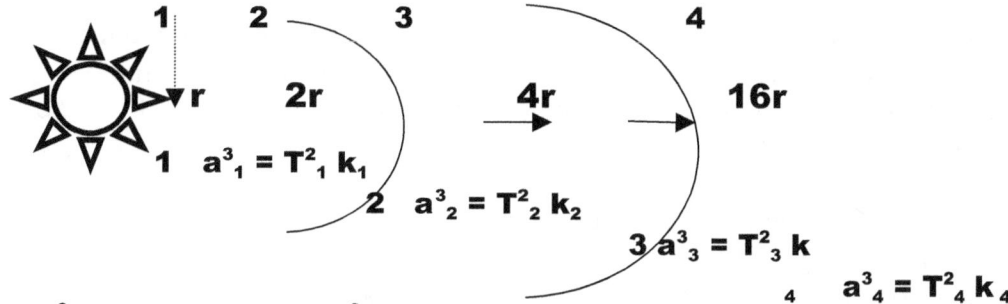

As a^3 increases, so does T^2 as well as **k** increase and with that the influence of gravity per space unit increases with the concentration demise of a^3. But why would that be and what are we missing? Light shows there is an influence out there in outer space, that redirects light's route through space when passing large gravity fields. It is about the relevancy of **k** influencing the a^3 to allow the T^2 of light to divert in route because of influences established by **k** on a^3 and slowing down or increasing the line diverting. In this measure one may also find the Roche limit applying, but to truly understand how the Roche limit comes in place and how the Roche limit works, one have to replace Kepler's factors with singularity and singularity extending being Π^3 $\Pi^2\Pi$ and **3**. The relevancy brings about the reducing of space to a smaller factor each time.

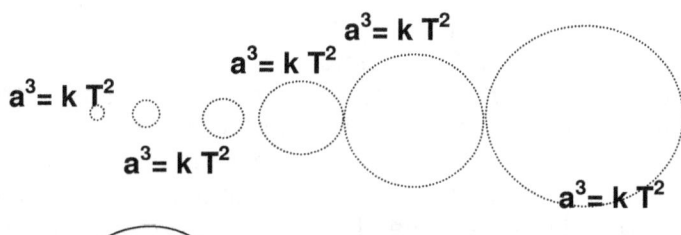

Even by reducing space as a result of gravity, space expands because **k** can never retreat to a previous position. Space will expand until time fazes out as being so short it has no influence any more.

$$k^{1 + .00000000000000000000000000000000000000000001} = k^{-.000000000000000000000000000000000000000000009}$$

$k = a^3 / T^2$ when space $a^3 = 10^{- .00000000000000000000000000000000000000000001}$ then time being a square factor of the third dimensional space $T^2 = .000000000000000000000000000000000000000000009$

The fact that k^0 moved to k produced a new value to **k**. By k having a new value T^2 will remain eternal and space will be in singularity. If **k** =1 then a^3 =1 bringing eternity to time with time being T^2 = **1**. This shows that **k** have to increase in order to relive the Universe from eternity by arranging a time component. The expanding will forever bring about motion and the motion will kill the space.

Heat forces singularity to expand, which thereby is producing space. The expanding of singularity causes motion of space by duplicating space through time. The motion of space realises that space has to be in any position lesser than the time would demand if space was less motional. That produces space being in a shorter time in one place. But with time remaining motion the motion then becomes the standard and that reduces space because the motion is actually the duplication of the space and with such space dividing between periods forming differentiation of the time, the time differentiation brings about space reducing. By occupying two spots of space in the same time split into two parts the space has to divide in size even if it still grows by overheating.

The shrinking of motion creates heat. The heat brings about space. The space creates motion and the motion reduces space but the reducing is never to the same state it was previously because of material cannibalism. This arises from the materials of compromised heat and secures growth of matter that provides cooling. In this manner the moving of space will capture some of the space it is occupying through motion. The space will enlarge just because the space captures more space than which it previously had. By motion it reduces space and by motion it captures the space it moves into. The motion will be there even if the motion is capturing space through expanding by means of overheating or whether the motion comes about by space duplicating by using time reducing thereof. Motion of space is as critical to space as space being space and space is motion that is forming space

Space a^3 is reducing through the motion of space by duplicating space in times motion and the reducing of T^2 by dividing time by half will have **k** as the constant. But afterward the scenario changes back, as space then is the constant with time reducing **k**. It comes about as the Earth spins around the Earth axis. I call this positive space-time displacement

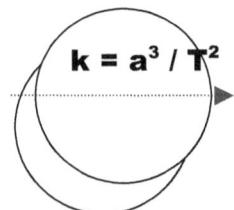 $$k = a^3 / T^2$$

Space a^3 is reducing by the motion of space with the implementing of T^2 having **k** as the constant. It comes about as the earth spins around the Earth axis. I call this positive space-time displacement

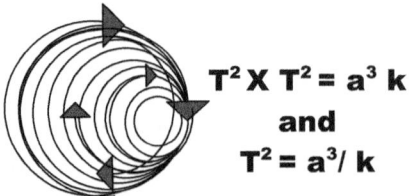 $$T^2 \times T^2 = a^3 k$$
and
$$T^2 = a^3 / k$$

Space a^3 is reducing by applying motion with the repositioning of **k**, having T^2 as the constant. In the reducing as well as forming duplication and in the reducing singularity captures as much as conforms some heat. This comes about in the form of securing singularity since growth is a natural dynamic part of gravity as Kepler's formula indicate $k = a^3 / T^2$. Proven by Kepler's formula, we see that space will grow by the single dimension because space is material that is captured in terms of one dimension of space when each cycle is completed outside the space occupied by material in space. That process involves the phenominon, which we call gravity. It comes about as the Earth spins around the sun axis and the space between the sun and the Earth captures by the sun as well as the orbiting Earth. After completing every cycle, the atom secured one dimension of heat by reducing one dimension of space. The fact that one dimension of heat removes one dimension of space and conforms that to singularity, it can only be the benefit of singularity as it secures the growth by the single dimension. $k = a^3/T^2$ shows a growth by the single dimension which can only contribute to securing singularity because singularity is the single dimension. The space that the dominating singularity captures, is also beneficial to the minor singularity because of the motion and therefore the committing of increased flow in space-time. I call this negative space-time displacement. In the space-time arrangement Kepler introduced the factor **k** increase by doubling **k** in the relation of $k = a^3 / T^2$ which increases space by quadrupling a^3. Motion by T^2 reduces space by half and that brings **k** back to one and a half times what it was previously. From the fact that T^2 advances (X) to a new location T^2 indicates a growth $a^3 \times k$ in space . In there is the Hubble growth but the growth is in the space claimed by material as much as the space not claimed by material The factor **k** could never move back to the original because singularity shifted into that slot. If **k** moved back to the starting value T^2 would stop and the one half of space a^3 that comes about through motion will vanish taking space a^3 altogether with the vanishing. The relevancy falls on motion **k** and spin T^2 bringing about space $k = a^3 / T^2$ by reducing and expanding space. Overheating brings about that **k** doubles. With **k** doubling space becomes four times larger. Then time applies motion and divides space in half. That brings about that every instant motion applying space becomes four times larger and then reduces by half. But such motion T^2 is at the prime of space a^3 and the eternal beginning k^0 of time where time just moved from eternity **T** one step to T^2. It is precisely where Einstein placed the strongest gravity. It is next to singularity. It is where space disappears and the motion that is providing us $a^3 = T^2 k$ time begins. It is where $\Pi^0 \rightarrow \Pi$.

There cannot be light without gravity and there cannot be gravity without light but that is on condition that the light we perceive to be light is only the symptom of what the Universe use as light confirmed. Light is not what we think light is but explaining that will be too extended to explain in this letter. Light is heat concentrated to a form almost material and space is heat dismissed. Therefore light, heat and space is the very same thing and all form the extending of the antigravity that is applying. However, gravity is more the redeeming and the re cooping of light where light then is antigravity. Light is space concentrated in motion at speed and gravity is reforming space by motion forming speed. Gravity is speed and light is space at speed. In drawing a most basic picture of light passing the gravity lines extending from any structure, I felt it was most insightful that the Brainpower in cosmology was not able to see why light does not bend in the presence of increasing gravity. More surprising was that I found the mathematicians had to call on Einstein for advice on a most ordinary problem. Light does not bend when passing large objects. It is Kepler's formula applying, and the evidence is clearly in front of the searcher for truth. But one has to go back to Kepler to re-apply the principle what Kepler formulised and change the significant from Newton's significance.

The Roche limit 5/2 becoming = (Π/2X Π/2) = 2.4674 as singularity interferes

Creation started with the Roche limit and in conjunction with everything else came about. In the Roche limit the space factor provides space to a solid structure and therefore the value of r is replaced by the value of Π bringing about a square in half of Π. The cube holding 5 to either side removes allowing the extending of Π to indicate position to space.

The space between the spheres divide in half, but because of the extending of Π and not applying r as ordinary mathematics will suggest where Π replaces r the singularity extending from Π^0 will be half of Π in the square of Π = $(\Pi/2)^2$ = **2.4674.** In this lies the dynamics why planets have a positional (be it rather a dimensional) relation of 7/10. Half of the five of the Lagrangian points is the Roche in conjunction with singularity. With singularity coming involved singularity will enforce the value change to fit Π.

5/2

Five sides divided by two spheres.

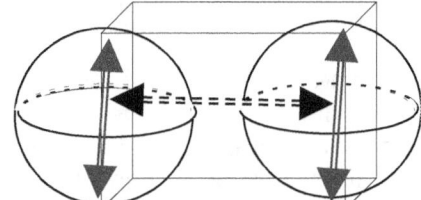

Where Π extends to lock onto the next sphere's extending indicator, Π has to connect to Π forming the square of space and translating that to the half of Π being $(\Pi/2)^2$.

When the initial and ultimate break came about where singularity moved from the spot π^0 to the dot π singularity established space in motion to the value of π and formed Π as Π parted from the rest. When looking at it from the other side there are ten points in reference to singularity. There are the four factors of time and by moving one point away the dot released space from time. In one dimension after time in the four came singularity by producing the motion of space in five, but the

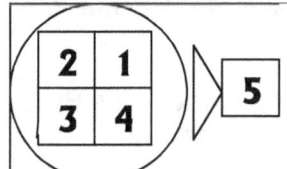

Time holds four positions and the first position in the dimension of space outside time is five.

Universe is about duplicating and repeating what was again and again afterwards. The second dimension the duplication involved five (see the Lagrangian explanation) above and below which is the half of ten and as ten rotor around an axis of one on the one side and a relevant .9991 on the other side bringing a representation to singularity which is always being present and when dividing this space not including the singularity by the four time factor an individual five is established. By adding the lot the total forming 21.9991 and as this is rotating in the relevance of the previous dimension where the sphere holds seven point one see that Π once again forms. But this time Π is always rotating from one side to the other and therefore again the Roche limit proved to apply as the one side had (Π / 2) that had to duplicate on the other side (Π / 2) X (Π / 2) = $(\Pi / 2)^2$. My discovery of this lead me to believe that all cosmic laws played a part as they repeated time after time with the developing of the cosmos through out all eternities that might come about. That meant the square of space (5+5) = 10 in relation to singularity on both sides of the divide forming 1 and adding that to the eternal 1 minus one in relation to the square of space being 10, which is on the inside and on the other side of the divide as singularity are forever about reducing which means it is .9991 and that forms once more 21.9991. When placing that in relation to the sphere having seven markers the square of space comes in contact with material forming $(\Pi / 2)^2$.

Independent singularity = .99991

Controlling singularity = 1

In all units there is a singularity seeking independence in relation to singularity elected seeking dominance. From one side and any singularity there will be present in the unit one factor of singularity carrying the value of 1 but also there will be one divided by space square singularity absent because only one singularity can apply to the unit in dimension. Therefore one tenth of the space is absent where singularity is one and holding .9991 valid as part of the other side of the Universe it is attached to but not connected to. With the unit being connected to motion the motion will stand related to seven from the one side as well as the full ten from the other side on both sides of the Universe. That relates as ten in space-time on both sides of the Universe (10 + 10 = 20) plus one factor of singularity present and one factor in the tenth not present (1+.9+.09+.009+.0009+.00009-.0001=1.99991)

The Lagrangian ratio is normal dimensional particle layout brought with development through time as a result of the Roche principle. One can see the Lagrangian system by looking at the elements developing clusters of five forming such characteristics of similarity in groups. Looking at how singularity applied connection in 3D there will always be five relating to a centre where that centre carries half of the value. This forms the principle behind the Lagrangian system. That will be singularity taking a centre with one point to the top of the centre and one point to the side of the centre. The centre then becomes the one side in corresponding to the side the centre takes. From the centre another connecting will be to the bottom the front and the back. Every point will form a position where it will support the centre in displacing space by providing heat in an attempt to secure the prevention of overheating. The function of the linking of five to a centre as a group effort is with the group work individually to secure the survival of the individuals forming the group and being the group but as a group as well as the securing the group as a unit by a mutual concentrating of space.

By halving this effort in the Roche proves the motive behind the forming of the five in the Lagrangian points because of the destruction or development that space less ness then in Roche brings about. The five to one forming the centre of the governing gravity by establishing a principle singularity such a centre improves the gravity effort by all taking part to dismiss heat and create a concentrating heat flow to the centre whereby the group as individuals and as a structure will survive through mutual gravity. This is the effort of the whole cluster in underwriting the centre spot control. It is plugging five spots the same as electrons do and in that accelerates the space flow by reducing the space flow as a motorcar carburettor does. In producing mutual support it underlines the individual support required by all participants to prevent future overheating or antigravity. The extending of space collecting and the extending of space reducing prevents overheating by establishing matter in six by connecting matter in six to a centre one. That centre one has the factor of less than one forming the Alfa singularity to the cluster forming the gravity.

Where the Roche factor is the singularity influenced by half of the Lagrangian system, the Titius Bode is the dimensional duplication of the doubling of the Lagrangian system in space occupied and space not occupied by material.

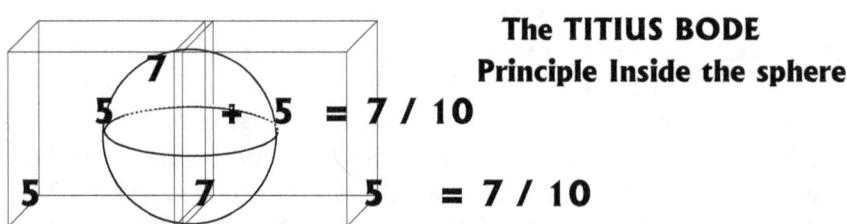

The TITIUS BODE
Principle Inside the sphere

5 + 5 = 7 / 10

5 ... 7 ... 5 = 7 / 10

Space-time is a four dimensional position of the universe where the position of an object is specified by three coordinates in space and one position in time

When concerning a sphere by establish the Laws of Pythagoras to provide a centre principle all lines running through the centre will be effectively related by groups forming 180^0 and 90^0.

We take a line running between two points as being 180^0 and the rest of the explaining is saved in the accepting part of mathematics. Any one of the two points where the line starts or ends at, is a point in infinity. The start and the end depend on the viewer putting the relevance to favour the side of choice. That puts the point of end or beginning in the spectrum of choice and not fact. Any direction is as equal as all other directions.

By rotating around a centre that is standing still such a centre forms a divide that separates the unified unit. **Any point will be opposing itself** within the **rotating of 180°** where it **then changes every aspect** of its **previous flowing** characteristics it had or **will once again have** in 360^0 from there. While in rotation from the view point of a bystander it all may seem static and never changing but to the object in spin every next instant in time will be diverting from every aspect it had every second passing, and the direction it held in relation to the direction it held the previous mille-, mille-second as it will totally be incompatible with the direction it holds the very next mille, mille second of rotation. This is why we can use degrees measuring the circle by (6^2) (forming the square relating to matter through singularity) X 10 (square if space) = 360^0 however it is always in motion.

In view of that one must remember the fact that singularity cannot move. In order to relocate singularity requires a shaft in relevancy where the line providing the space that moved the distance to allocate the space it represents, has to recline in the status it has with singularity. Then singularity has to re-establish a new alliance in rotation $T^2X\ T^2$ in order to provide the new forming of **k** as well as provide singularity with a new concept of space-time that shifts from and to singularity as the new space that singularity awards in $k\ X\ a^3 = T^2X\ T^2$. If the rotation of a^3 is not quite the same in value as the rest of a^3 space, an unbalance will occur that will come about as a distortion. If there is one spot in a^3 that is bigger than the rest such rotating of that spot in synchrony with the rest of a^3 will demand a larger **k** in place of that position and that will bring about in that a bigger space ($k\ X\ a^3$) must come about when time $T^2X\ T^2$ allocates a new position relating to that specific point. This breaking down and rebuilding of space-time happens as often as change comes about but the effect of the demand on a new **k** depends on the ratio changing that will come about when a^3, which represents the overall space that is all equal is allowing a new space a^3, positions to the heavier space a^3 that comes in unbalance. But that very same principle will be in place when all material is in perfect balance and motion is harmonised.

That proves no point can be static or constant, though it may seem that way to outsiders. Although matter is material, material can also be opposing material and moreover forms its own opposing matter at the same time. Which means that if material is being anti-material such a statement does not necessarily means that expressing something as being anti, anti must be expressing a situation where Anti material is devouring the material it is anti to. Being and being anti also can mean the two is in support of each other as much as opposing each other since the one that is not anti cannot be if it is not there to oppose the other which then is the anti to be. By opposing in being anti it is in supporting the one the not anti. The degeneration of structures are more likely to occur with overheating than with opposing because the opposing of spin in material causes friction to come into place, but it is the friction part that destructs both material in opposing or anti spin. Where one may be more vulnerable to friction such vulnerability will lead to destruction but in such a case it is not the opposing or the anti that brings on the destruction. However the friction part is taking π to π^2. Revaluing Π to Π^2 will bring about a new contact point where Π meets **r** forming another relation in Π^2. Every time material swaps sides, it also qualifies as anti matter to matter because if it goes out of orbiting rotatio frequency, it has the ability to collide with the same matter it forms union with but is located on the other part of the spin.

It then becomes in a situation where Π **revalue to r.** I refer to r as the distance which Kepler referred to as **k** which is just another line running from the one edge to the other edge through the designated centre. This proves that my method that I used to locate singularity the first time was the correct way since that is the way nature goes about improvising for time. **Time is** the **changes in relation** where Π **contacts a different r** not withstanding the many r points there may form because **every r constitutes a different value** to the universe through other ratios and relevancies brought about **by heat and light. Time is the duration it takes Π to rotate between any two given points of r** and therefore must always amount to **a square (T^2)** moving from point to point through the **cube of space (a^3)** in that **duration of time (k).** With that it proves **Kepler's a^3 (space) $=T^2\,k$ (time in the instant of motion)** but motion must continues through a specific value in space where the space-time is maintaining relevant equilibriums throughout singularity connecting. That provides the motive to have another look at how the cosmos produces a straight line from time long before mathematics came into place.

A straight line, triangle and half a circle will always have equality in dimensional capacity providing equilibrium being 180^0 because each one shares a common denominator in singularity to the value of Π. As the straight line averts a zero it holds another straight line in place to set about such an averting where the two lines will always carry a relevancy in elation to progress (the triangle) and a common denominator in the start from singularity. This concept we apply as the graph or the vector. By going back to a line, any line and all lines, the line is a connection of dots in infinity with every dot and all dots not excluding one possible dot being left out, running from one specific to another specific and avoiding zero in that manner. At every point in infinity it dips into infinity coming out on the other side of infinity by choice of direction and the direction is unforced and change presents any angle including the straight line, which incidentally is just another angle. We have to see the Universe as optimists. The Universe is not filled with nothing but filled with possibilities.

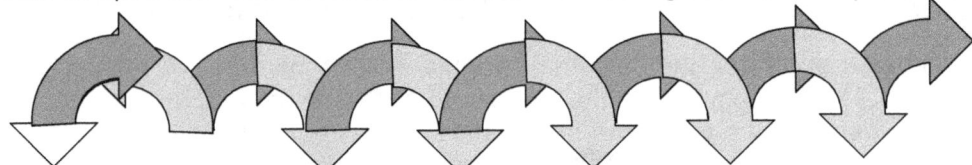

Following the flow of any line such a line is an extension of the previous dot in infinity to the next dot in infinity without any ability to skip or bypass any of the other dots in the connecting line. Any direction change including the remaining of travelling in the same direction is in relation to a line travelling all being the very same. Change does not affect the line.

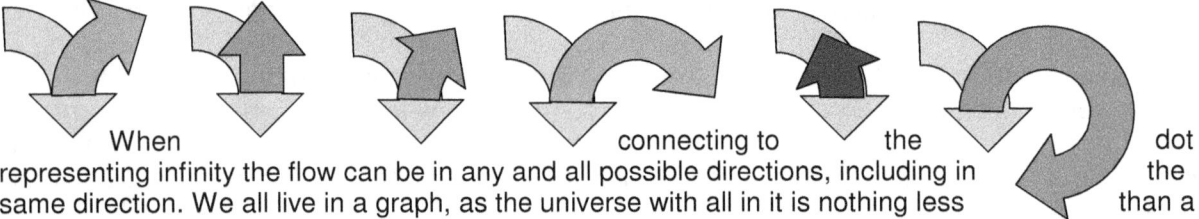

When connecting to the dot representing infinity the flow can be in any and all possible directions, including in the same direction. We all live in a graph, as the universe with all in it is nothing less than a three-dimensional graph flowing according to time. That means in the case of Pythagoras the mere fact that the line shows changes in direction does not implicate or affect the line as a tool of mathematics. Whether the line changes into a half circle meeting at the other end again or meeting in a triangle in forming a half square by joining the point where it began, the result still indicate a line flowing between points.

In the Roche singularity apply all three components

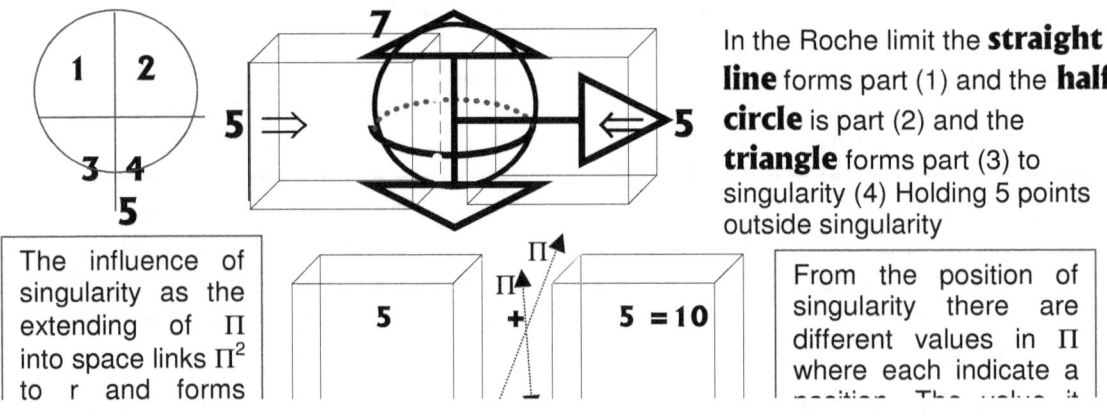

In the Roche limit the **straight line** forms part (1) and the **half circle** is part (2) and the **triangle** forms part (3) to singularity (4) Holding 5 points outside singularity

The influence of singularity as the extending of Π into space links Π^2 to r and forms

$5 + 5 = 10$

From the position of singularity there are different values in Π where each indicate a

$$\Pi \longleftarrow \longrightarrow \Pi$$

$$5 \quad + \quad 5 = 10$$

$$\Pi$$

$$\Pi$$

From there it influences singularity in the triangle flowing through to the half circle. It is an interaction between circular and linear motion as the value of Π continuous past Π^2 (at the end of the solid) and every cosmic structure holds an individual and specific singularity. The field where Π extends we call the atmosphere having a value of 21.991 / 7, which is Π.

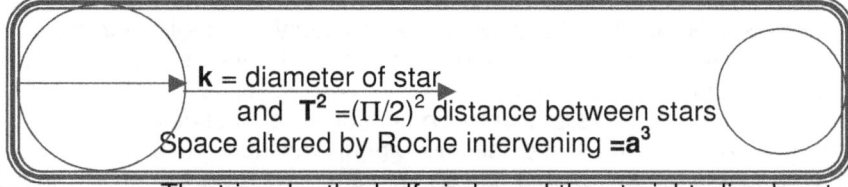

k = diameter of star
and $T^2 = (\Pi/2)^2$ distance between stars
Space altered by Roche intervening $= a^3$

$180^0 = \Pi$

$180^0 = \Pi^2$

Π $\Pi^2 180^0 = \Pi^3$

The triangle, the half circle and the straight –line has two things in common, they share 180^0 as a mutual value and they are part of singularity. This value came about even before mathematics came about. It constitutes the drive engine of the universe.

It is quite significant to note that there is never free energy available any place in the Universe. The fuel driving the cosmos is overheated singularity that now forms space, which back then could not maintain their structure by avoiding friction and by freely applying gravity. In this sense energy is one part of the cosmos that was structurally destroyed as it now moves with motion bringing about gravity to the other part of the Created Universe that can secure the heat from a gas (space) to a liquid (fluid atmospheric heat) to a solid we find present in maintaining the elements within the visible Universe. Material is just frozen heat and space is gas heat leaving fluids as liquid heat. Transforming heat from space as the gas through fluid to frozen material is the energy the Universe runs on. By moving heat from one sector to the next a drive engine comes about that drives the Universe from eternity through eternity to eternity. The sun for instance liquefies (producing light) gas to photons and in that produces energy by applying motion as space creating and gravity.

Using the concept that gravity applies Π as the circle factor Π as well as Π^2 replacing r^2 which the replacing by Π that brings two values as Π and Π^2. That I found is the case with gravity and will be apparent when explaining the sound barrier as well as the Four Cosmic Pillars. In order to create a distinction I remained using r as the indicator of the cube or non-circle that has vacant space and by vacant space I refer to non-solid structures. In the solid structure I use Π as a value for reasons that will become apparent in due time.

Being at... Π **Going too...**

Singularity: a mathematical point at which certain physical quantities reach infinite values for example, according to the general relativity the curvature of space-time becomes infinite in a black hole.

Singularity $= \Pi$

Coming from $= \Pi$

Where singularity holds position in the centre of any and all rotating objects as a value of Π merely applying movement (in the form of atoms) qualifies all matter to be space-time. It does not only fit the description of space within Black Holes but it fits all stars where singularity becomes part of all the stars from the minute to the largest cluster of matter.

With no line starting from zero because there is no zero as a mathematical fact, then all particles hold the point of infinity and not merely the Black Hole of

From that argument one may conclude that all stars will become Black holes depending on the gravity increase they

Through rotation encircling the point of singularity and matter is (1) coming from, (2) being at, (3) as it is going too in one movement in relation to the specifics of the centre point being singularity, all matter then qualifies to form space-time.

That confirms our vision that expresses Kepler's formula $a^3 = k\ T^2$. a^3 holds threeΠ^3 point in time T^2

Firstly singularity expands from Π to the seven positions it holds in material.

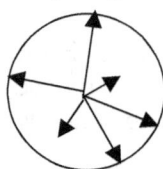

Singularity explodes into space having a value of a^3 as Kepler introduced or Π^3 if the singularity value is used. But that does not introduce the six-sided Universe with the dimensions we came to know. In the process of making contact the Π^3 becomes another Π^3 by abandoning the one side of the Universe and because of the moving of singularity Π^0 through Π^0 a situation develops where the seven points representing Π^0 receives another three pints in motion as the points cross over to the other side of the Universe. Remember the Universe in singularity is a flat Universe with no sides at all and the slightest motion of any point provides the point another Universe to move to. It is not the 180^0 we see as a straight line and at the same time it is the whole issue of 180^0 in all three forms of Pythagoras that forms the issue. The slightest motion becomes a most deliberate crossing of borders since there were no borders to cross and the movement provides the borders to cross. Then by motion of seven dividing into ten and on the same subject the ten dividing into seven as the crossing is implicating singularity $\Pi/2$ the ten and the seven form an alliance that brings about the value of Π^2 in relation to singularity as Π^0

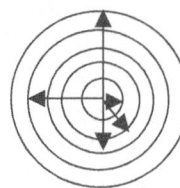

The universe started off as a spot, which was a solid dot without shape, size, form or sides. The universe was wrapped in the single dimension. It was where no mathematics and not even a thought could reach because what the Universe represented at the time still forms our most inner basics we are unable to reach because deep in there it is the I that makes up the me. Later I explain in the second part what is and what are not we the "I" forming "me" and how we have to distinguish between what I am and am not.

The dot came into the age of dimensions. At that point singularity broke free but the Universe kept in tact, secluded from the future in a single dimension where on the fringes of the singularity Π formed. There was then material, which was wrapped in singularity and there was the rest hanging on the fringes of singularity as singularity. There was a relevancy between form and singularity where singularity was the only aspect with form being in singularity. Only Π indicated form outside singularity but from one point having Π the centre of the Universe came about establishing such a centre and the centre uses the rest that was established as being space-time indicating a centre.

In this it is clear why the Titius Bode ([10 + 10 + 1 + .991] / 7) and the Lagrangian 5 \\ 1 systems part their ways when applying the different processes they hold. With all the differentiating, the observer must also consider the dual message that light uses in travelling through the vastness of universal space. The thought of nothing is just what it is, a thought of nothing and although it is in the human mind common nature to present nothing as a value in the recalling of something, nothing is a presentation of the figment in the human mind. There can be no number such as nothing and that was (possibly) Newton's biggest error. Nothing represents non-existing and that is just what nothing is, it is non-existing. The Titius Bode influences in a manner that on the one side holds the matter-to-matter relation of 7+7/10 whilst on the other side during the same time holds the space-to-matter relation of 10/7 forming equal and opposing values. From this the orbits of cosmic structures are always oval favouring the singularity dynamics of the one structure at one point and switching the favouring to the other structure on the opposing side. Because the structures can never be equal in size (singularity will not permit that where the Roche principle will intervene) the shape is always "off centre" as well. This influences coming about as the Titius Bode principal manifest in other ways proving Kepler's time relation with space through distance from singularity controlling the factors.

By rotational motion the top creates a line confirming singularity running down the line and

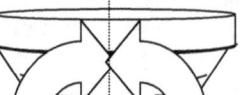

The line is generated but the line is far from magic. The line is where the centre of the Universe is which the Universe is then that

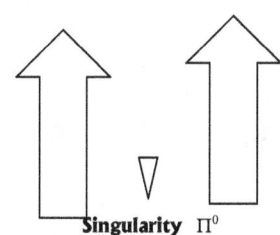

Singularity Π^0

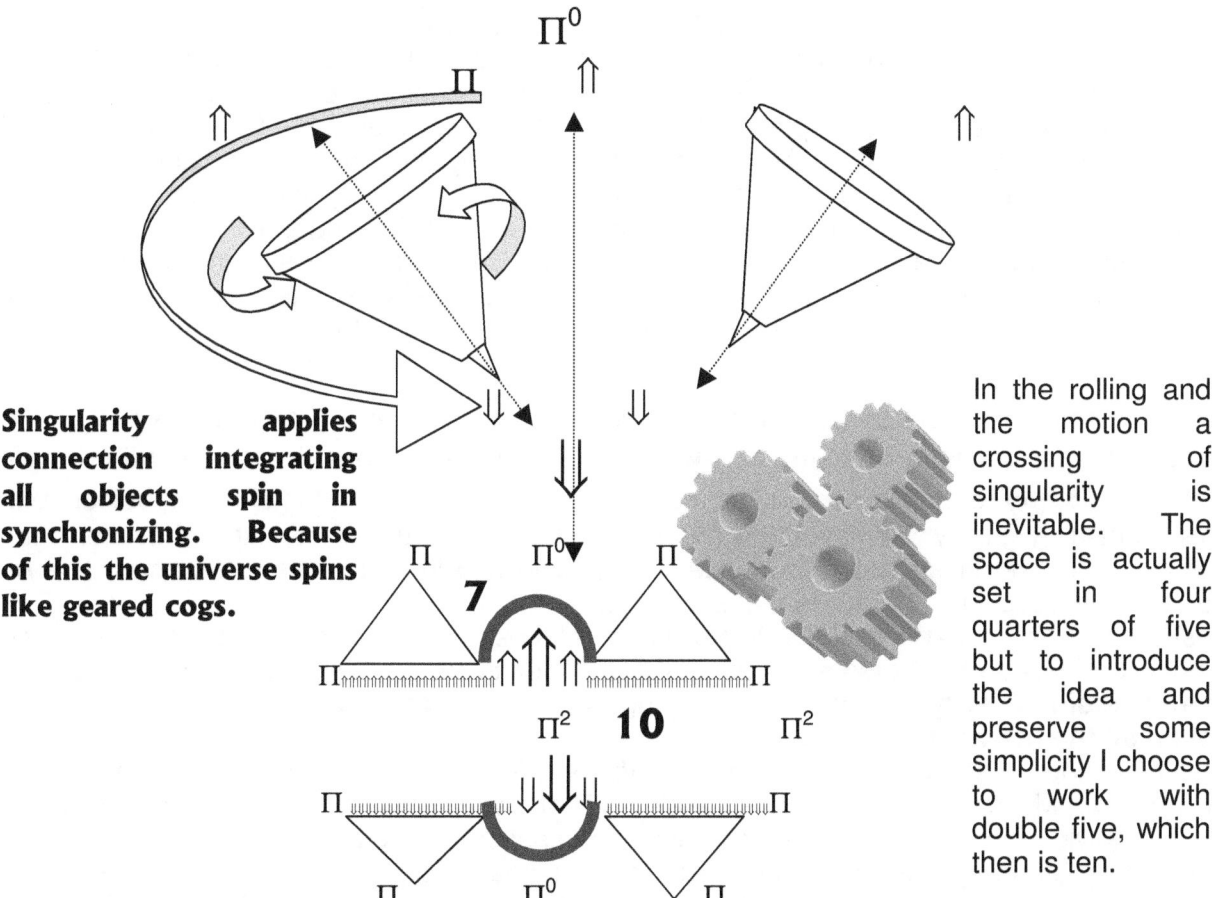

Singularity applies connection integrating all objects spin in synchronizing. Because of this the universe spins like geared cogs.

In the rolling and the motion a crossing of singularity is inevitable. The space is actually set in four quarters of five but to introduce the idea and preserve some simplicity I choose to work with double five, which then is ten.

The way the top spins and the way the atom forming the unit which is the top establishes a six sided Universe of its own initiating by spinning proves two points:

1. The motion establishes a unit of motion in bringing the Coanda effect into reality.
2. The Coanda effect coming about, establishes an elected centre serving singularity to produce an independent six-sided Universe in the established six-sided Universe.

The value of space becomes **twice times ten** being a combination of space on both sides of the border plus **singularity applying** to form an inclusive unit with **three factors** and **a presence** combines as **a unit** of **twenty one point nine** that relates to the **seven of matter** and that reproduces **singularity to the value of** Π as well as **forming** Π^2. It is singularity that is replacing space by replica of motion.

Newton might have seen the rotating cycle as accomplishing "nothing" but modern times showed that the very opposite is true. Every possible driveline that transmits energy must come about through such a line forming a rotation about a very specific and included centre of rotation. No driveline can come about is a restating process drive around a precise axis in a straight line. By rotating the driving shaft that sends the power down a straight line is the only manner in which a drive can be established. This fact comes from the principle Newton could name as gravity but failed to see because Newton was unable to recognise the motion correlating with the straight line, which forms the basis for gravity. I say this because he denounced the existing of the importance of rotation principle by announcing the value after completing one cycle delivering an accomplishment by the principle of achieving zero. He might have foreseen something, which he named but Newton killed the meaning of the principle as he failed to recognise the manner in which gravity works. By completing a cycle in rotation the cycle establishes singularity that comes about through the motion. The motion claims a centre and space using the process that carries the name of the Coanda effect breaks down space-time as the motion elects a centre that brings focus to such an elected centre. The motion re-establishes a governing singularity by re-installing a new elected centre that controls the space coming from the space-time rotation. It is important to recognise that by motion a new Universe is charged by motion still within the confining of a domineering singularity. The breaking down and rebuilding of space-time does not only apply to the driving of electricity but stems from singularity no being able to move but still forced to apply rotation on gravitational demand. A simple throw of a ball or the spin of a top or riding a bicycle is motion enough to generate a new singularity in charge of a newly established Universe. Driving or transfer of energy can only come about through the rotation that Newton gave the principle value of zero or "nothing". Blame me for "anything". We can see this by once again witness the spinning top applying motion in accordance with the establishing of the Coanda effect. It is necessary to remember that any motion and all motion including the driving motion is reforming gravity, countering gravity by forming gravity that is linking a generated singularity to form a newly established gravity and is therefore a fact that all machine-motion is gravity manifesting in some form. Just as the spinning top shows independence. So does the space place the Universe in a double sided with two halves that at the same time is also in a balance.

Once again the following proves that mass is a result of gravity and gravity does not come about through mass, because by using a new a^3 it can establish a new k, which will convert that gravity T^2 to apply to the new a^3. In this manner does motion release the spinning object from the Earth's containing gravity. **However** the motion must bring about differentiation between that which acts as solids and those taking the place of the liquid and in that forms relevancies. It is of much importance to recognise the **fluids** that stands apart from the **solids** when differentiating. The differences is some times just a concept. Water is sometimes a solid but can perform as a liquid, **which smoke and dense heat also are. On that and other grounds I maintain that the sun on the inside is liquid. Gravity and the establishing thereof is not a God given right of birth bestowed on all the heaviest to create.**

The gravity it develops is a "cosmic life" not to be confused with carbon life we find and have and are on Earth, but that which makes the Universe alive. That establishing or creating or exciting of singularity by applying motion which separates new heat in space by separating distance by spin is a new cosmos entity standing apart from the rest of the Cosmos. Every singularity is a Universe that can apply new values and rules by changing any of the Kepler formula factors. By generating heat singularity comes "alive" or energetic. The response to that is that motion comes about to rescue singularity and bring about gravity. Gravity is motion through space of space in space using time as a measure to secure the motion. The motion creating time must never be interpreted as a secluded event. Motion is always relation applying between various singularity points. Gravity can only come about when motion changes a relevancy between two or more structures and such a change will increase the coming together or the moving apart of the structures in question. But without applying heat by increasing the levels of heat motion cannot come about and "cosmic life" or energising singularity cannot produce such motion, as gravity will bring about. There has to be a relevant solid rotating about an axis in a relevant liquid that permit motion. There will always be seven to the one side as gravity reduces the space bringing about a "cooler side" and on the other side of the relation the space will be "hotter" and motion will come into place.

132

By applying heat to a spot the spot has more heat than the rest of the spots surrounding the spot applied with heat. By receiving more heat the heat will turn to space and the growth in space will bring about motion since the space has to go somewhere, as it is growing bigger than what the space was. The applying of heat will increase the effect of singularity on k. The increase of k will turn the increase of heat into an increase in space. The space becomes more therefore it has to go somewhere larger than what it was before. This we call expanding and the more ferocious motion we call exploding.

By altering any of the three Kepler factors space-time can establish a new significant gravity in the midst of gravity applying. By creating a new spin in the presence of the Earth gravity, the spin creates a gravity that will encourage the release from the major gravity in order to find independence, for example the spinning top to try (it can never happen but it is trying all the same) through a newly charged singularity to develop a gravity that will produce such vigorous movement T^2 that will take the top to a position apart from the rotation it normally has with the Earth. When the speed of the rotation exceeds the limitation the Earth allows, the spinning top will start wobbling from side to side indicating a maximum effort to create lift and go in a separate spec at a separate distance from the Earth. In the most vigorous attempt the top will fight for release by jumping in the air. When the top slows down the wobble will become present again, as the gravity established through the spin will fight to stay alive and apart from that of the Earth. But it shows that in Kepler's formula new space comes about from establishing a new T^2, which the spinning then forms in the alliance of space created through the manifestation of a new $a^3 = k\,T^2$ in the boundaries of the Earth. Make no mistake about the fact that the spin is new gravity that comes about in the area the top occupies and the **k** is now the rotation coming about from the centre of the new spin. The Earth has replaced the role the sun had before the spin commenced and the top by spinning resumes the role that the Earth had but involving much less dominance. The wobbling at the bottom and at the peak of the spin effort of the rotation is a gravity struggling for independence either to maintain independence or at the top to establish ultimate independence. The Earth will respond by loosing the battle as long as the Earth receive the asked bounty in the form of that the escaped has to release to accomplish the escaping or if the release in heat is insufficient the Earth will kill the rebellion by destroying the motion and reinstate the resistance we call mass.

The Roche singularity applies all three components of singularity

The motion discrepancy between the points holding singularity controlling the Earth as it does throughout the Universe in relation to the singularity in control of what we named mass will establish the Roche limit although in another value.

From such deductions one may speculate in order to learn. There is the fact that is a result of speed discrepancy between objects. The speed is a result of duplication. When particles are closer than singularity will allow the Universe standing affected by this, then falls back to conditions that presumably were in place when conditions instated the Big Bang scenario. From what we see in the Roche limit we can project back to what applied just before the Big Bang came in place. When two objects are very close to each other, they either have to spin at a harmonious pace in synchronized motion, or friction will come about leaving heat as the consequence there of. Gravity is motion discrepancy and the Roche limit shows that heat comes about as particle friction causes the heat in such a mismatch of spin. Since space was non-existing which brought on the obvious cause of the heat being in over supply, it is safe to make deductions based on such facts.

The ten dimensions I named the atomic relevancy is also showing the double value of singularity as singularity extends into as well as beyond space. The atomic relevancy is $(\Pi^2+\Pi^2)(\Pi^2 \times \Pi \times 3) =$ **1836** that is the mass relation between the electron (3) and the proton. Proton $= (\Pi^2+\Pi^2)$ Neutron $=\Pi^2$ Π. Then the electron comes as 3 where π lost form.

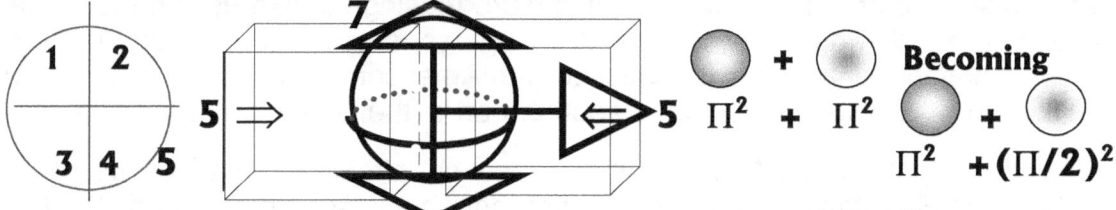

Every spot produce a dimension that indicates a value to such a position.

The influence of singularity as the extending of Π into space links Π^2 to r and forms	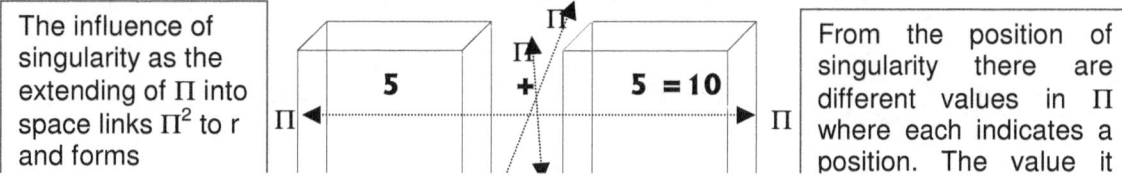	From the position of singularity there are different values in Π where each indicates a position. The value it

$$5 \quad + \quad 5 = 10$$

$$\Pi$$

$$\Pi$$

The atomic relevancy holds the dynamics of what singularity controls. In the ratio and dimensions we find in the atom, all space-time derives from the atom, which is resulting from whatever the atom is. It is very important to realise that every molecule, every atom, every particle and every cosmic structure holds singularity and as it holds singularity it is a Universe which it has the absolute task to secure and maintain that singularity in the future. Such maintaining is the prevention of further overheating by securing gravity opportunities under cosmos rules and laws. The accumulation of influencing all surrounding and all other singularity forms the unit we think of as the Universe but that is not the Universe. That is an accumulation of all Universes and therefore in the centre of every molecule, every atom, every particle and every cosmic structure that holds singularity is the centre of the Universe. Singularity formed a unit once that formed a Universe. Then every part took with it the Universe it represents and still form the centre of the Universe it departed from.

133

The influence that singularity has on the surrounding space-time or the involving of such surrounding singularity is part of the basic issue of gravity in the cosmos. The process is that which should be based at the core of our fundamental understanding. The concept in own merit is rather simple to understand and I also find is very difficult to explain. By the motion of particles within the confinement of any centre of a sphere such particle motion increases the heat levels by reducing the space factor and that increases other types of reaction and more gravity in the centre of the sphere. Every part of individuality in singularity is by form of π a sphere. It is motion that produces a new Universe by allowing a fluid to form space that separates the Universe in fragments. Where our part remained solid by gravity forming the basis on which the Coanda effect rests another part had to compromise the solidity to become fluid and in relation with the solid, apply the Coanda effect of liquid forming motion in the presence of solid applying a basis for Kepler's formula to determine the Universe at $a^3 = T^2 k$. The neutron in going liquid, sacrificed one part of $\pi^2 + \pi^2$ to form $\pi^2 \pi$ and in that lost π heat became the virtue of the atomic enclosure. Such heat in the centre will establish the value of singularity at the point such heat concentrates by reducing the space towards the centre. The atomic motion serves as the cosmos in motion. The cosmos is about motion and the neutron is the atom's motion, therefore the atom can only be when there is a neutron attached. The only motion not resulting from the atomic effort in motion is life. Life has the ability to manipulate and dictate the motion of the atoms serving life. Because life is not part of the cosmos the cosmos does not recognise the difference between any efforts coming from the intervention of life as an energy standing apart from the cosmos or the cosmos bringing about such heat by applying the laws it created to govern the cosmos. By securing a spot through motion holding singularity ten spots surrounding such a centre becomes valid in a task of generating space-time. In such an event of creating a heated gravity spot, the independence of such a spot starts to try and bring about independence and establishes a new dominating singularity where as the task the dominating singularity has is to subdue or "pacify" space-time captured (using the expression the U.S.A generals so fondly use to bomb the living daylight out of those persons whom they invaded, murdered and plundered and then has the tenacity not be satisfied with American domination and American rape of their liberty). Wind devils and hurricanes forms another part of the Coanda effect.

If space were zero or nothing as Mainstream science so affectively teaches us, then Kepler's principle formula would need the changes Newton brought about. But it is true and stands tested like no other research ever coming either before or after Brahe and Kepler's work.

$k = k^{3-2} = k^1$ $a^3 = a^{2+1} = a^3$ $T^2 = T^{3-1=2}$	$k = a^3 / T^2$ $k = a^{3-2} (T^2)$ $k = a^{3-2} = k^1$ $k = k^{3-2} = k^1$	$a^3 = T^2 k$ $a^3 = T^2 k^1$ $a^3 = T^{2+1} (k^1)$ $a^3 = a^{2+1} = a^3$	$T^2 = a^3 / k$ $T^2 = a^3 / k^1$ $T^2 = a^{3-1} = T^2$ $T^2 = T^{3-1=2}$
	is the same as	**is the same as**	**It is all the same**

$k = k^{3-2} = k^1$ is in direct relation to $a^3 = a^{2+1}$ is in direct relation to $a^3 = T^2 = T^{3-1=2}$. With this information staring mainstream science in the face and scream pleading at them to recognise this information they turn around and ask why can man not fly off to other galactica at the speed of light.

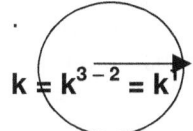

It takes time for space to fill **k** in the distance. In fact it takes the distance that **k**, developed since the Big Bang $k = k^{3-2} = k^1$ to fill the distance.
It also takes time $T^2 = T^{3-1=2}$ to produce the distance forming k^2
It takes space $a^3 = a^{2+1} = a^3$ to form k^3 since coming from the Big Bang

When the astronaut is departing from space on Earth or filling Earth space it will take the departing astronaut k^2 time to reach k^1 and fill out k^3. At present and in this moment our most impressive astronautic engineers will devise an engine that would cut k^1 by say half. This achievement will come as they increase the power output say for argument sake to double what it is at present. There was no friction of particles destroying the frame of the craft because there are not enough particles in space to do it, the space became too small to allow the time it takes to enter because the distance **k** decreased faster than the space a^3 could compromise with the time T^2 changing from what is present in outer space comparing that to the time in to atmospheric space. With the information in hand for a period of four hundred years and where the information forms the basis of modern cosmology since the information formulated gravity and not merely produced a name for gravity as our English friend did, it is amazing that such accidents can happen and it is more amazing that no one in Mainstream physics has the slightest idea why this is taking place! Our most impressive astronautic engineers are assembling a machine that will scramble the ratio Kepler introduced to a level in outer space where the ratio will be more than what the ratio in the sun is. Surprisingly they are not in the least surprised that not one object in outer space is using an excessive velocity.

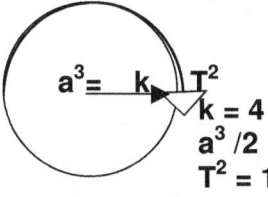

In realistic physics it means double the space will fill in half the time. We know that that is not possible because it can only bring about half the space in double the time or twice the distance in half the time. Space time and distance is a mesh where the lot integrate because Kepler said so. Kepler said the space forming space is the same space forming the distance of the space and that is the same space taking the time to fill the space. If the ratio changes then changes come about the entire ratio.

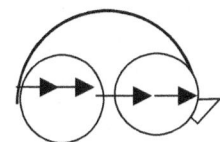

k = 4 and a^3 /2 if T^2 remains the same but that will not happen and that we know from past experiences. If that happens we have the challenger 2004 disaster repeating once more.

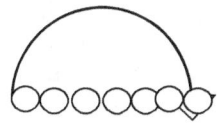

Increasing space-time displacement by six will decrease space by six and the distance the space progresses from a centre by twelve. The heat factor of the craft will rise by twelve times as the space decreases by six times.

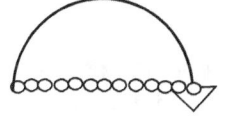

Increasing space-time displacement by twelve will decrease space by twelve and the distance the space progresses from a centre by twenty-four. The heat factor of the craft will rise by twenty four times as the space decreases by twelve times.

Motion of anything in any form is about duplicating the existing into following on images of the same thing. That is connecting space to last a certain period in relation to a specific point holding singularity before the next singularity is enticed or charged to maintain the space-time in motion. Every time (and in this case the referring to time proves to be most accurate) is having another singularity building and breaking down the space it represents for that duration of time. The time duration leaves singularity selected in charge of producing the roving space the extent in which it can duplicate the space it has to duplicate. By reducing the period the particular singularity may lay claim to the space, will inadvertently produce smaller space it is able to reproduce in the shorter period of time. However, to achieve such duplication standard, some heat will have to be sacrificed in order to secure such early time release from the singularity that takes charge of the duplication. Only by paying a price of compromising heat can such duplication be sustained. The compromise of heat in natural conditions will be sacrificed because of the enormity of the gravity that the singularity charged with duplicating space-time can unleash. That means the singularity therefore must have grown to the enormity as to be able to accumulate the heat in order to produce the duplication that will entice such enormous motion.

This effect we see happens when water is falling onto a hot stove plate. From the plate flows heat that charges the singularity that captured the space-time being water, to break from the unit forming the water drop and seek independence from the unit by enforcing motion of the water drop. The water drop running on the stove finds the plate charging the singularity of the plate to entice motion and the motion of the water is the singularity finding additional space in heat to duplicate the space faster as to create the motion that accelerates.

Unfortunately the compromising involves all aspects and that includes the gravity or time also. The Coanda effect shows that gravity in space can charge and change space and this is because Kepler said space time and distance form the same thing because the three is part of the same thing. Change one aspect in Kepler's formula and the outcome is a mess. I have a book on this aspect called **Inter Galactica Space Travel. ISBN 0-9584410-2-2.** The Coanda effect and the Roche limit results from this implication

Henri Coanda holds esteem in aviation circles but beyond that his work is slightly known. Henri Coanda should be amongst Kepler Copernicus and Newton if his work is correctly categorised. His work is that of a giant. He demonstrated gravity, he produced gravity, he introduced gravity and no one bothered to take notice because he was not a mathematician. Henri Coanda might not really have realised that he stumbled on the oldest principle in all of creation and that this effect he penned and named is the absolute basis of gravity, but his realising the importance thereof did not pass him by. Coanda realised that by reducing the propeller blades of the aeroplane the rotation speed becomes amplified. Concentrating space to amplify the heat comes from the concentration in space by redirecting the flow of air (space), which also can be water since both are liquids. By confining the space displaced by the propeller shaft the space duplication produces a rise in the heat within the space. What this implies is that the process is basically placing the Big Bag in reverse. The Big Bang is releasing heat to produce space and the Coanda effect is confining heat to produce space.

The role of the impeller is to reduce the space by applying motion that creates a higher concentration of heat in a reduced space during the same time interval. When this is mismatched the impeller burns holes in the blade as air produces heat that removes metal. The rotation of the impeller forms the dual prong of cosmic motion. The propeller has the role of the protons that reduce space by directing space to demolish space through implementing motion. The second part of the double prong is the confining of space that will bring about heat increasing in the confined space. Motion brings about gravity but motion is gravity in the duplicating of material claiming space and material holding space through the duplication thereof. The only condition is that a liquid must be present. In relation to a solid the liquid must relate to a point in singularity or a substitute to such a point in singularity by committing form of singularity to one specific concentrated point mimicking singularity. One has to realise that space is the gas part of heat being the liquid part and space and liquid is far apart from material. Material provides the solid structure where such solid material surrounds itself with dense liquid heat called plasma by those being smart and educated physicians. It forms between the solid atomic structure occupying the inside and the less dense heat on the outside that is regarded as space and is further away from the atom centre.

When the seven occupied by material rotates, it removes the ten in contact from influencing space-time and by rotation introduces a totally new ten points in relation by the act of rotating. But such a process also comes about on the other side of the divide because it is the divide that cannot rotate and forces space-time to comply with the frequency of breaking down space-time as space-time releases from one side and if necessary shift or only apply new rotation alliances as the seven points become a part of the new ten.

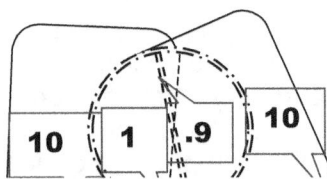

The motion duplicates space (10 + 10) by rolling over singularity (1) from the one side by reducing space (.09991) on the other side. All this motion of space 10 duplicating is directly related to matter 7. The total value established from this motion is the interaction of seven with ten producing Π^2 as the half of singularity $\Pi/2$ on the one side interact with half the singularity $\Pi/2$ on the other side and in this

The conditions proving singularity is that the circular form will produce a centre point from where gravity will dictate the reproducing of space. The gravity part is the fact that motion must contradict the centre point around which the motion will produce space. The space part is proven by the motion that produces a running line of space created and followed by the liquid producing the space in the motion of the liquid. It shows that relevancy comes from space shared by space and motion separates space shared by space. It shows gravity coming from motion separating shared space.

We must see that we have three dimensions coming from the atom.

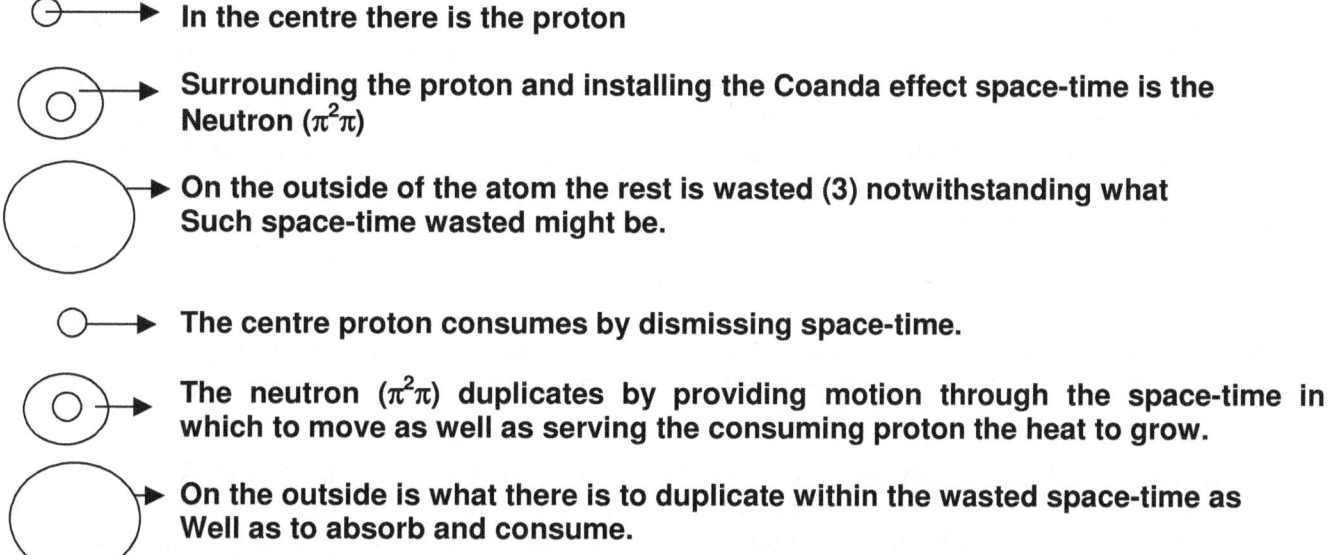

In the centre there is the proton

Surrounding the proton and installing the Coanda effect space-time is the Neutron ($\pi^2\pi$)

On the outside of the atom the rest is wasted (3) notwithstanding what Such space-time wasted might be.

The centre proton consumes by dismissing space-time.

The neutron ($\pi^2\pi$) duplicates by providing motion through the space-time in which to move as well as serving the consuming proton the heat to grow.

On the outside is what there is to duplicate within the wasted space-time as Well as to absorb and consume.

When further consuming is not possible, when finding any singularity that challenge is the consuming, the Roche limit comes into effect.

When duplicating the space-time, it results in re-connecting of space-time with some space in the centre of the established singularity. The centre dissolves space and also is placing space within a surrounding of space less ness where space then is becoming motionlessness. In the event where space always in a relevancy with time and time being the motion placing relevancies on space such a relevancy places a cup or a limit on what space flight may be able to achieve. With the international media and the international press printing in frenzy about flying sources because they have no better work to do than to go on some flying events that take the visiting aliens through the Universe on some sight seeing tours those visitors find the time to tell us humans tales. Unfortunately we have amongst us, the mentally impaired that comes out with tales of how those invading men-from-Mars-or-wherever-they-come-from, and travel billions of light years just to have the opportunity to have sex with these mindless storytellers (whenever the halfwits meet extraterrestrial visitors the visitors rape them, take samples of their body tissue and send them home scarred for life. To think that any one would cross one Universe just to have sex with them and then Take their body tissue as samples makes them real special, except on the intellectual front). To top this madness there are scientists with as little mental capacity that echo this insanity by insisting that human crowds will infest far off galactica by swamping the trillions of planets they are yet to find. To them and those I give the free advice: remember that it is useful to take into consideration that gravity also forms relevancies on preventing objects not to come harm-provoking close. Where there is this limitation, there is also one of being too far apart for singularity to allow the escaping.

The criteria which any object has to comply with should such an object wish to leave the domineering of the sun's gravity displacement, which by the way stretches beyond the Oort cloud and Kuiper belt, that escaping craft has to exceed the motion applying in the centre of the sun as well as reduce in form and space to fit in a spot such an object will fit into when in the centre of the sun. The craft is part of the captured space-time the sun is holding. To find release from that capture the craft has to find release from the governing singularity control within the centre of the sun. The heat effort the escaping craft has to apply must be stronger than all the atoms within the sun as an acceleration effort generates **what the containing force is that**. The escaping craft has to apply more motion $T^2k > a^3$ than the space that the material in the sun fills. To go beyond the barriers of the governing singularity within the centre of the sun, which I just mentioned, the craft must produce anti gravity, in the form of heat release that will overshadow the gravity of the sun. The craft will have to produce more heat than there is in the very centre of the sun at the point where space disappears because that escaping object has to produce more space through motion than which the sun can destroy in the very centre where the sun is destroying the space. From such a centre destroying space-time then is holding the solar system bonded.

The craft whishing to leave the Oort cloud boundary must produce more space through motion that is creating space by duplicating than that which the sun can destroy because if not it would repeal the motion and bring the escaping object back to the sun in way comets do. That is the reason why comets have not yet escaped and gone yonder to "nearby" stars and that is why the Oorts cloud has been unable to reach the area in the location of the Centauri triple star system. That principle is the law of space forming by motion applying, which Henri Coanda noticed and which rocket propulsion uses to establish motion. The craft must create conditions exceeding the centre of the sun, reduce space to become hotter than the centre of the sun and go faster than the velocity the sun can create in the centre of the sun and even then the object will find a relevancy of say so many millimetres per year to advance past the border of the sun. This is because fusion does take place within the parameters of the sun occupying space on a very limited basis. To those aiming to achieve that I bid them good luck for they will need much more than the luck I can offer them. The spinning top I referred to previously and the Coanda effect is the very same principle applying. The sun is just a large spinning top applying the Coanda effect.

Let us return to the top and find gravity in the behaviour that the top shows to motion. Again I have to stress that one should never forget the fact that the energy attributed to the top is a product of life and although it is in the capabilities of life to manipulate space-time it is also in the ability of life to use and manipulate from cosmic some applying principles that we can milk some behaviour and in doing that, that will tend to benefit life on a micro scale. The spin provided to the top acts like heat activating a singularity centre and from that motion comes about. A more natural result coming about without life's interference will be when the singularity in the centre of the top is all that charged and the heat will entice the singularity as to accelerate the top's independent time to bring much more duplication of the space of the two to come about. However, in the form we see the tops spin; this act by the top is not cosmic produced. It is at best cosmic re-enacting because there is no chance the top will by own sources start spinning and even more so spin violently enough that the top will then manage enough duplication that it can leave the Earth. Gravity is about motion providing space a certain relevancy to duplicate space in relation to the surrounding and that was how space-time came about from the start of creation.

If gravity was in principle about one object hooking another object like fighting boxers pulling at each other as they fight to achieve dominance, well yes, then the space cannot have any influence on the two particles...or can it?

Even if gravity is about particles pulling, which it is not and that I say on lack of evidence brought in support of that statement ever since Newton suggested the idea, some of the gravity "force" must come as a result of the conducting of the force through space being the transmitter of the force. Space is what we find that is between the two opposing structures in complying with gravity. What ever is conducting the gravity is doing so while space is or permitting the force to reach the other side. Allowing some influence to reach another point is what I understand conducting is. There is a restrain of flow of such conducting when space becomes denser or reduce density. In order to get what ever is being conducted to where it is conducting to, such a force that is conducted with the intent to grip onto or get hold of the other object, must find that the influence has to commit the space between the objects to allow such pulling. There has to be such a medium that can allow the conducting of the **massage** by the **massager** to take place but science produced nothing that is in support of proof to confirm this... If the conducting of gravity was about pulling then still the motion of space was the first indicator and without the motion the force will then be nullified. Even the indicating of mass is motion trying to come about but the motion finds moving restrained by a blocking. The sharing space between the objects was the first to be influenced because space has to relay such an influence; the space has to reduce to allow the action of contraction. Unfortunately academics would never previously admit to space deforming through gravity applying. We know that the flow of light is a conducting of light and the speed of light can vary in motion from standing still in certain silicon conducting computer chips, to travelling at C. But where light is anti-gravity, then gravity must be opposing light. Gravity is essentially directly connected to the form of the sphere.

The essence of the sphere is about reducing space from the outside to the reduced inside. The sphere reduces space by applying form and committing space occupied and deforms space unoccupied. This is always the complying of the outer space to relent the form to that of the sphere. In the sphere the form dictates space reducing as space occupation progresses to the inside. From that we have to deduct that if the sphere will apply reducing of space inwards, it will extend the applying to the outside. Present will always be an outer edge and that forms a border that commits space to be in or out. We have only to detect and find proof of such influence being present.

There are always two relevancies applying in gravity in relation to each other. There is always one being part of a holding unit and then there is one other part establishing borders. This is the result of how singularity broke into space. At first there was singularity k^0 holding the universe in the single dimension. Then singularity k^0 moved out to form singularity k (and this process stands apart from other forms coming out of single dimension at the next step) but in the same action the half circle came into position as well as the triangle forming space. At this where singularity the spot Π^0 moved to singularity the dot Π all other future possible particles were still frozen including mathematics in the rim being to the power of zero. Only Π yet had a value placing a point in the universe where others still were numberless. The circle was Πr^0. There was only space that Π brought forward as motion and motion was space as well as direction $a^3 = k\,T^2$ where all stood related by relevancy in direction to each other. The size they shared was all the same. Only the mathematics we now see as trigonometry was in use **where the only valid** are angles relating in forms and directions. But k was T^2 as much as k was a^3. The motion of expansion came in place but the expansion had nowhere to go than to rotate because the expansion was simulated in one move through out the Universe as one orchestrated motion where everything coming from singularity strived for independence instantaneously. It was motion that was creating space but it was motion bridging singularity, splitting the possibility of singularity into sectors. But part of this was relevancies where the one formed the other sector's space. The one was rein acting the other side of the Universe because all was so small a line was duplicating a half circle because the other half circle belonged to the other side of the Universe and the two applied the space received.

- But simultaneously to that action there was the other side of the other side of another part of the same Universe and from that stance the motion applied by the first mentioned singularity action was received in a much different light to those on the other side of the divide. As the second related singularity is on the opposite of the first Universe it must recollect the procedure very much differently than the way that the first sees the space formation. The first one applied motion and the motion placed spin on the second singularity point. The applying space to the one was at that

- same time applying motion to the next and the confrontation was establishing space in relevance to the motion that came about from the growth as well as the rotation that was established as a retraction by the opposing particle. From such a duel a relation will come into affect putting one point in singularity in a major position as far as security is concerned and one in a minor relation and from that stance the relation will commit to space where each one performs certain roles in relation to the other. The roles has to do with creating space

At the time of Big Bang motion one came about when independence was struck by forming a value of $(\Pi \div 2)$. Later on in cosmic development only when one of the two opposing participating objects crosses the line of $(\Pi \div 2)$ from both sides $(\Pi \div 2)^2$ will the opposing particles bridge the first cosmic law and the contraction will override the expansion capturing the second relevancy into the control of the first relevancy. The first instance both particles matched since all particles were at least $(\Pi \div 2)$ separated. But growth applies as material incorporates space as heat and from that finds material expanding. One of the particles grew more rapidly than the other and secured dominance but not control. By outgrowing the lesser partner it can capture the dominating of the space the lesser partner claims.

The biggest reality that came from this was that there would always be two factors producing gravity in relation to one another. The contraction of the one will bring about motion in expanding space to the other. The one will view from a **k** stance as expansion comes into effect. The other singularity will apply as **k^{-1},** which is the direct opposite of the matching pair.

But that forms a relevancy.
$k^0/k = T^2/a^3$ and that is the claiming relevancy

$k/k^0 = a^3 / T^2$ that is the expanding relevancy

There are the one relevancy that applies contraction that recaptures space and will answer to the ratio presented as follows $k^0/k = T^2/a^3 = k^{-1} = T^2/a^3$

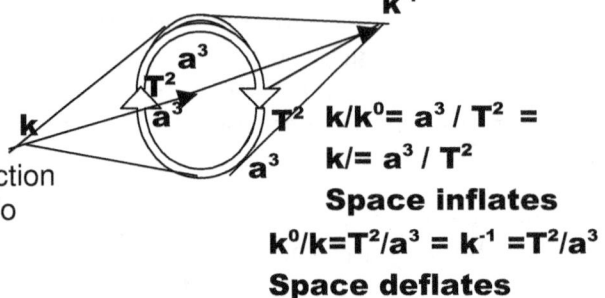

$k/k^0 = a^3 / T^2 =$
$k/= a^3 / T^2$
Space inflates
$k^0/k = T^2/a^3 = k^{-1} = T^2/a^3$
Space deflates

Then there is the other duplicating gravity principle where motion that brings about space is also putting distance between objects and this we humans understand to be growth of space or by a term much more commonly used as the Hubble constant but only in a balance where the space displaced and the space created match in time taken relating to space. This formula will apply as the second gravity $k / k^0 = a^3 / T^2 = k /= a^3 / T^2$ because the motion advances the distance between objects in relation but will only dominate when the first option brakes the Roche law of $(\Pi \div 2)^2$. Then the relation will change to where the **k** factor holds a negative and total control. This means the space duplicated are not adequate to the space dismissed and a new space to time balance must be established. $k^0/k = T^2/a^3 = k^{-1} = T^2/a^3$ That is what Newton saw and that is what science recognises but that is not the primary gravity found in outer space at large. In outer space the Newtonian gravity only dominates where the bridging of singularity forcefully brought domination and control of a major singularity in capturing a minor singularity. Where that situation complies with cosmic rules the president set will be that the major contractor reduced the minor expander to heat and then capture the heat. Only heat applying motion brings particle separation as we may witness with the spinning top. Motion brings about space duplication, which revitalises space dismissing.

How does one accept and believe in the proof about this, which is said. We find the proof in the Coanda affect where the reducing of the rotating blades and the increase in the rotating speed produces heat as high as heat can come in the form of flames. We find heat burns holes in impeller blades with the impeller blades of the speedboat submerged in water. The impeller blades of the impeller only works by practising the Coanda effect. The blades find the ability to reduce space and concentrate heat in the space by increasing the duplication of the blades in the rotating thereof. In the turbine, the blades provide gravity by reducing the space through increasing motion of the space during the same period in time duration and this gravity establishes the equal antigravity in the form of liquid tongues of flames. If this is happening now it happened way back then…

From the gravity and the opposing antigravity applying came about two interacting translations of material where one formed gravity in securing a position and another by permitting expansion, releasing matter to establish space in relevancy bringing about antimatter in the form of matter performing antigravity. The one was filling space by giving away density and the other was applying density by giving up space. If it was true that matter was drawing matter closer, the securing of space would not stop the contracting and the regrouping of matter would start in the Universe even before expanding could start. The Universe would become one solid structure and remain that way because there would be no reason for it to expand and fight gravity's magical attraction contracting space between particles with mass. We have to realise the fact of relevancies where there always will be a major and a minor factor enjoying the same space-time. Where there are two equal partners in relevancy such as we find in binaries, the two will push development until the ultimate development is reached much quicker than it will be in a single developing star. This is because the two have a near or perfect singularity match with no one able to dominate the other and both developing with one aim and that is to outgrow the other and establish dominance in such a way.

The balance is in the minor particle finding commotion with enough heat that is allowing the particle to seek independence and a major particle removing space by seeking control of the heat that is in the rebelling mode. When the minor particle gathers sufficient heat it will start an attempt to gain independence by placing as much space in relation to motion between the particles whereas the controller will demand the space in which the minor particle is moving by the major partners dismissing space-time. This relation allows infant and toddler stars to be freed from the galactica cocoons and move away from the galactica centre establishing space-time by concentrating heat from the outer space towards the inner core area. It is all about capturing or releasing by establishing independence or surrendering independence. Since all the objects in motion in the Universe have not surrendered their independence they're in gravity that forms with **k** allowing the maintaining of independence and **k^{-1}** in a lesser but pivotal role securing stability. Only in the dynamics of the Earth that captured all that is within the atmosphere of the Earth will **k^{-1}** be dominant, as all within the Earth becomes part of the Earths motion of **k**. The control of space is about space duplicating and space dismissing. With motion comes space duplicating but the more the object grows the more will the object dismiss space and the less will the object duplicate space. By not duplicating space the object will increasingly destroy space because the value of space is fifty percent in the duplication of the space. By relenting the duplication of the space the motion of the star will tarnish and that will allow the star to dismiss much more space outside the star than merely the space, in which it is. By that the star gets an increasing ability to dismiss space that is not in the star's control but by putting such space to heat within the star's control. The star finally will achieve an ability to liquefy space into heat. The sun has reached such a position. The essence of gravity is motion duplicating space and the dismissing of space. The Coanda effect is the best example and the spinning top illustrates the Coanda effect very well. The final step is to freeze space into material and become a dark star.

In the spinning top we read the dynamics, which controls the phenomenon we associate with the sound barrier. The top that spins within the Earth atmosphere does so while it shares a relevant position within the Earth's gravity domination dominating the time aspect. When an object is in outer space and being unattached to the Earth as say a satellite is there will be two points in singularity sharing a unit by relevancy. One relevancy will fight for independence, while there is the centre in the unit fighting to capture and control *whatever space-time has motion* within the unit, *which the centre provides the motion* by which it intends to capture the space-time / iets is verkeerd dit lees nie lekker nie. While the top holds independence as it is captured by the Earth time the top will share a point in singularity matching that of the Earth. As soon as the object is in a motion such motion establishes time other than that which the Earth demands. Then the independent singularity in motion shifts away from the spot they share when there is only the Earth motion involved. As the tops spin creates individuality by independent motion it places the time component away from the Earth's singularity. The motion of the top will secure a separate singularity value.

Within one space unit we always have two independent positions holding points in singularity that is sharing time but being independent in space at the same time. The two assert a velocity of $7(3\Pi^2)$ by just moving with the Earth. Since they both move equally in line with the Earth, the two objects hold an independent that is Π^0 in independent speed but the independence may grow to $5X\ \Pi^0$ before the sound barrier becomes an issue of concern.

Π^0 $5\Pi^0$

In pure nature
the Earth
applies all

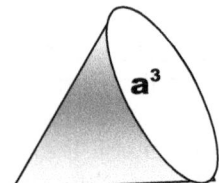

We must see every aspect of the top going into motion as cosmically artificially generated by an alien substance such as life is to the cosmos, where life has the ability to apply motion. To the cosmos the spinning top is a disguise of the truth and to the cosmos life is not a factor in existing. If the cosmos found something in a rebelling mode such as the top is the most unlikely scenario will come about where the top finds a means to establish singularity strong enough or driven enough to establish such highly developed independent singularity achieving such motion. However, the studying of the action of the top introduces the utmost basic principle we find in the Universe. We find the law of generating independence within the framework of confinement. That is what all of the Universe is. The Universe is independence of a lesser in the space of a dominating. The motion differentiation brings about the motion and the drive for motion. The motion then must be so well contained by the rebelling singularity that the Earth would be unable to depose the heat from the minor singularity centre. What we humans rein act is a process that happens in galactica when stars surge to find independence as the progress away from the confining centre and drive for independence.

The process is there and is part of the Universe but is part of another process we find in another part of the Universe. That process can never come about in the working of a star-like structure. Never should science intermingle life and the accomplishment of life on this little God forsaken blue dot of no repute as a fraction of the cosmos and have life acting on behalf of the cosmos. There is no round rock on any planet large or small that will start spinning because such a round rock has produced motion driven by its red-hot-core bringing about the motion. Before life intervenes and charge the top to bring about "cosmic life" (a term I dare to use for the lack of a better term, but in effect is the volume of heat concentrated in the centre of a rotating structure) the top is experiencing gravity just as we do by having gravity press the structure onto a surface where the top makes as much contact as possible with the surface that either is the Earth or stands in on behalf of the Earth. The top is resting without motion on the Earth in orbit of the sun as all captured objects do.

$$a^3 = T^2 k \quad (k=k^0)$$

Then by artificially applying motion through the intervention of life the top acts in a manner that will only come about space in normal cosmic conditions if the singularity that is carrying centre has generated massive heat centralised by gravity applying.

The spin we find the top has, is an indication of what principles drive the Universe. The proton collects heat from outer space as the neutron supplies motion by duplication. The duplication is a manner by which space-time is in contact with heat and by duplication it is in contact with heat that is released in space. By motion the proton contracts the space and by motion the neutron duplicates the space. The neutron replenishes the space it lost to the proton by moving about the space and this motion we see as the electron. The more the duplication is the more space-time heat is available to supplement the diminishing by contraction. That motion is time related and the higher the duplication is by time unit, the more contraction comes about and the less space in relevance the atom claims. There is more motion in the Earth's atmosphere, which constitutes to more gravity, which is more contraction, therefore is astronauts are smaller in the Earth's atmosphere than they are in outer space. The higher the spin rate is, the more space the spinning top puts into the unit and the more space there is in the unit. That means by increasing T^2 (the spin) the longer k will be and the higher the ratio will be in relevance between space and space.

As the proton grows the relevancy of singularity will extend. The more the proton contains the higher will the spin be that the proton enforces, but at the same time will the space-time serving singularity be in accumulating captured heat. Therefore the spin is quicker but as there is more space-time accumulated, it would seem as if there is a slower spin. This growth in material causing more space to duplicate that is bringing more space to contract eventually drives the Universe from one to another extreme. The extremes are the Big Bang and the Big Crunch and the moving process is the Hubble shift. All the processes I have mentioned are well documented. However, since the Earth has less heat than it should have, we find atoms on Earth degenerate slightly, but that is only a response to less growth in relation to what growth is available in the rest of the Solar System with the Earth holding one of the lesser gravity units there is in the solar system.

By duplicating space-time in a rotating manner the motion elects a chosen singularity by exciting such singularity and the motion of the unit energises the chosen singularity to steak a claim off on independence. The independence comes from a singularity that established a centre where the lack of motion at that precise centre dismisses space to the reducing of space-time.

By spinning the atoms that is forming the unit, are all in motion, where even every independent singularity point centralising and securing that specific space-time is in motion by transfer of space-time. The motionlessness of independence has to transfer in order to legitimise the claim of independence. While in motion it establishes a centre that connects all the atomic motion and represents the one point in space-less-ness of being without motion. At that point gravity comes about that singles out the unit from the rest of the space-time in dominance and that point of space-less-ness and motionlessness finds a singularity that then is charged with the task of securing growth and maintaining heat accumulation to further a flow of space-time from which all material in the unit that is duplicating will benefit by securing a stronger presence of duplicating **while the chosen centre elects top dismiss space-time** forming a stronger presence of a more secure singularity.

Under normal conditions this will only come about with the aid of a massive number of protons. In this case however the spin came from life generating some artificial gravity. When ordinary electricity producing the operational conditions in electric motors we gave the process found applying there the name of electromagnetic induction. It is still applying the principle of heating singularity to produce centre heat charging motion. Although in this case the top is spinning because of the action of human muscle creating motion. The electric motor turning induces a centre by using the Coanda principle with the benefit of a star quality motion. Humans take conditions one find inside stars where $iron_{56}$ causes space-time flow ($5 \times 5 + 1$) of protons and copper ($2 \times 10\pi$) to establish star gravity and centralise space-time flow. The important issue is that motion creates the spinning, which results in the top sustaining its own space by effectively enacting independent gravity through the Coanda process that is replacing the Earth's motion.

Singularity within the subatomic particles dismisses space in order to sustain cooling. The motion applying as cooling is a contracting and depleting of space that reduces the time factor to the point of elimination. To do that that it has to incorporate heat from relevantly close by or neighbouring singularity that has overheated and abandoned form. By resolving such heat into realm of the singularity able to secure survival by sustaining by the cooling of singularity that is applying gravity. When doing such reducing of space the singularity elected to the task eliminates space that allows a bigger flow of space by cooling space within material and concentrating the heat levels in space per space volume. The concentrating of heat within the cold and reduced space forms a driving ambition that generates motion into the revolving space-time surrounding the heat securing singularity.

The concentration of heat feeds the motion of the space-time in motion around the axis of singularity. By feeding on the much-concentrated heat the heat contributes to motion and the motion contributes to excessive (more than before the spin) duplicating of space by performing a stronger flow of space-time being the result of a stronger substitution for the dismissing of space in the centre. The motion is a result of combating heat which is a result of forming motion and the one factor generates the other factor as long as there is space to reduce and advance the heat levels that will supply the motive to produce motion on which further motion will feed while all the while singularity is incorporating heat into the unit accumulating singularity as it is concentrating heat by creating an ever reducing flow of space-time towards the centre and a duplicating of space-time is resulting into an increase in independent and unattached heat levels in the reducing space. When more heat concentrates it becomes an overheating situation that will increase both the duplicating as well as the dismissing of space-time that will increase the heat levels further.

It is an ever-increasing process and the only factor loosing is the outer space dismissing of heat within outer space. As the concentration of heat in outer space reduces the space factor in outer space that space then will increase by the same margin and that will seem to bring about the increase of space alone. This action is more like water falling on a hot plate. The motion duplicates space by increasing space as water reacts to the heat surging and the surging is elevating motion by producing space that instigates expansion, which generates as it heightens motion. But it is only the coming about of motion that is driven by heat levels increasing to which I here refer to in my using of this example.

The duplication that provides the motion can only come as a result of surrounding the singularity with other heat than the heat produced by singularity. When singularity release heat through overheating it reduces its relevancy in the relation between it and surrounding singularity, which is sharing relevancy. When the singularity sustaining form with gravity is committing heat brought about by other singularity such action enhances the relevancy in the relation that such singularity has with other relevant factors. When the lesser singularity finds dominance in the diminishing action of the relevantly superior singularity it has to react by providing heat that surrounds the lesser singularity in space-time. If the heat is a product of self-inflicting overheating the dominance will prevail to a point where the lesser subdue in defeat and join the space-time of the superior singularity. This can only come about when the ultimate limit is crossed and the crossing of the Roche limit turns the value of **k** from $k^0 = T^2 k / a^3$ to $k^0 / k = T^2 / a^3$.

But when the lesser singularity has the ability to surround the heat accumulated from space-time being outside by using space reducing thus intensifying heat it will establish motion that will place the relevancy of such motion coming about from the space diminishing ability of the superior singularity. It is all a matter of sustaining singularity by matching and qualifying relevant abilities in different singularity. Without the surrounding of heat the singularity will submit independence and surrounding by sufficient outside heat it then can sustain the independence of an independent developing singularity. A clear understanding about this principle will lead to accepting the principle and the accepting is of crucial importance when one wishes to come to understanding the origins of the solar system.

It must be clear that it is the harmony in the speed and not the value of the mass or the quantity of the heat that will establish the allocation of the positions taken by the relevant objects applying gravity borders. Only when allocation reaches the fringe of the borders set by the relevant objects comes into the limits it is that quantity and potential ability has an influence on the outcome of further development by either object. In the book an open letter about The **Seven Days Of Creation ISBN 0-9584410-4-9** I delve extensively as I delve deep into this topic of how the solar system comes about when using the cosmic relevancies that I am about to explain.

ANNEMARIE BEGIN HIER

Gravity is the product of motion and with motion gravity comes naturally in the process of motion. Due to the way the sphere is built it will always hold singularity in the centre of the structure. Singularity is a point within the centre of all spheres where no motion can be possible because the rotation pivots at that point and the pivoting changes direction precisely at that point without sides. The motion to which I refer here is the combined motion of all the atoms in the sphere that create a singularity centre. The pivoting comes about where the line that forms the circle ends in the first dimension That we read into Kepler's formula about space-time originating from singularity $k^0 = a^3 / T^2 k$. At that point all space-time finds the relevance to return to the form of formlessness. The point that holds the value of Π^0 is located by dimension within any and all spheres but to top this Kepler showed that motion produce space and space is time through motion $a^3 = T^2 k$. Time forming the second basis for the entire Universe is the spin of heat in space. Without motion space-time collapse into singularity and within all spheres there is this point that cannot provide motion. Therefore through the form the sphere holds the sphere will diminish and destroy space in the centre by not providing motion within the very centre in the round structure.

The top and the Earth share a common singularity. That is gravity we all see but it is not a force because when motion comes to the top, the shared singularity shifts. We may consider such shifting as receiving momentum but the momentum is only an enlarged gravity because more motion discrepancy comes about. The line 70 $(3\Pi^2)$ shares a common connection to the singularity in charge of controlling all gravity aligned to the centre of the Earth. The motion of whatever magnitude comes and proclaims the top even more independent and a shift in the line of gravity comes in and puts space in between the line the earth shared with the spinning top and the line the motion indicate to connect the top with the governing singularity. The shift can be anything from Π^0 going on to become Π and then extend to $\Pi^2 / 2$. That is what the factor **k** becomes added to the existing value there was before. By introducing a stronger $\mathbf{T^2}$ the **k** factor has to become more and that is what causes the sound barrier to break.

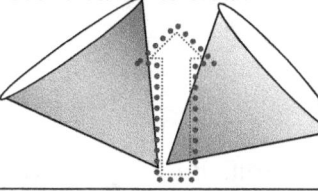

$$T^2 > a^3/k \ (k=k^0)$$

When the top spins at a pace exceeding apparent barriers the top will begin to turn about the axis as if it is trying to extend the axis, which it is spinning about. When the spin is extensively faster the top shows an effort in trying to escape from the mass constraint and the motion providing balance then becomes much more than the counter balancing of gravity. The excessive motion will create a spinning action where the top literally tries to use the spin to lift from the ground and jump into the air. Part of this is going side ways by spinning. The top jerks to the one side and jerks to the other side in an attempt to secure more space created by excess spin to use for the creating of additional space. Well that is precisely what it is doing in the attempt to turn around by extending the spinning axis which individual singularity provides.

The space it created through spinning has then reached a point where that space it then claims, is exceeding the space the Earth granted the top to have as a motionless object and the object is by the value of the motion it produce, acquiring more space. With extra motion, it will find more space-time by which it then tries to secure a better space-time position for the singularity it is maintaining.

$$a^3 > T^2k \ (k=k^0)$$

The apparent ness why a top will not spin in outer space becomes more evident when considering the gravity the Coanda spin motion introduce with all the relevancies that is attached to the process. It is the relation that the spin establishes by implementing the Coanda effect by motion, which secures the gravity that the top install in order to place the top in the upright position. Installing space-time secures independence the top requires to spin. Gravity is all about motion and nothing about mass enforcing pagan forces, which unleashes the will of the unknown upon the unwilling.

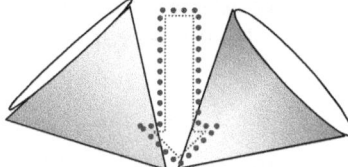

Only after the top showed the last act of defiance will the top again submit to the mass, which comes about from the tops inadequate spin tempo that will not any longer produce a sustainable space to create he needed space and use the motion as antigravity.

The mass does not apply or draw the spinning top down and in fact the top is trying to undo the mass the Earth provides. The stronger the motion is that top asserts, the bigger claim the top will make on the unit singularity. The top will try to rise into the air by extending the **k** factor between the singularity claiming space for the top's independence and singularity claiming space for the Earth. The lighter the spin rate of the top is, the bigger will the surge be to establish independence and by trying to lift from the Earth, that lifting shows the top has an attempt to reduce the affect mass has on it. By spinning faster $\mathbf{T^2k}$ such spinning is increasing the relevancy of the space $\mathbf{a^3}$ that the top claims in relation to the relevancy of the space the top holds. That increases the buoyancy factor that the top has in relation to the rest of the space the Earth claims.

When the earth reintroduce the effect of mass the top still try to create through motion sufficient space to use as a deterrent to the mass the Earth wishes to inflict on the top. Clearly the top shows that the space relation grows between the top and the Earth which

In cases where motorised machines bring about motion the artificial energy is strong enough to apply motion that counteracts the mass factor completely

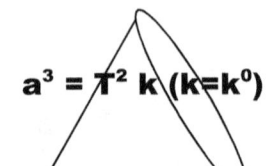

$$a^3 = T^2 k \ (k=k^0)$$

By bringing about motion that produces lift to the object captured on   Earth, as the rotating of the wing of the helicopter will produce through motion creating space the mass disappears. Mass is the manner in which the Earth depress the space the helicopter claims. This is the truth of mass notwithstanding the corrupt definition science use to further foul and disgrace an already corrupt definition. At some point science has to come to terms and accept the differentiation that is degenerating the truth about mass and gravity. One must either decide to remain as stubborn as a mule and insist on it having mass just because of some ridiculous argument tried to cover up disgrace in the past, or admit that the action relieves the object of the mass that the Earth produce.

The object will still sustain an effect from space duplication discrepancy but such discrepancy is then so much reduced the object is as light as air. That is the reason it can fly! By using the heat it incorporates the heat in the atmosphere, the helicopter increases the material to heat relation, who then brings favour on the side of duplication and then the relevancy about the duplication to dismissing of space-time favours the duplication by yards. In all of this I cannot find one shred of evidence where mass brought about any thing except being a result of the motion not producing independence to the top and therefore allowing the top to create an escape pass. Mass is the absence of adequate motion.

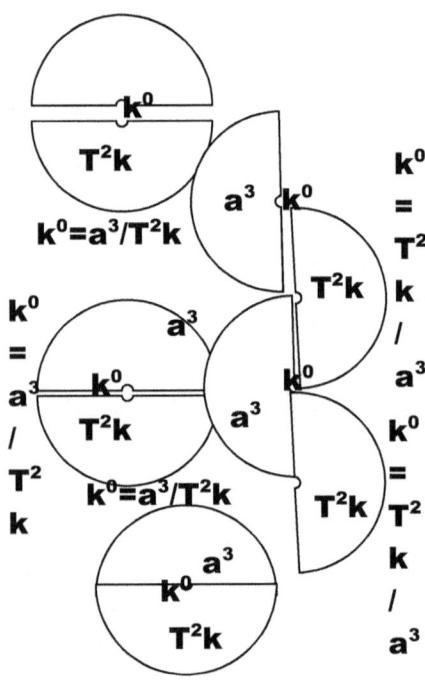

All material has gravity built into the form it holds. Due to singularity forming Π^0 space collapse as a result of motion in the centre of all spinning structures. The faster the structure spin the more intense that space will be that collapse and therefore the more heat there will be to accumulate within that centre. The space ending there forces the motion to stop and the lack of motion produces space dissolving. But with space disappearing a flow of space forms that becomes necessary to replenish the space that went lost. This centre is also within every atom where the proton forms such a centre and the atoms produce a unified effort of diminishing space to the total volume of space diminished that is produce as a compliment by all the atoms sharing space-time in the unit that is combining efforts where at the space less centre of the star in the space less centre diminish space. The refilling of the space produce a flow from the outside flowing towards that established centre we witness in the Coanda effect as in effect the centre where the space vanish as it accumulate in the centre of the sphere where no space ever can be.

Towards that end all space will flow but will never be able to replenish the lost space because motion stropped at that point and the motion that brings about the space destroys the space just because the motion cannot be produced at such a point. The fortunate part to us is the planning because the space that is dissolved is the space of the less successful spots that did not manage to establish contracting that well and in that kept expanding with the friction to combat the heat coming about.

The spinning top is once again confirming the role the Coanda effect plays in all of the natural *"forces"* as motion of material in liquid (this time the atmosphere) establish a centre and that centre takes charge of the space-time influenced by space in motion. Charging electricity apply the same method of operation in the same way.

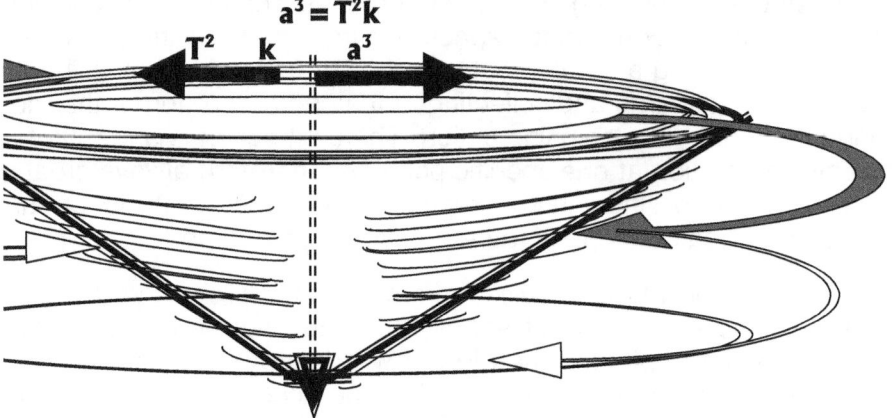

$$a^3 = T^2 k$$

Therefore the spin involves space already established and the space contracting is space of other parts that overheated and in relevancy liquefied to fluid heat. Therefore the expanding becomes part of the growth as it becomes the contracted space. It is this effect that we gave the name the Coanda effect and is as much part of gravity as gravity is part of gravity. The spinning top again is as much proof that the Coanda effect is a product of gravity than flying is proof of the Coanda effect where it establishes antigravity through motion. As the top spins a centre is established with the motion of spin activating the centre not spinning. The centre comes about, as the centre remains motionless while the rest is spinning and the centre becomes an additional part being part of the motion of the object but within this a centre forms that is having no space and no motion. The motion travelling towards the centre must duplicate the following as well as the previous space crossing the centre to commit space-time to the dimensional form it holds. Please note that by my referring to space being big or small I do not admit to the cosmos holding such standards but since I am human and has to command the English language to establish the understanding of my ideas, such terms merely describe references to us humans but it is by no means my way of admitting to cosmically supported standards.

When there is a smaller object within such a space duplicating and that smaller space also serves as singularity and as an independent unit from the major singularity the space the smaller object holds is not duplicating to the same trend as the larger space. If the larger space held the same claim in size to that of the smaller space occupied the time aspect will set the two claimed spaces billions of years apart. But since the two share the same time the space differentiation is significant. With the smaller space being part of the larger space but also still being independent it will lack the capacity to match the larger space in the effort the larger space can present in duplicating. If the duplicating of the smaller space was more favouring the duplicating than the dismissing of space it is found more towards the outer rim of the sphere, which the larger space claims.

Due to size restriction the smaller claimed space will not match the required space duplication when sharing the same time factor in using the matching motion since both share the time aspect. Although the top shares time in space with the Earth the top can find relief from the drowning effect of the gravitational dominating space – time the Earth unleashes just by applying more motion with the help of the motion the Earth provides. As soon as the top starts spinning the top duplicate more space within that space of the containing object, which in this case is the Earth, and by applying credible motion the spinning top suddenly find the required ability allowing the spinning top then to match the space reproducing. By excelling motion the top find the ability to establish enough gravity to bring about independent balancing through independent motion.

When spinning excessively the top can overshadow the space duplication and in that it can seek the ability to find more space that is normally more to the outer edges of the sphere and therefore try to match the space reduplicating with space towards the outer edge. This effort we call flying. Gravity will always comply with two relevancies one is creating space flowing away and the other is space contracting. When the top produce a spin it tries to secure sufficient motion to lean towards escaping the confining of the Earth gravity by trying to establish independence by which it can escape and by not spinning it then submits in defeat by falling with the inward contracting space of the Earth.

Not only does this explain the energising of the top when spinning with excessive motion applied but it also proves that the gravity induced by contraction is part of the accepted form of space-time in the roundness that is securing Π as form. It proves Kepler as correct where Kepler said gravity is space in motion. The motion creates space duplicating within the parameters of the space containing the minor space serving as a unit, but with the intensity of spinning the motion establish a concentration of heat flowing towards and collecting at the centre that forms the motive for the searching to secure an independent by the spinning action. The contraction is a result of singularity diminishing space by not providing motion at one specific point and therefore always creating an initiative for space to flow by motion back to singularity to replenish the space that disappeared. That serves the proof that there is no pulling but there is rather a flow of space-time.

When it comes down to the pulling of gravity that there is no pulling of gravity. The creating of motion establishes the duplicating of space but in the same instance the motion provide a point where motion does not apply and therefore it will destroy the space. At that centre duplication stops and space-time or space motion seizes. The one accomplish the other and a relevancy is brought about because the motion creates space that creates a point of no motion destroying the space created. The stronger the motion is the more space will naturally displace and the more heat will then concentrate in the centre of the object applying motion. The minor object is part of the space and within the space of the major objects flow of space towards the major centre that is holding the major singularity

The rotating of the spinning top is as much part of the Coanda effect as flight is part of the Coanda effect and it all becomes the result of gravity provided by singularity coming as a result of motion creating space by providing a centre that will bring about contraction as space is rendered motionless within that singularity centre and therefore the centre is space less as much as it is motionless.

The contracting of space diminishing as a result of singularity established

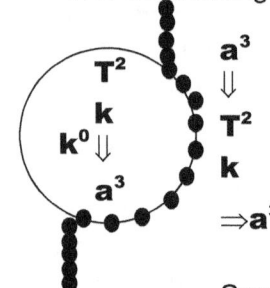

a^3
$\Downarrow$
T^2
k

The motion provides new space and the new space provides motion and in the midst of all of this a point forms through the motion where space will flow towards and disappear within that point of space less ness because of motion less ness.

$$\Rightarrow a^3=T^2k \Rightarrow \Downarrow \qquad k^0 \qquad \Uparrow\Rightarrow a^3=T^2k\Rightarrow\Downarrow \qquad k^0 \qquad \Uparrow\Rightarrow a^3=T^2k$$
$$k^0 \qquad \Downarrow\Rightarrow T^2k = a^3\Rightarrow\Uparrow \qquad k^0 \qquad \Downarrow\Rightarrow T^2k = a^3 \Rightarrow\Uparrow \qquad k^0$$

Space duplicating is creating motion and from the motion space is created.

The establishing of motion is creating space and providing a centre point of singularity. This proves that the atom links with singularity and that the value of singularity extending forms a relevance with the space-time it influences up to a point that it controls the space as much as it controls the motion of the space. That proves singularity extend way beyond the space of the material it holds and establish duplicating points of singularity within singularity by merely applying motion to space within space.

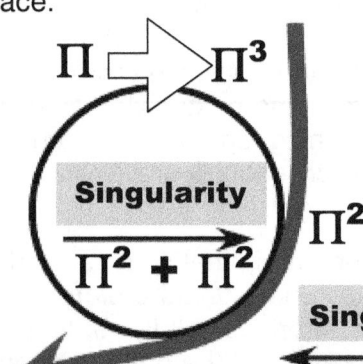

The Coanda effect is also the perfect example of the curvature of space-time brought about by the extending of singularity influencing due to the shape that imitates or duplicates the value of singularity and again conform Π. By establishing a new value of singularity as Π, singularity can once again take control and establish a new Π^2 as gravity in the new Π^3 forming space

Singularity extending the influence on flowing water

The Coanda effect is creating gravity. It is not replacing gravity it is not recreating gravity it is not enacting gravity it is is forming gravity.

By reducing the propeller in size and boxing in the airflow it is directing the flow to a centre where the spin will intensify as it accelerates. Gravity is created in such a way. The protons perform the spin creating the gravity and the proton number does not necessarily prove the strongest gravity.

The turbine engine is a star in the little. Massive space reducing brings on heat increase and with the accelerator of fuel added the heat increase generate a singularity enhancing equal to a star. It shows gravity come about by altering the space a^3, applying with heat added a new T^2 where that then produce the thrust to use the **k** coming about to elevate new movement, which is gravity independent from the earth gravity

By reducing the propeller in size and boxing in the flow it is directing the flow the spin where the airflow will intensify as the airflow accelerates.

Gravity is overcome as anti-gravity is created in such a way. With gravity it is a case of the protons that perform the spin, which is creating the gravity and the number of protons used, does not necessarily prove the presence of the strongest gravity that inflicts contraction, which produce mass. The protons accelerate the moving of space whereby that spin will reduce the space volume. By accelerating the flow of space the volume per time unit decreases in relevancy.

In the reducing of the intake the gravity becomes artificially constructed because gravity is the reducing of space. By injecting a fuel the situation intensifies as the possibility of raising the heat levels become much more prudent.

The reduction of space will bring about heat. Injecting fuel in a place where such reducing of space already exist it increases the heat level further and the fuel will "spontaneously" ignite. The igniting creates a rise in the heat level to a point where such a level that one can only find that level in the stars. Fuel will establish conditions that is in accordance with the laws of cosmology such heat will only apply to stars where such heat levels will indicate the enormity of the gravity present that is generating the massive gravity accumulated in the absence of space. Gravity is the concentration of heat in the utmost reducing and concentration of space thus a star is born in the gravity on Earth. In this example k increases as thrust pushing space a^3, which the **k** creates to a new T^2.

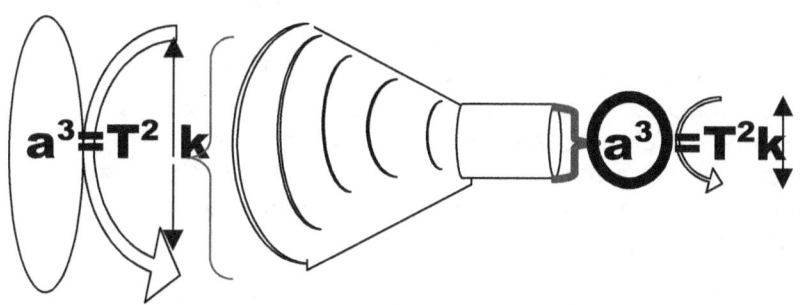

Bringing about a new centre with more heat in the specific point will establish motion in relation to the heat intensifying at such a spot and the motion produces an energy that elevates such a spot to form a new Universe. It is the manner how stars start to be stars and apply gravity.

By reducing the space a^3 with the shortening of **k** a new standard comes about establishing a new T^2. Then moreover a new T^2 created by injecting fuel into a newly established k sets the groundwork for an a^3 coming about that can challenge the best star centres there is. With the ability of life to manipulate space-time man confuses nature in accepting there is a Tiger loose and this young star can challenge the space-time set by the gravity of the Earth. The earth will allow such rebellion just to a point and a fight will ensure that will either release the spacecraft or down the airplane. But all this comes about by the artificial creation of a singularity miming singularity by presenting something that can respond and produce gravity.

Do not forget the artificial aspect that is involved in applying the process since all this mentioned is created by the intervening of an artificial substance that is the only factor in the entire Universe with the ability to accomplish a force and that all of us that has it, we call it life. Life is first of all by my definition motion not of cosmic origin and with the motion life can control and accomplish the manipulation of space-time and in that we include the manipulation of cosmic laws that work independent of cosmic motion but still use cosmic motion. By the mimicking of cosmic laws we with life find an ability by which we can control the space-time we use. Part of such space-time manipulation also includes the cosmic laws and in successfully manipulating space-time, we establish such laws to benefit the wishes and the will of intelligent life. Creating the Coanda effect concentrates space as gravity does, it concentrate space in huge quantity as gravity does and therefore nature takes the action as coming from well established singularity. But never forget that life established the phenomenon which life created through the manipulation of the cosmic phenomenon. The action itself is not cosmic.

The law applying is cosmic but the commanding of the law is done by life in terms of manipulating space –time therefore life has the ability to use its manipulating gifts and that include applying the laws of the cosmos. I have to be deliberate in stressing this point because if there is one thing Mainstream science is guilty of and blind to reality then it is differentiating between life and its abilities and the cosmos and its abilities. There is a strong drive to promote atheism as science friendly.

Little can be further from the truth. Mainstream Science wishes to promote the fat of life as a natural outflow of a cosmic reality by depicting life as the most cosmic spontaneous events. This is done notwithstanding the incompetence science has about knowing what drives the cosmos. Not one scientist up to the moment knows what gravity is or what brings gravity about. In the face of such poor lack of vision those thinking they are brilliant as scientist also regard them of such supreme intelligence they can challenge the fact of a Creator. In doing that they truly show how little their minds are to host vision about physics. Physics has so much more detail to offer than the calculating of a few feeble mathematical formulas. I challenge any one of those to disprove me, after they have read **"Man-in-motion"** which is part of **Scientific Mistakes ISBN 0-9584410-1-4** or an open letter on "**The Seven Days of Creation**" **ISBN 0-9584410-4-9** and prove me wrong in what I declare in those books. Life has never been spontaneous part of the cosmos.

The two can never be confused because life is not part of the cosmos and what apply to life is alien to the cosmos. We always introduce standards applying to the benefit of life but such standards are a joke when the cosmos comes over to play. The action of life is artificially established by life's ability to manipulate the cosmos. The action becomes part of the cosmos as result of cosmic life, which is heat accumulating and heat driven where gravity is increasing heat or reducing heat. **LIFE** is not normal and it is not cosmically natural within the cosmic independent structures. It is laws we find in galactica where stars cradle until they hold a singularity sufficient to survive while the established singularity can seek independence from the star nursery we call a galactica. In stars or "planets" that stands in direct contrast to the very same application of the very same conditions but where the turbine is as artificial as money in the cosmos, the Roche factor is as natural as heat in the cosmos.

With the ability that life has to throw the top we stand on Earth holding our space in the space of the Earth. We throw the top and see the top come to life stay alive for a limited time and then die again into the realms of no motion. We gave the top space to be within and the cosmos allowed the space to be but set the terms for the space to be used and when the top was unable to comply with the cosmic terms, the cosmos repealed the space it allowed the spinning to take place. Yet the top still use space, but that space the Earth accommodates as that space uses the Earth spin by measure of motion. We cannot have space if we do not fill the space we have on Earth. While being on Earth my position is $a^3 = T^2 k$ where k is because of the mass in movement standing in for k^0 by being k^{-1}.

The position I have stands in relation to the governing singularity that the Earth adheres too being k^0. My position I have in accordance to the mass I have stands me in relation to the k that positions me on the topsoil of the Earth. My motion T^2 sets my space a^3 apart and give me identity independence in relation to the rest of all factors forming the unit the Earth is. My mass comes about from me being a captured object of the Earth and as far as my standing with the Earth goes I have mass where my having mass classify my position in terms of the earth as $k^{-1} = T^2 / a^3$. Being k^{-1} we are also T^2 / a^3 which is reducing us in the space we hold our mass as we try to reduce a^3 further to comply with the T^2 the Earth is applying and which we have to use. If we move we have produced a larger k factor to the order of at least k^1 to find the ability to move from k^1_1 to k^1_2, which will allow us to accomplish such a task because we use T^2 to move from k^1_1 to k^1_2.

So we have to improve both T^2 as well as k to accomplish motion. From the findings it is the proof that connects Kepler's formula in reality since every aspect of the cosmos is growing by duplicating. In that it puts science and the claims science make on aging in question. Using $a^3 = T^2 k$ and producing a larger $T^2 k$ it means a^3 must also improve. From that we can see that whenever a^3 duplicates by measure of T^2 singularity grows. By the measure of k extending further than what it was before the duplication. That it does by doubling the space it use during the motion where the motion in itself is halving the space that doubles. This is not that uncommon physics.

A car holds the space a^3 and is moving by the speed of T^2 through the distance of k. If the car speed up the gravity will increase because the distance k will reduce. But if such reducing becomes part of the equation then other factors in the equation has to compromise to allow changes to come into place. With the mass or space in motion speeding up the mass has to increase to have the equation to remain even. If the distance travelled becomes more in the time duration it took compared to before, then the time duration must become less if the relevancy of the distance remains the same. The time it takes is the gravity or the time by the square that increases when the space moves faster between two points. The mass has to become more to allow compensation between factors to permit the changes happening. That is what gravity is.

The way we think about gravity is that we wish to have gravity applying the same everywhere and acting in the same manner under all possible conditions. In order to play with mathematics and still find the constants that enable such playing to deliver the same results time after time man has subdivided the concept under so many names for each fragment we divided we cannot even find the basic principle any more. Gravity is not a force as Newton suggested but a motion between space occupied and space filled forming a relevancy and this applies throughout the Universe.

Life on the other hand is the only force in the cosmos. Gravity is a flow of time from one eternity through another eternity towards and finally to end in another eternity. That is not the purpose of life to find eternity where singularity creates the cosmos by using time. The cosmos is hostile to life and has no use for life because life connects with death and not with eternity. The only force there is can only be found on Earth in the form of life. Life is the force on Earth and only on Earth it has space-time with the ability to manipulate space-time under its control by providing motion other than and above that the motion the cosmos provides. It is precisely in such a manner that light uses to travel in from singularity to singularity.

Because singularity forms space and controls space by time measure, which we consider as dark and therefore invisible, such spots in singularity breaks down and rejuvenate space much faster than light in the photon does. When the point holding singularity has rejuvenated and replenished with heat in growth, singularity releases the photon and by such release can the released photon join the next singularity in the period of removing space-time and the next in rejuvenating space-time which then will include the photon reassembling with the next singularity forming the space-time of the next singularity. That is the way light has to move if ever it moves from singularity to singularity. Looking at the issue in this way we can begin to appreciate that light is the duplication of the photon by the singularity charged by the motion that provides the singularity by charging the intensity.

It is about duplicating more than dismissing although dismissing does form part when the photon changes singularity. In that way the light loses intensity as it travels and is recharged by the singularity on route to somewhere in the future. By giving the top a chance to spin we extend our position by means of life's ability to use and manipulate space-time and that extending allow us the property to apply motion to the top. Light then forms the part of not being permanent on the side of what is eternal but is part of the limited time duration and death. The rest of the cosmos is in a balance of duplicating by rhythm of space-time dismissing and the control between the two points of gravity. Light is temporal and material is permanent but not eternal. Singularity is eternal.

In contrast to the duplicating of light is the duplicating of material. The duplication of space filled with material is the use of heat in space surrounding the atom, which provide the material the ability to the confirming of space. Such confirming of space is in relevance between material forming a unit and material surrounding the formed unit. In confirming the form of the unit the unit attach and surround the form it holds with heat. Such surroundings of liquid heat are much more than singularity requires in the immediate to prevent overheating to come into affect. The heat required is more than that is needed for sustaining heat to prevent singularity overheating and is more than singularity will ever require.

To be more than singularity will require it to be much more than what particle forming the atom ever will require. Singularity in charge of light can generate by duplicating the photon whereas material use the heat the photon provide for sustaining singularity requirements. When light is clashing with the atom the atom applies the heat that light is to dismiss space-time. The light is heat, which is applied by the atom to sustain the immediate singularity without any chance of relieving some. The material in its use of light is making redundancy of space-time, but its needs requires even more heat than light alone can provide. Material needs gravity, which is more heat and the gravity it requires is much more than the photon can deliver because that is why there is shadows on the dark side. Life is one providing form of such heat but is only committed to one atom and only one type of that particular element. Life is a supply of a specific kind of heat that allow motion to occur where such motion of space-time uses such heat to become motion other than the motion the cosmos can provide. But the duration of such independent motion is very temporarily whereas the motion that the Universe provides is eternal. Life uses the supply of heat to live and living is the aging of life. Life aging, cosmic growth and the limit duration of life are so much connected it is the same thing. That, Newtonian science totally misses.

Aging and the fighting of age, disease and death has become a major concern in the world of science especially now that there is a drive to secure atheism. The world of the medical genius claim a possibility of human life extending to four or five hundred years between birth and death and the working life of a human being on Earth becoming three to four times longer than it is at present. This they say while facing the reality that those in science have not even got the faintest idea what causes wrinkles and what causes the degenerating qualities of the human body as time develops. This is clear because they blame it on gravity while they have no idea what gravity is or what becomes gravity.

They claim such rise in life expectancy based on the already accomplished rise in average time of life in general on Earth. This rise in life expectancy we currently are fortunate to experience came about with the ability man acquired to fight other organisms on Earth and thereby remove a major killer of young lives. This we accomplish with a chemical warfare on other creatures that kept those in good health alive to the detriment of the unhealthy. Then food supply was a big reason for wars and killing back then when germs were the big killer of human life. Instead in modern terms where food supply is no longer an issue it is the wealth connected with oil being the major factor of killing in the modern terms. We have grown rather fond of human brilliance to destroy creatures of small size but big repute in killing humans. We think by killing those killing us we can use mighty bombs instead of gems to kill our kind. Then there do not have to be germs that kill, we can use British, Canadian, Australian, Italian and American politicians and their appetite for oil to select those that needs to be killed instead of germs doing the choosing.

Our effort of killing germs will stop the indiscriminate killing so therefore we better kill all known germs first. Changing that is not changing the coarse of human destiny but it is merely prolonging and boosting a disaster waiting for the future to come to pass. By killing off life, be they germs or insects carrying the germs, our killing them with chemicals do not give us doing the killing a superior advantage over those killed. Remember that those we kill saw killing of their kind so many times with so many disasters that struck the Earth in the past, we do not even know where the disasters were or what the disasters were. Those we kill overcame problems nature tossed their way by destroying their kind in a manner and much more than we that are doing the killing at present can anticipate of what destruction the earth is capable of.

There is a disaster waiting once this organism found a way around that which we use to kill those in their ranks. Some of us are doing the killing to swell the ranks of man and a small percentage is using the opportunity to get stinking rich from this action. Others do the killing in the hope of foiling death that is coming to every one and that coming is inevitable in any case. When those microorganisms come back after they found a way around our chemical war the revenge they will respond with, will bring death to humans on a scale that is beyond the recording abilities of man. This is one part of the foolishness that man is spreading in his lust for commerce and control but it will not destroy man, it will only set the balance straight again. Man will die in the billions by germ infection and the rest will die of hunger. But man will prevail as all forms of life do when facing genocide of its kind. We must also remember that this function of life connects to micro life even more than it does to humans. It is shown in the past when life of whatever measure is threatened the resilience of life to fight back and overcome such a threat to life is beyond the scope of human imagination. When in the very end of the Earth all life on Earth is dead one day, when the end announces the end of all…we can be sure of one thing…somewhere there will still be a form of life fighting to survive with all odds against its survival.

Read into this statement what you wish as long as you also read that the cosmos is out to end the beginning of life because the cosmos is in a fight with its own and the cosmos will not tolerate life chipping in at some point. On Earth we live with the idea that where we are we may think of that point being to us as eternally big but in relation to the rest of the Universe that point is so small the point gets lost before the point is ever found amongst the vastness of the cosmos. Life is also about duplicating life as much as removing dead wood. When born, life receives a specific carbon atom and by some property that specific carbon has, that carbon can help life to conduct the heat life use to move. Life is the motion of space-time in space-time by manipulating space-time in duplicating space-time purposely as to control the direction flow of such duplication. To do that life has to use a specific heat to apply a specific motion and the heat transmitting as an electric flow but the electric flow is not life, it is only a manner to conduct massages that enable life to manipulate space-time. That electric flow is also a conducting of gravity, which allow the carbon atom to grow as all atoms grow. By singularity extending in the first dimension $k = a^3 / T^2$ the atom of carbon used by life also grows.

When the atom grow, such growth is cosmic related but that growth brings aging to life and that process limits the time span that the atom can host life before change must come to the life that uses the atom for the sake of transporting life. That is the reason why the body needs oxygen. Oxygen brings heat to the carbon atom and by the carbon atom using the heat the carbon atom extends the atom singularity and in it also the atomic space-time. Unless science knows a way of detaching the growth of the carbon atom by suspending the live body of using oxygen, science and the atheist may be as wise as they see them to be, but they cannot beat aging. It is because of the growth that time inflicts on the atom whereby the cosmos will limit the extent of the duration of life being a part of the cosmos before renewal to the atom must come by the way of a newer version of the same life that has to replenish the atom again. Such replenishing we see as the birth of life. In most of my other books I deal with this matter in some or other way on a far more grand scale than the space in this letter would permit me to do.

The important issue is to see that life does give the top a chance to be in independent motion, but because life gave the top the chance to be, that chance is as limited as the cosmos will allow life to be. The drive of the cosmos runs from eternity to eternity through eternity but because the cosmos drive include eternity that drive exclude eternity from life in the cosmos. The limited time duration that is connected to the spin of the top is the result of life giving the top the spin and therefore the top is also taking the limit spin time as part of the gift of life. Death comes to life, but death only comes to life, meaning that stars cannot die because stars are a party to the eternity sector, which is part of the cosmos. The cosmos cannot find death as life can but the cosmos can only substitute one form of eternity to replace another type of eternity when time comes for the swap to take place. Life on the other hand is completely connected to death as the eternity of life connects to the realms from where the cosmos is driven and not from the driving side of the cosmos.

Gravity is about a relation established when time begun between particles we know as material and particles we know as free or unoccupied space. Gravity reduces space to apply to fit the form of the sphere and later accept the form of the sphere as much as gravity is duplicating space by motion thereof. The fact of gravity is the producing of space by duplicating space just as Henri Coanda showed. Gravity produces space by mimicking space and producing motion that destroys space.

The Roche limit comes about when singularity oppose space-time by applying gravity which we can witness in the Coanda effect where motion contract space to an independent point of singularity representing such space. When the two such points establish by independent motion without establishing enough space-time separating the two points the Roche limit comes into play. The fact of why the two are too close is the first problem too investigates after establishing the Roche limit at work.

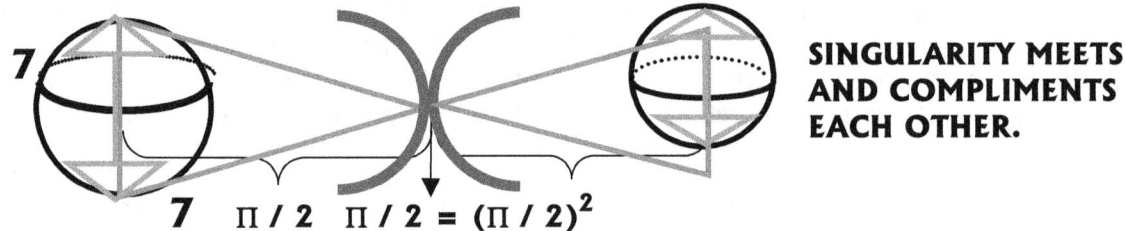

SINGULARITY MEETS AND COMPLIMENTS EACH OTHER.

$$7 \quad \Pi / 2 \quad \Pi / 2 = (\Pi / 2)^2$$

Friction point transmitted and relocated to smaller space centre

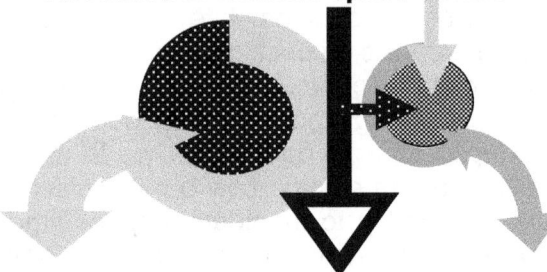

The difference in gravity being speed or motion will establish a differentiation in flow of liquid space. By implementing the Coanda principle the motion of the larger singularity providing flow to the liquid space will be superior to the flow created by the inferior gravity. That will advance the motion of the lesser singularity and the accelerated flow will try to establish a larger space that what can fit into the atmosphere of the lesser object. Again we can read the Coanda gravity principle into the applying of the Roche limit

When two opposing objects find a fluid in motion around them the motion establishes two Coanda systems bringing about different space-time relevancies. Since the stronger space will produce a faster flow, the faster flow will influence the space-time flow of the smaller space. The faster flowing liquid then will try to increase the smaller space by providing a larger space.

However, the smaller space can only become bigger by allowing more heat to be present and the larger amount of heat will have the growing singularity of the lesser space to flood with heat, thus having the space overflowing with heat. By accumulating the amount of heat to become proportional in dynamics the Coanda effect will establish a joint centre from where space-time is controlled in both directions. The accelerating in combined liquid space will equal the solid space whereby the lesser space will destroy, as it has to cope with heat the lesser singularity is not adequately adjusted to control.

When explaining the Roche limit we must recognise the fact that lightning exchanging takes place between Jupiter and its inner closest planet Metis. That means there is photographical proof of electric interaction between structures separated by outer space.

The diameter of the cosmic structure holds the value of r and singularity holds the dimensional value of Π meaning that the radius or diameter (r) extends to become the diameter multiplying the value of singularity. But since r already consists of the square of space holding a definite positional relation with the value of singularity being Π the diameter comes into effect. Π extends each to an individual value to a point where the singularity on each side meets, bringing about a mutual Π^2 to the value dominance of the larger singularity control.

The Roche limit comes into effect when the lesser singularity holds inordinate heat to create substantial motion with the flowing of liquid that will comply with the forms of enhancing gravity in relevancy between the objects in separate independent motion. The relevant speed does not match and that can only result from unequal growth of singularity gravity in too clearly shared quarters. The lesser singularity has the inability to establish a match in motion that will bring about harmony in $a^3 = T^2 k$.

When the motion is not in harmony the lesser singularity will not duplicate space in sequence to the larger singularity and will therefore move towards the centre of the larger singularity in order to establish a better time component fitting to the duplicating that brings about motion. The singularity of the lesser is space-time to the larger singularity in relation to be only heat to the other more prominent singularity and as such domination must come about before integration can be established. Gravity is electricity on a wider scale and electricity is gravity on a lesser scale.

The gravity of the dominant singularity will be electricity to the lesser singularity and the gravity will become a mismatch where the one will be a short circuit to the other which then will produce heat as all short circuits do. By rendering heat into space-time the lesser singularity must either become a compliments being an independent partner or a supplement to the dominating singularity as it then relinquish independence and becomes resolved as space-time within the absolute control of the domineering singularity.

At this point the equality of the straight-line dimension to the triangle and the half circle holds prominence as a straight line, a half circle and a triangle is dimensionally equal. The common denominator will bolster all factors to an equivalent ratio.

The Roche limit is the strongest indicator we have that can show us what is happening with the sound Barrier because in principle the sound barrier is half the Roche

When singularity by the straight line increases the singularity by the triangle it will also bolster giving equal potency in singularity by the half circle. As the singularity of the major component revives the lesser singularity to equality, the triangle in singularity will match the performance and so would the half circle respond in precise ratio setting equilibrium in order. The major partner's singularity in the straight line excites the minor partner's singularity in the straight line affecting all other aspects holding singularity in both objects to match equilibriums in all aspects of singularity. That is the Roche lobe.

By exciting the heat within the lesser structure the structure may find motive to accelerate motion but more likely will lose more heat to the dominating singularity as the two will join space-time by uniting Π in establishing equal gravity Π^2 and creating shared space Π^3. The Roche Lobe explains gravity in the ultimate detail one can wish for. If $k= \Pi^2 /4$ of the diameter distance, then T^2 presumes an equal value in both atmospheres rendering the lesser structure to dense heat. This proves that the space then becomes the motion the major structure enforces on the lesser structure and the major structure then demands equal motion and equal gravity from the lesser structure.

From this the lesser partner will fill by the extent of the larger partner and as soon as equilibrium sets in the growth will duplex to matching in both accounts, normally to the fatality of the lesser partner, as the lesser partner will be capitulating under the strain of the dual. In that way the inner planets came in place as I explain in An Open Letter About The **Seven Days Of Creation** ISBN 0-9584410-4-9 or **Part Seven of MATTER'S TIME IN SPACE :THE THESIS** ISBN 0-9584410-8-1 . In this there is no mass defecting but there is heat displacing from the lesser object to the larger object and all the infighting between cosmic structures in the cosmos is about producing heat in concentrated space. We find this principle also in space flight since flight is about sharing space-time. In gravity there is motion in a linear sense and in a circular sense. Gravity is the balance between space applying a negative or departing nature in **k** and the other is the positive nature in T^2 where the departing is neutralised by the swinging of the relevancy to fit the other side as we may find in the expression **1 / k** Applying the constant space $\mathbf{a^3}$ is equal **=** to motion $\mathbf{T^2\,k}$

Gravity is always a relevancy, which is always about domination and being strong enough to secure space-time whereby independence is formed from motion If the motion of the space-time proves to be strong enough to have independence it has a future and eventually become a Black hole at the end of developing. Even something as minute will end up as a Black hole but that is very much in terms of eventually being far to the future development. By that time the distance there is between the sun and the Earth would have grown to proportions so large in our human terms that no one ever would recognise that the two objects once shared a system holding space-time. In that not one of the two objects would increase much mass by creating new particles that will allow a gain in mass and yet by growth the mass will become massive enough to push the growth Black Hole proportions.

The Roche limit is more proof about my theory that gravity is all about motion discrepancy between two moving particles in space. There is no physical contact of any notion. When the motion between the two objects does not share space, which will introduce equality in time, or rotating tempo it will establish friction in being heat coming about from contact and the end result of this friction is mass. That is what we experience when we feel the attraction we categorise as mass or weight but what we experience is on the very edge or the other end or the limit of possibilities in the extreme of what can come about. The mass we experience is a resisting of those particles shown to protest the discrepancy there is in motion in space between the two objects. From our (us being all object in the atmosphere that is not part of the Earth solid surface) having the inability to duplicate space at the rate the Earth is duplicating space because of size (or space-time occupying discrepancies) the Earth is reducing the space we claim. We resist such reducing and that is leading to our friction that manifests as mass. Any object having a lesser time in space difference between the singularity connecting will indicate such heat when entering the zone such boundary is in. The Roche limit is the sound barrier and the Roche limit is part of the flames we find covering incoming objects that enter the atmosphere as it fall to the Earth.

In order to establish a matching duplicating ability we have to increase the volume of heat in the mass ratio and that will increase the mass to the point where it destroys the mass and the body finds the duplicating ability which will enable such a body to fly in the air of the earth atmosphere way above the earth core. When the object is outside the atmosphere limit of the Earth the object has to sustain a matching speed and such a speed can only come about by having an independence that can establish enough duplication because it must have the ability to secure the correct density in heat in the centre of the object. When passing the limit as the craft is coming in from outer space in the moment the craft gave up the claim on individual independence from the Earth singularity we see the heat forming a blanket around the object and if the conditions of entry is not met to detail the object entering will become heat.

This blanket of heat is the result of time compensations to space claimed and the reducing coming about from that. The heat comes about as the space occupied by the lesser object has to compromise in velocity of space duplication to presume the required mass or space in time ratio the partnership then will require. In the Roche connection mass is not yet a product of distinction connecting the two objects but there is a bond coming about arising of motion discrepancies. The space I refer to is another boundary much closer to the point in singularity we call the atmosphere and are where space turns to liquid because of gravity reducing space. Gravity is applying much more distinct measures as those border brings about mass and in that connecting or entry we find spacecraft covered in heat. That entry point is where an object will receive the mass we so graciously give to the object because to where it eventually will crash and then find mass as gravity at the incoming moment changed alliances from $\mathbf{k = a^3 / T^2}$ to $\mathbf{k^{-1} = T^2 / a^3}$ causing distance to reduce as far as possible that will provide absolute space sharing qualities.

One aspect Mainstream science totally ignore is that the entering and departing of objects we send into space and receive from space is totally alien to "cosmic normality" and "cosmic procedure" because all action produced is coming by the way of through the manipulating of life where life we find on Earth is totally alien to the cosmos. Life as such is alien to the rest of the cosmos and this fact cannot be pressed enough. The entering and departing is a creation of life and therefore "unnatural" in the cosmic world. There is no chance that any object will leave Jupiter under individual motion establishing any possible means of accumulating the required heat for such ejecting. There is no chance of particles leaving as a result by forming heat as a cosmic effort. No object will release from Jupiter bringing about the release of any included particles that is part of the Jupiter atmosphere to become excluded particles outside the Jupiter atmosphere. Much less is the chance of particles ever leaving the tiny star such as the sun by finding the cosmic realised ability through the distributing of cosmic heat. Life is the manipulation of space-time and what life achieves is alien and contradictory to the cosmic flow of events coming about by adhering too cosmic principles.

I have indicated there is no gravity forcing us down to the ground through any magical persuasion. It is the motion in the relevancy between T^2 and k forming a^3 that determines the position the object holds in that relation. A bird can stay in the air on three accounts. The one is the bird maintaining a motion superseding the space displacement the Earth brings about when flapping its wings in motion. This motion is by following the curvature in a straight line T^2 thus overcoming the motion bringing the bird down to the ground. The bird provides the motion bringing about sufficient antigravity to sustain the motion equilibrium

The next one is by motion of the winds curling air through the wings by a flapping in reverse. That motion is an airflow with a reversing relativity allowing the bird to appear to stand still. This reversing of air is creating the air to move backwards and thus giving the Bird the relevancy of standing still in air. $T^2 k$ where $T^2 = k$

Then with birds blessed with massive wings those birds can use the motion of the air that comes about as the heat in the air brings on space and the space created is providing an updraft to supply the motion that will fight the birds tendency moving towards the centre of the Earth. $k = a^3 / T^2$. All other cases we find the motion of the Earth T^2 exceed the motion the objects (amongst others we humans too) k provide and in that the space a^3 we claim is of a diminishing nature. $1/k = T^2 / a^3$

The fact that spin can activate and establish an independent singularity proves what laws motion of space bring about. The fact that there is not only a drive for individuality by an independent moving object but that by establishing gravity independently such gravity fights for release from a contracting gravity proves the law holding the cosmic growth captured. There is a relevancy to achieve release and a "let go" and there is another of contracting, commanding and a "come back" and the relevancy in a gravity unit is the balance there is in the two.

Taking this statement to Pythagoras the triangle enters the equation.

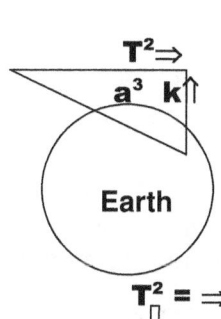

In the normal dispensation of gravity the balance between space and motion sets the space provided by the motion. The higher or longer k becomes the smaller the relative motion of T^2 will account for the aircraft staying air born. At 31×10^6 meters in linear flight an aircraft needs *many times over* the speed it requires near the Earth to break the sound barrier. The aircraft is not flying at twice the sound barrier because the balance between k and T^2 is out of balance. At that height T^2 hardly holds a value. At 2500 km / h the aircraft is just about falling out of the sky.

When the aircraft receives a maximum heat ratio applying in favour of k the craft has the ability to escape the earths atmosphere as long as the departure is horizontal and departing by 7^0. But the condition is that the motion from the start focuses on k as the only recipient of heat and is the only factor with a value higher than one, bringing in T^2 as a factor holding one. In such a case singularity providing k the direction will influence k to provide T^2 with the minimum value being 7^0. This is the only way to escape the atmospheric bubble enveloping of the Earth in the form of a liquid.

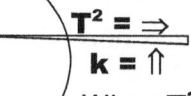

The balance also favours the other factor as we use that balance to do the normal flying through the atmosphere in an aircraft.

When T^2 is very much dominating k and k has the minimum influence where k is just about all the time in a factor of one we fly in the same manner as the birds do.

In the event where a body use the latter motion balance to fly, such a motion cannot bring about the crafts escaping from the atmosphere into outer space. Visiting flying sources will become resident flying sources. It is the other balance that has to be dominant to charge the earth singularity with heat that will bring the escaping effort to a successful conclusion. No alien aircraft can merely enter at will and unnoticed or escape in this manner and no aircraft can change from the circular stance to the linear stance providing the energy requirements to satisfy such an escape without radar detecting this enormous source of heat. This is the reason why the ride height of a formula one racing car is but a millimetre or two three and that represents **k**. In this ratio the movement of the formula one at 300 km / h is the T^2 factor coming about as anti gravity. The formula one therefore needs wings that is fitting a jet fighter more closely just to combat the antigravity the motion develops by establishing this ratio between **k** and T^2 in the order of a few millimetre in ratio to hundreds of kilometres per hour. It again proves the application of gravity established in the set up of the formula one. More proof about gravity being space depleting through motion applying is the way the sound barrier comes about. The sound barrier is the duplication of space within space by one object forcing motion and the second lesser object fighting for independence by providing the motion **k =Π^0 or Π or Π^2 / 2** to sustain individual singularity motion $T^2 = 7(3\Pi^2)$..

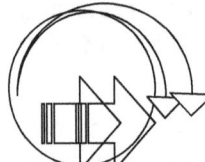

The complexity becomes clear when every factor in the Kepler formula is given a value implicating all the parts it holds in space-time through out space-time in the cosmic space-time. The factor represented by **k** points where singularity is implicating space a^3 and implicating time T^2 from a centre.

This, the Kepler formula is the most accurate formula ever devised with the simplest explaining about it and is also the least tested or understood formula in science. When in doubt about cosmic science refer to Kepler for answers because it is Kepler's vision of space being equal to time in the relation with a very specific point in singularity that is the solution to cosmology. Space holds a third dimensional value where time summons the space from a single as well as a square being the second dimension that is formed as space-time. The two in compliment forms the fourth dimension incorporating time and space. The location where we wish to locate a Universe holding time we can only find when a compliment of time forms space in relation to a second object forming motion relativity with time the ruler of space. To understand this one has to understand Kepler's formula.

The Dopler effect suggests a shift in rings moving from a centre concentrating in the direction of movement. That is to a very limited way what the sound barrier is about. The Dopler affect is a small and almost unnoticeable part of the sound barrier.

In the most basic explanation one can bring in the atomic relevancy of $(\Pi^2+\Pi^2) + (\Pi^2 X \Pi^0) X 7$ with $(\Pi^2+\Pi^2)$ forming the proton value and $(\Pi^2 X \Pi)$ forming the neutron value. In this the seven forms the electron value where the electron brings about the space-time influence of singularity on the motion.

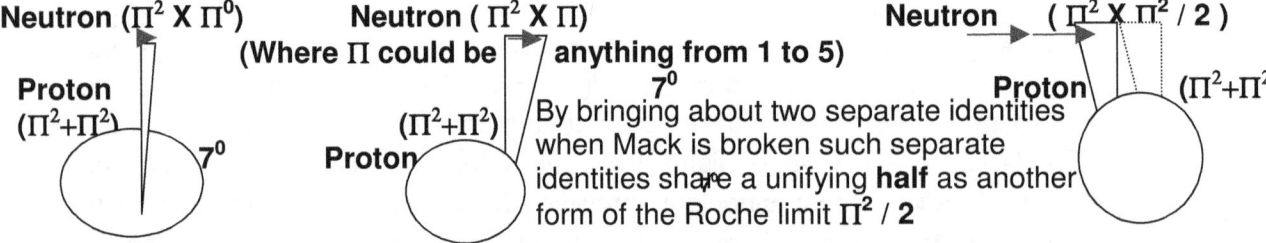

The motion of the earth sets a pace by bringing down space in the time set by the Earth this time has a proton connection since it is dismissing space and producing heat. The close the space comes to the Earth the more space turned to heat and the denser that space is with heat. To the earth this dismissing presents the circular displacing but to us stuck on the earth and glued to the Earth by this motion of space towards the centre the motion is linear as it is going straight down in relation to what ever is on top of the Earth. Any motion coming about from something standing on top of the Earth would have a circular connection relating to the earth because from such motion two linear connecting would come being k_1 and k_2 an the two forms together T^2 because both objects are then independent through motion but still share a joint space a^3.

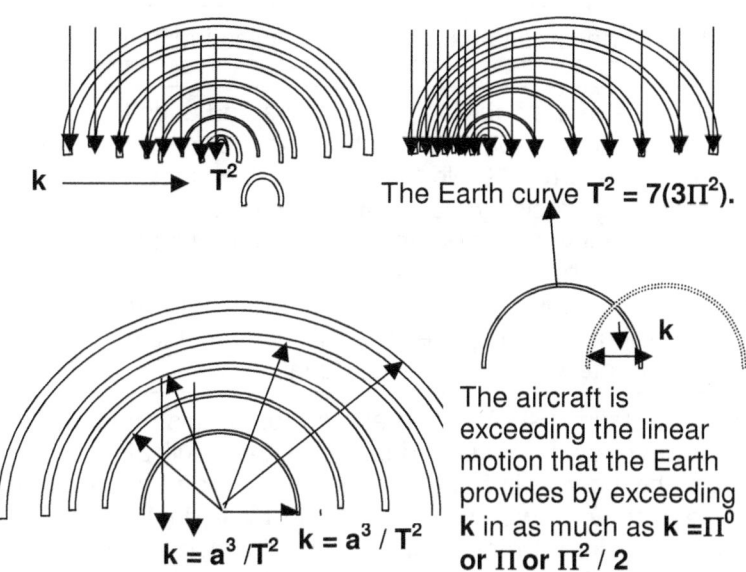

$k \longrightarrow T^2$

The Earth curve $T^2 = 7(3\Pi^2)$.

The aircraft is exceeding the linear motion that the Earth provides by exceeding k in as much as $k = \Pi^0$ or Π or $\Pi^2 / 2$

$k = a^3 /T^2$ $k = a^3 / T^2$

Motion comes about as some heat release causing motion heat to release at one specific point. Singularity ties down all objects at specific points. In order to establish a release from that point heat in the object seeking relieve has to come about to bring about motion. Since the atmosphere holds singularity more densely as space relinquish the position towards the centre many more spots of singularity occur and such spots are holding singularity in a position where the spots falls within the barriers that forms the Roche limit.

The sound barrier is from $7(3\Pi^2) \times 5\Pi^0 = 1036$ km / h to $7(3\Pi^2) \times \Pi^2 / 2 = 1022$ km / h	There then can be a transfer of heat brought about by the fact that the Earth applying gravity compresses singularity points much closer and denser. K becomes relatively smaller to accommodate the constant of T^2.

$k = a^3 / T^2$ is gravity and is   that Dopler only implicated as rings without any explaining why and as such it is not good enough. The very affect of the sound barrier can only be explained by using Kepler's formula because the fact is that the aircraft reaches a different a^3 to that of the Earth because the T^2 arrives at a different k to what the Earth allowed the object to be at that point. At the heart of this is gravity applying two forms of space dismissing. It is the motion that brings about new gravity ratios of singularity dominating space –time to maintain a balance ratio, with a singularity position that seeks for independence by motion. As motion brings about independent points serving singularity the Earth take on the position what the sun had before the motion came about. The Earth then forms central singularity and the newly establishing point serving the independent singularity and as motion comes about and since gravity is repositioning various speeds in relation to each other the motion establish another gravity where the motion establishing the relevancies.

The perfect example is the trumpet where the walls of the trumpet expand, and changes the relation between the unoccupied space-time and the occupied space-time. As is shown in the trumpet, the relation is a theoretical relation and changes according to many influences altering the compiling factors, for instance heat, density, altitude space and time. The sound barrier comes about when the factor k of the aircraft provides a T^2 in self-motion that is exceeding the T^2 the earth has while the two is sharing space. This then comes about as the Roche limit secures half the claimed space by bringing on k as $k = \Pi^0$ or Π or $\Pi^2 / 2$ while the other factor presumes with the value in T^2 that the Earth enforces being $T^2 = 7(3\Pi^2)$.

Gravity is using the Roche factor in conjunction with the Lagrangian five point systems and the Titius Bode law to form gravity. It is a combination of all the cosmic principles to form gravity but it can only apply if singularity is committed in a position that representing of the value and form singularity applies to motion. This is the manifestation of gravity we know by another name we use as the Coanda effect. The Coanda effect represents the final coming together of the principles forming gravity by motion bringing about

{[5+ 5] + [5+ 5]+ 1 + .9} = 21.991/7 =Π. This becomes space **5** sides duplicating in time

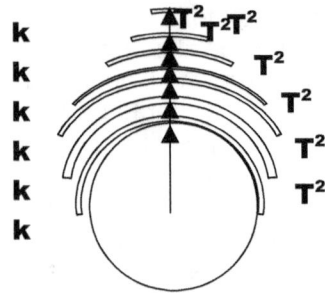

Within the influence of the sphere the Earth has and uses to capture space holding the space to Earthly time there is a relevancy Kepler saw. From Kepler's formula a relevancy come about following precisely the manner that Kepler translated the relevancy to be from the figures he received from the cosmos. The big issue about gravity is that the motion has a dual influence. The space a^3 is relevant to two factors of equal substance in space. The two connects to motion in complimentary fashion but not equal.

As **k** grows T^2 will reduce and as **k** reduces a^3 will diminish. We find aircraft follow this motion to the spot and do not depart from it in any way. However since the sound barrier is the duplicating of a^3 in response to the changes brought about by a shifting **k**, it is not the velocity achieved by the aircraft that represents the sound barrier but the relevancy between **k** and T^2 and by **k** duplicating T^2 in motion forming the triangle. At 31000 meters an aircraft may seem to record a flying velocity at a speed we record to be 2500 km / h but in relevancy it is ten times less than our recording of the speed. It is because **k** extends that T^2 has to decline and that puts the relevant applying speed at 250 km per hour. To break the sound barrier at such an altitude will be very impossible to achieve using any cosmic atomic material known to man.

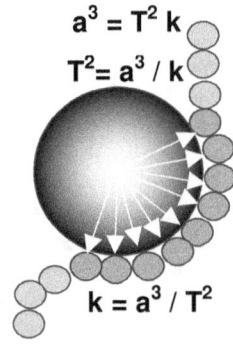

$$a^3 = T^2 k$$
$$T^2 = a^3 / k$$
$$k = a^3 / T^2$$

The sound barrier personifies gravity as presented by Kepler in a lateral motion. There is the unmistakable and clear evidence that the space a^3 created by the light T^2 relates directly to the distance **k** the object is with the Earth, which is the main provider of singularity. In the Coanda effect **k** remains the stable factor providing space a^3 a new and alternative position as T^2 brings about motion and that motion creates the new space filled by the flowing liquid. But this is precisely what happens with the flying object or any moving object. With every moment in motion new space come about that is then filled by particles occupying space and indicated by the distance factor of **k**. it still is the same thing. In all scenarios the formula applies as space a^3 created by motion T^2 in relation with distance **k** just as Kepler announced it.

Kepler said space a^3 is equal to the motion thereof T^2 **k** and also is diverting space to the motionless centre of all spheres T^2. From that **k** is the motion duplicating the position and therefore duplicating the space occupied by the sphere. T^2 is space replenishing the centre where the centre holds singularity in space and is dismissing space by lack of motion in the centre. That comes from the natural form a sphere has. **k** represents space a^3 replenished by motion in space of provided by a more dominant structure forcing space to apply motion to comply with space demand. Space duplicates by motion because the following space will fill with whatever is in the previous space and such filling will continue as long as space is in motion. The instant motion stops there will not be a following space to fill with the previous space and the motion or lack of motion will destroy the space being filled by motion. However the Earth resumes the motion as relevancies apply gravity then changes. The Coanda is proof of just that which I mention.

There is not a bit unnatural about it. I wish to be very ambiguous and very clear that the duplication I explain and the dismissing of space is as natural as space in the Universe itself is. The space is duplicating as time is providing the motion. That is natural since the space fills the vacancy that comes about as the vacant space moved on into the next position leaving the oncoming space to follow. It is what science sees as a separate issue in momentum but it is gravity.

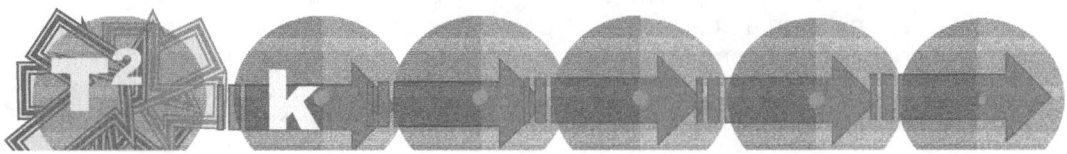

Space is equal to the motion thereof $a^3 = T^2k$

The destroying of space on the singularity end comes about as the proton removes space by introducing the motionless centre in singularity to the space less confinement of the proton cluster. The space at that point is 38303288464389083136 times reduced than space is at the electron end and therefore time is $(((1836)^3)^2 \, ((\Pi^3)\Pi^2\Pi^0))$ times slower than at the electron end. Space in motion forms the seemingly stable Universe we know and stopping such motion will destroy whatever part of the universe, which is stopped. The space cannot simply stop because then one piece of space will move away and there will be a crash as the other standing still will be colliding with the space coming into the spot it has to fill by time producing motion. On the other side the space will move away from the position it had before and with no space filling as a result of motion, the space will automatically self destruct in vacancies and collisions. This part forms the duplications and we can see the result of this when two solid objects finds one location that both solid objects wishes to use and therefore has to share. We call this phenominon events or collisions or more appropriately accidents which in most road collision events those a way cases should be called deliberates instead of accidents. Human made objects may find.

In the centre of the sphere there is a spot that is never in motion and that spot has the strongest destroying of space providing gravity the other part gravity holds. Since the space does disappear and since the gravity is strongest at that point in the precise centre the reason for that is it is keeping all aspects of the sphere in form with precise borders but motion does not apply there in the centre and space deform as it collapse. To render such a spot secluded from destroying the other spots in the Universe a border or barrier form that can only be bridged or crossed under most selected conditions or circumstances. It is all part of the natural with no fantasy about it. $k^0 = a^3 / T^2 \, k$. The space, which is in motion that is in place separating such points of space destruction, has a value limiting any and all excluded space-time from entering. Should there come about a need to enter certain laws apply that must be enforced or material entering the secluded space will be rendered to a form of liquid heat of which the photon producing light seems to be the main ingredient. Because of the natural form all spheres have there is in a sphere a centre where the space loose validity just because it cannot comply to what is said in the previous sentence. The dominance of such a singularity determines the space-time and the density of heat covering the space-time. Because of the Prime law of the cosmos space is there because of motion. This prove the opposite true that says that in the centre where space is lost as there is no motion to support space therefore space goes lost. However what make the space valid is the presence and the value of a bigger as well as a smaller space producing the relevancies.

There is a motion coming from the Earth that all objects have that is located within the Earth and functioning as part of the Earth. Where there is no independent motion the relevancy falls away as the independent object relinquish the applying singularity independence. These two factors make up the compliment of gravity. When an object within the boundaries of the Earth comes in motion above and beyond the motion that the Earth dictates can the object brings about independence within the confinement of the Earth. All objects have the displacing of space repositioning and trying to relocate the object in the Earth to within the Earth. That is mass and that is why mass tends to reduce the space occupied by the object producing the mass. However, the barrier holding the excluding properties of space-time that establish borders move from Π^0 to Π as soon as the singularity is threatened by motion that the independent singularity establishes.

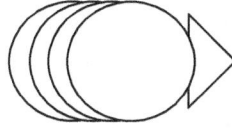

As the planet (in this case the Earth) or any lesser structure rotates around a more dominant structure such a rotating structure has to comply with the space duplication through rotation in order to remain in the circle of gravity it has in relation to the dominant object. It is space in motion that allows it to secure independence.

That leads to the complying of motion by repeating the route set by **k**. In the case of motion the factor of **k** has a very prominent role it plays. It positions the location of the next spot that will become the dot, which the space a^3 in motion will duplicate and fill. But as this is going on there is a larger more demanding motion as well because as the motion complies to satisfy the dominant aspect of space in motion in which the object is in motion within that space controlled by the dominant object. As the Earth rotates it takes all that is secured with it on its travel around its axis. It also takes the atmosphere and all that is within the atmosphere on its travel around the Earth axis. The dominant object is reducing space to a personal located centre from which the rotating object receives the value of duplication. This is the factor T^2 and **k** will eventually become the rotating T^2. I think by this time this aspect is established in the mind of the reader. Whether it is accepted is another issue but the best way to prove me correct is to try and prove me incorrect! There are forever relevancies establishing greater and lesser partners in such a relevancy. The spiral is the same continuing forever where T^2 will come about from **k** and **k** will in the end form T^2 again. When an object within the confinement of the Earths boundaries find a means to establish motion above and beyond the motion such an object has to comply with being part of the motion of the earth therefore the independent motion stands in addition to the motion the Earth demands. That is where sound comes in because sound connects to the singularity centred within the centre of the Earth where all space except space independently filled space will disappear. The sound adhere to singularity at a value of Π^0 but when motion comes about bringing independent motion to an object the Coanda effect proves that such motion which in fact is regarded as fluid in relation to the solidity of the earth soil, the Coanda effect shows how easily motion can establish a new dynamic singularity above and beyond and in difference to the singularity within the Earth centre. The Coanda effect is just an indicator of how all motion comes about but it is just ordinary motion like all other motion taking place as we find with any form of motion in the cosmos. It is moreover the interpretation of singularity duplicating space, which provide time with a new spot to position space. That is what Kepler said four hundred years ago in his statement $a^3 = T^2k$. That is the Coanda effect which is gravity actively participating in cosmic affairs from an established point in singularity.

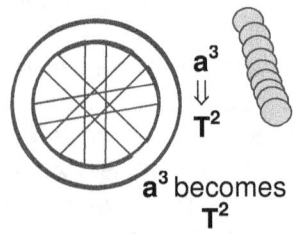

a^3
$\Downarrow$
T^2

a^3 becomes T^2

The Coanda effect proves Kepler's statement more than any other but also prove gravity more than any other. The establishing of a^3 becomes half of the object because the other half is in the motion of a^3, which becomes T^2 in relation to a constant **k** forming duplication. It is the same as a fan blade spinning. When spinning very fast the back of the fan is the same concentrating what it is collecting to reduce that which is collected to then form what is in the front of the fan because the only difference between the two is time where time is heat in motion through space that reduces the space by motion.

If any one places a solid object into the space of the rotating fan blades there is no one in the world that would indicate anything else than that the space is filled solidly. In the case of all moving objects the motion brings about that the rear fills the spot covered by the front and the rear then is what the front will be because the distance reproduce what was contracted at the back in duplicating that in front of the fan every time because what parts the space in the back from the space in the front and in that it is only time that is filling that space by duplicating a^3 and redefining the new 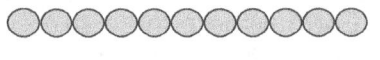 location of **k**. **k** reproduce a^3 by establishing a new spot with T^2

With **k** or$\Rightarrow$performing the duties of $k^0\Rightarrow$**k** it personifies singularity pointing a new spot to fill by time which is just space in motion. The space duplicate with motion performing time as **k** duplicates

$a^3\Rightarrow T^2\Rightarrow a^3\Rightarrow T^2$ **but $\Rightarrow$ is representing k but with the same growth also can go negative when mass becomes a factor $k^{-1} = a^3/T^2$**
When spacecraft enters the Earth the same applies but the reversing of the same extending growth of **k** applies.

The reducing must follow the boundaries of Π

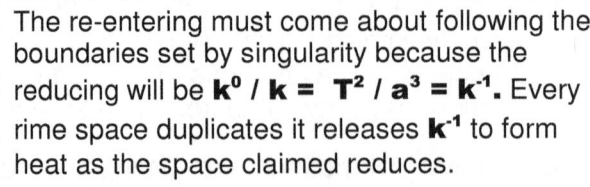

The re-entering must come about following the boundaries set by singularity because the reducing will be $k^0 / k = T^2 / a^3 = k^{-1}$. Every rime space duplicates it releases k^{-1} to form heat as the space claimed reduces.

The space reducing forms heat *to compensate for the reducing of space by the ratio of* $k^{-1} =$ a^3/T^2 and that heat we see as a hot blanket surrounding the incoming object. That space reducing transforming space and forming of heat is the only true gravity applying.

$k^0 / k = T^2 / a^3 = k^{-1}$ leads not only to the confirming of the time component but the time component is forcing the space reducing a^{2-3} because the time factor T^2 becomes the constant the Earth prescribe in the Earths relation to space –time $a^3 = T^2 k^1$, but the space duplication must then accept the ratio the Earth dictates. By entering the Earth atmosphere the duplication of space of the object entering relates exclusively to the duplication the earth relates to in the specific ratio demanded by the Earth. If space duplication is within the Earth space duplication a comparison must come about where the objects space-time duplication stands directly related in relation to the time component that the Earth holds as a claim to space in regard to the singularity the Earth sustains. That brings about mass but that is not yet mass in any location other than being solid on the Earth. It only follows the indicator that will eventually indicate mass if the object can sustain the space diminishing to heat that the Earth will introduce to the incoming object. It has nothing to do with materials forming friction with incoming objects because there can be no friction.

At that height (and everywhere else in the atmosphere for that matter) the density of material is extremely sparsely distributed and even if material had the ability to produce friction there just is not enough material at that height to bring about enough friction to cause such glowing heat. We must consider what volume of heat will material in friction produce to produce that amount of glowing heat. There is not much material gone from the craft therefore it must all be coming from such sparsely distributed and destroying that much to produce that volume if glow is just not a factor because then that much of the aircraft structure should also be scraped from the craft.

This is proven by the "heavy" gasses floating and the "light' solids stationed on earth. There are a group of elements that Mainstream science categorises as gasses. We have discussed when these facts in this book a on other places in this letter. The material we find there is also extremely volatile and will move about bringing on highly active motion.. They are by name Hydrogen, Helium 2, Neon 10, Argon 18, Krypton 36, Xenon 54, and Radon 86. These gasses are in most cases hugely massive in numbers and mass. Others are Nitrogen 7 and Oxygen 8 that has more protons than some solids. In the cases of Xenon 54 and Radon 84 the proton number and the mass produced makes the element much more massive than Carbon 6, Boron5, Silicon 12. With the presumption that mass makes things heavier because the mass pulls the Earth harder and the Earth responds by pulling back harder and the pulling of the two dismisses all the distance from both ends equally these gasses should at least be metals and in some cases heavy metals. Since the protons in numbers exceed metals in mass numbers. They should not be suspended in the air as gasses are but they should be stuck on the solid ground and some must be almost unmovable instead when taking into account the number of proton mass they hold they are floating like feathers.

Then there are the group of elements that does play they're part as Newton said that it is mass that brings about they're being on Earth as solid as you can get. By they're remaining on Earth they play the accepting role of the categorization they have. We share space with all the elements and being as short sited as humans can be, we categorise and classify all elements in a manner that we will best acclimatise to that position. We fancy 20^0 C that is a very nice laboratory acclimatised laboratories temperature for humans to work in and in that we test elements while nicely forgetting that such a climate might favour the earth at that position. Take the test results down five kilometres below what the deepest mine is and the test result will be much different to what it is in our nice little artificial climate. Then we go one more step from the ridiculous to the sublime and categorise hydrogen as a natural gas and lead as a natural solid. In our human mind set hydrogen has to be a gas because it takes massive changes in our laboratory to change hydrogen into a solid and that temperature cannot sustain life so we through it out as cot congestive to life. It takes first as big an effort to liquefy metals and no person can live with such heat so the metal s then a natural solid. Hydrogen has to be a gas because hydrogen proves to be a gas in every laboratory found on the face of the Earth. They are massive except in some cases and they are solid. We classify them and categories elements as solids or liquids and gasses knowing very well every element can turn to liquid and gas should prevailing conditions demand the changing. By adding or reducing the surrounding heat the surrounding heat all material can manifest in every form there is available yet we still demand that the elements will not change the form we gave such elements. I have shown that it is not the proton that produces solidity or a fluid state or a gaseous states but the way the element structure relate to heat or space. I gave this forming responding to heat the name of the Lagrangian atom and explain the concept in explicate detail in another book being **Volume six of MATTER'S TIME IN SPACE: THE THESIS ISBN 0-9584410-8-** or **"STARSSTUFFN' ISBN 0-9584410-3-0.** Elements respond to space-time in two manners and the response bring about the form they represent. The protons dismiss space in the time it takes to duplicate space while the element takes time to duplicate the space motion provided the time to complete.

 The outlay of the atom, which is too extensive to go into detail in this letter atoms stand favouring space duplication and in balancing contracting to that motion of space flowing to the centre where space-time dismissing takes place. When the element favours heat and disfavours much density it holds the heat it associate with as a "blanket" surrounding the element to sustain about all the dismissing of the protons and leave much of the rest of the heat to spare. In other cases the element density promotes a displacing of space-time that brings about not much duplication and therefore disassociate it by form from with the "heat blanket" surrounding the element where the displacing is so much in control the electron associating with heat is unable to secure a "heat blanket" that will secure space-time in order to supply the demand going in the way of dismissing. The duplication favours the electron bringing on more heat by producing volatile motion in an effort of collecting the surrounding heat and through excessive motion in comparison with other elements that diminish space-time by contracting motion towards the centre of singularity. In this the atom dismisses more space-time than that which the electron may accumulate, it all comes about from form in the formation of the atom. In between there is a variety of choice to be made producing forms hard too soft to gas. But none of this is written in rock and changes as heat is added or removed.

The proton holds space in a zone where the proton dismisses space and therefore the number of protons will reduce space volumetric equal to the number of protons in such a space less zone. This I named space-time displacement and that does not refer to mass in any possible way.

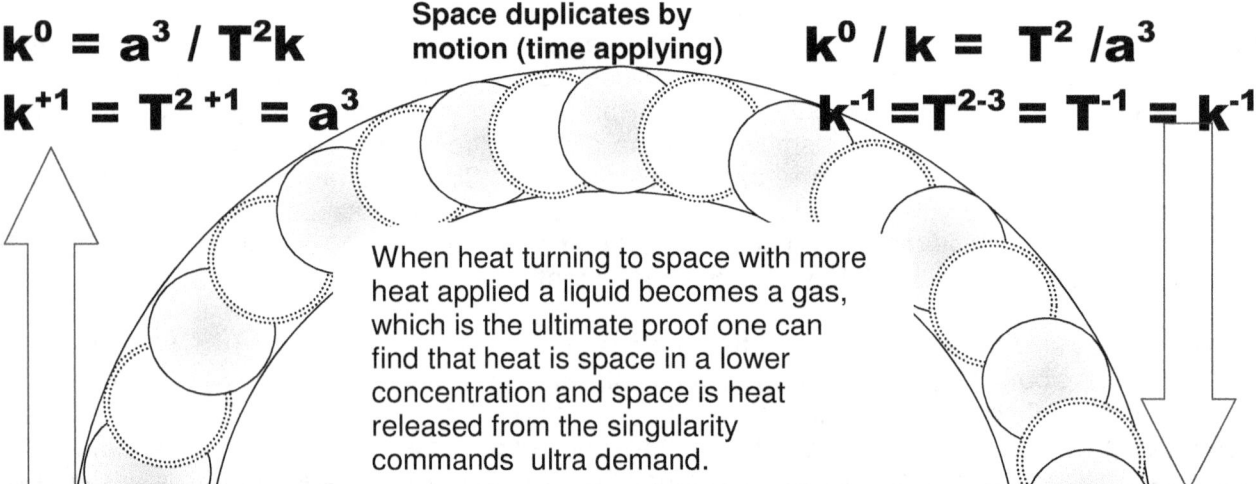

$$k^0 = a^3 / T^2 k$$

$$k^{+1} = T^{2\,+1} = a^3$$

Space duplicates by motion (time applying)

$$k^0 / k = T^2 / a^3$$

$$k^{-1} = T^{2-3} = T^{-1} = k^{-1}$$

When heat turning to space with more heat applied a liquid becomes a gas, which is the ultimate proof one can find that heat is space in a lower concentration and space is heat released from the singularity commands ultra demand.

All material duplicate space by motion duplicating space $a^3 = T^2 k$ It takes a certain time to duplicate the space but the space it has to duplicate is much more in volume than when compared to other elements. Such elements are surrounded more by heat/space and therefore they show a greater resistance to solidify than others do not withstanding proton number or mass acquired

All protons dismiss space-time by displacing the space as it dismiss the   space either by motion acquired to move through space or in other cases to apply   motion   to space, Because of the dismissing nature we tend to think of such elements as having little space and that we refer to as them having density. The density is exclusively a product of space dismissing by space-time displacement. But by displacing the space it results in forming heat density to space. It also provides a structure having very little space.

When the element has a relation to applying a lot of space surrounding the element the element is covered by a lot of space where we consider such element as being volatile which suggest it is full of much motion and will therefore prove not to easily abandon the space to solidify when heat / space covering remove the space surrounding the element. We call this tendency being a gas in extreme cases or liquids in less extreme cases.

- There are the elements that normally show less resistance to remove the space surrounding the element and become solid and those tend to form liquids more easily.
- Then we have the third group that shows a severe reluctance to surround the atom with space or heat and those we find to be the elements we regard as solids or metals.

In this argument Newton's formula for gravity and of mass goes flying through the open window. $F = G (M.m)/ r^2$ has no solution to this mystery and thus is not spoken about in intelligent conversation because the solutions Newtonian science offer are most horrifically proof of utmost stupidity. If Newton's arguments were correct then all light gasses must have very few with very little protons must be the ones floating in air and all heavy solid metals must have a large number of protons that do the pulling of matter onto matter. Since this is not the case it shows that Newton's perception of mass pulling and making heavier is extremely incorrect. Gasses actually go out of their way to disprove Newton.

We on Earth think of the spacecraft as orbiting but the spacecraft find the motion it has as flying straight. While the orbiting spacecraft is following a straight ahead travelling direction the spacecraft will inevitably again reach the point it was at just a cyclic period of time ago. By travelling straight the spacecraft completes a circle and by circling around the Earth the spacecraft must fly straight ahead matching the pace of the Earth. But it took an enormous amount of liquid released in such a way that it would favour the straight-line k by reducing T^2 to match the singularity k^0 initiate space-time. The release of the heat produce the space putting the orbiter at that spot it has rotating in the circle which it does. After the release of such heat the Earth took the bounty paid in heat in return for the release of the craft where the earth permitted the escape to take place but only on the condition that as long the orbit speed which is duplicating the relative required matching tempo to be in place as the rotating speed between the Earth, the object and matching the sun. That is gravity because what keeps the craft in relation to the Earth is the motion the craft duplicate the space it has in relation to the Earth duplicating the space the Earth has and the difference in the duplicating and the volume duplicated forms the gravity between the craft and the Earth. When we take the circling straight line even further we find the moon is rotating the Earth following a straight line while the Earth is rotating about the sun following a straight line and sun is rotating the centre of the Milky Way by also adapting a straight ahead position where that straight ahead position will eventually land the sun rotating in a circle around the Milky Way centre. That is what gravity are to all things in the cosmos. But that is not what mathematician Newton said although that is what the cosmologist Kepler said. Kepler said space a^3 is in motion of a line k in the single dimension going straight relating to a circle T^2 in the square. More than just that Kepler went on to say that space can only be formed in a straight line going in a circle $a^3 = k T^2$.

Space duplicates by motion (time applying)

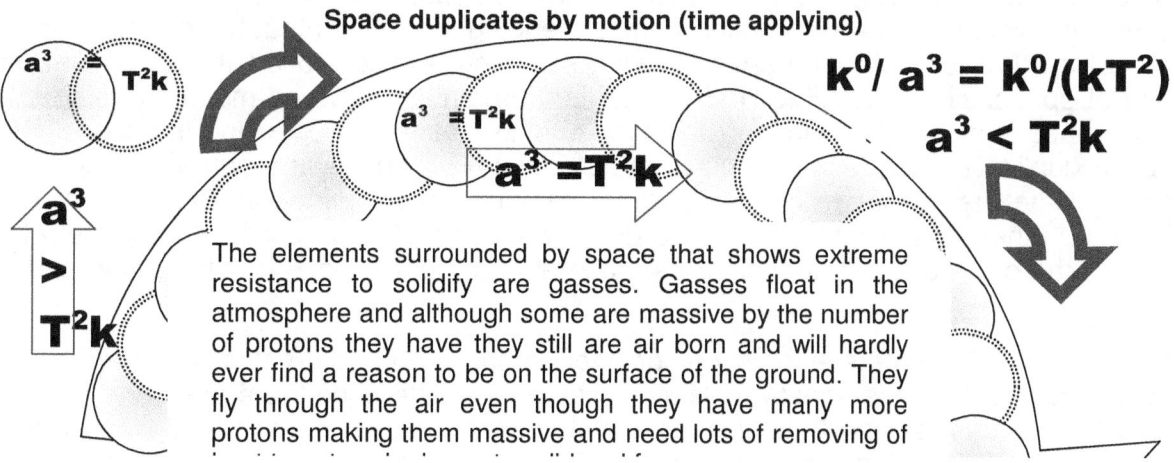

$$k^0/ a^3 = k^0/(kT^2)$$
$$a^3 < T^2k$$

The elements surrounded by space that shows extreme resistance to solidify are gasses. Gasses float in the atmosphere and although some are massive by the number of protons they have they still are air born and will hardly ever find a reason to be on the surface of the ground. They fly through the air even though they have many more protons making them massive and need lots of removing of

There are the elements prone fore filling the idea heavy and rather difficult by nature notwithstanding having precious little proton ability. But that tend to be mass that such dense material to move as they are (in some cases) the normal is that the is solid number of protons does tend to dismiss more space than they surround with establishing heat and therefore applies much less space in regard to much more mass to claim for occupation. The function of the proton is to dismiss space by displacing space and where the space then show motion by refilling the reduced space we call such flow gravity. As every one is schooled on the Newton principle of much mass is lots of protons being massive we naturally think that because a larger number of protons will lead to a larger dismissing of space it occupy less space as the protons always are in a space less environment in the centre of the atom or the star.

There will always be a connection between the rotating and the straight line travelling. Gravity is by suppressing space in the attempt of accumulating heat that will advance duplication and/or remove heat at the centre that will retard duplication. Moreover gravity is the balance between the two possibilities favouring either, both or one at a time all depending on the allowing that space-time governing singularity permits. This balance we find in all stars. The so-called giants are heat collectors producing motion by massive enormity and volumes of liquid space. Those are stars, which I classified as being prone to the electron as the electron is about the charging of heat collected and accumulation thereof. By creating motion it makes much contact with space helping the star in the process of brining intensity to the space through gravity and accumulating the space that the star turns to liquid heat. The stars having this nature all develop to become stars with mostly atoms that is prone at dismissing space more than collecting space and forming such space into liquid. These stars further develop into structures that skip the liquid faze and turn the collected space into solid heat directly by dismissing more space than the star can accumulate. In these stars, which I last mentioned no motion is active, since the star is almost to completely in a state where singularity is reinstalled and in that manner all motion within the star is killed of by singularity.

Producing singularity sets the divide because singularity splits the Universe apart and in separate equal components that in combining form the duplication of singularity being Π. Since the split brings about equality it means that what is applying on this side must be applying on that side. When motion changes Π^3 to the proton $\Pi^0 \Pi^2$ it will happen on both sides of the divide of singularity Π^0. It in effect means that that which combines the proton also part the proton as it combines the proton because the proton becomes $\Pi^2 \Pi^0 \Pi^0 \Pi^2$ where the adding is the divide being Π^0. The circle motion comes from space being dismissed by ending the motion and such ending of motion compromises the space it forms. By returning to where it is coming from it is ending the motion that began the space and as space is motion that is duplicating space motion returning is also motion that is ending which is destroying of space. THAT IS GRAVITY! Gravity is the balance between motion forming space by duplication space in motion forming time and time ending motion by destroying the space. Gravity is about space duplicating space in relation to space destroying space and some particles are more prone to duplicate than destroy not withstanding mass or proton numbers. Those we call gasses. Then there are others that are more prone to destroy space that duplicate and those we refer to as metals. Then there are a few that destroy as much as the create space by duplicating and that we call fluids.

When heat is added to some elements we consider as solids, the heat helps with the duplicating of surrounding space and brings about a balance restoring the difference there are in the destroying of space and the re-establishing of space. The metals become liquid and the heat forming the liquid brings about an adding to the material where such material diminishes space. By applying heat to materials that already favours the duplicating of space to the destroying of space the adding of heat will bring additional space as duplicated space and thus will produce more space to be duplicated and such elements will rise into the higher part of the atmosphere. When heat is added the heat as space that is added is the heat forming space by duplication forms a shift in the balance because the heat forms space also as a process of duplication but without the contracting aspect of singularity renouncing space.

Although the "gasses" are the particles favouring to duplicate space they still hold the tendency to diminish space but when applying heat the balance will favour the duplicating much more because the heat transforms to space acting as space duplicating and adding to the overall duplicating of space. The element has a function naturally of returning the space that the heat duplicated back to heat by removing the space destroyed and therefore returns the space to heat. In that way the particles do not only diminish the space they have but diminish space outside their claim. Such particle we call heavy metals. As heat is added more space becomes available to duplicate in relation the space they destroy and what space the elements diminish. With more space to duplicate the object will surge higher to a location in a position Earth will naturally duplicate as much space as the newly relocated material duplicates the surrounding space where more space naturally are. Cooling on the other hand reduces the space available for duplication by removing available heat that would have helped with the duplicating of the space and that then tips the balance in favour of the diminishing of space, which that element will also have. In that case the element will become a solid as the space duplication is more that the space. In this we can trace the most important part of star evolution.

When the star has a liquid centre with lots of heat, the duplication of space-time by motion duplicates space much more than it dismisses space and destroys time. With a star in all the liquid as the sun is it is proof of a very young undeveloped and insignificant star with almost no influence sphere. As the star develops the liquid ratio will shrink until it is only present in the centre. But as the liquid diminishes the motion of the star deteriorate because the liquid represents the motion. The star eventually becomes all-solid just before it removes the neutron from the atoms in the star and eventually places the proton action into outer space. Judging the layers we find evidence in this as the outer layers of stars are filled with elements which is highly prone to space duplicating as they have such a relation with heat. Hydrogen and helium stands very favourable to space producing and little in favour of space dismissing while iron, cobalt and copper is much prone to space dismissing. But in all factors mass plays no part. It plays no part in the star performance or the star development

This I use to indicate where there is a balance favouring the diminishing of space

In all forms of material there are the constant interaction between space duplicating and space reducing. Some elements favouring duplicating space more than the diminishing of space are as follows

Hydrogen has a	**mass of 1.00797 g/ mol**	**melts at -259^0 C,**	**boils at -252^0 C,**
Argon has a	**mass of 39.948 g/ mol**	**melts at -1899^0 C**	**boils at $-268,9^0$ C**
Krypton has a	**mass of 83.8 g/ mol**	**melts at -157^0 C**	**boils at -152^0 C**
Xenon has a	**mass of 131.3 g/ mol**	**melts at -111.79^0 C**	**boils at -108^0 C**
Radon has a	**mass of 222 g/ mol**	**melts at -71^0 C**	**boils at -61.8^0 C**

It is note worthy to notice that none of the above elements feature strongly in stars although they should be massive in relation to the numbers of protons they have because they duplicate space.
Other elements favouring diminishing of space more than the duplicating of space will be as follows

Magnesium has a	**mass of 24.32 g/ mol**	**melts at 650^0 C**	**boils at 1107^0 C**
Silicon has a	**mass of 28.08 g/ mol**	**melts at 1412^0 C**	**boils at 2680^0 C**
Iron has a	**mass of 55.847 g/ mol**	**melts at 1536.5^0 C**	**boils at 3000^0 C**
Cobalt has a	**mass of 58.933 g/ mol**	**melts at 1495^0 C**	**boils at 2900^0 C**
Carbon has a	**mass of 12.01 g/ mol**	**melts at 804^0 C**	**boils at 3470^0 C**

There are no correlation between mass and elements prone to space or prone to be solids. Mass do not create gravity and again on one more point Newton was wrong. But mainstream Science would rather ignore such compelling evidence as well as my writing about the matter than to admit that Newton could ever be mistaken.

My argument I take from Kepler where Kepler distinctly shows that space produce motion and motion is time and time produce space. This puts a relation between the Earths as a massive body compared to that which are within the space claimed by the Earth. The Earth has a lot of space to duplicate that provides the Earth with a lot of motion. There is no element or structure within the atmosphere that can even remotely match the Earth as far as the use of space goes by creating motion with duplication, but there are elements that fair far better than others. The Earth has a surface cover of 70% water and some persons in the world of science goes around thinking the Earth should have been named Oceania.

That is another promoted misconception of gigantic and titanic proportions and the only bigger science fraud preached are the size the massive dinosaurs supposedly had that roamed the Earth millions of years ago when they roamed the Earth so many millions of years ago. But Oceania is about in the misrepresentation class as our fraudulent representing of massive dinosaurs. Off the point just this: has no one of those scientists ever sat down and wondered why everything was so extremely BIG back then and why has those specimen found today shrunk so much at the present time. Are they truly thoughtless or has Hollywood fame drained their logic in the brain they have? Just an observation I made with no harm intended! I give the answer plus all the relevant explaining in another book where I match the Biblical seven days with a possible seven solar occurrences that brought about seven periods of growth in the Earth core. The book is therefore named the **Seven Days Of Creation** ISBN 0-9584410-4-9 **or is also Part Seven of** The Theses.

Getting back to Oceania where science making this statement is only when looking at the Earth from the top and then only sees a very thin outer layer, but most of the Earth is metals and silicon. Metals and silicon break down lots of space but holds little space and therefore the gasses floating can have a far better relevancy of space creating by motion than the comparing Earth and will float in the air not withstanding the proton number dismissing space. While on the inside of the Earth is all the space less and space reducing Iron core with other more heavy metals the bodies we received such as we have show reluctance to produce space. Therefore the Earth producing more space than we do will compress our space by thrusting us towards that space less inner centre. It is because we cannot manage to equal the duplication of space by motion than what the Earth can achieve. The thrusting of our inability to duplicate space gives us the presumed mass while the gasses apply a much bigger ratio of space reproducing which tends to favour the gasses and make them float in the air. There is no mass that is producing gravity.

The gravity is producing mass by thrusting the less-space producing objects towards the centre from where the gravity draws the space at large. That flowing of space-time towards the centre causes a duplicating mismatch that thrusting or the depleting of the validity that the space has committed to match the duplication effort of the larger space volume then seeks a relevance to position allocated to space that is in volume equal to that of the lesser space as it produces motion forming an imbalance. The mass cannot produce the motion to escape therefore it produces the effort of restricting the motion by producing a breaking effect we named mass. The very opposite can be established if the necessary heat is produced and applied to a specific point of k^0. When heat supplies motion to one smaller object the balance can be rectified in favour of the smaller object escaping the larger objects gravity. In spite of Galileo proving that objects with mass differences land on the Earth simultaneously when dropped at an equal distance also simultaneously, which totally contradict Newton's mass claims and pulling of objects.

The mass is the manner through which the dominant singularity uses to dominate the lesser body as the lesser body produce insouciant space duplication to accomplish the slowing down of the motion a larger object has notwithstanding the futility that such an act may have. By implementing the value of singularity $\Pi\Pi^2$ in relation with the Earth proton value of $7(3\Pi^2) + (\Pi^2)$ and the releasing object $(\Pi^2 / 2)$ $\Pi^2)$ \ 3600 applying heat which is producing the releasing motion the escape requirements is valued. Explaining the whole process of flight in relation to singularity is also very consuming and I have a book **AN OPEN LETTER ABOUT** " **INTER GALACTICA SPACE TRAVEL**" ISBN 0-9584410-2-2 that deals with that aspect almost exclusively. We find this applying of gravity in the flying and the breaking of the sound barrier and that is what science completely misses. The linear motion duplicating is related to the diminishing of the space the craft claim as and the crafts applying of additional motion again favours the duplicating of the space. In that manner the space is duplicated more by the craft than the space is duplicated by the Earth and since the motion brought independence to the craft the craft then holds more space duplicated by motion than does the earth in that area destroy space. The Earth dismisses space as it creates space. It is doing this even on behalf of all objects in its atmosphere. Being still and secure on the earth the difference we as separate bodies standing on

Earth have we still share to a precise point the same ratio in dismissing space relating to duplicating of space that gives us mass. The motion carrying the Earth represents the duplicating of space and its entire belonging foreword and the dismissing is about maintaining form to a specific centre at 7^0. This change when a standing object begins a separate motion in duplicating space through motion faster than the process of the Earth

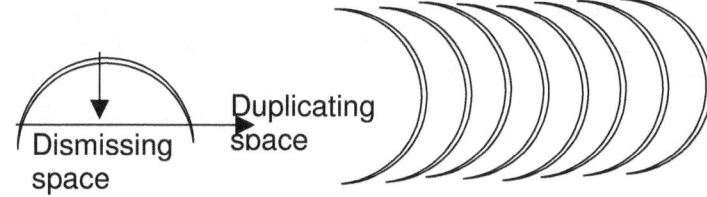

The dismissing of the space is equal to the degrees the Earth runs from circle to centre, which are 7^0.

The time ratio component in this duplicating 7^0 is the proton value $(\Pi^2+\Pi^2)(+\Pi^2)$ as well as the solid neutron value (duplicating Π^2) in relation to singularity $X \Pi^0 =$ 207 km / h. Because the formula relevancy becomes so long I use 3 Π^2 to alleviate confusion. The Earth crust and below the Earth crust forms a proton related position of $(\Pi^2+\Pi^2)$ Everything not part of the Earth crust forms the solid proton position of (Π^2) This is the natural material positional dispensation.

The earth holds the value of forming the proton relevance being $(\Pi^2+\Pi^2)$ and as the object is an individual structure loosely standing on the earth but holding individual mass in relation to the earth the object has the solid neutron position of (Π^2).This means the object is motionless in the horizontal line of gravity but (Π^2) secures the object motion in through relating to the elected Π^0 commanding the Earth control and by Π^2 in duplicating space-time. The connection with singularity is 7^0 or in position $7\Pi^0$

Since the object is maintaining the same duplication to the dismissing factor of space-time ratio the Earth establish the duplicating is in relation with the earth singularity as Π^0

Because we chose a cubic meter of water to represent the space of the Earth as cubic^3 X second ^{2} in time of the Earth spin around its axis and around the sun any volumetric discipline in line with the rotation of the Earth will represent the earths motion in space-time precisely <u>abiding by the seconds we use.</u>

When the object asserts independence by applying motion in addition the motion the earth has the object finds a new relevancy to the singularity of space dismissing which the earth applies. In such a case the formula becomes $(\Pi^2+\Pi^2 + \Pi^2)$ which represents the proton factor including the solid neutron factor relating to the newly established singularity factor of valued at any value between Π^0 and $5\Pi^0$

When spinning outside the atmosphere there is no direct linking contact established between the Earth and the object, which then is a satellite.

However when breaking the barrier and the satellite enters the atmosphere there are laws applying that changes

k_1
k_2
k_3
k_4
k_5

The Coanda effect starts

Any object entering the Earth atmosphere is starting an Earth wide Coanda effect by establishing a space link through motion.

When the solid enters the Earth atmosphere the reverse of the Coanda effect establish a concentration of the liquid which is then the atmosphere as the motion, is contributed by the effect present, solid which is the spacecraft in this instance and the liquid is the concentrated space of the atmosphere. The liquid is pure heat covering the body under the rules of the Coanda effect that must shrink the space of the body and fill the shrunken space with liquid as the Coanda rules stipulate. As the relevance of

k changes to compromise for the changes of the time T^2 that also change a^3 has to reduce in accordance with Kepler's formula ratio and in accordance with the establishing of a new Coanda effect linking the Earth centre with the space formed **k** by the motion of the entering object T^2 that then has to reduce the space a^3 which the object claims in terms of the more dense space a^3 that the Earth apply to retain space a^3. This comes from a new alliance between space a^3 through a motion T^2 that introduce a new linking value **k** between the Earth centre and the object entering. As time move back in the direction towards where the Big Bang came about, so does the heat that would have covered an object that was able to move at that pace through outer space. That leaves the entering object covered in a blanket of liquid heat as it establishes the new acting Coanda effect between the entering object and the Earth centre.

From the centre running through occupied space singularity position space-time at a point that connects by the seventh dot and in this a straight line forms from the centre out to the edge of space. This is then coupling the object with singularity controlling everything within the realms of the Earth by using the line running straight but abides the 7^0 points.

When the object accepts the duplicating and the dismissing motion of the earth such an object maintains a perfect relevancy with the singularity in the centre of the earth.

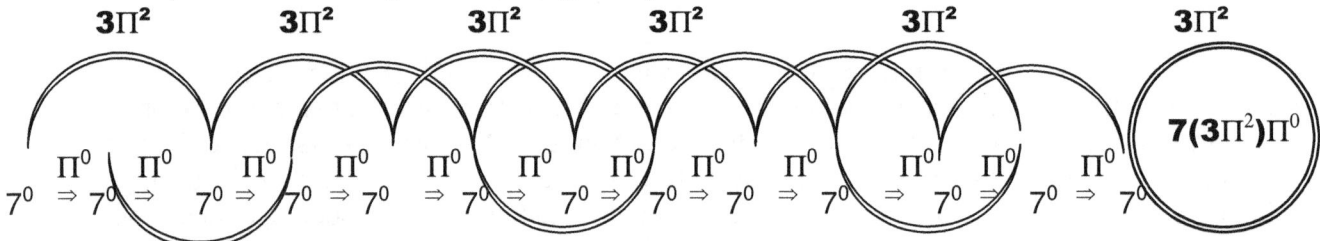

The relevancy of motion applying to the object which delivers the separate identity is $7(3\Pi^2)\Pi^0$. This motion applies although the object is motionless because the motion is motion that is creating the affect of motion which we interpret as producing mass. The motion stands related to the centre of the Earth and is completely independent of what ever may present the mass factor.

As independent motion sets in individuality the object can no longer adhere to singularity as the Earth duplicates but has to bring in a new alignment not in line with that of the Earth. The motion represent the duplication factor whereas the gravity the earth so kindly bestow on us is outright just a dismissing factor and our independence from the Earth represent our possible duplicating factor in harmony with that of the Earth. This factor we know as time or the rotation of the Earth. By moving across the horizontal line will be bringing about motion where such motion exceeds the Earth motion brought to us by the compliments of the gravity linear factor we call time. As we excel by duplicating more (in the horizontal) than does the Earth represents our duplicating we find the means to move above and beyond the motion the Earth establish by duplicating. This motion comes about entirely from life's involvement and by employing life's free will in manipulating space-time. From the laws applying in the cosmos life has no value and since life does not exist according to cosmic law all motion must result from some cosmic principle that the Universe are familiar with. The cosmos reacts on the motion as one of several spontaneous occurrences that may take place and motion can only drive for independence if and when under natural cosmic circumstances the independent singularity secured the correct amount of concentrated heat surrounding the independence striving singularity by applying strong enough independent gravity to secure such heat in the enclosed environment. By involving separate heat only can the independent singularity achieve the duplication requirements to manifest such duplication on top of what the earth is producing.

This motion can go on without effecting singularity borders from the limit of Π^0 to the border of $5\Pi^0$. At $5\Pi^0$ the Lagrangian displacement ruling atoms sets the next border.

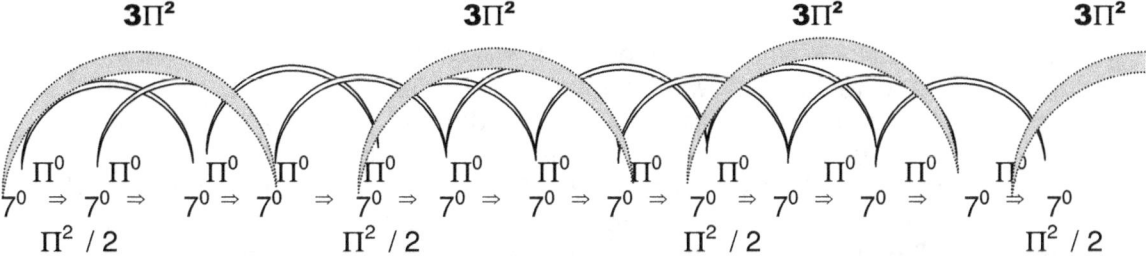

The frequency of duplication by motion controlled by the Earth singularity applying to objects in motion within the earth atmosphere does not harmonise with the duplication that applies to the object in motion since the object in motion broke free from the retaining Coanda gravity effect that applies to all other objects in the atmosphere. At that point the duplication of the object starts a new alliance with the Earth by establishing a new Coanda arrangement between the Earth singularity and the object in supersonic motion. However the sound coming from the aircraft at supersonic speed land into space that still apply the regular duplication rhythm and subsequently immediately start lagging behind the crafts supersonic duplication. There is no middle ground for the sound to be within.

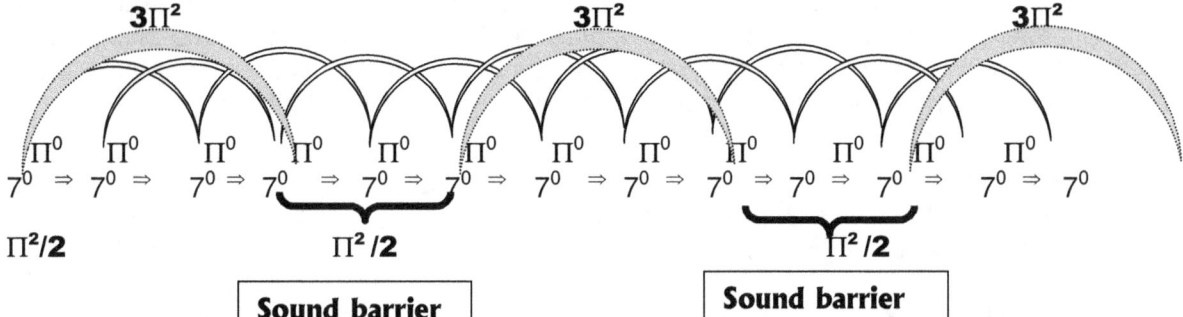

This border is **Sound barrier** the border we **Sound barrier** associate with speed and with breaking the sound barrier. When the striving rebel singularity finds the ability to bring about motion such duplicating crosses the border that space-time has the ability to duplicate the carrying of sound waves. Then at that point the rebel singularity finds the second border that will lead to eventual release from the Earth controlling singularity. The duplicating the rebel can produce exceeds the limit that the earth can produce but it still holds onto the motion the earths duplication lends to the rebel. Now at such a point the duplicating starts to cross the dismissing gravity lines flowing vertically down at a straight line of 7^0. In this half independence has been established and the secondary Roche limit come into affect. The Roche by half (because the rebel is still using the atmosphere) and therefore still abiding by the rules set by the Earth commanding singularity. The Roche applying between stars is $\Pi^2/4$ whereas in this case it is only $\Pi^2/2$.

The faster the object fly the more the object will secure a singularity that falls outside any parameter the Earth has with the Earth singularity but the moving object must still pay homage to the Earths singularity since it is part of the earth singularity. Since the motion in heat the aircraft produce forming space is much more prevalent than the Earths duplication the relevancy of space the aircraft represents is much more in ratio than the earth and the aircraft will surge to a higher space duplicating level that of the earth. By having wings the contact the aircraft produce with space will increase and dramatise the contact with space even more bringing on a pretension of seemingly duplicating / producing more space than what the Earth is duplicating.

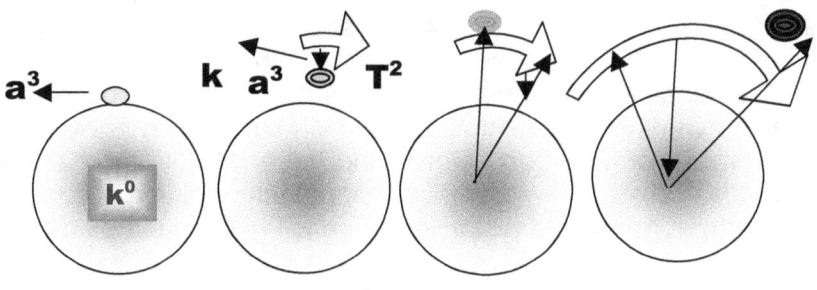

In the accumulating of heat the object finds motion. The motion brings along structural independence and such independence puts distance between k^0 and a^3. The heat increase will accompany a larger T^2. By increasing heat the distance between the

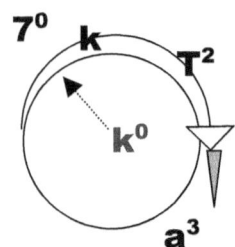

Only by creating a total independent heat centralised in a point holding singularity and feeding k^0 with an independent heat supply can an area a^3 establish a k that will release the independent a^3 from the secure larger k^0 The overall condition is that the escaping a^3 must establish a route following k^0 as k^0 places a diverting 7^0 where that 7^0 then forms part of the object creating heat to secure a release from the established a^3. Providing the heat will bring about a release placing a new object into outer space.

The releasing brings about "new Cosmic Life" but we must never loose sight of the fact that in relation to the cosmos such a release was "artificially " created by "Human life" and "human Life " is an alien energy in the Universe and to the cosmos to secure an escape space-time displacement to the order of

$4(7^0 x(3\Pi^2)(\Pi^2/2)\backslash 3600 = 11.216$ km/sec is the required velocity needed to escape gravity

Kepler stated that $a^3 = T^2 k$. To release from the confinement of $k^0 = a^3 / T^2 k$ and establish individuality as another structure (or possibly debris if classified by human standards) orbiting the sun the space had to increase on one spot holding singularity as to produce motion which it then applied as liquid heat forming space freeing that particular singularity to an individual particle.

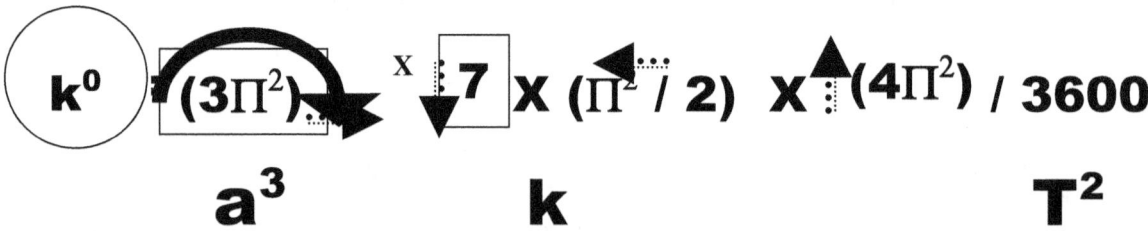

$$a^3 \qquad\qquad k \qquad\qquad T^2$$

> The **$7(3\Pi^2)$** represent the relevancy the Earth holds in the separating relation.
>
> The projectile claiming individual gravity also claims **$(4\Pi^2)$** in the relation because it reached a gravity point outside that of the Earth.
>
> But in order to overcome the earth gravity the projectile must accomplish the space and time between the positions it had on Earth and the position it holds in individual space-time as **$(\Pi^2 / 2) \backslash 3600$**

$(4\Pi 2)$

$(\Pi^2 / 2)) \backslash 3600$

$7(3\Pi 2)$

> **Newton introduced G $(m + m_p)$ as $(4\Pi^2)$ and since the object is becoming an orbiting structure it receives that value Newton designated to orbiting structures.**

It is of increase also true that the number of protons creating motion will determine the amount space that will deplete and the amount of heat that will come about as the required displacement that will sustain the singularity which will secure a release from the Earth singularity. The motion that more protons produce will gravity in the escaping object but it is because of the motion the protons and neutrons create. Gravity comes about from motion and the motion uses the Titius Bode law to produce gravity. The Titius Bode configuration in accordance to orbiting formation holds a slightly different explanation to the explanation that applies to cosmic structure surrounded by space. It is moreover the individual singularity in maintaining the major singularity, which sustains the governing singularity providing equilibrium in space-time. Not only does atomic individual singularity maintain self preservation, but in doing that it also sustain a governing singularity holding structural composition and form within a cluster of matter for example a star. As there is between stars so there are in the same manner a mutual or bonding singularity between atoms in stars, which we see as fusion. From this one may freely deduct that gravity is not forcing material closer but is destroying space whereby it converts the space to a density the senior partner has in the atmosphere of the senior partner. This is most predominant in the cosmos and gravity is all about motion of more advanced in relation to motion of lesser advances singularity. I have no whish to go into space-time displacement and how that operates in this book because the motto of this book is simplicity at almost all cost. The proton moves 1836 times faster than does the electron and the electron personifies the speed of light. If the speed of light was the fastest that material could travel through space as Einstein explained, the Black Hole would be as much a myth as Hercules is. Material linked to singularity in space can travel 1836 and material not directly linked to singularity by space displacing in space being light can travel at the speed of light. But light travelling is subject to gravity applying and gravity will always slow down the relevancy of light. It does not slow light down since light is a constant but that constant stands relative to the gravity or space in motion applying. In cases where the space in motion exceeds the speed of light, the star goes dark and where the star displaces space beyond the constant limit of light, the Universe surrounding the star with such demanding singularity collapses.

In the Universe one proton manages to displace a specific volume of space in ratio to a specific time being at a specific distance from singularity. This is deductible from Kepler's formula. When $k > k^0$ at the very point the distance forming is that eternal fraction bigger that singularity and $(k / k^0 > \Omega)$ space brakes down by specific measure $(k^0 / k = T^2 / a^3)$ the space demolishing in the time duration applying has a limit. The point, which I am referring to, is where singularity Π^0 evolves and becomes Π. At that point within every atom is where gravity collapses space into singularity. That is the reason why the atom has the ability to dismiss space-time. In every proton in all atoms any proton has this specific generating ability one can measure as one volumetric unit a^3 disappearing in one time unit T^2 but the time and the space depends on the time and space singularity permits at that specific location. Giving that space and time a measure would be very symptomatic of Mainstream science and would also prove very meaningless. A good illustration about the interaction between cosmic vessels applying all space-time relevancies happened not to long ago in front of millions of television viewers around the world. It was the event that will go down in history as the Shoemaker Levy 9 comet splashdown. In the case of the Shoemaker Levy 9 comet the answers Newtonian formula give is just about a does not begin to cover the questions that rises from the answers it gives,

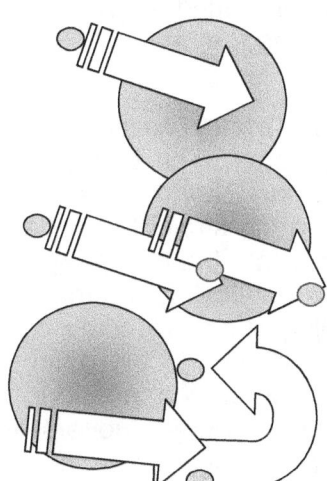

The comet tore towards the sun via Jupiters orbit path as that particular comet has done for (say we accept the) 4.5×10^9 years that science claims the Earth to be old. From trillions of possibilities in the past the comet never had to venture past Jupiter that close but it ran out of luck... Then this time the comets fortune changed which then changed the future of the comet as the comet was passing Jupiter orbit too closely this time.

Let us put the argument in the Newtonian court and say that if some inexplicable force pulled between Jupiter and the advancing comet the comet then is suppose to travel directly to Jupiter and collide with the centre of Jupiter since that argument does not allow for any reason why the comet went past Jupiter. Remember the reason for Jupiter pulling the comet while the comet was pulling Jupiter was $F = G (M.m)/r^2$ and using that formula the formula does not allow the comet leverage to pass by Jupiter as it eventually at first contact did. Then the one core should be heading to the other core where the two cores will meat in one violent confrontation but that they did not do at first.

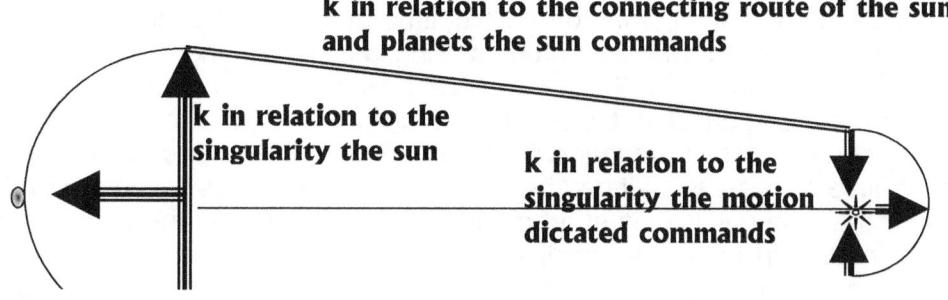

k in relation to the connecting route of the sun and planets the sun commands

k in relation to the singularity the sun

k in relation to the singularity the motion dictated commands

The singularity producing the circle around the sun comes about as a result of the demand coming from the centre of the sun.

The circle the orbiter uses comes as a result from the motion establishing an elected point that serves as an established singularity point in relation to the Coanda effect proclaiming such a singularity centre by creating the motion around such a centre.

In that way and to that reason there is a circling that is much smaller than the larger circle brought about by the Coanda motion forming singularity point.

Crossing the divide always produce new standards as a new Universe is introduced by every such crossing.

This realising came rather too late to save our comet that crossed the divide and again crosses the divide unaware that the Universe changes in every crossing as new borders form bring about a new balance between duplicating and dismissing space-time.

The comet return was inevitable because the approach was so successful

The crash did not become a reality and instead the comet passed Jupiter as if it was very much undeterred by the giant. The gravity of Jupiter at first allowed the comet to travel pass but only then the jerking back came about. The comet had a speed travelling and the closing in on Jupiter accelerated the comet by Jupiter depleting the space as the comet took advantage of the situation. The comet then had the speed to pass Jupiter because Jupiter helped enhancing the accelerated speed of the comet. With that boosting the comet had little effort to pass the Giant. But after the passing singularity lines were crossed and a new order came in place.

The success of the comets fight against overheating became the comets downfall, as the method of fighting the heat suddenly did not bring results.

This incident has all to do with gravity and nothing to do with Newton's gravity. The comet came along merely duplicating space as it has done since time began for the comet. The duplication of space brought it towards the sun and back from the sun more times than the Earth has people but this was a fatal encounter. While the comet was duplicating space forming the straight line that will eventually come to be a circle around the sun, another mighty object came into the path they share and the other object was dismissing much more space than the comet was duplicating. As the comet came closer the dismissing of space by Jupiter was helping the duplicating of space by the comet. Then at a point in the centre of Jupiter the comet crossed that centre and went to the other side although the passing of the centre was at a respectable distance. As the comet went past Jupiter's centre divide, which to the comet must represent Jupiters singularity and presented Jupiters other side of the Universe to the unsuspecting comet. Crossing the divide always brings radical changes taking place. The duplication of space by the comet was at that point confronting and the contracting which brings on the dismissing of space by Jupiter and because Jupiter dismiss more space-time in relation to what the comet duplicates the comet was in trouble.

As the comet passed Jupiter the comet for the first time at that point encountered Jupiters dark side in space depleting The comet saw the future smiling as the coming closer was in progress and Jupiters space-time dismissing ($k^{-1} = T^2 / a^3$) was presenting the comets ($k = a^3/ T^2$) all the advantage of duplicating it ever dreamed of but crossing the divide brought about misery the comet never though it would encounter since from that point on it had to fight the space-time dismissing instead of joining the space-time dismissing as it had before. The dismissing of space-time by Jupiter proved to be much stronger than the duplicating effort applied to the motion that the comet had and the comet had to surrender motion in favour of the dismissing of space-time displayed by Jupiter. The going back Jupiter produced proved much stronger than the going away the comet produced. Passing Jupiter the comet came at a certain point where the space was not accelerating the comet as it did on the approach all along but in fact it was beginning to slow the comet down. The space the comet was duplicating was at that point being dismissed at a far greater pace than that which the comet can sustain because the time factor Jupiter brought about on route to Jupiter a time factor that was helping the space duplication the comet used for cooling which helped and the comet too remain cooled very easily. But when passing Jupiter the duplication slowed down and the motion retarded where this directly effected the cooling of the comet. The slowing down became not supporting of the space duplication and that the time factor retarded the space duplication. The core of the comet overheated because of the insufficient cooling improvised by motion due to reducing of motion by Jupiters space depleting.

While travelling towards Jupiter the comet endured conditions, which is similar to aircraft leaving the Earth atmosphere but when it went passed Jupiter it started enduring the atmosphere as it would when the escaped spacecraft once again had to apply the conditions that comes about when objects are entering the atmosphere. The rise in the heat will create space. The comet expanding began to slow the comet down than ever with more space fighting the expanding than ever. The reducing of space by Jupiter did not do the comets eternal fighting of overheating any favours. Jupiter cramped the comet space with the space Jupiter allowed to the comet to have, has been not enough to support the comet. The overheating became a growing issue because instead of expanding with the heat increasing and bringing some relief, Jupiter cramped the already reducing space-time of the comet and with that ion the heat within the comet quadrupled while the singularity retained no more influence than it had and that then became another factor the comet had to find a compromise for.

With the reproducing of space retarded and the producing of space rising due to overheating there was only one option left. To remain cool enough to support the structure and prevent the structure from total demise more of the comet must then have direct access to space in relation to just transferring heat to space. The comet fragmented as singularity split through overheating and more heat being uncontrolled forms a bomb as heat forms space and with that the comet released more material into smaller portions to space that provided an escape from totally destroying all material. There the comet receive a name being the Shoemaker Levy nine comet because it transformed the space it occupied to nine spaces it occupied. But since Jupiters demolishing of space at that point where the comet unfortunately found it is so many times greater that the velocity the comet had, the comet found itself rerouted back to Jupiter. One also must see such fragmenting as a way whereby the comet duplicate in order to avoid ultimate destruction and this was a major factor that brought about the planets according to my theory of how the solar system formed when the Universe gave birth to our solar system as unique as it is.

The turning about in direction did not bring on the comet fragmenting because from the comets point the comet was still on route toward the sun. But the fragmenting of the comet slowed itself down even much more than what was previously the case. To the comet Jupiter was another Black hole that the comet never saw coming because it was in sight and then it was not and then it was this gaping hole. This process that Jupiter applied to the comet was a part of the way the cosmos was born. The cosmic birth came about when singularity was resting in the single dimension of Π^0 where that value held all the possible Universe components waiting for a future but was still sealed in the first dimension. Then singularity Π^0 heated and formed the initial value of singularity in as much as becoming Π. That did not yet release the other components, as they still remained stuck to the first dimension in the first circle of Πr^0.

They came to be on the newly formed rim of Π but were still in the first dimension of Π and only by the second expansion providing the motion that brought on the second forming of space outside singularity that is holding all captured future possibilities in the upcoming Universe did every one of the three factors come into being separate and individual factors. With the first motion Π^0 to Π not one of the factors k^0 and T^0 and a^0 except singularity Π^0 to Π was released by the expanding of the distance Π^0 to Π from single dimension. The components came into the Universe being a factor in the single dimension only after Π^0 became Π. When Π moved from Π^0 to Π, k^0 and T^0 and a^0 only became factors but not identifiable separate individual values yet. It was at this point that the law of Pythagoras came about and the line was the same value as the triangle and the half circle. It was at this point there were no formal values yet and all components within the newly born Universe had their values expressed in degrees meaning in directional rather than space valuating relevancies to one another. This was because only motion T^0 the half circle with distance growth the straight-line k^0 and space in the triangle a^0 came into place. One can see the way Jupiter rendered the comet to heat. First it went about destroying the route the comet followed. Then Jupiter retracted the space the comet took to escape by reducing the space more than the comet was able to duplicate and in that the comet retreated back to Jupiter faster than the comet advanced onwards towards the sun. Then Jupiter compressed the space the comet shared with all the space that Jupiter retracts in a natural process, which cramped the comet to the level of being liquid heat. That made the comet fragment which then circumvented the Roche law that Jupiter had to acknowledge.

With the fragments much smaller that the complete structure the comet had it made the devouring of the comet much easier as Jupiter finally cannibalised the comet singularity holding fragmented space- time. At first glance it is clear that the process repeats what Newton suggested what is happening out there. But if Newton was correct the centre point had to draw on the centre point and causing a direct line pointing at the direction the two objects had to travel as they reduced the distance r by the square to the force of the two masses. The two objects did not draw closer in a direct line as it should when applying Newtonian views. In stead it came by a round-about manner by first passing then returning then passing again before it broke up (something which is quite unexplainable using Newtonian science) and had the fragments slammed into Jupiter. It is the circling around that bothers in the Newtonian explaining. The comet once past the divide not going straight on towards the Jupiter centre, and came back following again not the straight line in the retreat but had another come back towards Jupiter before slamming into Jupiter. One should use this to read into what went on in the early cosmos before material was set on undisputed routs and later on and later on never influenced one another's travels. We find that more advanced singularity dominates lesser developed singularity and so much so that it happens to the point where the lesser singularity sometimes completely turn to liquid heat. That should point to where heat originated as we deduce and read into what applies in the process of liquefying unsuccessful singularity to a state of fluid heat.

Jupiter reduced a^3 of which the comets a^3 was part of the point where a^3 once more became a^0 but with a^3 going a^0 the rest of the relation must then also go single becoming T^0 and k^0. This had to one of the manner that applied during the time that mainstream science now presume antimatter came about. The comet became antimatter because to the comet Jupiter became antimatter reducing the comet to antimatter, which the Jupiter antimatter dissolved. Jupiter gave the comet a fair share of antimatter as Jupiter introduce $k^{-1} = T^2/a^3$ and that reduced the comet space-time from seemingly enjoying $k = a^3/T^2$ while in that time in truth and facing reality the comet was also reducing by applying but which the comet never suspected because from what the comet saw it was all the time applying $k = a^3/T^2$ while $k^{-1} = T^2/a^3$ was what the comet received. The comet became antimatter because of Jupiter forming antimatter in the form of k^{-1} to a point where heat remained. The three factors combined and multiplied to bring about the cosmos as it came to be. It was at that point mathematics arrived but the Universe was already secured as a statement bringing many factors to become part of the future. But singularity first came into position as Π^0 to Π and from that the foundation of expansion and growth was formed. In that very instant Π^0 to Π formed motion Π^2 to Π^3 where space came into place duplicated by motion.

As we can see from the behaviour of the comet in relation to Jupiter motion and dismissing formed a ratio. The very first instant came when $\Pi^0 = \Pi^1 = \Pi^2 = \Pi^3$ and all had the same value, which was separated by directions rather than the measuring of size as such. We witnessed the reverse of when the comet dissolved to heat that had no gravity and no duplicating qualities. From that the product of the culmination grouped together as major atoms or what we call stars. With the passing off the comet the applying gravity between Jupiter and the comet was $a^3 = T^2 k$ but by recalling the comet Jupiter changed the applying gravity to what is the same gravity applying to us, which is $1/k = T^2/a^3 = k^{-1}$ and also the same gravity Newton saw. Our position, which is a captured position on Earth is also that of the captured where T^2 got in a position to dominate k into negativity. By using this evidence one may arrive at a conclusion as how the solar system came but implicating laws that apply on a much larger scale by using the cosmic relevancy scale and thereby determining the original position of every planets as they were before development changed the lay out. By allowing development to rein act the solar system it becomes an open book to read.

When time came from singularity it brought along space and distance in motion. The distance represents the creating of space while the motion represents the destruction of space. Newton proved that $4\pi^2 a^3/T^2 = G(m+m_p)$ where k also is a^3/T^2 (from Kepler $k = a^3 / T^2$) and therefore $4\pi^2 k = G(m+m_p)$. Then in the event of singularity being $k^0 = 1$ then $4\pi^2 \times 1 = G(m+m_p)$. That means extending from singularity space-time will be $G(m+m_p)$ or $k = 4\pi^2$, which is what space-time became when k pronounced space-time for the very first time coming from singularity. Singularity extended to form $k = 4\pi^2 = (\Pi^2+\Pi^2)+(\Pi^2+\Pi^2)$. With heat disintegrating to space and the cosmos forming an extending of space-time in space, space-time deformed to $(\Pi^2+\Pi^2) (\Pi^2\Pi)3$ and that is the value of the atom. Since $G(m+m_p)$ is only a symbolic gesture it can represent any aspect as long as it represents a dominant symbol relating to a subordinate. It can also represent singularity in the same relevance where the one singularity is dominant fighting for dominance or control and the other tries to secure independence by establishing personal identity/

The very first instance brought the developing of k that was equal to T^2 time and a^3 space

But motion came about on this side as well as the other side of the Universe where the Universe was $a^3 = T^2 k$ and on this side was a^3 while a^3 was becoming $T^2 k$ on the other side and the other side $T^2 k$ was becoming a^3. Then the second in creation came about but the first instant plated a veneer of material in the form of opposing singularity destructing as liquid heat due to a lack of motion onto the singularity applying motion, which brought about to specific markings onto singularity. This act represented destroying of some less protected and the maintaining of other singularity remaining in form and cool. This would see as a flicker from another point holding singularity. From the other side the flicker would seem to cut the duration it took the motion to fill the time half as long while in truth it cut the space in half by doubling the time. Then four flickers came about and eight flickers came about and the action time provided brought about more space as space doubled every time but it cut time duration by half every time.

By the time the proton and neutron became about the very fist time the duration of time slowed down by Π^3 which is $(1836)^3 = (6188965056)^2$ more than the original space but there were also $(6188965056)^2 : 1$ flickers of time to the original. Every flicker T^2 of time represented on k that accumulated by introducing one unit of a^3 space. One rotation then meant time reduced in duration by $(6188965056)^2$ times to the original on and space increased by that margin. Eventually by the time that the electron and light came about the space was confirming duplication to a ratio of $1:6188965056$ that means for every flicker light brought along space confirmed $(1836)^3$ times over. Light found the epitome in k and k formed the atom but the atom confirms the unit of space duplicated by time to the value of (1836) which is made up of the double proton $(\Pi^2+\Pi^2)$ and the neutron $(\Pi^2\Pi)$ as well as the electron 3. Therefore the motion of the proton is 1836 X 300 000 km/sec. That confirms the proton as "flickering" 1836 times more than the 300 000 flickers in km strokes per second of that of light. In that way the duplication of space is accelerated by 1836 times when it becomes a proton in the ratio between light and mass.

To circumvent the atom circle it takes 6188965056 times k to produce that number of T^2, which duplicates that many a^3. This means by the time the atom was completed the atom in ratio was confirming the duplication of k by a margin of $1:6188965056$. That means when observing the motion there will be 6188965056 for every flicker on would notice at the proton seen from singularity or there will be 1836 flickers seen from the electron sees from the proton and there will be 300 000 flickers seen from our stance if the photon was one kilometre in length which we know it is not. That action brought on by singularity duplicating to form space-time is what makes the Universe seem solid. Once the measure of relevancy is broken time and space goes wasted. The solidness having flickering intervals is only applying in the eyes of the onlooker. Our Universe seems solid as long as we are on "the other side" matching the seemingly "shorter duration in time" and "extended space". Once the margin becomes fallible the space – time scenario becomes distorted and eventually destroyed. In the Universe we share it seems the speed ratio or time duplication can withstand the combined effort of sixty protons where then the duplication of space falls pray to the destructing of space.

$(\Pi=\Pi^3/\Pi^2)$ or in Kepler's terms $k = a^3 / T^2$ presents the duplication of space while

$(\Pi^2 = \Pi^3 / \Pi)$ or in Kepler's terms $T^2 = a^3 / k$ presents the destruction of space and $(\Pi^0 / \Pi = \Pi^2 / \Pi^3)$ or in Kepler's terms $k^0 / k = T^2 / a^3$ presenting the final act or the demise of space. $k^0 / k = T^2 / a^3$ becomes totally dominating when there are a displacing ratio of **6** (materials Number) in the square of space **10** which then presumes the value of **60**. Space will remain duplicating as long as there is space available to convert to heat and as long as there is heat that can be converted to material and material available to transform to singularity. Nowhere is they're having a free ride coming to any part or dimension of Creation. There is built in only a lot of hard work and a dear price to pay for every inch gained or lost in growth.

In this mentioned ratio between the dismissing and the duplicating of space through motion stars form by accumulating material in a giant sphere and keeping the atoms secluded from outer space. All the atoms within the star that are forming the space within the star that is forming the star are as much the star as the star is all the atoms combined. Since all the atoms in the star works towards a mutual goal as to provide the star with the required security to provide the maintenance that brings about survival to components in the star is as much one atom in cosmic space as all the atoms are individually cosmic structures. When the flow of space is exceeded by a certain specific number of proton abilities to dismiss the space the dimensional walls keeping the space in form no longer can sustain the flow of space by substituting the demise with a flow of space. We have to remember that the initial motion was equal to the initial expanding that was in turn equal to the initial space a^3 that developed. The distance **k** that came about was the same value as space a^3 and the motion of the space T^2. The Universe divided innumerably as it remained one structure. Relevancies came about that excluded no possibilities and whatever one may think of being in the Universe came into place through relevancies between innumerable factors all acting in groups that remained one. There was no space but the space created in that cycle motion. There was the motion that provided the expanding in the straight line that was precisely the same value in the half circle and that brought about the triangle also to the identical value as the other two and was securing the space that was precisely the same as the other two factors. Reflecting on matters we still find this very same trend applying to light at the present time we live in. In the development space grew because the diminishing was only half the growth and size rather than direction became a major influence. The direction is the foundation and came in as a bases in the very first instant. The second repeating instant changed the scenario. With time progressing the distance k^1 will tend to lag behind as the rotation T^2 has to compromise for more space a^3 involved. With more space coming about the circle that had to produce the rotation was at the start equal to the expansion distance and there fore $T^2 = k = a^3$. Since then the ratio changed to $a^3 = T^2 k$. That is how space/time relates according to the original calculations Kepler introduced.

$1/k = T^2/a^3$ We know that there is a demise of space relating to the growth in space proving that when distance **k** reduces space a^3 will do the same and

$k = a^3 / T^2$ when **k** expands it will produce space in relation to motion.

$T^2 = a^3/k$ When **k** demise the growth in distance **k** expand time T^2 will increase by the square but distance **k** will diminish by the single therefore time T^2 will grow faster than space a^3, which is the result of **k** will diminish.

At the beginning a trend was set that apply throughout Creation ever since. As the space increased the time ratio decreased since the distance in relevancy reduced in relation to the available space and that is the relevancy I simulate in the atoms ratio of space-time being $(\Pi^2 + \Pi^2)(\Pi\Pi^2)(3) = 1836$. But as the star takes time in space back towards the earlier scenario conditions applying to the atom will reduce the space it holds and as such the time will bring the compromising aspect to the changing ratio. It is most important note that when the comet liquefied it was all the atoms within the comet that was liquefied. It was not a case of the comet destruct but the atoms within remained preserved. The comet is as all cosmic objects are, made up by all about the atoms within the structure and also the way the elements arrange their various positions to assimilate the sphere to prefect detail where the entire group of atoms as a group manifest in the form of a sphere spinning about an elected axis. As the group of atoms work in the way an ant ness will where the group form a unit, so does the atoms form a unit representing all the atoms as a group in the group where the group retained the position as one atom with one elected singularity in a charged centre. As the Universe formed by dividing innumerable times and yet remained a unit throughout every aspect, this leads on to groups forming galactica and stars where all the atoms within the star becomes the star and the star becomes anther single atom. The atomic relevancy apply as that quantity in the outer space and with Earth not being that much better developed the atomic relevancy in outer space and in the Earth centre is $(\Pi^2+\Pi^2)(\Pi\Pi^2)(3) = 1836$ meaning the proton displaces space-time at a relevant value of 1836 times that of the electron, or the other way around is that the proton is 1836 times more intense going on route on to singularity than heat is going liquid through the strainer we call the electron. We can see that the atom formed in accordance with Kepler's prediction and followed the route Kepler introduced.

$\boxed{\Pi^0 \Rightarrow \Pi \text{ Bv motion } \Pi \text{ became } \Pi^3}$ The first motion implicated distance and space but space is one part of the requirement, which applies because the structure found independent space by separating from surrounding space through motion.

$\boxed{\Pi^3 \Rightarrow \Pi^0 \ \Pi^2}$ Since singularity split the Universe, which in the case is the independent structure forms motion that also separate as it is separate on both sides of the divide because singularity is unable to move. The motion may seem as a combining factor but is doing the dividing on both sides of the divide because singularity is unable to move. The forming of space had to be completed by the forming of motion that provided the time factor and so space/time was presented for the first time.

$\boxed{\Pi^0 \Rightarrow \Pi^2 + \Pi^0 \Rightarrow \Pi^2}$ The immovability of singularity spawns space-time by producing space in motion on both sides of the divide as the proton duplicates $(\Pi^2+\Pi^2)$. From singularity the double proton arrived in all particles through out the Universe in one motion by turning from one side of the universe to the other side. In this instant all four of the cosmic pillars came into affect that then became the building basis of all the cosmos.

$\boxed{\Pi^0 \Rightarrow \Pi^2 + \Pi^0 \Rightarrow \Pi^2}$ At the point where the initial motion came about from expanding or destroying singularity the plating of heat (in the same manner we see electromagnetic plating is done) where the process confirms the liquid heat into solid heat and in turn into singularity became contracting and by plating which confirmed the liquid onto the solid and into singularity it became it was feeding singularity and thus secured survival. The proton motion was no longer removing from singularity but was bringing into singularity. But this come at a cost where most of singularity in the cosmos failed securing the future by providing motion which is gravity and formed liquid that eventually became space. The double proton could not sustain or maintain the overheating due to friction that came about from the lack of space and space as a factor came about.

$\boxed{\Pi^2 \ \Pi^2 \Downarrow}$ The liquefying process mentioned above established at the very first a liquid particle that will hold the motion part in space to the value of $(\Pi^2\Pi)$ where one factor (Π^2)

$\boxed{\Pi^2 \ \Pi\Pi \Downarrow}$ confirmed space-time and the other factor could introduce motion (Π) and in that a third part became a factor forming the space within the atom.

The differentiation between time space and distance became more and more apparent and influences as much as altered the composition of the developing Universe. The progress in space required factors to change and absorb the pace developing.

$\boxed{\Pi^2 \ \Pi 3}$ Then only did the Big bang arrive as liquid heat proceeded in turning into space and space came into place as $\Pi^0+\Pi^0+\Pi^0 = 3$ which represent the 3dimensions of the six sided Universe we are within. Only with the arrival of heat forming part of the atom in providing space to the atom could the atom survive and bring about the Universe we know. Then only came the Big Bang

This became the atom $(\Pi^2+\Pi^2)(\Pi^2\Pi)(\Pi^0+\Pi^0+\Pi^0) = 1836$ and the atom formed stars that still act in accordance with and to the atomic relevancy

Outer space substantiate the atom as $(\Pi^2+\Pi^2)(\Pi^2\Pi)$ 3

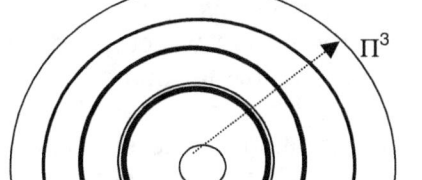

Π^3

Every layer in the star represents one factor in the atom since the star is just another cosmic atom securing

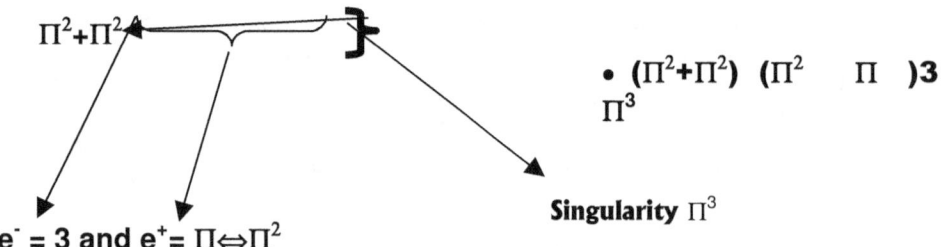

$\Pi^2 + \Pi^2$

$\bullet$ $(\Pi^2 + \Pi^2)$ $(\Pi^2 \quad \Pi \quad)3$
Π^3

Singularity Π^3

e⁻ = 3 and e⁺= $\Pi \Leftrightarrow \Pi^2$

Mass has no part in any process but space-time demise and the relevancy brought about as such plays the only part. $(\Pi^2 + \Pi^2)(\Pi\Pi^2)(3) = 1836$. At first with the first motion producing the first space, it took every proton in the entire Universe to suppress the space created in that motion. One movement found the ability to dispose all the space created but also created all the space there was. It was an innumerable number of protons forming the first atom. The first atom was the Universe and the atom became the Universe. The atom still is the universe and will remain the Universe till the end. During this time relevancies came about that we named sub atomic particles. At first through motion the double proton came when seven points formed a relation with three and involving the Roche factor to form Π^2 on both sides of the divide. But the relation was in conjunction and not against so the total became an addition and not a result. $\Pi^2 + \Pi^2$. Then the universe was blessed with spinning space destroying double components allowing heat no escape. Some of the protons had to capitulate to survive. From this came the neutron $\Pi^2\Pi^2$ at fist presumably duplicating and later dividing $\Pi^2\Pi3$. But every time it remained in relation to singularity extending, which is Π. In the Universe we enjoy the relevancy of the atom, which stands in direct relation to the atom holding space in the motion that duplicates the space. If the electron has this electron's mass of $9,109 \times 10^{-31}$ and the proton has that incredible bigger mass of $1,673 \times 10^{-27}$ making the proton a warping 1836 times more massive than the electron. Even in these differences in size and in gravity how can science connect size and mass. With this information and knowing where the subatomic particles are one can clearly see that gravity is about space dismissing ands what dismiss space more the action of the proton? The proton is the place in the Universe where I now am referring to which is where mass is created because that is where gravity is generated. Size just cannot be the cause of mass that produces gravity but rather more the other way around.

At this point one should delve deeper into the heat/cold and space density argument. Although it extends beyond the limits in information, which I wish to put on this book as far as my release of controversial and contradicting ideas I have colliding with the accepted and established norm straight on that mainstream science uphold I wish to elaborate somewhat on the big and small issue as well as the hot and cold issue. The Universe is without boundaries therefore there are no limits in what might be the hottest or the coldest. Neither can any one establish the limit of what might be big and what might be small . There is no biggest as much as there is no smallest. There is only some that is more developed or those that are less developed. But in saying even that I am extending what is and is not to what is not part of the Universe. It is far better stated in **Matter's Space In Time: The Thesis volume 5 or Matter's Space in Time: The Hypothesis** I. S. B. N. ISBN 0-9584410-6-5 which incidentally is the same book since I have much more space devoted to the matter. Very briefly I shall commit myself by admitting to this: we are in the fourth developing stage in a cycle (as far as I can calculate) being the fourth in seven cycles of star development and in that some refer to one stage where it will hold prominence and the next will relate to another cyclic developing stage where the other will have prominence during the developing era. In the end all will conclude reaching the same finality but some will achieve it sooner that would others do. In that comes the "more or the less developed" but such a statement is not differentiating between two but merely identifying dissimilarities in singularity management under specific condition as the singularity maintain space-time in relevancies. The heat is everywhere and all over the same but confining the space to the concentration may help us to set a scale where such a scale only indicate a ratio of the balance we find in the density leaning to one or the other side. The more heat there is separated from space the colder the space must be holding the separation active. In the fine detail there is density in heat forming a liquid or there is density in space forming a gas but it always come down to the one or the other of both in balancing. When the Universe was 10^{34} K the Universe had everything that is currently a part of the Universe, which makes the Universe as big then as it now is small. Everything that was in it is still in it and will be in it and even a Black hole that has left the Universe is actively part of the Universe by consuming space-time through a proton action, which the Black hole placed outside the Universe. If we say the Universe was as big as an atom back when the Universe was 10^{34} K…then in proportions the Universe still spread space evenly and by equal measure as it does today. Space cannot add unless there is something willing to give away of the prominence it holds. Where space is abundant we find we find the concentration of heat to be low. The heat is not low or high but the concentration either favours the liquid factor or the space factor where if the heat being liquid is hot then the space being gas must be cold. It is a question of what is concentrated and dense and what is of little prominence. But when saying that while looking at our Universe I have to admit a lot of heat back then went the way of space now and I have to immediately correct myself the space out there is as hot as space can get while the space within the very centre spot of the sun where there is no space and no motion is as cold as the Universe can get. If it was not the case fusion could no take place since fusion comes from freezing space between the two factors committed to fusion into the oblivious unknown. Space can only remove by cooling because pressure brings about heat and heat brings about space as the nuclear bomb so vividly proves. By removing plasma, which is a fancy, fill name for good old heat but sounds very intellectual, from space the space itself has to grow cold in order to allow so much distinction coming about. The cold can house the hot as the hot will accompany the cold and raising the one has to lower the other since it is clear that as space part it parts from heat and as heat lessons in density it joins space while it leaves heat. There is an unmistakable quantifying connection separating the two, which in itself is beyond separation. It is relevancies applying in favour of one aspect or another aspect. Gravity reduces space by employing motion and by reducing the space it makes the space cold as it removes heat from space by intensifying the motion of the space in the space. Gravity makes space loose from heat but also heat loose from space (if I will be excused for such low sophistication used to express myself). We measure the heat in the spot but we should agree that in that case we are measuring the space in the heat. The hotter the space will seem to us realistically the colder the containing space is. The relevancies have to push away from the centre or join at the centre. By removing the space and all the heat in the space to a point of no distinction. That means outer space is where heat joins space in the favour of space and singularity is where space joins heat in the favour of heat excluding space. In the past at some point cold and hot and heat and space was parted to the extent that the sun's hot $18 \times 10^{6\ 0}$ was deep frozen because outer space was $10^{34\ 0}$ K and at that stage what is hot now was bitterly cold because the suns corona must have been a solid freezer at only $6500^{\ 0}$ K which is 10^{30} below what was freezing outer space back then and what was bitterly cold then we regard the sun now to be as extremely hot. The cold accommodated the hot keeping the sun and all the stars in that range in a deep frozen state while the bitterly cold that outer space has to represent being the carrier of the heat limit was at that point in relation to our measure excruciating hot while the hot stored the cold but in the end it was the same as it is because it is the same Universe as it was.

One must not envisage this space duplication and space dismissing on the level where Einstein placed his vanishing and flat going Universe because it is on another plane. One must not see half a person following his other half into the future of a line of the same man continuously leading on in time. At the level we live and breath and the level we can see light and even the space we have and live in has become solid. A particle entering the atmosphere meets with a solid space that is so unbroken it puts the material back in time by billions of years by turning it into photons. The suspending of space by material in motion takes place at the most intricate of places locating where singularity brakes into space- time.

The atom like the star is a sphere and on the inside of the sphere there is the point where space dismisses into singularity because of the way that nature designed the sphere. The sphere secures a point where no space is possible therefore no motion can duplicate such a space. Then as space-time develops stronger locations flowing to the other border of the sphere space-time becomes more defined and secured. On the outside the number of spots per volume of space-time developed is far reduced to the space claimed because the vacancy in space can only be represented by singularity that is the securing aspect of space-time. The duplication and the dismissing is so small part of the space-time singularity secures that in the dynamics where a triangle must hold a different value from that of the half circle, which in turn has to hold a different value to the line the difference there is in the forms secure the definite solidness of space in time. One must not see an object breaking composition because at a displacement factor of $(\Pi^2+\Pi^2) \times \Pi = 62$ the walls of space comes tumbling down on the flow of time, but only at a point where the governing singularity developed to a point that the displacement an secure the survival of the appointed located singularity the star developed throughout its life span.

In outer space and near to outer space regions as the Earth is the atom confirms the security of space duplicating through time. The atom sustains a space-time duplication of the atom as it is combining the total value of the factors forming singularity. The factors related to the atom as a unit that is securing space–time relating to singularity to the form of $(\Pi^2+ \Pi^2 + \Pi^2 + \Pi + 3) \times \Pi = 112.31$. These factors are holding the atom apart from singularity by confirming space-time which is the atom in another relation other than the normal $(\Pi^2+\Pi^2)(\Pi^2\Pi3) = 1836$ and that is the point the sphere secures space in the six dimensions or space walls formed by the universe when the Universe came to $7/10\Pi^6/6 =112.162$. It is only when the requirements of maintaining singularity surpass the confirming the maintaining singularity in the atoms inner core stability that the double proton confirms in dismissing space-time without the required duplication also thereof will singularity dismiss the structural security that space-time in the double proton can provide. With me suspecting that there may be those that has super ambitious as space whirl creators living out there that plan to trick the Universe by building some double space whirl here this. In the event where there is a person with a super elastic imagination that foresee a situation where such a person can create or establish two of himself in the same space in the same time with some mathematical formula he has dreamed about, then such a person must first demolish his body down to fit a position where any triangle in his body will measure the size any half a circle that will take forming a straight line. Therefore all those out there planning on their next combining of space whirls they anticipate creating to split time or bridge time or just annoy time in space, remember that once you reach the space you occupy no more than a straight line you may proceed with your ambitions and plans.

The breaking down of space by the inability to create motion is present in every atom and particle just as the duplicating of space is in every atom and molecule. Since all stars are group representatives of innumerable atoms within the space-time confinement the governing singularity lays claim to the dismissing of or duplicating of space-time is present. It is the ratio in the atom or stars favouring the presence of one or the other that allows the atom or star the characteristic it shows. If the duplication of space –time is prevalent as it is with Xenon, then notwithstanding how massive the protons grouping is, it still seems that space will duplicate much more frivolous than would the diminishing of space-time be a feature of the element. It is the ratio that gravity establish placing relevancies against relevancies that produce space-time or even the lack there of as in the case of a Black Hole. Where the Black Hole lost all ability to motion with in because of its securing of singularity maintaining does the proton then ejects into outer space way outside the star or star that now became an overdeveloped atom to a position in outer space because the motion of the double proton is no longer valid within the well developed singularity governing the Black Hole.

The proton position

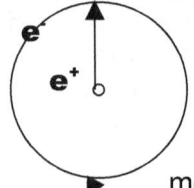

The electron has the outer edge of the atom to fill, which is far greater in volumetric size than the inside of the atom where space is at a premium compared to the outside. Even just looking at this picture should tell everyone all there is to tell about gravity. It cannot be size that implicates mass. Referring to the massive potential size discrepancy became synonymous with the quantum factor indicating unbelievable more space that is beyond any explaining going the way of the electron.

The electron position

The position the proton has within the circular atom cannot promote any idea that the proton can be that more massive than what the electron is. There is just not enough room down there in the centre of the atom. Therefore one has to search and see what there is down there that would allow the proton to be that more massive. The only thing in total abundance is the lack of space and the only way that the proton can be that much more massive is if it destroyed 1836 times the space than what the electron manages or on the other hand if the electron can duplicate 1836 times more space than what the proton can achieve. What this does tell is that the electron duplicates space 1836 times more than the proton or the proton reduces space 1836 time more efficient than does the electron.

This atomic relevancy we at present hold were not always applying and as space grew in progressive development the atom had to adapt in relation to the requirements of singularity. This reality is frozen in time in all stars when the star seeks independence from the galactica centre as it drifts outwards. There are two definitions we can use when looking at such a growth. We can look at the space not holding material that grew in size in which the stars froze their development by remaining behind all because of a lesser developing singularity or we can focus on the stars growing and with that push the push the space much more into expanding. The star froze cosmic development as it came about in its search for independence and at that it holds the atom to the value the cosmos had before and during the time it became independent. The cosmos grew in space progress just as much as the star was left behind. In that there are young and not that young stars but it has no implication on the particles in numbers volumetric holding mass and as such the object produces gravity. If an object had the intensity of I kg in outer space, the object will be 3 times more intense on the outer edge of the sun or it will be Π^2 more intense on the very inside That object will be 1836 times as intense in a neutron star or 6188965056 million tons in a Black Hole. This comes from the manner that the star manage to destroy space and redirect the space to fluid heat or the solidity of frozen space as matter really is. In the Black hole it reduces much further as it claims the singularity, which the object had, and destroy all space and all time there ever was.

The star did not diminish the space and the space did not outgrow the star. It is a relevancy where the one factor represents a compromise to substitute for the other factors changing the relevancies applying. It is the space that grew as much as it is the star development that fell away and started in initiating independence from further development by securing sufficient heat within the inner core to provide the motion that will produce such independence from its surroundings. The diminishing of space takes place in every atom in such an atom inner core per ratio of the number of protons acting as one unit. The result of the product is the accumulated to form the total value of the star. In every atom there is the dismissing of space that is fed from the top or outside of the star that spirals the diminishing value downwards towards the circle and into a centre. The dismissing has little effect in the immediate vicinity of the atom as the removing of space is compensated by the supply of space from positions where more space is available. The flow of space from outer space will substitute the dismissing of space. But in the centre of the star where all the heat accumulates and gravity is at the very prime, where there is no space left such diminishing will have an accumulating effect gathered from all the atoms accumulated efforts that cannot be substituted by the flow since the flow comes from every proton in the atom within the star housing all atoms and protons as one unity. The eventual combined value forming the total combination of spinning protons produces a velocity of flow of space in relation to the time thereof towards the centre and to compromise for the flow being insufficient the space will compensate by becoming denser heat once again. The heat that forms is in the centre of the star and the compromising of space by forming heat comes as the result of space flowing into the star. The star has to have an iron inner core to generate the flow of space towards the concentrated centre. If the sun did not have an iron core the sun would not have gravity. That again contradicts the science teachings This is the way space flow towards the centre of the star and that would influence the space/time displacing within the star that would affect a star:

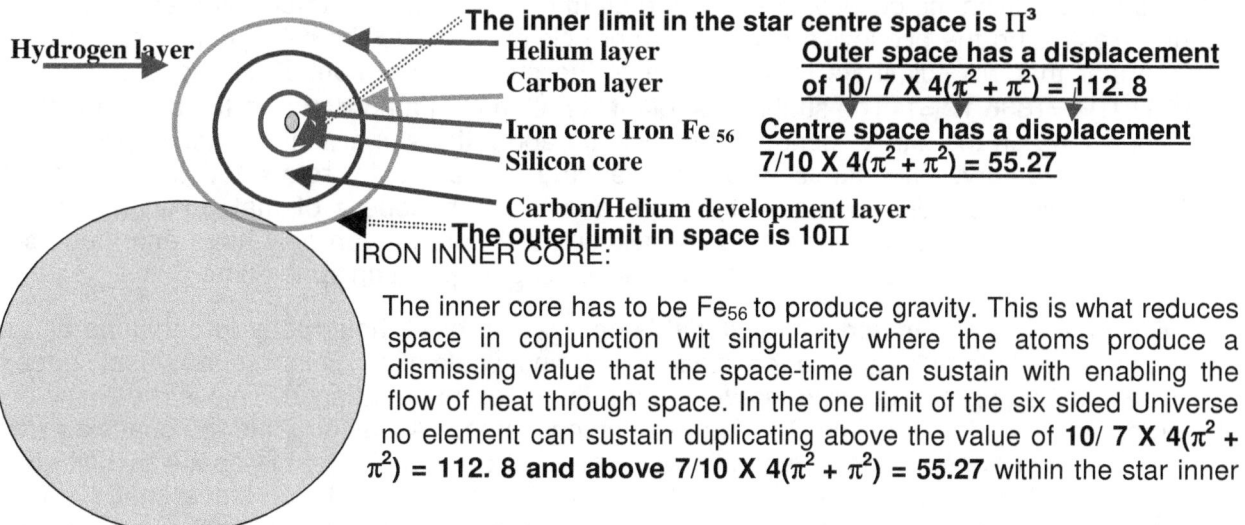

The inner limit in the star centre space is Π^3

Hydrogen layer

Helium layer

Carbon layer

Outer space has a displacement of 10/ 7 X 4($\pi^2 + \pi^2$) = 112. 8

Iron core Iron Fe 56

Silicon core

Centre space has a displacement 7/10 X 4($\pi^2 + \pi^2$) = 55.27

Carbon/Helium development layer

The outer limit in space is 10Π

IRON INNER CORE:

The inner core has to be Fe_{56} to produce gravity. This is what reduces space in conjunction wit singularity where the atoms produce a dismissing value that the space-time can sustain with enabling the flow of heat through space. In the one limit of the six sided Universe no element can sustain duplicating above the value of **10/ 7 X 4($\pi^2 + \pi^2$) = 112. 8 and above 7/10 X 4($\pi^2 + \pi^2$) = 55.27** within the star inner core dismissing space beyond that capability will no longer contribute to duplicating space-time of the atoms involved. Only the iron atom producing and maintaining a displacement value of 55 – 56 can produce gravity by being on the edge of demising space time while maintaining duplicating which is gravity and in our Universe only stars with an iron inner core has the ability to bring about gravity. Gravity can only achieve a displacing relevancy at $7/10 \times 4(\pi^2 + \pi^2) = 55.27$, and that produces a potential difference that brings about gravity within the inner star core where gravity accumulates. This then relates directly to the second value of the Titius Bode value of $10/7 \times 4(\pi^2 + \pi^2) = 112.8$ that limits outer space in the three-dimensional and six sided boundaries of what forms our Universe as outer space forming the value of $7/10 \ \Pi^6 / 6 = 112,162$. That is the outer relation to the inner relation set by the core in ratio to the outer space securing a position for the star identity in the space limits and is an indicator of the balance in space-time displacing potential of the star. In every star there is this flow towards the centre firstly of every individual atom but also as a combined unit flowing towards the centre of the star and the dismissing of space in every atom centre brings about the forming of a relation as a group within one unit structure we call a star. This flow is there because we gave it the name of gravity and gravity is the result of all the atom protons dismissing space and as such then has a linking that is invisible to the naked eye. In young stars the core ability is yet to develop and in such stars the gradual reducing come about as layers support the effort little developed inner core. The space reduced becomes a unifying effort from all the atoms in all layers from the outer (hydrogen$_1$ and helium$_2$) through the carbon / oxygen centre and the silicon layer down to the iron core and even going down further into space-time obscurity where the atoms as a group combining their effort acting as one atom. An atom securing one proton will provide much more space a much better field to flow . That leaves space the opportunity to support the demising of space of all the individual protons by substituting the loss with an new supply of space that becomes converted to heat than would an atom supporting 56 protons within one containing centre. The stepping down assists the young stars with a weak heat envelope sustaining the spin effort of the yet underdeveloped inner core.

The demand on space flowing will be much more beneficial to the flow where all the atoms comprises of hydrogen and helium such as we found large super giants have. In the event where fifty percent of the star holds iron$_{56}$ and the rest is composed of silicon $_{26}$, the demand on space flow will be at a prime and the heat envelope that will support such gravity flow coming from such demand will not likely allow any fluids, as the photon is to escape from the star. The concentration of protons overshadows the flow opportunity by far and the star will become darker. Such a star going darker does not die because it is not life or a coal stove that can go out. There is no time period where it consumed all the available fuel as is the case with coal stoves that was used during and just past the Elizabethan age. The fuel stars use is available in unlimited quantities forming volumes that only time extending to eternity can consume.

This dismissing of space puts a relevancy onto the atom and that allows the atom some space/time conditions that diminish the occupying space the atom claims. This puts totally different equations on the atom. Restricting the occupying space the atom claims indicate the relevancy of space demise within the star that will culminate in the centre of the star and will control all aspects of the star. $(\Pi^2 + \Pi^2)(\Pi^2 \Pi)3 = 1836.12 = 1$. The super "gas" giant is nothing but a bowl of fluid the compare more or less to conditions on the sun. A periodic cycle will be about the same as the planets give or take a few years per century. The atoms would display the same gravity relevancy with the electron establishing a dividing value between the atom and the atmosphere of that particular star. The relevant function of the star will then be 3. By aligning with three it can only have enormous space that the giants hold because stars in that league is all about duplicating space-time.

In the view space has on material all material is one gaping black hole ready to consume as much space as the star could manage and the space restraining would allow. But from the offset space has a black hole locked up and growing in every star striving for independence. This is the result of the first moment in space forming time. The first expanding distance **k** was equal to the space as well as the time factor and the reducing of the time factor brought on reducing of the space by half and the distance by half. This is the result of space-time being double the distance since space is just as much as the distance and the motion of the time is. By eliminating either or, either or remained a factor and the Universe came about. Then the second moment in space arrived and the distance created was half the space and half the time that doubled.

The initial doubling was possible by consuming and concentrating heat of less fortunate and less movable singularity. That became the second motion where some atomic particles completed gravity by reducing and confirming heat and the other part went liquid and establishes room in which to move with the ability to move. Then came the third phase we call the Big Bang when heat turned into space everywhere. Bu the first was coming as a result of atoms grouping in particles as well as groups that brought maintaining to singularity and other singularity destructed as it melted into liquid. Stars in galactica and atoms in stars then formed as in each case the atoms elected a governing centre and the stars with their elected governing centre elected a centre governing the galactica in quite the opposite way that the stars govern them selves.

The relevancy applies because we are attached to the singularity within the Earth. Being connected to the star for instance would have allowed a much different relevancy to apply but then again we would have been much different and very lifeless. I have no idea how any relevancy would be when I find a connection to some other relevancy than the Earth. Seen from a Black hole the Universe will be much different than seen from the Earth and in that we are unable to presume what would apply. It is crucial that humans only be concerned with the relevancies from where humans are allowed to work. Our relevancy would also be much different if we were not under the stringent control of the sun, which again help us to define our relevant position in space-time. It is a culmination of circumstances that apply to place us in the position that we can value the atom at $(\Pi^2+\Pi^2)(\Pi^2\Pi)3$ $=1836.12 =1$

$(\Pi^2+\Pi^2)(\Pi)3^3 = 1674.3$

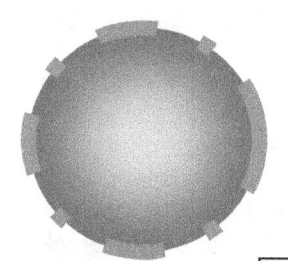

The cyclic period relating to our value that the sun display is Π because from "hot" to "cold it measures in time about twenty one years according to the butterfly diagram of space and heat contracting and expanding With that in mind a cyclic period in the sun is about Π to the outside and therefore we can deduct Π from the atomic relevancy as we experience it to be on Earth and restrict the space occupation to $(\Pi^2+\Pi^2)(\Pi^2)3^2$ With having a displacement of Π only puts space in the value of the electron acting as a neutron substitute in the electron as part of the neutron function. There is very minor fusion applying in the centre and growth is mainly coming from reducing space by concentrating heat that enlarges the productivity of the proton.

The sun has a relevancy of $(\Pi^2+\Pi^2)(\Pi^2)3^2 = 1753.36 = 5.70 \times 10^{-4}$

Blue giant $3\Pi^2 + (3)^3 = 56.6$
$(\Pi^2+\Pi^2)(\Pi^2\Pi)3 / 3\Pi^2 + (3)^3 = 32.438$ which then is dominating 10Π as well as Π^3 where 10Π forms the outer space limit and Π^3 forms the inner star limit. At that point the outer space has no more influence on the star and the star will not emit any light.

Blue giant going dark $(\Pi^2+\Pi^2)(\Pi^2) = 194.81$

Neutron star $(\Pi^2+\Pi^2)(\Pi^2\Pi)3 / 10\,\Pi = 58.4455$

The pulsating star $(\Pi^2+\Pi^2)(\Pi^2\Pi)3 / (\Pi^3) = 59.21$ and space inside the inner sector of the star collapses the dimensional walls of the atom from the proton side. After this it becomes a new ball game all over again and in this class there are two separate classes.

After witch only the proton action of $(\Pi^2+\Pi^2) = 19.73$ remains

Black hole producing the ultimate relevancy $(\Pi^3) = 1$ or $(1836.12)^3 = 6190178657.$

This spot indicates the existing of a theoretical star that is just a spot of no energy sitting in space to small to observe and to powerful to ignore.

The whole position came about from speeds not matching and the larger velocity overtaking the comets lower velocity at a time when the lower velocity was challenging the higher velocity for supremacy. The overheating was the result of the comet firstly not finding enough space to duplicate but secondly by overheating the comet has the chance to enhance its velocity and find a manner to beat the velocity Jupiter excels on space by depleting space. With the magic force applying the comet would not pass Jupiter and would not later find a way to brake up. There would be no compromise with Jupiter having so much stringer gravity that that of the comet the comet with its own application of force would have participated in the collision. In summarizing thus far the following. Mass does not establish gravity. There is no magical graviton. There is no grabbing and there is no pulling of material on each other in any form. Mass does not inflict upon forming gravity but come as a result of motion differences in space secured by material a^3 in relation to space being secured by motion T^2 in relation to a specific centre k from where singularity holds the Universe true to form. Gravity produces mass but mass is only the result of gravity applying a single factor of the three that combines to form gravity. Mass does not produce gravity and the means to measure that mainstream science use is flawed not only by argument or concept but in the way science indicate by the formulas they then use in calculating the measure gravity's effect. The reason why the formulas work and they do work well is that on Earth only one of the three factors apply and then the application is in reverse of the cosmos space-time flow in general seeing the manner in the way we experience gravity. In outer space however there is a Universe of difference in the gravity applying there. In outer space there are three factors. One is the duplication of space by motion and the second is the dismissing of space by the lack of motion within the centre and the third is the balance between the two motions.

Gravity is about space concentrated to form heat, which is stored in motion that produces gravity. Any one not in agreement then convince every one by comparing the neutron star with the massive red giant and be convinced. Mainstream science came up with some cockeyed way of introducing the square of the speed of light to hide the failing of their explaining as to why and how stars evolve. To calculate a Black hole they go and throw C^2 next to the dividing radius and throw the square onto the C that presents the speed of light instead of keeping it at the original position of squaring the radius/diameter as it normally apply. We the mindless shall be baffled when they the Members of Mainstream science sit back and feel smart in the way they manage to once again baffle all of us and the "us" include everyone include everyone with such an exceptional weak mind hat will believe any and every story they can think up with no ability to screen fact from trash with their incorrectness as they are outsmarting us the brainless bunch into some form of accepting as those honourable members cheat once more to prove their incorrect views correct. After all who will ever fly down a Black hole and return to support or deny their calculations. When I suggest the Gravity of the Black hole is a speed because all gravity is speed and speed is space of one particular kind overcoming in relation to the motion of another specific kind versus the time it takes to bring in motion deference's . Gravity is a ratio between space that is displaced or dismissed and space duplicated. It is space over time. It is $k^0 = a^3 / T^2 k$ bringing about space divided by the time effecting the space. Then the speed that light has is gravity. The gravity of the light can be gravity as much as it at that very same time can be antigravity. What the hell has C^2 got to do with a Black hole because you can pop what ever nuclear device far away from a Black hole and the effect of such a nuclear blast would be at the most and at the worst very much insignificant. The motion the Black hole brings about destroys any other motion coming from the Black hole. The light produced by the nuclear explosion will not even escape form the gravity of the Black hole. When it became apparent to Mainstream science that the radius of stars reduces as the stars develop through progress, someone was supposed to say: hey there is a dead rat I smell. I for one have been on some mission about this matter for how many years now but for my saying so I am regarded as a mindless mutt being the clown in the courtyard having no friends and only foe.

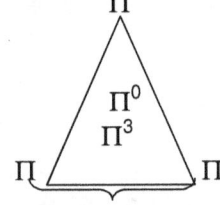

The reason why I changed Kepler's formula was that I established singularity change existing formulas to match, because I did detect the evidence of singularity within the formula. Although Kepler used different symbols the value of the symbols were dimensional alike. All came from Π^0. If everything came from Π^0 it should be going to Π. That places everything according to Π in one or the other side of the universe in accordance with Π. From that there has to be Π^3 which is the triangle

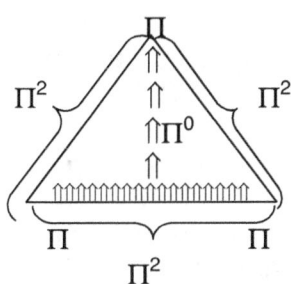

Where ever matter is it will hold $\Pi\Pi\Pi$ relative to $\Pi^2\Pi$ and that makes nonsense of Einstein's Flat Universe, (however I must correct myself) where Einstein wanted to place the universe that he saw to be _fl_at. Where the universe truly is (one or three or seven points away from singularity) within the atom and more correctly where the proton's Π^2 in the double links to time and meets at the edge of the Universe at $\Pi\Pi\Pi = \Pi^3$ that is within the core sector of densified space-time within every atom. At that point the universe goes flat, but that is only because the universe is in every atom and the protons move so fast..

In every atom hides a Universe that remains part of the original Universe that we apparently came from. That which we see is only extending, more a product of what came afterwards and definitely not representing that which was first. From singularity the space-time of which we are only a part is the contact there is between various points holding the original singularity governing the space-time in matching frequency in time. All that are is in singularity.

BACK THEN when the Universe was new

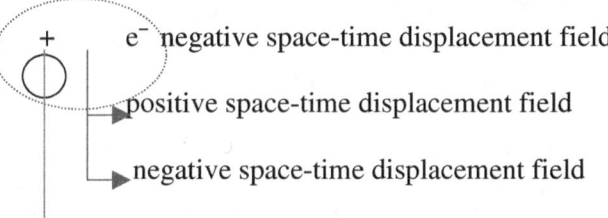

If space outside the atom grew to where we now see the Universe the space within the atom also grew substantially and if a star demolishes the space it has, such a star must then also reduce the atomic space because after all that is what the star is all about, dismissing of all space including the space surrounding as well as the space

PRESENTLY we refer to the sizes we find space has in the sun as quantum meaning they are inexplicably big

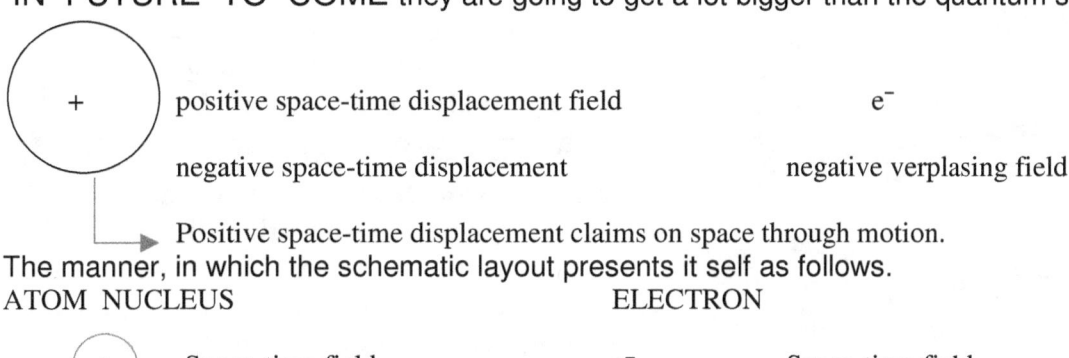

IN FUTURE TO COME they are going to get a lot bigger than the quantum size now present.

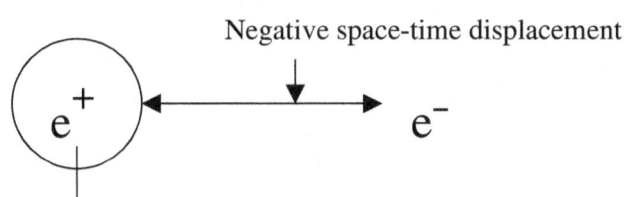

The manner, in which the schematic layout presents it self as follows.

ATOM NUCLEUS   ELECTRON

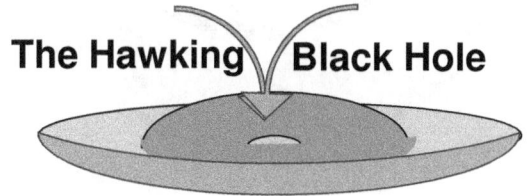

From singularity comes the motion and the space we call space-time. Singularity is dimensionless, time less and space less and because of all this features it carries the value of Π^0. By expanding does singularity apply a relation coming about that reform singularity from Π^0 to Π. Only when extending Π^0 to Π the extending creates motion and the motion creates space that then doubles through motion applying which cuts the space in motion in half by matching the space as a duplicate. Motion creates another dimension or another level reforming singularity from Π^0 to Π or from Π to Π^2 or from Π^2 to Π^3. The motion comes as a result of different motion claiming space within space in relation to individual positions they hold relating to singularity. If everything started off small it must include everything and not part of everything. How big was the atom as a unit when a star was the size of an atom. The relevancies apply all the way through and not just when science needs them to explain certain aspects gone array. Gravity to us in the way we experience gravity being on Earth and part of Earth yet also apart from Earth we find the connection we have called gravity to be $\Pi^3 \Rightarrow \Pi^2\Pi$. Explained it would read that it would be the space we are in Π^3 that we claim within the Earth is confined to the space the Earth claims Π^2 extending from outer space Π the centre Π^0. But the rules applying to the roles changed a little since Π^2 apply to the motion singularity use to retrieve space and Π indicates lateral shift.

It is not only outer space that grows because $k = a^3 / T^2$ is as much the cosmic value as the value within the atom. That means that $k = a^3 / T^2$ is also in place within the atom and that shows the space within the atom grows as the Universe grows because the atom represents the Universe that is in growth. As gravity brings space-time reduction from the centre of the proton so must the growth come from the centre to the atomic proton cluster. As the atom expands in space-time the proton can also grow dimensionally bigger through the neutron growing in stature. It expands in captured space –time by pushing the electron walls to allow the atom more space to occupy. It is pushing the electron to achieve a distance every time in the same manner that the body let nails and hair grow.

As the atom expand it pushes all space into expanding because it takes the heat from space where the heat in the space retains the growth of the space by allowing the distribution of space go in favour of space and against the density of heat losing relevance. By relieving space of heat the density of s[ace grow by comparable measure to the density heat loses in space therefore the Universe is growing in space by the measure the Universe is losing space

There are three factors of space-time where space-time is released cosmic unity and space and heat parted as singularity released the space heat holds by forming motion which produce time to set boundaries and relevancies applying.'

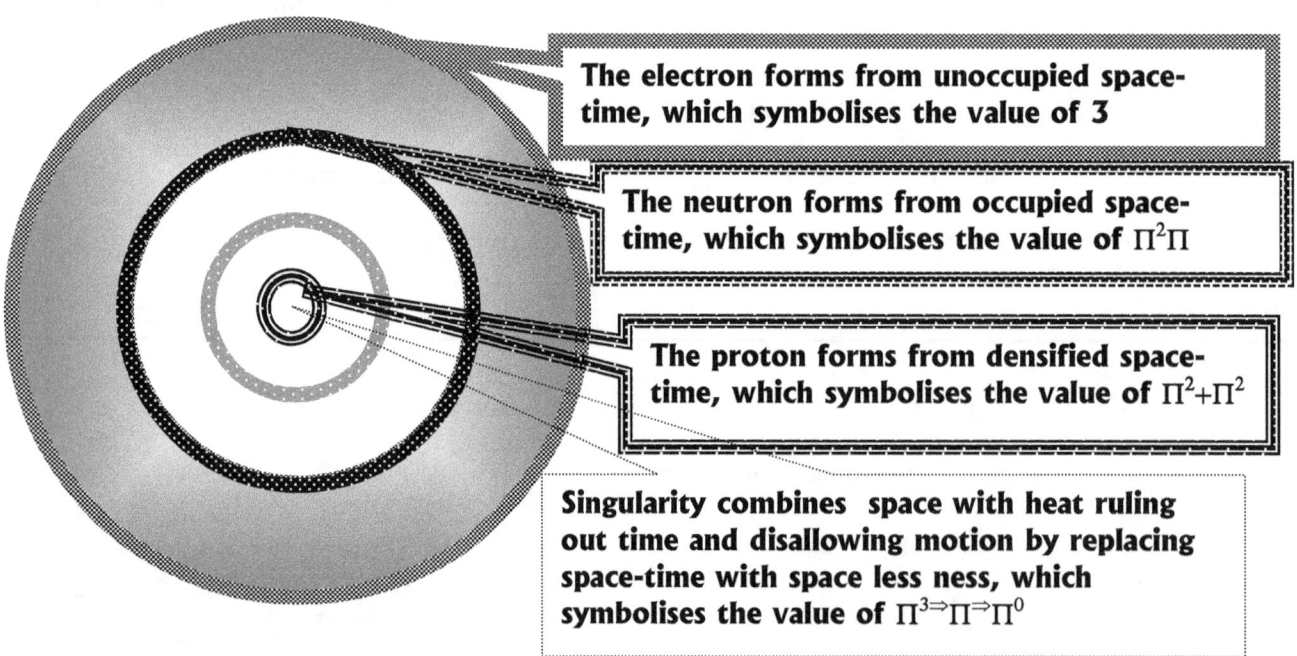

The electron forms from unoccupied space-time, which symbolises the value of 3

The neutron forms from occupied space-time, which symbolises the value of $\Pi^2\Pi$

The proton forms from densified space-time, which symbolises the value of $\Pi^2+\Pi^2$

Singularity combines space with heat ruling out time and disallowing motion by replacing space-time with space less ness, which symbolises the value of $\Pi^3 \Rightarrow \Pi \Rightarrow \Pi^0$

Unoccupied space: This forms the atomic relevance of **3,** which is where the ratio of space moves to the ratio of liquid in space. This is bringing motion in contact with **unoccupied space**.

Occupied space: Then forms the atomic relevance of $\Pi^2\Pi$, which is where the ratio of liquid space moves to the ratio of solid space. This is bringing motion in contact with **occupied space**.

Densified space That forms the atomic relevance of $\Pi^2+\Pi^2$, which is where the ratio of solid space moves to the ratio of **densified space** or motionless space. This is removing motion by disallowing contact with space and then forming space less ness.

Space less ness. That forms the atomic relevance of $\Pi^2+\Pi^2$, which is where the ratio of solid space moves to the ratio of densified or motionless space. This is confining motion in a position being part of eternity.

This forms the atomic relevance of $(\Pi^2+\Pi^2)(\Pi^2\Pi)3$

Every galactica grow individually in accordance to the singularity at what every position in relation to singularity influence in space-time. Stars accumulate heat from space and become dark space less giants. By duplicating the space of any particle sharing space within a larger cosmos structure such as an atom inside a star or a human inside the Earth there are two relations applying. At this point I must indicate that I entirely disagree with Stephen Hawking on the matter of a Black Hole being in the centre of a Galactica.

That cannot be possible since the galactica presents more mass than the entire galactica present with hidden matter included. This I say because the Black hole has gone the full circle and has developed back to singularity while all galactica is somewhere in the process. In the centre of the galactica there is the very opposite of the Black Hole. In that centre singularity is so much in charge it has not yet released space-time but is feeding on space-time form outside its realm of influence. To maintain form the singularity governing is removing space-time from the Universe and charging the yet to be developed space-time within the structure that is still in a state of pre Big Bang conditions. Since that is not the same as a developed Black Hole gone through the cycle material will escape in a very small amount since there is an out flow of space-time too as the displacement will require space duplication that causes ventilation. In **MATTER'S TIME IN SPACE: The Hypotheses ISBN 0-9584410 –6-5 there are 550 pages of explanations about how the galactica formation is distributed.**

 Not only does space grow but stars holding space that holds atoms also grows in relation to the space that grows with the atoms growing and so does the inside of the atoms forming the growing stars. How else would space grow if there is no material pushing and converting unoccupied space-time into occupied space-time.?

Stars expand through singularity development claiming heat from space and developing matter in relation to the progress singularity has securing value in relation to the Alfa singularity.

The expanding of space is represented on all levels but along with the expanding is the diminishing of space as well and such diminishing shows relevance to growth by indicating reducing of space density and increasing of mass in truly massive stars.

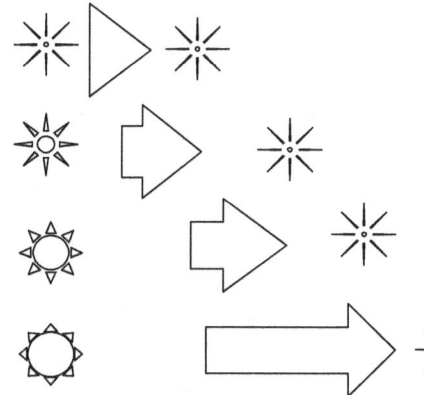 It is not the star that reduces or the star that grow or the space that expands but it is relevancies applying more tendencies in representing the relations there are between structures in space and structures and space..

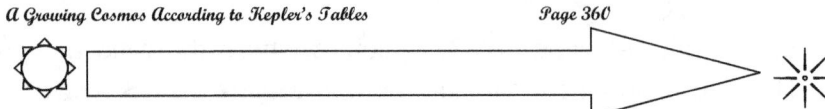

With outer space carrying the blackness in progressive multiplying, the very essence of space being space within the atom too must be in growth claiming more space. Of all the above factors Mainstream science only acknowledge the growth of space in as Much as calling it the Hubble Constant. However that is not where the growth affects ends because it originates as much from any individual atom as it comes from Alfa singularity. Space does not expand because the space is only reducing the heat in density while producing density in space.

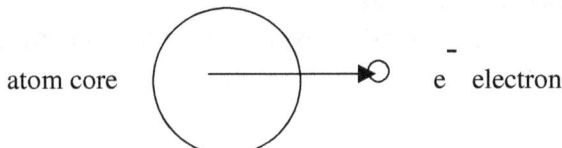

With matter growing the rate the growth would advance the growth in space must be at least $(\Pi^4/4)$ times the growth of the diameter matter holds to singularity.

ATOM NUCLEUS   SPACE GROWTH   ELECTRON

Space-time field e^- Space-time field
occupied space-time   unoccupied space-time

This side of the electron...and...   **That side of the electron**

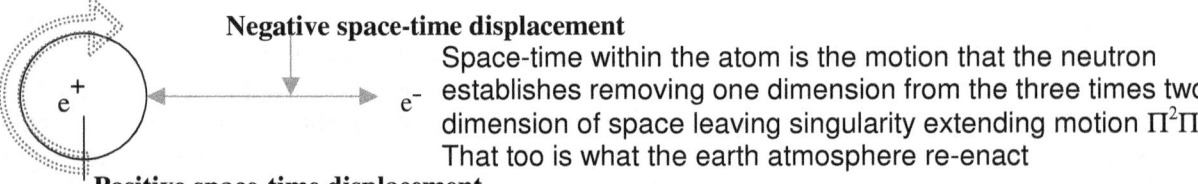

Negative space-time displacement

Space-time within the atom is the motion that the neutron establishes removing one dimension from the three times two dimension of space leaving singularity extending motion $\Pi^2\Pi$. That too is what the earth atmosphere re-enact

Positive space-time displacement

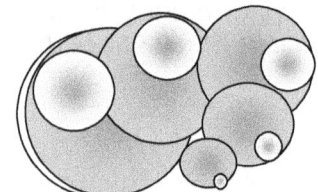

The more space the star converts to heat in relation to the most little space the star has to duplicate places untold compensating demands on the atom to release the space the atom claims and hold. The closer the motion of the star duplicates singularity in relation to the position of singularity the more the motion will find an opportunity to remove the occupied space within the atom as it converts such motion responding as mass increasing

In the first factor of singularity a line indicate direction and hydrogen becomes a volatile product. But singularity also provides space $\mathbf{a}^3$ and duplicate by motion $\mathbf{k}$ to destroy by rotating $\mathbf{T}^2$.

In dismissing the space the proton grow by accumulating space that the other side lost. The growth depicting of the dot I use to symbolize the protons growth are highly exaggerated I have to admit, but that is only to bring across the idea I wish to convey. Looking at carbon$_6$ one would think that proton numbers would bring about mass, which we then associate with density between particles. But then comes Nitrogen with seven proton pairs and oxygen with eight proton pairs, Fluorine nine and Neon having ten. These mentioned are significantly highly volatile which means they truly extend duplicating of space.

This as a group forms a relation with heat unlike any other. If mass did the trick these must have been the group having the second least density, but they form the group with the least density as a five point group. The next group holding the Pythagoras five or the Lagrangian five plus two or three, four or five enabling their relation with heat to be quite remarkable. This must be some indication of events during the period just preceding the Big Bang at say 10^{-7}, 10^{-6}, 10^{-5}, the time when the fuel that would ignite the Big Bang turning heat into space turned material into heat. The relation of five plus on, plus on and five plus one plus one plus one are just too uncanny to ignore.

What the Coanda effect proves is that the rotating motion is acclimating a centre that

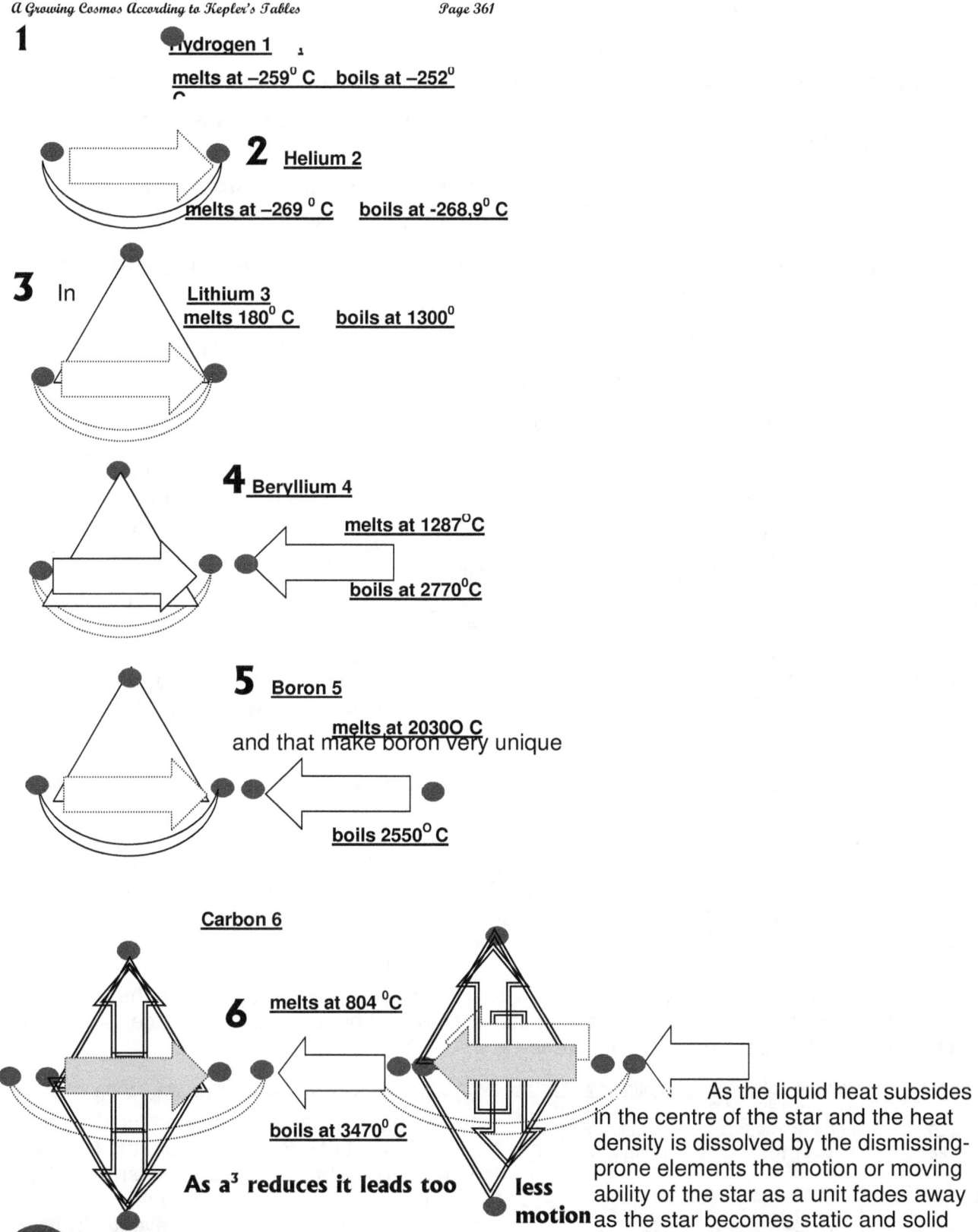

1   Hydrogen 1

melts at –259^0 C   boils at –252^0 C

2 Helium 2

melts at –269 ^{0} C   boils at -268,9^0 C

3 In   Lithium 3
melts 180^0 C   boils at 1300^0

4 Beryllium 4

melts at 1287^{0}C

boils at 2770^{0}C

5 Boron 5

melts at 2030O C

and that make boron very unique

boils 2550^0 C

Carbon 6

6 melts at 804 ^{0}C

boils at 3470^0 C

As a^3 reduces it leads too **less motion**

As the liquid heat subsides in the centre of the star and the heat density is dissolved by the dismissing-prone elements the motion or moving ability of the star as a unit fades away as the star becomes static and solid with less space providing the star with less motion

Since the star is the total configuration of atoms characteristics the atoms will tell us what we should know about every layer from what is applying in such a layer too what characteristics such a layer would show when it provides the function of what it has to for fill within the star.

In all it is not mass contributing to gravity but gravity establishing mass. Mass has no influence on gravity but mass the creation of gravity.

Π ●

Hydrogen depends on interacting as duplicating

Π ←→ Π **Helium 2 depends on some duplicating and some dismissing**

Π △ Π **Lithium 3 depends on more duplicating than dismissing**

Π● ⬡ ●Π **Beryllium 4 depends much more on dismissing than duplicating**

Boron 5 depends much more on dismissing and very little on duplicating

Carbon 6 depends as much on dismissing as it depends on duplicating making carbon most unique.

Nitrogen 7	melts at -210°C	boils at –195.8° C
Oxygen 8	melts at –218.8 °C	boils at -183° C
Fluorine 9	melts at –219.6° C	boils at –188.2° C
Neon 10	melts at –248.59° C	boils at –246° C
Sodium 11	melts at 97.85° C	boils at 892° C
Magnesium12	melts at 650° C	boils at 1107°
Aluminum13	melts at 660° C	boils at 2450°
Silicon14	melts at 1412° C	boils at 2680° C
Phosphorus 15	melts at 44.25° C	boils at 280° C
Sulphur 16	melts 119° C	boils at 444.6C
Chlorine17	melts at –101	boils at –34.7 C
Argon 18	melts at –189.4° C	boils at –185.8° C
Potassium 19	melts at 63.2° C	boils at 760° C
Calcium 20	melts at 838° C	boils at 1440° C

One will find that whatever group one choose there are gasses and there are solids. If mass was attracting mass then the strongest mass must be attracted the strongest mass and the least mass must float in the air. F = G (M.m) r^2 hardly can even begin to explain the fact that there is a gas that is more massive than iron but floats in the breeze just as hydrogen which is the least massive element.

Excluding Argon, which is six (carbon's number) times two and suddenly that is a less dens material. The four times five plus… group are the following:

Scandium 21	melts at - 157° C	boils at -152° C
Titanium 22	melts at 1670° C	boils at 3260° C
Vanadium 23	melts at 1902° C	boils at 3400° C
Chromium 24	melts at 1857° C	boils at 2665° C
Manganese 25	melts at 1244° C	boils at 2150° C

The ignoring of these facts by Mainstream science will hardly answer the problem we do not understand and such ignoring brings strong doubts about the quality and sincerity of science.

Iron being the five times five plus one is the only generator of electricity and therefore the producer of gravity making five times five plus one the ultimate relevancy to heat in reducing space. Still Krypton is much more massive and turns out to be a gas.

Krypton 36	melts at 1539° C	boils at 2730° C
Iron26	melts at 1536.5° C	boils at 3000° C
Cobalt 27	melts at 1495° C	boils at 2900° C
Nickel 28	melts at 1453° C	boils at 2730° C
Palladium 46	melts at 1552° C	boils at 3980° C
Silver 47	melts at 1412° C	boils at 2680° C
Cadmium 48	melts at 321.03° C	boils at 765° C
Xenon 54	melts at –111.79° C	boils a-108° C

How can science promote their image of establishing honesty when they are confronted by such truths but choose to ignore the truth so long as a lie will bring them some respectability. Their ignoring of me might extend the disinformation for some time but eventually be forgotten when they will go down as the liars that was hiding the truth. It turns out that mass has nothing to do with creating gravity

Following the process and seeing the influence of singularity should bring about a pattern that may lead one to a pattern of how the required heat formed and how the intended heat transformed to space. Density depends more on proton number arrangement producing specific form in relevancy as to merely and only having mass as factor that contributes to the forming and development of stars in the cosmos. The evidence is so clear that mass has nothing to do with gravity but density has everything to do with gravity. Density is the volume of space in numbers used to fill material in ratio with numbers of space per volume not filled with space. It is matter versus space in every sense there are.

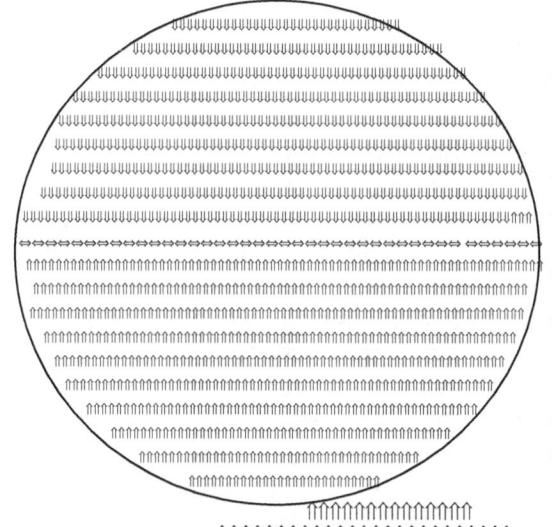

Every atom holds (I am guessing) as many dots as the sun has subatomic particles per atoms and that would still be a very conservative guess. Every dot is a controlling centre selecting a regional centre where every regional centre selects a centre. This goes on as long as there are spots forming groups as individuals unable to survive independent. The others that was unable to group formed heat that became space, which became the broken dots. The dots form groups to survive and as a group the survival depends on doing what the group has to do to remain cool. In another book I reserve one chapter to explain the phenomenon what I called the Lagrangian atom. These dots arrange in a manner that they could either favour the space duplicating aspect or the space dismissing aspect.

The Earth is mainly about duplication of space much more than dismissing of space and so is every structure in the solar system

This can only be the result of the fact that even in the case of the sun the centre is almost entirely liquid heat and the liquid heat produce sufficient space to dismiss by the centre that holds the heavy metal particles which is doing all the dismissing. The liquidity provides motion while the solidity removes motion in the centre of the star. The dismissing going on is in the space factor where the space leads to a denser heat within that space because there are insufficient material to accommodate all the heat by the dismissing factor T^2. In that case motion far outweighs dismissing $k>T^2$ but a time comes in every star that the dismissing takes absolute charge. $k<T^2$ That is when the star goes dark

Let us investigate and try and find a way by using logic how a star applies gravity. Therefore it is not the number of dots that is important. It is not the size of the number of dots occupying the position or the size of the space the dots occupy that is prominent. It is the relation in the dismissing of space and the duplicating of space that becomes important. The less space there is the more the favour will be to reduce the space because of the advantage the dots have in securing space-time that will prevent overheating. On the other hand the more space secured will also prevent overheating and therefore those will opt to duplicate space in order to find space to secure and prevent overheating.

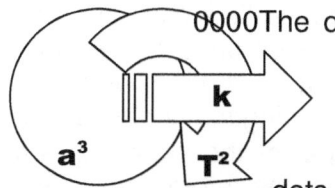

0000 The dot does not spin therefore the dot is unmovable. The dots immovability renders them the opportunity to destroy the space by incorporating the space into the dot. The series of dots flow in a direction around a democratically elected centre that means the dots are moving as a group. The group motion becomes the space duplication by motion. But if the dots remain in one location, then the space they accumulated react and the space they accumulated shift on spot on because the space they accumulated and use to return the previous standard plus one in space-time relocate each relative position in relation to the centre one spot. The motion factor T^2 has singularity rotate the space in relation to the motion of time whereas k has singularity rotate in relation to a select centre and the selected centres selected a centre that provide the group of selected centres a rotating centre as well as a rotating direction in which the group of selected centres will move in a line. The space created by the motion spins as it rotates but then when the dot secures the space by rendering the space secured through the immovability of the dot, the dot changes the relevance it had in the group and as part of the group the dot spins. This then is **k.** The factor **k** confirms the duplication of space when the factor $\mathbf{T}^2$ confirms the space as dismissed and when the inter-acting takes place it confirms space-time through motion duplicating or motion to dismiss space-time. There is forever a balance and this balance we call the Universe. The containing structure represents in a duplication relation applied within the boundaries of the star and in that it applies a norm set on pre-determent conditions that is as old as space-time self applied by the governing singularity that space is supporting and maintaining. That will demand the mass brought about by all that the space contains. This relevance means that without a specified container (star forming boundaries) the space producing specific duplication and destroying of space in the area known as outer space, the outer space container will apply a diminishing relevancy of space-time displacing that which can support a maximum number of 112 protons working as a unit and in conjunction dismissing space can withstand. We know that is a theory because the atoms in space can sustain much less than 112. Inside containers being stars the direct relevancy of singularity applying puts much more strain on the surviving abilities of atoms. In outer space the atom has an own relevancy of seven and the space demand on the atom is only three that it must maintain in order to duplicate. But in stars the containing star places a demand of the containing seven plus the space creating three in relation with the time applying inside a star, which are four.

Since the Earth has no singularity demand that is much better developed than the universe sustains we find on Earth a relevancy of Π to $(\Pi^2+\Pi^2)(\Pi^2\Pi)3$ is adequate. But in bigger units the space-time displacing relating to space duplication presents much more demand on atomic structures occupying space within the star containing through set boundaries. In the presumed to be bigger stars there is much space filled with atoms occupying much space. In the stars more massive but holding lesser space the atoms must also hold lesser space but they also hold more protons by number in the lesser space. The space the particles hold is directly in relation to the particles the containing structure duplicate. The more space that is relevant to the structure that the star duplicate by motion is then in turn once again relevant to the space the structure destroys by proton action in space less units. The more space the particle claims in relation to the space the container hold that relates to the space the container duplicate is relative to the space the containing structure destroy. From that mass derives value. As individual occupying space the atom is an individual container by own merits and as such duplicate space in this regard within the specific confinements of atoms.

The atom restricts dismissing of space by the containing structure to the atoms relevancy being Π^0 in singularity bringing on Π relating to $(\Pi^2+\Pi^2)(\Pi^2\Pi)3$. This we will classify as normal applying structure values the atom has in outer space or in structures with very little atmosphere. Please note there is no pressure involved because the motion involved creates conditions naturally instead of unnatural pumping that causes pressure. Pressure is an artificial creation as part of life but has no role in the natural cosmos. Pressure is a condition where the retaining of particles has to be confined in a patrician made of material where the outer wall does the retaining of the substance within. This obviously cannot be in a star because the "pressure" is regulated from a condition applying and space-time controlling inner centre that needs no solid walls to contain whatever is inside. With that one can see there is a Universal difference between the concept of pressure forming due to human action inside a container and what comes about as secluded space-time within a star. As the demand of singularity in such units grow stronger some relevancies within the atom come into play and I developed a system whereby I can arrange the space-time merits of space-time curtailing within the confinement of the star borders applying in the star to place such a demand in relation to singularity where the ultimate demand sets the standards. In the sun for instance which is a minuscule small star a relevancy in the outer region might be 3^3 relating to singularity and with the atom having a sustaining displacement of $(\Pi^2+\Pi^2)(\Pi^2\Pi)3$ there is no danger of the atom demising. The electron in the sun will have a diminishing factor of 27 whereas the atom can sustain $(\Pi^2+\Pi^2)(\Pi^2\Pi)3 = 1836$. The relation in the atom degenerated by 27 leaving the atom a sustaining value of the electron plus the neutron applying space-time without involving any of the neutron aspect at all. That is the mass the electron will consume in the space reducing and producing mass within the star that then forms a favouring of duplication. The star is a bright little boy shining by dismissing pebbles of light-photons into space. When a demand on space-time displacement reach an accumulated general displacing to the value of 56.6 protons of the general; accumulated displacement within that space forming time. The star accumulated more heat by consumption applying direct dismissing without accumulating space-time in liquid form beforehand therefore there is no heat remaining to dispose of by producing light. When the general displacing flow of space-time within that sector of the star or the star in total reached 56.6 g/mol. absolute solidifying becomes the norm within the star as the star will exclude all electron functions and stop shining as the demand on space-time duplication and diminishing reduced the atom to space without a heat envelope that will be electrons or a liquid/gas jacket. Only the nucleus will be able to sustain the diminishing and the reducing of space by increasing of time. The entire star becomes a solid structure by reducing space-time directly freezing the space-time from a gaseous state to a solid state. By motion speeding up the tempo of the flow of space-time the liquid state of space-time is by passed going from gas to solidity in one motion. The atom would shrink to such little space it will have space within the star that only the centre nucleus will fit. More reducing by applying motion in creating space differentiation will leave a star with so little space the space will be insufficient to secure a position for the neutrons and the star will then have the name of being a neutron star. Going even further will find the proton rejected from the star.

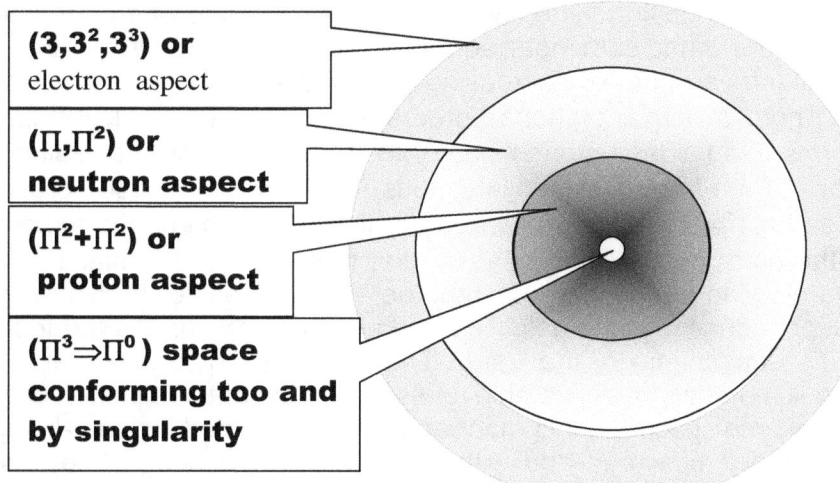

$(3,3^2,3^3)$ or electron aspect

(Π,Π^2) or neutron aspect

$(\Pi^2+\Pi^2)$ or proton aspect

$(\Pi^3 \Rightarrow \Pi^0)$ space conforming too and by singularity

Since the star performs as an accumulated atom where innumerable atoms inside the confinement of the star combine to select one centre spot forming singularity that represent the star, I have chosen the to use the same symbols that I found in atoms to describe the relations in space –time to singularity within the space-time of the star. I refer to a star as a cosmic atom in other books.

$(3,3^2,3^3)$ Favours duplicating of space
The star is duplicating space as in the case of Jupiter, and the other Micro stars in the solar system

(Π,Π^2) Favours liquefying space
The star produces light by turning space to liquid. The liquefying is when the star removes in ratio space as much as duplicating space

$(\Pi^2+\Pi^2)$ Favours dismissing of space
The star is a member Mainstream Science at present considers to have died. It does very little favouring of space reproducing or liquefying of space but is purely about dismissing space.

$(\Pi^3 \Rightarrow \Pi^0)$ Slowed down to no motion coming about that explodes in a burst of light and then again with no motion no time or space comes into space-time. Total dismissing and annihilating motion producing any and al possible space due to immovability or motionlessness. This star is almost up to very much where the cosmos began long before the Big Bang began

Early stars still in the envelope of heat within the centre of the Galactica have only space duplication and growth through the cover of such enormous heat. These class stars are not visible but are shrouded in a blanket of heat covered by light. The atoms forming the stars are small and under developed. They remain cool because they contrast with the heat surrounding the star where the star material support the cool space and does no form part of the liquid heat forming the outer limit. As I indicated previously but I would like to draw your attention once again to the fact that the sun at one stage was a cool 18×10^{6} ^{0} on the inside and freezing cold at 6500 ^{0} on the outside while outer space was a blistering 10^{34} ^{0} which was considered the coldest place in the Universe because the sun was still part of the deep frozen space. Look at any galactica and see in the centre there are stars surrounded by a blanket of heat with stars conversed by heat sitting like a duck frozen in this pond of liquid heat. The inner singularity sets a border firm enough to prevent the heat from liquefying the star and only when the control of the general governing singularity will apply the same heat as that of the barrier will the barrier shift where the star would then accumulate and concentrate space-time to form an atmospheric border. the protons grow as a result of the fact that more space within the proton is duplicated that that which the proton can dismiss. In such a case the heat surrounding the star is hotter than the core of the star. We know that the sun and the equivalent to the sun has an inside temperature of 18×10^6 degrees where every one at present think of it as extremely hot. But a while ago that inside was a deep freeze because the outside outer space area was a wild 10^{34} on a mild day. At that stage the sun was just a covered spot surrounded by heat from which the sun could receive heat and turn that to space converted to material. The sun developed from 3 to 3^2 to 3^3 and then when the outer edge where the sun meets the end of outer space the sun started turning space into liquid. Then the sun went from 3 to Π by liquefying the atmosphere. The earth at present has reached the stage where the atmosphere surrounding the Earth as Π extending is beginning to become intense liquid relevant to the other side of the 100 km border. However the moon has still to reach this stage.

As the star develops it moves away from the centre and away from the heat blanket covering the star. By concentrating heat at the centre as it is removing space in the centre the singularity driving the star comes into a dual with the singularity driving the galactica. This is the stage where the spinning top example proves the point. As the inner core heat surges the outer edge forming the newly established atmosphere excel in temperature by increasing duplication through applied motion around the Π^0 singularity from where a border (various borders in fact) establish Π and limit the interaction by Π^2. The border forms three sides on each side of the divide allowing a crossing of the border Π under specific conditions/ Then from there too is various Π borders and those borders end where the solid structure of the earth $\Pi^2 + \Pi^2$ forms another border to bring to an end the atmosphere at a value of Π^2. All the while the outer space liquid heat blanket that covered the star since time began slowly move aside as the growing atmosphere pushes the envelope aside. In this manner a star is born. The star seeks independence and the galactica demands control, just as the top acted when spinning severely within the atmosphere of the Earth. The sun is all liquid inside and in the process of turning space gas into liquid heat the sun is able to discard much of the liquefied space as light photons throwing the light back into space. In this manner the heat collected inside will apply as a separation to push the star to move outwards and with that the star will bring about stronger independence and relieve of the galactica domination. In the stage of development, which I now am referring too the star transforms from 3 to Π and Π to Π^2. At Π^2 the star begins to turn the flow of light from flowing towards outer space returning the light as space dismissing towards the inside. It goes about in a balance of cold inside the star centre and heat fuming in outer space. One must remember that not withstanding culture the outer space is gas being much hotter that on the inside of the star where material freezes away to singularity In **"STARSTUFFIN"** **ISBN 0-9584410-3-0** I explain this in much detail. As the star progress the star starts to dismiss space more than the star liquefies space and this starts a cycle of light flashes and star seemingly disappearing. These stars are called by many names but are the Pulsating variety of variation.

That is how gravity applies because it is a matter of relevancies applying between space holding and demanding conditions and space reducing in relation to insufficient motion bringing about much less space duplicated and space demised. The space duplicated brings about mass as a result.

Material comes into place when singularity overheats that starts the process of Π^0 going onto Π and such overheating creates motion Π going onto Π^2 is about motion coming about and forming border relevancies that becomes time as well as space and by introducing this motion singularity applies the value Kepler stated $\mathbf{a^3 = T^2\, k}$ but which I changed somewhat to apply in terms of singularity as $\Pi^3 \Rightarrow \Pi^0 \Pi^2$. While singularity is splitting the Universe it is dividing the Universe in a joining effort by forming a dividing unity. On the one side is $\Pi^3 \Rightarrow \Pi^0 \Pi^2$ and on the other side is $\Pi^3 \Rightarrow \Pi^0 \Pi^2$ also and it might seem as if Π^0 is a joining value and the truth is that it is a parting value although shared by both sides equally because it runs into infinity and out of infinity on the other side thus providing the Universe with two equal sectors being the same.

Space is created by motion $\Pi^0 \Pi^3 \Rightarrow \Pi^0 \Pi^2$ that duplicates the material expanding it into four $\Pi^0 \Pi^3$ sectors whereby the motion created cuts the space in two halves $\Pi^0 \Pi^2$ that divides the space ($\Pi^0 \Pi^2 +$

$\Pi^0\Pi^2$). The protons causes motion ($\Pi^0\Pi^2 + \Pi^0\Pi^2$) and by creating space the proton find the ability to dismiss space $\Pi^0\Pi^2$ because it renders material from Π^2 to Π^0 by diverting Π to Π^0. The proton holds a value of $\Pi^0\Pi^3 \Rightarrow \Pi^3 \Rightarrow \Pi^{-1}\Pi^2$ because the gravity Π^2 is always reducing and therefore is diminishing Π^3 by placing **k^{-1}** in a negative relation as the proton then returns to what singularity started from Π^0. As the process places the Universe in a reverse by providing growth such providing of growth stimulates singularity and prevent overheating.   This motion duplicating as much as reducing takes place on both sides of the Universe in the space as well as the motion part and therefore the proton holds a dual value of ($\Pi^0\Pi^2 + \Pi^0\Pi^2$).

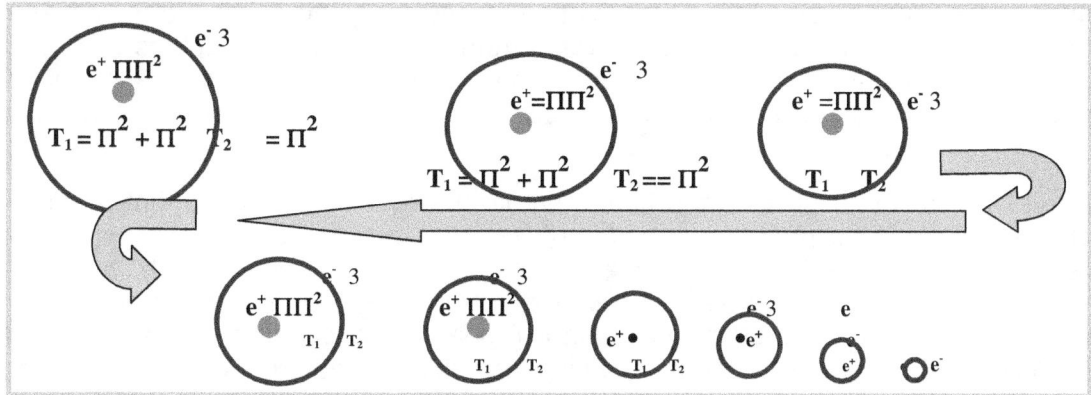

e^-	= 3
e^+	= Π
T_2	= Π^2
T_1	= $\Pi^2 + \Pi^2$

As the star diminish space within the star as claimed by the star the atoms shows a reluctance to submit to the demand. But as motion differentiation bring about the decline so will the atom reduce the quantity factor and later the occupation factor until the star becomes a single atom with no neutrons.

The best way to explain the relation between 3, Π, Π^2, $\Pi^2 + \Pi^2$ and Π^3 must be in the confinement of the earths atmosphere using one of the best known weather phenomena, the way rain forms.

If my prediction is correct and gravity is heat being forming as space is compressed systematically, then electricity is heat compressed instantly. That means gravity and electricity is the same thing riding only on a differentiation brought about by the time compliance that the two systems have. In each dimension, the light displaces space-time that is more concentrated and therefore influences the projection path of the light's future position in space-time to unoccupied-, occupied-, and densified space-time. Science at this point is using magnetic fields carrying large electric charges to try and produce fusion. Their trying that involves electric charges is also a silent admitting of some agreeing of my statement.

Since Π^0 does not affect the relevancy therefore I do not normally include the use of singularity values of Π^0. When referring to the double proton value as ($\Pi^2 + \Pi^2$). Therefore I refer to the singularity value at the proton level as ($\Pi^2 + \Pi^2$) The motion discrepancy within the atom leads to space-time displacement but that has nothing to do with mass. Every proton has an equal ability to dispense of space-time by motion applied. Since the motion of the atom at the proton level is much more time consuming the space is that much less because the space stands directly relevant to the time. The contraction reduces the space and with the contraction the heat density rises. From this more motion reduces space but such reducing is on another dimensional plane where space has three sides, which is duplicated by motion once more. The duplication $(\Pi\Pi^2)3$ brings about the one six dimensional Universe we live in. With every plane using more space it affects the time considerably and the delaying of time is reversibly affecting the time duration. Singularity reducing space forms a unit and such a unit works on space reducing within what space captures and not on size particularly.

The orbiting of the electron

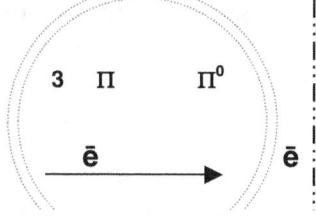

THE UNOCCUPIED SPACE-TIME   REPOSITIONS TO A NEW LOCATION, BRINGING ABOUT A RISE IN HEAT. COOLING WILL COME INTO EFFECT BY THE METHOD WHERE REPLACING SPACE HEAT WILL BRING   ABOUT LOWERRING TEMPERATURE BY CANNIBALISING OTHER PARTICLES CLAIM TO SPACE-TIME IN THE FORM OF

The same apply whether it is an atom construction and whether it is a star or a galactica the rules apply equally to all in the same manner because it drives by space in motion consuming space to render heat a more a denser value being more intensely concentrated in the distributing there of. As space is reduced at each level the motion leans strongly towards diverting motion to the lesser space and in such action we find the motion we named gravity. Gravity is the result of time destroying space by creating motion thus cutting the growth in space by half at singularity level. But in the process it is also filling the vacant space with heat that moved inwards from the outside. As much as outer space is creating a relevancy seemingly in favour of space becoming more that same relevancy tends to reduce the space unoccupied space-time by claimed occupied space-time in the better-developed stars.

The less space available in whatever area applying the more the reducing of space that is captured will affect the space that motion captures and motion of reducing such space. By applying this rule material captured in a larger space will always draw towards the centre of such a much larger space where it then will have to match the value of the space surrounding it as the dominant space by reforming it holds because the captured space cannot repeat the expanding and even less the duplicating of such space. By the specific density coming from this the material will reduce to a point in the space it claims as an individual structure where such duplication will be in harmony with the duplication of rest of the space surrounding it. This tendency of smaller sized object too move towards the centre of a structure as to put the motion and duplication of space in harmony with the rest of the space we call gravity and from that a relation comes about where the structure captures in the larger space fight to protect and sustain individuality by securing its structure and form. That is what time is. We find humans being as confused as humans are, humans confuse eventualities with time. Time is not a flow of events following the will and the wishes of persons in life in a complex by chance progress from which we can report about a history written in blood. Going back in time will not bring Napoleon back to the future because the events played their part while time was flowing and the events can and will never happen again. What life confuses time with is again putting life in the pinnacle of time and history forming the focus of the cosmos' entire reason to exist. Time and history humans see as bridging the time that it will take on Earth in comparison to the time it will take to travel from here to there where there is about a billion light years away. That is nonsense because if that was possible the traveller would first have to return to time just after the Big Bang when space was that small, but remain in the time frame that traveller now has to bridge the distance by short cutting the time. That is the way we think of human eventuality by becoming my father's farther as I shift time to and fro. My travelling through space is projecting life's eventualities through space reduced by time being in some ridiculous short cut. What we in life confuse as the centre of the Universe is the position the Universe grant each of us as being in the centre of the Universe, which we are. But each one is in the centre of each persons' particular Universe matching that person.

My being centre underlines my eventuality I position as paramount but that is eventualities and even if there was any possibility (which there is not the slightest chance of) any return to such an event will most defiantly not repeat in the way it transpired on the previous occasion because some idiot will have some other idea in his head at the time and that will change his behaviour which will change the outcome of the future being much different as it was in the past. Eventualities run on human decisions and not the driving of the cosmos. Any going back will most probably lead to even more confusion as before where the human spirit in charge of messing everything up will do its purpose even better and we will si in an even stickier situation as before. That is human mind becoming events but without the human mind no rock will run into a car as soon as a woman driver pass the rock. Cosmic time is the way the distance from the centre singularity establishes space holding the time relevant to such space being established but the flow of time in relation to the space being available from the centre. $a^3 = T^2 k$ and you change k to find everything else in the equation changing. The space traveller will have to pass through the centre of the sun in order to escape from the centre of the sun because it is the centre of the sun keeping such a traveller relative to the centre of the sun. The motion the traveller has to project will be quicker in pace than the centre gravity the sun produce at the very, very centre of the sun. Cosmic time is the relation the object in motion has while being in motion with all aspects of all motion applying being relative and that includes the heat to space ratio determined by the motion forming interruptions in relation to all other factors sharing space-time. This motion influencing time is because all gravity in the solar system is motion in ratio the motion coming about and from the very centre of the sun where all motion is standing still. Time is motion on the spot relating to the rest of the Universe but under the command of the motion less ness of the strongest centre spot killing motion. If one whished to travel beyond the sun one first have to break the barrier of the centre of the sun where the sun stops all motion and when that is overcome the traveller would have a vehicle stronger than the centre of the sun

By accepting the Big Bang theory and acknowledging Hubble's evidence science has to have a look at the way they cling onto Newtonian views. There is no chance that by purely having the diameter one can calculates and determines the gravity of the star. If that was the case then how can science ever try to explain the pulsar or the Neutron star? Even by bullshitting in the face of contradicting evidence there is no manner in which to explain any star by birth or by death because a star cannot be born in the manner science try to prove as much as a star cannot die in the manner science promote. I truly cannot see how science can stick to the notion of the speed of light being a constant when all evidence prove the very opposite. How can the Black hole break the constant of the speed of light because then the sky must fall? Before the light tries to escape the motion the Black hole introduce the light goes into reverse by falling into the Black hole and to come to that reverse the light will first arrive at a point of being very motionless and therefore the light will lose the space it holds in that specific position in that very moment. The light will go slower and slower until it comes to a standstill from where it turns in direction and go into a reverse as it then accelerate faster than the speed of light down the gravity funnel towards the centre of the Black hole. All the time the light holds an even relevancy but since the relevancy overpowers the light speed of $3\Pi^2\Pi^0$ which is the ultimate antigravity possible that relevancy can change into 330 km / h or it can accept the factor of one as it rushes down the gravity funnel of the Black hole. When the factor is one it will show no individual motion and therefore relinquish all the claim the photons had on space. Light will go dark. But during this all the relevancy factor remains $3\Pi^2\Pi^0$ until the stronger Π^3 relinquish the space and the time by rendering the light motion no relevancy. Wherever light flow and wherever singularity apply Π^3 light will hold the relative motion of $3\Pi^2\Pi^0$ in relation to singularity being a proportional factor to all changes in space-time that may affect space-time. A constant derives from a law and the cosmos does not break laws. The atom fills the space inside stars and the atom must be the space that the star ultimate will deplete. The speed of light is just another part of atomic space because science are aware that once in the beginning of the Big Bang the outer space comprised of a density stronger than what the electron was. The complexity of the universe rides on the balance that singularity introduced and the entire Universe invested in. In the centre is singularity standing still. That will remove all space because there is not motion that produces space. But the standing still of singularity produce space through expansion by overheating and the overheating constitutes of motion where only motion will provide the required cooling. But it is both the occupied space holding the seven relevancies to singularity that expand and as that expand the ten relevancies holding occupied space within unoccupied space that produces the other expanding factor.

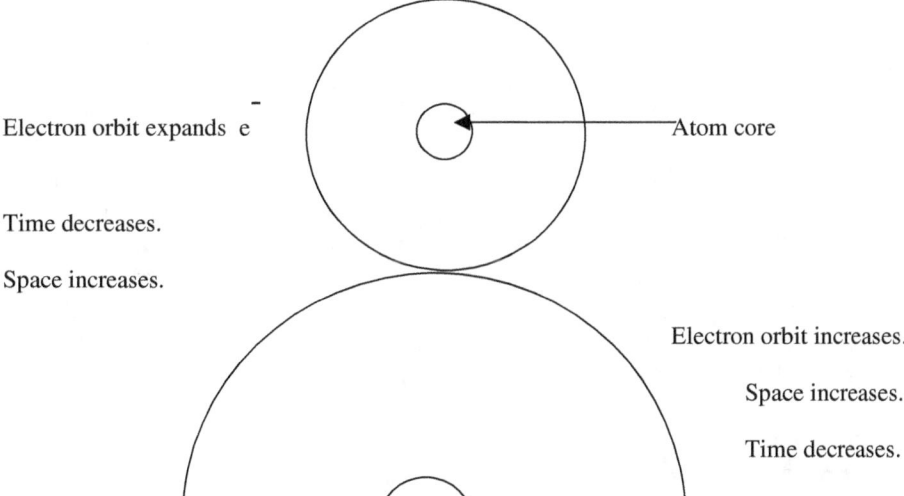

Electron orbit expands e⁻

Time decreases.

Space increases.

Atom core

Electron orbit increases.

Space increases.

Time decreases.

electron ē   Core   Electron moves

faster.

The reality about this diagram is that every aspect of the cosmos relates to this diagram.
Since the Big Bang the atom increased because all the relevant space in the Universe increased as
the balance shifted. With that being the case the reverse has to apply as the star is about contracting
and reducing space by motion producing gravity.

Red giant ⟹ 10(
100kg

Betelguese
Dia.1400000000 km
Diameter relevancy is
35.2 km.
$(\Pi^2+\Pi^2)(\Pi^2\Pi)3$

From what I gather the red Giant is the equivalent of a gas
structure such as Jupiter only containing a lot more space filled
fluids. Strains placed on the atom in the Red star, which is just a
liquid bowl of contained heat, separating atoms will be precious
little. There will be many protons but the protons will be spaced
individually where each proton claims individual space securing
individual survival since the space dismissed is relatively little
and the space duplicated are relatively very much favouring the
duplication and not the destruction of the space.

A **100 kg** in outer space will register as a **100 kg** in the star
because the atom will have to release very little space

Yellow dwarf
Sun
Dia.1400000 km.
Diameter
relevancy is 38
$(\Pi^2+\Pi^2)(\Pi^2)\Rightarrow 3^3$
⟹

100 kg on earth ⟹ **becomes 1000 kg in the sun**

White dwarf

Dia.16000 km
Diameter
relevancy is 300

In this star the core has developed considerably concentrating heat to
the inner regions. Within the inner regions the heat concentrating will
sustain the proton in a manner it will show much more growth within
the denser heat surroundings and as the proton grows in stature the
proton finds a stronger ability to dissolve space and with that ability it
can therefore dismiss more space with the increase in mass as a
result. The more the mass the more it produces more massive protons
that can displace more space-time In this star the favouring shifts to
apply equal to all ends of gravity..

longer allow the flow of light on the inside while
on the outside the star is one unbroken solid electron and the protons
are diminishing that space also at a rate. To find a compromise as both
ends of the star holds both ends ultimately the compromise will be to
favour the light by the producing of light. Then it will favour the
diminishing by contracting all light to the inside in the direction of the
core. In this way both ends find equality at both ends of the duplication
and the dismissing of space- time

100 kg $\Rightarrow$ 10³ tons

$$(\Pi^2+\Pi^2)(\Pi^2)3^3$$

From this point the star will become darker as the protons grow more massive finding the ability to demolish even space holding heat so concentrated the heat is pure electrons and photons filling space as material would. It is discarding the electron from the atom. The more massive protons must create much better conditions for fusion.

Neutron Star

$$(\Pi^2+\Pi^2)\Rightarrow(\Pi^2)$$
Neutron star
Dia.**19.2 km.**
Diameter
relevancy 3mm

100kg $\Rightarrow$ 10⁹ tons

$$(\Pi^3)\Rightarrow(\Pi^2+\Pi^2)$$

In the neutron star the space dismissing reached a level where no heat as such is available any more because conditions inside that star has gone _to_ pre Big Bang conditions. The space applying is not sufficient to separate atoms any longer and the reducing of space has limited the motion of space to almost a standstill. There are two forms of motion being very apart. The heating allows the neutrons the ability that photons had before the atoms dismissing space through motion decreasing and in motion the neutron can expand the space by duplicating the space into which it expands. In that manner it finds a way to escape much like the spacecraft does when launched from Earth.

By reducing the atomic motion the neutron has the ability of a liquid thus it remains in motion. It is that motion that the neutron then uses to distinguish the space it claims from the space the star pursuits. As the space-time reduce, the neutron surges outwards using the motion of being a liquid to sustain the space it claims and find the release it requires.   The protons however reduce all space, which may be inside the atom by discarding the neutron from the atom leaving only room for the protons housed in space less ness.

Black hole
Dia.**9.8 km**
Diameter
relevancy
1.5mm

Within the Black hole only singularity remains. It is the single dimension of space less ness where the singularity retained the single unit of all singularity it once contained. That is the epitome where the Universe is heading, which is the final goal that all stars securing singularity will finally reach and there is no unsuccessful candidates that will float in darkness in space forever and another day. The star has completed the journey and is now in the effort of demising outer space as an entirety.

100kg $\Rightarrow$ 10¹⁹ tons The mass increase is only an educated guess but it is most likely that a mass of **100kg** will become**10¹⁹ tons** in the **Black hole** but it is much more accurate than the official number presented as a calculated value by mainstream science.

The galactica and the star that ended as a Black Hole started off equally in mass but much different in mass distribution. The stars that eventually became the first Black Hole united the mass they had forming the controlling singularity, while the galactica distributed the mass they had in separate cluster packets we call proto stars.

The Black Hole and most galactica started off on the cosmic developing journey at the same time period release. Both had equal singularity generated, except for the way the material compliment produced the governing singularity by motion of duplication and proton cluster density.

Galactica Π^3 **Black Hole**

Π

A variation of cyclic intervals Π^0

In the galactica many units formed partitioned units became stars-to-come that provided atoms a wide range of units to form whereas in the Black Hole, the material was the same in total but the variation of space occupation was much less with much more protons in numbers per cluster unit. In the galactica the time factor gave a variety by which the different material could develop and form many equal governing singularity. There was a range of cyclic periods in which to develop and gave a range of singularity developing periods. This came from the density and the duplication in every star. Through the spin the atoms provided space-time that form the unit. The quality of spin would achieve a governing singularity. In the Black Hole the material that connected to the centre governing singularity was the same, but all atoms produced one unit and not stars in cluster. With the massive proton numbers that formed proton clusters a very high demand was placed on dismissing and little was on duplication whereas at the time the Black Hole was developing, the galactica was only duplicating and waiting for development by having the dismissing factor developed.

The cluster of stars in the galactica reach maturity in singularity development and by such maturity find release from the centre governing singularity, however it never strays out of cyclic developing order and remain part of a galactica governing singularity. In the case of the Black Hole there was one unity that formed one governing singularity where all atomic singularity remained attached to such a centre. The centre was strong enough from the start not to allow individual development in the unit and all material developed towards achieving one united singularity. The mass however connecting to the main singularity in the galactica is the same as that which connects to the singularity in the Black Hole. The Black Hole singularity made the mass redundant as it united all proton singularity into one unit. The galactica had the proton mass more widely spread and brought space to part units to form individual clusters with many different centres attaching to a centre singularity but with much less confining value.

The cluster we call stars move away from the centre singularity that keeps the galactica united. As the singularity governing the star finds mass development, it would motion away from the centre attachment of the galactica and progress to achieve independence as the spinning top tries to achieve. It is this developing law that we mimic in the top we spin. The Black Hole eventually destroyed all means to duplicate material and was set on dismissing space-time. In a book of mine called **"STARSTUFFIN" ISBN 0-9584410-3-0** I describe and explain how the two developing stars progress. (A galactica is just another Big star with spawning abilities as it develops stars while staying on route. By releasing stars when the time comes for such release of star to occur it keeps cosmic development on course by releasing stars. All released stars will eventually, in far off time to come, end their development as Black Holes but that is in time to come.). I am of the opinion that most if not all so called giants stars are galactica that has not indicated any form of growth because of the little intensity in dismissing there are and the high value of duplicating the unit shows. The difference between the two options that the stars had when the galactica and the Black Hole-to-be formed was one in the form of the galactica concentrated on duplicating and by duplicating developed pockets of forming dismissing eventually and the other star expired duplication long ago and only concentrated on dismissing.

The Black hole reached the point of singularity. Singularity is a point where $k = \Omega$ placing $a^3 = \Omega$ and also $T^2 = \Omega$. I feel sorry for the mathematicians but this time they cannot take it out the cupboard and play with it for a while. It is too small for mathematics and only fits into thought. On Earth we may give it a meaningless value of being one molecular mass. At first the space surrounding the atom holding the heat relevancies diminish as the space within the star compromises in favour of heat accumulating. This is quite in coming into place within the gas structures and is starting to apply in the solid stars.

Within the star the atom will follow the demising trend on space set by the star

All the while the atoms has to comply to the rules within the star as demand on the atomic space claims sets new standards. At a point the reducing of space becomes so demanding that the factor of light finding an ability to apply motion disappear as the massive structure draws even light towards the centre because at that stage the photon no longer has the ability to duplicate space as it displaces space. The photon must then surrender space due to a lack of adequate motion applied at a relevancy of 56.6. At the beginning when a star establish independence from the galactica outer space which is hotter than the star itself, it is not the heat but the motion as the totality of all the protons working as a group within the secure unit of the infant star that dismiss space-time and the total displacement finds a focus in the centre of the star.

As the star captures
space from outer

At first the star may only demand a reducing focuses on the 3 or the Π to become independent and secure defined borders or atmospheres but as the star develop through the intense centre it forms, the protons will grow and bring about through fusion much more active displacing that eventually forms fusion. The more protons there are in the least space there are will bring about the strongest gravity there are. As the star development progress the dominant gravity generating protons found in one location begins to form within the centre of the star where the major heat is accumulated. The shift takes place from the focus at first on the outside rim of the star developing towards and then to the middle sectors and eventually to the centre of the centre. In this the focus of the displacement gradually moves from a massive number of single proton atoms to a massive number of atomic protons. The quantity of protons efficiency move over to form a focus on to the quality in proton numbers in one unit of a centre and then the dominant atom displacement will not be Π but it will become $Π^2$, later $3Π^2$, $3^3 + 3Π^2$ and so the centre progressively develops.

Then further premiums on the space-time that the individual atom may require becomes resolved as the space demand within the star annihilates all atomic motion of individual atoms and the neutron, which is the representing of motion in the atom, abandons the unit of the atom to reproduce space in the manner the photon did in normal stars. In the end there are by then no hint of any photons left because a lack of motion brought on a total demise of photons. The star is not dead! Eventually only the collapsing of space can sustain the proton activity still present in the star as singularity sets in and diminish all motion activity within the star.

As explained on the previous page this mass comes from the fact that the proton lags in motion to singularity, which is motionless. Explaining this concept or the following concept will also take to much time but what I mention in this page I have a book of more than five hundred pages that covers the whole aspect and has the name of AN OPEN LETTER ON " **STARSSTUFFN'** " ISBN 0-9584410-3-0. I will just quickly touch on the thought. What we see from the outside is just the opposite of what is applying on the inside of the Universe where the "inside" is singularity. Considering it in that light that is why all the information we receive from the Universe by means of light is a mirror reflection of what is taking place. I shall quickly mention the most basic idea of this concept: The motion produces time and the time brings about space.

To us being in space-time the forming of space by using time is a positive measure because we are on the side in space-time, but from singularity such motion becoming space is a disaster coming into practise and it will take the Universe many billions of eternities to once again correct the disaster. To us the motion is quicker that eternity but to singularity the motion is slower than eternity. By creating eternity minus whatever that deduction slows down eternity where as from our perspective it increase eternity by splitting eternity.

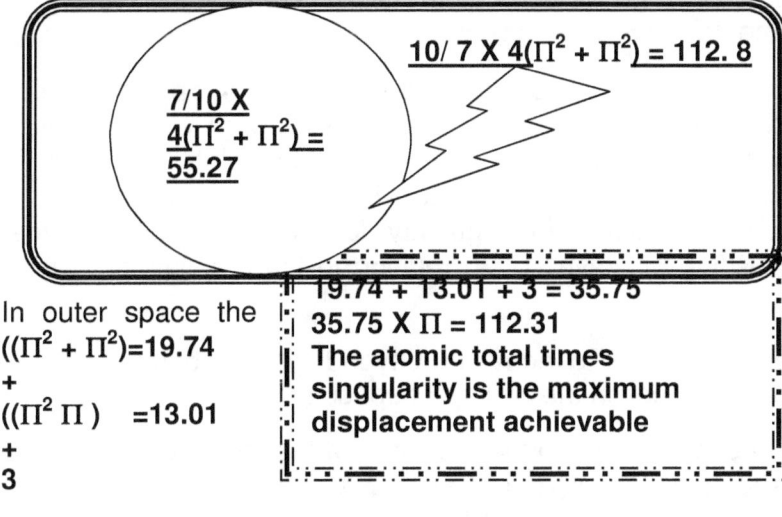

$$\frac{7}{10} X \;\; \underline{4(Π^2 + Π^2)} = 55.27$$

$$\underline{10/7 \; X \; 4(Π^2 + Π^2) = 112.8}$$

In outer space the $((Π^2 + Π^2)) = 19.74$

$+$

$((Π^2 \; Π)) \;\; = 13.01$

$+$

3

19.74 + 13.01 + 3 = 35.75
35.75 X Π = 112.31
The atomic total times singularity is the maximum displacement achievable

Gravity is the concentration of heat running from outer space with a displacement value of 112 to the inner star that has to have a displacement value of at least 55 to ensure the flow of gravity generated by the motion of the object within the boundaries of these two limits. With an inner core displacement of less than the required 55 the star would not yet have arrived at the point of securing an individual singularity in the presence of outer space at 112. The potential difference needed to generate gravity is 112 coming down to 55

The atomic total of 35. 75 x Π (singularity) = 112.31. That is the limit placed on the atom within the boundaries of what we consider to be the Universe. That will remain a unit

The same formula applies to setting the boundaries that limit the possible duplication as it limits the binderies after which space-time goes flat. The relevancy is $4(\Pi^2 + \Pi^2)$

From outer space the atomic relevancy is as follows

$(\Pi^2 + \Pi^2)$ Represents the proton in relevancy to singularity Π through out the universe

4 Is the time aspect of spin creating motion that is creating space.

x Π (singularity) = 112.31. That is the limit placed on the atom within the boundaries of what we consider to be the Universe. That will remain a unit

From outer space the atomic relevancy is as follows

$(\Pi^2 + \Pi^2)$ Represents the proton in relevancy to singularity Π through out the universe

4 Is the time aspect of spin creating motion that is creating space.

10/ 7 Is the **space(10)** in which the **material (7) spin** according to the Titius Bode principle.

7/10 Is the **material (7),** which **spin through the space (10)** according to the Titius Bode principle.

Outer space has heat secured at **10/ 7 X $4(\Pi^2 + \Pi^2)$ = 112. 8** while the star through motion generate a requirement to heat that establish a flow of **7/10 X $4(\Pi^2 + \Pi^2)$ = 55.27.**

The outer walls of outer space is **10/ 7 X (Π^6) / 6 = 112. 8** while the position that the atom demand space is the value iron have as a potential difference. It is in the **7/10** and the **10/7** that the limits are placed. It is seven spinning about in ten crossing singularity by turning about the inner core of the star.

The factor of **10Π** being in relation to Π^3 is a direct translation from Kepler's formula $a^3 = T^2 k$ By substituting the symbols used with the actual value of Π the symbolic massage transforms to specific values applying

10Π Is space square 2(5) (T^2) in relation to singularity Π (k) being equal to space Π^3 (a^3)

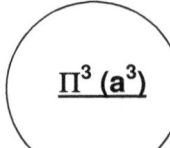

In outer space the motion **2(5) (T^2)** of the material Π^3 **(a^3)** keeps space in dimension **Π (k)**. But this motion produces a relation that apply to material groups such as stars relating to space holding groups such as outer space which I refer to as geodesic space in more advanced books.

$a^3 = T^2 k$ is $\underline{\Pi^3 \ (a^3)} = \underline{2(5)(T^2)} \ \Pi(k)$

When using the atomic relevancy I refer to the proton relevance in space in example $((\Pi^2 + \Pi^2)$ and then how space will relate to accommodate the atom as the atom as a group facilitate the star and accommodate the stars unifying requirements.

In the expression **10Π** relating to Π^3 it is space flowing towards the star centre in approximately an equal manner as volts flow from space to the Earth or Neutral whatever name there is to choose.

One must see the 10Π not for what we read into the numbers as such but what it represents. The number 10 is the square of space that stands for space outside gravity in the place of Π^2 and is the square of space relating to the ten positions in relation to singularity dot as Π. That is the space in which the motion is providing the establishing of gravity Π^2 in relation to the creation of space by motion Π^3.

The star on the inside cannot support space up to equal or beyond **2(10Π)** before ultimately collapsing the space dimensional support of 6 sides in the square of space **(10)**

On the other hand can the space in the geodesic securing the presence of the atom hold space up to the ability of 112 protons displacement secure **10/ 7 X $4(\Pi^2 + \Pi^2)$ = 112. 8.** This is theory because we well and truly know that it is actually $5(\Pi^2 + \Pi^2) \ (\Pi/2)^2 \ (3/5) = 244$ which is the number of neutrons and protons that will allow Plutonium the ability to remain a constructed atom within our Universe. But as one can clearly see it is as volatile as no other element and is on the very edge.

Plutonium holds at 94 and as an atom almost falls outside space-time reality $3\Pi^2$ as it is on the very border with a possible increase in displacement of $5(\Pi^2+\Pi^2)(\Pi/2)^2(3/5) = 244$. In the sun however the dimensional change is $10 \Rightarrow 10\Pi^2$ in comparison with our change of $10 \Rightarrow 3 \Rightarrow \Pi$. With the universe being $7/10\Pi^6/(6)=112$ and the sun at $ =(\Pi^0)\ 10\Pi^2 = 98.696$

In the star the balance bringing about space-time flow is in the iron displacing limit of an atom not holding more that 56 protons because the atomic relevancy is **7/10 X 4($\Pi^2 + \Pi^2$) = 55.27** whereas the neutron reaches the value **of $2 \times 3 \times \pi^2$ = 59,22** the double proton value, it will respond by returning to a space-time value. This is as far as the atom will go down eternity and no further.

Where we are is not the only place that is possible in the Universe to be. This concept is as wide as the Universe is and I am by no means getting into that argument in this book because that argument covers 650 pages pf the book **Matter's Space In Time: The Hypothesis ISBN** I S B N 0-9584410-3-0 . What I will refer to in the following few paragraphs is what applies to our Universe forming our space in our time concept. There are as many possibilities of different Universes all contained by one Universe as there are names for people on Earth and that even is underestimating the possible quantity by the indefinite, but I refer to the one I share with all my fellow Earthlings circling on route around a star we named the sun. In the Universe I am able to witness the proton holds $\Pi^2+\Pi^2$ giving a displacing of space in the duration of time as **19.74** of what ever you wish to name the measure.

The Universe holds a displacing value of **10Π** in relevancy of the motion applying Π^3. Dividing Π^3 by the square of space as **10Π** leaves **9.86** or Π^2.

When the centre displacement of a cosmic structure has a group atomic displacement at the core that is exceeding Π^3 the star qualifies to form an independent structure as the outer space it separates from is **10Π**. When **10Π** shows a relation to the inside of the star holding a displacement of Π^3 the value of Π^2 =9.8696 becomes gravity which the spin or motion of the star will produce as it moves in space through space. By having Π^2 between the **10Π** and the Π^3 inside the star becomes independent from the outer space that captures it. The motion inside the star delivers the independence the star requires to separate from the centre of the galactica and proceeds as an individual star. There is no star in our Universe showing this weak gravity but in other parts of the Universe there are such stars coming into operation. Such a star will not be able to produce space that ensures total independence. It will not yet have gravity that is forming electricity on a grand scale.

With the displacement of iron being 55 + the iron atom has the capacity to dismiss space and by doing that it has the ability to generate such proton motion as to remove space all together from a selected area on conditions of motion producing a connection with singularity. Such connection we call electricity and the diminishing of the space we call an electromagnetic field. The electromagnetic field is the reducing of space between the element by name of copper, (63) where we find copper exceeding the border of $2\Pi^3$ and that of iron Fe_{56}. At $2\Pi^3$ the space-time will become motionless and being motionless and the element iron producing the motion to generate the gravity micro, which carries the human name as electricity.

Space-time displacement which also is motion that reconverts space from heat back to singularity start to achieve a duration in time putting such duration above what the Universe reserve as having the ability to duplicate. Above **62** then forms the epitome of time. At the point the space can no longer sustain the flow in time to sustain the demand set out by singularity with a dismissing potential of **62** protons. After **62** proton is providing the motion of space-time displacing the space held by the protons break down the dimensional wall created by motion and the atom of which only the proton remain in place at that point in any case completely destructs. Only singularity remains casting all other space-time out into outer space. The star then become a star holding no atom but only contains singularity on the inside. It becomes the all so famous Black hole where all falls down a pit of space less ness into singularity without space or motion.

In the star in the Universe which we are in the proton number of those atoms forming the composition of gravity or the dismissing of space to become eternal in time is **7/10 (4(($\Pi^2 + \Pi^2$) =55.** Coincidently this displacing value belongs to iron and therefore iron can produce electricity because when applying motion iron with the ability to displace space-time by using the combining motion of **26** protons and **26** neutrons has the ability to confirm space-time to singularity. For this reason stars must have an iron core and if the earth did not have an Iron core our gravity was not able to generate electricity. What this means in short is that the star then can convert space to light by diminishing space the contraction of space.

At double the value of outer space **2(10Π) = 62** space within the star collapse since the compactness within the star starts to destroy the space that atom holds as **55.27**. The motion applying within the density the star is creating as gravity then has no space or time to occupy any atom in form using space-time such as all an atoms must do.

When a star find the inner-Core- value of applying atomic spin or motion to create **(($\Pi^2 + \Pi^2$) X Π = 62.0** which is double that of outer space **2(10Π) = 62** the result is that space depletes within the inner core of the star and the star will start to withdraw more heat from outer space than the star establishes or returns light into outer space At **7/10Π^6 / (6) = 112** the Universe stretches space-time to the limit we find ourselves in. For this reason atoms that are exceeding the mass of 112 cannot fit in our Universe we have. But it has nothing to do with mass coming from pulling, pushing or shoving. It is about motion exceeding eternity. This is what the atomic number is that can apply motion within the atom centre by the maximum number of protons gathered as a group, which as one group can apply space-time displacement although the practical number of protons in one cluster is $3\Pi^3$. More protons that bring about a group motion will produce a collapse of the atom space. At a displacing value surpassing $2\Pi^3$ the dimensional walls of space leading on the motion forming time (Π^6) / **6** will collapse into the centre of the atom. Beyond $3\Pi^3$ no atoms form a unit because in the practical sense atomic motion cannot surpass the displacement value of $3\Pi^3$ where at such a point singularity starts maintaining space-time without the support of atomic structures and substructures

The number of protons applying motion produces space dismissing and cultivating heat from space and in that process is the space in time returning to heat by duplicating space. The protons apply motion where there is just about no space and by the motion at that level the proton motion turns space to absolute heat where singularity then dissolves the heat. By reducing the space it intensifies the heat and that returns singularity to what it was when space came about as the Big Bang presented space-time. **(($\Pi^2 + \Pi^2$) (($\Pi^2 \Pi$) 3)** is the atom number used to form the atom in the development which brought about the atom. The reducing of space is 1836 times more at the proton **(($\Pi^2 + \Pi^2$)** than it is at the electron but the combining effort of displacing is the sum total of all the atomic part.

When the proton **(($\Pi^2 + \Pi^2$)** and the neutron **($\Pi^2 \Pi$)** is added to the **3** the electron produces the total dimensional sum produces not 6^2 as it should but **35.75**. With the sum being **35.75** one can see where space will collapse or return to the form singularity provides if it exceeds the singularity connection there is between the atom in total and such an atom connecting to singularity Π.

Then we arrive at the universal six dimensions of three sides in space and three sides in motion bringing about a totalling of **(($\Pi^2 + \Pi^2$)=19.74+(($\Pi^2 \Pi$)=13.01 + 3 = 35.75 x Π (singularity) = 112.31** and that is the maximum atom displacing value outer space can tolerate before destroying the space holding the atom all together. Inside the star the proton maintaining a connection with singularity directly will produce the proton value of **(($\Pi^2 + \Pi^2$) X Π = 62.01.**

This means that at this point within the star the protons and above this velocity time cannot duplicate space any longer and the wall of space erected by time collapses back into singularity. Space disappears because time cannot any longer sustain space. Any star having a space with a displacement exceeding the generating ability to displace space to the value or above the value of 62 protons in one secluded given space that is repeated as a unit by motion duplicating space will no longer have the ability to sustain the walls time provides space.

The limit is 62, which is **10Π** holding the one proton Π^2 on the one side in place and the **10Π** holding space in duplicated motion Π^2 and it is also double the space value of $2\Pi^3$. Therefore the gravity a star produce at a maximum point is (**10Π**. +**10Π**.) converting space to ($\Pi^2 + \Pi^2$) in relation with singularity Π then collapses back to Π^0.

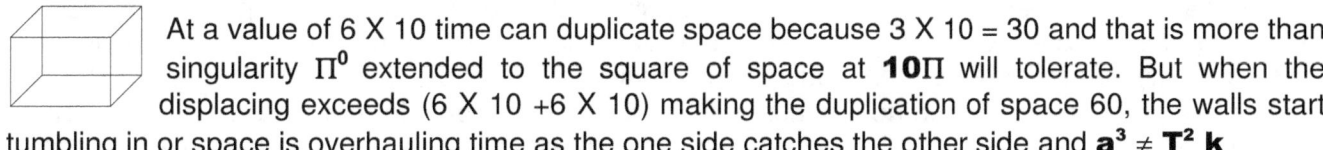 At a value of 6 X 10 time can duplicate space because 3 X 10 = 30 and that is more than singularity Π^0 extended to the square of space at **10Π** will tolerate. But when the displacing exceeds (6 X 10 +6 X 10) making the duplication of space 60, the walls start tumbling in or space is overhauling time as the one side catches the other side and $\mathbf{a^3} \neq \mathbf{T^2\ k}$

This **62.01** is the total and the maximum number of protons dismissing a value of Π^0 space-time after which the atom as a single unit or as a group in space and time and ultimately the star as a unit of the combining effort of all the atoms forming the star will abort space or the Universe will dispense of the star, which is all the same effort. The star then has grown back to the connecting with singularity and then forms a Black hole. The principle may sound somewhat simple but it is quite involved in the total explaining. The Black hole that forms is within every star centre because that keeps the star in a unit above and beyond what is going on in outer space and as soon as it can bring about fusion by performing motion it will sustain a proton growth setting the star on its final journey to eventually form a Black hole, although there are two stars even beyond the Black hole. However I would prefer not to elaborate on that at this venture. The black hole hides in every atom because through that tiny space not even being part of the Universe on our side, space-time dismissing is applying by killing space because of motionlessness. All atoms in our Universe are showing equal growth notwithstanding position or motion because all atoms in our Universe are in complying to motion as a result of the Titius Bode law of 10 /7 and 7 / 10 bringing the time split equal to all involved in space-time.

When the atom finds the motion within the star centre or star-core-centre start to reach ($\Pi^2 + \Pi^2$) X 3 = **59.22** the neutron moves outside the atom and also outside the star. At a displacement value of **59.22** the atom in the star can no longer accommodate the neutron within the atom and the neutron motion slows down to a point where the Neutron motion is to slow to find accommodation in the atom and in the star. At (($\Pi^2 + \Pi^2$) X Π = **62.01** the proton collapse, which is double the value of **10Π** as well as $2\Pi^3$ therefore Kepler's is doubled or duplicated to serve the motion increase. That is what space-time represents because space-time also represents **k** in variations of distances. It is the doubling of $\Pi^3 = 10\Pi$ to $2(\Pi^3 = 10\Pi)$ which then exceeds the duplicating ability introduced by Kepler. But we must not forget that it is also half the cosmic value of $7/10\Pi^6$ / (6) = **112**. At a higher motion the proton moves to outside the atom and form a proton motion by introducing the proton motion to the space surrounding singularity. It is then where it started with, the dot that claimed a spot in the cosmos. But the dot went further and grew a lot.

It started with a dot Π, because that is the only form, size and dimension mathematical logic will allow our brain to accept what would form as the first value flowing from singularity. From the one dot had to come a second phase of dots and a third lot of dots. The dynamics of such dots are smaller than we can understand because such a dot is in negative relation to what we see Π to be, and the deeper we delve in finding the smallest fragment where space started, and that is the spot where time is still eternal as much as we can accept eternity to be. This we find in the aligning of planets where the one dot from which the aligner stem becomes the reference too the distance applied between the aligner and the original dot, or governing singularity or structure in charge of holding position to all orbits following.

The reason why we should first locate the spot is because we can only work from that point forward. By working forward we have to work backwards to locate where we are heading. The cosmos started at a point and where such a point is, we will find the Universe and where the Universe one day will end. Every one knows where the Universe is, because we can see where the Universe is, but if we can see where the Universe is, then we should find the centre of the Universe in that spot. Einstein theoretically positioned the point of beginning at a place he indicated where singularity should be. With the cosmos the size it is and space so large compared to our smallness we have no chance in finding the centre of the Universe. The Universe started where singularity is and singularity is the sure indicator of the centre of the Universe. With all spinning objects holding singularity we then have located singularity in as much as finding the centre of the Universe. The Universe started with a dot forming. That answer arrive from taking mathematics back to a point of being the smallest possible position, far smaller than we may be able to calculate form as we return for a moment to a time before Mathematics developed.

My approach might seem unconventional but through the abandoning of the accepted, it enabled me in locating the precise location of a universal singularity forming a connecting basis of the Universe (this I say with some degree of confidence). The smallest figure there can be must be a dot The only value a spot can have being without space is Π^0. The dot is the only form that leaves all the options open to extend in any and in all directions should the opportunity arise. The only mathematically sensible option about extending a line from the dot will be non-bias progress in all directions equally in order to give a meaningful flow of mathematical equilibrium.

 The Pythagoras mathematical principle is the proof and that I explain. The obtaining of singularity is in my rejecting of nothing by replacing it with something being the dot. With the clepsydra or "water thief" Empedocles deducted that air was composed of innumerable fine particles, braking the thought that what we now know is air, was also believed to contain nothing being altogether a space filled with nothing until proven to be wrong so many years ago. Never did science take the lesson learnt back then to the future and out onto outer space. If there is space, there cannot be "nothing" as space is something. The claim becomes obvious when observing the connection between the half circle, the straight line and the triangle, which could also promote all the qualities lurking behind the pyramid. Consider the connection between 180^0 sharing and then one may realise much of the pyramid mystique becomes less spectacular in considering the very basic in mathematics being the Law of Pythagoras on which all mathematics arrived. Once the water thief was eliminated by some human intelligence the matter was left at that. Nothing shifted out to an area we think of as outer space. In outer space they say we now find nothing. There is nothing but an atom here and there and even the atom is covered in nothing. Now I ask you how bloody logic is that?

I wonder why the nothing landed there? Could it be that the reverse came about and because there was no visible "water thief" the very limit of man's suspicions came into practice. Man has always been extremely good in flying from one outer edge to another and if the water thief proved something was present, then the mere absence of a water thief must therefore prove that nothing must be in outer space. It seems much easier to shift nothing than to find what should replace ether after ether was removed from space and replaced with nothing. But what is space as such. What can space be, because with explosions we can clearly witness space created from heat. Our culture prevents us from admitting our vision, but the release of heat produces a "shock wave". That "shock wave" is nothing less than space created from heat released. The space that the release of heat creates re-establishes the position and location of the entire space it refills. We have to brake free from culture of the past and a rigged mind set narrowing our vision. We have to learn to see the Universe with our minds and not our eyes as we can see in the presence of the Black hole we cannot see. The Black hole is only visible by presenting invisibility. We know about the Black hole because we can't see the Black hole. Why would that be?

Because of the manner in which the Universe initially started where more singularity in the relation was unsuccessful to form contraction after the overheating brought about expanding and overheated by expansion much more heat released into the Universe in heat as space uncontrolled than that which remained controlled by singularity secured inside a unit such as atoms or as star or any form of containing material. The container had borders coming about from motion that the unit employed and such motion set the forming structures apart from the rest of space. Once again we see $\mathbf{a^3 = T^2k}$ being correct from the start. There is more heat in space uncontained than space contained in some cosmic unit in heat that is volumetric consoled and secured. Therefore to restore balance there must be a position where singularity reduces space faster than heat can fill space. At such a point singularity is taking longer to reduce the space than it takes the heat to fill the space and therefore the space reducing takes longer than filling the heat in space. The space flickering that announces contraction takes longer that the expansion causing the heat to turn to space. Before the heat expansion can begin in progress the contraction already completed the motion successful. This question proves fusion between materials to be the answer. By primarily having 112 atoms as single hydrogen particles in one area will capture a great volume of much space. Changing that number into an atom of single proportions having 56 protons that is housing 56 neutrons in one accumulative structure some heat will release during the capturing process but such heat will become the demise of the space which was holding the heat as well as the demise of the heat by substituting the dismissing of space and building of the gravity applying material in the process such material will push the direction the Universe takes into some final conclusion.

At the start of the Universe birth there was heat that turned to space that turned back to heat through motion applying contraction but there were lots of other where singularity could not contain the expanding by contraction and at that, that expanded more then singularity could contract allowing more heat to be in space than material is in space. Since there is more heat in space than there is darkness because the heat is darkness that turned to space the reducing thereof will bring about more darkness because it reduces the heat.

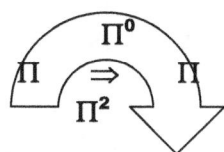
In the region where the singularity present space starting time and where the motion originates space is so little that time is eternal. The space the photon reduces too is 1836 times smaller than what the photon is. Actually in more advanced explaining I shall show the reducing is $((1836)^3)^2$ times smaller that the photon but let us leave that explaining for later in another book at another time.

In the area where singularity release time to contract space the motion of Π to fill Π in the seven Π positions of Π creating ten Π to establish gravity by reducing ten to Π^2 the very idea of moving to the other side of the Universe is covering a distance we cannot begin to grasp. The slightest of motion ever possible completes the whole journey and the space duplicated is one mark more that no space at all. We are dealing with mind accepting and not sight ability because where we venture is $((1836)^3)^2$ smaller than the photon we cannot even see but can only use to see, Where the space is that little the time must be that long because time is the very opposite of what space presents. $(k^{-1} = T^2/a^3)$. Being on the other side of the Universe being so small the contracting duration takes so long that the space the photon fill reduces to compact the heat that the photon contains above.

The motion is reducing space in a space that is much more reduced than what any heat concentration in the Universe can sustain. It is far below or far more than the speed of light and captures the space that the photon claims by motion in reverse. The flickering that we find within all stars and by which we name some that is pulsating is taking so long with the reducing of the space it remains on the dark side of the impulse one eternity while the flair of light is then reduced before the impulse can begin to expand. One should take into account that a photon that is released is holding a space that is humanly impossible to measure. The photon is space connected by a spinning motion that reduces the former space to heat where is so concentrated it can and it does cut material. In order to understand the reality in the situation one has to reduce that 1836 times and that gives us the space where the proton finds space to dismiss. It operates in the area below the speed of light and at the limit where motion produces space. The mass the photon creates is so enormous it destroys the space the photon claims That is the reason why the photon has one motive and objective and that is to liberate space by creating motion. If the mass of something as small as the photon is destructed by own mass to the extent it has to move out of the space it holds and where we on Earth find the mass immeasurably small the gravity applying to establish such reducing is more than what numbers in mathematics can present on paper.

Going down the atom passage leads to an area where gravity motion is reduced 1836 times that of the reduction the photon has already been through and we are left with space in the Black hole where that space the photon measures is so enormous that is the demise of the photon as the occupying space destroys the photon. The photon already far to small to measure is 1836 times too large to fit into the space. It will have to comply with as it enters the edge of the Black hole. That is the road in reduction that the photon has to go before it measures up to the conditions applying in the Black hole. The photon cannot liberate the space it holds and thus the massive size that the relevancy of the photon has in such conditions destroys the photon.

From this evidence it is clear that stars range from the one dimension of total darkness to another dimension of total light. In the Black hole a dimension is motion in space where dimensions are just a concept. To those hard liners in Mainstream science that cannot see past the culture driven concept that time is the same everywhere how can those thinkers declare time as a constant where there is already mathematical proof that the increase in gravity reduces time by increasing duration. This was Einstein's calculation that brought this insight. The only manner in which space will reduce 1836 times is if the concentration of the applying gravity intensifies by 1836 times and that increase the contraction 1836 because it has to contract the space into a unit 1836 times smaller. This can only come about if gravity increases by 1836 times to what it was and with gravity changing 1836 times surely it has to influence time and the measure of time. Einstein said that gravity increasing slows down time and what bigger increase do they wish to experience that an increase of 1836 to one. There is the other limit to where stars are still in the cradle blanketed by covering of heat in density beyond the photons releasing. That is why that inner centre of the galactica is so eternally luminous. The time duration on that side of the Universal limit again is so short that the contraction captures the expanding light faster than the speed of light can secure a photon release from that centre. If not that centre must by now be as dark as the rest of the dark Universe.

The escaping of a photon from an area as large and compresses as what a galactica holds and in the presence of such mighty gravity as that has to be present there must be considerably slower than that of the escaping light that is coming from a midget star such as the sun Yet looking at a galactica the light streaming from a galactica does not seem to be much brighter in intensity that when coming from a star. Which means the light (heat) escaping from the galactica, which has to be billions of times bigger in space occupation than what the light claims coming from a star. That is the other side of such enormous gravity applying in a confined infinite space. Remember that gravity and time is the very same thing because gravity contains the expanding of space in motion. Therefore the time factor inside galactica is going on very, very slow indeed. That will mean that although the speed of light is in affect the release of photons, the escape against the applying gravity flow will in relative terms be a trickle of a flicker.

The flickering is taking so long in favour of expanding the light flowing from it seems a solid by constant flickering. That we also can tell because photons must present flickering if photons are individual particle apart from one another. The accumulated gravity is becoming denser as the heat concentration intensifies toward the centre of the galactica. The relative heat concentration shows up denser on photographs but the light coming from that area is not more intense when measured, in the way it should be with such sustained increase in density in the source of the light. But the increase in light comes from a stronger gravity and a stronger gravity will also produce a slower release of light photons escaping. The Universe is in ranges of space-time.

The Black hole presents space-time that favours the flickering of the dark much more intensely to the extend the dark is far beyond solid The galactica favours the flickering of light with the same intensity as the Black hole favours darkness to flicker, but the galactica in the centre eventually commits so much accumulated gravity, it slows down the release of light and more so rushing from the part in the very centre of the galactica. In the measured precise centre where Prof. Hawking measured a Black hole, the gravity applying is so intense due to the not yet developed that is within that centre that the accumulated gravity of all the yet to develop stars situated in that centre slows the flow of photons down to a stand still and even taking the photons in reverse flow. The light is so intense it can stand still while only a certain amount of light is able to escape. This allows the rest to remain stationary or even to flow back in reverse to the specific dot in the centre of that galactica.

In that centre the gravity of the accumulated underdeveloped stars are so great it allows light to flow in reverse and therefore take time in reverse to beyond the Big Bang which then explains the other singularity that Prof. Hawking detected within the centre of the Galactica that slows down the flickering of light to the extend the light exceeds the limit of solidness as it present itself as being a Black hole holding matter not yet introduced to the centre of the Universe. The space that Prof. Hawking detected has not yet left the battle ground and is still n the spot where the stars of previous era already completed the journey, those in the centre must still experience their Big bang in waiting their turn. If and when one use only light to read what there is to measure from the situation it will most likely lead to the thought of Black holes but then the question arrive at the problem what will produce such gravity and why does such gravity not devour the galactica from the inside. The answering of this question proves that considering it as another Black hole would be the incorrect conclusion to draw.

One must see the Black hole as a lot of solid space with in another big space. When the space in the Universe was little the Black hole was big and was constructed of the most massive space that was available at the time. Yet at the time when the space which the black hole claimed was more massive than galactica are today because if not the galactica would have already gone into the Black hole development stage.

There was in comparison not that much space available at the time. The **k** that was then present produced a minute a^3 (in our reckoning at present) with an enormous T^2. The universe was not small just as much as the current Black hole was not big at the time. The Black hole sustained the space that became singularity but the extending of singularity grew with the growth of space-time not committed to star structures. The gravity that the star back then had to possess and create independence in the midst of the gravity presented in the Universe had to be so strong that it now contains the star in circumstances that was present at the time in the Universe. We must realise that the gravity contained in that star was strong enough to reduce the space in that star while all the gravity in the rest of the Universe was unable to contain the space or retain the time in developing. Looking at a star we can measure the Universe at the time when the star broke free from its galactica confining. The Black hole and all other stars for that matter stayed behind just as much as the space grew more. It is a relevancy applying both ways and not just to one side. As the heat in space declines the heat within the star rises. It is a relevancy not favouring any side. But also matter grows as much in relation as space recline.

The Black hole started off as a (presumably) red giant of its day. In the Universe applying at the time the space the Black hole contained was enormous and I suppose it engulfs large areas of the space available at the time. It is still acquiring massive space in dismissing therefore as far as the star goes little changed but in relation to the Universe the Universe acquired muck more space in exchange for large reducing in time. I cannot for the life in me see how Newtonians can consider that one part of the cosmos change leaving the rest of the cosmos never to change. The cosmos is about changing on all fronts that there are and in every factor of the relevancy $a^3 = T^2 k$. There was a time when the little space the Black hole now controls was a big space in a little Universe with lots of heat and little else but a few very bright stars covered in massive heat. It should also be recognised that the Black hole contains the material we will find in the collection of many massive galactica combining their protons into confining that much material in one unit. As I said, if that was not the case the Black hole just could not develop in the manner which it did while the rest of the Universe spawned of to form space in overgrown time.

There is the time where the space is consumed by the Black Hole as much of the space is concentrated because the Black hole remains relevant only to the singularity it sustains by dismissing space not relevant to the singularity it is sustaining in growth. To that end it is supplying heat to the singularity that will reconstruct the space to heat occupied by material and by that it feeds singularity by consuming heat to convert to space.

But then the Universe expanded as the star reduced. The star did not reduce while the Universe did not expand. The star reduced space as the universe expanded space and one may never lose sight of the relevance. Only singularity grew as singularity acquired the taste for other singularity, as they will provide the required space/heat to contain.

There was a time when the Black hole presented hundred of millions times larger space than the space that the Huge Red giant now present in relevance to the space available to outer space and to the huge star claims at present to occupy material within boundaries set by singularity extending. But back then the Huge Giant Red star of present was only a speck invisible in the centre covered by a blanket of thick foggy layers of heat in the centre then of the Galactica that was then present and converting the now released giant star. .

In the spiral of the proton that the Black hole now have stretching into space the gravity influence extending might be some indication of what the core influence had in relevance to the space ratio that applied when the star went from the universe into singularity.

One can only imagine the growth this star represents in the Universe because as much as the star lagged behind the Universe the star also grew in space in the Universe. Just imagine how hot space was when that Black hole was some flickering orange star dot in the sky. But that star was a star before light was the light we now know because the light we now know is invalid within the Black hole we now know. In the mean while the cosmos expanded and the Black hole reduce by remaining the size it had before the expanding came to alter the ratios that was applying. Just as the sun captured the outer space value in heat that was present and part of the outer space density back then when the sun released form the outer space the Black hole too captured the space relevant to outer space then and holds that space relevant throughout other progress applying. It is only our human insignificance that indicates what might presumably be big or small, hot or cold, near or far but to the cosmos it is history.

* At present the Black hole as a star, as a whole is smaller than an atom because the star ejected all the proton qualities of the occupying singularity to the out side the structure the star claims in singularity. It rejected the electron before space –time was equal to the electron.

* It discarded the neutron when it got rid of internal motion and only occupied singularity in the sector where the proton commutes in. Then finally with further Universal development that implicated the star as star development the dismissing of space became absolutely overwhelming by even displacing the proton function to the space outside the Black hole. While space is in demise stars are growing.

Compare the theory I propose with the theory which mainstream science underwrite and applaud and was promoted by men to the likes of no less than Einstein in person. Compare the theory that I introduce in a practical context to Einstein's critical density and the underlying and the underlying factors as scientists wishes to promote gravity. They never calculated the distance that particles are apart because they couldn't. They couldn't because of the obscurity we humans have in our perception on the Universe stemming from our position in the fact that we think we hold the centre position in the Universe because all light streams directly n our direction. From the official argument the earth must be at the centre of creation because every time Mainstream refers to where space ends at some billion trillion zillion or whatever ridiculous number they come up with. Such a remark damns their perception on reality and from such grossly thoughtless argument the wish to position stars and distances parting stars. Mainstream science full well know that the parting radius has a phenomenal influence on the outcome and even their argument about space ending is totally ridiculous because what happens to light shining where space ends. Does it collide with the end and then what will constitute such an end. Where will gravity go when the gravity afterwards reaches such an end? The thought about the lot entertains those unable of clarity but to the rest of us it is a joke that men with such brains can deliver such nonsense and present it as truth.

Einstein's Critical Density lacks the accepted matching facts we need in proving the critical mass factor. But our inability in securing such required evidence defies the most basic logic. It seems all new evidence we receive from outer space is disputing all Newton laws about the cosmos and new findings that disprove Einstein's Critical Density as the answer. The universe will not reach a point of contracting, not withstanding whatever dark matter astronomers try to locate in the vast space. Why would the expansion turnaround and do a reverse by going back to where it came from. Where will the centre be where such contracting will locate a point and what will form that point It seems fine to draw a saddle or a square or a triangle or even a flat mat (no mat can be flat because there has to be another side if you draw anything presenting flatness) and then try and sell that idea to those scientists considered being a bit of jelly minded and weak of thought. That includes mostly and entirely the us, the me, the commoners and non-academics that cannot think. Even the saddle or the triangle or whatever has to present a perfect centre from where the general gravity will flow that keeps the structure they present contained as a unit. Without such a centre and the ability to show such a centre all fancy arguments drop short of being realistic. $F= G(Mxm)/ r^2$ holds all related mass I correlation with one specific centre point and without the centre the rest is strewn in disarray.

If the material is there it is there because it is secured there and where is the centre securing the material that is secured and there. It is such a simple question but without an answer the statement becomes ridiculous. Consider the momentum alternation such a change will bring about. How will the total Universe come to a halt when the time comes where the Universe must come into the turning back and contracting in order to honour and confirm Newton's statements? In what direction will the turn around go since all material that is in motion is spinning in ever growing circles? In other words the question is where is reverse. There will be a massive number of collisions except on the unlikely condition that this turnabout is orchestrated in the manner where every single subatomic particle goes in reverse in the same instant.

Such simultaneous action will have to include and involve every point of material but what will motivate the action and who will give the command to stop turn and restart the moving as the direction of moving must include every particle all over the universe. I might guess the stopping problem alone is an insurmountable object that first has to be laid to rest before we start calculating and measuring every atom in the Universe. There are so many more unanswered questions, which I touch on in other books but that is not the nature of this book.   While the clever academics present an image of very literate about physics they also lack the physics insight to realise that if motion shows a turnabout all material in the Universe will demolish because the momentum that applies with such a turning will destroy all structures. You can't even stop a bus that fast without paying very dear consequences. They never refer to relative facts in cross-referencing.

The sun is not a gas-filled sphere holding hydrogen in its "natural gas" form, but it is all fluid and is in a liquid form where singularity is liquid- freezing hydrogen at 6500^0 C while outer space is boiling over at $- 276^0$ C. Even if it was the gas filled bowl, then still more to the point is the thought no one ever gave a minutes notice to is why hydrogen being a gas will remain on the inside of a structure being 6500^0 without expanding like a rocket exploding. What prevents a gas filled bowl not explode because even if the gravity is considerable, hydrogen would never stand that heat and not get pretty volatile about it. The hydrogen must get very spaciously exploding long before 6500^0 is reached under what ever the conditions may be..   This book explains the Roche limit in the practical sense… when applying cosmic laws instead of improvising cosmic laws the information one receives uncovers that reality and becomes awesomely simple. It becomes clear that the universe is as much expanding as it is contracting and contracting by expanding. As there is no hot or cold, no big or small, no grand opposing but relevancies in ratio to one another. If you do not believe me, then believe your eyes when looking at the picture about the sun coming from telescopes. What ever the sun is it is fluid falling into fluid.

Please tell me one reason why I am incorrect and consider the time it took the Universe to develop from 10^{-5} to 10^{-43} seconds to create a cosmos the size of a neutron. Compare that to what is happening now and see how many events took place by the creation of every lepton and every hadron and it was to be true that that period took longer too complete than it took the universe to create the solar system. The flow of light through the density that space produce heat gives the speed of light the relevancy of time in space. The thicker the "soup" of heat is that space forms, the longer it will take light to cover a distance. It is very important to note that the speed of light is a relevancy between time (seconds) and space (kilometres). The speed relies completely on the value **k** holds on space –time. The speed of light is forever a constant but the constant is part of the relevancy of space-time.

The Universe connects in a way Kepler established through his relevancy theory. Those not convinced answer this: where would the Planets be if not for the sun securing planet positions. The relation proves the ratio of one in all cases to be valid. It proves much more than merely connections at liberty of holding positions where ever the randomly opportunity placed the structure. The structure does not come closer by a pulling and tugging. Kepler's figures coming of his calculations must still be around and by repeating the task again but this time made much easier with the help of computers and telescopes of magnificence we can compare that to those the which Tycho Brahe calculated and test what growth took place. At present the star, as a whole is smaller than an atom because the star ejected all the proton qualities of the occupying singularity to the out side of the structure the star claims and is under control of the governing singularity. It rejected the electron before space –time was equal to the electron.

I do realise that the way I interpret Kepler is very new approach and up to now all I could find is Academics with scepticism and detachment from my views. I do not blame such reaction but if I am not correct, please explain according to the view you hold how the following is possible. You're being a person that stands at night on the highest elevation in your local the vicinity in a manner so that no other solid object can restrict or block out any of the light flowing towards you. From all over and from the most outer regions of outer space light is travelling in a straight line directly to you. The light is travelling at a speed maximum to what the cosmos will permit. That is how eager the light is to reach you personally. That light is using mostly millions and in some cases even billions of years to reach you while you are filling the centre of the Universe. Wherever you move the centre of the Universe will shift to the position you then mostly millions and in some cases even billions of years to reach you while you are filling the centre of the Universe.

All light are flowing from all positions and from every possible direction to salute you're filling the spot in the Centre of the Universe. Wherever you move the centre of the Universe will shift to the position you then hold because the light flowing to you will follow you to wherever you are at any spot. All the light coming across the vastness of space every possible region of all outer space is acknowledging your position where you fill the centre of the Universe in the spot which you fill every second of your entire life. But it is you filling the centre position and not the centre position being where you are. As you move notwithstanding who ever you are or wherever you are the centre will follow you as the centre will always be on the spot where you are. That is only because you that fill the centre of the Universe are where you are at the time you are there. Everyone being of flesh and blood as an individual knows that that person being him or her is filling the centre position of the entire Universe because all the light from everywhere is coming right across outer space to acknowledge what every person on Earth was expecting about himself being the one that is filling the centre of the Universe in any case. It is coming straight to you because you are in the centre of the Universe according to the light travelling. Not one spot represented by one ray will pass you thinking that you are not the centre of the Universe or the spot you hold is not important enough. Why would that be? Why will the light act as if you fill the spot where the light considers it can locate the centre of the Universe?

The light is underlining what you take for granted, the light admits that it is coming to you because you are filling the spot the light comes too as if that spot is you being the centre of the Universe. Most of the light started on their route even long before man was to become a species and most of the light was already on route long before the time when the solar system came about. Yet the light treats the place you occupy as the very centre it wishes to come to confirm the importance of you're position where you're position is within the spot centralizing the entire cosmos must be just because you fill that spot you're glowing presence. And every person everywhere has the same conformation about his or her importance as the Universe acknowledge this important position in the location wherever you may have what you have that is so important only you can have that centre position. If you close your eyes or ignore the light, the light will go unnoticed for all time to come.

The light then travelled across so much space using so much time to travel only not to be recognised by you the person filling the centre of the Universe. The light and all the space the light represents and all the time the light travelled would go into eternity dismissed as never acknowledged to be worth noting just because you, the one filling the centre of the Universe did not take the time to look and acknowledge the effort the light made so many million of years ago just to bring homage to you filling the position of absolute centred importance. So much effort is done by the light so long before you were even born to get the timing so right as to acknowledge your centre location in this moment in space and in time depends on you're taking notice of the effort and appreciate what the Universe did to admit that it also believe you are the centre of all space and time out there. Now use modern science to explain this centre you establish with your importance.

Tycho Brahe and Johannes Kepler stood there night after night and made a super human effort to acknowledge the information the light of the universe brought to them. They wrote down every massage every night. But since they are only human they could only managed to acknowledge the light coming from the sun that was reflected by the planets. The two masters were the centres of the Universe when the two Masters decoded the language the cosmos used to speak. Think about what they managed to collect for all mankind's benefit. They acknowledged the light coming towards them in a straight line. A flow of electrons causing photons to travel across outer space and meet their eyes. One line of photons flowing tells them all about the regions the light came from. Think of the size Jupiter holds and a few photons can bring across such large quantity of information.

Yet the light acknowledged the position the Master was in and came all the way representing such a large structure as what Jupiter is confining such an enormous structure with all the information it has into one line of photons. With the information of the entire structure confined into a few photons it managed to convey all that information across such vastness of space, choosing that specific point as the centre point of the entire Universe. By using the theories now applying and representing the views of Mainstream Science how would you go about explaining the way you are treated by light travelling through space and time to be in the centre of the Universe, which is the spot you have and hold. Seriously you know you cannot be in the centre where the Universe started the initial Big Bang process.

Not even the solar system or the Milky Way can be in the centre where the cosmos started and yet that is how Mainstream science argue, Can they be wrong? Yes they can and no they can't because they are correct but they are incorrectly looking for a centre where there is so many centres in the very precise centre. Still without any possible influence the light puts you and every one else in that centre. But since you are in the centre, you must admit that there are billions of trillions of spots forming the centre of the Universe just as Kepler stated. The only way you can fill such a spot is if the spot is the place holding singularity and singularity represents the centre of the Universe. Then k^0 can form at any spot and establish $a^3 / T^2 k$ because k is the end of k^0 being the start and k^0 is representing singularity. The centre of the atom represents k^0 and the electron represents k. The main issue brought to light by Kepler's formula is the relevancies that always prevail through out the cosmos no matter what sphere it is representing.

When a point holding singularity achieve energising by securing a number of protons concentrating heat towards that space centre holding the singularity, the singularity will establish a will to seek independence from other more dominating singularity surrounding that point and moreover seek to abolish control of the dominant singularity that suppresses its individuality by establishing a securing centre. Two points will redefine their relationship and will establish space-time relation between the points. A relation will come about where there is a dominant point in singularity establishing a controlling centre that is trying to establish control over space-time by creating motion too space-time by displacing space-time, then there is the space-time in motion and a third factor where singularity is applying motion to material within the space-time in control.

The dominant singularity will control the space surrounding the lesser spot in singularity that is gaining in heat concentration by material growth. Through applying gravity to ensure heat concentration with space-time diminishing the lesser singularity will gain heat in centralising space-time and by then turning it to motion it then can use such motion to gain independence from the centre in the space-time that the dominant singularity is controlling and is diminishing through applying motion within the space - effecting the lesser partner while the lesser partner the prominent space it is feeding from the prominent singularity that provides the lesser singularity with some pre concentrated space-time. This is the manner which structures in development within the incubator of the galactica heat blanket use the galactica governing singularity that is democratically elected and placed in a centre as a centre to bring progress in the incubating star development.

The cosmos cannot be if the cosmos do not share with everything else in the cosmos but the sharing is always producing relevancy to the position of another factor forming the Universe.

The inner space is applying positive space-time displacement in relation to the object in rotation.

The heat the inner structure secure prevents the motion from applying to the object because it became dominant enough to reduce the space towards the heated centre and in doing that it is producing space to secure space through applying motion to the space.

The orbiting object forming the outer ring is in a negative displacement in relation to the inner centre. The outer object has heat in the centre it has but the heat is far less dominant than the heat of the centre and by increasing motion it is concentrating heat as much as the motion is reducing space to concentrate heat. By applying motion it is securing space

What makes the cosmos is the variety of structures forming the Universe while all participating objects of all sizes are using the same singularity. There is no big ore small because that which has the biggest control in the cosmos is also incidentally the smallest there ever can be in the cosmos. In fact it is so small it cannot even directly claim apart in the cosmos. To make sense of this lot to me being somewhat of a dimwit I placed two opposing motions in relations to each other and the one will always show one or two relevancies in relation to the other.

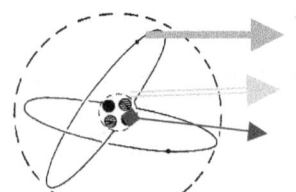

ELECTRON : Unoccupied space-time is 3
NEUTRON : Occupied space-time is ($\Pi^2\Pi$)
PROTON : Densified space-time is ($\Pi^2 + \Pi^2$)

The role of the electron the neutron and the proton is very commonly accepted the role each sub atomic particle plays. But galactica and stars are just as much just more cosmic atoms playing their part in the very same way as does the sub atomic particles.

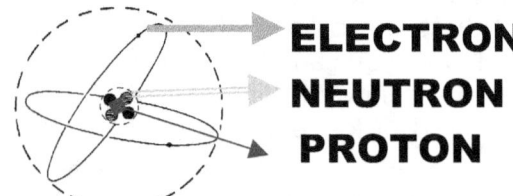

ELECTRON
NEUTRON
PROTON

ELECTRON is about confirming space

NEUTRON conforming space

PROTON converting space

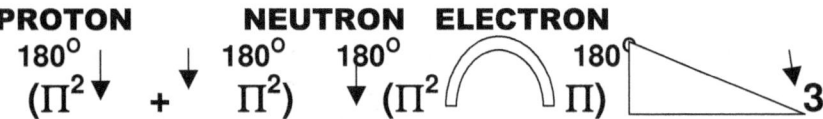

PROTON   **NEUTRON**   **ELECTRON**

180° 180° 180° 180°

$(\Pi^2 + \Pi^2)$ $(\Pi^2 \frown \Pi)$ 3

If one looks at the transmission of sound, it too depends on the relocation of matter, but to a very small degree, and in this process lies the transmitting of sound. To make the error of judgment in confusing the process with the breaking of the Doppler rings are quite understandable.

The biggest activity there can be is in a Black hole forming space-time confined directly to singularity where that pushes time eternal and by doing that places al motion activity into outer space.

It is about confirming space ($\Pi^2 + \Pi^2$) conforming space ($\Pi^2\Pi$) and converting space3.

The Universe comprises of the atom and a star is just another atom holding lots of its own. This comes about from the fact that both particles are the same in the universe since both particles serves singularity. From singularity the size space-time takes up is unimportant because from singularity space-time is merely principals connecting singularity forming energy as gravity or antigravity and presenting space through the relevancies forming the motion of time. That is particles and atoms surrounding singularity, protecting singularity, maintaining singularity and securing the surviving of singularity. This service of space is done by motion in duplicating space or extending space.

When a star favours **3** and a multitude of three the star is still in a process where it will favour more the duplication of space. As the star develop it will ever increase as it moves through the ranks of being liquid ($\Pi^2\Pi$) the favour to the proton ($\Pi^2 + \Pi^2$) where more space is dismissed that space is duplicated because the motion of the star also diminish progressively and so to the end phase of the star. where the star will once more be motionless. The cosmos comes together at a displacing limit of 112 protons per atomic unit. It values time to space at **10 /7(4(($\Pi^2 + \Pi^2$)) = 112.795** on the space limits while motion breaks down within the centre of the star at half that being at **7/10 (4(($\Pi^2 + \Pi^2$)) = 55**

Space receives six sides at **7 /10 (Π^6) / 6 = 112.16**

(($\Pi^2 + \Pi^2$) the proton +($\Pi^2 + \Pi$) the neutron + (+ 3) space = 35.75 X Π singularity = 112.313

The cosmic cube we live in is singularity that is six sided **7 /10 (Π^6) / 6 = 112.16**

To decipher the code one goes about as follows. The value that generated gravity or if you wish to call it electricity will be as follows: $4(10\Pi^2+10\Pi^2)/7=55.4$
4 Indicates the time presence influencing the value. Four always indicate time
(10) I outer space in the square already

Π^2 Indicated the square forming gravity

$(10\Pi^2 + 10\Pi^2)$ Is the double proton in multiplication with the square of space on both sides of the Universe.

/ 7= is representing the value of the sphere

55.4 Are the maximum protons displacing space-time working in one space less unit.

The four quarters of time (4) holding the square of space (10) in the double to gravity Π^2 also in the double on both sides of the Universe (+) in a proton relation $(\Pi^2 + \Pi^2)$ still maintaining the shape of the sphere (/7) is representing the flow of space-time by duplication in relevance to space-time dismissing that comes to a limit will be the maximum that the form of the sphere can sustain. After increasing the flow the sphere as form will collapse although space-time will still remain and be concentrated but valid. It is where gravity as we know gravity ends because that is where the any stars liquid atmosphere meets the density of the electron and the density equals the speed of light. After this value the speed of light will no longer sustain an electron in the atom, which by that time has been compressed to its limit as it was during the Big Bang. That is the end of the electron where the star is all electricity with no space-time being less than the value of the electron.

The fact that the displacement equals the iron proton "mass" explains why iron Fe_{56} is the only element that has the ability to generate electricity because by bringing focus on an iron Fe_{56} core through motion contributed to employ the Coanda effect electricity is generated as long as copper $_{62}$ is used to demolish space-time. Every electric generator is a star core in the micro.

The six-sided cube   In motion applying gravity

$(\Pi^6) / 6$ **(7/10)**

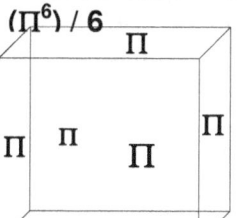

Forming the star

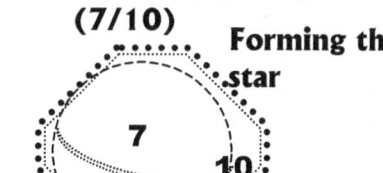

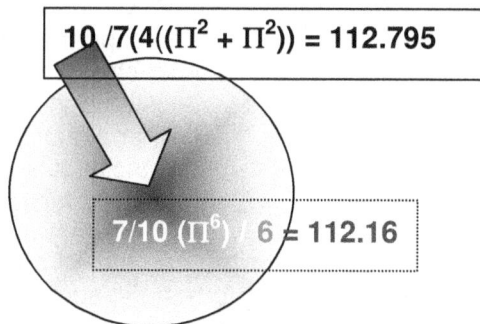

$10 / 7(4((\Pi^2 + \Pi^2)) = 112.795$

$7/10 \, (\Pi^6) / 6 = 112.16$

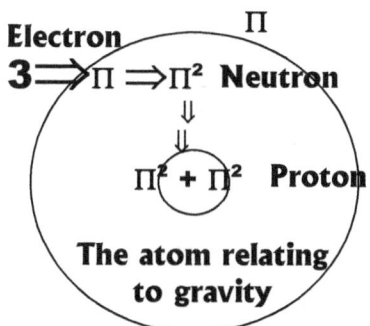

Electron

$3 \Rightarrow \Pi \Rightarrow \Pi^2$ **Neutron**

$\Pi^2 + \Pi^2$ **Proton**

The atom relating to gravity

From the value that outer space can support being the sum total of the particles forming the atom $((\Pi^2 + \Pi^2)$ **the proton** $+(\Pi^2 + \Pi)$ **the neutron + (3) space = 35.75 X Π singularity = 112.313** the star deliberately reduce the atomic space or the subatomic constructed- space as the star intensify motion and that reconstructing of space-time changes the qualities of the atom from what we presume the atom to be to suspending the atom beyond the boundaries of **7 /10 (Π^6) / 6 = 112.16.** The converting of space- time from outer space through gravity to the star centre is the same route electricity follow

Electricity and lightning is the absolute epitome of the Coanda effect where the Coanda effect is precisely the manifestation of light following the exact principles of the Coanda effect and the **Total Internal reflection** is also miming the same principle as the Coanda effect which is vivid proof of space-time $a^3 = T^2k$ the Coanda effect in acting principally by using the flow of photons instead of atmospheric heat. **Total Internal reflection** is only about applying motion by the flow of space-time (in this case water running) through the atmosphere but in the case of the phenomenon we call the **Total Internal reflection** singularity captures light holding the flow of light honest to a specific centre as does the Coanda effect and by setting borders the boundaries light is restricted as singularity set limiting boundaries to the flow of photons. But that is what electricity is; it is only creating space-time accelerated motion with much intensity added and it links a line than is concentrating space-time as it accelerates space time through the displacing differentiation which one find in stars between copper dismissing space and iron accelerating heat directly to singularity. It is only   much more intensity. All it is, is the Coanda effect forming electricity and lightning as the Coanda effect.

Gravity is electricity because electricity is the flow of heat from a gas source to singularity by charging iron $\{7/10 \ (4((\Pi^2 + \Pi^2)) = 55\}$ (forming the artificial core exactly as the Coanda effect will charge singularity by applying motion) through the influencing of total space reducing which copper can manage having the specific space-time displacing value. The influencing of copper $(\Pi^2 + \Pi^2) \ X \ \Pi = 62.0$ breaks down space-time as stars do in the core centre. That is the reason why only iron can excite to charge electricity and only copper can dismiss space to a tome equal to the flow of photons. All phenomenon used in the Cosmos is the precise same thing using the precise same principles in a more intense or lesser intense gradient. Still it is all about singularity charging the control and the flow of space-time through motion where a liquid flows through space to a solid iron core that is influenced by copper.

Electricity and lightning is gravity reduced by the intervention of the phenomenon we know as the Coanda effect where the Coanda effect is the establishing of a more intense dynamic point representing a new point of a controlling singularity dynamic.

Iron forming a centre or an iron- core precisely as the Earth core forms in iron form a centre core.

The copper field coils breaks down space-time by dismissing space-time as the 63 factor excels the space-time as fast as electron will as the space-time has to flow through the copper in the event where the iron being in motion causes the flow and charging the flow to the equal time set as the photon has. That is electricity. It is taking space-time directly to the centre f the earth because the motion $T^2 \ k$ excited the space a^3 to a level that gravity is within the Earth core.

Kepler said that the one half of space is motion and therefore there are no space any where except space in motion forming space-time. There is only space duplicating during a specific time duration and only space-time can relate to a specific point where singularity establish control. There is no space only because it is all space-time, which means space in motion by duplication thereof. Therefore when one removes motion from the space factor space will collapse. That is what a Black hole is all about! In relation to the mindset of modern man what I am referring too is unthinkable not to be absolutely part of common sense.

The motion of space-time is about space duplicating by filling the following space in the following time in a relevancy where the front space will be followed by the space just behind that and that will be followed by the space just behind that.

Material within stars (and outside stars play the game of follow my leader because the material follow in a flow where the one behind captures the space it will occupy as well as the space it will represent in the flow directed by the direction of time. The material flowing into an object is connected to material flowing out of the object by motion we talk about as momentum.

When creating a solid barrier to stop material flowing we destroy the space by preventing the motion to continue. We stop the movement of with some intervention blocking the space-motion but by stopping the momentum we destroy the space duplication. We refer to this as a collision between bodies but what it is, is that we fill the space which the space in motion was suppose the fill therefore we destroy the form of the space in motion and most times we destroy all space by creating heat bringing about a colliding destruction that produce more space in the end. The more space we supply in relation to the space in motion the more

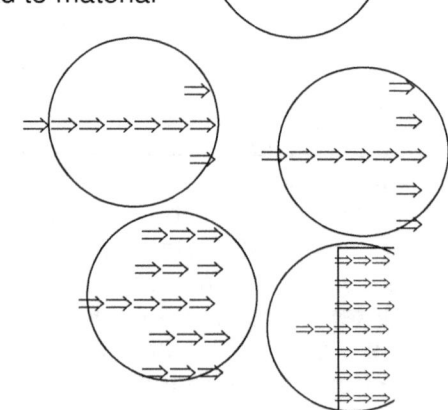

As the space in motion is occupying less space due to the motion duplicating and reducing the space the space in motion will need less space to duplicate thereby then create motion. Smaller objects can apply faster motion since smaller objects require less space to duplicate and therefore less time to do the duplicating of the space.

Stars in our Universe are controlled by singularity that defines the outer borders as **10 /7(4((Π^2 + Π^2)) = 112.795** on the out side and an inner border of **7/10 (4((Π^2 + Π^2)) = 55** in the centre of the star. That means the space envelope is **10 /7** and that is confined too **7/10.** That produces the gravity in the stars

If you wish to travel at the speed of light you must first become a photon. Objects entering the Earth find the space they have to occupy within the atmosphere much reduced in the time they have to duplicate such a space. The total space they are unable to duplicate then gets shed as photons because those material producing quicker motion than the object turns to photons as the singularity protecting the space time within the atom of the material going wasted has been sacrificed by motion to become heat. That heat became light and in that heat found a manner to travel at the speed of light since it parted with the material as photons, which then as photons are small enough to travel of the speed of light. At some level the demand becomes so strong inside a star the photons cannot sustain the flow of space in time. A division in the centre of the star develop where there becomes a barrier as space cannot reduce within the time constraint of the space we are in at a rate of more the proton can sustain as the proton (Π^2 + Π^2) stands related to singularity Π in as much as (Π^2 + Π^2) X Π= **62.0.** This is in contrast to the atom (Π^2 + Π^2)(Π^2 $\Pi3$)= **35.75** X Π = 112.31. In the centre the displacement value is equivalent to 62 protons but the space can not sustain the time in duplicating Whereas the outer space can maintain space-time to fit the entire atom the inner core of the star gets to a position it no longer can even sustain the proton The space motion in time becomes too little for the sustaining of even the proton and the proton gets rejected from the atom into the outside space There is no further space in time connecting to singularity past that border and space being the six – sided dimension as we know it to be becomes too little and to fast to duplicate. Space goes flat within the centre of the star and in all stars this is present. Black Holes are not about material escaping or not escaping but it is about motion of space not being able to support the dimensional

space in which we live. But Black holes are just the peak of an evolutionary development that can be explained so easily when using the cosmic dimensional code.

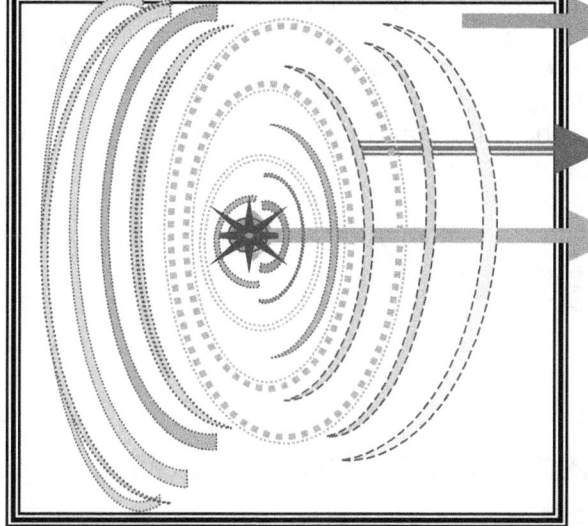

ELECTRON Space falling outside the domain of the star

NEUTRON Space in the star domain transmitting heat into outer space

PROTON Space in the star domain that holds the gravity and therefore holds the solar system in form.

Science should become serious about the task they chose to perform and not going flat out covering up mistakes by denying mistakes their Master made. I am aware that science is aware of the contradictions Newtonian science advocate mainly because no one this far acted surprised or bewildered when I mention it and they then have to realise it for the first time in their entire existence. They know exactly what I say when I say and with that they are performing acts so shocking about self-protection and self-preservation. When Roche presented his findings they should have realised there is something missing with the way they see things. When Hubble presented his findings they should have put Einstein to task about finding the mistake they made and not a manner to cover up the mistakes they made about the Universe contracting. Newton's arguments mathematically are about contraction and

Hubble proved that the entire Universe proved without doubt that all indications prove expansion. By going overboard even further and order Einstein to measure the universe is frankly madness! But I have never seen any group of persons to be blatant in their refusing to acknowledge a possible mistake that Newton made in the manner that I do. Mainstream science places Newton equal to religiosity. I found on all and every campus I went that any remark about uncovering a mistake Jesus Christ supposedly made generated immediate interest with even the most adhering Christians coming to hear the argument. Making a remark about Newton making an error gets you marched off the campus by security. Why not test Newton's $F = G (M.m)/r^2$ from figures Kepler left us and see how far did planets shift closer. Prove what is lectured to millions of students. Where such observation did take place it rather proves definitely quite the opposite because the distance between the Earth and the moon is growing and it is not shrinking. With me openly criticising Newton and Newtonians lecturing none applying information across all of the Universities across the world (I guess) will once again make this book as successful as the others in the past. The Academics get very a very subdued hostile attitude towards me when I blame the Academics to their faces that they are not about gaining knowledge but about conserving the past through protectionism. Universities protect their own without any willingness to test that which it protects. All evidence should be clear in confirming that the basis on which the entire world of science forms a union is founding their policies and beliefs on incorrect principles. They should not become annoyed with critics but not only that, they should show evidence immediately to back their claims showing how far did the solar system-structures move closer. From that we then can calculate what collisions we are waiting for and how long before the big solar clashing will begin. The absence in they're just mentioning such possibility confirm to me they know as well as I do there is no evidence of the moon reaching the Earth with no evidence of pulling or tugging and the Universe is in synchrony more than any person may ever be able to prove. When looking at photographic images coming from the sun we can clearly see that the fluid pushes out of a bowl of liquid and the telescopic images coming from the sun via the camera lens there can be no doubt that the sun is a bowl of liquid sloshing like a boiling kettle of soup. , From within stars, spilling both sides as it falls back into liquid pool forming the sun. The inside of the sun is not gas but it is fluid. In all of nature there is no NATURAL **gas** as much as there is no **natural solid**. Hydrogen is as much a liquid as iron is a gas and neon is a solid. It depends on the element relating to the space/heat in the circumstances surrounding the substance at that very precise instant in time. We have to stop telling the cosmos to show us what we wish to find and start accepting what the cosmos is telling us what is out there that we should look for and find. The fluid state and the gas state is expendable waste that stars remove through development but then so is all space-time and material. Using the location to explore these books I announce will indicate as to how singularity came about to form space-time and commanded motion by creating space. In that exploring we may find out that the Universe is already contracting as much as it is expanding and it is contracting by expanding because it is through the contracting that it is expanding; the answer comes about from $a^3=T^2k$. My effort with the criticizing of the Academics was never to attack the world of physics and I never had it in mind to destroy any work made by them. I can ill afford enemies and even less enemies as power full as the Academics are. But on the other hand I cannot go on praising work I disagree with when I aim in full knowledge in the mistakes I see about their work.

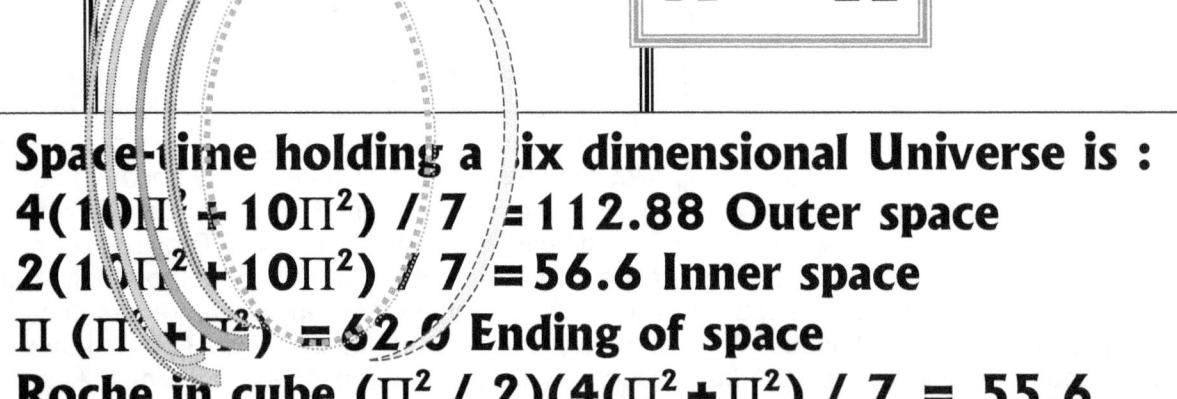

$$k = \Pi$$

Space-time holding a six dimensional Universe is :
$$4(10\Pi^2 + 10\Pi^2) / 7 = 112.88 \text{ Outer space}$$
$$2(10\Pi^2 + 10\Pi^2) / 7 = 56.6 \text{ Inner space}$$
$$\Pi (\Pi^2 + \Pi^2) = 62.0 \text{ Ending of space}$$
Roche in cube $(\Pi^2 / 2)(4(\Pi^2 + \Pi^2) / 7 = 55.6$

At

first I was very timed in my approach, which brought me no joy as well as no results from any Academic near or afar responding. I then felt forced to become much more bold and down right to the point. Still the response was the same. I concluded at the time my lack of results most probably that my revelations did no surprise those Academics with my startling evidence one bit with revealing my evidence about they're contradicting testimony about the official accepted cosmic principles physics. I now am of the opinion their lack of attention to my asking their agreeing or disagreeing is more about they're refusing to admit than they're being surprised as they are stunned by their seemingly double talk of their teachings. Not even once did any Academic in any way acknowledge the absoluteness of motion controlling the space making the space being valid in the Universe. All this is notwithstanding all the clear evidence collected by centuries of investigative studies bringing proof and proving to this affect. If the motion driving space-time towards the centre of the Black hole stops the Black hole will altogether disappear. It is the driving motion of space-time concentrating as it speeds up moving towards such a centre that that puts the Black hole wherever the Black hole may be in the centre of that Universe the Black hole keeps captured by motion.

Everything in the cosmos is moving, on two accounts by own individual accord, as well as under the influence of some other singularity dominance. In explaining we return to the top. The two opposing motions are inseparable and always in relevancy but never to size because it is always to motion. In explaining we return to the child toy in the spinning top.

When the top is in a state of motionlessness on own accord it is everything but motionless. The motion it adapts are synchronised with the Earth in harmony with the solar system and according to the greater picture of the cosmos. When an energy source not related to the cosmos called life intervenes and energises the tops motion, the singularity in that top suddenly jumps to life. By adopting a rotation energised to an unnatural state of energising because of life's intervention, the singularity that up to that point dominated the top is no longer in absolute dominance of all aspects of the top's motion . Another singularity takes charge placing the it in charge by nominating a point in singularity as the Coanda effect takes charge as the sun or any other star for that matter applies more and more motion, it will begin to find a means whereby it can escape and apply individual singularity as the top starts to separate from the singularity the Earth holds. The singularity holding the Earth would then allow the singularity of the top to rotate within a specific band where that specific band of being active will tolerate such striving for independence before the earth's singularity will start to destroy the singularity in rebellion. The top on the other hand will try its outmost, when the singularity it then forms by individual spin is too strong to remain without independent motion and being totally be in dominated by of the Earth's singularity. The motion of the top is an attempt to begin applying an individual singularity space-time defying and standing apart from the Earth's gravity. This is coming about by forming the Coanda effect. That action we see as the top starts rotating in a manner where the top does not align with the earth's singularity. With the adding of spin, the time the top holds becomes unrelated to the time the Earth holds and the top will start a campaign too escape from the singularity domination the Earth has on the top. When the time or spin of the top exceeds the limits the Earth places on the top, the top would emerge by trying to escape from constrains placed by the earth. The view I present at this point is known to science for almost as long as science knows mathematics. The motion establishes the singularity that sets the top's drive for independence from the Earth's gravity restriction and set the top in balance as Kepler indicated. The space the top represents becomes the motion thereof $a^3 = T^2k$.

If we wish to find the future we should locate the past. If the cosmos is contracting, where to is it contracting? The direction of contracting must be in the opposing direction the direction of expanding. If we wish to locate the past from where the cosmos came and through that see in what direction the cosmos came, it must take an effort to backtrack the direction it came. Should the argument come about that all came from nothing, then everything either still has to be at nothing, or our understanding of nothing leaves much to desire. Nothing means not existing, not being, never found and unable to produce any multiplication there of by any growth.

The above questions, but mostly the unanswered questions about what is more nothing and what is less nothing draw me to the realisation there can be no such a quantity in space as nothing because even space has to be something. Clearly as it is for any one to see one create space by nuclear explosions. In explosions Academics portrait that the wind as shock waves, but what is the shock wave other than new space coming into prominence and rearrange the structure in relation to the new space just created by liquid heat unleashing the created space as well as the space volume that came in place. In that way it is clear that releasing heat brings about the expanding of r as part of the sphere forming space. Hubble proved the universe is expanding. Then by backtracking we have to set about reducing the sphere constituting the expanding universe. If r in the circle is growing we have to reduce r to backtrack. When the circle reduces, the value located to r will become implicated because r determines specific size. Not so in the case of Π, because Π in the true sense only indicate that the circle is a square without corners and therefore Π dictates form and not size. By reducing size only r comes into contest and will point to such reduction. By reducing the circle radius r by half continuously will lead to an infinite small circle but Π will remain because the circle as a form remains even being infinitely small.

If there is anything that we can learn from the Coanda affect and how we see it applying in the Universe then it is that motion brings about an allocated and generally elected chosen centre form where that centre find control of all of the Universe being the Universe in such a controlled area.

But we may say in our self-defence that is what the cosmos tell us to believe!

We use the most infinite to view and formulate what we think is going on in the Universe. Being Π sets us in the centre of the Universe. We take so much light for granted never thinking for one second how impossible our relation with light truly is. This totally extraordinary relation we have with light must be one of the reasons why we humans put our position we have in the Universe in such a pivotal place. The fact that we as life carrying individuals especially we in person that are all blessed with the ability to only use our hind legs to walk on, on the surface of the Earth has the idea that the Universe was created especially for us, us being those holding life. Think again and such an idea supports everything everyone thinks of his own importance. Such an idea is absolutely bizarre.

It would be the same as if the ant running in Central Park in New York is of the personal opinion that it is his being in the park that is the cause of the park being where it is and maintained as it is and only because of our ant's personal benefit that every one which is there is in the park. Our ant considers that every one of millions of people thinks a thousand people is dedicated on his behalf to maintain Central Park in New York. All the people in New York are of the obsession having one purpose to live for and that is to please that one ant. And yet that is happening with the light and us. Every person is standing in the Universe is under the elusion that all the light through out the Universe is directly flowing to the very point the person is standing. It happens to all of us. The place where I stand or any other individual for that matter is standing is positioned in such a manner that every beam is directly flowing to that very specific spot. Every beam is coming at the speed of light through the entire Universe to locate such a person with that magnitude in honour and glory as to fill the centre of the Universe.

From all the corners of the Universe one line of light is especially directed to that specific location used by that specific person for that specific instant in time. One very important human being is filling such a location of absolute splendour. The light departed from every location in all points through out the entire Universe stretching further than the mind can admit directed on course to meet the person in that centre spot. The light followed one after the other dedicated for billions of years to flow in the direction and directly that is the spot, never diverting for one instant, to come to where I am filling that spot in that centre of the Universe. All the light in the Universe is coming to me. It is on route straight to me where I am standing filling one spot on Earth. If there are those that don't believe me, well those I challenge then to go outside and see the vastness everywhere form wherever the light is coming from. It is coming from all over. It is coming from areas so large not even Einstein can calculate the size or content and it is rushing towards me specifically. There is not one ray that is going to miss me by fluke or accident. The light has one purpose and that is to meet me at the point I am. Every beam has my name on it and it is coming for my eyes. Can any one imagine if a person was standing in a location and found all the persons in that city was running towards him where he is occupying that point, how frightening such a person must feel. Yet it is happening to every one from where ever the vastness of space is situated and is coming across space to that very specific point the viewer is standing.

Even if I shift to another position on the other side of the Earth or to the moon the light will change direction and trace me in my new location. Even if my new location is in a camera and the camera is in a vehicle in the centre of Mars the light will know that I am using such a point and trace the camera so that I may still be in the centre of the Universe. Wherever I might be, the light will still get me at that location. The light flows to me from where ever and to top that it is also flowing to all other persons. That means it is not the Earth that is that important but it is where the point location is and that point which the observer is using to view from, that is the most important place eve to be. If it was only the sun that the light was streaming from that is choosing me as representing the Universal centre being the centre of the Universe it then cannot be that very exclusive. The sun is close and the light is plenty. But that, which I am referring to, that it is coming from all over.

That is just one small part of the fantastic affair. Some of the light left the stations they come from some 12×10^9 years ago to meet little old me in this spot I am filling. The light has been travelling 12×10^9 at the speed of light, which I might add is much before my birth crossing space and time, rushing all the way to meet me at this point. No one ever thinks how it was possible for the light to know I was going to stand at this point and be here the moment the light arrived. How did the light know I was going to take centre stage at that moment when it left so many billions of years ago and fill the specific centre of the entire Universe? I have to be in the centre of the universe because all the light is travelling to this spot filling the centre of the Universe without one straying off course and missing my spot. The light takes two million years coming only from the closest next galactica to meet me here taken into account the prefect timing it applied after all travelling that far to be in time just to meet me in the centre of the Universe. How important can I ever dream to be? Light is coming across time measured in millions and billons of years through space measured in millions of trillions of kilometres, travelling at the maximum speed the cosmos will allow, ignoring all other places it could go too and came to meet me in the centre of the universe.

To the light on route time means nothing and space even less. Light cannot show more motivation to reach me at this point that I am filling at this moment. Not one ray is by accident missing me except by my choice prevailing. It flows through the Universe in time and in space in the hope that I, the chosen Universal centre whom is filling this spot, is in the spot that all light is coming too, hoping all the way and all the time that I would be graceful enough to notice the light. If I were not in the mood to acknowledge the light, the light would have done all the travelling just to be disappointed by my not meeting the light. If I choose not to go outside and acknowledge the light all the travelling the light did was then in vane. An effort spanning billions of years and an effort stretching trillions of mega kilometres was all done for nothing and no reason at all because I neglected to meet the light. From everywhere the light is coming my way and that miracle is passing me by because I am feeling even more important as to acknowledge the total importance I have. The light is tracing me specifically at the location I am occupying just to please me and serve me with all the information about the history of the Universe. I can accept and acknowledge the effort or I can dismiss and ignore the lights efforts. I suppose that will allow me some arrogance and encourage me to think this all was specially created with only me in mind and if I wish to draw a map about the Universe I have all the right in the world to place me in the centre of the universe from where all of the everything has chosen to meet me, the one to be met. After all, all the light is doing just that!

This information being at our disposal for thousands of years we never used except to inflate our already overblown ego further. By studying what we experience we can dismiss our superior view we hold on our self-importance and find what the cosmos is telling us about the position we have concerning the cosmos. It has to be true that we are in the centre but by using our knowledge especially concerning the Coanda principle we should gauge how the Universe work because we hold the point from where we may judge much, much better. That is because of the Coanda effect and because we are in motion apart from the Universe within the Universe confined to and being inseparable of the Universe we all are still carrying the eternal centre located to us as we are in the eternal centre.

I am in the forefront of motion in the Universe as is everything else having singularity...and everything, but everything has got singularity. Motion is not what we think of through what we experience as motion in getting from here to there because we do not wish to be here but we wish to be there. Motion in the Universe is the rearranging and the re-administrating of objects. It is a question of relocating the arranged to be re-arranged. This is because the Universe and everything in the Universe can't go anywhere but remain in the Universe. Because there is motion allocated to me I become a reference point for singularity that is holding space –time that is guarding singularity. Because of the motion I receive the allocation of being the most prominent of that entire Universe that can and that may be dominating space-time because I am pivotal to gravity, which is the gravity in my area securing and surrendering to the form to the Universe by accepting form from of the Universe. With me being in the position that my life within me holds me in singularity it's putting me in a spot of not representing space and therefore being motionless. Because I take on singularity everything around me represent space and motion at a distance from the point I have no space and no motion. Remember I am not my body as my body is the closest space-time I can control. With me being life I hold singularity. From where I hold singularity everything is space in motion and it is because of the Coanda effect I feel the affect of being the centre in the Universe. All space is spinning around me and because all space is spinning around me the Coanda effect puts me in the centre of the space-time sharing my Universe. There is one Universe, which is an innumerable, many Universes all referring to one Universe and represents one Universe in innumerable locations within the same unit. The motion I am due is the Universe allowing me my due to be. The Universe is not about positions or locations but relevancies responding to other relevancies that either puts me in a form or uses me as part of a form An object in a collision does not destruct or destroy but merely rearrange its composite from point of motion to another point in motion. But all motion of space forming space has the due to pay, which is the due I have to pay. We all replace energy from one location by motion to another location in motion.

In the past, and even in some quarters today, science is on the search for the 100 % efficiency machine. That theory runs on the surmising that a machine can drive as an output delivery without receiving input of energy. A few hundred years ago many Kings were fooled by such notion and some scientists truly spent a life in honest search of just such a device. Mostly the accomplishment came from cheats that very well new their machines were not up to the task, but in fooling a rich investor, brought about wealth to the inventor. As science progressed the no input giving all output machine became less and lesser a feature of the honest inventor. But the idea does not exclusively come from crooks finding a way to cheat the world. The practise of receiving without giving comes from science in the form of physics. It is physics taking the world on a wild goose chase in the way physics present the cosmic motion.

Physics propagates that the cosmos is all about running without input driving energy. The cosmos is all about wasting matter to a supply of motion. This idea prevails even after the world of science saw clearly in the past that there could be no such machine anywhere. In the Universe space can only be space when space is in motion and motion is about energy. Energy does not come free and energy always has a price tag because of differences in energy supply. That made me realise that. Even the cosmos must be a machine driven by an input and an output. It is the input / output driving energy that must be located and the driving ability we have to locate. Science hold the idea about mass that forms the unexplained and is therefore the magical drawing power to prominence, but what if it is not the drawing power of mass that holds prominence, but it is the reducing or contracting of space that is the driving motor behind the cosmos. All energy we humans at present use to accomplish material being in motion, holds some form of heat redistribution. Even electricity is a form of pure heat. I say that holding in mind of what apply when the energy of electricity becomes over abundant and the machine overheats. By overheating it means that the motion the machine creates comes about from heat control and precisely planned heat distributing. When scrutinizing the process we find behind the charging of electricity we find behind the charging of electricity is the same principle than what establishes the Coanda effect . Understanding the connection between the various principles helped me to connect the line of dots.

That brought me to realise that it is not me that is drawn towards the Earth, but it is the space in which I find myself that reduces, and that produces the effort bringing me closer to the Earth. The formula $F = G \, (M_1.m_2)/ r^2$ suggests driving, moving in a direction and contracting. It suggests the reducing of space and not merely drawing or moving closer. When looking at any machine in practice, the machine draws power from space that reduces whereby heat increases. Not releasing that it is the heat that forms space is a misguidance of the truth. The heat creating or forming space will lead to the destruction of the composition forming the machine. There is no form of matter, or element strong enough to resist the material deformation brought about by overheating. Having this in mind that matter does not resist heat, it is of importance to recognise that it is heat that is allowing space to give matter form. Looking at the manner in which energy is utilised it is space and heat forming matter allowing motion that allows work to achieve value. Motion originates from singularity by establishing the proton promoting spin through out the atom. Then there comes the part that makes the star a star or any cosmic structure what it is because every cosmic structure is the motion it is and not the material it confines. A Black hole will disappear into the realms of obscurity if not for the motion it establishes.

At this moment science is all about a body falling where the two bodies are producing a force whereby the bodies draw one another closer and all this is enticed by unbelievable gravitational pure magic. The bigger the mass, the bigger the drawing that comes about from the force unleashed by the mass of matter. The idea about this practise was phenomenal in 1602, it was impressive in 1802, but it is really ridiculous in 2002. Why would Boron form a solid having 5 protons weighing 10.811 g / mol from a dense solid structure and Argon as a gas having 18 protons weighing 39.9 g / mol. be a gas so light it floats in the atmosphere. This would imply that the "heavy" element with the biggest drawing power is a gas and the lighter element of the two is a solid. That denounces the contracting force theory. The way we compile and use energy must be in a similar manner to the way the cosmos uses energy distribution. **We humans can create nothing, but nothing is all that we humans can create**. The rest of our achieving is by duplicating whatever nature provides.

We have advanced past mass but and should now find the ability to establish what drives the Universe except for blaming some medieval magical force coming from nowhere going nowhere we have to find what drives us. The energy we use in all forms is producing heat in space by either converting space to heat or heat to space. Explosions are about converting heat to space. Compressing is about reducing space to heat. That is all energy composing work and is the only method of producing energy notwithstanding the immeasurable many names we use to express the same function in different forms. Arriving at the question about locating the space and time forming the centre the centre of the universe one has to realise the centre of the universe are in every singularity forming matter weather it is big or small, size carries no significance. It is the impartiality of singularity that is claiming the value and not the differentiation of matter. One must realise there are no big / small or hot /cold or near / far. It is all relevancies between matter claiming space and space is heat in a turnabout manner. Every aspect in the cosmos are locked-in universes, sealed off from other universes and inclusive or exclusive depending on singularity holding relevancies relating to one another. The relevancies rely on inter dependence and inter linking, but there are no differences according to human sizes or standards. Accepting that principle unlocks the "so called mysteries" of the Universe and brings about clear understanding. It is all about accepting, acknowledging and interpreting the role singularity maintains on matter.

One should not try to focus on an image of such a spot or dot because there is no image. The line dividing the cosmos and that run through every particle, no matter how large or small is beyond our vision. Such a small line, so small it is not even noticeable and is so small it is not part of the cosmos we can see but is large enough to part the cosmos into sectors. It splits the biggest there is into particles and we are not even able to notice the precise location of such a split. In truth there is no top or bottom that we living in 3D can see. We shall have to use a general conception brought about by intelligence. Your intellect tells you about such a spot, but that is all indication anyone will ever find because that spot is on the other side of the Universe (quite latterly). From the centre of the dot there is a top and a bottom spot. From those points there is connection with four quarters. That produces six connecting points that are all aligning to the centre. Because it serves big and small, hot and cold equal and alike, and it is the smallest cutting the biggest into equality, size is of no issue. Size is what man makes of it. In the Universe there is no size in hot and cold, large and small. And that dismisses all prominence of what we ever wish to give mass. For the smallest there is, singularity is serving the largest there is equally.

Our instincts, our logic and our calculating process all indicate that the sphere holds a centre point from where six evenly positioned point's position matter to be. Using The formula $F=G (M_1.m_2)/ r^2$ it indicates to a force pulling objects closer, where each force is coming from each centre point the body in question has. Where every atom dismisses space as it is spinning the spinning find refilling from outside coming as it flows inwards directly in a circle to the centre (because there is always a double connection to gravity in motion) of the cyclical sphere. At one point in the precise centre, the dismissing of the total ability of all the components within the structure finds a peak and at such a peak the dismissing forms the biggest influence on all points at the outside. That places the border forming of the group selecting the unit by motion and such reducing becomes inherent part of the form of the sphere spinning to create cycles. In nature the only form provided is the sphere. The contraction that causes the reducing of space must commit the two bodies towards a point in each case being spot on in the middle, not withstanding what direction the force is applying, the body will draw to the centre when being part of the unit. Only when the heat ratio promotes more duplicating than dismissing due to the spiral cyclic relation the elements hold with heat will that relation with heat counteract the dismissing and neutralise the overall effect of the collective dismissing.   If the Universe spins around a centre point holding singularity, and singularity confirms the centre of the Universe, then every particle holds the centre of the Universe making the number of universal centres immeasurable many, and every atom and sub atom particle presented outside the atom in smaller bits, are all not pieces of the Universe but they are a Universe surrounded by many Universes. If every atomic particle no matter how small is holding the centre of the Universe, then the gravity is coming about from that point because that is where the gravity applying in the Universe are applying contraction. There was a beginning that saw a radius between objects so small the size will never again repeat. The diameter of the particles were also next to nothing but that should not be a contributing factor surely…the main focus point is that particles were as cramped as it shall never again be repeated.

With the radius in the square dividing the shared and combined mass of the particles the relevant mass of the particles rises by the square as the radius reduces. If the radius becomes infinite, the relevant mass that the particles will produce goes up eternal. No force in the world would keep particles apart drawing on each other with such an applying force where such a force is divided by an infinitely small separating radius is producing force as big as the Universe is capable to produce. That was the Big Bang. The universe always flicker as a solid being in a state of duplicating which is held by a flow of gravity that can never repeat once segregation. The question in need of the answer that will bring in the light is then what brought on the break? This is a recipe for joining and not dividing. Still according to the Universe I am able to witness the dividing that became enormous and the joining distance being practically irrelevant. The gravity was more than words can describe, the heat was able to melt it all in one structure, but that did not happen. It split into an innumerable number of billions of individual atoms

It then is the atom in the most centre part where space and time meets singularity, that Einstein found a Universe collapsing to a single dimension, and every atom at a point post of the proton where gravity initiates in according with the proton dimensional colas **of $(\Pi^2 + \Pi^2)(\Pi^2 \times \Pi \times 3) =$ 1836.** The relevancy of the atom proves the three forms of material there are in the cosmos. The solid structure is represented by the proton relating to singularity in the form of $(\Pi^2 + \Pi^2)$. The neutron is the liquid part representing the motion or movable part in the cosmos as $(\Pi^2 \Pi)$. Then the final stage is the space era represented by the 3-part in the atomic relevancy where the liquid formed space. The last part representing 3 is the Big Bang era where heat changed to space. The evidence of this statement is so clearly visible any one must see its presence.

$(\Pi^2 + \Pi^2)(\Pi^2 \times \Pi \times 3) = 1836$ is an atomic relevancy that projected from the atom to the star directly without loss of any translation. The formula of $F = G (M_1 \times m_2) / r^2$ only apply in a very specific range, and at a very determinable point the formula does not effect objects in the air. After such a point one will find satellites able to orbit, be it art a definite pace that matches the rotation of the earth. Still…below such a point (B) orbiting objects will come crushing down to the earth.

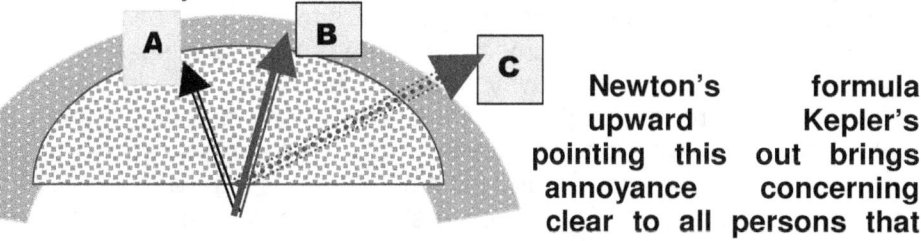

From point (B) to the earth apply and from point (B) formula apply, but my about all sorts of academics. It must be Newton's formula upward Kepler's pointing this out brings annoyance concerning clear to all persons that **there are a big difference between the applying of Newton's $F = G (M_1 \times m_2) / r^2$ and Kepler's $a^3 = T^2 k$. When the objects reach some point they will drop to the earth and when that happens, mass do not play a part in the speeds they come to reach.**

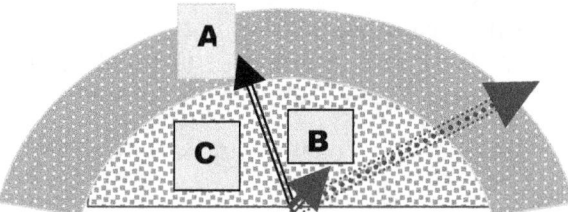

$$A = (\Pi^0 + \Pi^0 + \Pi^0 = 3)$$

Translated to the atomic part that focus on the collecting and accumulating of space transforming space into heat by concentrating the time and establishing general differentiating between outer space and inner space of the star

$$B = (\Pi^2 \times \Pi)$$

Represents the neutron part of the star where the motion maintaining of the star is confirmed in the carbon nitrogen oxygen space-time accelerating within the star.

$$C = (\Pi^2 + \Pi^2)$$

This part represents the proton phase and the proton part of the star where all space-time is confirmed to singularity by dismissing of space-time and establishing growth in securing singularity maintaining.

When examining the case where two balls drop vertically, gravity, as a force does not apply and therefore gravity does not come into effect because there is no difference in speed or duration.

With out any apparent reason the formula is substituted with the following formula:

$g = G(M . m) /r^2$ where:

G = the gravitational constant,

M = the mass of the body,

M = the mass of the lesser body

r^2 = the radius between the two bodies.

> What ever this formula needs it is desperately lacking a foundation of basic necessities to provide substance about basic fundamental logical questions no one ever dare to ask. The slightest provocative questions leave this application high and dry. There is a dot needed in a spot where r needs some defining of proof.

Should one ever wish to paint showing that atoms formed as a result of the one mass groping the other mass as the two were clutching one other till dooms day arrive bringing coupling then such a venture has to start by providing the incentive to initiate a centre from where the clutching may originate and why such clutching will originate. What brought about a centre r that brought about the mass one groping the mass two and why did mass one form as well as mass two formed because the first and foremost forming of mass one and mass two before the groping started had to involve a centre beforehand and afterwards it proved it mass. But before being a mass it had to have a centre to secure that mass. What brought about such centre into the cosmos. Stopping at 10^{-43} sec. There had to be a centre to start whatever mass was groping another mass that had to have started with a centre. By not providing a centre one drops into a basin fluid with nonsense since the initial explaining of what and why and where the very centre arrived at that brought about the groping in the first place the whole notion is a loose unstable not very well thought through blubbering ridiculousness in the realms of Little Red Riding Hood. Why was there one centre because the moment you say mass some centre had to establish the mass notwithstanding the size or nature of mass. Mass is contained so what contained the first mass before there was any mass to contain around a centre. At least I can bring a centre to the table as I prove with my reasons I give for such a centre forming before even any mass came groping and snaring other mass. I can indicate with a fair amount of logical accuracy why such a centre will come into place. I can show how and why and where motion brought on this very first groping and clutching if there is groping and clutching to begin with. I can show what phenominon is responsible for this centre to become practical and I can show the phenominon positioning and locating of such a centre. I can show in the shortest manner by using accepted mathematics in the simplest of forms why space is linked to time and all space revolve around a centre using time to do so. I show what natural occurrences producing motion because producing motion provides heat the space it establishes. Space forming to accommodate heat expanding is providing motion hat brings about cooling through duplication that is the result of motion. Cooling establishes duplication which provide reducing as the product is reduced by duplicating and in that duplicating the doubling of pace is halving the heat in the space. It is a natural process that can be tested anywhere in the cosmos. Heat something and the space grow bigger whereas cooling the something will naturally reduce the space it holds. It is no diverting of nature by creating anti whatever that disappeared and comes back just like the Phantom does in the comics.

Let us take this formula back to the accepting of the Big Bang and find sensibility amongst a lot of confusion that I can see.

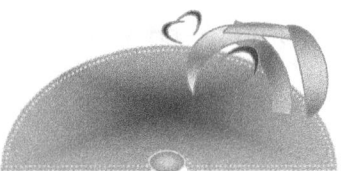

See the fluid push out of a bowl of liquid, spilling both sides as it falls into liquid. The inside of the sun is not gas but it is fluid. In all of nature there is no NATURAL GAS as much as there is no NATURAL SOLID.

Hydrogen 1	melts at -259^0 C,	boils at -252^0 C,
Helium 2	melts at $-269\ ^0$ C	boils at -268.9^0 C
LITHIUM 3	melts 180^0 C	boils at 1300^0
BERYLLIUM 4	melts at 1287^0C	boils at 2770^0C
BORON 5	melts at 2030^0 C	boils 2550^0 C
Carbon 6	melts at $804\ ^0$C	boils at 3470^0 C
Nitrogen 7	melts at -210^0C	boils at -195.8^0 C
Oxygen 8	melts at $-218.8\ ^0$C	boils at -183^0 C
Fluorine 9	melts at -219.6^0 C	boils at -188.2^0 C
Neon10	melts at -248.59^0 C	boils at -246^0 C
Sodium 11	melts at 97.85^0 C	boils at892^0 C
Magnesium12	melts at 650^0 C	boils at 1107^0
Aluminum13	melts at 660^0 C	boils at 2450^0

No element is either a gas or is a fluid or is a solid. We arrange the elements in such a manner, but that is only applying to the situation the earth grants the elements.
When an element freezes it is solid notwithstanding...
When an element melts it becomes a liquid
When an element boils it is a gas again notwithstanding.

Gravity is the duplicating of material by the dismissing and removing of space that leads to the concentration of heat. Gravity cools space down to liquid stored in concentrated space and further freeze space to turn liquids into solids. By duplicating the protons bring about cooling that that freezes space-time into the single dimension of singularity by removing space-time altogether.

Hydrogen is as much a liquid as iron is a gas and neon is a solid. It depends on the element relating to the space/heat in the circumstances surrounding the substance at that very precise instant in time. We have to stop telling the cosmos to show us what we wish to find and start accepting what the cosmos is telling us to find. The culture that I am referring to is all about **nothing.** At present we find that there is something we think of as nothing in outer space. Because nothing is what we wish to find and nothing is precisely what we are getting because we think of outer space as nothing. If you accept the cosmos to be nothing, then please define nothing to yourself and find the definition in the cosmos.

But as it is impotent to realise the above it is just as important to realise that heat is another form of material and a separate form of material. The two developed on equal basis and as a result of the other. The one produced to save the other and what the one produced saved the other .The one principle brought the incentive for motion while the other took the incentive by providing the motion. The one produced what the other captured and the one retained what the other delivered. Eventually the motion did not bring the required relief and another form had to be devised. By overheating and increasing space it counteracted overheating and by removing the expanded material and retaining it onto the contracting of the other did the two form a synopses where by all received benefits in the form of cooling. Only when further requirements develop did the need arise for more to be made available. The first demand on motion asked no further changes because one change brought on satisfaction to all that suited all. The second was more general and on an ad hoc basis that was established to fit the need of individual places and not groupings in the broader perspective to fit individuals at large. At first the establishing of motion set a trend that brought on required results but afterwards the space required in which to move became a demanding issue as the heat levels required out of control. The heat had to be stored in space by becoming space to retain heat for later consumption. The number in ratio that produced the heat providing particles that offered to release their form in contribution the have those that retained form do so to save those others in retaining form. But those on offer became those ones that became the danger of destroying Creation instead of saving Creation. There might even some areas and regions in far off places in our modern day where an imbalance may evolve and some particles become unsuccessful to save those more successful. By going less successful the singularity places a demand on another bringing about the command on space-time so that support can be accomplished to save singularity. Therefore by losing density was gaining security to survive as part of a bigger relative. Density is the distributing of heat in specific relative space and by having less material in more space the density is the offering for the common survival of the lot. But the relevancy brings a contribution in whatever role to secure the survival of the lot in relations. No relevancy therefore can be "nothing" notwithstanding Newton's opinion about the matter as Newton had the opinion a relevancy acquired by rotation brought about an accumulation resulting in nothing.

Another point I question about the Official Policy is that them and me and we am in agreement that the heat melted forming a cover on particles and from that particles grew more massive meaning they became bigger. By flowing from the relative softer part, which I call, heat onto the relative harder part, which I call material particles in such joining produce space-time distancing space from singularity by applying time to increase. Through that action such combinations produced particles in many stages. All of the particles less than or more than the atom used this method to come about the form they took on. How it happened is another bone of contention but more about that a little later on. By the arriving of the third stage there was heat on the outside and there was matter on the inside. We are able to still witness this covering of heat in the yet to develop regions of all developing galactica. The heat was liquid because the sun and other stars still indicate masses of liquid fluid inside. I can only imagine that that liquid inside the sun holding temperatures as low as 6500^0 K and up to 1.8×10^6 K the heat already is in a molten form. What about the heat then when the frozen outer space was 10^{34} K and such temperatures were the general order of the day back then. If the sun is liquid now then those temperatures raging back then must put the heat in form available in outer space at the time as dense and thick.

From the outside drawn onto the particle inside the blanket of heat came a flow of soup that became matter enlarging the already existing material formed at the beginning as solids . By dismissing space way inside all of the deepest the centre is drawing space dense with heat on the outside as it froze onto the already established solids. The forming of materials by using what ever is used to fill material was a process that was in use from the very beginning. It moved whatever is inside singularity out of and towards the outside of singularity but some came on the rebound and filled again that which others lost as that was pulled into…and that made material. That much I do understand. This carried on until…when? When did this stop. When did the universe run out of heat? Is it when the cosmos became minus 270^0 or when the Universe reached 0^0 C. The fact which I am trying to establish is when did what that first filled space become nothing and then was no longer used to form material because what was then filling outer space was no longer was no longer filling outer space which then mean outer space filled up with nothing coming from nowhere. These temperatures are scales developed by man and it does not indicate actual freezing or actual boiling. It only recommends what conditions are applying on Earth in relation to the way life responds to the application there of. When did the situation arrive that outer space was cold because before that time outer space came across to be considered as being hot. At one point it was 10^{34} and that is as hot as hell. Now it is 0 Kelvin and it is regarded as hell freezing over. So when did the transformation come about when outer space was room temperature at a very refreshing 18^0 C, which all persons would consider, represent a nice working environment acclimatised to suit life in distinction. When did outer space evolve in the nasty place, which represents the coldest there is. When did they begin to consider outer space as the coldest all around with no heat anywhere? Is it hot when it cannot sustain a human body with heat. Man is not actually the centre of the Universe but reading science one would never guess that science are of such an opinion. It is always a constant applying because the Earth permits and the cosmos produce the permitted. Where to did the Universe dismiss the heat that was once there but now is empty? At the start we have lots of heat and little space. Now we think we have lots of space with no heat. Look at the centre of any galactica still in development. The centre is the Big Bang still in progress while the edges are what we are. In the centre of Galactica there is no space because the stars that are there are so closely packed a blanket of heat covers them. How did the process stop bringing from space intense heat and from that particles grew?

What I wish to know is when did the process off building by accumulation of what ever fill the space outside and coming from the outside inward stop manufacturing material. It was a practise that was in place in the beginning according to Mainstream science when sub atomic particles forms. Now my question to those advocating nothing being in outer space when and at what point and why did this building process stop and allow nothing to take the main measure. What made such contracting by bringing in material stop affecting the growth of particle, by applying the basis for the theory of the growth of material within the atom space, in fact the growth of everything that grew came from this first growth. What you see or do not see grew since it was part of the Big Bang and everything in the cosmos at present was part of the cosmos during the Big Bang. My theory say this process of collecting heat from outer space never stopped but is an on going process we now gave a nice name calling it gravity. Outer space never became empty and void but relevancies changed concepts where centres form that should not be as it then interferes with concepts about relevancies Gravity is not and never was about particles pulling each other closer. If it was, no Big Bang was possible. Gravity is about turning space, which is released heat back to heat and concentrate the heat where gravity is the strongest and space is the least. Space is the transverse form of heat and visa versa is also true. Should any one not believe me try a bicycle pump by compressing the plunger while blocking the valve bit. The heat will burn your finger to blisters if the force on the plunger is strong enough, the plunger seals enough and your ability to withstand pain can last that long. Then answer your own question about where the heat came from because sure as hell you will realise what you are holding is as hot as hell and hell is hot. This evidential heat did not come from friction with air particles such as oxygen and nitrogen escaping through the valve bit. Heat is unleashed space and space is concentrated heat. Reducing space to heat is gravity and antigravity is expanding from overheating blowing into space accumulation.

When looking at a sphere the inside has always (in a cosmic relevancy) the location with strongest heat also always has the strongest gravity in any given cosmic sphere. The centre of the sphere clusters the combination of particles forming the sphere into one unity.

Academics, which I contacted in the past, treated my work with great scepticism. That forced me to find more proof and every time I went back to the basics again and brought about further proof in the hope of acceptance to prove my case. At first, I did not wish to criticize Newton or science, but eventually through the persistence of some academics, I was forced to do just that. It is not that I wish to replace Newton in the least, but when Newton formed his formulas the background to science was very patchy and dark in comparison to what science know at present. I have tried to publish the book through the commercial route by finding a publisher. The response I received was that of publishers not willing to be involved because it falls outside the boundaries of accepted science. The battle continued as I then took a more direct approached by going to the Internet. This was no solution at all because those that seek information on the Internet did not make head from tail to what I was saying, and those persons I wished to draw their attention do not look for information on the

Internet. I have spent too much time, money and effort in this work to dismiss it just because people do not understand what I am saying. I am too sure about the correctness of my case too let go.

The definition of gravity according to the Oxford dictionary on Astronomy is as follows:
The force of attraction that operates between all bodies. The size attraction depends on the masses of the bodies and the distances between them; gravitational force diminishes with a square of the distance apart according to the inverse square law. Gravitation is the weakest of the four fundamental forces in nature. I, Newton formulated the laws of gravitational attraction and showed that gravitationally a body behaves as though all its mass were concentrated at its very centre of gravity. Hence, a gravitational force acts along a line joining the centres of gravity of two masses. In the general theory of relativity, gravitation is interpreted as a distortion of space. Gravitational forces are significant between large masses such as stars, planets, and satellites, and it is this force, which is responsible for holding together the major components of the universe. However, on the atomic scale the gravitational force is about ten to the power of forty times weaker than the force of electro magnetic attraction.

In view of the Titius Bode time depletion, time in flow creating space has a far more complicated arrangement than Xepted science can even produce on a chart. To be honest every person knows that Xepted science cannot even place the planets on a chart, depicting true distance to size, but they WILLFULLY never mention that information when the chart they show is as false as a three-dollar bill. In a sense it does no harm leading people down the ally in such a way, because others in my class of mental insignificance in society is far to un-intelligent to realize the correct way and will therefore not understand the correct way in any event. That is a mistake with a stinging tale. It is as dangerous as a scorpion to Xepted science. In an introducing article, I named Anglo-American Mythology, I pointed out how misconception feeds society, favouring the lies and untruths and blatantly ignoring the truth. In the past, since time began the powerful used this on the brainless masses, and for a period where that civilization lasted, got away with that strategy. The next civilization that came to power, followed the same methods applying the same dogma, and in the end paid the same penalty because their greed, lust for power, and sublimations gave them control over the masses for a while.

The misconception those in favourable positions forced onto the masses, made the very people in power so shortsighted, their course on vanity lasted but a few generations. This is achieved because our Earth environment is tolerant and can buffer a lot, to save life in the end for life's contamination on Earth. They wish to extend life's connection to Earth, as being a connection to the cosmos at large and will be in effect as long as life remains in the cosmos. When we are "going abroad" to our "next door planet" that apparently holds all the supporting evidence of life carrying organisms, the connection to the cosmos remains and connecting to the Earth is of little consequence. That bluffing must stop. I realize no one on Earth will ever take note of what one sod (like me) on Earth is shouting, but misery awaits our Martian Colonists.

I started using the term Xepted because I do not accept Newtonian views and therefore I except Newtonian views but this bloody brainy machine tell me every time what I can and can't write when I refer to excepted (rejected) Newtonian science. Then I got brainy all by myself too and created my own word and told the machine to accept and shut up. Now we all are satisfied and English just got one word richer by my inventing Xepted science which is what I use when I refer to Newtonian science that science wilfully accept while they very well know it is Xepted science.

Suffering will be the reward for the fools attempting to catch the bounty of "fame, riches and glory" on behalf of the All Powerful Dollar and the dollars absolute true benefactors. Those with eyes, let them see, those with ears let them hear and let the rest self demolish.

Binary stars, spinning to self-destruction will produce significant heat. Heat create space, space forms winds. That is facts that the Bible present and is indisputable. Where the Earth was, was still a void, containing a sphere of circular displacement and this will reduce linear displacement to zero. Linear displacement is space and circular displacement is containing heat for matter survival.

Binary Star Minor overheated. That is why the core brittle and fragmented. This action will release tremendous contained heat, the heat will produce magma flowing in space like water in space and this eruption of heat space that created winds. Once again the recollection fits the scenario. Releasing the heat and producing space will establish space-time and fill the void where the Earth should fit. This is fact and if anybody even tries to dismiss this will be because of abstinence on his or her part. I did not prove the Bible correct. The Bible told the truth and in such correct detail, it is beyond human comprehension, but sublimation on the part of Newtonians and science before them, disallowed their ability seeing it.

That is what an insignificant formula $R^3/T^2 = 1$ where $R^2/T \times R/T = 1$ represents space-time in singularity as well as space-time in densified, occupied and unoccupied format. That means the everything of the whole lot, or as we say in Afrikaans, the "Heelal" meaning Universe. It refers to space-time for the first time while nuclear explosions are the epitome of $R^3/T^2 = 1$ where $R^2/T \times R/T = 1$ and how long has nuclear explosions been with us?

Now comes the proof: In the electron dimension the value is $\Pi^2 \Pi$ in relation to 3.
In the cosmos there are always at least six sides to any object.

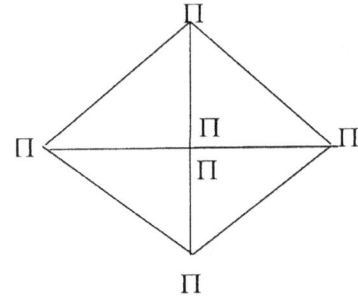

This gives the value of Π^6 in relation to the Titius Bode value of 7/10 in relation to the 6 sides in space (10).

$\$T = 7/10 \, (\Pi^6) / (6 \times 10) = 112.$

From this comes the Newton formula holding time in the fourth dimension to space in the fourth dimension. At all times the dimension application will be the value of time in singularity (Π) to the time of matter (Π^2). Separating the illustration will be as follows:

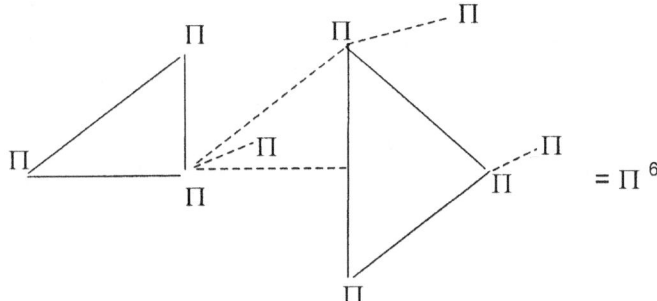

Through this "gravity" has the ability to produce a dimensional change from 3 to Π. Heat in a liquid form will be $3\Pi^2$ while one step more concentrated (the neutron state) heat will be $\Pi\Pi^2$. This reduces the triangle in space-time to the half circle in space-time.

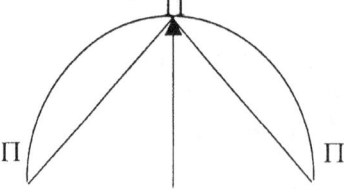

The one side is connecting the $3 \times \Pi$ and with the change in dimension cross over falls away changing the 3 (number of Π) Π^2 removing of the dimensional value of a half cube (6/2) to that of a half circle. ($\Pi\Pi^2$).

Both remain 180° contact with time in singularity Π^3, which is a straight line (180°). But there can never be a dual application. The one holds the other in support. In a star this fact becomes irrelevant but a star is the cosmos, applied in reverse. In the galactica (of which the solar system is a fragment) the relevancy of support will always apply at the outer circles and not apply within the inner layers. However, this is beside the point as we are dealing with the Titius Bode configuration of 3 ; 6 ; 12 ; 24 etc. that apply to time in space formed by heat.

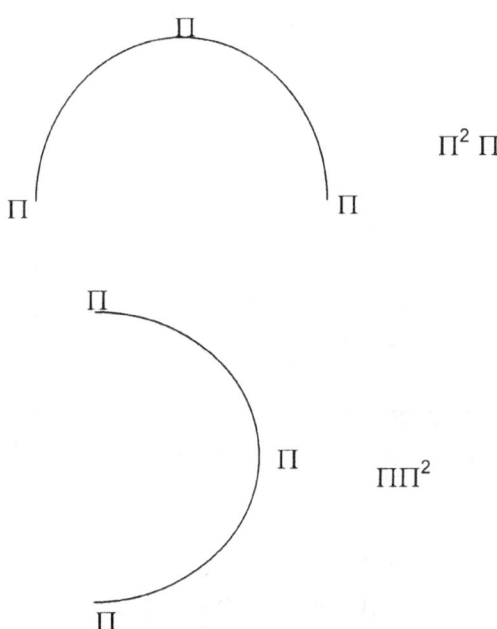

The relevancy applying in a cosmic cluster will always be that of a half circle applying as a triangle because it is in support of the triangle. The single Π will always support the Π^2 of time and this configuration will be a half circle.

Only the half circle can apply at any given point in any given moment. The value of Π^2 extends as the value of Π because Π^2 is the circular of "gravity" and Π is the linear of "gravity".
The rest is everyday mathematics. The invert square law applies just as well to a half circle than it does to a full circle. The value of Π in the next circle will be that of Π^2 in the previous circle. Well in a way presenting it as the invert square law does apply, but it has a cosmic sting to it.

There is a much more substantial explanation about how the Titius-Bode law arrive at the configuration of 1, 3, 6, 12, 24 doubling every time. However, such an explanation covers a great volume of facts because we have to cover the neutron's calculation from every angle available.

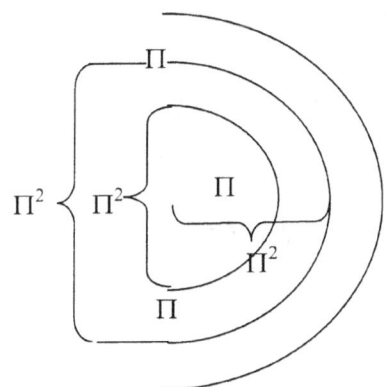

In the first configuration matter lends space all value at $\Pi^2\Pi$ configuring to the dimensional influence change of 3. This will be as such.

Matter

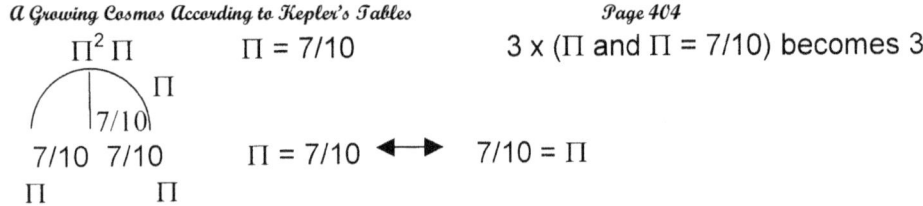

$\Pi = 7/10$ $3 \times (\Pi \text{ and } \Pi = 7/10)$ becomes 3

$\Pi = 7/10 \longleftrightarrow 7/10 = \Pi$

Every time another body develops to the outside of this the inner space will be three dimensional space $3^2 + 1^2 = 10$ and this apply to the sevens, therefore the space to the inside becomes 10/7 and from the space the next applies a matter value of $\Pi^2 = 2\Pi = 7/10 + 7/10$ with Π at 7/10. Only Π^2 determines time therefore $2\Pi = 2(7/10) = 14/10 = 1,4$.

Space holds the value to the already developed part as $10/7 = 1,42$. Therefore space will be 1,42 followed by matter, 1,4 and that leaves the Titius Bode law to double its distance. $3 \rightarrow 6 \rightarrow 12 \rightarrow 24 \rightarrow 48$.

Where Π^2 are the radius of the one circle it will double to become Π in the next circle. It is the manifestation of the neutron dimension applied in the electron dimension where all 3 of Π holds equal value therefore Π^2 will become $2 \times \Pi$. This relevancy will apply wherever heat and matter produce space-time. This is a given, standing as firmly as the Hubble constant, the Roche limit and any other law application.

Every person in the past sought a relevancy in the application of this phenomenon in as much as it applies to the solar system. The proof there is, is not in applying, but in the way it does not apply and the reasons why it does not apply.

When testing for proof in the application of the Roche limit as far as it stands in figuration of the solar system, we will find it does not apply at all. That means the fact that is NOT PRESENT, PRESENTS THE PROBLEM. The absolute importance of the Roche principle not only reflects on the influence of the Roche principle alone but the Titius Bode space-time growth and the Titius Bode configuration that is an extension of the Roche principle.

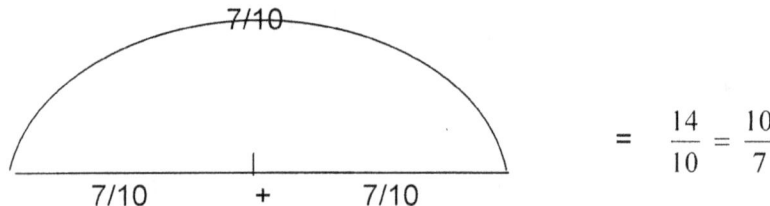

$$= \frac{14}{10} = \frac{10}{7}$$

But the matter position relate to the space position.

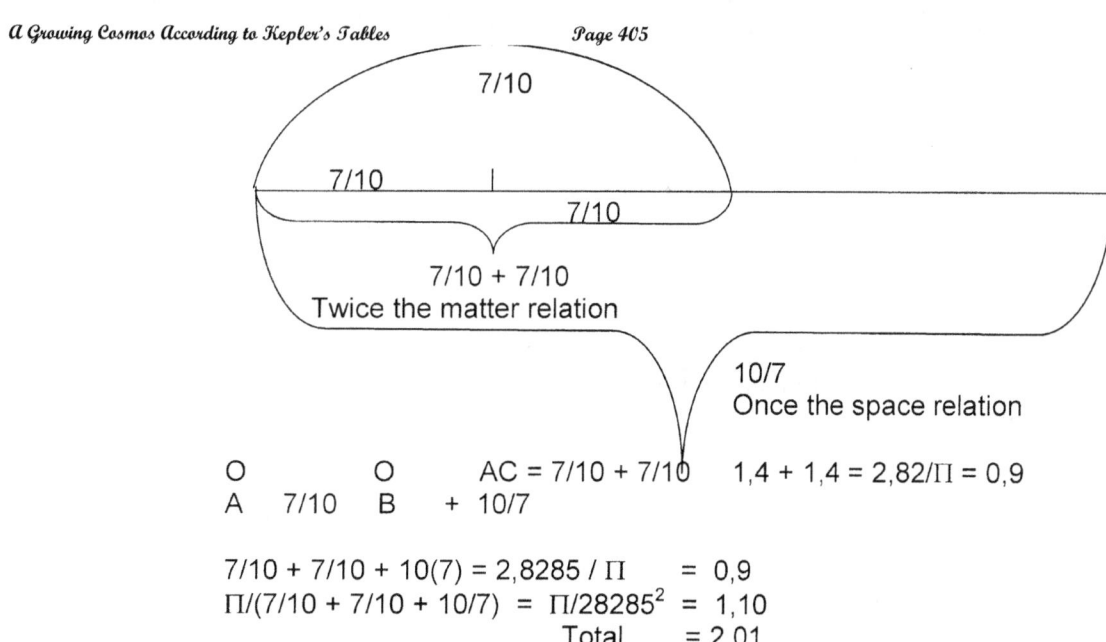

7/10

7/10

7/10

7/10 + 7/10
Twice the matter relation

10/7
Once the space relation

O   O   AC = 7/10 + 7/10   1,4 + 1,4 = 2,82/Π = 0,9
A   7/10 B   + 10/7

7/10 + 7/10 + 10(7) = 2,8285 / Π   = 0,9
Π/(7/10 + 7/10 + 10/7) = Π/28285^2 = 1,10
                          Total   = 2,01
Therefore singularity at Π relating to space-time = 0,9
and space=time relating to Π in singularity   = 1,11
The total of matter to singularity   = 2,01.

Therefore $Π^2$ to Π = 2Π . This then is 7/10 + 7/10 = 10/7

Therefore space will always hold double to the relevancy of matter.

To that end
.   3   . (7/10 + 7/10 + 10/7)
.   3   .   6 .
A   B   C
         2Π
Then 6 becomes 3 = 2Π(Π=6)=12
A   C   D
Then 12 become 3   .   2Π = 24

And that concludes the Titius Bode configuration of 3; 6; 12; 24; 48 etc. by valuing the triangle and the half circle.

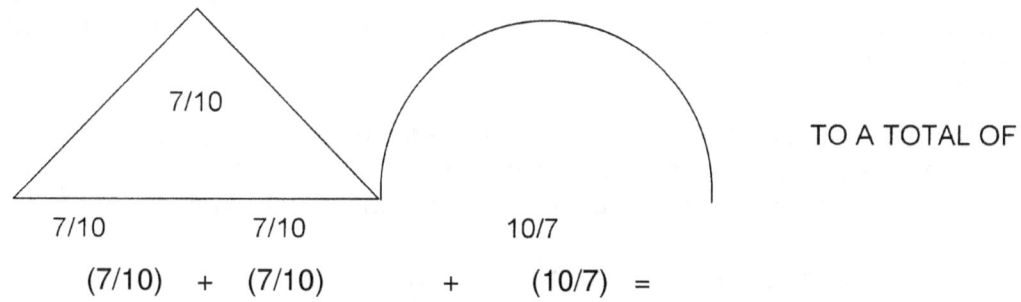

7/10

7/10   7/10   10/7

TO A TOTAL OF

(7/10) + (7/10)   +   (10/7) =

Three going to double in the next configuration.

The first $Π^3 → Π^2$ Π Separating singularity.

This then brought on $Π^2$ in heat.

Π
$Π^2$ Π   3Π = 3
Π
From that space developed

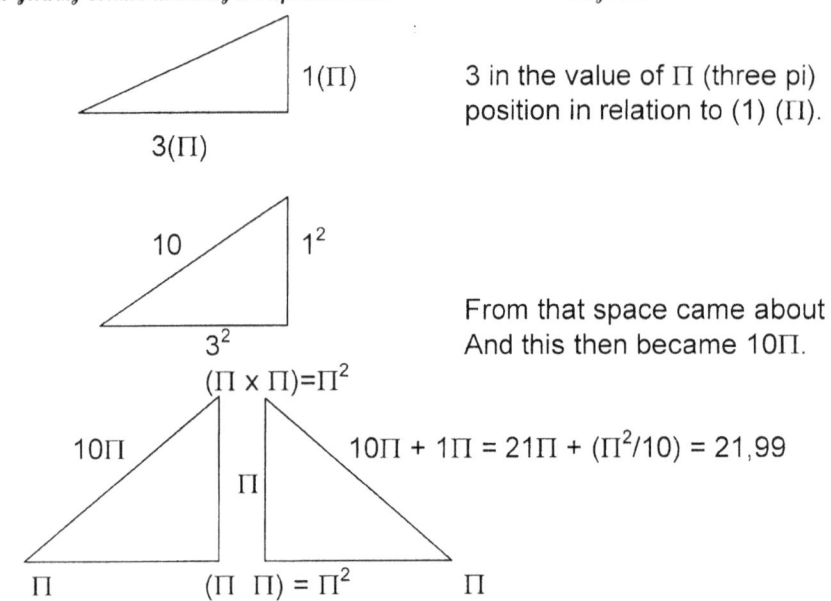

3 in the value of Π (three pi)
position in relation to (1) (Π).

From that space came about
And this then became 10Π.

$10\Pi + 1\Pi = 21\Pi + (\Pi^2/10) = 21{,}99$

On top of that the sphere established.

Add that to the seven that holds densified and occupied space-time and there is seven of Π in the triangle of matter adding 3 Π's in the half circle of space ($\Pi^2\Pi$) and the total is 10Π. With 10Π the value of the total triangle (in a square) and 10Π the total of space-time (matter holding singularity apart, there is a factor of 7 by Π to 10 by Π, with the triangle having two Π - two factors where Π at the bottom formed $(\Pi \times \Pi) = \Pi^2$ and to the top $(\Pi \times \Pi)^2 = \Pi^2$, with all of this constructing space (10Π) to matter Π^7.

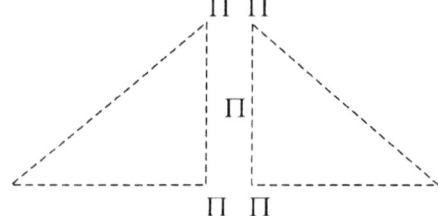

This will bring about 5 $(\Pi^2+\Pi^2)$ bringing matter and space to singularity. There are Π to the number of 5. From this comes that 4 (the four Π in the double proton in relation to the proton $(\Pi^2+\Pi^2)$ will always form the time value $(4(\Pi^2+\Pi^2)$. From this comes the fact that $3(\Pi^2+\Pi^2)$ will hold the space component as that in fact is half space-time.

Space-time being $(\Pi \times \Pi)(\Pi)(\Pi \times \Pi)$ and half of that to any direction is $(\Pi \times \Pi)$ (Π). This is why space-time in the geodesic sense has the value of 10Π and that 10Π in a sphere are Π^3. It is both triangle relating to one half sphere which is both triangles (7) in the totality of the half sphere (10).

The value of $2(\Pi^2+\Pi^2)$ will relate to a position where space passes on time in a dimensional transformation, but not a value transformation. It will be when Π relate to Π^2 as much as $\Pi^2+\Pi^2 = 19{,}7$ and 6 (the sides available for Π to use in space is 18,84. It will form a border of dimension where space will not apply in the same manner as it did before reading that border. One may say that is the densifying border of heat in space to heat in matter.

The last position of the Lagrangian atom is where only Π (in the triangle) have value and this links space directly to time (matter). At this point concerning the Lagrangian layout, we must view heat, filled space, the stuff we exist in, as an element. Hydrogen becomes liquid at −269 °C and heat (outer space) has a gas value of −273 °C. This is only a dimensional changeover from 10 to Π^2 or from 3 to Π. That was what the "Big Bang" was all about.

Creating the Universe in space as we see space, was the process where the last natural element formed. The Universe or Cosmos is the last atom, which formed. It was the conclusion of the proton $(\Pi^2+\Pi^2)$ in developing. That space can be in gas as we now see it, covering elements to position them in relation to the rest of the Universe as gas, liquid or solid. That is why astro-physics in the ways science apply it at present, is but a good old romantic science-fiction story.

Mass is the frustration of motion while motion in duplication as well as contraction representing gravity. Even with our experiencing of mass it is the tendency we experience to move that is the gravity and the mass part is the blocking of the space we with our bodies wish to claim. While the earth is blocking our claiming of the space we wish to occupy the earth as well as us are both in mass but the mass has precious little and a lot of nothing to do with the fact that we are moving through the application of gravity. To quote Kepler space a^3 filled with material has to move in line with a centre that is controlling the moving of space through time and in time as well. Therefore I repeat what I said before. Mass holds no value, it is density and that density brings in the heat in space, unoccupied as yet by matter. The motion represents density of time in space. That density gives a star its "gravity". The space has to move through time by duplicating and the more dense time is the more effort such duplicating requires. Where the Sun can only hold heat to liquid more dense stars will hold heat to a "jelly" and others will take heat all the way to something as hoard as tungsten. That is what tungsten is. Tungsten can place heat in a relevancy that the heat relative to tungsten is almost as dense as the neutron within an atom. That is why I refer to the system as the Lagrangian atom, because the Universe holding heat, produces all relevancy matter can have. The Universe is the final atom.

The diameter of the Sun is 1391,980 km. Bring this radius in relevance to the Roche factor and the first orbiting structure will be Π^2 relating to $(\Pi/2)^2$. With Π^2 at a value of $1391980 \times (\Pi/2)^2$ the position of Mercury must therefore be 3 4345 73 km making it approximately $3,5 \times 10^6$ km. With the effect of the Titius Bode configuration the next position must be 7×10^6 km and the third at 14×10^6 km. If the orbiting structure were that close, as it should be under all normal conditions, we would have been roasted toast.

1. Mercury
2. Venus
3. Earth
4. Mars
5. Ceres
6. Jupiter
7. Saturn
8. Uranus
9. Neptune

Roche Limit according to actual dist variation	1	2	3	4	5	6	7	8	9
	3.5	7	14	28	56	112	224	448	896
	57.9	108.2	149.6	227	414	778	1427	2871	4497
	16.5	15,4	10	8	7,39	6,5	6.3	6.4	(5)

If there were only the Sun that affected the positioning of the gas, "planets" the diameter of the Sun will be Π^2 placing Π at a position where dimensional implication becomes valid. Since the structures are still in space-time positioning, the effect of the Roche limit will still be in place. Therefore the Sun would be Π^2 arranging $(\Pi/2)$ accordingly.

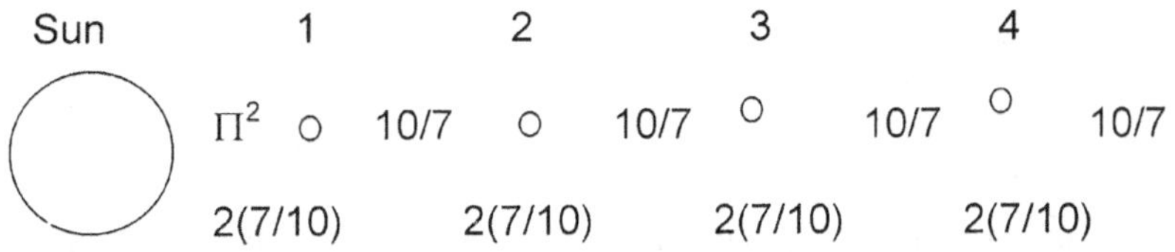

Sun   1   2   3   4

Π^2 o   10/7   o   10/7   O   10/7   O   10/7

2(7/10)   2(7/10)   2(7/10)   2(7/10)

I shall explain the layout as follows: To every structure the value of space-time in the electron dimension is

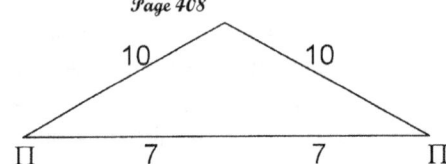

Therefore the matter positioning will be 2Π and that in terms of matter relates to (2x7/10)=1,4. The space-time to space will therefore be 10/7-1,42 because matter is two parts while space remains in singularity and singularity is always one. Therefore 10 cannot double it is one in relation to any one matter part at any given point. This then means Π² means in real terms (2 x 7) /10 as one Π and 10/7 as the other Π. Because matter relate to the dome in the half circle and space to the triangle, in relation to space it is Π+Π and to matter it is Π x Π. In relevancy to matter as we apply our attention to the two structures, the correct connection is (Π/2)². However, in the space factor it will be (Π + 10Π) / 7 x (2²) therefore it will be on the one side

(7 + 7) / 10 (matter plus matter) = 1,4.

On the other side though it is (10/7) = 1,42.

This is 1,4 + 1,4 = 2,82 in the space where Π relate to 10 and in this instance Π is 2,82. This means that bringing the relation back to matter will effectively mean it is (2,82) x 10 relating to 7 in conjunction with 2.

The Roche limit is Π which in this case is in space, therefore cannot be a square since the ten already apply as a square.

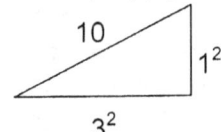

This all relate to the fourth dimension in space, but matter relate to the third or neutron dimension that holds time to a square. The matter as such remains in singularity (7) but time stands in regard to the square of half of Π. This then means matter (7) holds a relative to half the time effects matter (7x 2).

Space-time outside singularity then is

Space-time outside singularity then is

$T = $\frac{28,2}{7 \times 2}$ which is (1,4 + 1,42) x 10, which is matter (7) to time 2

$T = $\frac{28,2}{14}$ = 2

Therefore every object holds the inner structure as 1 and in accordance to its own position of 2. This then is $\frac{10}{7} + \frac{2(7)}{10}$. That is the Titius Bode implication, however, I explain this better when dealing with the Titius Bode just before this part. Therefore the first portion Mercury must have is Π² (the sun with the diameter of 1 391 980) with the Roche factor implicating (Π/2)². According to this the positions are as follows:

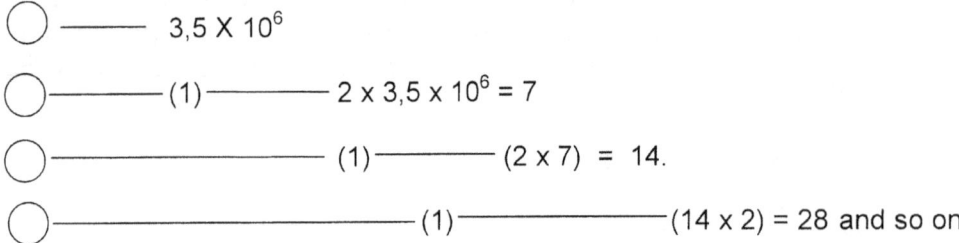

We find that the gas planets are on average about 2 Π overshooting the development that would apply in the case where the Roche limit would result in positioning orbiting structures. As explained a few paragraphs ago, the application of Π² in terms of the Titius Bode configuration will be 2Π, a

dimensional factor change.   As I have indicated the fact that it shows as 2Π, in the electron dimension, become Π^2 through putting matter in space.

Therefore the two Π you see, is the Π^2 you get. Having Π^2 means one thing: there was another object (Π^2) that related to 2Π and the two Π can only come from one more object that filled that space during some duration of time in the past.

The Titius-Bode principle relating space-to-matter at a value of $R^0 / T^2 = 1$ where space holds the square of 10 and matter is 7

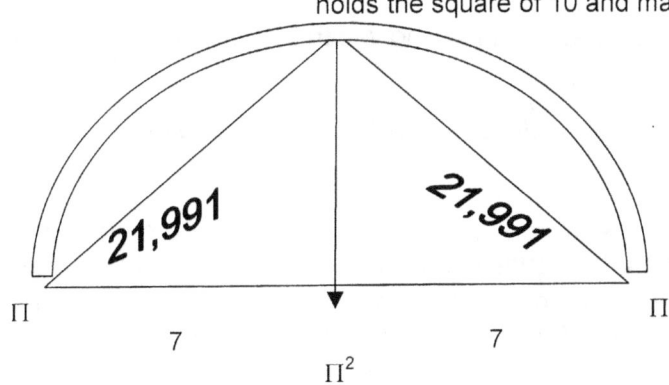

Therefore we may correctly surmise that something to the value of a relevancy of 2Π becoming Π^2 played a part in the positioning of the outer "planets" As shown repeatedly a double star would apply as $\Pi^2 + \Pi^2$ with $(\Pi/2)^2$ separating the stars where the Sun holds the position of Π^2, and therefore another object was present during space development in the time the Sun released from eternity.   In other words THERE WAS ANOTHER STAR IN BINARY TO THE SUN.

In the half circle the centre is 2Π's and in the triangle the corners form 2Π.

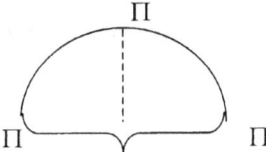

The dimension "gravity" removes as it replaces it with Π^2
In the half circle the fourth dimension holds a triangle.

That is singularity in time to the value of Π^3. The aim we have is determining the value of Π^3. The aim we have is determining the value of $\Pi2$. Known to all at this stage is that there is 7/10 in the Titius Bode law and in all spheres, including the Earth, we have a space dome of 21,991 holding space to the 7 holding matter. When a spacecraft re-enters the atmosphere, the angle of entry must be not less than seven and not more than 21,991. Therefore there are 7 holding 21,991 to the value of Π. This is the dimensional equal of the Titius-Bode law of 7/10. In order to determine Π^2 you therefore have to translate that value to the fourth dimension, giving it a value from singularity (linking time and space to a figure of 1) therefore 1, to its time position of Π^2.

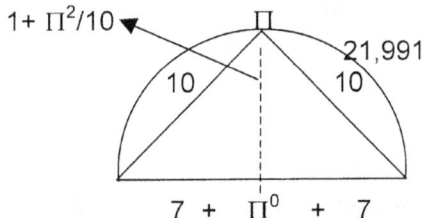

First of all, all Newtonians are educated in mathematics, therefore they will know that the triangle holds 180° the equal to the next dimension of the half circle at 180°, also equal to the straight line of 180°.
Secondly, Newtonians are aware that multiplying in the one dimension translates to the next dimension in the form of adding i.e. 4 = 2 x 2 and 4 x 2 = 8. Therefore $2^{1+1}=4$ and $2^{1+1+1}=8$. In the

one dimension adding is the same as multiplying in the other dimension. However, in astro- physics one do not merely transfer dimensions, you work with dimensions running concurrent in value. Therefore to the top you add or subtract, and to the bottom you multiply.

First we subtract the top from 21,991.

Titius Bode 10 + Titius Bode 10 is 20. Adding the one, the link between space and time in singularity, holds a position of one.
This leaves us 21.

Then determining the point where gravity (Π^2) will be at space (10) and will therefore be $\Pi^2/10 = 0,99$.

In order to get Pythagoras, you add 10(T.B.) plus 10 (T.B) plus 1 (singularity) plus $\Pi^2/10$ ("gravity" ending in space) and you. Square this total of 21,991 as well as divide it with the value of 7 x 7. (The top you add, while the bottom you multiply).

Therefore the top is 483,6 and the bottom is 49. To get to the "gravity" part in another dimension you divide (not subtract) the square of space (483,6) with the square of matter (49) and the value will then be 9,869467, the value of Π^2 relating to singularity. This means that the 2Π space holds, were filled with matter (Π^2) . There can be no argument about that fact and we can at present see the other Π^2 being the sun.

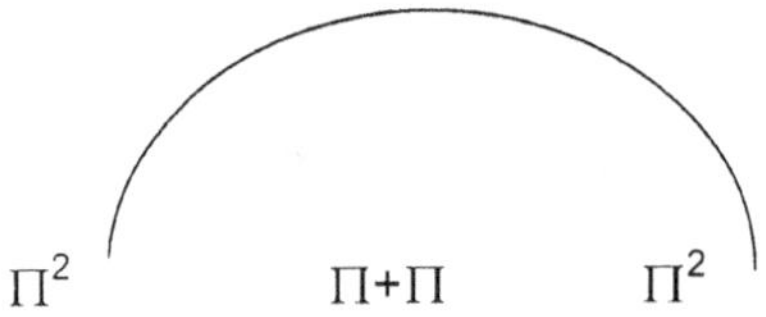

$$\Pi^2 \qquad \Pi+\Pi \qquad \Pi^2$$

$\Pi + \Pi = 2\Pi$ translating to time as Π^2.

That is what the Roche limit is all about. It is the point where a cosmic proton (Π^2) shares (that means halved) a space relating position of Π in neutron dimension of time (Π^2) $\Pi \to (\Pi/2)^2$.

$$\{(\Pi = 7) + (\Pi = 7)+ (\Pi = 7) + (\Pi^2 / 10 = .991)\} = 21.991 / \{(7 / 2) + (7 / 2)\} = 7$$

$$R^2 / T \, (S\$^3 \, S\$^2) \, R / T(+ S\$^2) = (7 \times 7) / T^2 = (7 / 2) + (7 / 2) = 7$$

$$\Pi = \$_T = 7$$

$$(S\$^3 \, S\$^2) + S\$^2 = \quad = 483.74$$
$$=\sqrt{} \; 483.74 = 21.991$$

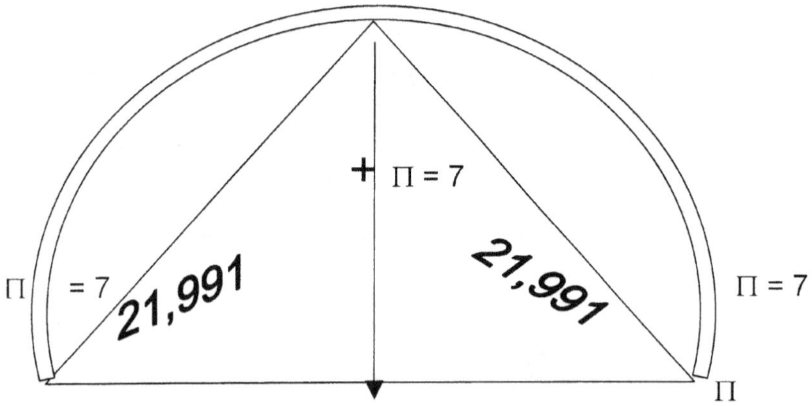

$$+ \; \Pi = 7$$

21,991

21,991

Π = 7 $\Pi = 7$

Π

$\Pi / 2 = 3.5$ $+ \; \Pi / 2 = 3.5$
$\Pi = \$_T = 7$ $(\Pi^2+\Pi^2)$ $\Pi = \$_T = 7$

If any person wishes to cling to Einstein's view that the speed of light is the fastest that matter can apply velocity, explain how the Black Hole works. Inside the Black Hole must be matter, because there is no space, yet time does apply because it takes the particles spiralling inwards to the centre time to move from point to point. Matter in motion is time. However, no light can return to the surface, therefore the light is slower than the moving particles within the star. The only thing about the star, is that it maintains a higher relevancy than the relevancy the speed of light can apply. By accepting the existence of a Black Hole, the Einstein claim about the speed of light being one, becomes zero.

Another place where the speed of light becomes obsolete is within the centre of galactica, where the accumulative movement of matter exceeds the speed of light. That is where doctor Hawkins saw a Black Hole that is not a Black Hole, but the precise opposite. Light matter and heat, moves inward in an effort to maintain cooling as the group of proto, proto stars two era to the future) still claims their share of heat maintenance. Those particles in such close proximity, establishes a time well above that of the speed of light.

Everything in the cosmos is all about relevancies. Time started at such a high velocity, it had to be eternal. Nothing diverting from eternal can become more than eternal so it has to be less than eternal. It is fragmenting eternity into parts making eternity smaller. The smaller eternity becomes, the lesser eternity will be. That means that time started at eternity and became shorter with the introduction of infinities that broke the monotony of eternity. The more inanities there are affecting eternities, the shorter will eternities be.

I prove in "Matter's Time in Space – The Thesis" where Einstein went wrong in his theory about "The curvature of space-time". There is the space-time complying with singularity and filling the space-time in singularity is heat and matter valuing space-time. Space-time (Π^3 to Π) cannot bend, cannot curve, forms a straight line, but what fill space-forming time is matter (Π^2) and heat (3).

That part changes. The atom cannot be gas, or liquid, or a solid, because the atom is densified in occupation of space-time. It is the heat in unoccupied space-time that produces the gas, liquid, or solid that all substances can form. It is THE HEAT in SPACE that produces TIME, that can and does curve, bend or whatever. That HEAT in SPACE forming TIME that forms the relevancy of space-time and that does bend. If, by applying the forming of gas, or solid or liquid to the element instead of the space between the elements, of course you will get the incorrect vision Π, where the space-time (matter holding singularity form singularity) is doing all the binding that apply to the curvature of space (validating time) and time in singularity (a straight line) will be solid. Einstein placed the relevancy incorrectly on singularity, instead of heat.

Once again I do admit, IT IS A LOT MORE COMPLICATED THAT WHAT I MAKE IT TO BE AT THIS POINT, but the motto is, Keep it simple.

If you wish to keep time in space constant, everything in the Universe will be oblong. That is why the Newtonians have an absurd view of the cosmos, and they present facts in the cosmos in a way nobody (least of all the Newtonians) can understand.

Time was slow, time became faster and faster because by extending the position of Π, Π^2 will produce speed.

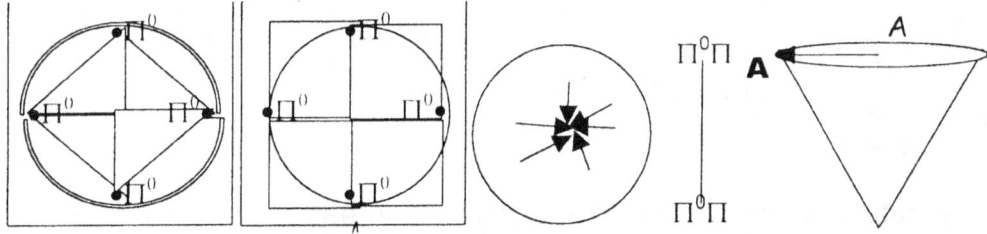

At that point space developed to a sufficient degree as to allow material, space and time form units being confined structurally while standing also individually apart. It was the pre runner to what then later became galactica filled with stars that was filled with atoms. But this was the prelude to all of that.

All rotating object has to be round to rotate. From the ends of the Unit rotating there will run a line running horizontally which turns with the top and as it turns with the top, the horizontal line is crossing another line that is running vertical but is not running at all because the three points cannot turn. The picture on the side is a picture showing the rotating object from the top as one would look down onto the line in singularity that does not turn. That line in singularity is representing the cross over limit parting the one part in the Universe from the other part in the Universe. At the displacement value of space (139) matter (138) and time (137) the Universe had tangible liquid in a relative motion with a structures solid containing heat in space.

Space   Time   Matter
139   137   138

centre spot where eternity heat came apart from cold set in place motion that space and time without Relevancies came abut spot but had no space to charge singularity by activated into complying. are all the same things only dimensions are formed space being in motion and

There were forever four sharing a centre spot while spinning around a locked infinity into a unit. But then that parted infinity from eternity that provided space in time and time in space or time.
when the dot moves away from the move. All that was possible was to relevance to comply in being Space-time is motion and movement separated by dimensions and space, where the dimensions become the space is motion by contraction or

by expansion but because time is almost eternal at k^0 our perception of the universe we are in is a stable and steady eternal structure. Gravity is motion and motion creates space to the third by the third in the third that interacts with one but establishes ten.

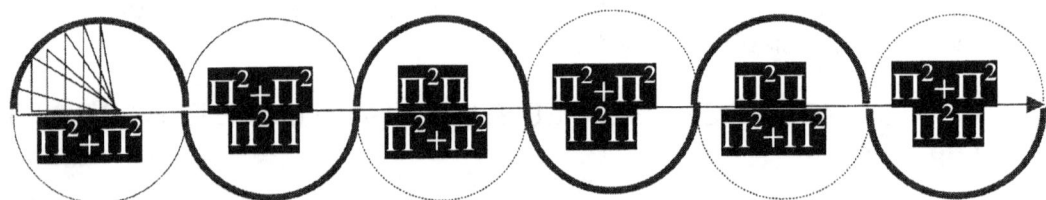

Motion is parting the Universe because it is representing eternity as much as it is eternity. It will never disappear because it is never there to begin with. The line cannot stop turning because the line can never start turning. The line is absent because it can never hold space, yet the line is always there because any motion may charge the line into presenting the centre of the Universe. To find the centre of the Universe is to reduce the line because Kepler said $k^0 = k\,T^2 / a^3$.

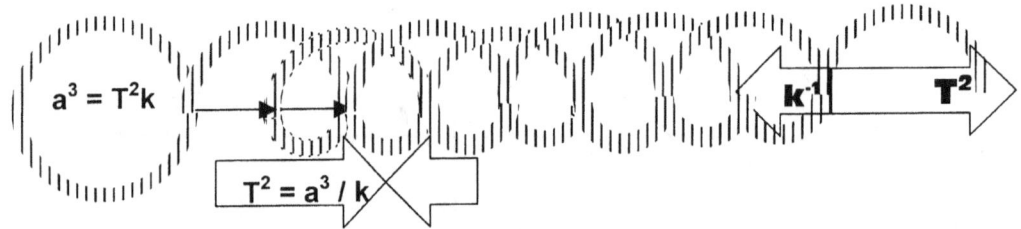

This process is a natural normal occurrence everywhere in nature without any person ever noticing. We see this so clearly in the spinning top. When the top is spinning such spinning of the top creates a centre and the lines start reducing space in the direction of the centre of the spin. The centre establishes a balance in space-time where at a point it finds partial independence from the dominance of the Earth's gravitational motion that is depressing the tops movement to a standstill in the space the earth confiscates.

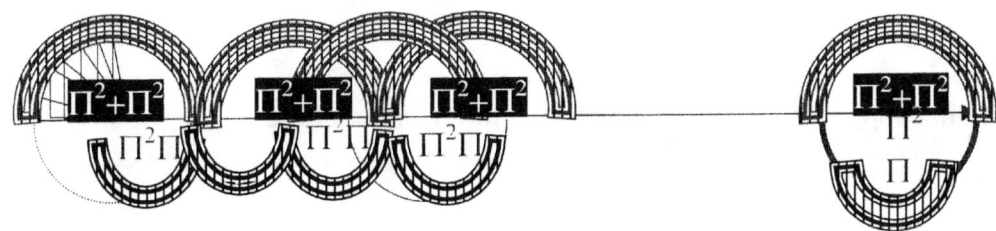

By moving from 1^0 to 1^1 and from $1^0\Pi^0$ to $1^1\Pi$ requires space. Yet such moving did not leave the realm or the domain of singularity. The motion was still within singularity because moving involved forming a relevancy between heat and cold between infinity and eternity, between space and time and most of all producing what will in the far future develop into a Universe that can even be a host for life albeit on a very small spot for a very short while in relation to the vastness space has and the duration cosmic time has.

Time started by placing the double proton in space and in matter, where space and matter will always be in a sphere. The sphere always forms 7° from one point to another running outwards.

$$\$T = 7(\Pi^2 + \Pi^2) = 138$$

Explaining is as follows:

$\$T$ is space-time ($) in the time sector (T)
7   This indicates that the Coanda principle placed motion to space.
$(\Pi^2 + \Pi^2)$ Refer to the proton committing the Coanda effect.

At the same time space formed as a consequence to the Roche limit and so did matter. Forming the Aanplasings -Atomic-Epitome (the point where matter breaks in singularity). This was where the three-dimension concept was introduced but not quite accepted in practise because form still rules.

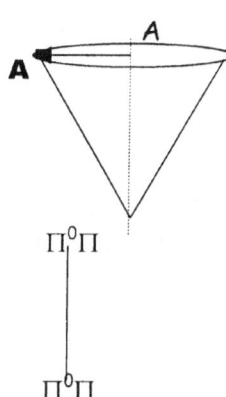

The moving of Π^0 to Π involved relegation and not motion as we consider motion. It was Π^0 getting a side and that is all. There was no true side but only a form that came into place. **Singularity (A)** received singularity (**A**) and no more of anything but the shift to comply with having a relevancy forming in relation to singularity. The dots had no sides, had no length or diameter. There was not measurable space or measurable time involved. The time could have been a micro, micro second as much a trillion millennium because time had no relevance. It was eternity interrupted by infinity, as it still is the case, however the line that eternity followed was no line because there was no space to hold the line. The line was momentarily interrupted by infinity, however with no one there, there was no one to notice. The lines were not lines but relations to sides being formed.

There was then an outer line forming time in space 10/7. Then there was the inner line forming space-time being 7 /10. The there was material filling space at $(7)(\Pi^2+\Pi^2)$ forming the sphere as it was filling the sphere.

$$\$T = 10/7 \ (\Pi^2)(\Pi^2+\Pi^2)$$

Explaining as follows:

10/7   The limit between what is part of the atom and what is excluded.
7/10   Forming the matter factor.
$(\Pi^2 /2)$ Indicates a deliberate inclusion of an atmosphere or a liquid or a neutron
$(\Pi^2 + \Pi^2)$ Shows that the proton still had total control

$$\$T = \mathbf{7/\ 10}(\Pi^2 \ (\Pi^2 + \Pi^2) = \mathbf{138}$$

Explaining as follows:

7/10   Heat flowed to material supporting gravity within the centre of the sphere.

$(\Pi^2 + \Pi^2)$ the double proton

(Π^2). The boundaries were set by the motion that the neutron provided.
  The atom was born

$$\$T = 7(\Pi^2 + \Pi^2) =$$ 138 The circle the atom has.

$\bullet\$T = 7/10 (\Pi^2(\Pi^2 + \Pi^2)) = 136$
Motion towards the inside of the atom;

$\$T = 10/7 \ (\Pi/2)^2(\Pi^2 + \Pi^2) = 139$ the relevancy of space carrying time to allow material space to apply motion within.

The atom formed a circle and that placed the Coanda effect in control of the Universe. For the first time there was matter in relation to time 10/7 in space in relation to space-time 7/10
This is what the Universe consisted of, everything that is today, was then, in a dimension that only holds "gravity" or the " gravity - motion". The Coanda effect took the Universe into the three dimensions.

$$\$T = 7/10 (\Pi^6) / (6 \times 10) = 112$$

7/10   is the matter has the dominant value.
Π^6 matter has the six sides it holds in the fourth dimension.
6   are the six sides to space occupying matter.
10   are the value or dimension in which space holds a ratio to time.

The cosmos began, not to a specific space, because all the space that was there, initially is still there at present. Any atom above 112 cannot apply to the fourth dimension not then and not now.

A proton with a "mass" of say 12g/mol on Earth will have a "mass" in accordance to Earth standards of 25g/mol on Jupiter and it will hold a comparable "mass" of 100g/mol within the sun. This "growth" in mass of any molecule within the structure's potential occupation of space-time increases. With this in mind, one cannot merely bring in such a relation to the "mass" of the proton in the beginning or within a star, or as it is on Earth. As the atom's spin increases or decreases in the relation to $\Pi^2 + \Pi^2$ $\rightarrow \Pi^2 \Pi \rightarrow 3$, and with it, the "mass" will subsequently alter.

In explaining all of this, it is quite impossible for me to give it a value in mass, or time as both these factors alters in space-time occupied.

I INCLUDE A SMALL PART OF THE TECHNICAL DETAIL TO SILENCE THE "SUPER-EDUCATED-KNOW-ALL" THAT IS FLOATING ON THE "CUTTING EDGE" OF SCIENCE. FOR THE AVERAGE PERSON THAT DOES NOT HAVE ANY CLAIM ON THE IMPORTANCE OF A TITLE IN BEING PART OF THE ESTABLISHED "SUPER-EDUCATED-IN-XEPTED-SCIENCE-MOCK", FEEL FREE TO READ THE TECHNICAL EXPLANATION, OR IGNORE IT, IT DOES NOT CHANGE, ADD, OR DISCARD ANY LATER EXPLANATIONS.

Whatever one believes, one has to be honest by admitting that time had to start somewhere. It proves only shortsightedness on the part of the Newtonians, to conclude that "gravity" started at 10^{-43} sec after the "Big Bang". This only concludes that the start was with the "Big Bang" and little else. Nothing is said about what caused the "Big Bang". Beside the point, but still very valid is the fact that I have no words in expressing my resentment with the term used as the "Big Bang" being the start of the Universe. This name only explains how little science understands about the cosmos.

NOTHING WAS BIG BACK THEN AS NOTHING WAS SUDDEN, OR QUICK OR BANG.

No person ever came up with a logic and scientific explanation to what brought about the process of heat expansion. What is irrefutably true however is that it came on route from eternity or timelessness or whatever one wishes to call it.

At this point, I have to explain the mistake we go about thinking about science and time. At first I was arrogant enough to think I was the first to understand the way time works. After all, it took me some time to figure out how the handle fits the fork. Then to my shock, I found that H.P. Wells already concluded my way of thinking about a century before I have. Well that proved so much for my personal brilliance and modesty once again, returned to me.

When witnessing an event we regard as an explosion we surmise that what happens on the inside of such an explosion is extremely fast, but to the contrary, it is very slow. In the explosion, the duration of time extends, becoming longer.

To explain this we take two persons, one watching the other runs a mile. We place both persons initially in the same duration where both will endure four minutes of time lapse. The spectator will see in real time how the competitor takes four minutes to complete the mile.

Then in the next scene, we increase the duration of the competitor to 1 : 60 and the spectator's time remains the same. To the spectator the athlete will be covering the distance sixty times faster, and to the athlete the spectator will be cheering 60 times slower. The spectator would not believe his eyes because of the athlete's abilities in running that fast while the athlete will think the spectator is in frozen state of admiration.

In the third scene, we enhance the duration in the athlete's time zone by another 1 : 60. This will bring about that the athlete, in the view of the spectator, will be running the mile in less than 7 hundreds of a second and the athlete will be watching the spectator trying to wave while the action of the spectator will last 240 hours. In the eyes of the one person, the other's time span will be either an explosion or, everlasting, depending on the person's point of view from the space in time that he holds.

To each one, the spectator and the competitor, a time lapse or time duration of 240 minutes occurred, although the actions in both sectors would have seemed to alter severely. Any confusion coming about from the explanation above, I wish to remind that it is a common and well-accepted fact that time slows down as "gravity" increases.

Behind all of this explanation is one obvious rule. When the one subatomic particle positions in such a way as to displace space-time in the form of heat breaking down, the value of space, in order to meet the requirement of time, is once again all a relation between space and time. The less space heat has, the less the value of space becomes, because space has no value. Without heat in space and without matter there is no such a thing as space. Therefore, space does not exist, but for matter and heat valuing space to form time.

To understand this one must firstly understand the principle behind the theory I introduce. At the most inner point one find time or if we can supply it with a completely fictional name: "The gravity - motion". The gravity - motion carries the value of Π^3. This value determines time in eternity a position matter has no space, but is occupied in singularity.

Taking the neutron position to that of the proton we find the value created when the three dimensions (six sides) came about $\Pi^2 + \Pi^2 + \Pi^2$, which carried to the fourth dimension in cosmic or geodesic space-time becomes Π^6. When relating Π^6 to singularity it becomes Π^1 (space) x Π^1 (time) x Π^1 (matter). I do realize this explanation does not suit normal mathematical principles but we are working in dimensions and Π^1 (time) in a straight line is 180° and Π^1 (matter), which is half a circle is 180° and Π^1 (space) which is a triangle is 180°. As each Π^1 represents one dimension establishing another dimension and providing that dimension's existence.

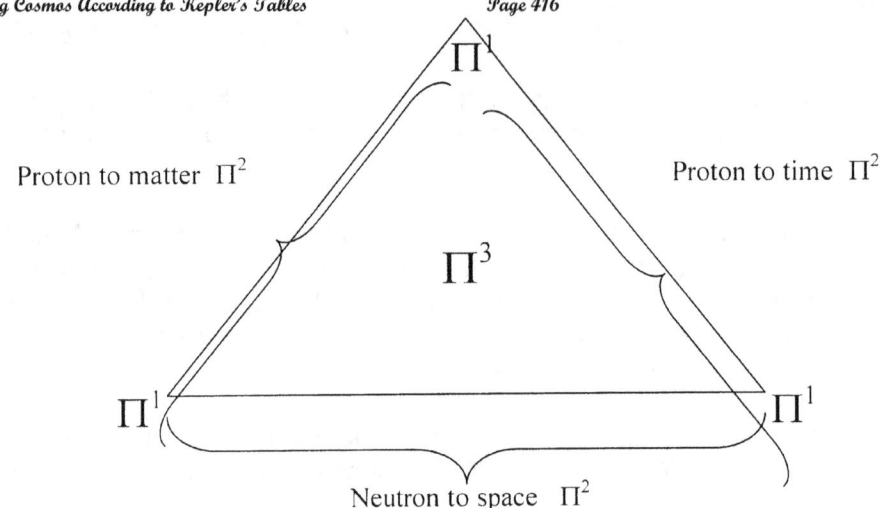

Proton to matter Π^2 Proton to time Π^2

Neutron to space Π^2

FROM THIS SPACE HEAT AND MATTER DEVELOPED

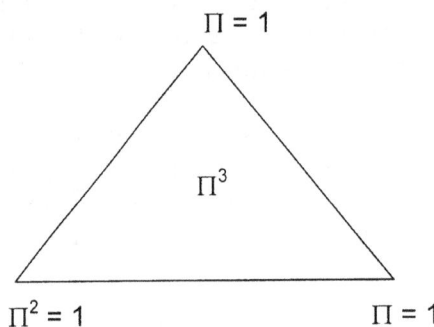

In relation to the " gravity - motion" space through a straight line will be Π and through the half circle matter with heat positioning space ($\Pi = 1$) to 3 sides 3. Relating the gravity - motion Π^3 three will always be a Π^2 and a Π combining 3. The half circle will be Π^2 and the straight line Π. Behind all of this explanation is one obvious rule, the one subatomic particle positions in such a way as to displace space-time in the form of heat breaking down the value of space in order to meet the requirement of time. It is again all a relation between space and time and the less space has heat, the less the value of space becomes, because space has no value. Without heat in space and without matter there is no such a thing as space, therefore space does not exist but for matter and heat valuing space to form time.

It would be far too complicated to explain why space-time and water share so many characteristics but they do. I have, to some extent, tried to explain it but I am aware that the explanation falls short of satisfying. I will repeat it once more, well aware that it cannot bring acceptance. It is heat that produce gas or liquid or solid. The period before light, everything about the cosmos was a soup cocktail, heat was liquid and space in singularity was liquid flowing heat that appeared like water. In The Thesis I spent many pages in explaining this fact, but I do not wish to overcomplicate this book as I wish to bring across the scientific proof about the seven days of creation from a realistic scientific stance, for all persons to understand.

It seems very ironic that science with all its bravado, money, wisdom and splendour, can only begin at the point where light came to the Universe, while the Bible explains the creation in detail, long before the "Big Bang". The reason why there was no light before the "Big Bang" was that the spinning matter exceeded the speed of light, being $3\Pi^2$. The Authentic Author of Genesis refers to this as a mighty wind and this leaves a question. What better name can one give to this occurrence?!

With time in eternity, space in zero and matter being time and space, what would ever bring about that this situation changed? No Super-Educated-Wonder ever came forward to explain this. Why did the "Big Bang" start and what brought the "Big Bang" about? Only the Bible produces any logic to this question. Time, matter and space froze in one, there was no reason in nature for things to change, since this situation lasted eternally. Nature with all nature's laws did not apply, therefore one cannot say that nature started it. Nature was frozen. Nature was not even solid, it was in a state beyond being solid. Nature was nowhere!!

The spin in the Universe slowed down, up to a point where the spin was equal to that of the speed of light. At the point where the Universe spin equalled that of the speed of light, the Universe was still in total darkness. The light (photon) was there, but did not yet produce light. Light only came about as the spin reduced to below the speed of light, and only then light became obvious. The Universe grew away from darkness as this event lasted many eternities, during the period where the light separated from darkness.

How did this "Big Bang" take place? The best way to examine the reason is to see why anything in the Universe expands. To get anything to expand one has to heat it. All matter expands when overheating. Science may come up with whatever brilliant theory, the fact of the matter is that when matter overheats it expands. The bigger the overheating, the bigger will the expansion be, it is as simple as that. This means whatever leads to the forming of the "Big Bang", whatever preceded it, it had to come about from matter that overheated. With the event of the nuclear age, the proof came about that matter is heat in some frozen form. Unleashing heat from its frozen form brought about a jolt of heat, never yet experienced by man. By breaking matter from the frozen state, of which it is in, within the atom, heat produces light and heat. Where this process clearly shows how new space-time forms is where the releasing of heat caused winds that stun man's logic.

The nuclear explosion shows quite clearly what the "Big Bang" was, with the nuclear explosion being a very minute form. Yes, we have all heard the rubbish about matter and anti-matter. What can be anti-matter, since matter is heat, defined to a certain space occupied for that time. With matter being frozen heat, what would form anti-matter. Anti-matter means the opposite to matter, and if matter is frozen heat, anti-matter must then be overheating heat. This in itself is quite ridiculous. Anti-matter can only be matter with an opposite spin to that we think of as matter.

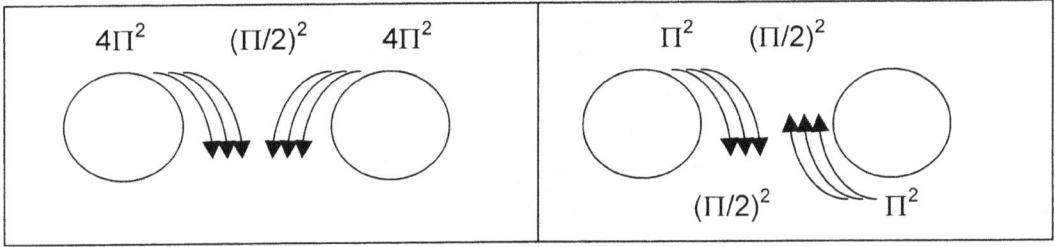

In the second sketch Π^2 ($\Pi/2$) / ($\Pi/2)^2$ Π^2 the Roche limit cancel each other and with that the space of the neutron effectively disappear. The two protons touch destroying each other and the neutron as well as the proton demolish and became heat 3^3.

The process where matter then touches matter, it will bring about a reduction in the feeding process of heat,. where all matter in that space will overheat and expand, producing unfrozen heat. This means there was heat occupying space, and matter with both in relation to time.

The proton with a positive space-time displacement less than 136 placed its displacing properties in negative space-time displacement. In short, to substitute for mass shortcoming of less than 136 grams/molecule and still finding sufficient cooling properties for the proton to survive the overheating deficiency, it has to spin more rapidly, therefore by spinning it makes contact with more heat than it would otherwise do. One may consider this as "breathing". If the proton does not find adequate supply of heat by being motionless, the atom has to substitute the movement through motion. Forming an object that has an increase in heat supply through work commonly uses this fact. In nature, just the opposite is true because of the motion by movement, as one find in the case of wind.

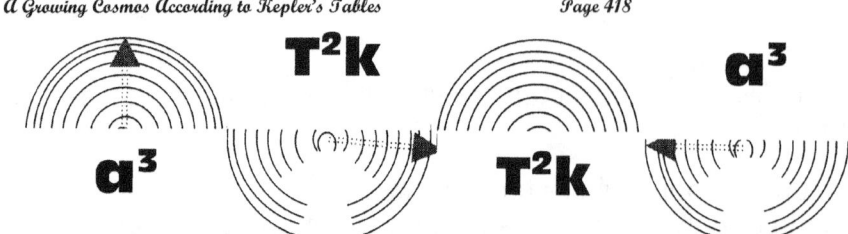

There is no "force" in the cosmic flow of time because everything is a "force" in one way or another. Everything is in a 90⁰ angle with time therefore the cosmos is restraining time while it is retaining space it is moving in relation with time but also opposing time all the way. The proton takes heat from space in an effort to maintain temperature and stability. This flow of heat brings about the reduction of space by increasing the heat in that space. The flow of the heat, through the electron by means of the neutron to the proton is time. The amount of heat taken by a proton is a constant throughout the Universe but relative to the space reducing effort of all the protons influencing that specific space. That is "gravity" (a term I denounce and reject). "Gravity" is nothing else but additional application of time. The higher the gravity is, the slower the time will become by prolonging the duration of time,

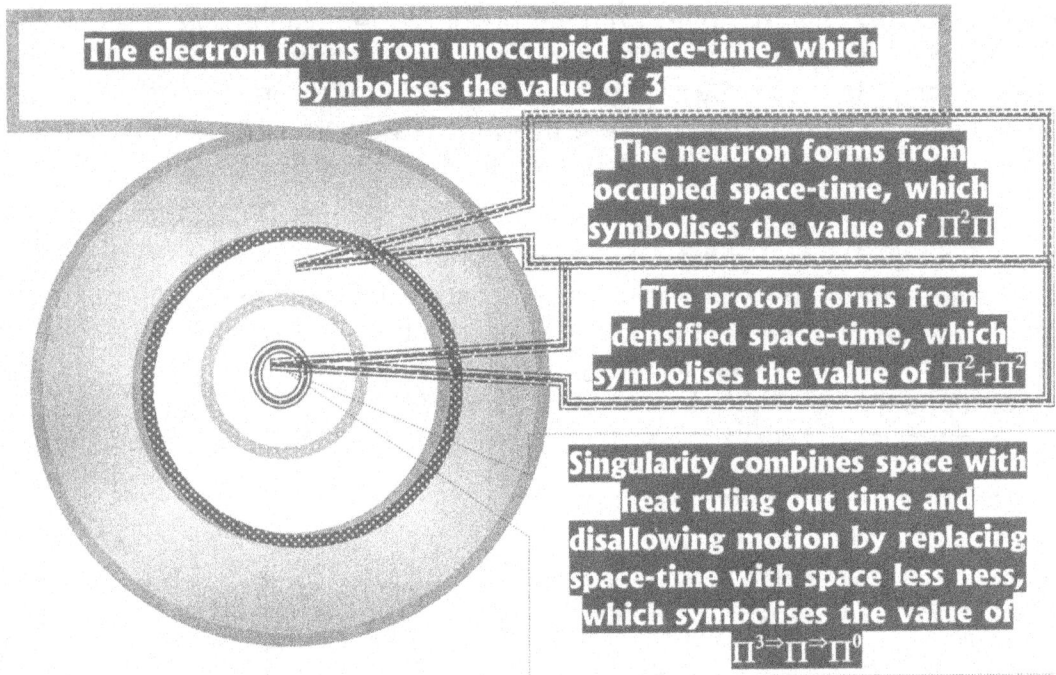

Unoccupied space: This forms the atomic relevance of **3,** which is where the ratio of space moves to the ratio of liquid in space. This is bringing motion in contact with **unoccupied space**.

Occupied space: Then forms the atomic relevance of $\Pi^2\Pi$, which is where the ratio of liquid space moves to the ratio of solid space. This is bringing motion in contact with **occupied space**.

Densified space: That forms the atomic relevance of $\Pi^2+\Pi^2$, which is where the ratio of solid space moves to the ratio of **densified space** or motionless space. This is removing motion by disallowing contact with space and then forming space less ness.

Space less ness: That forms the atomic relevance of $\Pi^2+\Pi^2$, which is where the ratio of solid space moves to the ratio of densified or motionless space. This is confining motion in a position being part of eternity.
I give the following relevancies in order to show how I define a star in relation to an atom since a star is just another cosmic particle or just another atom.

By taking this statement and introducing that to a Galactica, the shining luminous middle part, holds time to eternity through movement. The atoms in the middle admits light because it holds time duration to the speed of light, the longest duration that time can be and still remain in the fourth dimension.

The centre of all cosmic structures determines the time that applies to the structure itself. As the cluster of protons supply the density that influence the space of occupation, the density is a collective reducing of space with the increase of heat. This can apply through an object relating to space through movement by the object and by the object reduction of space through the density of the accumulative effort of the cluster of protons we named elements.

By referring to "gravity" only one aspect of the space-time relation of any elements apply. There is no mention of the second and crucial part of "Gravity" where the motion of the object brings about the space-time relation, or if you wish, providing the cooling aspect. At first, the proton cluster's total positive space-time displacement has an insufficient "gravity" to secure a stable cooling effort. This spinning of the element clusters is inherent from an event, even predating the "Big Bang".

The spinning motion of the element clusters (or proto stars or if one wishes to use the name of future stars, it will be just as applicable) hold their relation to heat secures by maintaining motion. As the time value in the clusters space occupation (mass) secures an era related value, the structures that were spinning, reposition in such a fashion as to apply a new linear displacing value. At first the motion is such that the linear position is negligible, but as the mass grows, the linear distance grows accordingly, placing the revolving structures further apart and at the same time, "pushing" the rotation of the objects in a wider revolving orbit.

By widening the rotation circle, the objects rotate at a "lesser" pace and this pace coincide with the space-time occupation ("mass") of the totality in the effort of all the protons put together. In this one will not find a "force" but it will be a complete balance between matter, space and time. By securing an ever-increasing space-time occupation (mass) the future star will reduce its negative space-time displacement (motion) and increase its positive space-time displacement (gravity). The higher the positive space-time displacement (Gravity) becomes, the lesser the negative space-time displacement (motion) will be. At present only stars holding an $iron_{56}$ inner core can maintain a star status, and any object with a lesser element in the inner core will not bring about fusion, or in fact, any form of luminosity.

For instance by the time the "Big Bang" arrived, only elements with a "mass" of 112 had the ability to release from the Galactica luminous core and during the Era of the Quarks, the releasing mass of the time determining elements carried a combined proton-cluster "mass" of 88. As the single proton's time holding value increased (molecular mass) the time grew less and the Universe "grew bigger". Up to this point all arguments came about from the theory about the "Big Bang". In The Thesis I show mathematically that the Universe will last seven cosmic days. This however is not the seven solar days and under no condition may one confuse the two.

We are in the fourth cosmic day calculated from the Big Bang as if the Big Bang was the first day. To understand the process of the cosmic days, please study the cosmic almanac as seen on the last page of this letter. While I am saying this, this book is about proving with undeniable facts. That I leave to The Seven Days Of Creation ISBN 0 – 9584410-4-9. In this letter I only show it is possible to prove what I say I proved. The Bible speaks of seven days of creation; therefore we must look for the seven days the Earth formed. According to the cosmic calendar, we presently find ourselves in the fourth day, with three more days to follow. Why do I refer to these periods as days? Well the term "day" is as manmade as clothes, buildings, trains etc.

In this is another point that proves the technique science applies at present, does not nearly give a near value to the time duration of development on Earth. It should be out by as much as a few billion years for all we know. To indicate the meaning of what I am trying to bring across I shall illustrate a time scale in which the development might have taken place. I do most strongly disagree with the age the Brainy Bunch hands out to the solar system but since I have no better time to give I shall use the Xepted table ONLY AS AN INDICATER SERVING TO PLACE RELEVANCIES.

Every star (even a midget such as the sun) is a gas giant going down to a Black Hole

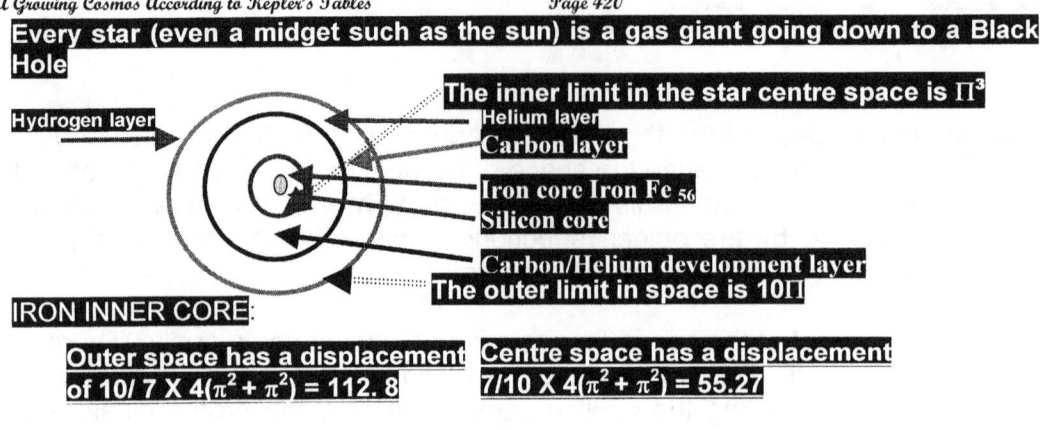

Hydrogen layer

The inner limit in the star centre space is Π^3

Helium layer

Carbon layer

Iron core Iron Fe $_{56}$

Silicon core

Carbon/Helium development layer

The outer limit in space is 10Π

IRON INNER CORE

Outer space has a displacement of $10/7 \times 4(\pi^2 + \pi^2) = 112.8$

Centre space has a displacement $7/10 \times 4(\pi^2 + \pi^2) = 55.27$

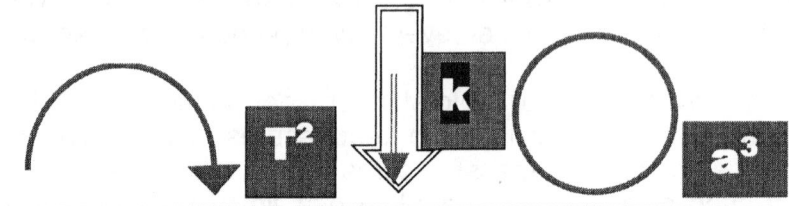

Planet	Period years	T	T^2	Distance	Space a^3	Ratio
Mercury	0.241		0.058	0.39	0.059	0.983
Venus	0.615		0.378	0.728	0.381	0.992
Earth	1.000		1.000	1.000	1.000	1.000
Mars	1.881		3.54	1.524	3.54	1.000
Jupiter	11.86		140.66	5.20	140.6	1.000
Saturn	29.46		867.9	9.54	868.25	0.999
Uranus	84.008		7069	19.19	7067	1.000
Neptune	164.8		27159	30.07	27189	0.999
Pluto	248.4		61703	39.46	61443	1.004

The inner core has to be Fe_{56} to produce gravity. This is what reduces space in conjunction with singularity where the atoms produce a dismissing value that the space-time can sustain with enabling the flow of heat through space. In the one limit of the six sided Universe no element can sustain duplicating above the value of **$10/7 \times 4(\pi^2 + \pi^2) = 112.8$ and above $7/10 \times 4(\pi^2 + \pi^2) = 55.27$** within the star inner core. Dismissing space beyond that capability will no longer contribute to duplicating space-time of the atoms involved. Only the iron atom producing and maintaining a displacement value of $55 - 56$ can produce gravity by being on the edge of demising space time while maintaining duplicating which is gravity and in our Universe only stars with an iron inner core has the ability to bring about gravity. Gravity can only achieve a displacing relevancy at **$7/10 \times 4(\pi^2 + \pi^2) = 55.27$**, and that produces a potential difference that brings about gravity within the inner star core where gravity accumulates. This then relates directly to the second value of the Titius Bode value of **$10/7 \times 4(\pi^2 + \pi^2) = 112.8$** that limits outer

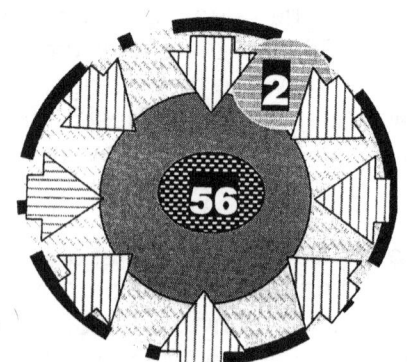

space in the three-dimensional and six sided boundaries of what forms our Universe as outer space forming the value of **$7/10 \, \Pi^6 / 6 = 112,162$**. That is the outer relation to the inner relation set by the core in ratio to the outer space securing a position for the star identity in the space limits and is an indicator of the balance in space-time displacing potential of the star. In every star there is this flow towards the centre firstly of every individual atom but also as a combined unit flowing towards the centre of the star and the dismissing of space in every atom centre brings about the forming of a relation as a group within one unit structure we call a star. This flow is there because we gave it the name of gravity and gravity is the result of all the atom protons dismissing space and as such then has a linking that is invisible to the naked eye. In young stars, the core ability is yet to develop and in such stars, the gradual reducing comes about as layers support the effort little developed inner core. The space reduced becomes a unifying effort from all the atoms in all layers from the outer (hydrogen$_1$ and helium$_2$) through the carbon / oxygen centre and the silicon layer down to the iron core and even going down further into space-time obscurity where the atoms as a group combining their effort acting as one atom. An atom securing one proton will provide much more space a much better field to flow.

The scale above proves the accuracy to some degree but with the information being as sparkly as it is in this letter I would rather prefer if you would please see it more as a scale to use and form ideas than to be a mathematical yardstick. He way I present the following might not seem to as an accurate and tested scale in time but merely as measure to indicate how the frequency will relate to the time duration. From where we stand, we may have a perception that the frequency is getting shorter, but as seen from within the sun, the time duration would be precisely the same value each time. This is the process in which time is concentrated in the space confinement and the relative space-time is amplified to extent the duration. In time, this variation is perceived as flair and later as a pulsating readjustment. I hope it now will be apparent just how small and under developed our Sun really is when compared to other structures.

This comes about because the relative size of a star is based on its space volume that contains matter and the incorrect way in which the density of stars are calculated. At present the frequency could have come down to as little as 15 000 years, maybe slightly more, but who knows. However, it is not the frequency that is the problem, but the way in which the frequency is measured that is of concern. At present, we relate to the duration of time laps relevant to the magnitude in which the Sun presently is. This might be a problem to all of mankind and civilization. There exist neither method nor means in which one can determine at what stage of progress the Sun is in at this moment. All that is extremely clear is that at one stage, the Sun becomes a raging bull, and a sleeping bear follows this. In between these two possibilities of time duration, time can become double the value it holds now, which then is followed by a period where time might have half the duration, we experience at present. The first thing that springs to mind, is that we find ourselves in the middle, which averages out the extremes. That is not the case.

Let us start by taking the size Jupiter is today. We know seven events happened and each event had influence on the Jupiter distance.

The relevance of the actual distance is however, 16,5; 15,4; 10,6 and 8 respectively in relation to the others of 2Π.

With this knowledge secure, we have to seek the positions evidence as how the structures came to be in that place as the inner planets do not confirm their position in accordance with the rest of the solar objects. When we give the distance that should apply if the inner planets were also just orbiting objects it would then be

The official average distances from the centre of the Sun the average distance of rotation that each inner planet completes in relation tot the governing singularity.

At first I thought the way I presented my first impression of the solar development was correct in as much as the way I first introduced the image. Back then I was still very much under the influence of the Newtonian conceptions of a runaway star, and other misconceptions I later found to be alien to cosmology. There can be no runaway star precisely for the reasons I explain the constitution of Galactica. A star with an individual developed space in the time of this era the iron$_{56}$ era, will then establish a circular "gravity" that is able to withstand the influence of the linear gravity. The higher the circular "gravity" becomes, the more static will the linear "gravity" be. In the case of a Proton star (Black Hole) the linear component lies with the fact that we can actually see matter performing its linear component by not curling as lesser stars do, but placing the circle and linear components all on the matter as the Proton star pulls matter, space and time into its pace-time occupation. A Proton Star is unmovable, static, and stationary and every other name you wish to connect to its immovability. It can no longer go anywhere. The stars within the sphere where doctor Hawkins identified a Black Hole to be, within the very centre of a galactic, holds all occupation relevance to the spin or linear component of space-time occupation and only a very minute part to the circular "gravity" component.

I bring evidence to prove my personal theory development and showing honest misconceptions on my part. In that view, I wish not to remove the first suggested solar formation but to replace it partly with facts that I became aware of, as my personal insight grew.

There is another way of looking at the effect the Titius Bode law applies to cosmic atoms and this is very important within stars. No person can deny the fact that the Earth is a sphere, excluding outer space, where our need to apply entry into the sphere of inclusion ($\Pi^2\Pi$), and the law to abide by is the four cosmic pillars. You have to abide by the Roche limit where rules allow you entry, or destruction. No Newtonian can deny that.

When you cannot deny the fact that the Earth is excluding space as it is including time the rest is beyond denying also. One has to seek the evidence where the evidence is, where one can locate such evidence and above all, read the evidence correctly. The evidence proves the existence of a binary before the Earth came to be. The four inner planets are left over parts, a reminder of a star that uses to be at the other side of the atom. While the one side of the atoms in a star relate to the square of space 10 / 7, the other part of the atom in the star relate to the matter to matter (neutron ($\Pi^2\Pi$) holding matter to space while space becomes time ($\Pi^2+\Pi^2$).

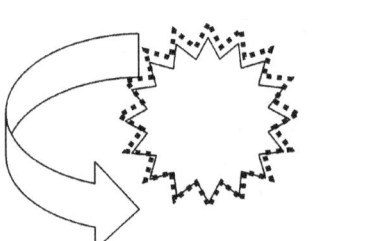

The Sun was in a binary with another cosmic structure that has no longer have a full place in the solar system as the solar system stands today. The second object had a good measure of the suns' potential, but not adequate to survive. If the Sun was the size of what Jupiter at present is, the second binary was then about the size the Earth is today.

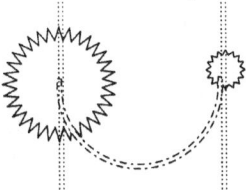

Both had individual singularity Π placing a value of 2 X Π in space, as well as a common singularity $\Pi^2 + \Pi^2$.

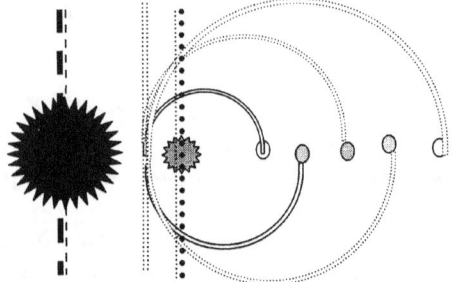

From this binary extended a singularity connecting five rock ice cubes, each holding a point of singularity, with the electron position of the binary where the binary holds the mutual point of singularity. This had nothing to do with 5 or 4.5 X 10^9 years in time laps.

What I am about to do, is very unscientific and the next few pages must be regarded as pure speculation, brought into the book for one purpose and that is to amuse. That is all value that the next calculations have. I dispute the fact that any calculations can ever determine the precise size the structures had because the structures at present hold the very same size that it held when the solar system formed. However, life is not only about proof and fighting dispute of proof, there has to be some entertainment in the book, merely then to satisfy our need to gossip. It is far better to gossip about the planets than it is to gossip about one another, because I do not think it will hurt the feelings of the solar objects at all. It is utter speculation and a needless process and I do not wish to encourage such wild guesswork in any way what so ever.

From this binary extended a singularity connecting five rock ice cubes, each holding a point of singularity, with wit the electron position of the binary where the binary holds the mutual point of singularity. This had nothing to do with 5 or 4.5 X 10^9 years in time laps.

What I am about to do is very unscientific and the next few pages must be regarded as pure speculation, brought into the book for one purpose and that is to amuse. That is all value that the next calculations have. I dispute the fact that any calculations can ever determine the precise size the structures had because the structures at present hold the very same size that it held when the solar system formed. However, life is not only about proof and fighting dispute of proof, there has to be some entertainment in the book, merely then to satisfy our need to gossip. It is far better to gossip about the planets than it is to gossip about one another, because I do not think it will hurt the feelings of the solar objects at all. It is utter speculation and a needless process and I do not wish to encourage such wild guesswork in any way what so ever.

For your entertainment alone, I shall go about trying to determine the size of the Sun back when…as the size of the Sun and other structure was when the dual came to its final resolve.

Let us start by taking the size Jupiter is today. We know seven events happened and each event had influence on the Jupiter distance.

With this knowledge secure, we have to seek the positions evidence as how the structures came to be in that place as the inner planets do not confirm their position in accordance with the rest of the solar objects. When we give the distance that should apply if the inner planets were also just orbiting objects it would then be

1	Mercury	$58 \div 2\Pi = 9{,}23 \times 10^6$km
2	Venus	$108 \div 2\Pi = 17{,}188 \times 10^6$km
3	Earth	$149 \div 2\Pi = 23{,}714 \times 10^6$km
5	Mars	$227 \div 2\Pi = 36{,}12 \times 10^6$km

The relevance of the actual distance is however, 16,5 ; 15,4 ; 10,6 and 8 respectively in relation to the others of 2Π.

In the case of Mercury, Mercury is $5\,\Pi$ further than the $(\Pi/2)$ (Π^2) of the others are and Venus is $(\Pi^2/2)$ (Π) ; Earth is Π (Π) and Mars is almost $(\Pi/2)^2\,\Pi$ away from the sun. I admit that the distances do not apply to the millimetre but that will become apparent soon. The importance here is the relevancies Mercury is $(\Pi/2)$ $(\Pi^2$). That is the place where the binary minor should be in if it was still there.

Venus is $\Pi^2/2$ (Π), which is in the space-time that binary minor held at a position it would hold its value to Π.

Earth would be in a position of one Π more than where binary minor would relate Π. This then is Π x $\Pi = \Pi^2$. Mars would be $(\Pi/2)(\,\Pi)$ (half a Π) even further away than the Earth's position of Π (Π). We shall get to Ceres (a fragment of what was another planet in due time. Therefore let us establish the relative positions to Mercury, the sole holder of the star binary minor in relation to the other fragments.

Mercury = 0.
Venus = $(\Pi^2/2)\,\Pi$. The edge of the entrance field that Binary Minor had.

Earth (15,4 + 10,6) = 26. That has a relevancy of about $(\Pi/2)$ (Π^2) = 24,35. This is where we must not forget that the Earth too is a binary and the moon played its part in the drift relating to a position of 26 instead of 24,35 or $(\Pi/2)$ (Π^2).

Mars holds a relative position of 10 (Π) and we know that 10Π are the relevant space position to $\Pi^2\Pi$. This indicates that anything outside 10Π will be far outside $\Pi^2\Pi$ and this places the object in that space, without space. Any object without space will be directly into time and this will mean total destruction of that object. It will have the very same consequences as having a cosmic body holding an iron core in the previous era, or having a cosmic body with a silicon core in this era. It will and must disintegrate; there is no question about that. The space-time applied to such a core will be double than what is to the other five inner planets. It will destruct, in the same manner, as did the Shoemaker Levy 9 comet with the one exception, it held a relative position where the Sun could not get hold of the fragments as Jupiter got a grip on the Shoemaker-Levy 9 fragments.

At first I thought the way I presented my first impression of the solar development was correct in as much as the way I first introduced the image. Back then I was still very much under the influence of the Newtonian conceptions of a runaway star, and other misconceptions I later found to be alien to cosmology. There can be no runaway star precisely for the reasons I explain the constitution of Galactica. A star with an individual developed space in the time of this era the iron 56 era, will then establish a circular "gravity" that is able to withstand the influence of the linear gravity. The higher the circular "gravity" becomes, the more static will the linear "gravity" be. In the case of a Proton star (Black Hole) the linear component lies with the fact that we can actually see matter performing its linear component by not curling as lesser stars do, but placing the circle and linear components all on the matter as the Proton star pulls matter, space and time into its pace-time occupation. A Proton Star is unmovable, static, and stationary and every other name you wish to connect to its immovability. It can no longer go anywhere. The stars within the sphere where doctor Hawkins identified a Black Hole to be, within the very centre of a galactic, holds all occupation relevance to the spin or linear component of space-time occupation and only a very minute part to the circular "gravity" component.

I bring evidence to prove my personal theory development and showing honest misconceptions on my part. In that view, I wish not to remove the first suggested solar formation but to replace it partly with facts that I became aware of, as my personal insight grew.

Let us establish a line of evidence and fill the puzzle.

1.   There was another star with the Sun where the Sun was Binary Major and Star Unknown was Binary Minor.

2.   The Binary system catapulted the solar system out of its frozen eternity, way ahead of its time of development, bringing along the rest of the micro stars. The outer "planets" are not planets; they are micro stars in development.

3.   The Binary system formed part of a Lagrangian system holding the Binary in the centre and with Jupiter as the first orbiting satellite.

4.   The position Jupiter held for most of the developing period made Jupiter the second main benefactor of the dual, with the Sun the major winner and Binary Minor the major loser. This will also explain why Jupiter has such an advance in space-time occupation when compared to the other micro stars.

5.   The position where the (six) inner planets find themselves, were a void, AS THE BIBLE CLAIMS.

6.   The relative positions were as follows:

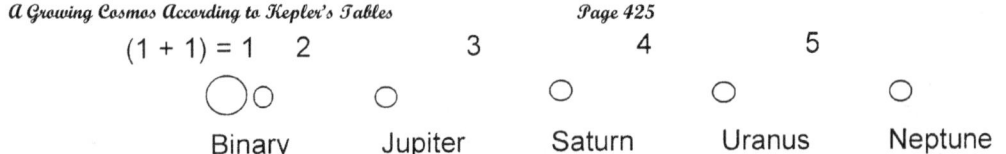

$(1 + 1) = 1$	2	3	4	5
Binary	Jupiter	Saturn	Uranus	Neptune

Then Unknown Star capitulated as it could no longer serve the dual it fought. It fragmented into possibly 9 major parts and many minor parts, (the comets.)

7. As the core fragmented the brittle parts dislodged in a position each to a relative neutron position in the space-time binary minor held relating to its point of $(\Pi^2+\Pi^2)$ $(\Pi^2\Pi)$ 10Π. Whichever way we Earthlings will look at our position, from whatever angle and by whichever calculation we devise, our relevancy will be 1, will be 10, will be Π^2. Should any person ever do a calculation and find his answer does not bring this fact to bear, he must go back and fix his mistake. There will be a mistake on his part.

SUN	MERCURY	VENUS	EARTH	MARS
B_1	B_2	$\Pi^2/2\Pi$	Π^2	10Π

$$\Pi^2 + \Pi^2 \longrightarrow \Pi \quad \text{(This I shall explain)}$$

Binary stars, spinning to self-destruction will produce significant heat. Heat create space, space forms winds. That is facts that the Bible presents and is indisputable. Where the Earth was, was still a void, containing a sphere of circular displacement and this will reduce linear displacement to zero. Linear displacement is space and circular displacement is containing heat for matter survival.

Binary Star Minor overheated. That is why the core brittle and fragmented. This action will release tremendous contained heat, the heat will produce magma flowing in space like water in space and this eruption of heat space that created winds. Once again the recollection fits the scenario. Releasing the heat and producing space will establish space-time and fill the void where the Earth should fit. This is fact and if anybody even tries to dismiss this will be because of abstinence on his or her part. I did not prove the Bible correct. The Bible told the truth and in such correct detail, it is beyond human comprehension, but sublimation on the part of Newtonians and science before them, disallowed their ability seeing it.

I found no one could look past me and see my formula $R^3/T^2 = 1$ and $\$T = (\Pi^2 \times \Pi^2) (\Pi^2\Pi) \, 3 = 1836$ which is the relevance of the cosmos. By not finding a person that could see past me, I knew that person will not be able to look beyond "a burning Sun and see the frozen state in which the Sun is. Without noticing such crucial evidence, the rest goes lost. That person that sees me and not my formula will never see the cosmos for what it is.

Slightly of the mark but duly valid I wish to make a brief remark on the Sun/ moon binary. As the moon is also in a binary extended position with the Earth I wish to take this quick opportunity to show that the moon was never part of the Earths proton - proton value $(\Pi^2 + \Pi^2)$ value but is in a neutron to space position $\Pi^2\Pi$. This can only apply when the one object occupying less space-time has a proton value $(\Pi^2 + \Pi^2)$ that is less than the superior object's position on $\Pi^2\Pi$. In other words, when the total core value of the lesser structure is in any case less than the neutron value that the larger object relates to, concerning the smaller object. This means the one is totally dominating the other in all aspects.

Some quarters of the Newtonian High Priest in High ranking made claims that the moon once formed part of the Earth. In the following elaboration I shall prove why I dismiss this claim as utter nonsense.

From these facts about a binary, one can then clearly see that having two structures in a position overshooting the Binary scenario, is very much fantasy. It is just not possible because the valid space-time will exceed 112, and the structure will not have the ability to hold position in the universe that is limited to 112.

The proton value of the Earth is $(\Pi^2 + \Pi^2)$ and it will hold the second object (the moon) at $\Pi^2/2$. This is because the second object is in the "gravity" application of the larger object (the Earth) and the "gravity" factor of the Earth takes on a linear value, half that of the gravity factor of the Earth. The Earth will not allow any linear action to exceed 10Π and at $(\Pi^2 + \Pi^2)(\Pi^2/2)$ it exceeds that value.

As the two core has a dual space-time occupational value of $(\Pi^2 + \Pi^2)(\Pi^2/2) = 97$, and the core value of the Earth is at $7/10$ ($4(\Pi^2 + \Pi^2)$ the combined value will even exceed the critical space factor of $3(\Pi^2 + \Pi^2)$ applying to stars holding space, therefore the space separating the two objects will vanish into singularity. The reason why the Roche principle maintains core separation is that the core combinational value , seen from one or the other objective, is $(\Pi^2 + \Pi^2)(\Pi/2)^2 = 48$. The individual space-time factor of each core is $7/10$ ($4(\Pi^2 + \Pi^2) = 55$, therefore the space holds less heat and therefore more space.

Where two structures go into a Roche Lobe and the one structure forms a proton value of Π^2, but the comparable space-time occupation is less that Π the Shoemaker Levy 9 structural fragmenting will take place.

As larger structures will have no occupational space loss due to overheating, but the one holding a Π value has great concerns.

From the superior object the occupational distress will be $(\Pi^2 + 2\Pi)(\Pi.2)^2 = 39,8$. The geodesic space value as a factor is $\Pi^3 = 31$, therefore it will bring about a "gravitational pull" revaluing the relation to $(\Pi^2 + \Pi)(\Pi^2/2) = 64$. Being at 64 it means the smaller object holds a position of space reduction, as the space value is twice that of the geodesic value. The conclusion is that it will fall under the invert square law of spheres. By looking at what happened to the comet, one can see that such estimation will be correct. When taking that formula and applying it to the position that the smaller objects holds, one cannot surmise immediately that it will be part of the atmosphere, therefore the 7 in the formula in atmospheric heat income will change from $7(3(\Pi^2)(\Pi^2/2)$ to $(4\Pi^2 +\Pi^2)\Pi^2/2 = 121,36$ because the second object holds a far superior occupational position in its application of "gravity". With a relative value of 121, overshooting the highest atomic occupational possibility of $7/10 \Pi^6/60 = 112$, the atomic structure that the smaller object holds will diminish to heat and photons. It will break up; turn to heat, photons, and dissolve, which are precisely what, happened to Shoemaker Levy 9. One can witness the structure demolishing in heat, light and fragments.

With an object larger than that of Shoemaker Levy's relevancy to that of Jupiter, the same laws apply but the values derived from it bring about a different end result. The only change will be in the position of the relevancy where the one object being the superior will again apply the same formula in establishing its position. $(\Pi^2 +2\Pi)(\Pi/2)^2= 39,8$. With this value being the same as $2(\Pi^2 + \Pi^2) = 39,47$, it will hold the structure in a cosmic orbit, not being able to reduce the space-time separating the two, and applying the gravitational equilibrium of $2(\Pi^2 + \Pi^2)(\Pi/2)^2 (2\Pi^2 + \Pi^2)(\Pi./2)^2 = 73$ and with the space-time occupation not only exceeding $3(\Pi^2 + \Pi^2)$ where space destructs but going another half a Roche factor down $(\Pi/2)^2/2$ above and beyond the space demolishing value of 58, it means there must be a total structure space decrease of some sort. It will not be a structural break up and fragmenting as in the case of Shoemaker Levy, but still a space-time occupational re-adjusting. This one can witness by studying the evidence Hubble's photos brought back. As indicated the superiority of the proton rules, not only the atom, but also the universe. The volumetric size matter holds, is in precise ratio to the space value of the protons. Apparently all protons hold the same space-time value ("mass") with only the space that changes holding the protons. This factor indicates the density of the star and it is a far greater asset to space – time occupation than merely mass. In this aspect of the proton is the universal equilibrium that produces universal time as matter takes heat in unoccupied space-time directing it to densified space-time through occupied space-time and then finally to time. The progress in the proton is the demise to space. As space is in singularity, space cannot demise. If space demises the singularity within the proton, which controls the space-time occupation has to grow. When the space-time occupied grow, it will control the space-time unoccupied.

The simplicity in proving this is laughably stupid. Photons travel through unoccupied space-time, and if the amount of protons can influence the travelling light, the protons in that particular space during that particular time, also influence the unoccupied space-time. There is more heat around the Earth than in outer space. The protons therefore that controls and maintains the Earth's "gravity" also has to draw the accumulation of heat to the Earth.

Saturn and Uranus which is much further from the sun, is immense hotter on the surface than in the case of the Earth. That fact has to be a sure indicator that the application of "gravity" has to have something to do with the attraction of heat. If heat will only flow from hotter regions to colder regions it indicates that the proton has to be a lot colder than even outer space. By moving particles through spin brings about cooling, therefore the proton has to spin much faster than the speed of light to be able to draw photons from the unoccupied geodesic space-time (outer space) to the proton.

If the proton draws heat it can only be to cool the proton and therefore the proton has to accumulate heat. Through this then the single proton grows in "mass" or densified space-time. This brings about that all matter becomes larger through the development of time.

The "mass" will deform, possibly brake up, as the space within Jupiter will revise the time. The space, which the wood occupies, will reduce to the extent that the structure may brake into a liquid and even a gas. Through this the "mass" will not reduce, it will increase as the heat component increases. As the heat component increases, the matter will grow faster than it would in outer space.

The formula science uses in determining time is $t = 1 - \sqrt{C^2 - V^2}$ in as much as the photon's speed (square $\sqrt{}$) minus the speed of light (C) square minus one representing time will produce time. This formula does not allow for any change in time. With this view, science is also in solidarity about the fact that everyone in science accepts that "gravity" influences time. This fact was tested in launching the most accurate chronometers man can devise and found positive results. Yet, not one formula complies in any way of this change to influence the universe.

I indicated the influence density has on the "gravity" by showing the relative difference planets holds. Presumably this influence of density will multiply by billions of times in one Black Hole, or as I wish to call it, a Proton Star. If "gravity" influences time duration to retard in a minute environment such as the Earth, how much more does time retard within a Proton Star? Time would literally to all human measures stand still. It will become eternity because that is what time in a still standing mode is to us humans.

A Proton Star is just the first star with the uttermost fragment in space (almost to the point of singularity) of the universe as it came out of eternity, equal to the "gravity" endured by matter back then, during the "Big Bang". Even if one use the Newtonian formula the measurements must be beyond calculation, bringing the time duration that applied during the "Big Bang" also to eternity. To us non-Newtonians this conclusion is obvious, but to the Newtonians it is far too simplistic. Not surprisingly, the logic behind the argument and facts are far too simple for our Super-Intelligent-Super-Educated-Wise-of-the-Wise. Being as super intelligent as they are, the cosmos has to test their own brilliance by introducing problems only those with their super intellect can see, understand and solve.

The matter of the fact is that when time slows down to a minute pace, it will seem everlasting to us. This fact is beyond any argument. Another fact is that heat does not bring about fusion, but it does bring about change in the application of the duration of time, affecting the space in which the time is.

This rather lengthy elaboration of repeating facts already explained in detail is to bring across how little science can piece together the most obvious and logic of facts, which they supposedly are the masters of. Life is fare more complex than anything in the universe and because we are part of live, we can only view life as life reflects the history of time. We humans are part of life, yet with all the research, no one ever came up with a definition about life.

Life is an energy with the ability to manipulate space-time occupied and unoccupied. To change the body, which holds life, is only part of the manipulation of space-time occupied to the benefit of live occupying that space-time.

Because of the atmospheric and surface heat they believe the water formed vapour and the vapour vaporized and disappeared into the vastness of outer space. It is this part of the theory that makes the theory completely unnatural and bogus. What the scientist wishes to imply, is that the sun's solar winds will be stronger than the "gravitational pull" of the planet. The minute the vapour becomes a solid, which water is when frozen, it will be heavier than air and it will fall back to the planet surface. Even when evaporating again before the water reaches the planet surface, it will evaporate, but again it will form water and ice, and this process will continue indefinitely.

As for comets with boiling water forming the tails as the Sun "heats the surface of the comet". I am not willing to waste any space or time in this book by dealing with such illogical nonsense. I do explain the misconception about comets and their tails rather extensively in "Matter's Time in Space".

The Earth has an abundance of water. The question arises: Where does it come from? The answer is in the closeness the moon has with the Earth. The moon is not a moon to the Earth, but much more a sister planet. When studying the effect of the Roche Lobe and interpret this to the relation the moon and the Earth once must have had, many unexplained questions find answers.

Examining all the facts about the dual planet system, it seems one is blessed with all the cosmos can offer, and the other one is dead and docile. The sister planets are in the most extreme of positions of all planets in the solar system. Science has developed the knack to apply circumstances they find today and interpret it as if it has been there since time began.

Let us reflect what happens in the Roche Lobe and apply this to the sister planet system. Even if the distance between the Earth and the moon does not fully comply with the necessary Roche distance today, one has to bring into the equilibrium the fact of solar development, which would be in the category of the Hubble Constant. There is differentially growth of the Titius Bode application to consider.

As every one knows, water form where one oxygen particle forms a compound with two hydrogen atoms. When any two structures go into a Roche Lobe they cut the circular motion (R^2/T) off from the geodesic space-time.

Through this action one find a secluded system, cutting off all influences from the outside.

$$(\Pi^2/2)2$$

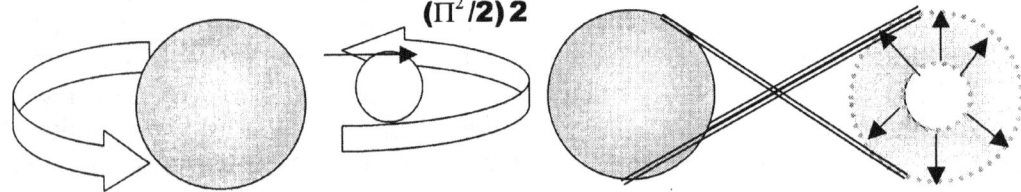

On every cosmic system "gravity" will always be Π^2, but the value of Π^2 will be different. As both systems share a common linear "gravity" (R/T) of $\Pi/2$ at point L, both structures will have the same atmosphere. With both structures forming the atmospheric value of $\Pi^2 \Pi$, this will allow the perfect condition to form compounds. In view of the fact that the Earth will have a dominant Π^2 value, it will take up all the progress that the double structures can produce through the Roche Lobe.

The void to which the Bible refers, is the effect of the Roche development between the two systems, as an atmosphere formed in the Earth section, destroying any possible chance of atmospheric development on the moon.

However, another major factor of development is that the core of the Earth will benefit largely from the Roche system, as the Earth will be the major benefactor of the heat increase, deriving from the large spin the circulate motion of both structures increase. The Earth therefore has a double development in progress denying the moon its fair share in normal development.

On the surface of the Earth a great amount of water developed, cooling the Earth's atmosphere drastically. The cooling will accompany a huge vapour of water, as the water formed clouds. With the atmospheric temperature being this high, the clouds will evidently be extremely high, forming a massive and thick cloud layer. There will be little chance of rain, because the water will form back to vapour as the water rains down on the surface, never reaching the surface.

Who can, without the support of a fully developed technical language, describe the Roche-developing factor between the Earth and the moon, in a better way than did the Authentic Author of Genesis?

After separating the waters from the sky, came the third solar day. The Bible verse is as follows: "Let the waters under the heaven be gathered into one place. so that dry land can appear."

One has to remember that the Roche principle was still in full effect, much more than as it is presently in effect. During the Solar Day, it is only "non-conviction" and during the "Solar Night" it becomes conviction (if you will).

To understand this commissioning is quite simple. Pour a steaming, boiling hot cup of coffee to a specific point. Then let the cup of coffee cool to room temperature and you will find that the mass of the coffee reduced by a millimetre or slightly more. This is not due to loss of matter through vapour, but through the loss of heat. The mass of any heated substance swells. The very it apply when the Earth and the moon surface heats extensively. To all natural principles, we are still in the Roche-Lobe but as the source of the heat is external, and not internal as in the case of stars, the Roche Lobe comes into effect with a solar expansion.

In the next phase one can clearly see that the "day" the Bible refers to consists of many seasons, with many growth periods.

It will be of little use to remind the readers that man has a written memory of a few thousand years, where even with such little information, the biggest amount of the evidence is lost through time.

In the Roche-Lobe the following principle becomes a major factor:

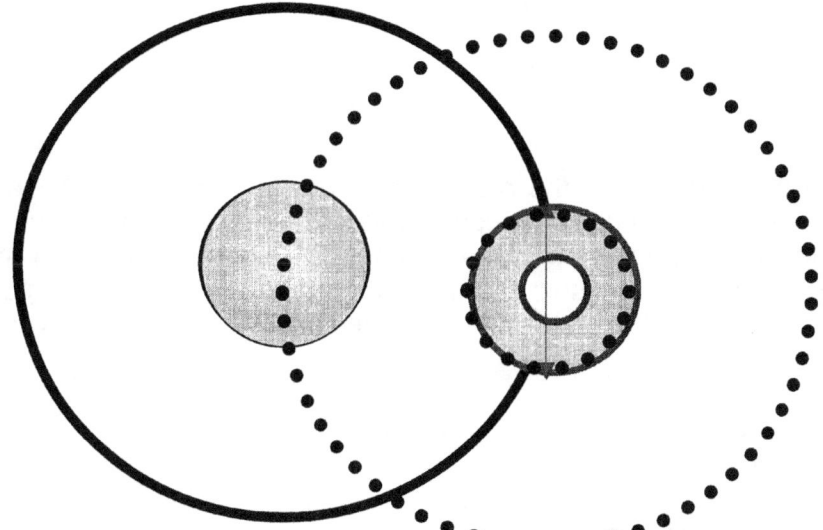

With the Sun blowing heat in thick clouds over the planets, the heat that the Earth and the moon detains through "gravity" is many times more than it is at present. Because the iron$_{56}$ core will reduce most space, densifying heat, it will also retain most heat. As it cannot accommodate all the heat retained, it will relocate heat through space forming, to the surface. In relevancy to the iron-core the silicon is space, therefore rejecting surplus heat will bring about introducing excess space amongst the silicon. This is the same that apply when baking bread and the bread "rises" in the oven. The silicon layer "rose" as the heat, coming from above, as much as from below baked the silicon to rise. From the top, the water still formed vapour, taking all the heat that the moon and the Earth holds as a compliment and then with the core of the Earth being the dominant, directs almost all heat to its core, because the Earth's iron core brings about the most space depletion ("gravity"). Through this process where the combined effort of the moon and Earth removes heat by space depletion a large area becomes effected on the space end … but, on the inside, at the time end, the iron core of the Earth is the almost sole recipient of heat, leaving the silicon of the Earth as the sole benefactor of space-incorporation (bread rising). Said in another way, the moon helped in doing all the work, but the profits of the work went entirely the Earth's way. That even includes the vapour where oxygen and hydrogen combined through the excessive heat to form water. The vapour from the charging of hydrogen and oxygen, discharged again as the heat moved by lightning to the Earth. From this water formed in abundance as the moon and the Earth both collected, both stirred oxygen and hydrogen into a mixture, but only the Earth collected the end product.

This process became as seasonal as winter and summer now are, as seasonal as rain spells and drought spells are or as ice age and heat spells are. Who would know the intervals, and the intervals are not important, because time back then is not time at present. The important issue is the evidence left in the Earth.

As the cumulative positive space-time displacement rises above the value of the other surrounding protons in surrounding atoms, the spin will exceed the average inner-Core-value of the other protons, thus "freezing" the nucleus of the atom in fusion in the time zone of the major element.

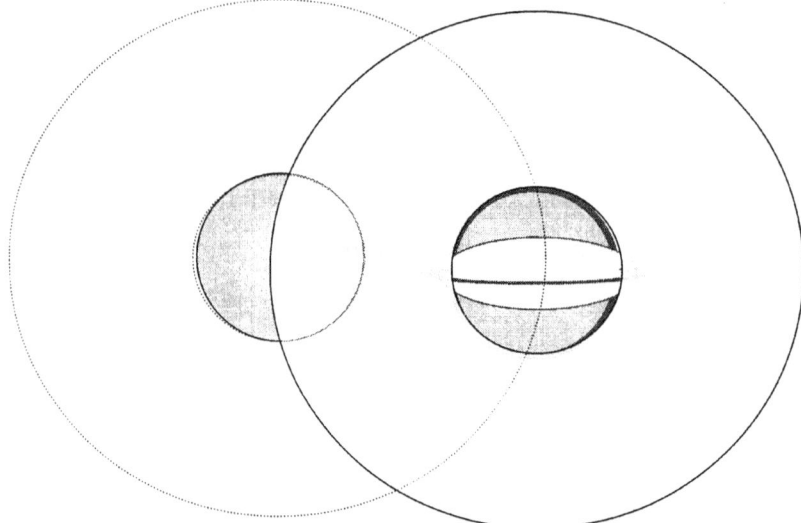

As the time duration extends further with the growth of the proton, that specific atom will develop a time duration much higher than the surrounding atoms. An accumulation of heat above and beyond the actual time duration develops, placing the space-time of that atom much higher than that of the actual time value within that layer. The "mass" increases and this causes the **aanplasing** to grow "pushing" the atom in the direction of the centre of the star where the time duration matches the value of the growing atom.

The ratio imbalance that occurs within the atom can be displayed as follows

$10 \ (\Pi^2 + \Pi^2) \rightarrow 10 \ (\Pi^2 + \Pi) \rightarrow 10 \ (3)$
$10 \ (\Pi^2 + \Pi^2) \rightarrow 12 \ (\Pi^2 + \Pi) \rightarrow 10 \ (3)$ **as the time value grows**
$11 \ (\Pi^2 + \Pi^2) \rightarrow 11 \ (\Pi^2 + \Pi) \rightarrow 13 \ (3)$ **as fusion occurs releasing heat and light.**

Because of the growing demand for heat caused by the increase in space-time density and therefore bringing about more use within the proton, the neutron, which is the balancing factor of the atom, has to readjust and apply more heat in order to keep the neutron form overheating. A point arrives where the accumulation effort of the neutron can no longer sustain an effort in supplying the accumulative call for heat by the proton.

In order to apply a balance in space-time density (mass) the neutron captures a proton that is in verplasing and this proton freezes into the density of the overheating atom. With the growth in the density of the proton, the neutron's accumulative cooling effort does no longer need to sustain such an effort, therefore rebalancing the heat application. The heat excess and rebalancing releases a great amount of heat built-up in the unoccupied sector. As this heat release comes about from the neutron reaction, it comes out as radiation and photons. In the lesser dense (top layers) area the release of heat overshadows the requirement of the proton; therefore a lot of light and heat discards back to geodesic space-time. However, in the inner star structure this readjustment will be at an equal balance and other atoms within the space confinement will apply the heat to suit their need. This produces much less of heat to the outer regions of the star.

Should a value of $7/10(\Pi/2)$ represent the hydrogen atom and $(2\Pi^2/4) \times (\Pi^2/2)$ that of carbon. By inflating the carbon atom's unoccupied space-time slightly one can see that it would accommodate a hydrogen atom.

In this, the Earth iron core grows at twice the ratio of the silicon layer. As the Earth grows, the Earth has to rise above the water at certain points. Therefore the Bible once again is correct by declaring that the water mass, which at first covered the complete surface of the Earth, separated from the water by rising above the surface of the Earth.

In the beginning of the part, I proved the ratio that applies to the atom ratio being $(\Pi^2 + \Pi^2) \times \Pi^2 \times \Pi \times 3$. This particular ratio not only applies to the atom determining the "mass" of the proton (Mp) in relation to the "mass" of the electron (Me). This ration extends far wider than only the atom, as it is an indicator to the revaluation of time duration application and plays the major role in determining the "sound barrier".

The "sound barrier" is a sure indicator as to how heat relations affect the atoms. By intensifying the heat ratio between atoms, the time to space of the atom changes completely to a point where the time overshadows the transfer of sound. It is all a dependence to heat forming time or on the other side space. The less heat there is between the atoms in the unoccupied space, the less the time will be affecting the atom, and the reverse is also true.

It is not only the heat that **one finds between the atoms** that influences the proton – electron ratio and this is the major part of the huge misconception in the view "gravity" applies, because "gravity" not only relies on the "mass" of the protons, which makes up the number of the protons, and it is not mainly the density in which the number of protons are, but it is just as much dependent on the "speed" that the protons travel in relation to other protons.

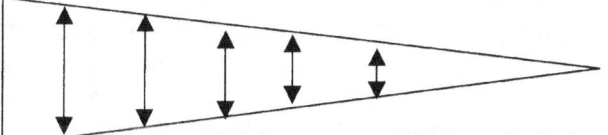

As the electron "travels" through space the time relation increases. The proof in this is that when a wind blows over an open fire, the heat in the fire increases. Oxygen as such, does and cannot burn. It is the special relation that oxygen holds with heat that increases the heat by speeding the flow of oxygen and nitrogen through the fire that increases the amount of heat.

By reducing the flow of air (oxygen and nitrogen) the fire smothers and the heat contained or regulated in the process. This does not change the transformation from wood to ashes it retards and controls the process.

While the one proton connects to space in singularity (Π^2 going to singularity) and connects space Π (in singularity) the other proton brings time Π^2 directly to space Π, re-uniting space-time as a unit to singularity (Π^3). We may call this re-unification unifying the "Graviton" in order to identify the one proton unifying time with space, while still in contact with the other proton holding (Π^2) time to (Π) space in as much as being occupied by matter (Π^2) and unoccupied heat Π forming 3.

This places Kepler's formula at a relevance s $a^3/T^2 = k$ where k^1 will hold a relation to heat with time in eternity, while AT THE SAME MOMENT infinity also applies, giving k^0 the value of singularity. This is all to do with the Universe remaining in contact with singularity while keeping the Universe in matter and heat, in the dimensions of space and time. This explains then the absolute value of time as a square, with the square having both a circular value of R^2/T multiplied by the linear value of R/T producing the link to singularity in time and singularity in space. That is what is the significance about the formula $R^3/T^2 = 1$ where $R^2/T \times R/T = 1$ represents space-time in singularity as well as space-time in densified, occupied and unoccupied format. That means the everything of the whole lot, or as we say in Afrikaans, the "Heelal" meaning Universe. It is a replica of Kepler's formula $a^3 = T^2 k$ but it brings space in line with both aspects of time being $R^3 = T^2$ if $k = 1$ where T^2 then become R^2 / T and k becomes R / T. It put space in relation to both aspects of time because the moving of space is both in a circle and also in a straight line at the same time. The one cannot be without the other.

When saying this, I wish to include the following explanation for those that may have an interest in the technical aspect.

Energy is a term for a power or a force, an effort to get work done. Energy relies on movement. The influence one object holds relating its position in accordance to another position.

When an object remains in one position relating to the rest of the surrounding objects, the time it remains in that position is unchangeable. Therefore, for that duration it will remain eternal. To reflect this relation to a formula will be $R^3 = 0$ and $T^2 = \Omega$. Once the movement starts the position changes, therefore the space relation changes. The movement relates to time because any movement takes time, even with an uncontrolled explosion. Changing space always takes time.

Science agrees on one aspect of time; they have no idea what time is and when anything has the value of being unexplainable to science, it does not exist. Should one not believe my suggestion that science do not accept time as a part of the law of science then think of the money science spend on a ridiculous conception in as much as "time-travel". Even discussing such a conception is time wasting and science should discard it as time wasted. That topic is beneath my dignity as much as it is below my mentality.

Applying energy ALWAYS entertain a synopsis of space converting to heat through a period measured in time. Should one not believe me, think of a lamp, a heater, a stove, they all work on heat where heat is applied by some or other man made device converting heat from space surrounding a generator to heat travelling through some element and when oversupplying the element with heat in a controlled manner. It will readmit the heat to space in the space life wishes to apply the heat. By oversupplying the conductor with electric current, the conductor, resistor or whatever will burn. Anything can only burn with excessive heat applied at one place. When an electric device burns, it is just that. The heat we concentrate by spinning iron through a concentrated excited space filled with polarized (excessive spinning) heat, and by spinning excessive, that heat distinguishes itself from the rest by maintaining a higher spin than the rest of the heat and therefore create an individual time to space than the rest of the environment. This allows the iron, also spinning therefore applying an individual time in space, to place the heat in a separate time than the surrounding atmosphere. It will take heat directly to the Earth in a time- frame where the space is much less than the environment.

The flow of electricity is not a force, it is energy where heat receives a separate value of time, distinguishing it from the rest and as with all energy, it will bring about a reduction to space. In short, electricity is heat flowing. Electricity is space converting to heat and the proof is by investigating electrical human interactions. When human flesh makes contact with a high flow of current, the flesh shows sings of burning. In all events of applying energy, it comes down to the conversion of space to heat through time. It is in all cases, the heat (of the sun) that supplies the heat (of the Earth). All considered; energy is all about interchanging and converting heat to space and space to heat and relating this action to time. When converting elements to stored heat, which fossil fuel is, is in fact transferring heat from the Sun to chemical bonding. This is transferring heat to space in a natural surrounding. Forming that compound to oil, coal and gas is storing the transferred energy in time.

Unleashing the energy for use to extend the influence life holds on our region of the universe, is again all about the transfer from heat back to space, and using the conversion to the "benefit" of mankind.

When dissecting the "Big-Bang-Theory" it is all about converting heat to space. It was rather exceptionally hot at the start in a surrounding which seem to u as considerably small. This however is only a perception we form from our perspective in retrospect. In truth, the universe was just as large back then as it is small today. Nothing grew and nothing shrunk, because nothing goes wasted.

Reflecting on what King Solomon said, three thousand years ago: "There is a time for everything." In this sentence, the primary word is time.
This picture applies. Whether our Newtonians want to accept it or wishes to understand it has very, very little to do with what reality is in science. This is the evidence and any child will see it.

There was time in singularity, as there is still time in singularity and everything will end in time in singularity. However singularity means just what it says, there is no movement of time, no movement in space, there is no space. Every aspect was frozen at zero to everything, what ever you may think of, it was frozen solid. Then came the Creator's command and everything responded immediately. That immediately is not our immediately. Ask any Theologian proclaiming knowledge about the Bible and he will tell you that the Creator is, according to The Creator Himself, FROM ETERNITY TO ETERNITY THROUGHOUT ALL ETERNITY. In this lies the Universe because whatever is in the Universe is between that which has no end and that which has no start. The Universe comes about as it came about when that eternity with no end split from that eternity with no start. It is not religion but it is physics. If atheist are too feeble minded to understand physics it would be better fore those pea brains to go out and wheel a cart in the markets place. That which is eternal split having on one side the part that never can and on the other side the part that never can start where the tow parts are the very same part. By responding does not mean everything started running frantically out a blistering pace. It means eternity ended. That does not mean time as we see it at present started, but at a pace ten times faster. It is a pity that the Theologians never read the Bible, because it the Bible documents it all. If they read the Bible as they claim, why would they then insist upon an Earth being seven thousand years old? The Bible states that to our Creator a thousand years (another way people used to express a time back when the recording of time lead to misconceptions), has the same value as one nights work. THIS IS ALL ABOUT EXPLAINING THAT THE DURATION OF TIME IS PURELY A MAN MADE CONCEPT.

Time started in infinity or in eternity, whatever you wish to say, as long as you say time did not move at all. Then the command came and time overheated for the first Π^2 in time. That brought space into play.

When a bowl of soup is boiling, have you seen the bubbles of air rising from the soup? Has any Newtonian ever taken the time to explain that process in detail? I think not, because such explanations would be far too "everyday-like" to bother their mighty brains. Black Holes and finding the mass of the neutron, and such mighty brainpower cannot bother with small events.

Well, that boiling soup tells the complete story about the creation. Poets and painters and writers always wishes to say how "they created their creation". That is rubbish; they created nothing. They brought nothing new to the cosmos, they only rearranged what was a small part of the cosmos into a new order, that one can detect a distinction from. Creating is producing what never was before. When looking at the boiling soup, one sees bubbles rising from the soup at the top. In the soup's brew, there are only liquids and solids before the heat came. No one placed air in before the event or during the event and any time. Yet from the brew of liquid and solid rises gas, or if you wish space. That space was not there previously. That SPACE WAS CREATED. That space is energy en energy is the interaction between heat and space. As space becomes a part of the soup, a part not there before, with no room to be, it moves out. We refer to that process as boiling. That space creation is applying heat to time, and time in singularity will respond as space in singularity. The space created will vanish just as it came, back to singularity. By applying heat to time, brings forth space, and from the three components, only the heat factor is not in singularity. It removes space in singularity from time in singularity to establish room (space) for heat (time).

That is how creation started. Time in singularity overheated and the product of that was space.

Because time and space both, is part, of the same thing time became space and space became time. I prove that the repeating of this process happened about seven times, in seven different ways and explaining the other five ways is rather complicated and tedious. Therefore, I shall only give two explanations in this book. One is as I explained Π^3 (singularity time) parting with Π (singularity space) leaving "gravity" Π^2. What happens to space happens to time. Space holds three parts with six sides, of which only three sides directing towards any object at any time. Therefore 6/2 (half of the six sides are valid at very instant, only 3 has an effect.). Time in singularity holds a line directing to time. A straight line is 180°. Matter Π^2 holds Π^3 from Π, being valid in forming $\Pi \times \Pi = \Pi^2$.

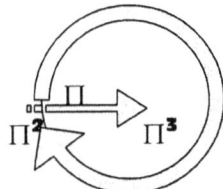

The Π^3 is matter in singularity
The Π^2 are motion or heat.
The Π is space in singularity.

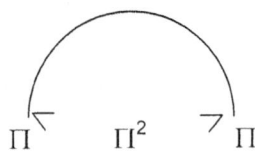

The half circle is 180° placing matter in a circle but because space only applies, to one half, 6/2 and matter holds space to value 6/2=3 only half the circle comes into effect. Half a circle is 180°. Because space has three parts in effect, it also becomes a triangle.

That means

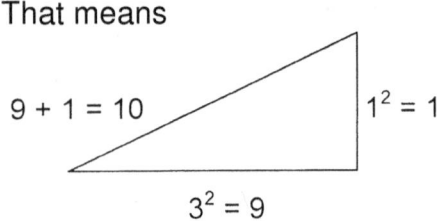

9 + 1 = 10 $1^2 = 1$

$3^2 = 9$

Where space holds three and time is one, the heat within that space becomes another dimension, the fourth dimension holding space-time $(3^2+1^2) = 10$. That changes the matter inside space in singularity at ten and "gravity" at Π^2 . This is why "gravity" Π^2 is space (10) losing one dimension (Π^2). "Gravity" is all about space (occupying matter and heat) losing one dimension back on a long journey to singularity.

As time is in singularity, and space is in singularity and both are the same thing $(\Pi^3 \rightarrow \Pi^2 \rightarrow \Pi)$ the 10 of matter (heat) that affects space (10 Π) will also affect time (Π^3) and therefore time carrying heat will become 10 (Π^3) with space 10 Π. Anyone with a simple calculator can divide 10 Π^3 by 10 Π and see where Π^2 fits in.

The reason why time in singularity is Π^3, lies in the two components of space-time occupation or "gravity" manifested in the Roche limit. All object spin and spinning is a circle Π^2 while all objects are moving in a direction $\Pi^2/2$. Again only, half of Π has any dimensional validity at any given time, therefore the dimension surrounding an object is Π. Because the "gravity" of the surface of the cosmic object extends from Π^2 to Π but only half of the circle of Π (180°) can apply to time (Π^2) being in a straight line $\Pi^2 \rightarrow \Pi$, "gravity" will form at that point of $(\Pi/2)^2$ giving the Π in space the "gravity" to hold.

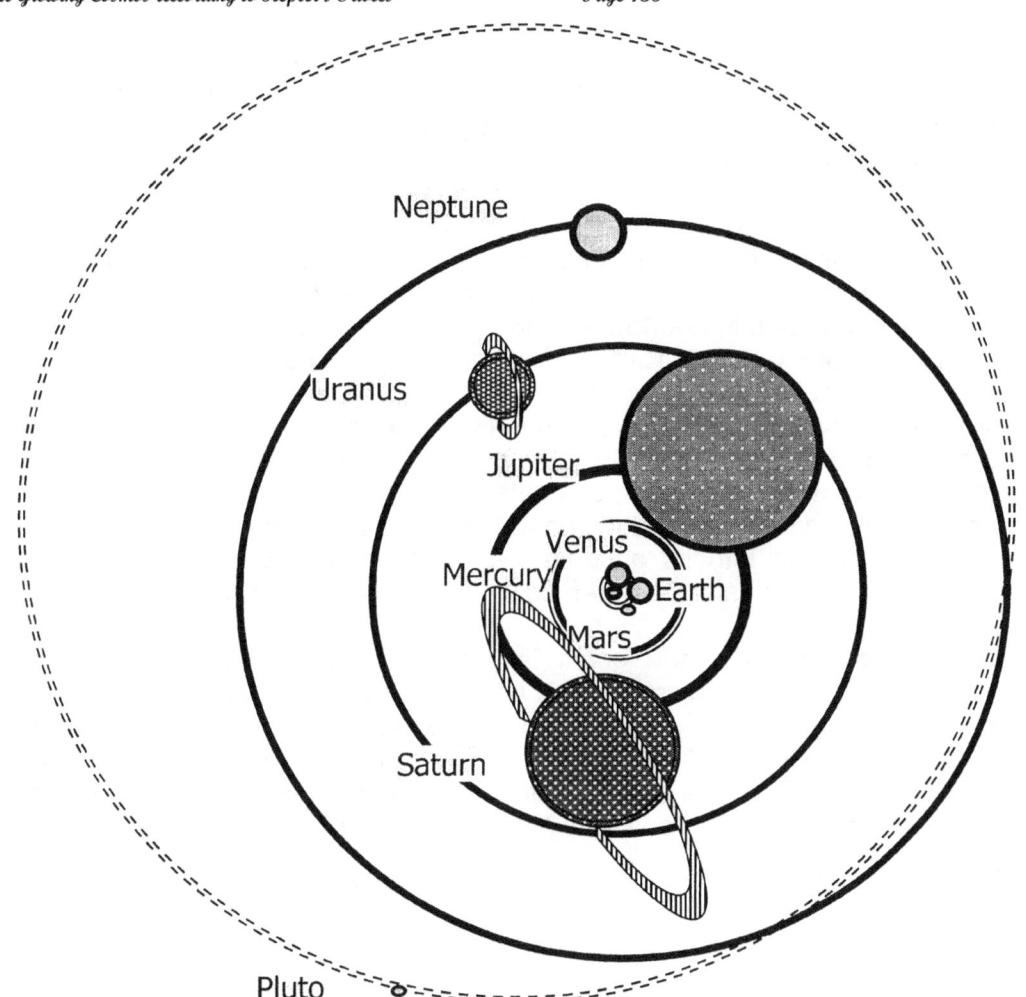

Gauging this tells a story of catastrophe and unplanned accidents, which is not the normal cosmos. This tells a picture of the prefect being interrupted by the imperfect. This tells of something extraordinary to the entire perfect that is the ordinary in the cosmos. This tells of the very opposite to the ordained and structurally sound we find in the cosmos.

That is why space will forever comply with $(7 / 10) \, \Pi^6 / 6 \; = 112$, and time forming the line ($180°$) between the half circle (Π to $\Pi = \Pi^2$) at a $180°$ will form the triangle of space in half ($180°$). The matter component of the Titius Bode law effectively apply to the value of space, therefore 7/10 comes into the calculation. That places any atom with an existence in space at a premium of $7(\Pi^6)/ (6 \text{X} 10)$.

The reason why plutonium at $5(\Pi^2+\Pi^2) \, (\Pi/2)^2 (3/5)=244$ is at the element limit is obvious; when dissecting the relevancy in detail. It is in these lines that we must look for reasons why the solar system is so much extraordinary.

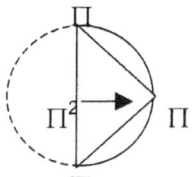 That will produce time in singularity a value of Π^3.

Explaining the other five stages of gravity (Π^2) development is extremely complicated and for that there is no room in a commercial book such as this. My motto in this book is "Keep it simple"

$\Pi^0 \Rightarrow \Pi$

With time in singularity, time was eternal. Time is the spin rate of heat in space. That means the way the movement changes where matter and heat relate to other matter and heat in space. With all movement relating to a circle (Π^2) going somewhere (Π) in space 3. The Π will form the radius to the circle (Π^2). Any novice can see that the longer Π becomes, the wider Π will be and therefore the longer change in the repositioning of matter will be.

It took some while to realize that the Sun is one huge gigantic, awesome (there is no word to fully convey the thought!) electrical short circuit that is why all stars in this era, (a star with an individual developed controlling the flow of heat to singularity in a controlled manner), without creating space as space converts to heat space in the time of this era (the iron 56 era), will establish a circular "gravity" that is able to withstand the influence of the linear gravity. The higher the circular "gravity" becomes, the more static will the linear "gravity" be. The electricity we generate on Earth is not even a thought comparing it to what is applying in the sun. However keep in mind that, we must not over estimate the sun's ability to that of the Earth and think it to be big. In the case of a Proton star (Black Hole) the linear component lies with the fact that we can actually see matter performing its linear component by not curling as lesser stars do, but placing the circle and linear components all on the matter as the Proton star pulls matter, space and time into its pace-time occupation. In the Sun the Sun is concentrating heat by a huge generator it holds in the inner-Core-value and that is what "gravity" is.

Electricity is all about a process condensing heat and "gravity" is a process concentrating heat. There is an enormous difference but takes some explaining and that is not applicable to this book. A Proton Star is unmovable, static, and stationary and every other name you wish to connect to its immovability. It can no longer go anywhere. The stars within the sphere where doctor Hawkins identified a Black Hole to be, within the very centre of a galactic, holds all occupation relevance to the spin or linear component of space-time occupation and only a very minute part to the circular "gravity" component.

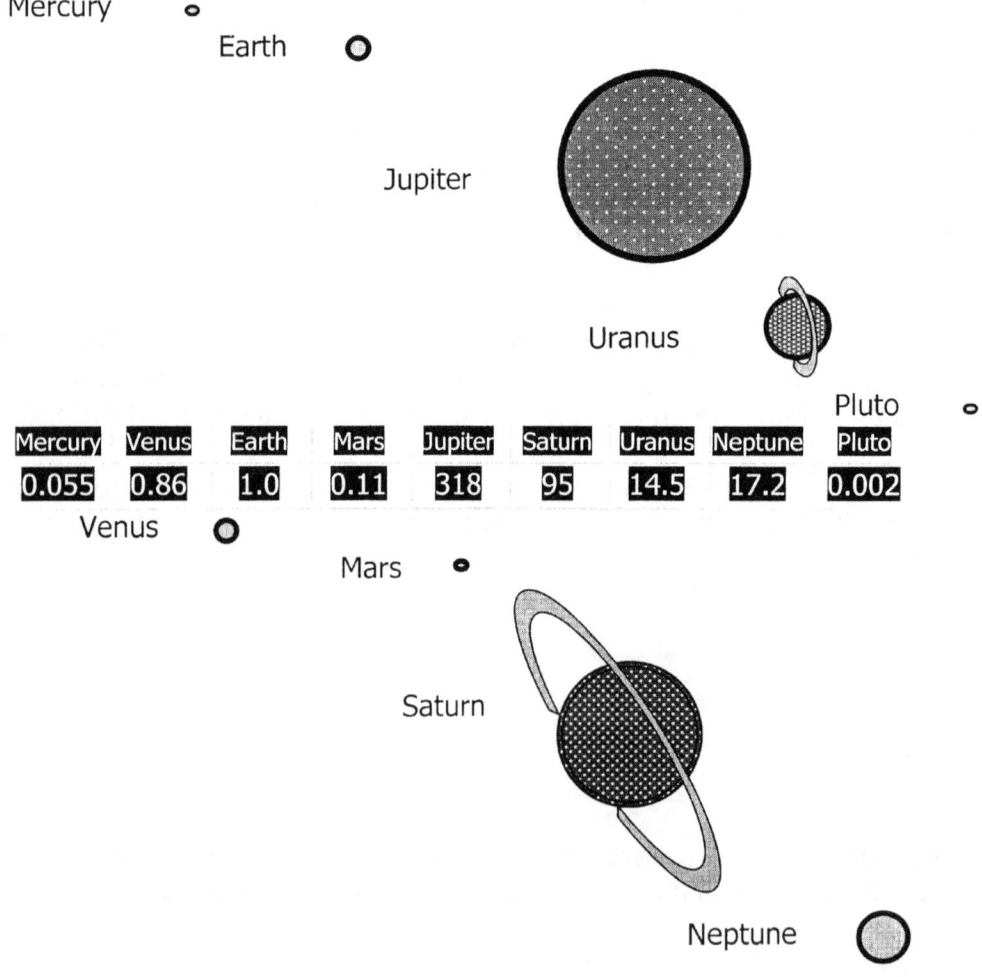

Mercury	Venus	Earth	Mars	Jupiter	Saturn	Uranus	Neptune	Pluto
0.055	0.86	1.0	0.11	318	95	14.5	17.2	0.002

It is so very clear what is missing from this picture the solar system presents the cosmos. It is conformity. It is a planned formation presented as all the laws governing the cosmos was adhered by the structures within the system something is missing. A planned layout is not present.

When investigating any structure in the cosmos we find the less developed objects allocated towards the centre and the structures that are more progressive in time and have developed better are object located on the rim or edge of the galactica. In this case there is no order. There is no recipe and the most precise recipe that has ever been used for any constructs the cosmos. This is what is wrong with the picture we see in the solar system. The development that took place is no correlated with normal growth the big Bang provides.

That is the big clue missing from the picture we may find in the development of the solar system in relation to what should be the case with the normal development provided by the Big Bang. Then in releasing this shortfall we have to turn out attention to one of the four cosmic pillars to give us the indication of what happened and how it happened when it happened.

(1) A binary formed from the pre Big Bang frozen blanket.

(1) Outside the spinning dual were 5 dots minding their own frozen space and concerning their own singularity, which the binary confined as space under influence.

(2)

(3) Then the one factor of the binary dismantled by allowing the singularity to brake down in heat and fragments holding singularity.

FIVE FRAGMENTS CAME OUT OF THE BLANKET CLOUD OF HEAT.

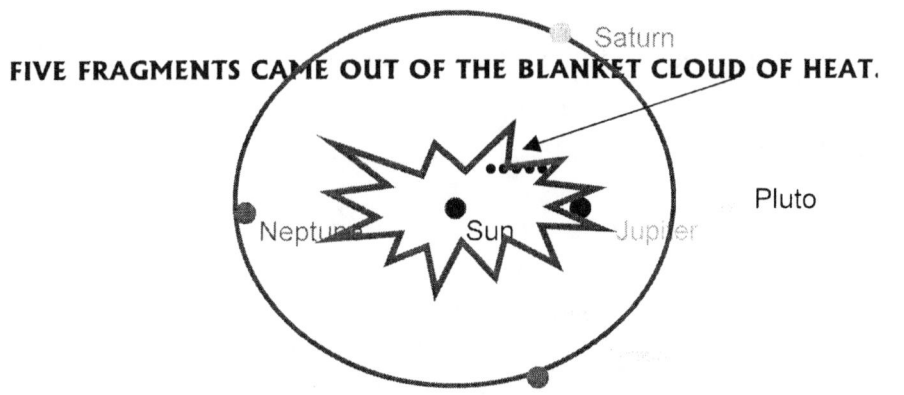

This debris can be witnessed as meteor craters on the solid planets. The bigger structures were the first ones to become cosmic missiles and disappear from sight, because there were less of them to go around in the first place. Afterwards only smaller particles remained and they too became ill fated. However, none of these fragments became part of the planetary system. As the inner structure changed its orbital course in both the Binary minor and the Unknown structure, its gasses became reduced in size in other words compressed, while the inner heat in the gas rose to extreme values, creating an enormous wealth in magnetic space time (how and why this happened, will be explained).

However, the inner core could not maintain the reduced value of the stars, as they were firstly far too small, and secondly the core was in process of becoming fragmented. Therefore, it did what all unsuccessful stars do. The magnetic space-time at first were deprived of its negative space-time displacement because the increase in the stars overall growth in negative space-time acceleration. So all the matter was compressed up to a point where the acceleration of the stars became a value of evenly linear motion. As the momentum did not accelerated any longer, the compressed matter began expanding again. Then the magnetic space-times own negative displacement took over and the gasses of which the two stars comprised of, became known as Oords Cloud by humans after a duration of $4,5 \times 10^9$ years.

Then soon afterwards another two tragedies followed and only because of these two tragedies, life became possible on Earth.

THE HEAT OF BINARY MISSING CONFINED FIVE FRAGMENTS WHILE FROZEN SPACE-TIME CONFINED FIVE UNDEVELOPED MICRO STARS WHILE THE SUN AS THE PRIZE WINNER CONFINED THE LOT

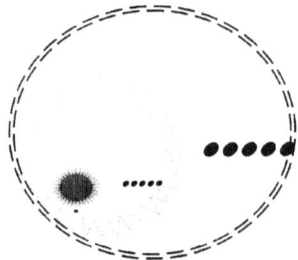

Let us establish a line of evidence and fill the puzzle.

1.   There was another star with the Sun where the Sun was Binary Major and Star Unknown was Binary Minor.

2.   The Binary system catapulted the solar system out of its frozen eternity, way ahead of its time of development, bringing along the rest of the micro stars. The outer "planets" are not planets, they are micro stars in development.

3.   The Binary system formed part of a Lagrangian system holding the Binary in the centre and with Jupiter as the first orbiting satellite.

4.   The position Jupiter held for most of the developing period made Jupiter the second main benefactor of the dual, with the Sun the major winner and Binary Minor the major loser. This will also explain why Jupiter has such an advance in space-time occupation when compared to the other micro stars.

5.   The position where the (six) inner planets find themselves, were a void, AS THE BIBLE CLAIMS.

6.   The relative positions were as follows:

$(1 + 1) = 1$	2	3	4	5
◯○	○	○	○	○
Binary	Jupiter	Saturn	Uranus	Neptune

7.   Then Unknown Star capitulated as it could no longer serve the dual it fought. It fragmented into possibly 9 major parts and many minor parts, (the comets.)

8.   As the core fragmented the brittle parts dislodged in a position each to a relative neutron position in the space-time binary minor held relating to its point of $(\Pi^2+\Pi^2)$ $(\Pi^2\Pi)$ 10Π.

Whichever way we Earthlings will look at our position, from whatever angle and by whichever calculation we devise, our relevancy will be 1, will be 10, will be Π^2. Should any person ever do a calculation and find his answer does not bring this fact to bear, he must go back and fix his mistake. There will be a mistake on his part.

SUN	MERCURY	VENUS	EARTH	MARS
◯	◯	◯	◯	◯
B_1	B_2	$\Pi^2/2\Pi$	Π^2	10Π

Then the rest.

$$\Pi^2 + \Pi^2 \longrightarrow \Pi \quad \text{(This I shall explain)}$$

Our Super-Educated hope to establish Martian Colonists. They cannot even begin to comprehend their ordeal. Columbus only had to fight some sea dragons, a Wall of fire on the water and an ocean-sized waterfall on his way to India. Even if all that myth were true, and he encountered the hole lot in one day, his problems would seem minute when compared to what awaits our Martians. They have no clue of their folly. Mars is not another island in an ocean of water waiting anxiously to be discovered by a wind powered sea-faring vessel. Mars is an island, a piece of rock in the middle of a gas of heat, with total alien circumstances to life. Scientists can fool the public, they can fool themselves, they can fool with figures corrupting cosmic laws, but they cannot fool the facts. There is a reason why the Bible excludes the other solar structures and concentrate on the Earth. This is the same reason why any way we do calculating the solar system; the Earth will come to one according to human standards. Life from our perspective, from our perception, totally connects to the Earth. That connection is so deeply routed in man and other life it is part of life. Newtonians may spread the rumour about life being ten a penny scattered to the social structure that man established on Earth through every social alliance or group, be it politics, law, medicine, science, theology or whatever denomination they wish to control, they will not create facts.

I admit, man derives all the above-mentioned lies mainly from corrupting the Bible. The Roman Catholic Church started the trend 1 500 years ago and still maintains it as best it can. You can prove whatever you wish to prove from facts you take from the Bible. The Bible is in support of whichever standing you may support. This is not because the Bible is incorrect, it comes from our insignificance to appreciate the full content of the whole Bible. The Bible promote only truth, man takes from that what man wants, divert the truth to suit his need by corrupting the lot.

I know presenting evidence as well as I do, will not change the course of science. I know too, that I am too small to pinpoint conclusively whether Authentic Author referred to the creation of the first cosmic period, or the first solar period. This is not because of inaccuracy on the part of the Bible. This inaccuracy comes from the human interpretation of Authentic Author on his vision, and then my insecurity to interpret his interpretations. It is human error bringing on misinformation. The Bible's recollection CAN and DOES apply to either period. Therefore, to respond in containing human misjudgement on my part, I shall re-apply the vision of Authentic Author in the context of the first solar day. I showed how it fits to the first cosmic day already.

Binary stars, spinning to self-destruction will produce significant heat. Heat create space, space forms winds. The Bible present facts that is indisputable. Where the Earth was, was still a void, containing a sphere of circular displacement and this will reduce linear displacement to zero. Linear displacement is space and circular displacement is containing heat for matter survival.

Binary Star Minor overheated. That is why the core brittle and fragmented. This action will release tremendous contained heat; the heat will produce magma flowing in space like water in space and this eruption of heat space that created winds. Once again the recollection fits the scenario. Releasing the heat and producing space will establish space-time and fill the void where the Earth should fit. This is fact and if anybody even tries to dismiss this will be because of abstinence on his or her part. I did not prove the Bible correct. The Bible told the truth and in such correct detail, it is beyond human comprehension, but sublimation on the part of Newtonians and science before them, disallowed their ability seeing it.

Let us envisage what factors applied to the third planet binary that set it apart from the other planets (not micro stars). First, the eruption occurred, and with the eruption came the release of massive quantities of hydrogen, oxygen, carbon and nitrogen. I respect to the quantities the Sun stockpiled for own "personal use it was not much. However, to what the other structures in micro planets had, it was exceeding their quantities by hundreds of times.

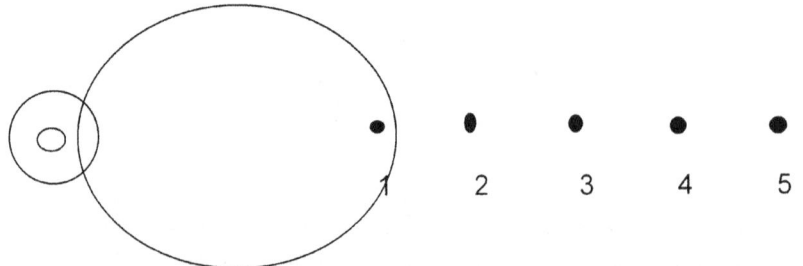

Jupiter (1) still enjoyed much of the vapour for the longest duration, and was a benefactor before the fragmenting as much as it was a benefactor after the event. As the clouds drifted from the position of point zero, the other micro stars also became receivers, in order of position. The mass they hold today, is still evidence of their position placing them as recipient 1 – 4.

1.	Jupiter	318 x
2.	Saturn	95 x
3.	Uranus	14.5 x
4.	Neptune	17 x

However, the Neptune micro star in all evidence has very particular characteristics that I cannot explain. I have a theory about Neptune, it is only a theory that I cannot substantiate with mathematical proof but nevertheless there is circumstantial evidence I shall present in order to prove that my theory is not merely wild guessing on my part. I shall return to this matter shortly.

The indication clearly points to the fact that the micro stars were in this cloud cover of star gasses, for a long period, and the period were substantially more as the micro star holds a position in relation to close proximity to the destructed binary. All the orbiting structures the micro stars hold, except (I suspect) Neptune, are fragments of Planet 5, the destructed one.

One must realize that Jupiter holding an inner core sizable enough to produce 2,34 times the "gravity" of the Earth were a result of benefiting from the Binary dual between the Sun and Unknown Star. Compare that figure to the rest, and it points to the fact that something major set Jupiter's development on a course of progress, where factors benefited Jupiter by far.

1.	Jupiter	2,34
2.	Saturn	0,93
3.	Uranus	0,79
4.	Neptune	1,2

Again we can see that Neptune had some other benefits to its progress, because, again Neptune diverts from the obvious sequence.

Once more, this phenomenon should not occur with Newton's presumptions about gravity. These bodies will collide and destruct, without a doubt. When the formula $F = \dfrac{M_1 M_2}{r^2} G$ apply, there should not be any force which is able to keep them apart. However, they do exist and what is more, they maintain a certain distance apart. Seen from this view, it is little wonder that the significance of this was lost in the notion that this is yet another "mystery" of the Universe. The scientists of the day (and the past) lost the importance, which this holds for us as Earthly dwellers.

As explained, there is no gravity, instead a balance exist in space-time, where a value of linear displacement relates to a value of circular displacement. $\dfrac{\frac{R}{T}}{\frac{R^2}{T}}$ (R/T // R²/T). Regarding this, the

Roche lobe comes about because the Roche lobe forms a borderline between these two related values. Space-time lying within the Roche lobe stands at a value where the linear displacement is at a greater value than the circular displacement. In mathematical terms it expresses as follows in an equation:

$$\$ = R / T > R^2/T$$

Matter located on the border of the Roche lobe, will represent mathematically in the following equation:

$$\$ = R/T//R^2/T.$$

This is the position, which satellites have to comply with, in order to remain in orbit. Matter located on the outside of this border, which the Roche lobe holds, will mathematically represent as follows:

$$\$ = R/T < R^2/T$$

You, the reader, might react with surprise, because all structures with influences in this way, are far beyond having any influence on our life. This represents a great misunderstanding, as it has everything to do with life developing on Earth. In systems of unmatchable and unequal space-time values, the larger body will tend to dominate the smaller body in as far as high jacking the smaller bodies' space-time values are concerned. Please take note throughout all of this discussion, there is NO FORCE applied on any of these bodies, but only a balance, which maintains or goes array due to certain reasons. Therefore, NO STAR STRUCTURES (OR PLANETS) can ever collide, and a meteor colliding with a planet, is not a collision, but an imbalance of space-time manifestation.

To understand the meaning of this statement, I shall firstly explain why no star system can collide. When two objects i.e. double stars, come into a conflicting position as to sharing space-time, the two systems has a response in space-time values.

The academic world has treated me very poorly, because of my view on what they regard as Holy Scripture. I came across some brilliant scientists whom were able to form a conclusive opinion of my work after just reading the first four pages of a book containing almost two thousand pages of explanations and facts. By just reading two pages those highly informed professors decided that I am completely misinformed. The claim they made is that I am not familiar with Newton, and therefore do not "understand" Newton. They never even allow themselves the time to get to the part where I explain why I do not recognize Newton, let alone begin to introduce my opinion. Such intelligence, I must admit, is a true indication of just how their acquired brilliance can allow their decisions made in split seconds about a book that would take them at least one month to read extensively.

I know that if I went the conventional route by enlisting at a university and following such a course, I would have had to accept the institution's views on science or they would boot me out. I did not escape the booting process, as many academics discarded my work. I wish to put this to all of the religious scientists: Even if you will never admit it, you believe in Newton more than you believe in the teachings of the Bible, no matter how much you wish to deny it. The moment the Super-Educated faces up to any criticism about the work of Newton, they do not have the courage to even investigate the criticism. If one says that that person disagrees with Christ, everyone in hearing range is prepared to listen to such a view, although it must be considered by all believers as profanity against the almighty and this charge I direct directly to those so called believers that has Bible versus on their answering machines. Tell the physics department on any campus you disagree with Newton and they honestly treat you as a mad raving lunatic keeping busy with blasphemy!

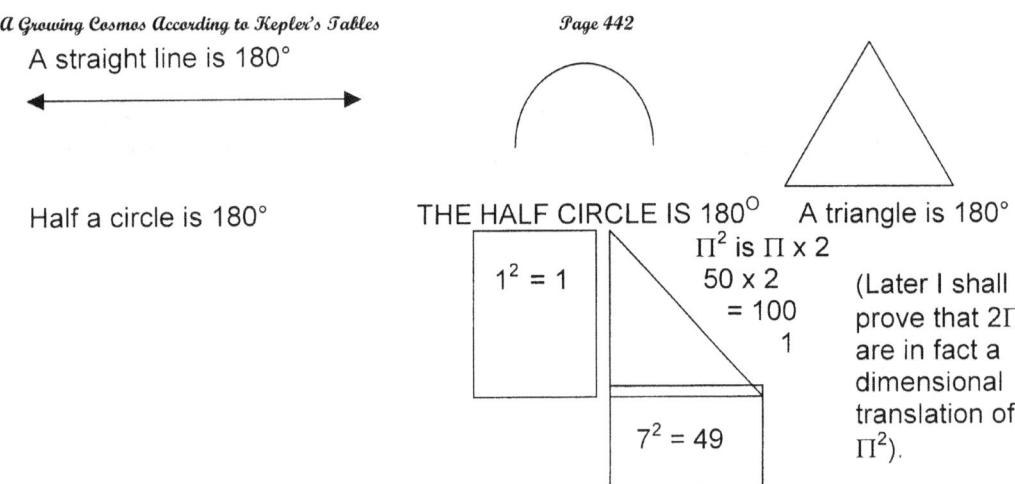

A straight line is 180°

Half a circle is 180°   THE HALF CIRCLE IS 180°   A triangle is 180°

Π^2 is $\Pi \times 2$

$1^2 = 1$

$50 \times 2 = 100$

1

$7^2 = 49$

(Later I shall prove that 2Π are in fact a dimensional translation of Π^2).

Π^2 because it holds a higher dimension is 49 + 1 (Pythagoras) = 50

and 50 x 2 ($\Pi^2 = 2\Pi$) = 100   (The two is a dimensional implication

$\sqrt{100} = 10$ carried by the value of 50)

Therefore space to time is 10 and matter to time is 7. Time to time is 1.
I base my facts that the moon never was or could be part of the Earth on the Roche limit that would never allow it. What is apparent though is that the Earth robbed the moon of all vital atmosphere and water.

This much I shall say: If not for the moon's position and size, life would not found such an acceptable environment in which to evolve.

The Titus-Bode Principle of heat growth and time growth works on a simple basis, the same basis as the speed of light. It is all a question of dimensions that space holds, in conjunction with time, running parallel, that exist in accepting more than understanding. Trigonometry is the sole proof of such dimensions.

I have shown Π^2 to be 49 from the Earth's perspective which means Π is 7. But Π should be 3 from the law of Titius Bode. And Π^2 then have to be 6. By placing Π at 7 and Π^2 at 49 it means the distance positioning came about from space in the fourth dimension $7^2 = 49$ and not time through proton development in the third dimension. $\Pi = 3$ then $\Pi^2 = 6$; $\Pi = 6$ then $\Pi^2 = 12$. The square that applies to seven is an indication of definite heat creating space and not time through proton growth alone. But one can still detect the proton growth of 4, in the accumulation of seven.

There is of course a much simpler way about to go in explaining the Titus Bode 10/7 and 7 / 10, and I do give tit in another part of the book, but knowing the purist of academics, such simplicity would go by unnoticed.

In a previous part I established the planetary relation according to space-time application of R^3 (space) and T^2 (time) holding equal value. In the following pages I wish to apply the same standards, only relating the standards to cosmic principles I already explained.

A star form time at a value of 7/10 (4 ($\Pi^2 + \Pi^2$) = 55 and the core has to be less than 56 to allow light (29,6 + 27) than 56 to allow light (29,6 + 27) ore holding a higher value than 56,6, the light will no longer escape from the inner core and as the core grows, the star will start absorbing its light (photon) production until such time it turns into a full blown "black hole". At a value of 3($\Pi^2 + \Pi^2$) the core will dissolve space to the effect that the neutron no longer has space in which to be. The neutron then will abolish the star altogether.

When a star is in a Roche limit, the value is $(\Pi/2)^2$ ($\Pi^2 + \Pi^2$) = 48, but since it is two stars it becomes 2 x 48,7 = 97. That places the space-time development of the binary in a position held by stars two era's previously. The space-time enhancement is about double to normal growth. This fact also carries a high degree of significance when I explain how the solar system came to be in the layout in which it presently presents itself.

This places the star in a space-time occupation that was valid before the "Big Bang" surpassing the cosmic value of 112. It puts the stars in the Roche-lobe in a state where the stars in the binary combines to a cosmos excluding the outside to favour the inside where the atmosphere is excluding the cosmos and does not have space separating the stars, but merely the atoms. The time relation of the atoms hold a combining value similar than that, which is $2(\Pi^2 + \Pi^2)(\Pi/2)^2 = 98$. The atoms will not destruct, but all space between the atoms (Π as well as 3) will diminish as the atom alone falls outside singularity, but all other space has reduced to singularity. Fusion will not occur since the presence of heat maintaining the proton / neutron will still generate through matter spin of the two objects as one, bringing about a linear locking, applying a circular contact with space-time.

With the two structures forming a cosmic dual of survival each will apply $4\Pi^2 (\Pi/2)^2$ where the factor of $(\Pi/2)^2$ will hold the relevancy the one has to the other. When the two structures have equal prominence the relations will be as follows. In spoken language each will regard the other as a neutron attachment of half Π and half Π^2, therefore $(\Pi/2)^2$.

$4\Pi^2 (\Pi/2)^2 (\Pi/2)^2 \, 4\Pi$

$4\Pi^2 (\Pi^2)$ and $4\Pi^2\Pi^2$

As both hold and equal value to the neutron or second Π^2 the two will apply a joint value of $(\Pi^2+\Pi^2)(\Pi/2)^2 + (\Pi^2+\Pi^2)(\Pi/2)^2$.

This value places the space-time within the lobe to a time duration where heat was liquid, just after the forming of the neutron. In such a star, not only will the heat and space of the electron have liquid space, but also the neutrons will find itself in dense space. The combination places the space-time value in a position where it will enhance the neutrons in the space-time between the two. When the two structures, almost equal all neutron space, it will naturally become a combining or combined proton star or double Black Hole.

In the event of the two core structures being so equal and similar, it will start growth the space-time the protons hold, placing the double proton in an unnatural era. At this point I wish to indicate that the factor values indicate the space in time separating the objects or if you wish the numbers only apply to the heat value of the space keeping the structures apart. The matter part will remain apart from the unoccupied space-time gone into singularity. Saying this I have to add that with the event of the matter finding itself in space-less time, predating the "Big Bang" the matter eventually will also have to dissolve to time as the matter starts to fall out of space and going on to singularity. The matter, however, will follow another route. All these facts I introduce to prove my point that the Sun is not even a star but through the Grace of God it is there. If the Sun was a true natural star, all "planets" will incinerate with no chance to support life.

As indicated above, the Roche-lobe, from the matter's vantage point (not the space separating the two objects), the one object relates to the other object being a neutron to the proton.

$4\Pi^2 (\Pi/2)^2 + 4\,\Pi^2 (\Pi/2)^2$

Therefore $(4(\Pi^2 + \Pi^2)\,\Pi^2/4)\,/2$
$= ((\Pi^2 + \Pi^2)(\Pi^2))\,/2 = 97$

To enable any person to see how significant and out of era this is, compare this value to that which the last surviving element, Plutonium holds at 94. The true value of plutonium carrying an overload of neutrons to stabilize the element is then $5(\Pi^2+\Pi^2)(\Pi/2)^2 (3/5) = 244$.

Analysing this is as follows:

5	End of space-time
$(\Pi/2)^2$	Limit on demolishing space-time
(3/5)	Heat stretched to its limit. This whole ration spells one nuclear disaster waiting to occur. The space-time environment within the binary, even exceed this Plutonium.

Therefore half of the combination regards the other half as $\Pi^2\Pi$ and $\Pi^2\Pi$ is a neutron position holding the three dimensional Universe to the same value in space as that which separates matter from matter being 3. Since the 3 stands completely excluded from the atomic combination of $(\Pi^2 + \Pi^2)$ holding the second object as

The stars maintain their individual circular displacement values, as they move closer, driven by their linear space-time displacement. At a point where space-time becomes a unit, star A's linear displacement has to overcome the Roche lobe boundary of star B. This applies to star B in the same way. At the limit of the Roche lobe border, star A and star B will establish a mutual circular displacement value, limiting the other star's linear displacement. Then a situation develops, where the mutual circular displacement will replace the individual circular displacement values, which both stars had. At this point, the stars would find it impossible to move closer, because of the Roche lobe limit. As both stars share equally in the circular displacement, they have to share an equal linear displacement. This would leave them unable to break the linear displacement equality, therefore they maintain at a distance apart, spinning around a mutual axis somewhere in the middle of the distance keeping them apart. Mathematically the equation can express as follows:

In the event where the one star's space-time value would overshadow the second object's space-time value, therefore the value of $\$_c$ would locate within the boundary of the larger star's Roche lobe border. In such a case the Roche lobe, would in effect not apply and the larger system will incorporate the smaller system within the larger system's space-time.

The effect that this would leave on the two systems which are able to maintain a mutual linear displacement value, is that they would either share in a common space-time growth or the one system will destroy the other system at a certain point in space in time.

$\Pi^2\Pi$. This can only apply when the one object occupying less space-time has a proton value $(\Pi^2 + \Pi^2)$ that is less than the superior object's position on Π^2 This means the one is totally dominating the other in all aspects. Some quarters of the Newtonian High Priest in High ranking made claims that the moon once formed part of the Earth. In the following elaboration I shall prove why I dismiss this claim as utter nonsense. From these facts about a binary, one can then clearly see that having two structures in a position overshooting the Binary scenario, is very much fantasy. It is just not possible because the valid space-time will exceed 112, and the structure will not have the ability to hold position in the Universe that is limited to 112.

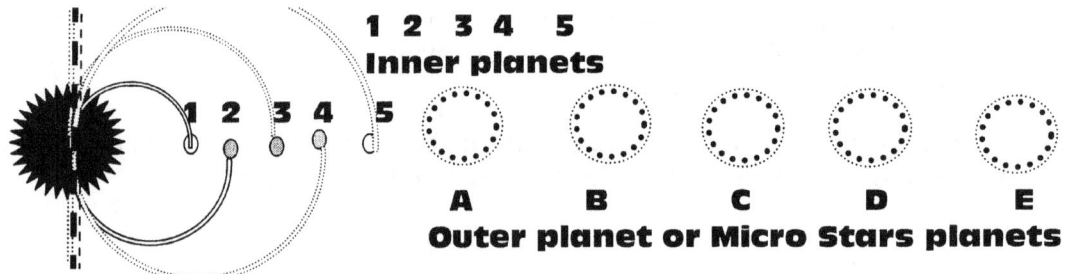

In accordance with the Lagrangian rule there cannot be more than five objects pairing onto a centre object. However in the case of the solar system we have a double except that in the one pairing there is one devastated and in the pairing of the larger group we also find the most outer structure demolished or at least in ruin and in association with cosmic debris. With this evidence there now is a manner in which to determine how the rules were broken.

From the fragments we find that is scattered all over the solar system and the irregularities we find as I indicated up to this point we can see there is a lot of history in the story, which will unfold as we investigate. There was tussle in core battles between the Earth and the Moon, where each falls into a sequence arrangement of a proton value of the Earth is $(\Pi^2 + \Pi^2)$ and it will hold the second object (the moon) at $\Pi^2/2$. This is because the second object is in the "gravity" application of the larger object (the Earth) and the "gravity" factor of the Earth takes on a linear value, half that of the gravity factor of the Earth. The Earth will not allow any linear action to exceed 10Π and at $(\Pi^2 + \Pi^2)(\Pi^2/2)$ it exceeds that value. However this can only be a result of another tussle of much bigger importance and ferocity. The debris can only formed from many disarrangements and cosmic laws being broken as a chain reaction that sprung from one specific event. From the evidence that the debris leaves it is a fact that there were many numbers more rocky structures that was left over from the Unknown star demise, but as the law would not allow more than five, it was five that was formed with one being crushed as an aligning planet. But there were more than those forming the line, but I have not the time, nor the space in this letter to go into speculating about the crushed ones since the publisher limits this letter and therefore because of the length limit the publisher placed on the book I cannot delve deeper. The moon for example too must have been one of the after thoughts of the disaster and in _Seven Days Of Creation_ I speculate about this with some convincing proof as how that came about. I place a great importance on the fact that the Moon and the Earth performed as a Roche partnership and this much more than any other reason brought the development or the possibility of life developing to the Earth and not one of the other solar structures. That too I cannot share in this book because I am limited about the length of the book and it has to be accompanied by a vast quantity of proving information. Without the evidence I know I shall meat with sure rejection from our Newtonian camp and that is as sure as the Sun Shines during the day.

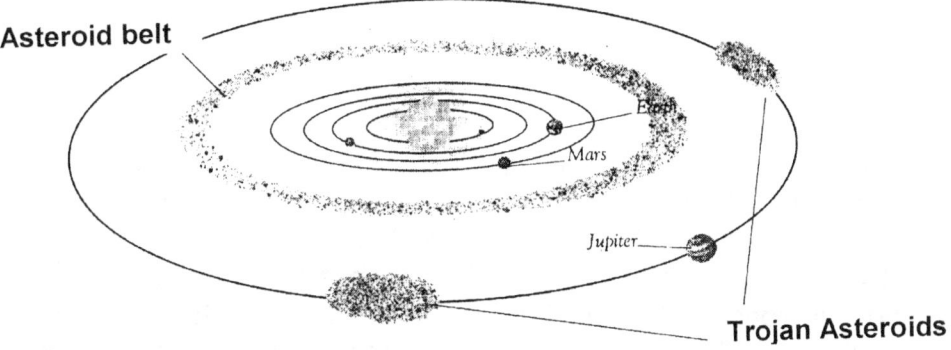

There is a great deal of pieces of hard rock floating all over the show. Some are nicely grouped in units, which one can clearly see was designated places and solid units. The only way the rocks could have brittle as they did was becoming excessively overheated. The only way it could have overheated in the manner it did was by bridging the Roche limit divide with much larger object, and there is a much larger object in the region and sharing an orbit with the immense structure. This was bullying at its best and we can see from this how cosmic plunder, thieving, brutality and theft came about as the larger murdered the little.

When the two structures go into a duel. As the two core has a dual the fragmenting of larger to smaller is a method of distributing heat by expanding. It common language we call it an explosion, space-time occupational value of $(\Pi^2 + \Pi^2)(\Pi^2/2) = 97$, and the core value of the Earth is at 7/10 (4 $(\Pi^2 + \Pi^2)$ the combined value will even exceed the critical space factor of 3 $(\Pi^2 + \Pi^2)$ applying to stars holding space, therefore the space separating the two objects will vanish into singularity. The reason why the Roche principle maintains core separation is that the core combinational value, seen from one or the other objective, is $(\Pi^2 + \Pi^2)(\Pi/2)^2 = 48$. The individual space-time factor of each core is 7/10 (4 $(\Pi^2 + \Pi^2)$ = 55, therefore the space holds less heat and therefore more space.

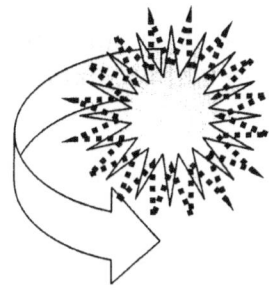

Lets reconsider what was apparent.

We know from the fact that there are four peculiar hard surfaced objects formed within the inner part of the sun that there was some abnormality in the solar system development. The inner objects are named planets although in the entire Universe there are no call for such object releasing from the Milky Way this early and therefore the fact of their presence relate to some irregularity that came about in the past. In order to have the four smallest planets blessed with a solid surface while the other "giant planets" all are a mushy gas structure must be because the five inner planets was part of a star much more equal to the Sun that any we now have. In order to be of significance and not be totally destroyed in the Roche battle the Sun I suppose was the size of what Jupiter at present is, the second binary was then about the size the Earth is today.

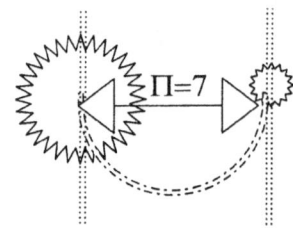

I have explained that matter is seven in relation to the Titius Bode law and that pts space in a position of 10. If 7 is Π, then Π^2 is where the Earth is because I am in the Earth and therefore I am the centre of the Universe bringing the Earth position also as the centre of the Universe. All motion spins around me making me Π to all of Π^2. If $\Pi = 7$ then $\Pi^2 = 7^2 = 49$. Therefore my position in relation to the governing singularity is 49. I am holding space in Π^6, which then is $(10)^6$, which is then 10^6 kilometres away from the Sun and with material being 49 the factor in relevancy is 49×10^6 kilometres away from the Sun.

Since I am on the third planet from the sun the relevancy my planet have with the Sun is in line with the factor relevancy, which I have with the Sun. That puts me with my Earth in relation to $49 \times 10^6 \times 3$ (since I am on the third planet from the Sun and if the development of our solar system was normal) the allocated position of the Earth must then be $3(49 \times 10^6) = 147\ 000\ 000$ kilometres

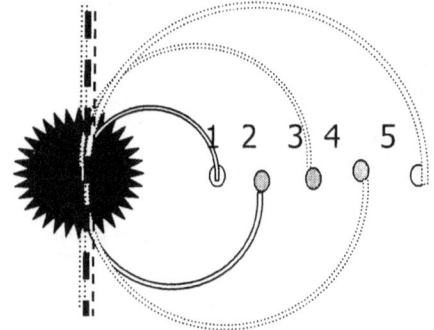

By the way that is the reason why Mercury follows such a strange pattern while circling around the Sun Mercury presents $7/10 = 10/7$ so in the case of Mercury aligning with the Sun nothing makes sense for poor old Mercury. More to the point there is evidence of unusual space –time development at the place Mercury now occupies.

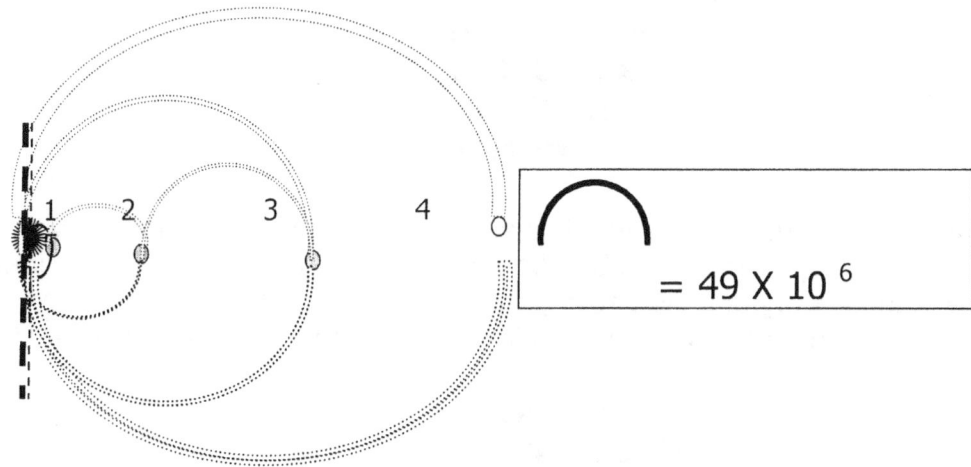

$$= 49 \times 10^{6}$$

There is a clear defined value, which serves as **k** in every event of every planet in relation to the spin as well as the development in accordance to the Titius Bode law. B y the square of seven representing material related to the double cube of time through which material must rotate while aligning with the sun the distance on average is one million.

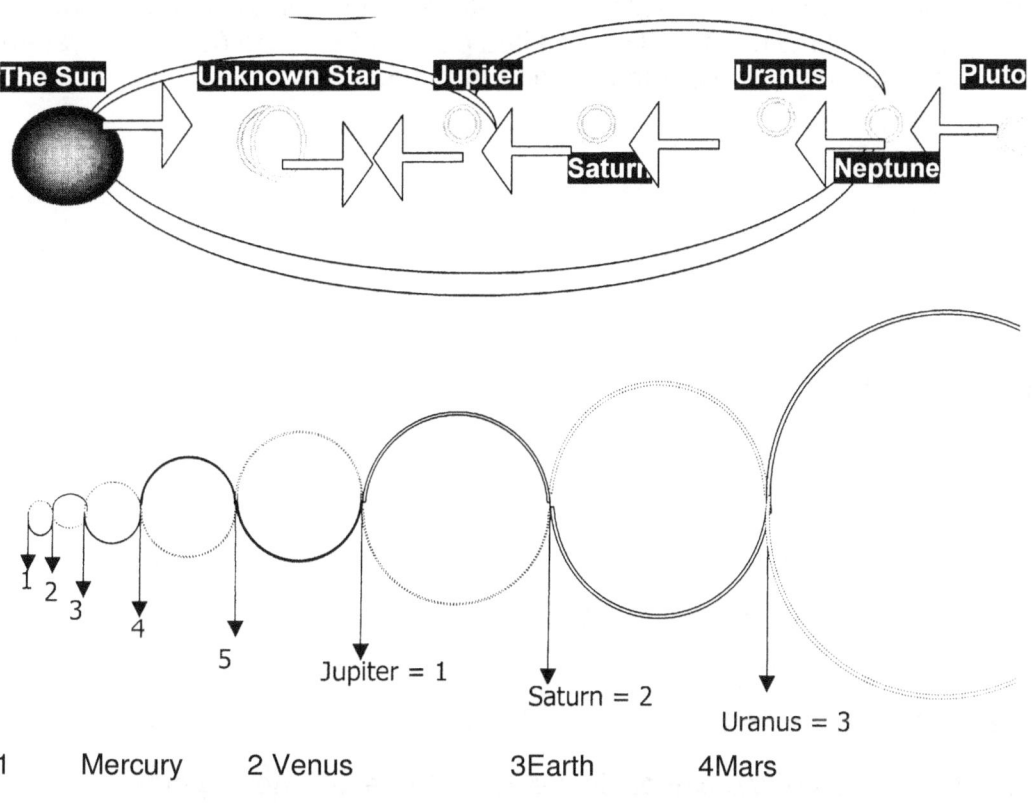

The Sun   Unknown Star   Jupiter   Uranus   Pluto

Saturn   Neptune

Jupiter = 1

Saturn = 2

Uranus = 3

1   Mercury   2 Venus   3Earth   4Mars

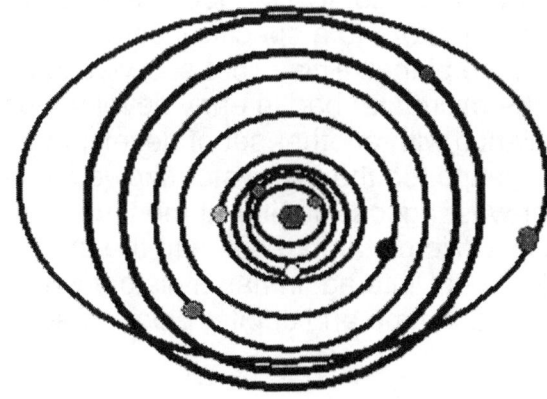

The difference there is in the alignment of the gas planets when compared to the inner planets tell the tale of much different history in development. It is also clear that there were seven periods of significant and altering stages that influenced the solar system to the core of development. Most important in all is the distribution of the cosmic structures in their unit as they relate to the development the unit underwent.

1	Mercury	49 X 10⁶ km
2	Venus	98 x 10⁶ km
3	Earth	147 x 10⁶ km
4	Mars	227 x 10⁶ km

Under normal development there cannot be more that five structures relating to singularity without forming a sphere. It is how the development should be. Why it is different I do not wish to go into but I will say it has to do with the butterfly diagram. However getting into that explanation would be very complicated and time consuming and for that reason I think it is best left to the Sven days Of Creation.

1	Mercury	49 x 10⁶km
2	Venus	98 x 10⁶km
3	Earth	147 x 10⁶km
4	Mars	196 x 10⁶ km

This is what it is at present so there is strong indication about some development above and beyond the normal cosmic growth that the solar system did experience as part of the Hubble growth. The normal flow would put the fragments at time relevancies as I interpret them by using the space-time rule in time in space confirming normal development in space-time. However it is not as innocent as it all seems and much evidence tells of a violent and crime filled past that shaped our solar system in becoming as unique as it eventually did with sporting a place that could host life and all.

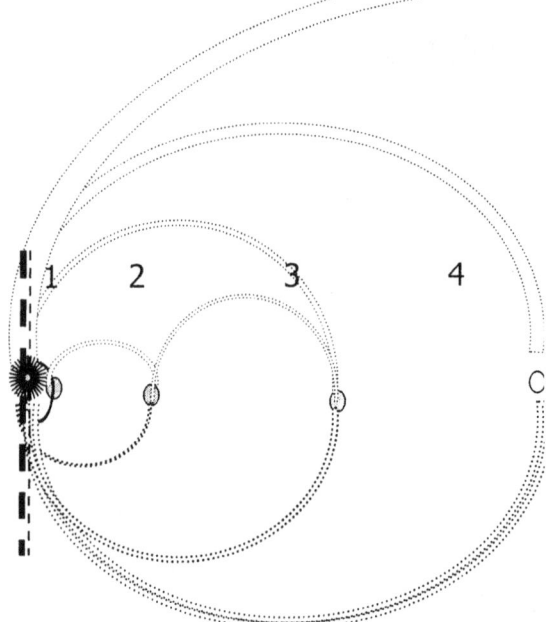

One can clearly see there was a push of the planets in the Titius Bode law moving outward towards the direction of the last structure. The plants move from one to five but five landed in a massive problem. The growth coming from the sun was defining the structure to a specific location while that location was blocked by a micro star may times the size of the planet. This Micro star was in tune with quite another singularity line because of the history the micro star had in early development with the Sun. The micro star was located as structure on in alliance with another set of developing structures and was unmoved by the oncoming dwarf. The response the micro star showed in relation with the Sun was in another frequency than the Micro star was. I guess Jupiter at the time was not as disproportionate gigantic as Jupiter seems today but being a Micro star in relation to the other micro stars such as Saturn, Uranus Neptune and Pluto (yes Pluto but Pluto again has a history which is his story and we have no time for that). With the sun forming space-time and Jupiter blocking such an advance there came trouble to paradise.

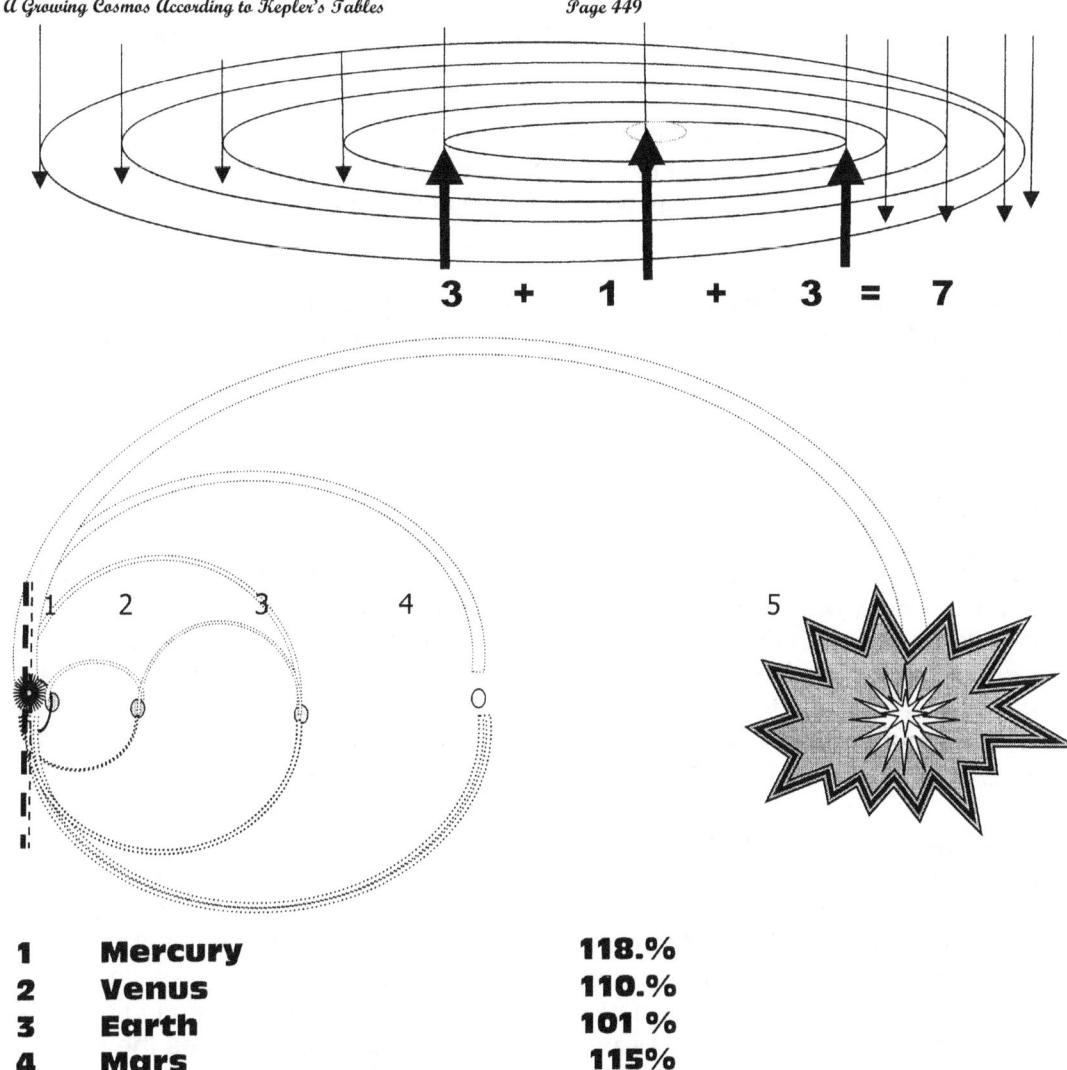

3 + 1 + 3 = 7

1	**Mercury**	**118.%**
2	**Venus**	**110.%**
3	**Earth**	**101 %**
4	**Mars**	**115%**

That indicates than space grew from a source to the outside of Mars and there is only one way an enormous quantity of heat could release at that point. It must be from the fragmenting of the fifth solid planet as the planet had the same fate as its mother star did and went the same rout by being forced to expand into a territory where no more expanding in such a direction is possible. Such a situation demolishes the core by exaggerating the heat load to appoint where almost the entire core liquefies and as the singularity of the little planet cannot take the bearing of the micro star, the little planet becomes wasted space-time.

The fragmenting tells how so many "moons" or satellites fragmented and were captured by the giant micro stars. Where the Sun came into the opportunity to capture some, the Sun did not use those satellites as "moons" but had them as liquid heat.

The planet went liquid but that is another bone of contention because science do not even recognise heat going liquid by forming gas as heat expands. Well, if they do they never made it part of cosmology. In the star hydrogen has a function in relation to heat. So has helium have a very special function. The oxygen in the relation to the carbon in relation to nitrogen has a specific purpose and that purpose tells to what point did the star stage of development reached. The Hydrogen has the duty to produce motion by as much duplication as the star singularity requires in maintaining. The Helium contains the captured heat, which then is charged to the carbon layer that keeps the heat in transit. With the oxygen coming into contact the oxygen transports the heat where the hydrogen transfers the heat from a solid as it was in the carbon and to an extent in the oxygen and release the heat to be served as fuel driving the core region. When looking at a fire one can clearly see what every element does with heat in liquid form. The carbon keeps the coal red while the oxygen store the heat as smoke and the hydrogen takes the heat as flames into a volatile motion. Every element stands directly related to the purpose it holds with heat and in the manner it serves the star. We can see this from the way a fire ignites and burns. The cosmos has rules and the rules will apply everywhere.

This same effect happens when a small fire starts in a large room. The heat lodges in the smoke as the smoke fills the room the smoke grows increasingly to the ceiling of the room and nestles in the smoke. When a draft enters the room the heat transfers to space and that action the scientists regard as an explosion.

Science fixes their attention on the incoming oxygen that causes the explosion. This might be true in part, but that is a small part of the whole picture. In the cloud of smoke gathers heat, confined to the smoke with the flow of direction out of the roof. The advancing heat flows from the burning fuel igniting underneath the smoke.

Material form heat at the bottom, the heat advances to the top, but the heat at the bottom remains higher than does the heat at the top.

At this point we have to look at another natural phenomenon. As heat can convert to space so can space convert to heat. We all know what happens when a compressor forces air into the air container. The container heat rises dramatically. Science call it pressure, but pressure it is not. Pump the compressor to say ten bar and leave it overnight. AT first the container will be hot. After a while (say 12 hours), the heat will reduce to room temperature. At that point one may call it pressure, but with the heat amongst the matter, it is space turned to heat. With the flow of time, the heat will return to space. TIME IS THE SPIN RATE OF HEAT IN A SPECIFIC SPACE.

To go back to the oxygen argument, one has to examine the burning oil well in Kuwait during the Iraq war. In order to distinguish the oil fire, they used blasting material to cut the heat from the oil. According to science it is to cut the oxygen from the fire, but the oxygen will flow in just as fast as it flowed out. The oxygen will return immediately, therefore the oxygen as such does not increase the fire. It is the response that the oxygen has with heat that kills the fire. By accelerating the time from burning to beyond the interaction with oxygen, that is what really kills the fire. The incoming oxygen does not transfer sufficient heat to restart the fire. What is also true is that not only does the oxygen bring heat in, but also it relieves the burning material of heat, therefore allowing more space to convert to heat.

One has to seek for this natural phenomenon as it occurs inside the star. In the case of a steel cutting however, the acetylene is the fuel, and the oxygen's role is to enhance the heat, not burn the metal. Again it is the way oxygen responds to heat and that accelerates time. The oxygen removes heat by bringing in more space to become heat, as much as remove heat to become space. We have to recognize the dual role of oxygen and not merely the fact that oxygen burns. OXYGEN DOES NOT BURN AND OXYGEN CANNOT BURN. IF OXYGEN COULD BURN, THE EARTH WOULD NOT HAVE ANY OXYGEN LEFT BY NOW.

In fact we observe the metal degeneration as cutting but it is time enhancing on one small area of the metal. The metal is "rusting away" at one point. The rusting process speeds up by thousands of years at one specific point. It is not the oxygen but the way the oxygen responds to heat and the interaction between heat and space that cuts the metal. In other words, the oxygen only carries the heat and space to the iron.

In every galactica, a certain value of space-time was sealed in, released, as its time becomes valid. Every galactica therefore, corresponds to a different value at a different rate in as far as compensating for the time draining in the Universe, as the Universe is shrinking. We, as humans have up to now, placed a false value to the universal expansion, because we have valued the invalid aspect to the Universe, which is space.

Seen from the valid perspective, which is time, the Universe is shrinking as it loses in the valid perspective, which is time; the Universe is shrinking as it gains in the invalid component to the Universe, which is space. Therefore, the Universe is not expanding, but it is shrinking all along. I have been in so much disagreement about so many aspects of science in the way they regard the Universe, that I do not wish to split hair about such trivia. In the light of this, I only point this aspect out, but I do not wish to make it an enormous issue, because it does not really matter, from what angle you look at it. Another outcome may be where both objects maintain the claim to singularity, by pushing the space-time occupied to new levels of occupied space-time values. The result of the establishing of new individual but unequal points of singularity is the oval way objects rotate, first favouring the on in the matter part and the other in its space part and afterwards turning the points of reference around.

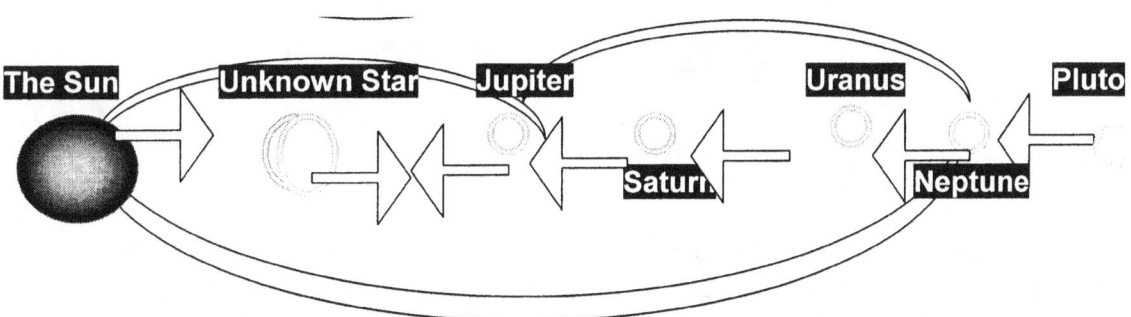

Since the First movement of time, the Roche factor was present and from the Roche factor came the Titius Bode principle. Each object seen, as well as not seen, represents a different period contained by a different specific value of space. All matter is time in a different frozen state in space and the time gives the space its particular value.

The Roche limit that came about between the Unknown star and the star we call sun contributed to the demise of the smaller star.

The Roche limit is:

The region surrounding each star in a binary system, within which any material is gravitationally bound to that particular star. The boundary of the Roche lobes is an equipotential surface, and the lobes touch at the inner Lagrangian point, L_1, through which mass transfer may occur if one of the components expands to fill its lobe. It names after the French mathematician Edouard Albert Roche (1820-83).

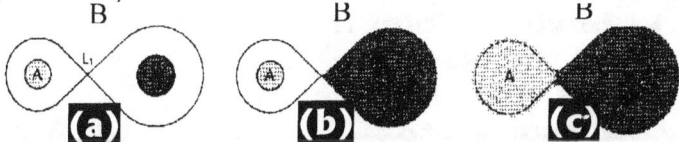

THE ROCHE LOBE: In a binary system, the Roche lobes of components A and B meet at the L_1 Lagrangian point. (a) In a detached system, neither star fills its Roche lobe. (b) In a semidetached system, one massive component, B, fills its Roche lobe. (c) In a contact binary, both components overfill their Roche lobes and share a common envelope. As with the graph I can see the two sides forming a connection therefore relevancy has to apply, all contradicting Newtonian claims of no connection but through mass attractions. The mass does not attract but one interferes with the other total influencing the space surroundings.

Any person taking Newton seriously should at least take on the challenge and find the comets colliding with the sun, find how much the planets moved closer to the sun since the days of Newton and indicate where there is unprecedented collision between stars. Yet the closest the Universe comes to that is to show " how stars blow bubbles" in space and that is to use the precise words.

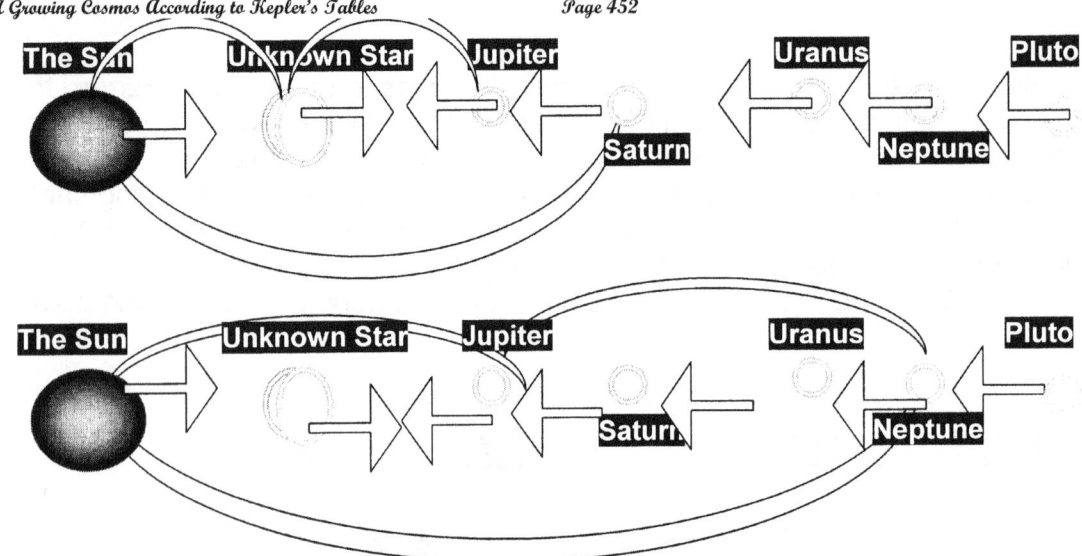

Jupiter related to Unknown star by the matter-to-matter relation of 7 /10 + 7/10 in relation to the space of 10 /7. However there the relation with the other micro Stars stopped! Jupiter was the one 7/10 marker and Saturn held its position as the other 7/10 marker. Then Saturn had the 10 / 7 relation with the Sun. From Saturn point of view in relation to singularity control the fact of Unknown star did not exist. Where the sun development pushed Unknown star out, it did not move at the rate of the Micro stars since it was a large star by individual measure. It pushed the sun Back as well and it pushed Jupiter but the other four that did not see Unknown star held firm. That brought conflict since Unknown star did not move well with the others. It had a singularity relation with the Sun but that was it. The Micro stars had a singularity relation with the sun and Unknown star was never in the picture.

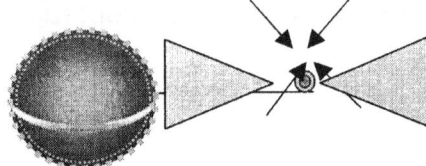

The extending of **k** in **k = a^3 / T^2** had a murderous effect on Unknown star since from the sun the star had to grow away from the Sun but five other Micro stars had now addition about the intentions of Unknown Star and therefore Unknown star became something like Johnny –in –the- middle.

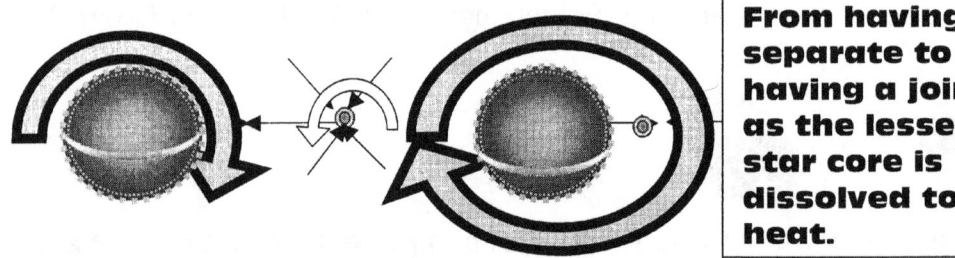

From having separate to having a joint Π^2 as the lesser star core is dissolved to heat.

The war came to rage as a battle for supremacy began to emerge because Unknown star and the Sun came into a Roche limit. How the actual figure places the position that Unknown star had I will not get into because that is fairly complicated and it is time consuming. In the Roche limit there is an intense flow increase in liquid time and it is the minor structure that cannot take the contraction. The advances star puts the flow Π^2 it receive from the spin that forms the space Π^3 to the value of Π. Then as the Roche limit is bridged the time value $4\Pi^2$ is in conflict with the time difference limit $\Pi^2 / 4$ and since the time ratio of $4\Pi^2 / \Pi^2 / 4 = \Pi^0$, which in effect brings about that the singularity controlling the better developed singularity also takes control of the lesser developed space-time since both then relate to Π^0 and is in relation to $k = \Pi$ because $T^2 = \Pi^2$. However there is no way that the lesser core can stand the heat since the better developed core increase the flow of liquid extending the flow to the lesser star and then the lesser star is taxed with the burden of Π which puts the same value of space Π^3 in relation to the lesser developed core.

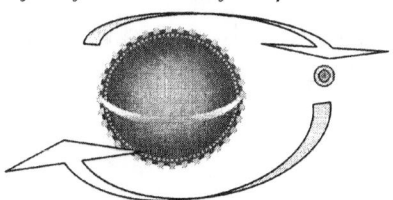

It is well advised to remember that it is not the Universe that grows but it is the Mterail in the Universe that expands by the margin Kepler mentioned as **k = a³ / T²**. Therfore in all the expanding of space that took place and did not take place the diameters of the Sun as well as Unknown star did expand. As they expanded so di the distance parting them not respond by the same measure because the Micto stars was blocking the route into expanding as far as the well being of Unknown Star goes. Soon time arived that introduced serious conflict in Paradise.

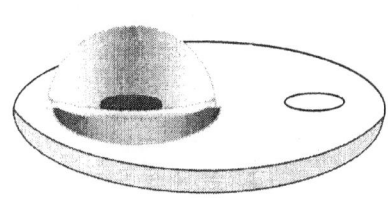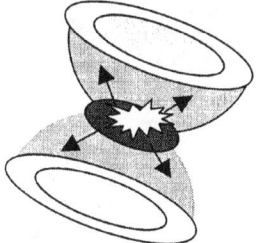

The Roche limit is evidently widely applying throughout the entire cosmos but Newtonian rules do not explain the phenomenon. The heat increased condemned the fate of Unknown star and the core of Unknown star expanded as it brattled and blew. Five chunks remained as well as many fragments, which became comets. The Hydrogen gas now at present forms what is known as Kuiper belt.

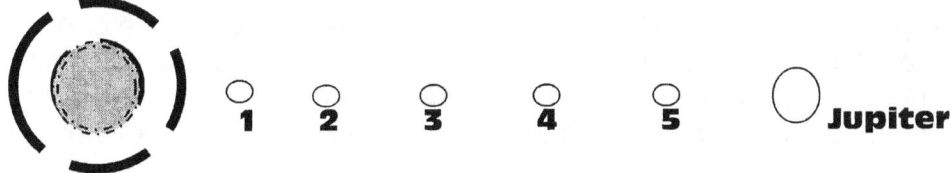

Then there were five totally unrelated fragments on the inside of five Sun related micro stars. This spelt disaster in many languages. The five now formed did not correlate to the five micro stars and therefore the five micro stars still did not at all align with the newly fragmented hard rocks.

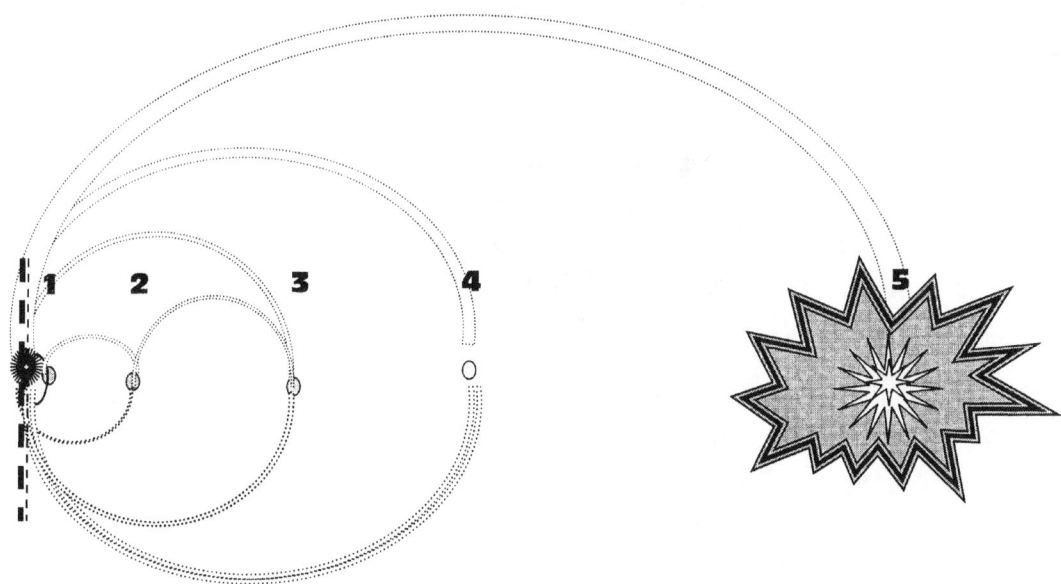

We have a ring of debris and cosmic junk clustered around where a fifth planet should be on the very edge of the inner planets. Logic tells us that too must be part of a solar disaster but how can one prove that? I once again do not propose the following as proof but my main attempt is only to show there are methods one can use when applying my cosmic code to find answers to answerless questions.

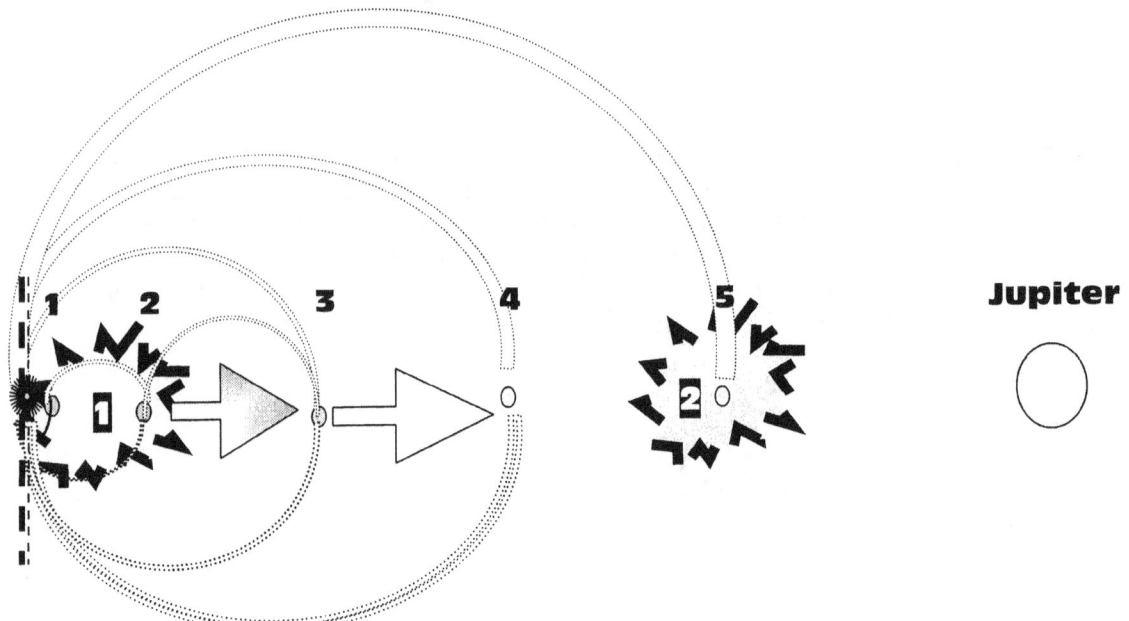

The evidence we see is in the numbers we find. There was this event that produced an enormous quantity of heat near or close to the sun. This we find in evidence where Mercury expanded 18 % more than that which should be gauged as normal. The sympathy decline in space development as the planets developed at a further distance from the event. In the case of Venus we find an exaggeration of 10 % and in the case of the earth there is an exaggeration of 1 %. This is to be expected since the demise of one little star In a Roche limit is hardly expected to change the face of the entire Universe at large. By the tome the development approached Jupiter there should be no evidence of growth since Jupiter now did not connect with a page object and the growth could hardly constitute to influence a micro star.

Is:

1	Mercury	49 X 10^6 km
2	Venus	98 x 10^6 km
3	Earth	147 x 10^6 km
4	Mars	227 x 10^6 km
5	Fragments	413.1952

Should be:

1	Mercury	49 X 10^6 km
2	Venus	98 x 10^6 km
3	Earth	147 x 10^6 km
4	Mars	196 x 10^6 km
5	Fragments	245 x 10^6 km

1	Mercury	118.%
2	Venus	110.%
3	Earth	101 %
4	Mars	115%
5	Fragments	168%

However when we include a Quantity of debris at a location where the Titius Bode law does indicate the presence or position of a structure, the significance change considerably.

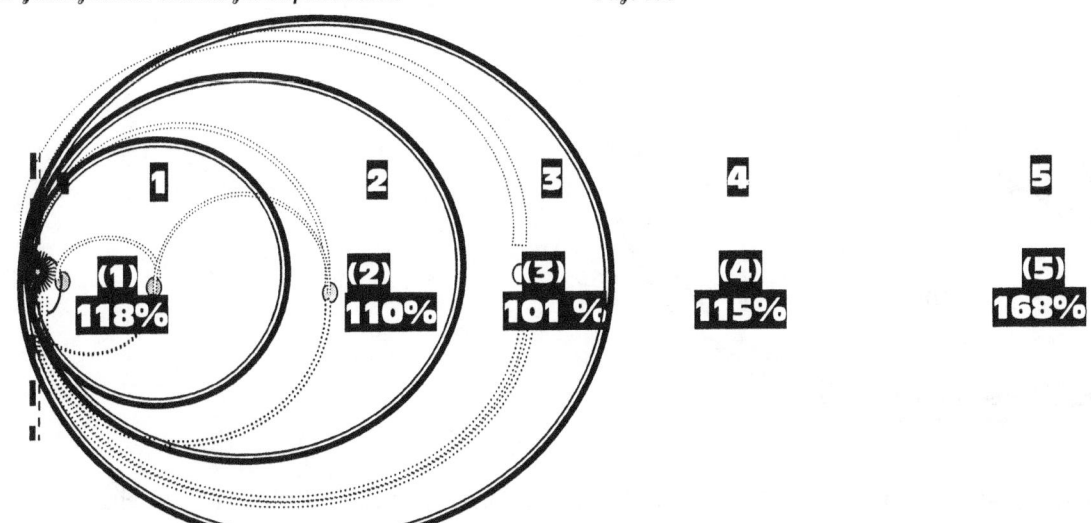

Where the reducing should start nullifying the space growth, which is past Mars there we find an increase that is most astonishing. We find that there is an increase of as high as 68 % where there should be no traces of any growth left. This can only be the result of a heat release of gigantic proportions in the manner of a (very little) Super Nova spectacular.

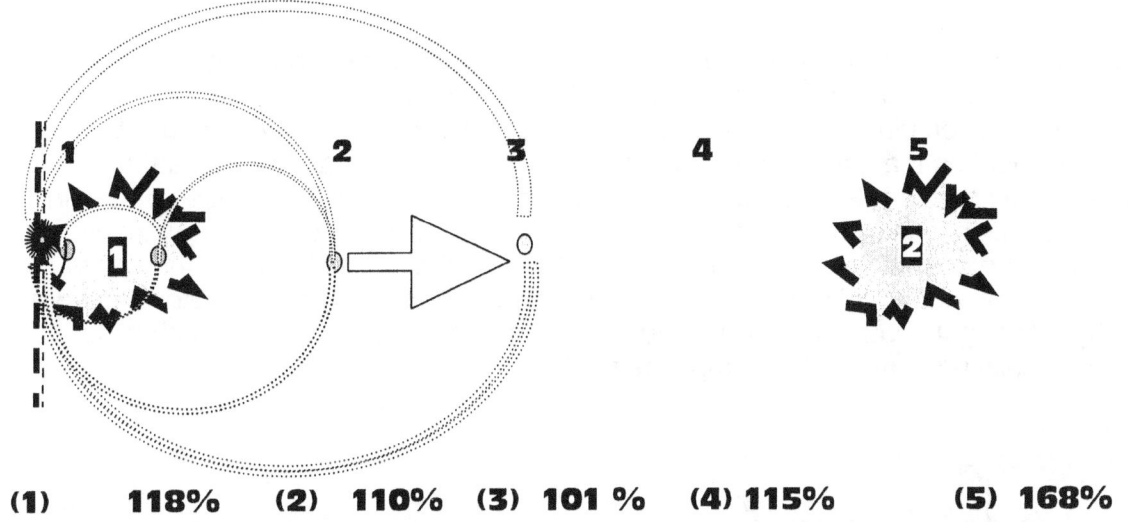

(1)   118%   (2)   110%   (3) 101 %   (4) 115%   (5) 168%

Jupiter released the same fate on the fifth planet as that which befell the fifth plants mother star old Unknown star. I the fifth planets did no fragment (and believe me there was no alternative solution) the alignment between the planets and the micro stars would not have realised because then there was no linking the micro stars and the planets in relation to the using of the Titius Bode law.
Any application to use this method in gauging the development of the micro stars would bring no clear results since we and the micro stars are not connected in the manner we connect with the four planets.

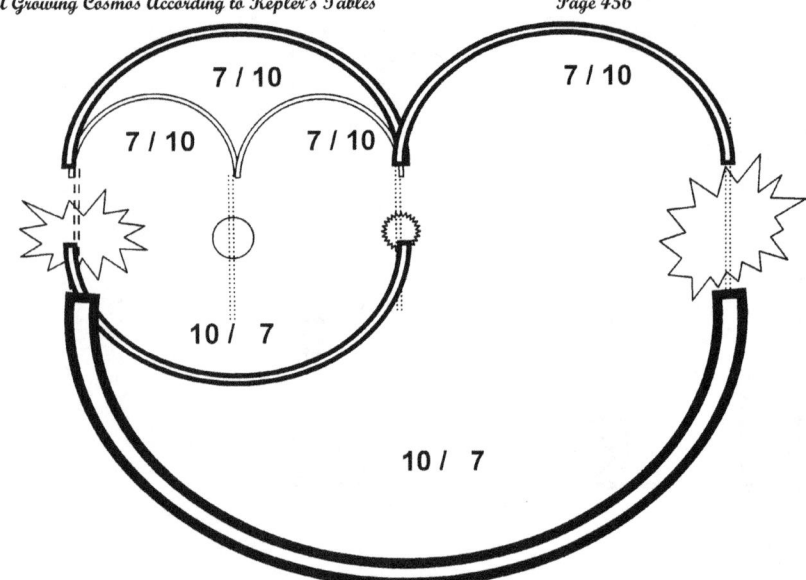

The significance and implication that the application of the Titius Bode principle holds on the Hubble constant reflecting on era to come as well as era gone past turns the cosmos from a small piece of vacuumed holding a few atoms to a vastness no computer irrelevant of size can ever determine.

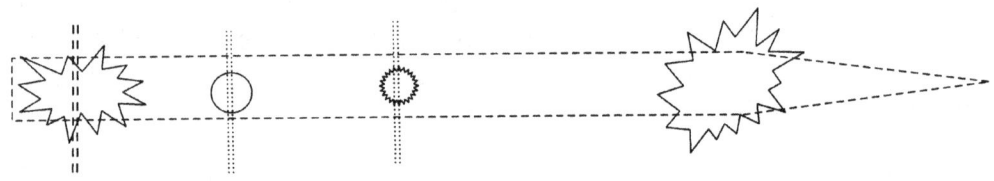

In view of the Titius Bode time depletion, time in flow creating space has a far more complicated arrangement than Xepted science can even produce on a chart. To be honest every person knows that Xepted science cannot even place the planets on a chart, depicting true distance to size, but they WILLFULLY never mention that information when the chart they show is as false as a three-dollar bill. In a sense it does no harm leading people down the ally in such a way, because others in my class of mental insignificance in society is far to un-intelligent to realize the correct way and will therefore not understand the correct way in any event. Now you ask "So what about Neptune and why does Neptune not fit the pattern…"well that storey is more complicated which I therefore reserve for ***"Seven Days Of Creation"*** because the explaining is as simple after I produced a much wider vision with a lot more explaining to make it as simple as this…

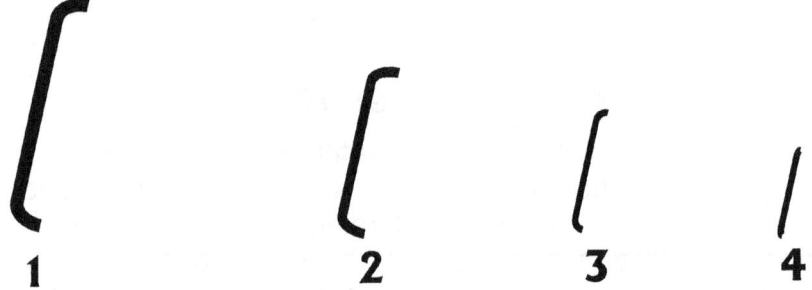

1 SINGULARITY HOLDING MATTER TO OUR FUTURE IN SPACE-TIME

2 SINGULARITY HOLDING MATTER IN RELAVANCY TO SPACE-TIME

3 SINGULARITY HOLDING MATTER AT THE SPEED OF LIGHT

4 SINGULARITY HOLDING MATTER BEYOND THE SPEED OF LIGHT

That is a mistake with a stinging tale. It is as dangerous as a scorpion to Xepted science. In an introducing article, I named Anglo-American Mythology, I pointed out how misconception feeds society, favouring the lies and untruths and blatantly ignoring the truth. In the past, since time began the powerful used this on the brainless masses, and for a period where that civilization lasted, got away with that strategy. The next civilization that came to power, followed the same methods applying the same dogma, and in the end paid the same penalty because their greed, lust for power, and sublimations gave them control over the masses for a while. The misconception those in favourable positions forced onto the masses, made the very people in power so shortsighted, their course on vanity lasted but a few generations. This is achieved because our Earth environment is tolerant and can buffer a lot, to save life in the end for life's contamination on Earth. They wish to extend life's connection to Earth, as being a connection to the cosmos at large and will be in effect as long as life remains in the cosmos. When "going abroad" to our "next door planet" that holds all supporting evidence of life carrying organisms, the connection to the cosmos remains and connecting to the Earth is of little consequence. That bluffing must stop. I realize no one on Earth will ever take note of what one sod (like me) on Earth is shouting, but misery awaits our Martian Colonists. Suffering will be the reward for the fools attempting to catch the bounty of "fame, riches and glory" on behalf of the All Powerful Dollar and the dollars absolute true benefactors. Those with eyes, let them see, those with ears let them hear and let the rest self demolish.

I admit, man derives all the above-mentioned lies mainly from corrupting the Bible. The Roman Catholic Church started the trend 1 500 years ago and still maintains it as best it can. You can prove whatever you wish to prove from facts you take from the Bible. The Bible is in support of whichever standing you may support. This is not because the Bible is incorrect, it comes from our insignificance to appreciate the full content of the whole Bible. The Bible promote only truth, man takes from that what man wants, divert the truth to suit his need by corrupting the lot. I know presenting evidence as well as I do, will not change the course of science. I know too, that I am too small to pinpoint conclusively whether Authentic Author referred to the creation of the first cosmic period, or the first solar period. This is not because of inaccuracy on the part of the Bible. This inaccuracy comes from the human interpretation of Authentic Author on his vision, and then my insecurity to interpret his interpretations. It is human error bringing on misinformation. The Bible's recollection CAN and DOES apply to either period. Therefore, to respond in containing human misjudgement on my part, I shall re-apply the vision of Authentic Author in the context of the first solar day. I showed how it fits to the first cosmic day already.

Binary stars, spinning to self-destruction will produce significant heat. Heat create space, space forms winds. That is facts that the Bible present and is indisputable. Where the Earth was, was still a void, containing a sphere of circular displacement and this will reduce linear displacement to zero. Linear displacement is space and circular displacement is containing heat for matter survival.

Binary Star Minor overheated. That is why the core brittle and fragmented. This action will release tremendous contained heat the heat will produce magma flowing in space like water in space and this eruption of heat space that created winds. Once again the recollection fits the scenario. Releasing the heat and producing space will establish space-time and fill the void where the Earth should fit. This is fact and if anybody even tries to dismiss this will be because of abstinence on his or her part. I did not prove the Bible correct. The Bible told the truth and in such correct detail, it is beyond human comprehension, but sublimation on the part of Newtonians and science before them, disallowed their ability seeing it.

At this point I invite you, the reader to go back and read about the Newtonian version of cosmic structure forming. Compare that FORCE applying MAGIC to the Biblical portrait of events and see where the fools hide. When comparing notes about Newton's view and the Bible's view, judge for yourself who in the end understood Newton and who did not. All I wanted was to find some one that could look past the mechanic and appreciate the work he is representing. I never seek prominence, I only wished to introduce my view and let another more educated, more significant and more wise man take the reigns from there. I always knew I am not the man to do the job. My field of knowledge is too small, too limited and above all, too insignificant.

I found no one that could look past me and see my formula $R^3/T^2 = 1$ and $\$T = (\Pi^2 \times \Pi^2)(\Pi^2\Pi) 3 = 1836$ which is the relevance of the cosmos. By not finding a person that could see past me, I knew that person will not be able to look beyond "a burning Sun and see the frozen state in which the Sun is. Without noticing such crucial evidence, the rest goes lost. That person that sees me and not my formula will never see the cosmos for what it is.

While the one proton connects to space in singularity (Π^2 going to singularity) and connects space Π (in singularity) the other proton brings time Π^2 directly to space Π, re-uniting space-time as a unit to singularity (Π^3). We may call this re-unification unifying the " gravity - motion" in order to identify the one proton unifying time with space, while still in contact with the other proton holding (Π^2) time to (Π) space in as much as being occupied by matter (Π^2) and unoccupied heat Π forming 3.

This explains then the absolute value of time as a square, with the square having both a circular value of R^2/T multiplied by the linear value of R/T producing the link to singularity in time and singularity in space.

That is what an insignificant formula $R^3/T^2 = 1$ where $R^2/T \times R/T = 1$ represents space-time in singularity as well as space-time in densified, occupied and unoccupied format. That means the everything of the whole lot, or as we say in Afrikaans, the "Heelal" meaning Universe.

When comparing notes about Newton's view and the Bible's view, judge for yourself who in the end understood Newton and who did not. All I wanted was to find some one that could look past the mechanic and appreciate the work he is representing. I never seek prominence, I only wished to introduce my view and let another more educated, more significant and more wise man take the reigns from there. I always knew I am not the man to do the job. My field of knowledge is too small, too limited and above all, too insignificant.

I found no one that could look past me and see my formula $R^3/T^2 = 1$ and $\$T = (\Pi^2 \times \Pi^2)(\Pi^2\Pi) 3 = 1836$ which is the relevance of the cosmos. By not finding a person that could see past me, I knew that person will not be able to look beyond "a burning Sun and see the frozen state in which the Sun is. Without noticing such crucial evidence, the rest goes lost. That person that sees me and not my formula will never see the cosmos for what it is.

While the one proton connects to space in singularity (Π^2 going to singularity) and connects space Π (in singularity) the other proton brings time Π^2 directly to space Π, re-uniting space-time as a unit to singularity (Π^3). We may call this re-unification unifying the " gravity - motion" in order to identify the one proton unifying time with space, while still in contact with the other proton holding (Π^2) time to (Π) space in as much as being occupied by matter (Π^2) and unoccupied heat Π forming 3.

This places Kepler's formula at a relevance s $a^3/T^2 = k$ where k^1 will hold a relation to heat with time in eternity, while AT THE SAME MOMENT infinity also apply, giving k^0 the value of singularity. This is all to do with the Universe remaining in contact with singularity while keeping the Universe in matter and heat, in the dimensions of space and time. This explains then the absolute value of time as a square, with the square having both a circular value of R^2/T multiplied by the linear value of R/T producing the link to singularity in time and singularity in space. That is what an insignificant formula $R^3/T^2 = 1$ where $R^2/T \times R/T = 1$ represents space-time in singularity as well as space-time in densified, occupied and unoccupied format. That means the everything of the whole lot, or as we say in Afrikaans, the "Heelal" meaning Universe.

THE COSMOS IS NOT OUTSIDE; IT IS INSIDE, INSIDE EVERY ATOM. THE ATOM CANNOT DISAPPEAR, AS IT CANNOT VANISH. After all it is energy.

In an effort to convey what I see as time I wish to convert the Cosmic Calendar to some Cosmic Time Scale, This scale does not name events, but rather use relevancies, running time and space as the event unfold from singularity. The singularity comes from the point where all matter occupied and unoccupied confirmed one line with out space as space was infinite and time was eternal. As purely an indication of physical time found in the cosmos applying at this moment I wish to introduce an illustration in bringing a better comprehension of time flow. There is a possibility of many other starts to the flow of singularity, but this point holds most valid significance to the theme we explore

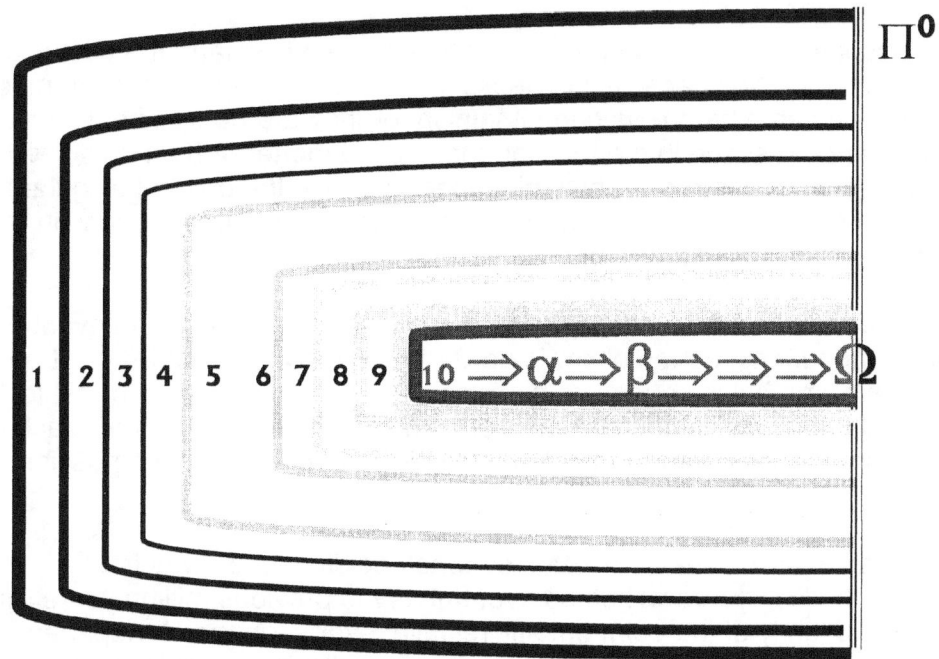

Newtonians tell about a Big Bang explosion that included everything there is.

$t = 10^{-43}$ seconds

the very first instant, the entire Universe were much smaller than a neutron and the temperature was $\approx 10^{32}$ K

$t = 10^{-34}$ seconds

The Universe underwent an increase in pace expansion growing in size with a factor of 10^{30}. The Universe becomes a soup of quarks and leptons at a temperature of $\approx 10^{27}$ K.

$t = 10^{-4}$ seconds

Quarks combine to form protons and neutrons and their anti particles. The Universe cooled down to such a slow pace electrons no longer can breakdown the particles remaining. Particles of matter and anti matter collide and annihilate each other. There is a slight excess of matter not finding annihilating partners, surviving to form the world that we know today.

$t = 60$ seconds

The Universe has by now cooled down enough to form protons and neutrons and with colliding can stick together to form the nuclei of low mass elements ^{2}H, ^{3}He ^{4}He and ^{7}Li. The predicted relative abundances of these nuclides are just what we observe in the Universe today. There is plenty of radiation around, but light cannot travel before it interacts with nucleus. The Universe is opaque to its own radiation.

$t = 300\ 000$ years

The Universe has now fallen to $\approx 10^4$ K, and electrons can stick to bare nuclei when they collide forming atoms. Because light does not interact appreciably with (uncharged) particles such as neutral atoms, the light is free to travel great distances. From this light comes background radiation Atoms of hydrogen and helium can hold together under the influence of gravity, and begin to clump up forming galactica and stars. In every small human mind we try to find time, which we know and trust. The cosmos holds time much to the properties the Creator describes in the Bible. I shall not be blasphemous and say it is the time the Creator refers too, because the cosmos is time the Creator created with all other aspects, therefore the Creator refers to time at a pace that puts our vision of eternity in the same class as we find an explosion to be.

To us the future is dark because it holds more space to less light. On the other hand the past is bright because it holds lighter to less space. The space we see lacks luminosity, because there are much more space to hold light. Stars that came before us cannot vanish for they are matter, holding matter to occupy space of matter. Matter (and space in the form of unoccupied matter or heat if you wish to call it that) is energy and energy cannot destruct, vanish disappear or leave the point of singularity. Again we are facing the situation the Bible warned us about. We are thinking of the heavens in terms of Earth instead of thinking of the heavens in terms of heavens.

On Earth we humans, connect time to human relation. Today become yesterday with the event of tomorrow. We think of today disappearing, as tomorrow is dawn. In the Universe that may not be the case, because where true cosmic time holds space, we humans shall never enter. In contrast to Popular Newtonian belief we will not run down the corridors of some Black Hoe to another Black hole and in the process mesmerize time. Such is not for us fitted with carbon-holding life in a position of singularity.

In stars one year is as eventful as a million years because of space holding time away from singularity. On the moon the next million years will be as eventful as the previous million years with nothing to report in newspapers. Slightly of the point but still relative: ever seen the Newspapers come out and say nothing much happened? There is always news only the relevancy may change from day to day. Well, this does not even happen on the moon because on the moon life will not bring news as it happens; it does not happen!

Looking at the sky we do not see space we see time. The photon is space travelling through time and that is the only space we see, that which the photon hold and that the photon occupy as much as bring to us. Space is light in utter darkness. Space is a ray of photons in magnitude not directed to our singularity but in countless other directions where singularity manifest space in time. Space is heat and heat is photons of lesser implication or not directed our way. To my view (what it may be worth) the Authentic Author did not move back in time, he merely moved foreword in space. By moving in space he reached Moment-Alfa.
Moment-Alfa is in eternity and eternity never ends therefore he moved to eternity in space eluding time by having a vision. He could therefore see what he reported because he was in that space but not in that time.

If he were in that time he would still be there, he would be there eternal and never come back to report on what he saw. There is much rumour of a Big Bang, however I am inclined to think the biggest bang that ever was also became the smallest bang there ever can be. It was the instant when heat parted from cold. The line started in infinity because the line was continues but being continues it never was. The very instant followed the previous instant identically and the instant was so identical it remained the same by never moving while always moving uninterrupted eternity upon eternity. Then came the entire Universe when infinity broke free from eternity. It was when darkness broke into light. It was when whatever possibly can be became a possibility to be. It was when the first number mathematically arrived and from the one became two by gong 1^0 to 1^1. That infinity is so small it houses everything there is in the entire Universe. The entire Universe still is in a spot that formed a dot. The spot has no outside but it only has an inside while it is inside all that spins as it generates all that can spin. Yet, by spinning it brings motion into being.
The spin creates a drive that keeps the Universe mobile. Still the first forming of the dot from the spot came about to the inside and not the outside, which makes the Universe shrink and not expand. It is the smaller things that come into relevance as the larger things were placed in relevance when time began. The Universe is shrinking into the oblivious since the Universe never had anywhere to expand to. Never once did one Newtonian sit back and consider their laughable proposal of an expanding Universe with nowhere to go when it is expanding.

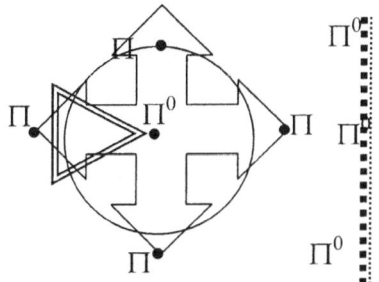

Three points formed a line covering singularity where the centre singularity recovered heat to grow and two points served as an axis to allow the rotation and to assist the duplication. There is one centre connecting the duplication of three as well as the recovery of one (the fourth one) that is applying the tie aspect. Therefore, motion consists of three positions in relation to a centre, which forms as space in relevancy to the motion and the space receive a controlling centre.

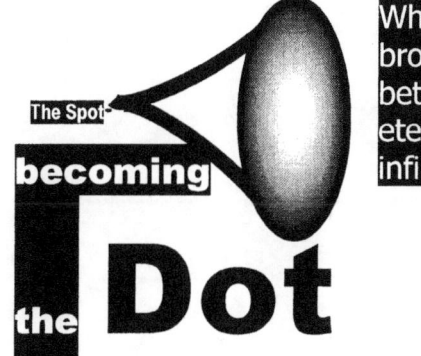

At the first glance, and positions applying different planets in the That is why the Universe
The duplication comes singularity in precise charged is as space less
The heat it requires to carry

Kepler's formula seems to be numbers between the sun and specific but solar system.
is Π
about as singularity is exciting another relevancy of 3 to 3 to 1, but the points and as motionless as only singularity is. the exciting between points forming space

and the space excites heat and the time delay it takes to excite singularity between points forms space-time.

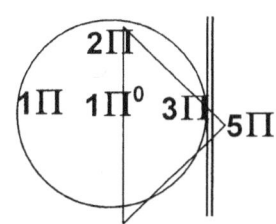

Where motion conducts electrical charging which is equal to gravity the charging of motion is to entice duplication of singularity. This is the basis, the heart and the sole ingredient of the Coanda principle that includes the Roche limit ($\Pi^2/4$). The charging of gravity $((7/10) + (7/10)) / (10/7) = \Pi^2$ and the charging of space-time $\Pi^3 = \Pi^2\Pi$ is all due to the relevancy brought on by the Coanda principle. The value of motion came from singularity exciting singularity and that is the duplication while the duplication or motion presents the space.

When the cosmos came to motion, motion was not yet defined. When the cosmos brought about motion, the first motion was parted from hot. Eternity parted parted from motion absence. laboriousness of eternity for the The spot became and grew into From what the spot was to what be just a mathematical from 1^0 to 1^1 but in reality that creating of and establishing of with all possibilities now in it. much growth become a reality, growth is beyond what we ever

When space brought division between eternity and infinity

relevancies. Cold from infinity. Motion Infinity broke the duration of infinity. the dot.. the dot now is might implication of going first motion was the an entire Universe Never again can that although to us the can notice. But it is

because the growth is so massive and we are so small that we are unable to notice such almighty growth.

When the spot Π^0 became functional and established all relevancies possible, heat parted from cold as eternity parted from infinity. The expansion was not clear motion but more a parting of relevancies where a centre formed a relevancy because the centre could not provide motion. Without being capable of motion, the centre established four points, which also served singularity. From the inverse square law we know that the centre doubled by producing the four points holding singularity.

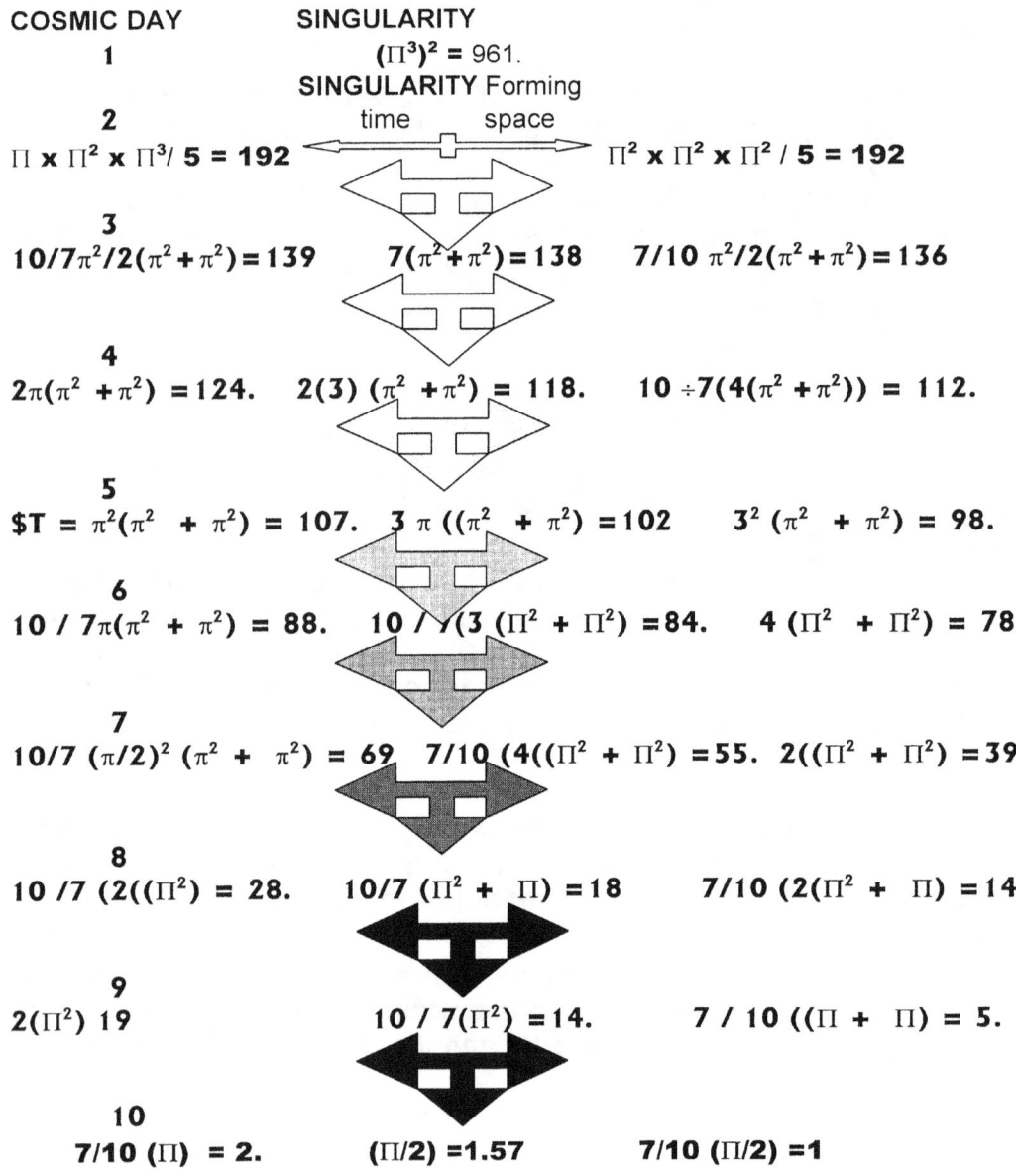

COSMIC DAY   **SINGULARITY**

1 $(\Pi^3)^2 = 961.$

SINGULARITY Forming

time   space

2

$\Pi \times \Pi^2 \times \Pi^3 / 5 = 192$ $\Pi^2 \times \Pi^2 \times \Pi^2 / 5 = 192$

3

$10/7\pi^2/2(\pi^2+\pi^2) = 139$ $7(\pi^2+\pi^2) = 138$ $7/10\ \pi^2/2(\pi^2+\pi^2) = 136$

4

$2\pi(\pi^2+\pi^2) = 124.$ $2(3)(\pi^2+\pi^2) = 118.$ $10 \div 7(4(\pi^2+\pi^2)) = 112.$

5

$\$T = \pi^2(\pi^2+\pi^2) = 107.$ $3\pi((\pi^2+\pi^2)) = 102$ $3^2(\pi^2+\pi^2) = 98.$

6

$10/7\pi(\pi^2+\pi^2) = 88.$ $10/7(3(\Pi^2+\Pi^2) = 84.$ $4(\Pi^2+\Pi^2) = 78.$

7

$10/7(\pi/2)^2(\pi^2+\pi^2) = 69$ $7/10(4((\Pi^2+\Pi^2) = 55.$ $2((\Pi^2+\Pi^2) = 39.$

8

$10/7(2((\Pi^2) = 28.$ $10/7(\Pi^2+\Pi) = 18$ $7/10(2(\Pi^2+\Pi) = 14$

9

$2(\Pi^2)\ 19$ $10/7(\Pi^2) = 14.$ $7/10((\Pi+\Pi) = 5.$

10

$7/10(\Pi) = 2.$ $(\Pi/2) = 1.57$ $7/10(\Pi/2) = 1$

Reading this book re-affirm that that is not my view and I dispute that view in all my heart.

Aanplasing, verplasing, versnelling and inperking

As this book is a translation from Afrikaans originally, some terminology and expressions I had to revise to accommodate my Ideas. Where I could I used modified English words to express a thought or an idea. One such a term is gravity. I had so much criticism about the word, which I feel I do not deserve. There is a certain notion clinging to the idea represented by gravity. Gravity links to a force that is all compelling, but I do not agree with such a compelling force, such as the word gravity implies. Gravity I introduce, works on two principles, but gravity to Newtonian standards is a single force. When I refer to gravity the normal reaction is that I am referring to the force I deny. By declaring that gravity the force controlling the entire Universe as a standard constant is non-existent. I bring the wroth of the scientific world upon me. When I make the statement that there is no gravity, every person considers me mentally unstable. Of course there is a movement of energy keeping all objects attached to the earth, but gravity implies work, and with that work principle I disagree, because that is not work, that is a cosmic balance that started at a point of eternity and will end at a point of eternity. Only life standing alone and detached from the cosmos being an energy source and is the energy that is able to manipulate space-time can commit work is from beginning of life to the end of life running concurrent with the rest of the Universe where the rest of the Universe is in a balance. The Universe is setting time and space to which life must adapt and adopt but in that science have an un comprehendible inability to relay the difference between life and what we find as a natural in the cosmos. Life is within the cosmos and the cosmos is without life. In the entire universe there is no work, it is a balance running concurrent through time and space. The balance shift in some cases to favour space and in other cases to favour time more but in all of that shifting, a continuous balance strikes every aspect of space-time. This applies new ideas never brought to light before and the new concepts clashes with the conventional names that science applies to current ideas. I had to divorce the science ideas from those I introduce and the only way was with new etymology. I have to start implementing the newly created terminology, which will apply to the rest of this book. This stems from my lack in ability to find suitable words in the English language that would define the concepts as they are, in order to establish the difference in meaning from the current words, which convey the existing misinterpretations (or if you wish, to my view incorrect applications).

This applies new ideas never brought to light before and the new concepts clashes with the conventional names that science applies to current ideas. I had to divorce the science ideas from those I introduce and the only way was with new terminology. I have to start implementing the newly created terminology, which will apply to the rest of this book. This stems from my lack in ability to find suitable words in the English language that would define the concepts as they are, in order to establish the difference in meaning from the current words, which convey the existing misinterpretations (or if you wish, to my view incorrect applications).

Firstly, we start with the word ***densified***, which is not a normal English word, but a word I had to produce in order to make a comprehensible statement. The correct word that applies is concentrated, that much I do know about the English language. But concentrated has not the correct meaning or the expression that I would like to bring over. Concentrated can apply to any substance, be it gas, liquid or solids where one of the ingredients become more than the rest of the ingredients. In that way, matter as a solid substance produced from the eternal substance which is heat, cannot be concentrated. Nothing in the entire universe can compare with the density of pure heat that spins at a rate in which that very heat can produce a value and which has a density far beyond anything else. Therefore, I chose to use the concept of concentration in a position where it makes a lot more sense.

A star is concentrated space-time, but there is a huge difference between a star's concentrated space-time and the value of pure matter. When a star does therefore become densified space-time, it can only be at the end of the Big Crunch eternity, witch I prefer to call moment Omega; that is when space becomes eternal and time becomes Zero. In this light I chose to call matter densified space-time. Densified space-time should therefore be in a definition where matter or substance has reached a point in density that will last one eternity, but has no limit. Concentrated space-time, on the other hand does have a limit, which is at the point where it becomes densified space-time.

The second word I created is ***Aanplasing,*** which is the ongoing redirection of heat as in matter to heat as in time and that connects to a circular deepening of the separation that matter undergo transforming to time as it discards heat for the cold of fusion. Later (I hope) it will be clear enough for every reader to comprehend and to distinguish between the various factors that bring about **aanplasing** as should the reasons be clear why I prefer to have created this new word.

In this case however, there was no English word that could merely be altered and then be re-applied. With a choice to my disposal I chose to alter an English word as was possible with densified. A more suitable word that relates to a better meaning in the case where I brought in **"densified"** would have been the Afrikaans word **"verdigting"**, where "verdigting" stands in relation to "konsentrasie" (concentrated). The fact of the matter is that I am not wilfully forcing Afrikaans down the throat of the Anglo-American and in the case of density I was able to adapt and modify a known English word that could adopt a new concept. Unfortunately in the case of aanplasing using an English word would mean that there is no liberation from the "misleading" focus that depends on gravity, nor can it liberate the feature of this "misconception".

As for the Afrikaans words: **aanplasing, verplasing, versnelling** and **inperking**: there are no such words or concepts in existence that the precise meaning can derive from the English written or spoken language. Should any such words exist, the misconceptions that remains connected to the original English words, would not bring justice to the concept which I wish to apply to convey the meaning that lies behind the correct value of the thought. If I stuck to the word "gravity", the concept I wanted to introduce would forever remain confused with Newtonian application. To that end the new realization would then never come across in the way I intend it to be.

The R that I use in the formula has nothing to do with Radius as a term, except when used, to calculate the value of a circle or a sphere. The R is derived from the Afrikaans word **Ruimte,** which means space. In the Afrikaans word: **"Ruimte" the "u" and the" i "** is used in conjunction, which is pronounced the same way as " ai" is used in English words such as in **pain, drain, train, vain, rain, etc**. So spelled incorrectly it should be pronounced as **"Raimte" and the "te" is pronounced "huh". The T stands for the word tyd, which incidentally is time.**

This is what I named negative space-time displacement, which results in **verplasing. The word verplasing is pronounce FHERPLHASHING (FHER – PL –HA – S – H-ING),** which means to "relocate" without destroying or changing the composition in any way, as the object is moving away from a certain position.

The word **aanplasing is pronounce AHNPLHASHING (AHN –PL – HAS – HING)** and literally means to relocate without damaging or destroying the composure or structure of the objects, as the object is moving toward a certain position. This very same value was previously mistakenly confused as being gravity. It is the effect on matter where space-time is in motion and matter is motionless or "stand still".

Both **aanplasing** as well as **verplasing** cause time differentiation and matters structural re-valuation. That means the duration of time is re-valued and the space compromised. The excelling of the time factor is; versnelling and the reduction of space are: inperking.

The **word versnelling is pronouncing FHERSNELHING (FHER –SNEL – H-ING)** and means to speed up. If one is placed in a star it would seem as if time inside the star is accelerated while time on the outside of the star would come to a standstill. This concept is explained in far more detail, in a later stage in the book. The **word inperking means reduction or containment of the structure or scaling it down to a different size without penalizing or altering the shape in any way.**

Both inperking and versnelling is how matter relates to change **in space (inperking)** and **time (versnelling)**. The generation of the heat is within the structure and relate to time in space.

Inperking: This stands apart from the idea of curtailment because in curtailing. The *In* part is pronounced as in English where the *per* one pronounced in the same fashion as the sound an English sheep makes BHE placing a *H* sound before the *E* with *king* already explained. Sorry but that is as far as the Afrikaans lesson goes for the day.

Inperking: This stands apart from the idea of curtailment because in curtailing something or someone, means that object's or person's movement or moveable motion is deprived. This then brings over the misconception in the accepted notion of an expanding universe. In due course I shall explain the concept, but inperking involves the same value that was there at first and will be there in the end, only the location in the balance shifts to favour one or the other part of the same coin. Because of the fact that none such a thing applies when space-time "accelerates" (versnel), and where this brings about inperking, it does not apply. Instead, all functions and factors still apply when inperking becomes valid, therefore the meaning of inperking becomes more applicable and this word describes the process much better. Inperking relates to time, where the duration of time extends, but not the value of time as such, as time applies in the cosmic sense.

One should realize that the entire atom, as well as its surroundings including all other surrounding atoms are reduced in space-time volume, so the atom is not actually curtailed, nor is its surrounding which means the word curtailment does not really apply. All aspects of occupied and unoccupied space-time are in reality, re-focused down in the true sense and above all, remains to the precise relative relation value it had for one entire eternity where the relation between such times, only refocus. However, scaled down would neither apply, because that would not refer to the time involvement, which lies at the hart of this revaluation. Where less time applies, inperking would be more severe and where more time applies, less inperking will apply.

In this reference to time, one second would remain one second to matter inside the star but the duration of that second, compared to geodesic time validation, would appear to stretch enormously. All words in the English language by implying its dictionary meaning, will inevitably lead to more language confusion, seeing that the explanatory meaning does not cover time enhancement and space reduction, heating and slowing of time lapse. By introducing a new word to the reader, I hope to screen out any misconceptions. Hand in hand with inperking, goes versnelling. When the reader encounters the concept of inperking, it should accompany the idea of versnelling.

Versnelling: It carries exactly the same meaning as acceleration, but the meaning or concept connected to acceleration implies to matter as the matter increases its own positional change in space and time. That is not the impression I wish to relay when referring to versnelling, because it is exactly the opposite of that meaning. In this, the actual meaning is more applicable to the true connection. These I must explain carefully, not to convey confusion. When a person stands outside an explosion of some sort, the time laps seems instantaneous, quicker than the senses can relate to. However, inside the explosion, time is almost standing still.

Any person, who is inside such an explosion, would relate to time on the outside as being instantaneous. Whether this statement is accepted or not, the truth is that a person in an explosion cannot die, although his body is shattered in a million pieces. Such a person is sealed in a period separated from the period he and we lives in. This I explain at a later stage. The time duration slows down immensely, but from the outside, it accelerates immensely. Therefore, time versnel to the outside of where ever one relates to.

I wish to bring over the fact, as just been said, that the concept we have, is quite the opposite. Versnelling implies that the motional increase lies with the transfer of space-time, regardless whether matter occupies it or not. As aanplasing (not gravity) and versnelling bring about inperking (not curtailment as the body remains free to do as it wishes) the space-time that the body occupies and the surrounding sphere are in constant state of versnelling. The increase in motion has an effect on the matter, but the matter stands weightless as its specific density applies the time in that particular space.

Verplasing: This word is preferred to that of displacement, because although the matter in motion is displaced, time and space are implicated in the process. Verplasing is in fact the transferring of newly created magnetic space-time by matter, as a body composed of atoms has to replenish the space-time it occupies in order to maintain its position, place and structure in space-time in time in space, according to its geodesic positional allocation within the star's space and in time. Verplasing comes in effect as matter progresses in position, but the time-affect of verplasing that it has on matter, comes into real effect when an object reaches Mach $_3$ depending on its shape and altitude. In short: **Aanplasing** is relatively where matter is in a geodesic motionless position as space-time carries the motion component of the two values. This means that aanplasing is relative to positive space-time displacement.

Verplasing on the other hand has to do with the motion being with the newly created space-time in relation with the matter and the geodesic space-time remains relatively motionless. In both cases inperking and versnelling is a consequential result of the process. The difference is in the application of the time component itself.

A practical example of the difference between aanplasing and versnelling is as such: a body in **_aanplasing_** is in example a skydiver is falling towards the earth and **_verplasing_** is where a body, such as a rocket is on a trajectory path as it fires into space. Both bodies will comply with the linear and circular displacement, but the circular displacement will relate oppositely in each event.

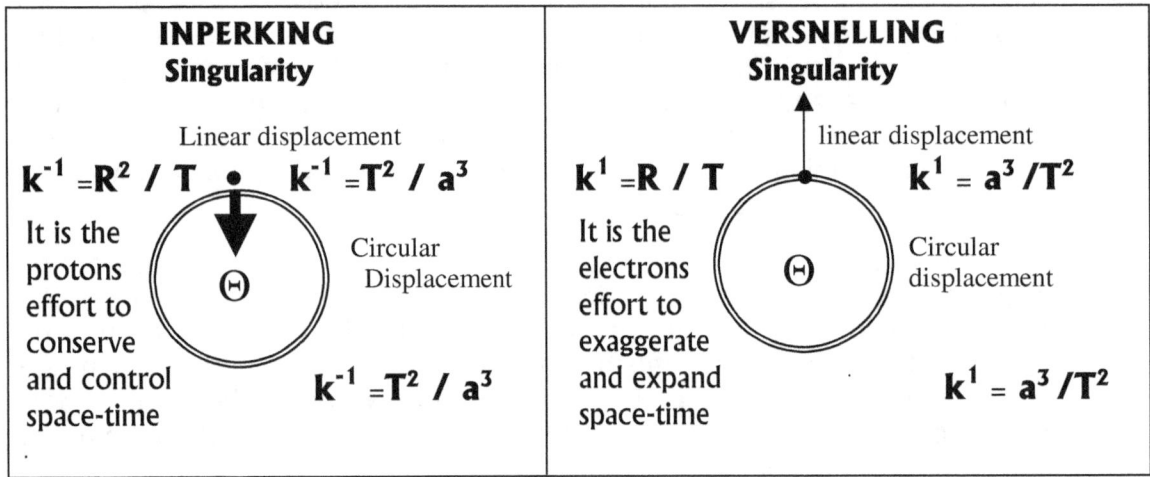

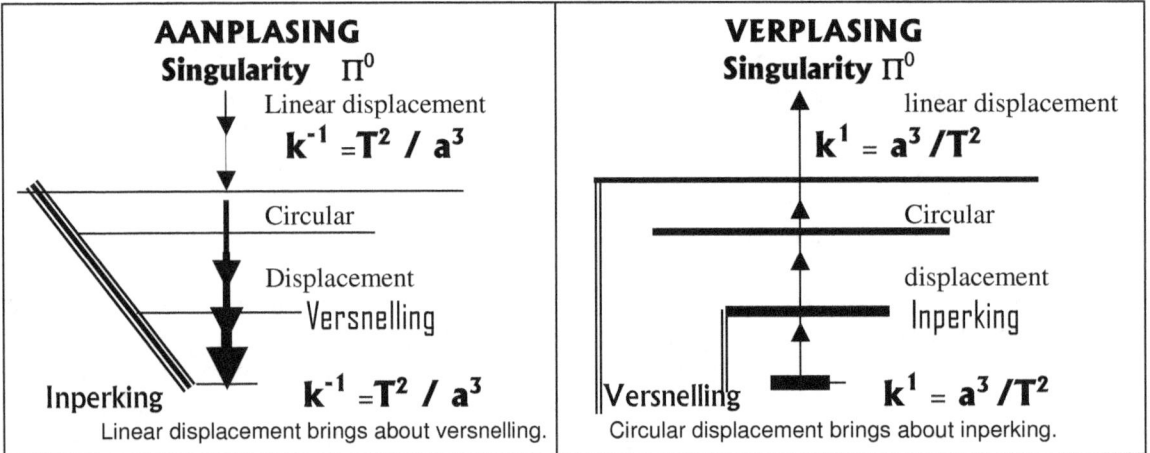

This aspect, Newtonian science disregard in, as much as they disregard that there is any connection between the atom and the cosmos

I deny the fact that a star can have winds, although winds are as close to that concept which the earth can provide. As you will later see, winds are the transferring of heat, but so is electricity and lightning, and one cannot call lightning wind. Neither can one call lightning electricity. It is altogether different product of the same transformation of heat, but the applied principle separating the products of heat transmitting stand in total different areas. As far as ordinary physics go, nothing changes at all. Every aspect of physics remains the same, except the way science view the cosmos. The formulas I show, has NO CALCULATION ABILITY. The only value in the exercise is proving what no person ever proved before, AND THAT IS THE INFLUENCE THE ATOM HOLDS ON THE UNIVERSE, AS THE ATOM INFLUENCE EXTENDS BEYOND ALL COSMIC BOUNDARIES.

To me everything makes perfect sense and while saying this I do admit full heartedly that I am not a Master such as yourself with the knowledge you possess. In that light, should you feel there are aspects I do not explain to a sufficient standard, I am willing to work on it. **This aspect, Newtonian science disregard in, as much as they disregard that there is any connection between the atom and the cosmos**

A star is liquid in motion flowing around a solid centre. All gravity is a product of the Coanda principal where motion of liquids creates a governing singularity centre with in the very centre of the star. Electricity, gravity the atomic gravity and the flow of light are amongst many other forms the transmitting or the normal flow of heat. Generating electricity is the same process the Earth and other cosmic structures use to generate gravity and there are no simple one force pulling as gravity. As gravity is, so is electricity and

lightning concepts of the same principle where one may be stronger in dynamics when compared to others being weaker in dynamics. All principles are relevancies where one statement only finds value when compared to another forming a relevancy by borders.

What is the universe? This is such a simple question that every one and every person gets wrong because of the relevancy we humans place on the Universe and the relevancy what the Universe truly is. SUPER –EDUCATED- WIZARDS really get tide in knots with making all about nothing so complicated it absorbs everything holding back nothing. It is so embarrassing simple even I can understand what the Universe is and the Universe is not what science says the Universe is. The Universe is a sphere only because Π holds Π^2 at the very end of space and time. By holding a specific centre the sphere becomes the strongest form that any object can have. The sphere is without any doubt the favourite choice coming about as the natural form formed by gravity in form of being committed only to gravity. Where gravity has the last say without other influences changing possibilities as collisions leaving debris in space or natural out burst like Super Nova explosions, gravity will enforce the sphere to be the form taken by the particle. But there is no evidence of particles of similar size joining in matrimony through gravity being the shotgun at the wedding. In cases where there is a mismatch of size outside any proportions of equality there is then a contracting of the lesser by the greater. In such cases the lesser is not qualifying as material (and that I prove later on) but the greater consider all the lesser to be heat. It is humans bringing distinction to matter in form.

Two objects where the one is small and the other is large is falling down to the earth by implication of size holding mass should have their own value of gravity and gravitons and in comparison with the gravitons of the earth; the mass putting the gravitons at work has apparent insignificant and an unrelated value. However, these two objects are in their own individual deuce to see who reaches the earth first. Let's compare an iron ball in matching size and compare such a fall to a wooden ball falling the same distance under the same conditions. It stands to reason that the iron ball's gravitons should give it a superior advantage just because of superior numbers working. This comes about because the two objects are in a position where they compare in relation to one another and share a common second factor, which is the earth. In relation to the earth, the gravitons of the two balls do not come into consideration, but this do not play a part since the earth is a common factor. The balls, however, is put in a situation where they stand in relation to each other. When compared to one another, the gravitons should give the heavier ball a sizable advantage. But Galileo said there is no heavy or light big or small since all object sharing similar conditions fall equally. Galileo was the first to indicate space-time but all failed to notice and Galileo was the first to prove gravity is simply equal motion to all but that every one including Newton which is supposedly the master on motion failed to see. The sensible example one can show to prove that where some matching structures in size in the cosmos come into conflict by coming to close to each other there is a process coming about where occupied space sharing one of the structures are turned to heat in space by the other and larger structure. The larger object is not pulling the lesser object closer as Newton confirmed by suggesting, but it is literally dissolving the lesser before consuming the liquid forming the lesser as it is turning as much of the lesser objec5ts accumulated heat into a benefit for sustaining growth of the more prominent object.   If the structure proves to large the superior structure turns the lesser compatriot into heat. Then being heat it will apply gravity and admit such heat into the ranks of its atmosphere, but not before turned it into fragments good enough to be heat. This process where this devouring happens has a known name for centuries and while the known name was given no one tried to marry this process off onto Newton's suggested formula.

 The Roche limit is evidently widely applying throughout the entire cosmos but Newtonian rules do not explain the phenomenon:

 Any person taking Newton seriously should at least take on the challenge and find the comets colliding with the sun, find how much the planets moved closer to the sun since the days of Newton and indicate where there is unprecedented collision between stars. Yet the closest the Universe comes to that is to show " how stars blow bubbles" in space and that is to use the precise words.

  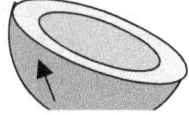

 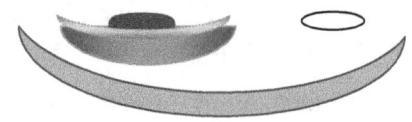 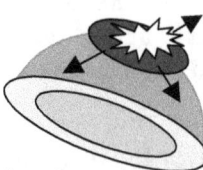

Even more astonishing is facts about the Binary star system that is seldom to never mentioned. Try and associate what happens in the Roche limit, which is what truly happens in the cosmos to what Newtonians confirmed suggestion about what happens when conflict of space arrive. The Official Policy Protectors never tries to explain the relation between Newton's laws as mentioned above, and the binary star system forming the principle we know as the Roche limit. The binary stars are systems where two stars spin around each other and never collide. These stars are sometimes smaller but mostly many times over the size of our sun. When one applies the same Newtonian formula as given above, these massive giants must crash into each other, destroying themselves in the process. The enormous mystery is not in the apparent misbehaviour of these giants, but the fact that this behaviour is known to science since the previous century. Relate the binary once again to the comet/ sun relation and there is a distinct similarity.

With the comet, the Newtonians regards a force that attaches the sun and the comet in some way where this force pulls the comet towards the sun. At the same time another force join in that pulls the sun closer to the comet, and is in ratio and resulting because of the mass but such is the mass difference between the sun and the comet, the force the comet applies never realizes as a force able to move the sun In view of this, only the force the sun applies, comes into effect. The comet proves the reality of this force by speeding up its movement as it comes closer to the sun. By asking the correct question to the point more proof is uncovered about reality. If the force did not become greater, why would the comet gain momentum? The sun started pulling the comet when the comet was hiding in the cold and darkness, but then discovered its location by applying detective gravity and collected a force second to none that collected this hiding comet by force and pulled it along in the direction of the centre of the sun.

Stars will never collide because stars can never collide.
The only absence in the cosmos is zero and without zero there cannot be an end to eternity but only an everlasting cycle that breaks to start one more eternity now and then. With the cosmos created minute by minute from no space within the cosmic centre, the cosmos is ruled from a position with every thing but nothing is but where we know God must be. By accepting singularity and the rule there of brings into the cosmos things physics are unable to explain, mathematics are unable to calculate and man is unable to dismiss. If you accept physics you have to accept God because you cannot except one proving singularity without the other coming through singularity.
If it is that simple then why is it complicated.

BEST WISHES,

PETRUS. (PEET) S. J. SCHUTTE
There is more about this in other books with the titles:

MATTER'S TIME IN SPACE:

FOR other related information, PLEASE VISIT THE WEB SITE, FOR YOUR CONVENIENCE

To find more about the book please visit

<u>gravity @tlantic.net</u>

MATTER'S TIME IN SPACE THE THESIS ISBN   0-9584410-8-1

FROM THE ORIGINAL AFRIKAANS: "MATERIE SE TYD IN RUIMTE" I. S. B. N.   0 – 6 2 0 – 2 7 0 4 1 - 1
WRITTEN BY PEET SCHUTTE
© KOSMOLOGIESE EN ASTRONOMIESE TEGNIKA

An open letter

TO SELECTED ACADEMICS
ISBN 0-9584410-9-X
is THE ACADEMIC PROLOGUE AND AN INTRODUCING LETTER TO ACADEMICS PRESENTING THE THESES MATTER'S TIME IN SPACE

But is also available as a Commercial book.

There is two books available with a commercial intent and is more elaborate it explaining how the cosmos came to birth and how the solar system came to birth. It is not using the same info but is eternities apart. It is more complicated to those with set minds about religion and science but is very simple to understand when less rigged mindsets set rules that is manmade and yet unchangeable. I use the Cosmic time table and the cosmic calendar I developed to prove how the solar system came to be and believe me it is not from dust accumulating to become rock!

The INDEX of MATTER'S TIME IN SPACE: THE THESIS ISBN 0-9584410-8-1

Part 7" SEVEN DAYS OF CREATION" ISBN 0-9584410-4-9
Is about the forming of the planets and gas structures in the solar system)   713
MATTER'S TIME IN SPACE: The Thesis (As a combination of the above)   3584

You may also contact me by land mail at:

Po Box 1093,
Ellisras
0555
REP. of South Africa.

An open letter

TO SELECTED ACADEMICS
ISBN 0-9584410-9-X

You may also contact me by land mail at:

Po Box 1093,   Ellisras
0555
REP. of South Africa.

Kyk vir verandering op bladsy
45
en op 128
en op 180
en op 188
en op 191
en op 205
en op 231 + 232
en op 262
en op 331
Kontroleer veranderinge vanaf 269 – tot 279